OBLIQUE SHOCK PROPERTIES: $\gamma = 1.4$

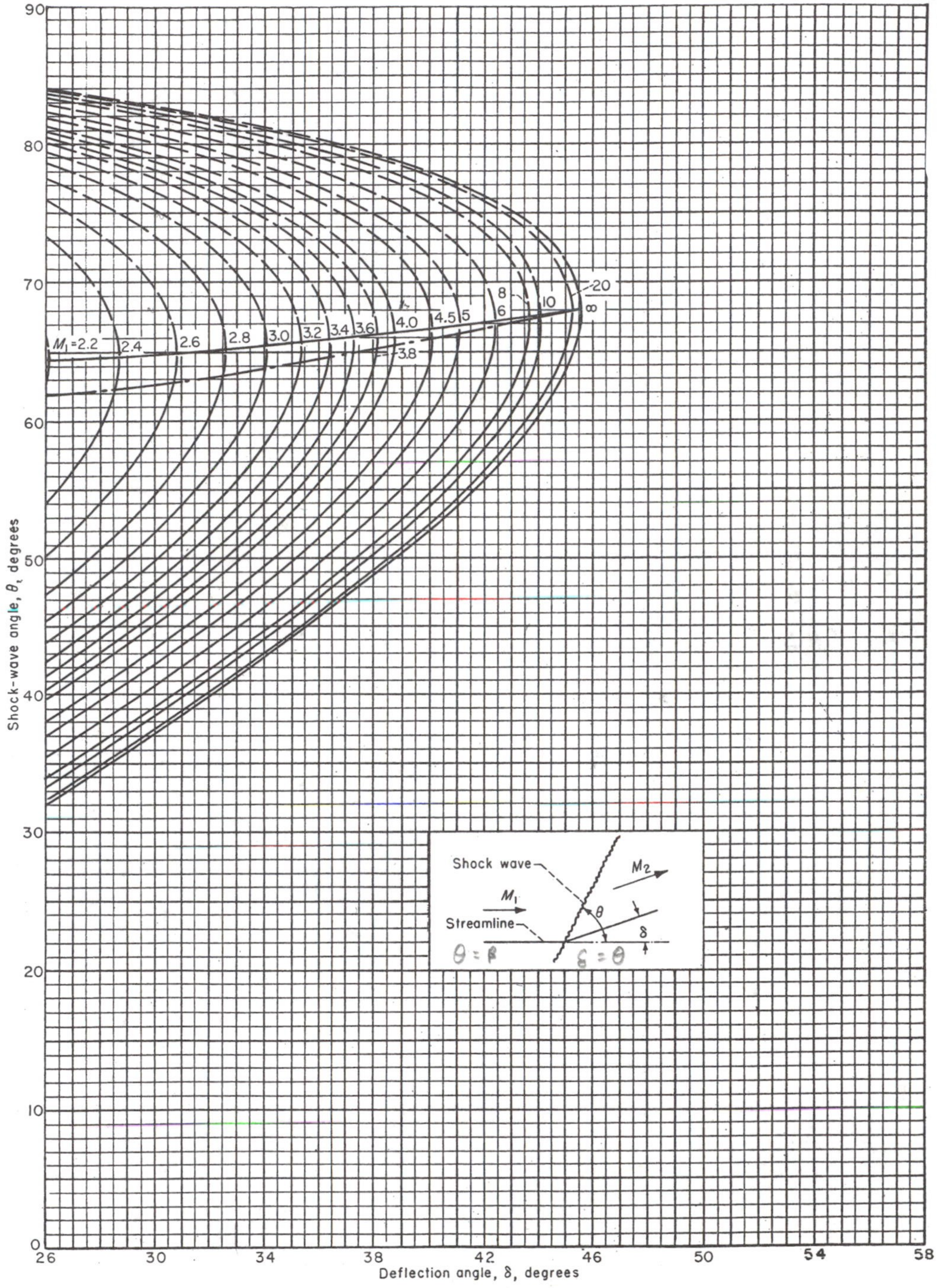
Shock-wave angle, θ_t, degrees
Deflection angle, δ, degrees
0
10
20
30
40
50
60
70
80
90
26
30
34
38
42
46
50
54
58
M_1=2.2
2.4
2.6
2.8
3.0
3.2
3.4
3.6
3.8
4.0
4.5
5
6
8
10
20
∞
Shock wave
M_1
M_2
Streamline
θ
δ

Modern Compressible Flow

With Historical Perspective

McGraw-Hill Series in Aeronautical and Aerospace Engineering

John D. Anderson Jr., *University of Maryland*
Consulting Editor

Anderson
Aircraft Performance and Design

Anderson
Computational Fluid Dynamics

Anderson
Fundamentals of Aerodynamics

Anderson
Introduction to Flight

Anderson
Modern Compressible Fluid Flow

Barber
Intermediate Mechanics of Materials

Borman
Combustion Engineering

Baruh
Analytical Dynamics

Budynas
Advanced Strength and Applied Stress Analysis

Curtis
Fundamentals of Aircraft Structural Analysis

D'Azzo and Houpis
Linear Control System Analysis and Design

Donaldson
Analysis of Aircraft Structures

Gibson
Principles of Composite Material Mechanics

Humble
Space Propulsion Analysis and Design

Hyer
Stress Analysis of Fiber-Reinforced Composite Materials

Kelly
Fundamentals of Mechanical Vibrations

Mattingly
Elements of Gas Turbine Propulsion

Meirovitch
Elements of Vibration

Meirovitch
Fundamentals of Vibrations

Nelson
Flight Stability and Automatic Control

Oosthuizen
Compressible Fluid Flow

Raven
Automatic Control Engineering

Schlichting
Boundary Layer Theory

Shames
Mechanics of Fluids

Turns
An Introduction to Combustion

Ugural
Stresses in Plates and Shells

Vu
Dynamics Systems: Modeling and Analysis

White
Viscous Fluid Flow

White
Fluid Mechanics

Wiesel
Spaceflight Dynamics

Modern Compressible Flow
With Historical Perspective

Third Edition

John D. Anderson, Jr.
Curator for Aerodynamics
National Air and Space Museum
Smithsonian Institution, and
Professor Emeritus of Aerospace Engineering
University of Maryland, College Park

McGraw Hill

Boston   Burr Ridge, IL   Dubuque, IA   Madison, WI   New York   San Francisco   St. Louis
Bangkok   Bogotá   Caracas   Kuala Lumpur   Lisbon   London   Madrid   Mexico City
Milan   Montreal   New Delhi   Santiago   Seoul   Singapore   Sydney   Taipei   Toronto

McGraw-Hill Higher Education

A Division of The **McGraw-Hill** Companies

MODERN COMPRESSIBLE FLOW: WITH HISTORICAL PERSPECTIVE
THIRD EDITION

Published by McGraw-Hill, a business unit of The McGraw-Hill Companies, Inc., 1221 Avenue of the Americas, New York, NY 10020. Copyright © 2003, 1990, 1982 by The McGraw-Hill Companies, Inc. All rights reserved. No part of this publication may be reproduced or distributed in any form or by any means, or stored in a database or retrieval system, without the prior written consent of The McGraw-Hill Companies, Inc., including, but not limited to, in any network or other electronic storage or transmission, or broadcast for distance learning.

Some ancillaries, including electronic and print components, may not be available to customers outside the United States.

This book is printed on acid-free paper.

International   2 3 4 5 6 7 8 9 0 FGR/FGR 0 9 8 7 6 5 4 3
Domestic   8 9 0 FGR/FGR 0 9

ISBN 978-0-07-242443-0
MHID 0-07-242443-5
ISBN 978-0-07-112161-3 (ISE)
MHID 0-07-112161-7 (ISE)

Publisher: *Elizabeth A. Jones*
Sponsoring editor: *Jonathan Plant*
Freelance developmental editor: *Regina Brooks*
Marketing manager: *Sarah Martin*
Senior project manager: *Kay J. Brimeyer*
Production supervisor: *Kara Kudronowicz*
Media project manager: *Jodi K. Banowetz*
Coordinator of freelance design: *David W. Hash*
Cover designer: *Rokusek Design*
Cover illustration: © *The Boeing Company*
Lead photo research coordinator: *Carrie K. Burger*
Compositor: *Interactive Composition Corporation*
Typeface: *10/12 Times Roman*
Printer: *Quebecor World Fairfield, PA*

Library of Congress Cataloging-in-Publication Data

Anderson, John David.
  Modern compressible flow : with historical perspective / John D. Anderson, Jr. — 3rd ed.
      p.   cm. — (McGraw-Hill series in aeronautical and aerospace engineering)
      Includes index.
      ISBN 0-07-242443-5 — ISBN 0-07-112161-7 (ISE)
      1. Fluid dynamics.   2. Gas dynamics.   I. Title.   II. Series.

  QA911 .A6   2003   2002067852
  629.132′323—dc21   CIP

INTERNATIONAL EDITION ISBN 0-07-112161-7
Copyright © 2003. Exclusive rights by The McGraw-Hill Companies, Inc., for manufacture and export. This book cannot be re-exported from the country to which it is sold by McGraw-Hill. The International Edition is not available in North America.

www.mhhe.com

John D. Anderson, Jr., was born in Lancaster, Pennsylvania, on October 1, 1937. He attended the University of Florida, graduating in 1959 with high honors and a bachelor of aeronautical engineering degree. From 1959 to 1962, he was a lieutenant and task scientist at the Aerospace Research Laboratory at Wright-Patterson Air Force Base. From 1962 to 1966, he attended the Ohio State University under the National Science Foundation and NASA Fellowships, graduating with a Ph.D. in aeronautical and astronautical engineering. In 1966, he joined the U.S. Naval Ordnance Laboratory as Chief of the Hypersonics Group. In 1973, he became Chairman of the Department of Aerospace Engineering at the University of Maryland, and since 1980 has been professor of Aerospace Engineering at the University of Maryland. In 1982, he was designated a Distinguished Scholar/Teacher by the University. During 1986–1987, while on sabbatical from the University, Dr. Anderson occupied the Charles Lindbergh Chair at the National Air and Space Museum of the Smithsonian Institution. He continued with the Air and Space Museum one day each week as their Special Assistant for Aerodynamics, doing research and writing on the history of aerodynamics. In addition to his position as professor of aerospace engineering, in 1993, he was made a full faculty member of the Committee for the History and Philosophy of Science and in 1996 an affiliate member of the History Department at the University of Maryland. In 1996, he became the Glenn L. Martin Distinguished Professor for Education in Aerospace Engineering. In 1999, he retired from the University of Maryland and was appointed Professor Emeritus. He is currently the Curator for Aerodynamics at the National Air and Space Museum, Smithsonian Institution.

Dr. Anderson has published eight books: *Gasdynamic Lasers: An Introduction,* Academic Press (1976), and under McGraw-Hill, *Introduction to Flight* (1978, 1984, 1989, 2000), *Modern Compressible Flow* (1982, 1990), *Fundamentals of Aerodynamics* (1984, 1991), *Hypersonic and High Temperature Gas Dynamics* (1989), *Computational Fluid Dynamics: The Basics with Applications* (1995), *Aircraft Performance and Design* (1999), and *A History of Aerodynamics and Its Impact on Flying Machines,* Cambridge University Press (1997 hardback, 1998 paperback). He is the author of over 120 papers on radiative gasdynamics, reentry aerothermodynamics, gasdynamic and chemical lasers, computational fluid dynamics, applied aerodynamics, hypersonic flow, and the history of aeronautics. Dr. Anderson is in *Who's Who in America.* He is a Fellow of the American Institute of Aeronautics and Astronautics (AIAA). He is also a fellow of the Royal Aeronautical Society, London. He is a member of Tau Beta Pi, Sigma Tau, Phi Kappa Phi, Phi Eta Sigma, The American Society for Engineering Education, the History of Science Society, and the Society for the History of Technology. In 1988, he was elected as Vice President of the AIAA for Education. In 1989, he was awarded the John Leland Atwood Award jointly by the American Society for Engineering Education and the American Institute of Aeronautics and Astronautics "for the lasting influence of his recent contributions to aerospace engineering education." In 1995, he was awarded the AIAA Pendray Aerospace Literature Award "for writing undergraduate and graduate textbooks in aerospace engineering which have received worldwide acclaim for their readability and clarity of presentation, including historical content." In 1996, he was elected Vice President of the AIAA for Publications. He has recently been honored by the AIAA with its 2000 von Karman Lectureship in Astronautics.

From 1987 to the present, Dr. Anderson has been the senior consulting editor on the McGraw-Hill Series in Aeronautical and Astronautical Engineering.

CONTENTS

PREFACE TO THE THIRD EDITION

The purpose of the third edition is the same as that of the earlier editions: to provide a teaching instrument, in the classroom or independently, for the study of compressible fluid flow, and at the same time to make this instrument *understandable* and *enjoyable* for the reader. As mentioned in the Preface to the First Edition, this book is intentionally written in a rather informal style in order to *talk* to the reader, to gain his or her interest, and to keep the reader absorbed from cover to cover. Indeed, all of the philosophical aspects of the first two editions, including the inclusion of a historical perspective, are carried over to the third edition.

The response to the first two editions from students, faculty, and practicing professionals has been overwhelmingly favorable. Therefore, for the third edition, all of the content of the second edition has been carried over virtually intact, with only minor changes made here and there for updating. The principal difference between the third and second editions is the addition of much *new* material, as follows:

1. Each chapter starts with a Preview Box, an educational tool that gives the reader an overall perspective of the nature and importance of the material to be discussed in that chapter. The Preview Boxes are designed to heighten the reader's interest in the chapter. Also, chapter roadmaps are provided to help the reader see the bigger picture, and to navigate through the mathematical and physical details buried in the chapter.

2. Increased emphasis has been placed on the physics associated with compressible flow, in order to enhance the fundamental nature of the material.

3. To expedite this physical understanding, a number of new illustrative worked examples have been added that explore the physics of compressible flow.

4. Because computational fluid dynamics (CFD) continues to take on a stronger role in various aspects of compressible flow, the flavor of CFD in the third edition has been strengthened. This is not a book on CFD, but CFD is discussed in a self-contained fashion to the extent necessary to enhance the fundamentals of compressible flow.

5. New homework problems have been added to the existing ones. There is a solutions manual for the problems available from McGraw-Hill for the use of the classroom instructor.

6. Consistent with all the new material, a number of new illustrations and photographs have been added.

This book is designed to be used in advanced undergraduate and first-year graduate courses in compressible flow. The chapters divide into three general categories, which the instructor can use to mold a course suitable to his or her needs:

1. Chapters 1–5 make up the core of a basic introduction to classical compressible flow, with the treatment of shock waves, expansion waves, and nozzle flows. The mathematics in these chapters is mainly algebra.

2. Chapters 6–10 deal with slightly more advanced aspects of classical compressible flow, with mathematics at the level of partial differential equations.

3. Chapters 11–17 cover more modern aspects of compressible flow, dealing with such features as the use of computational fluid dynamics to study more complex phenomena, and the general nature of high-temperature flows.

Taken in total, the book provides the twenty-first-century student with a balanced treatment of both the classical and modern aspects of compressible flow.

Special thanks are given to various people who have been responsible for the materialization of this third edition:

1. My students, as well as students and readers from all over the world, who have responded so enthusiastically to the first two editions, and who have provided the ultimate joy to the author of being an engineering educator.

2. My family, who provide the other ultimate joy of being a husband, father, and grandfather.

3. My colleagues at the University of Maryland, the National Air and Space Museum, and at many other academic and research institutions, as well as industry, around the world, who have helped to expand my horizons.

4. Susan Cunningham, who, as my scientific typist, has done an excellent job of preparing the additional manuscript.

Finally, compressible flow is an exciting subject—exciting to learn, exciting to teach, and exciting to write about. The purpose of this book is to excite the reader, and to make the study of compressible flow an enjoyable experience. So this author says—read on and *enjoy*.

John D. Anderson, Jr.

PREFACE TO THE FIRST EDITION

This book is designed to be a teaching instrument, in the classroom or independently, for the study of compressible fluid flow. It is intentionally written in a rather informal style in order to *talk* to the reader, to gain his or her interest, and to be absorbed from cover to cover. It is aimed primarily at senior undergraduate and first-year graduate students in aerospace engineering, mechanical engineering, and engineering mechanics; it has also been written for use by the practicing engineer who wants to obtain a cohesive picture of compressible flow from a modern perspective. In addition, because the principles and results of compressible flow permeate virtually all fields of physical science, this book should be useful to engineers in general, as well as to physicists and chemists.

This is a book on *modern* compressible flows. An extensive definition of the word "modern" in this context is given in Sec. 1.6. In essence, this book presents the fundamentals of classical compressible flow as they have evolved over the past two centuries, but with added emphasis on two new dimensions that have become so important over the past two decades, namely:

1. *Modern computational fluid dynamics*. The high-speed digital computer has revolutionized analytical fluid mechanics, and has made possible the solution of problems heretofore intractable. The teaching of compressible flow today must treat such numerical approaches as an integral part of the subject; this is one facet of the present book. For example, the reader will find lengthy discussions of finite-difference techniques, including the time-marching approach, which has worked miracles for some important applications.

2. *High-temperature flows*. Modern compressible flow problems frequently involve high-speed aerodynamics, combustion, and energy conversion, all of which can be dominated by the flow of high-temperature gases. Therefore, such high-temperature effects must be incorporated in any basic study of compressible flow; this is another facet of the present book. For example, the reader will find extensive presentations of both equilibrium and nonequilibrium flows, with application to some basic problems such as shock waves and nozzle flows.

In short, the modern compressible flow of today is a mutually supportive mixture of classical analysis along with computational techniques, with the treatment of high-temperature effects being almost routine. One purpose of this book is to provide an understanding of compressible flow from this modern point of view. Its intent is to interrelate the important aspects of classical compressible flow with the recent techniques of computational fluid dynamics and high-temperature gas dynamics. In this sense, the present treatment is somewhat unique; it represents a substantial departure from existing texts in classical compressible flow. However, at the same time, the classical fundamentals along with their important physical implications are discussed at length. Indeed, the first half of this book, as seen from a glance at the Table of Contents, is very classical in scope. Chapters 1 through 7, with selections from other chapters, constitute a solid, one-semester senior-level course. The second half of the book provides the "modern" color. The entire book constitutes a complete one-year course at the senior and first-year graduate levels.

Another unique aspect of this book is the inclusion of an historical perspective on compressible flow. It is the author's strong belief that an appreciation for the historical background and traditions associated with modern technology should be an integral part of engineering education. The vast majority of engineering professionals and students have little knowledge or appreciation of such history; the present book attempts to fill this vacuum. For example, such questions are addressed as who developed supersonic nozzles and under what circumstances, how did the modern equations of compressible fluid flow develop over the centuries, who were Bernoulli, Euler, Helmholtz, Rankine, Prandtl, Busemann, Glauert, etc., and what did they contribute to the modern science of compressible flow? In this vein, the present book continues the tradition established in one of the author's previous books (*Introduction to Flight: Its Engineering and History,* McGraw-Hill, New York, 1978) wherein historical notes are included with the technical material.

Homework problems are given at the end of most of the chapters. These problems are generally straightforward, and are designed to give the student a practical understanding of the material.

In order to keep the book to a reasonable and affordable length, the topics of transonic flow and viscous flow are not included. However, these are topics which are best studied after the fundamental material of this book is mastered.

This book is the product of teaching the first-year graduate course in compressible flow at the University of Maryland since 1973. Over the years, many students have urged the author to expand the class notes into a book. Such encouragement could not be ignored, and this book is the result. Therefore, it is dedicated in part to all my students, with whom it has been a joy to teach and work.

This book is also dedicated to my wife, Sarah-Allen, and my two daughters, Katherine and Elizabeth, who relinquished untold amounts of time with their husband and father. Their understanding is much appreciated, and to them I once again say hello. Also, hidden behind the scenes but ever so present are Edna Brothers and Sue Osborn, who typed the manuscript with such dedication. In addition, the author wishes to thank Dr. Richard Hallion, Curator of the National Air and Space Museum of the Smithsonian Institution, for his helpful comments and for continually opening the vast archives of the museum for the author's historical research. Finally, I wish to thank my many professional colleagues for stimulating discussions on compressible flow and what constitutes a modern approach to its teaching. Hopefully, this book is a reasonable answer.

John D. Anderson, Jr.

CHAPTER 1

Compressible Flow—Some History and Introductory Thoughts

It required an unhesitating boldness to undertake a venture so few thought could succeed, an almost exuberant enthusiasm to carry across the many obstacles and unknowns, but most of all a completely unprejudiced imagination in departing so drastically from the known way.

J. van Lonkhuyzen, 1951, in discussing the problems faced in designing the Bell XS-1, the first supersonic airplane

PREVIEW BOX

Modern life is fast-paced. We put a premium on moving fast from one place to another. For long-distance travel, flying is by far the fastest way to go. We fly in airplanes, which today are the result of an exponential growth in technology over the last 100 years. In 1930, airline passengers were lumbering along in the likes of the Fokker trimoter (Fig. 1.1), which cruised at about 100 mi/h. In this airplane, it took a total elapsed time of 36 hours to fly from New York to Los Angeles, including 11 stops along the way. By 1936, the new, streamlined Douglas DC-3 (Fig. 1.2) was flying passengers at 180 mi/h, taking 17 hours and 40 minutes from New York to Los Angeles, making three stops along the way. By 1955, the Douglas DC-7, the most advanced of the generation of reciprocating engine/propeller-driven transports (Fig. 1.3) made the same trip in 8 hours with no stops. However, this generation of airplane was quickly supplanted by the jet transport in 1958. Today, the modern Boeing 777 (Fig. 1.4) whisks us from New York to Los Angeles nonstop in about 5 hours, cruising at 0.83 the speed of sound. This airplane is powered by advanced, third-generation turbofan engines, such as the Pratt and Whitney 4000 turbofan shown in Fig. 1.5, each capable of producing up to 84,000 pounds of thrust.

Modern high-speed airplanes and the jet engines that power them are wonderful examples of the application of a branch of fluid dynamics called *compressible flow.* Indeed, look again at the Boeing 777 shown in Fig. 1.4 and the turbofan engine shown in Fig. 1.5—they are compressible flow personified. The principles of compressible flow dictate the external aerodynamic flow over the airplane. The internal flow through the turbofan—the inlet, compressor, combustion chamber, turbine, nozzle, and the fan—is all compressible flow. Indeed, jet engines are one of the best examples in modern technology of compressible flow machines.

Today we can transport ourselves at speeds faster than sound—supersonic speeds. The Anglo-French Concorde supersonic transport (Fig. 1.6) is such a vehicle. (A few years ago I had the opportunity to cross the Atlantic Ocean in the Concorde, taking off from New York's Kennedy Airport and arriving at London's Heathrow Airport just 3 hours and 15 minutes later—what a way to travel!) Supersonic flight is accompanied by shock waves generated in the air around the vehicle. Shock waves are an important aspect of compressible flow—they occur in almost all practical situations where supersonic flow exists. In this book, you will learn a lot about shock waves. When the Concorde flies overhead at supersonic speeds, a "sonic boom" is heard by those of us on the earth's surface. The sonic boom is a result of the shock waves emanating from the supersonic vehicle. Today, the environmental impact of the sonic boom limits the Concorde to supersonic speeds only over water. However, modern research is striving to find a way to design a "quiet" supersonic airplane. Perhaps some of the readers of this book will help to unlock such secrets in the future—maybe even pioneering the advent of practical hypersonic airplanes (more than five times the speed of sound). In my opinion, the future applications of compressible flow are boundless.

Compressible flow is the subject of this book. Within these pages you will discover the intellectual beauty and the powerful applications of compressible flow. You will learn to appreciate why modern airplanes are shaped the way they are, and to marvel at the wonderfully complex and interesting flow processes through a jet engine. You will learn about supersonic shock waves, and why in most cases we would like to do without them if we could. You will learn much more. You will learn the *fundamental* physical and mathematical aspects of compressible flow, which you can apply to any flow situation where the flow speeds exceed that of about 0.3 the speed of sound. In the modern world of aerospace and mechanical engineering, an understanding of the principles of compressible flow is essential. The purpose of this book is to help you learn, understand, and appreciate these fundamental principles, while at the same time giving you some insight as to how compressible flow is practiced in the modern engineering world (hence the word "modern" in the title of this book).

Compressible flow is a fun subject. This book is designed to convey this feeling. The format of the book and its conversational style are intended to provide a smooth and intelligible learning process. To help this, each chapter begins with a preview box and road map to help you see the bigger picture, and to navigate around

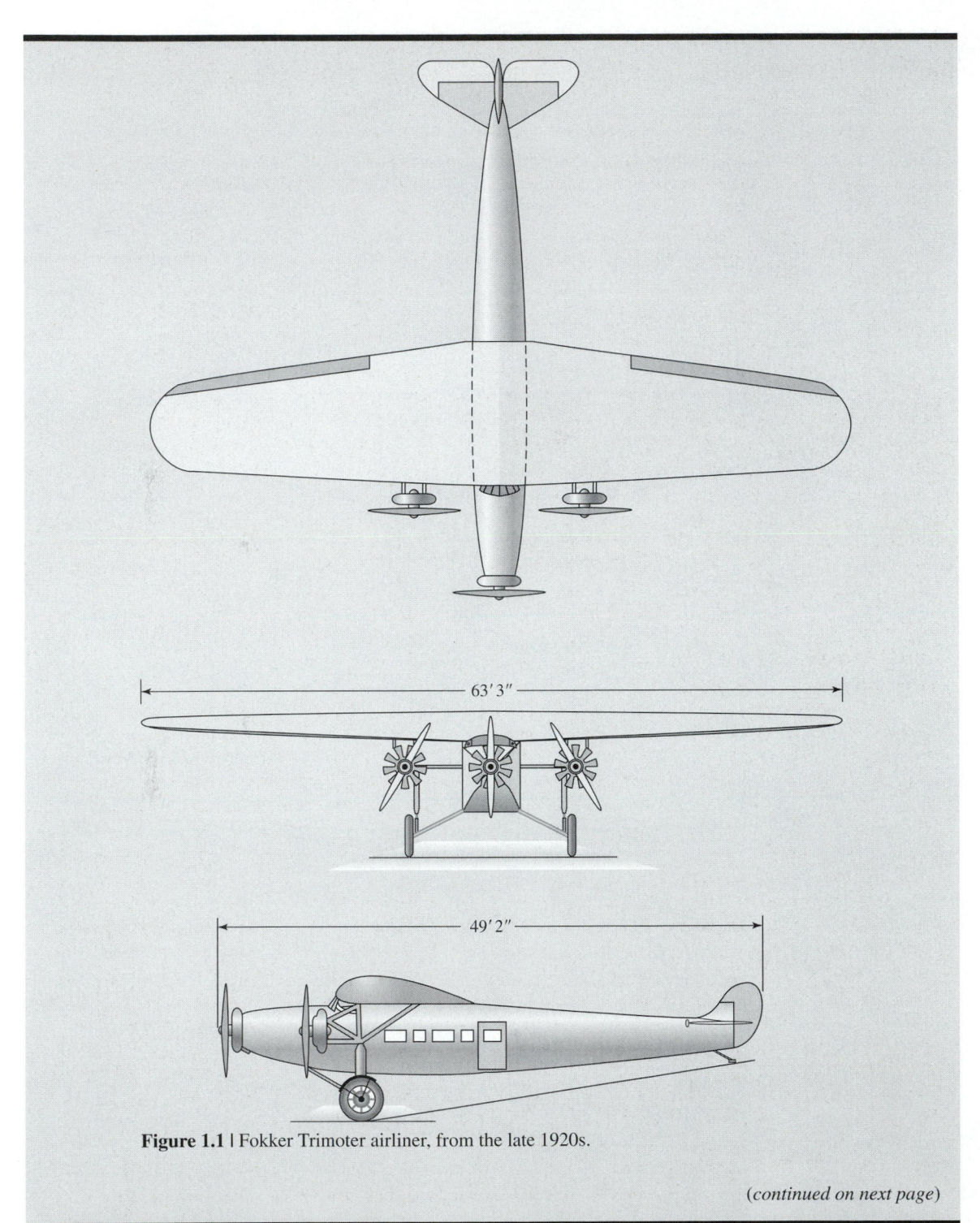

Figure 1.1 | Fokker Trimoter airliner, from the late 1920s.

(*continued on next page*)

(continued from page 3)

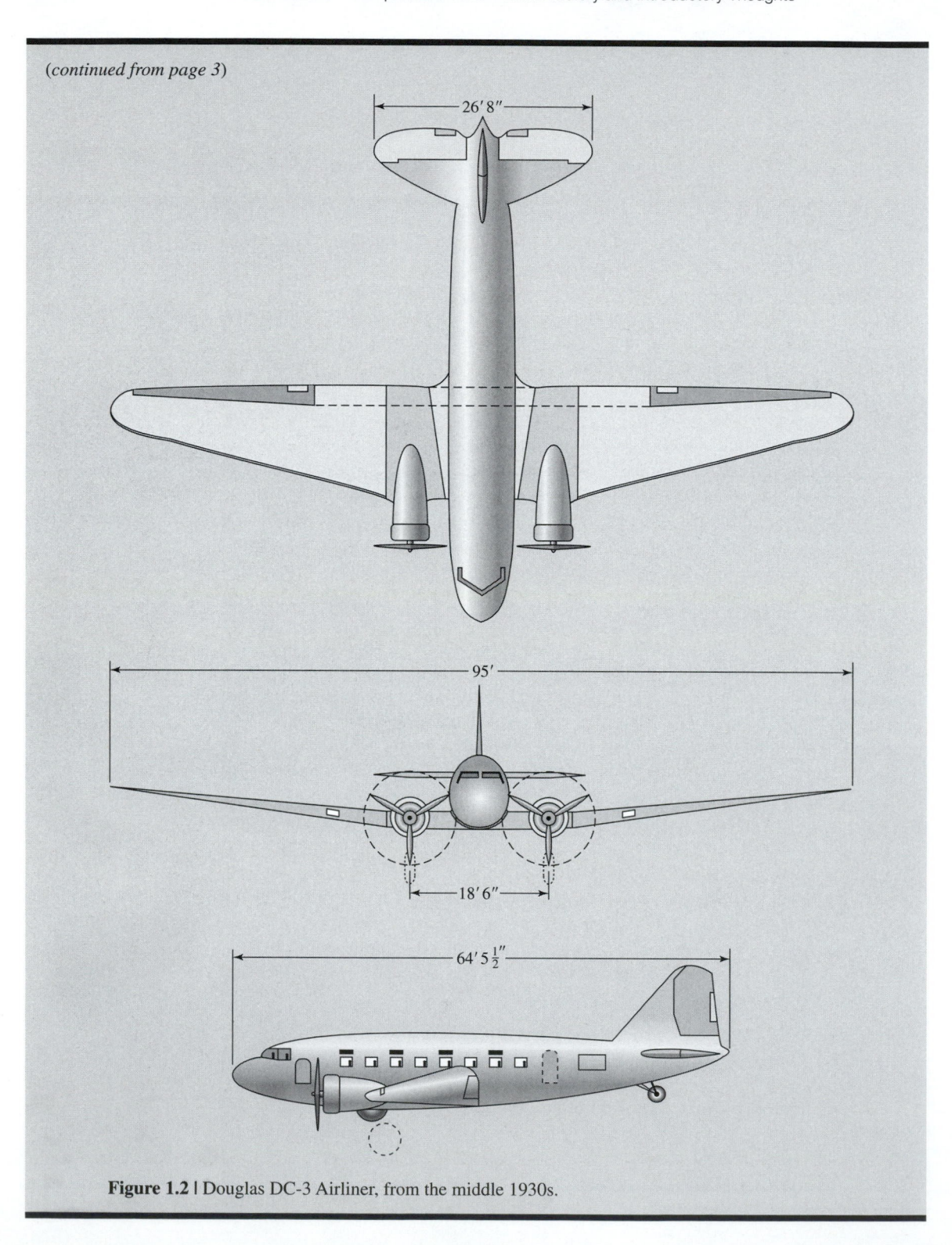

Figure 1.2 I Douglas DC-3 Airliner, from the middle 1930s.

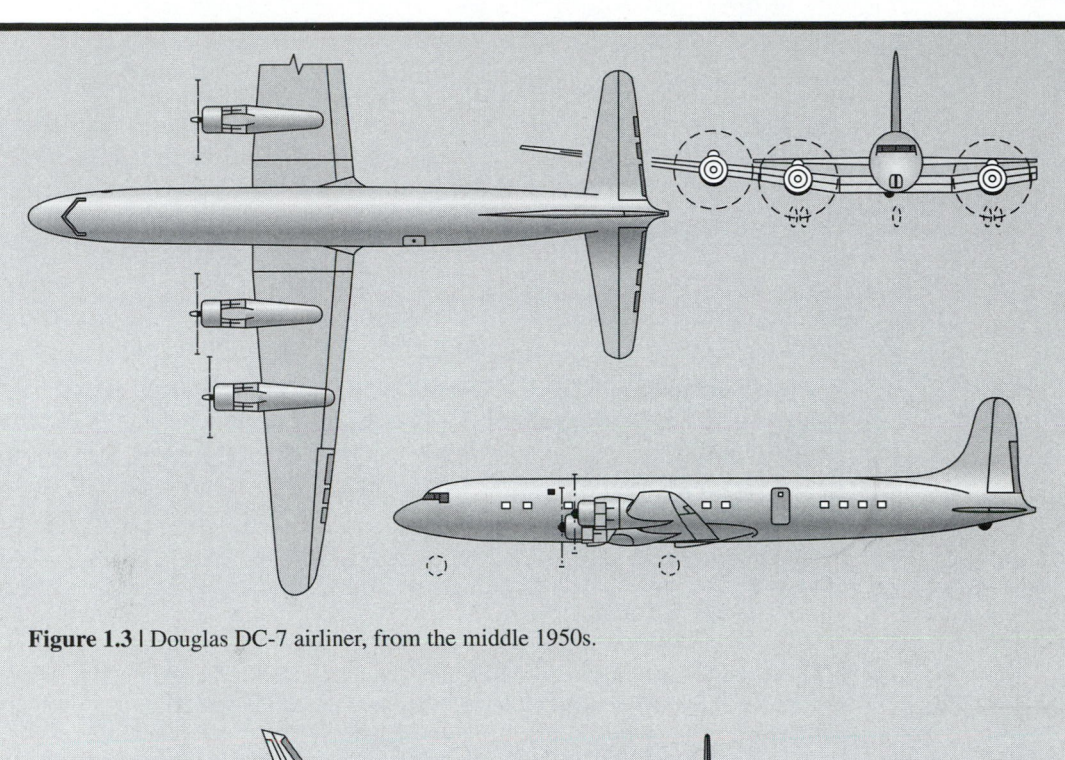

Figure 1.3 | Douglas DC-7 airliner, from the middle 1950s.

Figure 1.4 | Boeing 777 jet airliner, from the 1990s.

(*continued on next page*)

(continued from page 5)

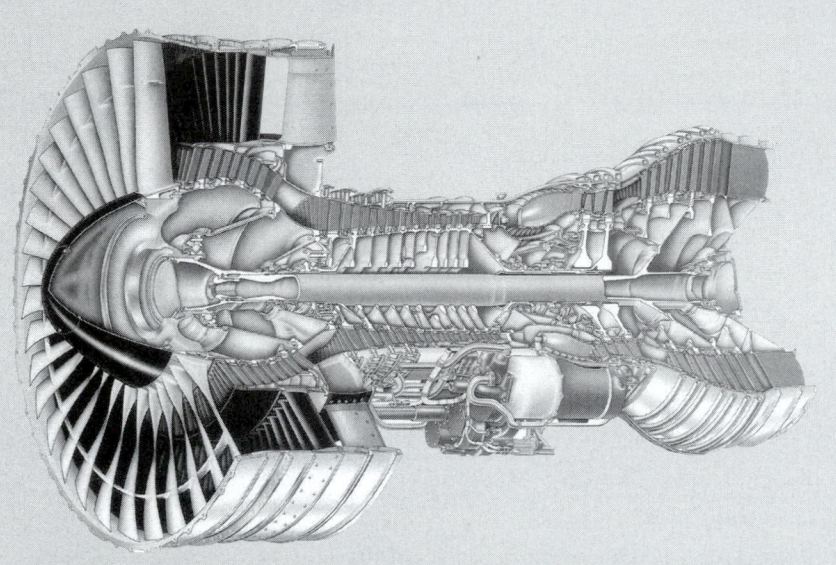

Figure 1.5 | Pratt and Whitney 4000 turbofan engine. Third generation turbofan for widebody transports. Produces up to 84,000 lb (329.2 kN) of thrust. Powers some versions of the Boeing 777 (see Fig. 1.4).

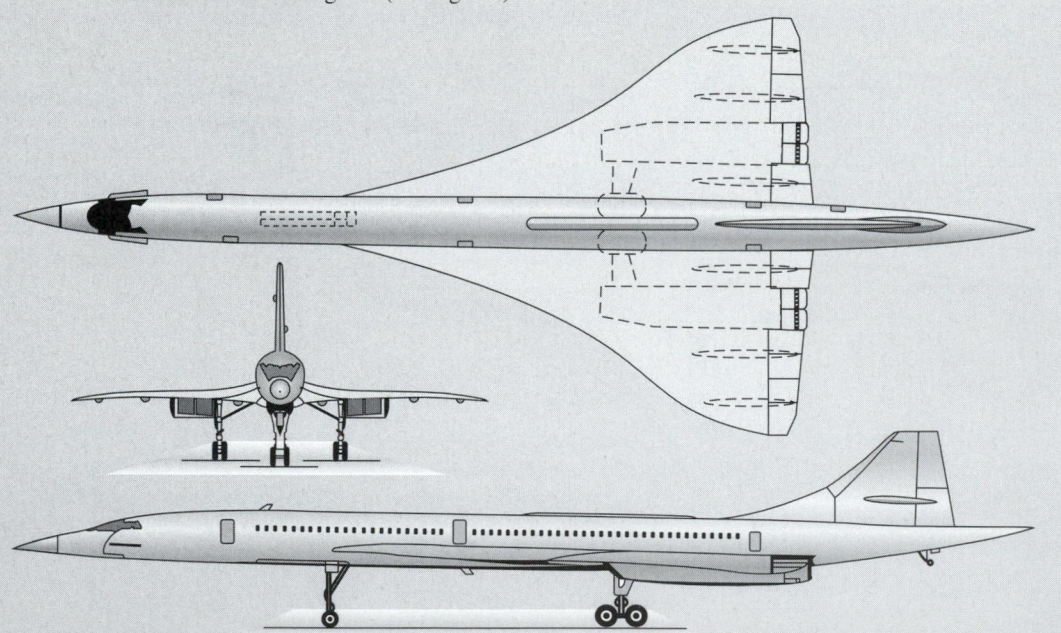

Figure 1.6 | The anglo-French Aerospatiale/BAC Concorde supersonic airliner.

some of the mathematical and physical details that are buried in the chapter. The road map for the entire book is given in Fig. 1.7. To help keep our equilibrium, we will periodically refer to Fig. 1.7 as we progress through the book. For now, let us just survey Fig. 1.7 for some general guidance. After an introduction to the subject and a brief review of thermodynamics (box 1 in Fig. 1.7), we derive the governing fundamental conservation equations (box 2). We first obtain these equations in integral form (box 3), which some people will argue is philosophically a more fundamental form of the equations than the differential form obtained later in box 7. Using just the integral form of the conservation equations, we will study one-dimensional flow (box 4), including normal shock waves, oblique shock, and expansion waves (box 5), and the quasi-one-dimensional flow through nozzles and diffusers, with applications to wind tunnels and rocket engines (box 6). All of these subjects can be studied by application of the integral form of the conservation equations, which usually reduce to algebraic equations for the application listed in boxes 4–6. Boxes 1–6 frequently constitute a basic "first course" in

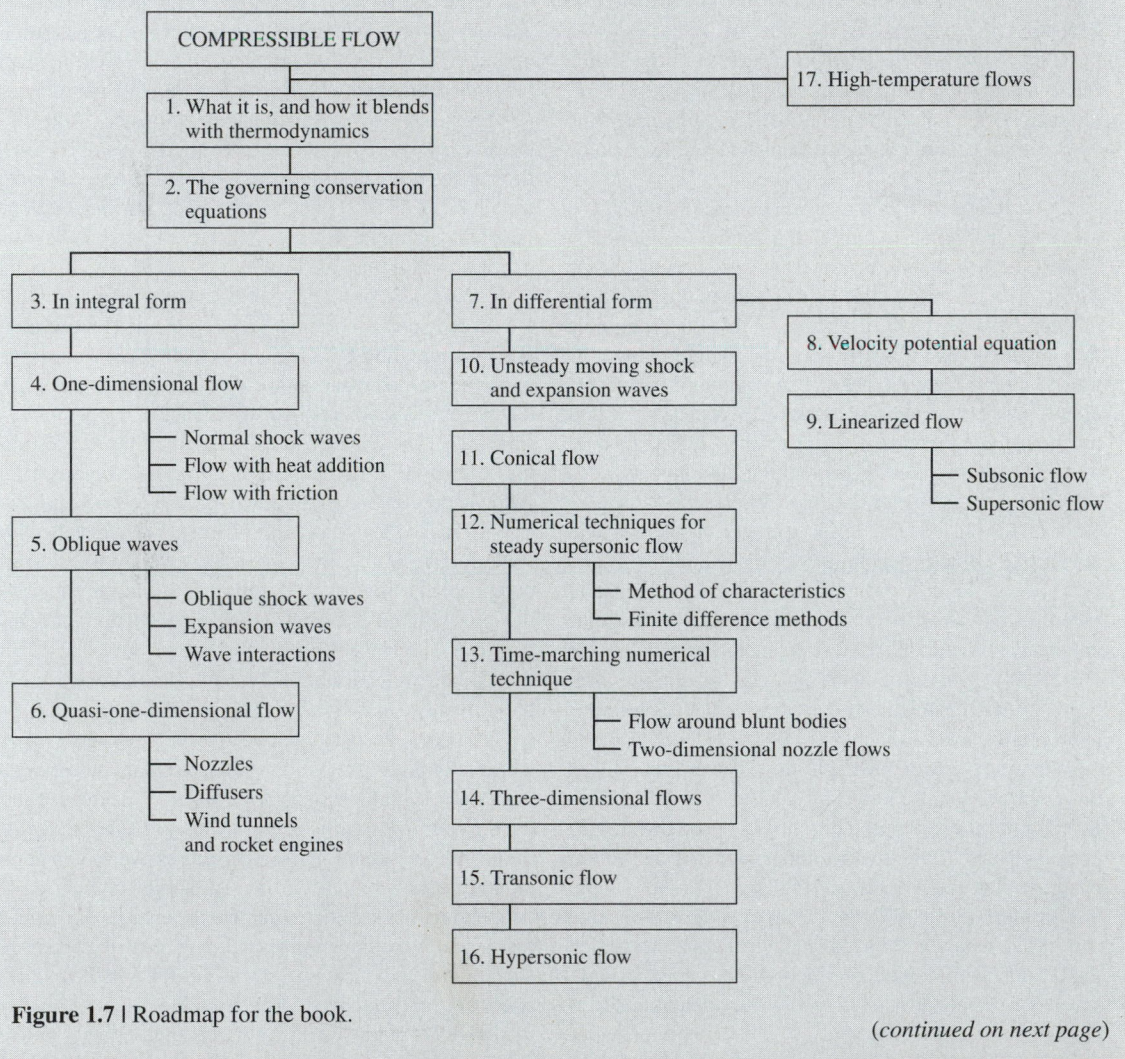

Figure 1.7 | Roadmap for the book.

(*continued on next page*)

(continued from page 7)

compressible flow, and the mathematics usually does not go beyond that of algebra. However, to deal with unsteady and/or multidimensional flows, we have to step to box 7 and obtain the governing conservation equations in differential form. They take the form of a system of coupled, highly nonlinear, partial differential equations. In some special cases for subsonic and supersonic flows, they can be linearized (boxes 8 and 9), leading to so-called "linearized flow." However, in most cases, we must cope with the nonlinear equations. The way we do this, and the fascinating physical phenomena we discover along the way, is told in boxes 10–16 dealing with unsteady flow, flow over cones, flows over supersonic blunt-nosed bodies, three-dimensional flows over bodies at an angle of attack to a uniform free stream, and the very special characteristics of transonic and hypersonic flows.

Our treatment of the material covered in boxes 4–6 and 8–16 in Fig. 1.7 assumes the gas to be calorically perfect, i.e., to have constant values of specific heats. This is valid as long as the temperature in the flow does not exceed about 1000 K. The vast bulk of compressible flow applications satisfy this criteria, including the flow around the Concorde when it is cruising at Mach 2. However, the flow over higher speed vehicles, as well as the flow through parts of a jet engine, will encounter temperatures high enough that the assumption of a calorically perfect gas is not valid. Witness the flow over parts of the Space Shuttle as it enters the atmosphere at Mach 25, where flow temperatures can be as high as 8000 K, and the flow through rocket engines where temperatures on the order of 4000 K or higher occur in the combustion chamber. At these temperatures, the flow is chemically reacting, and the analysis of compressible flow applications at these conditions must include the appropriate physical-chemical effects. Hence, to round out our study of compressible flow, toward the end of this book we identify, discuss, and analyze these high-temperature flow effects. This subject is somewhat self-contained and is relatively independent of the earlier chapters; for this reason in Fig. 1.7 we show high-temperature flows in box 17 in an adjunct position somewhat separate from the main structure. However, this is not to minimize its importance. In many high-speed flow applications today, high-temperature effects are very important. Any study of *modern* compressible flow must include box 17.

We note that all of the material in this book, boxes 1 through 17 in Fig. 1.7, assumes *inviscid* flow, i.e., flow with no friction, thermal conduction, or mass diffusion, except for the special case of one-dimensional flow with friction (box 4 in Fig. 1.7). Flows where the dissipative transport processes of friction, thermal conduction, and mass diffusion are important are called *viscous* flows. Viscous flow is a subject all by itself and is beyond the scope of this book. The assumption of inviscid flow may at first sound ideal and restrictive—flows in the real world are not so ideal. However, the important physics that dictates compressible flow, such as the propagation of pressure waves through the flow, is essentially an inviscid phenomena. Moreover, for the vast majority of compressible flow applications, the influence of the dissipative transport phenomena is limited to small regions, such as the boundary layer along a solid surface. Hence, the inviscid flows treated in this book are indeed very practical and apply to a vast majority of everyday applications of compressible flow.

All of this constitutes a preview for the material that is covered in this book—a broad, general view to give you a better, almost philosophical feeling for what compressible flow is about. As we continue, each chapter has its own preview box in order to enhance a broader understanding of the material in the chapter and to relate it to the general view. In this fashion, the detailed material in each chapter will more readily come to life for you.

In regard to the present chapter, we start out with some historical high-water marks in the application of compressible flow, and then discuss some introductory thoughts that are essential for our understanding of compressible flow in the subsequent chapters. For example, in this chapter we give a brief review of thermodynamics—but *only* those aspects of thermodynamics that relate directly to our subsequent discussions. Compressible flows are usually high-energy flows. Imagine that you are driving down the highway at 65 mph, and you stick your hand out the window; your hand will literally feel the energy of the 65-mph airstream, and it feels impressive. But 65 mph is really a low velocity in the scheme of compressible flow applications. Rather, imagine the energy you would feel if you were traveling at 650 mph, near the speed of sound, and you stick your hand out the window (definitely not recommended). You would feel a lot of energy in the flow. High-speed flows are high-energy flows. Thermodynamics is the study of energy changes and their effects on the properties of a system. Hence, compressible flow embraces thermodynamics. I know of no compressible flow problem that can be understood and solved without involving some aspect of thermodynamics. So that is why we start out with a review of thermodynamics.

The remainder of this chapter simply deals with other introductory thoughts necessary to provide you with smooth sailing through the rest of the book. I wish you a pleasant voyage.

1.1 | HISTORICAL HIGH-WATER MARKS

The year is 1893. In Chicago, the World Columbian Exposition has been opened by President Grover Cleveland. During the year, more than 27 million people will visit the 666-acre expanse of gleaming white buildings, specially constructed from a composite of plaster of paris and jute fiber to simulate white marble. Located adjacent to the newly endowed University of Chicago, the Exposition commemorates the discovery of America by Christopher Columbus 400 years earlier. Exhibitions related to engineering, architecture, and domestic and liberal arts, as well as collections of all modes of transportation, are scattered over 150 buildings. In the largest, the Manufacturer's and Liberal Arts Building, engineering exhibits from all over the world herald the rapid advance of technology that will soon reach explosive proportions in the twentieth century. Almost lost in this massive 31-acre building, under a roof of iron and glass, is a small machine of great importance. A single-stage steam turbine is being displayed by the Swedish engineer, Carl G. P. de Laval. The machine is less than 6 ft long; designed for marine use, it has two independent turbine wheels, one for forward motion and the other for the reverse direction. But what is novel about this device is that the turbine blades are driven by a stream of hot, high-pressure steam from a series of unique convergent-divergent nozzles. As sketched in Fig. 1.8, these nozzles, with their convergent-divergent shape representing a complete departure from previous engineering applications, feed a high-speed flow of steam to the blades of the turbine wheel. The deflection and consequent change in momentum of the steam as it flows past the turbine blades exerts an impulse that rotates the wheel to speeds previously unattainable—over 30,000 r/min. Little does de Laval realize that his convergent-divergent steam nozzle will open the door to the supersonic wind tunnels and rocket engines of the midtwentieth century.

The year is now 1947. The morning of October 14 dawns bright and beautiful over the Muroc Dry Lake, a large expanse of flat, hard lake bed in the Mojave Desert in California. Beginning at 6:00 A.M., teams of engineers and technicians at the Muroc Army Air Field ready a small rocket-powered airplane for flight. Painted orange and resembling a 50-caliber machine gun bullet mated to a pair of straight, stubby wings, the Bell XS-1 research vehicle is carefully installed in the bomb bay of a four-engine B-29 bomber of World War II vintage. At 10:00 A.M. the B-29 with its soon-to-be-historic cargo takes off and climbs to an altitude of 20,000 ft. In the cockpit of the XS-1 is Captain Charles (Chuck) Yeager, a veteran P-51 pilot from the European theater during the war. This morning Yeager is in pain from two broken ribs incurred during a horseback riding accident the previous weekend. However, not wishing to disrupt the events of the day, Yeager informs no one at Muroc about his

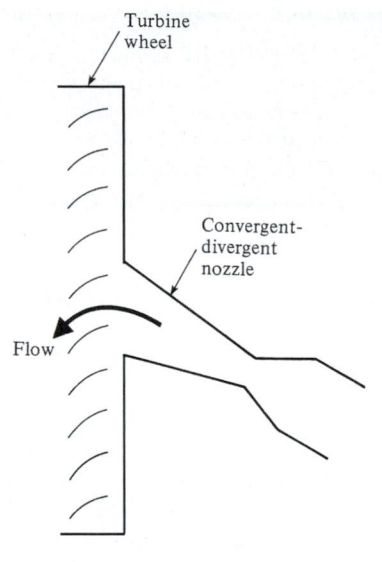

Figure 1.8 | Schematic of de Laval's turbine incorporating a convergent-divergent nozzle.

condition. At 10:26 A.M., at a speed of 250 mi/h (112 m/s), the brightly painted XS-1 drops free from the bomb bay of the B-29. Yeager fires his Reaction Motors XLR-11 rocket engine and, powered by 6000 lb of thrust, the sleek airplane accelerates and climbs rapidly. Trailing an exhaust jet of shock diamonds from the four convergent-divergent rocket nozzles of the engine, the XS-1 is soon flying faster than Mach 0.85, that speed beyond which there is no wind tunnel data on the problems of transonic flight in 1947. Entering this unknown regime, Yeager momentarily shuts down two of the four rocket chambers, and carefully tests the controls of the XS-1 as the Mach meter in the cockpit registers 0.95 and still increasing. Small shock waves are now dancing back and forth over the top surface of the wings. At an altitude of 40,000 ft, the XS-1 finally starts to level off, and Yeager fires one of the two shutdown rocket chambers. The Mach meter moves smoothly through 0.98, 0.99, to 1.02. Here, the meter hesitates, then jumps to 1.06. A stronger bow shock wave is now formed in the air ahead of the needlelike nose of the XS-1 as Yeager reaches a velocity of 700 mi/h, Mach 1.06, at 43,000 ft. The flight is smooth; there is no violent buffeting of the airplane and no loss of control as was feared by some engineers. At this moment, Chuck Yeager becomes the first pilot to successfully fly faster than the speed of sound, and the small but beautiful Bell XS-1, shown in Fig. 1.9, becomes the first successful supersonic airplane in the history of flight. (For more details, see Refs. 1 and 2 listed at the back of this book.)

Today, both de Laval's 10-hp turbine from the World Columbian Exhibition and the orange Bell XS-1 are part of the collection of the Smithsonian Institution of Washington, D.C., the former on display in the History of Technology Building and the latter hanging with distinction from the roof of the National Air and Space

Figure 1.9 | The Bell XS-1, first manned supersonic aircraft. (*Courtesy of the National Air and Space Museum.*)

Museum. What these two machines have in common is that, separated by more than half a century, they represent high-water marks in the engineering application of the principles of compressible flow—where the density of the flow is not constant. In both cases they represent marked departures from previous fluid dynamic practice and experience.

The engineering fluid dynamic problems of the eighteenth, nineteenth, and early twentieth centuries almost always involved either the flow of liquids or the low-speed flow of gases; for both cases the assumption of constant density is quite valid. Hence, the familiar Bernoulli's equation

$$p + \tfrac{1}{2}\rho V^2 = \text{const} \tag{1.1}$$

was invariably employed with success. However, with the advent of high-speed flows, exemplified by de Laval's convergent-divergent nozzle design and the supersonic flight of the Bell XS-1, the density can no longer be assumed constant throughout the flowfield. Indeed, for such flows the density can sometimes vary by orders of magnitude. Consequently, Eq. (1.1) no longer holds. In this light, such events were indeed a marked departure from previous experience in fluid dynamics.

This book deals exclusively with that "marked departure," i.e., it deals with *compressible flows,* in which the density is *not* constant. In modern engineering applications, such flows are the rule rather than the exception. A few important examples are the internal flows through rocket and gas turbine engines, high-speed subsonic, transonic, supersonic, and hypersonic wind tunnels, the external flow over modern airplanes designed to cruise faster than 0.3 of the speed of sound, and the flow inside the common internal combustion reciprocating engine. The purpose of this book is to develop the fundamental concepts of compressible flow, and to illustrate their use.

1.2 | DEFINITION OF COMPRESSIBLE FLOW

Compressible flow is routinely defined as *variable density flow;* this is in contrast to incompressible flow, where the density is assumed to be constant throughout. Obviously, in real life every flow of every fluid is compressible to some greater or lesser extent; hence, a truly constant density (incompressible) flow is a myth. However, as previously mentioned, for almost all liquid flows as well as for the flows of some gases under certain conditions, the density changes are so small that the assumption of constant density can be made with reasonable accuracy. In such cases, Bernoulli's equation, Eq. (1.1), can be applied with confidence. However, for the subject of this book—compressible flow—Eq. (1.1) does not hold, and for our purposes here, the reader should dismiss it from his or her thinking.

The simple definition of compressible flow as one in which the density is variable requires more elaboration. Consider a small element of fluid of volume v. The pressure exerted on the sides of the element by the neighboring fluid is p. Assume the pressure is now increased by an infinitesimal amount dp. The volume of the element will be correspondingly compressed by the amount dv. Since the volume is reduced, dv is a negative quantity. The compressibility of the fluid, τ, is defined as

$$\tau = -\frac{1}{v}\frac{dv}{dp} \tag{1.2}$$

Physically, the compressibility is the fractional change in volume of the fluid element per unit change in pressure. However, Eq. (1.2) is not sufficiently precise. We know from experience that when a gas is compressed (say in a bicycle pump), its temperature tends to increase, depending on the amount of heat transferred into or out of the gas through the boundaries of the system. Therefore, if the temperature of the fluid element is held constant (due to some heat transfer mechanism), then the *isothermal compressibility* is defined as

$$\tau_T = -\frac{1}{v}\left(\frac{\partial v}{\partial p}\right)_T \tag{1.3}$$

On the other hand, if no heat is added to or taken away from the fluid element (if the compression is adiabatic), and if no other dissipative transport mechanisms such as viscosity and diffusion are important (if the compression is reversible), then the compression of the fluid element takes place isentropically, and the *isentropic compressibility* is defined as

$$\tau_s = -\frac{1}{v}\left(\frac{\partial v}{\partial p}\right)_s \tag{1.4}$$

where the subscript s denotes that the partial derivative is taken at constant entropy. Compressibility is a property of the fluid. Liquids have very low values of compressibility (τ_T for water is 5×10^{-10} m^2/N at 1 atm) whereas gases have high compressibilities (τ_T for air is 10^{-5} m^2/N at 1 atm, more than four orders of magnitude larger than water). If the fluid element is assumed to have unit mass, v is the specific volume (volume per unit mass), and the density is $\rho = 1/v$. In terms of density, Eq. (1.2) becomes

$$\tau = \frac{1}{\rho}\frac{d\rho}{dp} \tag{1.5}$$

Therefore, whenever the fluid experiences a change in pressure, dp, the corresponding change in density will be $d\rho$, where from Eq. (1.5)

$$d\rho = \rho\tau\,dp \tag{1.6}$$

To this point, we have considered just the fluid itself, with compressibility being a property of the fluid. Now assume that the fluid is in motion. Such flows are initiated and maintained by forces on the fluid, usually created by, or at least accompanied by, changes in the pressure. In particular, we shall see that high-speed flows generally involve large pressure gradients. For a given change in pressure, dp, due to the flow, Eq. (1.6) demonstrates that the resulting change in density will be small for liquids (which have low values of τ), and large for gases (which have high values of τ). Therefore, for the flow of liquids, relatively large pressure gradients can create high velocities without much change in density. Hence, such flows are usually assumed to be incompressible, where ρ is constant. On the other hand, for the flow of gases with their attendant large values of τ, moderate to strong pressure gradients lead to substantial changes in the density via Eq. (1.6). At the same time, such pressure gradients create large velocity changes in the gas. Such flows are defined as *compressible flows,* where ρ is a variable.

We shall prove later that for gas velocities less than about 0.3 of the speed of sound, the associated pressure changes are small, and even though τ is large for gases, dp in Eq. (1.6) may still be small enough to dictate a small $d\rho$. For this reason, the low-speed flow of gases can be assumed to be incompressible. For example, the flight velocities of most airplanes from the time of the Wright brothers in 1903 to the beginning of World War II in 1939 were generally less than 250 mi/h (112 m/s), which is less than 0.3 of the speed of sound. As a result, the bulk of early aerodynamic literature treats incompressible flow. On the other hand, flow velocities higher than 0.3 of the speed of sound are associated with relatively large pressure changes, accompanied by correspondingly large changes in density. Hence, compressibility effects on airplane aerodynamics have been important since the advent of high-performance aircraft in the 1940s. Indeed, for the modern high-speed subsonic and supersonic aircraft of today, the older incompressible theories are wholly inadequate, and compressible flow analyses must be used.

In summary, in this book a compressible flow will be considered as one where the change in pressure, dp, over a characteristic length of the flow, multiplied by the compressibility via Eq. (1.6), results in a fractional change in density, $d\rho/\rho$, which is too large to be ignored. For most practical problems, if the density changes by 5 percent or more, the flow is considered to be compressible.

EXAMPLE 1.1

Consider the low-speed flow of air over an airplane wing at standard sea level conditions; the free-stream velocity far ahead of the wing is 100 mi/h. The flow accelerates over the wing, reaching a maximum velocity of 150 mi/h at some point on the wing. What is the percentage pressure change between this point and the free stream?

■ Solution

Since the airspeeds are relatively low, let us (for the first and *only* time in this book) assume incompressible flow, and use Bernoulli's equation for this problem. (See Ref. 1 for an elementary discussion of Bernoulli's equation, as well as Ref. 104 for a more detailed presentation of the role of this equation in the solution of incompressible flow. Here, we assume that the reader is familiar with Bernoulli's equation—its use and its limitations. If not, examine carefully the appropriate discussions in Refs. 1 and 104.) Let points 1 and 2 denote the free stream and wing points, respectively. Then, from Bernoulli's equation,

$$p_1 + \tfrac{1}{2}\rho V_1{}^2 = p_2 + \tfrac{1}{2}\rho V_2{}^2$$

or

$$p_1 - p_2 = \tfrac{1}{2}\rho\left(V_2{}^2 - V_1{}^2\right)$$

At standard sea level, $\rho = 0.002377$ slug/ft^3. Also, using the handy conversion that 60 mi/h = 88 ft/s, we have $V_1 = 100$ mi/h $= 147$ ft/s and $V_2 = 150$ mi/h $= 220$ ft/s. (Note that, as always in this book, we will use *consistent* units; for example, we will use either the English Engineering System, as in this problem, or the International System. See the footnote in Sec. 1.4 of this book, as well as Chap. 2 of Ref. 1. By using consistent units, *none* of our basic equations will ever contain conversion factors, such as q_c and J, as is found in some references.) With this information, we have

$$p_1 - p_2 = \tfrac{1}{2}\rho\left(V_2{}^2 - V_1{}^2\right)$$
$$= \tfrac{1}{2}(0.002377)[(220)^2 - (147)^2] = 31.8 \text{ lb/ft}^2$$

The fractional change in pressure referenced to the free-stream pressure, which at standard sea level is $p_1 = 2116$ lb/ft^2, is obtained as

$$\frac{p_1 - p_2}{p_1} = \frac{31.8}{2116} = 0.015$$

Therefore, the *percentage* change in pressure is 1.5 percent. In expanding over the wing surface, the pressure changes by *only* 1.5 percent. This is a case where, in Eq. (1.6), dp is small, and hence $d\rho$ is small. The purpose of this example is to demonstrate that, in low-speed flow problems, the *percentage* change in pressure is always small, and this, through Eq. (1.6), justifies the *assumption* of incompressible flow ($d\rho = 0$) for such flows. However, at high flow velocities, the change in pressure is not small, and the density must be treated as variable. This is the regime of compressible flow—the subject of this book. *Note:* Bernoulli's equation used in this example is good *only* for incompressible flow, therefore it will not appear again in any of our subsequent discussions. Experience has shown that, because it is one of the first equations usually encountered by students in the study of fluid dynamics, there is a tendency to use Bernoulli's equation for situations where it is not valid. Compressible flow is one such situation. Therefore, for our subsequent discussions in this book, remember *never* to invoke Bernoulli's equation.

1.3 | FLOW REGIMES

The age of successful manned flight began on December 17, 1903, when Orville and Wilbur Wright took to the air in their historic Flyer I, and soared over the windswept sand dunes of Kill Devil Hills in North Carolina. This age has continued to the present with modern, high-performance subsonic and supersonic airplanes, as well as the hypersonic atmospheric entry of space vehicles. In the twentieth century, manned flight has been a major impetus for the advancement of fluid dynamics in general, and compressible flow in particular. Hence, although the fundamentals of compressible flow are applied to a whole spectrum of modern engineering problems, their application to aerodynamics and propulsion geared to airplanes and missiles is frequently encountered.

In this vein, it is useful to illustrate different regimes of compressible flow by considering an aerodynamic body in a flowing gas, as sketched in Fig. 1.10. First, consider some definitions. Far upstream of the body, the flow is uniform with a *free-stream velocity* of V_∞. A *streamline* is a curve in the flowfield that is tangent to the local velocity vector **V** at every point along the curve. Figure 1.10 illustrates only a few of the infinite number of streamlines around a body. Consider an arbitrary point in the flowfield, where p, T, ρ, and **V** are the local pressure, temperature, density, and vector velocity at that point. All of these quantities are point properties and vary from one point to another in the flow. In Chap. 3, we will show the speed of sound a to be a thermodynamic property of the gas; hence a also varies from point to point in the flow. If a_∞ is the speed of sound in the uniform free stream, then the ratio V_∞/a_∞ defines the free-stream Mach number M_∞. Similarly, the local Mach number M is defined as $M = V/a$, and varies from point to point in the flowfield. Further physical significance of Mach number will be discussed in Chap. 3. In the present section, M simply will be used to define four different flow regimes in fluid dynamics, as discussed next.

1.3.1 Subsonic Flow

Consider the flow over an airfoil section as sketched in Fig. 1.10*a*. Here, the local Mach number is everywhere less than unity. Such a flow, where $M < 1$ at every point, and hence the flow velocity is everywhere less than the speed of sound, is defined as *subsonic flow*. This flow is characterized by smooth streamlines and continuously varying properties. Note that the initially straight and parallel streamlines in the free stream begin to deflect far upstream of the body, i.e., the flow is forewarned of the presence of the body. This is an important property of subsonic flow and will be discussed further in Chap. 4. Also, as the flow passes over the airfoil, the local velocity and Mach number on the top surface increase above their free-stream values. However, if M_∞ is sufficiently less than 1, the local Mach number everywhere will remain subsonic. For airfoils in common use, if $M_\infty \leq 0.8$, the flowfield is generally completely subsonic. Therefore, to the airplane aerodynamicist, the subsonic regime is loosely identified with a free stream where $M_\infty \leq 0.8$.

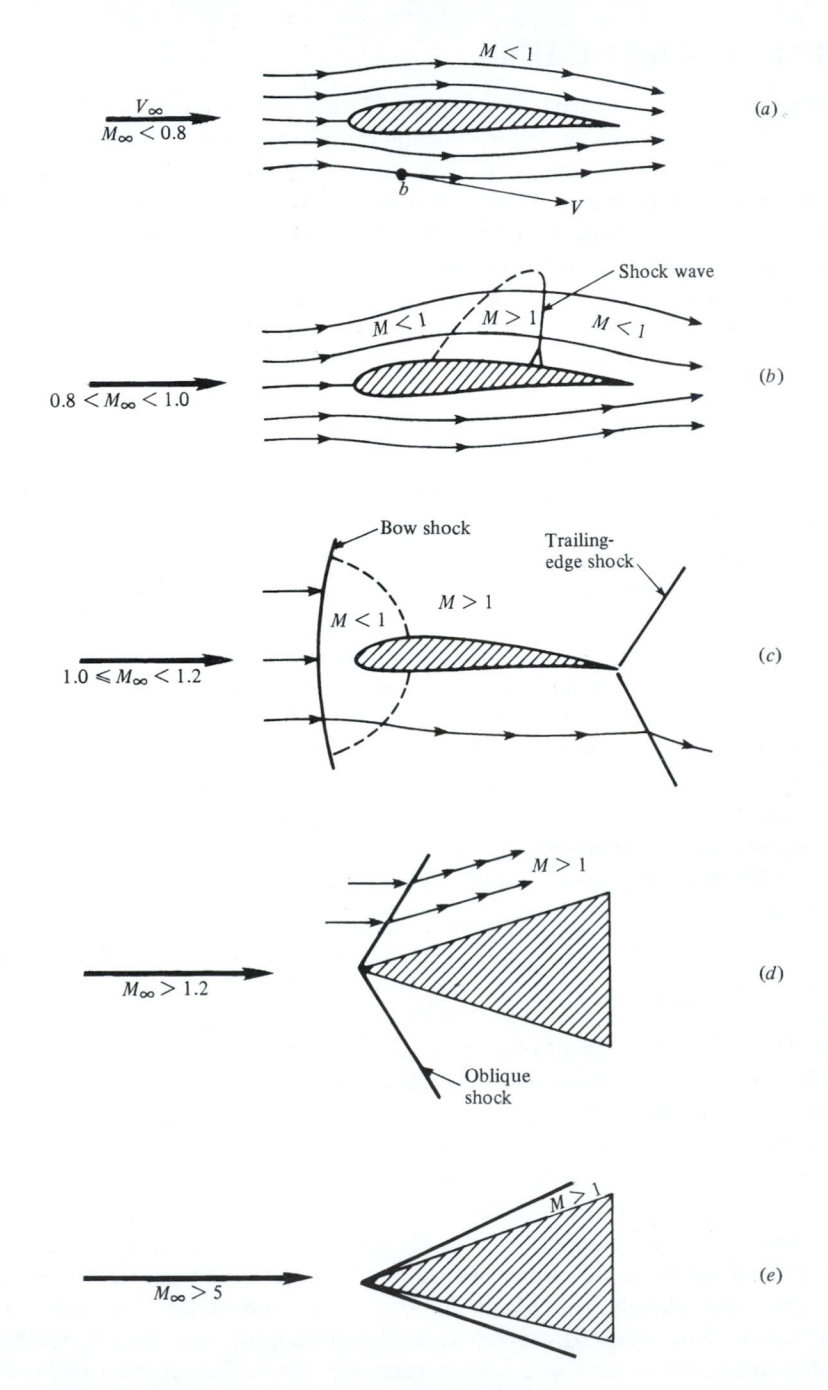

Figure 1.10 | Illustration of different regimes of flow.

1.3.2   Transonic Flow

If M_∞ remains subsonic, but is sufficiently near 1, the flow expansion over the top surface of the airfoil may result in locally supersonic regions, as sketched in Fig. 1.10b. Such a mixed region flow is defined as *transonic flow*. In Fig. 1.10b, M_∞ is less than 1 but high enough to produce a pocket of locally supersonic flow. In most cases, as sketched in Fig. 1.10b, this pocket terminates with a shock wave across which there is a discontinuous and sometimes rather severe change in flow properties. Shock waves will be discussed in Chap. 4. If M_∞ is increased to slightly above unity, this shock pattern will move to the trailing edge of the airfoil, and a second shock wave appears upstream of the leading edge. This second shock wave is called the *bow shock,* and is sketched in Fig. 1.10c. (Referring to Sec. 1.1, this is the type of flow pattern existing around the wing of the Bell XS-1 at the moment it was "breaking the sound barrier" at $M_\infty = 1.06$.) In front of the bow shock, the streamlines are straight and parallel, with a uniform supersonic free-stream Mach number. In passing through that part of the bow shock that is nearly normal to the free stream, the flow becomes subsonic. However, an extensive supersonic region again forms as the flow expands over the airfoil surface, and again terminates with a trailing-edge shock. Both flow patterns sketched in Figs. 1.10b and c are characterized by mixed regions of locally subsonic and supersonic flow. Such mixed flows are defined as *transonic flows,* and $0.8 \leq M_\infty \leq 1.2$ is loosely defined as the *transonic regime*. Transonic flow is discussed at length in Chap. 14.

1.3.3   Supersonic Flow

A flowfield where $M > 1$ everywhere is defined as *supersonic*. Consider the supersonic flow over the wedge-shaped body in Fig. 1.10d. A straight, oblique shock wave is attached to the sharp nose of the wedge. Across this shock wave, the streamline direction changes discontinuously. Ahead of the shock, the streamlines are straight, parallel, and horizontal; behind the shock they remain straight and parallel but in the direction of the wedge surface. Unlike the subsonic flow in Fig. 1.10a, the supersonic uniform free stream is not forewarned of the presence of the body until the shock wave is encountered. The flow is supersonic both upstream and (usually, but not always) downstream of the oblique shock wave. There are dramatic physical and mathematical differences between subsonic and supersonic flows, as will be discussed in subsequent chapters.

1.3.4   Hypersonic Flow

The temperature, pressure, and density of the flow increase almost explosively across the shock wave shown in Fig. 1.10d. As M_∞ is increased to higher supersonic speeds, these increases become more severe. At the same time, the oblique shock wave moves closer to the surface, as sketched in Fig. 1.10e. For values of $M_\infty > 5$, the shock wave is very close to the surface, and the flowfield between the shock and the body (the shock layer) becomes very hot—indeed, hot enough to dissociate or even ionize the gas. Aspects of such high-temperature chemically reacting flows are discussed in Chaps. 16 and 17. These effects—thin shock layers and hot, chemically reacting gases—add complexity to the analysis of such flows. For this reason, the flow regime for $M_\infty > 5$ is given a special label—*hypersonic flow*. The choice of $M_\infty = 5$ as a dividing point between supersonic and hypersonic flow is a rule of thumb. In reality, the special characteristics associated with hypersonic flow appear gradually as M_∞ is increased, and the Mach number at which they become important depends greatly on the shape of the body and the free-stream density. Hypersonic flow is the subject of Chap. 15.

It is interesting to note that incompressible flow is a special case of subsonic flow; namely, it is the limiting case where $M_\infty \to 0$. Since $M_\infty = V_\infty/a_\infty$, we have two possibilities:

$$M_\infty \to 0 \quad \text{because } V_\infty \to 0$$

$$M_\infty \to 0 \quad \text{because } a_\infty \to \infty$$

The former corresponds to no flow and is trivial. The latter states that the speed of sound in a truly incompressible flow would have to be infinitely large. This is compatible with Eq. (1.6), which states that, for a truly incompressible flow where $d\rho = 0$, τ must be zero, i.e., zero compressibility. We shall see in Chap. 3 that the speed of sound is inversely proportional to the square root of τ; hence $\tau = 0$ implies an infinite speed of sound.

There are other ways of classifying flowfields. For example, flows where the effects of viscosity, thermal conduction, and mass diffusion are important are called *viscous flows*. Such phenomena are dissipative effects that change the entropy of the flow, and are important in regions of large gradients of velocity, temperature, and chemical composition. Examples are boundary layer flows, flow in long pipes, and the thin shock layer on high-altitude hypersonic vehicles. Friction drag, flowfield separation, and heat transfer all involve viscous effects. Therefore, viscous flows are of major importance in the study of fluid dynamics. In contrast, flows in which viscosity, thermal conduction, and diffusion are ignored are called *inviscid flows*. At first glance, the assumption of inviscid flows may appear highly restrictive; however, there are a number of important applications that do not involve flows with large gradients, and that readily can be assumed to be inviscid. Examples are the large regions of flow over wings and bodies outside the thin boundary layer on the surface, flow through wind tunnels and rocket engine nozzles, and the flow over compressor and turbine blades for jet engines. Surface pressure distributions, as well as aerodynamic lift and moments on some bodies, can be accurately obtained by means of the assumption of inviscid flow. In this book, viscous effects will not be treated except in regard to their role in forming the internal structure and thickness of shock waves. That is, this book deals with compressible, *inviscid flows*.

Finally, we will always consider the gas to be a *continuum*. Clearly, a gas is composed of a large number of discrete atoms and/or molecules, all moving in a more or less random fashion, and frequently colliding with each other. This microscopic picture of a gas is essential to the understanding of the thermodynamic and chemical properties of a high-temperature gas, as described in Chaps. 16 and 17. However, in deriving the fundamental equations and concepts for fluid flows, we take advantage of the fact that a gas usually contains a large number of molecules (over 2×10^{19} molecules/cm^3 for air at normal room conditions), and hence on a macroscopic basis, the fluid behaves as if it were a continuous material. This continuum assumption is violated only when the mean distance an atom or molecule moves between collisions (the mean free path) is so large that it is the same order of magnitude as the characteristic dimension of the flow. This implies *low density, or rarefied flow*. The extreme situation, where the mean free path is much larger than the characteristic length and where virtually no molecular collisions take place in the flow, is called *free-molecular flow*. In this case, the flow is essentially a stream of remotely spaced particles. Low-density and free-molecular flows are rather special cases in the whole spectrum of fluid dynamics, occurring in flight only at very high altitudes (above 200,000 ft), and in special laboratory devices such as electron beams and low-pressure gas lasers. Such rarefied gas effects are beyond the scope of this book.

1.4 | A BRIEF REVIEW OF THERMODYNAMICS

The kinetic energy per unit mass, $V^2/2$, of a high-speed flow is large. As the flow moves over solid bodies or through ducts such as nozzles and diffusers, the local velocity, hence local kinetic energy, changes. In contrast to low-speed or incompressible flow, these energy changes are substantial enough to strongly interact with other properties of the flow. Because in most cases high-speed flow and compressible flow are synonymous, energy concepts play a major role in the study and understanding of compressible flow. In turn, the science of energy (and entropy) is *thermodynamics;* consequently, thermodynamics is an essential ingredient in the study of compressible flow.

This section gives a brief outline of thermodynamic concepts and relations necessary to our further discussions. This is in no way an exposition on thermodynamics; rather it is a review of only those fundamental ideas and equations which will be of direct use in subsequent chapters.

1.4.1   Perfect Gas

A gas is a collection of particles (molecules, atoms, ions, electrons, etc.) that are in more or less random motion. Due to the electronic structure of these particles, a force field pervades the space around them. The force field due to one particle reaches out and interacts with neighboring particles, and vice versa. Hence, these fields are called *intermolecular forces*. The intermolecular force varies with distance between particles; for most atoms and molecules it takes the form of a weak attractive force at large distance, changing quickly to a strong repelling force at close distance. In general, these intermolecular forces influence the motion of the particles; hence they also influence the thermodynamic properties of the gas, which are nothing more than the macroscopic ramification of the particle motion.

At the temperatures and pressures characteristic of many compressible flow applications, the gas particles are, on the average, widely separated. The average distance between particles is usually more than 10 molecular diameters, which corresponds to a very weak attractive force. As a result, for a large number of engineering applications, the effect of intermolecular forces on the gas properties is negligible. By definition, *a perfect gas is one in which intermolecular forces are neglected.* By ignoring intermolecular forces, the equation of state for a perfect gas can be derived from the theoretical concepts of modern statistical mechanics or kinetic theory. However, historically it was first synthesized from laboratory measurements by Robert Boyle in the seventeenth century, Jacques Charles in the eighteenth century, and Joseph Gay-Lussac and John Dalton around 1800. The empirical result which unfolded from these observations was

$$p\mathscr{V} = MRT \tag{1.7}$$

where p is pressure (N/m² or lb/ft²), $\mathscr{V}$ is the volume of the system (m³ or ft³), M is the mass of the system (kg or slug), R is the specific gas constant [J/(kg · K) or (ft · lb)/(slug · °R)], which is a different value for different gases, and T is the temperature (K or °R).[†] This equation of state can be written in many forms, most of which are summarized here for the reader's convenience. For example, if Eq. (1.7) is divided by the mass of the system,

$$\boxed{pv = RT} \tag{1.8}$$

where v is the specific volume (m³/kg or ft³/slug). Since the density $\rho = 1/v$, Eq. (1.8) becomes

$$\boxed{p = \rho RT} \tag{1.9}$$

Along another track that is particularly useful in chemically reacting systems, the early fundamental empirical observations also led to a form for the equation of state:

$$p\mathscr{V} = \mathscr{N}\mathscr{R}T \tag{1.10}$$

where $\mathscr{N}$ is the number of moles of gas in the system, and $\mathscr{R}$ is the universal gas constant, which is the same for all gases. Recall that a mole of a substance is that amount which contains a mass numerically equal to the molecular weight of the gas, and which is identified with the particular system of units being used, i.e., a kilogram-mole (kg · mol) or a slug-mole (slug · mol). For example, for pure diatomic oxygen (O_2), 1 kg · mol has a mass of 32 kg, whereas 1 slug · mol has a mass of 32 slug. Because the masses of different molecules are in the same ratio as their molecular weights, 1 mol of different gases always contains the same number of molecules, i.e., 1 kg · mol always contains 6.02×10^{26} molecules, independent of the species of the gas. Continuing with Eq. (1.10), dividing by the number of moles of the system yields

---

[†] Two sets of consistent units will be used throughout this book, the International System (SI) and the English Engineering System. In the SI system, the units of force, mass, length, time, and temperature are the newton (N), kilogram (kg), meter (m), second (s), and Kelvin (K), respectively; in the English Engineering System they are the pound (lb), slug, foot (ft), second (s), and Rankine (°R), respectively. The respective units of energy are joules (J) and foot-pounds (ft · lb).

$$p\mathcal{V}' = \mathcal{R}T \tag{1.11}$$

where $\mathcal{V}'$ is the molar volume [m³/(kg · mol) or ft³/(slug · mol)]. Of more use in gasdynamic problems is a form obtained by dividing Eq. (1.10) by the mass of the system:

$$\boxed{pv = \eta \mathcal{R}T} \tag{1.12}$$

where v is the specific volume as before, and η is the mole-mass ratio [(kg · mol)/kg and (slug · mol)/slug]. (Note that the kilograms and slugs in these units do not cancel, because the kilogram-mole and slug-mole are entities in themselves; the "kilogram" and "slug" are just identifiers on the mole.) Also, Eq. (1.10) can be divided by the system volume, yielding

$$p = C\mathcal{R}T \tag{1.13}$$

where C is the concentration [(kg · mol)/m³ or (slug · mol)/ft³].

Finally, the equation of state can be expressed in terms of particles. Let N_A be the number of particles in a mole (Avogadro's number, which for a kilogram-mole is 6.02×10^{26} particles). Multiplying and dividing Eq. (1.13) by N_A,

$$p = (N_A C) \left(\frac{\mathcal{R}}{N_A} \right) T \tag{1.14}$$

Examining the units, $N_A C$ is physically the number density (number of particles per unit volume), and $\mathcal{R}/N_A$ is the gas constant per particle, which is precisely the Boltzmann constant k. Hence, Eq. (1.14) becomes

$$p = nkT \tag{1.15}$$

where n denotes number density.

In summary, the reader will frequently encounter the different forms of the perfect gas equation of state just listed. However, do not be confused; they are all the same thing and it is wise to become familiar with them all. In this book, particular use will be made of Eqs. (1.8), (1.9), and (1.12). Also, do not be confused by the variety of gas constants. They are easily sorted out:

1. When the equation deals with moles, use the universal gas constant, which is the "gas constant per mole." It is the same for all gases, and equal to the following in the two systems of units:

$$\mathcal{R} = 8314 \text{ J/(kg · mol · K)}$$
$$\mathcal{R} = 4.97 \times 10^4 \text{ (ft · lb)/(slug · mol · °R)}$$

2. When the equation deals with mass, use the specific gas constant R, which is the "gas constant per unit mass." It is different for different gases, and is related to the universal gas constant, $R = \mathcal{R}/\mathcal{M}$, where $\mathcal{M}$ is the molecular weight. For air at standard conditions:

$$R = 287 \text{ J/(kg · K)}$$
$$R = 1716 \text{ (ft · lb)/(slug · °R)}$$

3. When the equation deals with particles, use the Boltzmann constant k, which is the "gas constant per particle":

$$k = 1.38 \times 10^{-23} \text{ J/K}$$

$$k = 0.565 \times 10^{-23} \text{ (ft} \cdot \text{lb)} /^{\circ}\text{R}$$

How accurate is the assumption of a perfect gas? It has been experimentally determined that, at low pressures (near 1 atm or less) and at high temperatures (standard temperature, 273 K, and above), the value pv/RT for most pure gases deviates from unity by less than 1 percent. However, at very cold temperatures and high pressures, the molecules of the gas are more closely packed together, and consequently intermolecular forces become more important. Under these conditions, the gas is defined as a *real gas*. In such cases, the perfect gas equation of state must be replaced by more accurate relations such as the van der Waals equation

$$\left(p + \frac{a}{v^2} \right) (v - b) = RT \tag{1.16}$$

where a and b are constants that depend on the type of gas. As a general rule of thumb, deviations from the perfect gas equation of state vary approximately as p/T^3. In the vast majority of gasdynamic applications, the temperatures and pressures are such that $p = \rho RT$ can be applied with confidence. Such will be the case throughout this book.

In the early 1950s, aerodynamicists were suddenly confronted with hypersonic entry vehicles at velocities as high as 26,000 ft/s (8 km/s). The shock layers about such vehicles were hot enough to cause chemical reactions in the airflow (dissociation, ionization, etc.). At that time, it became fashionable in the aerodynamic literature to denote such conditions as "real gas effects." However, in classical physical chemistry, a real gas is defined as one in which intermolecular forces are important, and the definition is completely divorced from the idea of chemical reactions. In the preceding paragraphs, we have followed such a classical definition. For a chemically reacting gas, as will be discussed at length in Chap. 16, most problems can be treated by assuming a mixture of perfect gases, where the relation $p = \rho RT$ still holds. However, because $R = \mathcal{R}/\mathcal{M}$ and $\mathcal{M}$ varies due to the chemical reactions, then R is a variable throughout the flow. It is preferable, therefore, *not* to identify such phenomena as " real gas effects," and this term will not be used in this book. Rather, we will deal with "chemically reacting mixtures of perfect gases," which are the subject of Chaps. 16 and 17.

EXAMPLE 1.2

A pressure vessel that has a volume of 10 m^3 is used to store high-pressure air for operating a supersonic wind tunnel. If the air pressure and temperature inside the vessel are 20 atm and 300 K, respectively, what is the mass of air stored in the vessel?

■ **Solution**

Recall that 1 atm $= 1.01 \times 10^5$ N/m^2. From Eq. (1.9)

$$\rho = \frac{p}{RT} = \frac{(20)(1.01 \times 10^5)}{(287)(300)} = 23.46 \text{ kg/m}^3$$

The total mass stored is then

$$M = \mathcal{V}\rho = (10)(23.46) = \boxed{234.6 \text{ kg}}$$

EXAMPLE 1.3

Calculate the isothermal compressibility for air at a pressure of 0.5 atm.

■ **Solution**

From Eq. (1.3)

$$\tau_T = -\frac{1}{v}\left(\frac{\partial v}{\partial p}\right)_T$$

From Eq. (1.8)

$$v = \frac{RT}{p}$$

Thus

$$\left(\frac{\partial v}{\partial p}\right)_T = -\frac{RT}{p^2}$$

Hence

$$\tau_T = -\frac{1}{v}\left(\frac{\partial v}{\partial p}\right)_T = -\left(\frac{p}{RT}\right)\left(-\frac{RT}{p^2}\right) = \frac{1}{p}$$

We see that the isothermal compressibility for a perfect gas is simply the reciprocal of the pressure:

$$\tau_T = \frac{1}{p} = \frac{1}{0.5} = \boxed{2 \text{ atm}^{-1}}$$

In terms of the International System of units, where $p = (0.5)(1.01 \times 10^5) = 5.05 \times 10^4$ N/m^2,

$$\tau_T = \boxed{1.98 \times 10^{-5} \text{m}^2/\text{N}}$$

In terms of the English Engineering System of units, where $p = (0.5)(2116) = 1058$ lb/ft^2,

$$\tau_T = \boxed{9.45 \times 10^{-4} \text{ ft}^2/\text{lb}}$$

1.4.2 Internal Energy and Enthalpy

Returning to our microscopic view of a gas as a collection of particles in random motion, the individual kinetic energy of each particle contributes to the overall energy of the gas. Moreover, if the particle is a molecule, its rotational and vibrational motions (see Chap. 16) also contribute to the gas energy. Finally, the motion of electrons in both atoms and molecules is a source of energy. This small sketch of atomic and molecular energies will be enlarged to a massive portrait in Chap. 16; it is sufficient to note here that the energy of a particle can consist of several different forms of motion. In turn, these energies, summed over all the particles of the gas, constitute the *internal energy, e*, of the gas. Moreover, if the particles of the gas (called the *system*) are rattling about in their state of "maximum disorder" (see again Chap. 16), the system of particles will be in *equilibrium*.

Return now to the macroscopic view of the gas as a continuum. Here, equilibrium is evidenced by no gradients in velocity, pressure, temperature, and chemical concentrations throughout the system, i.e., the system has uniform properties. For an equilibrium system of a real gas where intermolecular forces are important, and also for an equilibrium chemically reacting mixture of perfect gases, the internal energy is a function of both temperature and volume. Let e denote the specific internal energy (internal energy per unit mass). Then, the *enthalpy, h*, is defined, per unit mass, as $h = e + pv$, and we have

$$e = e(T, v)$$
$$h = h(T, p) \tag{1.17}$$

for both a real gas and a chemically reacting mixture of perfect gases.

If the gas is *not* chemically reacting, and if we ignore intermolecular forces, the resulting system is a *thermally perfect gas*, where internal energy and enthalpy are functions of temperature only, and where the specific heats at constant volume and pressure, c_v and c_p, are also functions of temperature only:

$$e = e(T)$$
$$h = h(T)$$
$$de = c_v \, dT$$
$$dh = c_p \, dT \tag{1.18}$$

The temperature variation of c_v and c_p is associated with the vibrational and electronic motion of the particles, as will be explained in Chap. 16.

Finally, if the specific heats are constant, the system is a *calorically perfect gas*, where

$$e = c_v T$$
$$h = c_p T \tag{1.19}$$

In Eq. (1.19), it has been assumed that $h = e = 0$ at $T = 0$.

In many compressible flow applications, the pressures and temperatures are moderate enough that the gas can be considered to be calorically perfect. Indeed, there is a large bulk of literature for flows with constant specific heats. For the first half of this book, a calorically perfect gas will be assumed. This is the case for atmospheric air at temperatures below 1000 K. However, at higher temperatures the vibrational motion of the O_2 and N_2 molecules in air becomes important, and the air becomes thermally perfect, with specific heats that vary with temperature. Finally, when the temperature exceeds 2500 K, the O_2 molecules begin to dissociate into O atoms, and the air becomes chemically reacting. Above 4000 K, the N_2 molecules begin to dissociate. For these chemically reacting cases, from Eqs. (1.17), e depends on both T and v, and h depends on both T and p. (Actually, in equilibrium thermodynamics, any state variable is uniquely determined by any two other state variables. However, it is convenient to associate T and v with e, and T and p with h.) Chapters 16 and 17 will discuss the thermodynamics and gasdynamics of both thermally perfect and chemically reacting gases.

Consistent with Eq. (1.9) and the definition of enthalpy is the relation

$$c_p - c_v = R \tag{1.20}$$

where the specific heats at constant pressure and constant volume are defined as

$$c_p = \left(\frac{\partial h}{\partial T}\right)_p$$

and

$$c_v = \left(\frac{\partial e}{\partial T}\right)_v$$

respectively. Equation (1.20) holds for a calorically perfect or a thermally perfect gas. It is *not* valid for either a chemically reacting or a real gas. Two useful forms of Eq. (1.20) can be simply obtained as follows. Divide Eq. (1.20) by c_p:

$$1 - \frac{c_v}{c_p} = \frac{R}{c_p} \tag{1.21}$$

Define $\gamma \equiv c_p/c_v$. For air at standard conditions, $\gamma = 1.4$. Then Eq. (1.21) becomes

$$1 - \frac{1}{\gamma} = \frac{R}{c_p}$$

Solving for c_p,

$$\boxed{c_p = \frac{\gamma R}{\gamma - 1}} \tag{1.22}$$

Similarly, by dividing Eq. (1.20) by c_v, we find that

$$\boxed{c_v = \frac{R}{\gamma - 1}} \tag{1.23}$$

Equations (1.22) and (1.23) hold for a thermally or calorically perfect gas; they will be useful in our subsequent treatment of compressible flow.

EXAMPLE 1.4

For the pressure vessel in Example 1.2, calculate the total internal energy of the gas stored in the vessel.

■ **Solution**

From Eq. (1.23)

$$c_v = \frac{R}{\gamma - 1} = \frac{287}{1.4 - 1} = 717.5 \ \text{J/kg} \cdot \text{K}$$

From Eq. (1.19)

$$e = c_v T = (717.5)(300) = 2.153 \times 10^5 \text{ J/kg}$$

From Example 1.2, we calculated the mass of air in the vessel to be 234.6 kg. Thus, the total internal energy is

$$E = Me = (234.6)(2.153 \times 10^5) = \boxed{5.05 \times 10^7 \text{ J}}$$

1.4.3   First Law of Thermodynamics

Consider a *system,* which is a fixed mass of gas separated from the surroundings by a flexible boundary. For the time being, assume the system is stationary, i.e., it has no directed kinetic energy. Let δq be an incremental amount of heat added to the system across the boundary (say by direct radiation or thermal conduction). Also, let δw denote the work done on the system by the surroundings (say by a displacement of the boundary, squeezing the volume of the system to a smaller value). Due to the molecular motion of the gas, the system has an internal energy e. (This is the specific internal energy if we assume a system of unit mass.) The heat added and work done on the system cause a change in energy, and since the system is stationary, this change in energy is simply de:

$$\boxed{\delta q + \delta w = de} \tag{1.24}$$

This is the *first law of thermodynamics;* it is an empirical result confirmed by laboratory and practical experience. In Eq. (1.24), e is a state variable. Hence, de is an exact differential, and its value depends only on the initial and final states of the system. In contrast, δq and δw depend on the process in going from the initial and final states.

For a given de, there are in general an infinite number of different ways (processes) by which heat can be added and work done on the system. We will be primarily concerned with three types of processes:

1. *Adiabatic process*—one in which no heat is added to or taken away from the system
2. *Reversible process*—one in which no dissipative phenomena occur, i.e., where the effects of viscosity, thermal conductivity, and mass diffusion are absent
3. *Isentropic process*—one which is both adiabatic and reversible

For a reversible process, it can be easily proved (see any good text on thermodynamics) that $\delta w = -p\,dv$, where dv is an incremental change in specific volume due to a displacement of the boundary of the system. Hence, Eq. (1.24) becomes

$$\delta q - p\,dv = de \tag{1.25}$$

If, in addition, this process is also adiabatic (hence isentropic), Eq. (1.25) leads to some extremely useful thermodynamic formulas. However, before obtaining these formulas, it is useful to review the concept of entropy.

1.4.4 Entropy and the Second Law of Thermodynamics

Consider a block of ice in contact with a red-hot plate of steel. Experience tells us that the ice will warm up (and probably melt) and the steel plate will cool down. However, Eq. (1.24) does not necessarily say this will happen. Indeed, the first law allows that the ice may get cooler and the steel plate hotter—just as long as energy is conserved during the process. Obviously, this does not happen; instead, nature imposes another condition on the process, a condition which tells us *in which direction* a process will take place. To ascertain the proper direction of a process, let us define a new state variable, the entropy, as

$$ds = \frac{\delta q_{rev}}{T}$$

where s is the entropy of the system, δq_{rev} is an incremental amount of heat added reversibly to the system, and T is the system temperature. Do not be confused by this definition. It defines a change in entropy in terms of a reversible addition of heat, δq_{rev}. However, entropy is a state variable, and it can be used in conjunction with any type of process, reversible or irreversible. The quantity δq_{rev} is just an artifice; an effective value of δq_{rev} can always be assigned to relate the initial and end points of an irreversible process, where the actual amount of heat added is δq. Indeed, an alternative and probably more lucid relation is

$$\boxed{ds = \frac{\delta q}{T} + ds_{irrev}} \tag{1.26}$$

Equation (1.26) applies in general; it states that the change in entropy during any incremental process is equal to the actual heat added divided by the temperature, $\delta q/T$, plus a contribution from the irreversible dissipative phenomena of viscosity, thermal conductivity, and mass diffusion occurring *within* the system, ds_{irrev}. These dissipative phenomena *always* increase the entropy:

$$\boxed{ds_{irrev} \geq 0} \tag{1.27}$$

The equal sign denotes a reversible process, where, by definition, the dissipative phenomena are absent. Hence, a combination of Eqs. (1.26) and (1.27) yields

$$\boxed{ds \geq \frac{\delta q}{T}} \tag{1.28}$$

Furthermore, if the process is adiabatic, $\delta q = 0$, and Eq. (1.28) becomes

$$\boxed{ds \geq 0} \tag{1.29}$$

Equations (1.28) and (1.29) are forms of the *second law of thermodynamics*. The second law tells us in what direction a process will take place. A process will proceed in a direction such that the entropy of the system plus surroundings always increases, or at best stays the same. In our example at the beginning of Section 1.4.4, consider the system to be both the ice and steel plate combined. The simultaneous heating of the ice and cooling of the plate yields a net increase in entropy for the system. On the other hand, the impossible situation of the ice getting cooler and the plate hotter would yield a net decrease in entropy, a situation forbidden by the second law. In summary, the concept of entropy in combination with the second law allows us to predict the *direction* that nature takes.

1.4.5 Calculation of Entropy

Consider again the first law in the form of Eq. (1.25). If we assume that the heat is reversible, and we use the definition of entropy in the form $\delta q_{\text{rev}} = T\,ds$, then Eq. (1.25) becomes

$$T\,ds - p\,dv = de$$

$$\boxed{T\,ds = de + p\,dv} \tag{1.30}$$

Another form can be obtained in terms of enthalpy. For example, by definition,

$$h = e + pv$$

Differentiating, we obtain

$$dh = de + p\,dv + v\,dp \tag{1.31}$$

Combining Eqs. (1.30) and (1.31), we have

$$\boxed{T\,ds = dh - v\,dp} \tag{1.32}$$

Equations (1.30) and (1.32) are important, and should be kept in mind as much as the original form of the first law, Eq. (1.24).

For a thermally perfect gas, from Eq. (1.18), we have $dh = c_p\,dT$. Substitution into Eq. (1.32) gives

$$ds = c_p\frac{dT}{T} - \frac{v\,dp}{T} \tag{1.33}$$

Substituting the perfect gas equation of state $pv = RT$ into Eq. (1.33), we have

$$ds = c_p\frac{dT}{T} - R\frac{dp}{p} \tag{1.34}$$

Integrating Eq. (1.34) between states 1 and 2,

$$s_2 - s_1 = \int_{T_1}^{T_2} c_p\frac{dT}{T} - R\ln\frac{p_2}{p_1} \tag{1.35}$$

Equation (1.35) holds for a thermally perfect gas. It can be evaluated if c_p is known as a function of T. If we further assume a calorically perfect gas, where c_p is constant, Eq. (1.35) yields

$$\boxed{s_2 - s_1 = c_p\ln\frac{T_2}{T_1} - R\ln\frac{p_2}{p_1}} \tag{1.36}$$

Similarly, starting with Eq. (1.30), and using $de = c_v \, dT$, the change in entropy can also be obtained as

$$s_2 - s_1 = c_v \ln \frac{T_2}{T_1} + R \ln \frac{v_2}{v_1} \qquad (1.37)$$

As an exercise, show this yourself. Equations (1.36) and (1.37) allow the calculation of the change in entropy between two states of a calorically perfect gas in terms of either the pressure and temperature, or the volume and temperature. Note that entropy is a function of *both* p and T, or v and T, even for the simplest case of a calorically perfect gas.

EXAMPLE 1.5

Consider the air in the pressure vessel in Example 1.2. Let us now heat the gas in the vessel. Enough heat is added to increase the temperature to 600 K. Calculate the change in entropy of the air inside the vessel.

■ Solution

The vessel has a constant volume; hence as the air temperature is increased, the pressure also increases. Let the subscripts 1 and 2 denote the conditions before and after heating, respectively. Then, from Eq. (1.8),

$$\frac{p_2}{p_1} = \frac{T_2}{T_1} = \frac{600}{300} = 2$$

In Example 1.4, we found that $c_v = 717.5$ J/kg · K. Thus, from Eq. (1.20)

$$c_p = c_v + R = 717.5 + 287 = 1004.5 \text{ J/kg} \cdot \text{K}$$

From Eq. (1.36)

$$s_2 - s_1 = c_p \ln \frac{T_2}{T_1} - R \ln \frac{p_2}{p_1}$$
$$= 1004.5 \ln 2 - 287 \ln 2 = 497.3 \text{ J/kg} \cdot \text{K}$$

From Example 1.2, the mass of air inside the vessel is 234.6 kg. Thus, the total entropy change is

$$S_2 - S_1 = M(s_2 - s_1) = (234.6)(497.3) = \boxed{1.167 \times 10^5 \text{ J/K}}$$

1.4.6 Isentropic Relations

An isentropic process was already defined as adiabatic and reversible. For an adiabatic process, $\delta q = 0$, and for a reversible process, $ds_{\text{irrev}} = 0$. Hence, from Eq. (1.26), an isentropic process is one in which $ds = 0$, i.e., *the entropy is constant*.

Important relations for an isentropic process can be obtained directly from Eqs. (1.36) and (1.37), setting $s_2 = s_1$. For example, from Eq. (1.36)

$$0 = c_p \ln \frac{T_2}{T_1} - R \ln \frac{p_2}{p_1}$$

$$\ln \frac{p_2}{p_1} = \frac{c_p}{R} \ln \frac{T_2}{T_1}$$

$$\frac{p_2}{p_1} = \left(\frac{T_2}{T_1}\right)^{c_p/R} \tag{1.38}$$

Recalling Eq. (1.22),

$$\frac{c_p}{R} = \frac{\gamma}{\gamma - 1}$$

and substituting into Eq. (1.38),

$$\frac{p_2}{p_1} = \left(\frac{T_2}{T_1}\right)^{\gamma/(\gamma - 1)} \tag{1.39}$$

Similarly, from Eq. (1.37)

$$0 = c_v \ln \frac{T_2}{T_1} + R \ln \frac{v_2}{v_1}$$

$$\ln \frac{v_2}{v_1} = -\frac{c_v}{R} \ln \frac{T_2}{T_1}$$

$$\frac{v_2}{v_1} = \left(\frac{T_2}{T_1}\right)^{-c_v/R} \tag{1.40}$$

From Eq. (1.23)

$$\frac{c_v}{R} = \frac{1}{\gamma - 1}$$

Substituting into Eq. (1.40), we have

$$\frac{v_2}{v_1} = \left(\frac{T_2}{T_1}\right)^{-1/(\gamma - 1)} \tag{1.41}$$

Recall that $\rho_2/\rho_1 = v_1/v_2$. Hence, from Eq. (1.41)

$$\frac{\rho_2}{\rho_1} = \left(\frac{T_2}{T_1}\right)^{1/(\gamma - 1)} \tag{1.42}$$

Summarizing Eqs. (1.39) and (1.42),

$$\boxed{\frac{p_2}{p_1} = \left(\frac{\rho_2}{\rho_1}\right)^{\gamma} = \left(\frac{T_2}{T_1}\right)^{\gamma/(\gamma - 1)}} \tag{1.43}$$

Equation (1.43) is important. It relates pressure, density, and temperature for an isentropic process, and is very frequently used in the analysis of compressible flows.

You might legitimately ask the questions *why* Eq. (1.43) is so important, and *why* it is frequently used. Indeed, at first thought the concept of an isentropic process itself may seem so restrictive—adiabatic as well as reversible—that one might expect it to find only limited applications. However, such is not the case. For example, consider the flows over an airfoil and through a rocket engine. In the regions adjacent to the airfoil surface and the rocket nozzle walls, a boundary layer is formed wherein the dissipative mechanisms of viscosity, thermal conduction, and diffusion are strong. Hence, the entropy increases within these boundary layers. On the other hand, consider the fluid elements outside the boundary layer, where dissipative effects are negligible. Moreover, no heat is being added or taken away from the fluid elements at these points—hence, the flow is adiabatic. As a result, the fluid elements outside the boundary layer are experiencing adiabatic and reversible processes—namely, isentropic flow. Moreover, the viscous boundary layers are usually thin, hence large regions of the flowfields are isentropic. Therefore, a study of isentropic flows is directly applicable to many types of practical flow problems. In turn, Eq. (1.43) is a powerful relation for such flows, valid for a calorically perfect gas.

This ends our brief review of thermodynamics. Its purpose has been to give a quick summary of ideas and equations that will be employed throughout our subsequent discussions of compressible flow. Aspects of the thermodynamics associated with a high-temperature chemically reacting gas will be developed as necessary in Chap. 16.

Consider the flow through a rocket engine nozzle. Assume that the gas flow through the nozzle is an isentropic expansion of a calorically perfect gas. In the combustion chamber, the gas which results from the combustion of the rocket fuel and oxidizer is at a pressure and temperature of 15 atm and 2500 K, respectively; the molecular weight and specific heat at constant pressure of the combustion gas are 12 and 4157 J/kg · K, respectively. The gas expands to supersonic speed through the nozzle, with a temperature of 1350 K at the nozzle exit. Calculate the pressure at the exit.

■ **Solution**

From our earlier discussion on the equation of state,

$$R = \frac{\mathscr{R}}{\mathscr{M}} = \frac{8314}{12} = 692.8 \text{ J/kg} \cdot \text{K}$$

From Eq. (1.20)

$$c_v = c_p - R = 4157 - 692.8 = 3464 \text{ J/kg} \cdot \text{K}$$

Thus

$$\gamma = \frac{c_p}{c_v} = \frac{4157}{3464} = 1.2$$

From Eq. (1.43), we have

$$\frac{p_2}{p_1} = \left(\frac{T_2}{T_1}\right)^{\gamma/(\gamma-1)} = \left(\frac{1350}{2500}\right)^{1.2/(1.2-1)} = 0.0248$$

$$p_2 = 0.025 p_1 = (0.0248)(15 \text{ atm}) = \boxed{0.372 \text{ atm}}$$

EXAMPLE 1.7

Calculate the isentropic compressibility for air at a pressure of 0.5 atm. Compare the result with that for the isothermal compressibility obtained in Example 1.3.

■ Solution

From Eq. (1.4), the isentropic compressibility is defined as

$$\tau_s = -\frac{1}{v}\left(\frac{\partial v}{\partial p}\right)_s$$

Since $v = 1/\rho$, we can write Eq. (1.4) as

$$\tau_s = \frac{1}{\rho}\left(\frac{\partial \rho}{\partial p}\right)_s \tag{E.1}$$

The variation between p and ρ for an isentropic process is given by Eq. (1.43)

$$\frac{p_2}{p_1} = \left(\frac{\rho_2}{\rho_1}\right)^{\gamma}$$

which is the same as writing

$$p = c\rho^{\gamma} \tag{E.2}$$

where c is a constant. From Eq. (E.2)

$$\left(\frac{\partial p}{\partial \rho}\right)_s = c\gamma\rho^{\gamma-1} = \frac{p}{\rho^{\gamma}}(\gamma\rho^{\gamma-1}) = \frac{\gamma p}{\rho} \tag{E.3}$$

From Eqs. (E.1) and (E.3),

$$\tau_s = \frac{1}{\rho}\left(\frac{\partial \rho}{\partial p}\right)_s = \frac{1}{\rho}\left(\frac{\partial p}{\partial \rho}\right)_s^{-1} = \frac{1}{\rho}\left(\frac{\gamma p}{\rho}\right)^{-1}$$

Hence,

$$\tau_s = \frac{1}{\gamma p} \tag{E.4}$$

Recall from Example 1.3 that $\tau_T = 1/p$. Hence,

$$\tau_s = \frac{\tau_T}{\gamma} \tag{E.5}$$

Note that τ_s is smaller than τ_T by the factor γ. From Example 1.3, we found that for $p = 0.5$ atm, $\tau_T = 1.98 \times 10^{-5} \text{ m}^2/\text{N}$. Hence, from Eq. (E.5)

$$\tau_s = \frac{1.98 \times 10^{-5}}{1.4} = \boxed{1.41 \times 10^{-5} \text{ m}^2/\text{N}}$$

1.5 | AERODYNAMIC FORCES ON A BODY

The history of fluid dynamics is dominated by the quest to predict forces on a body moving through a fluid—ships moving through water, and in the nineteenth and twentieth centuries, aircraft moving through air, to name just a few examples. Indeed, Newton's treatment of fluid flow in his *Principia* (1687) was oriented in part toward the prediction of forces on an inclined surface. The calculation of aerodynamic and hydrodynamic forces still remains a central thrust of modern fluid dynamics. This is especially true for compressible flow, which governs the aerodynamic lift and drag on high-speed subsonic, transonic, supersonic, and hypersonic airplanes, and missiles. Therefore, in several sections of this book, the fundamentals of compressible flow will be applied to the practical calculation of aerodynamic forces on high-speed bodies.

The mechanism by which nature transmits an aerodynamic force to a surface is straightforward. This force stems from only two basic sources: surface pressure and surface shear stress. Consider, for example, the airfoil of unit span sketched in Fig. 1.11. Let s be the distance measured along the surface of the airfoil from the nose. In general, the pressure p and shear stress τ are functions of s; $p = p(s)$ and $\tau = \tau(s)$. These pressure and shear stress distributions are the only means that nature has to communicate an aerodynamic force to the airfoil. To be more specific, consider an elemental surface area dS on which is exerted a pressure p acting normal to dS and a shear stress τ acting tangential to dS, as sketched in Fig. 1.11 Let $\mathbf{n}$ and $\mathbf{m}$ be unit vectors perpendicular and parallel, respectively, to the element dS, as shown in Fig. 1.11. For future discussion, it is convenient to define a vector $d\mathbf{S} \equiv \mathbf{n}\,dS$; hence $d\mathbf{S}$ is a vector normal to the surface with a magnitude dS. From Fig. 1.11, the elemental force $d\mathbf{F}$ acting on dS is then

$$d\mathbf{F} = -p\mathbf{n}\,dS + \tau\mathbf{m}\,dS = -p\,d\mathbf{S} + \tau\mathbf{m}\,dS \qquad (1.44)$$

Note from Fig. 1.11 that p acts toward the surface, whereas $d\mathbf{S} = \mathbf{n}\,dS$ is directed away from the surface. This is the reason for the minus sign in Eq. (1.44). The *total*

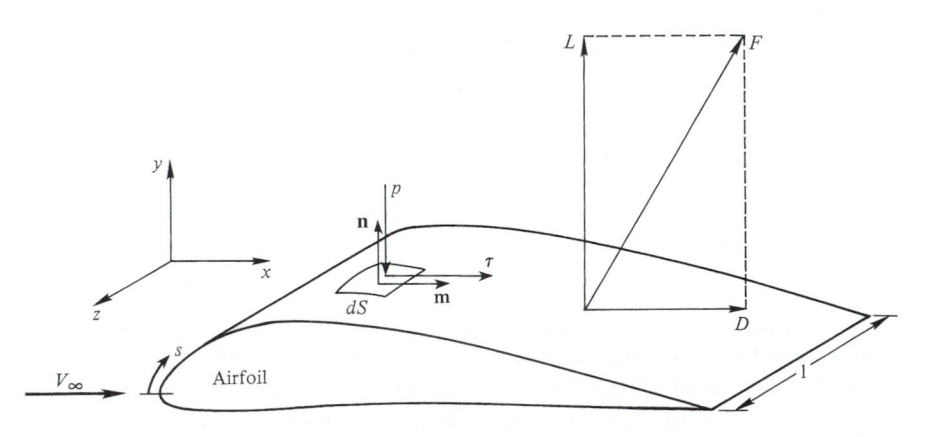

Figure 1.11 | Sources of aerodynamic force; resultant force and its resolution into lift and drag.

aerodynamic force **F** acting on the complete body is simply the sum of all the element forces acting on all the elemental areas. This can be expressed as a surface integral, using Eq. (1.44):

$$\mathbf{F} = \oiint d\mathbf{F} = -\oiint p \, d\mathbf{S} + \oiint \tau \mathbf{m} \, dS \tag{1.45}$$

On the right-hand side of Eq. (1.45), the first integral is the pressure force on the body, and the second is the shear, or friction force. The integrals are taken over the complete surface of the body.

Consider x, y, z orthogonal coordinates as shown in Fig. 1.11. Let x and y be parallel and perpendicular, respectively, to V_∞. If **F** is the net aerodynamic force from Eq. (1.45), then the lift L and drag D are defined as the components of **F** in the y and x directions, respectively. In aerodynamics, V_∞ is called the *relative wind,* and lift and drag are always defined as perpendicular and parallel, respectively, to the relative wind. For most practical aerodynamic shapes, L is generated mainly by the surface pressure distribution; the shear stress distribution generally makes only a small contribution. Hence from Eq. (1.45) and Fig. 1.11, the aerodynamic lift can be approximated by

$$L \approx y \text{ component of } \left[-\oiint p \, d\mathbf{S} \right] \tag{1.46}$$

With regard to drag, from Eq. (1.45) and Fig. 1.11,

$$D = x \text{ component of } \underbrace{\left[-\oiint p \, d\mathbf{S} \right]}_{\text{pressure drag}} + x \text{ component of } \underbrace{\left[\oiint \tau \mathbf{m} \, dS \right]}_{\text{skin-friction drag}} \tag{1.47}$$

In this book, inviscid flows are dealt with exclusively, as discussed in Sec. 1.3. For many bodies, the inviscid flow accurately determines the surface pressure distribution. For such bodies, the results of this book in conjunction with Eq. (1.46) allow a reasonable prediction of lift. On the other hand, drag is due both to pressure and shear stress distributions via Eq. (1.47). Since we will not be considering viscous flows, we will not be able to calculate skin friction drag. Moreover, the pressure drag in Eq. (1.47) is often influenced by flow separation from the body—also a viscous effect. Hence, the fundamentals of inviscid compressible flow do not lead to an accurate prediction of drag for many situations. However, for pressure drag on slender supersonic shapes due to shock waves, so-called *wave drag,* inviscid techniques are usually quite adequate, as we shall see in subsequent chapters.

EXAMPLE 1.8

A flat plate with a chord length of 3 ft and an infinite span (perpendicular to the page in Fig. 1.12) is immersed in a Mach 2 flow at standard sea level conditions at an angle of attack of $10°$. The pressure distribution over the plate is as follows: upper surface, $p_2 = \text{const} = 1132$ lb/ft^2; lower surface, $p_3 = \text{const} = 3568$ lb/ft^2. The local shear stress is given by

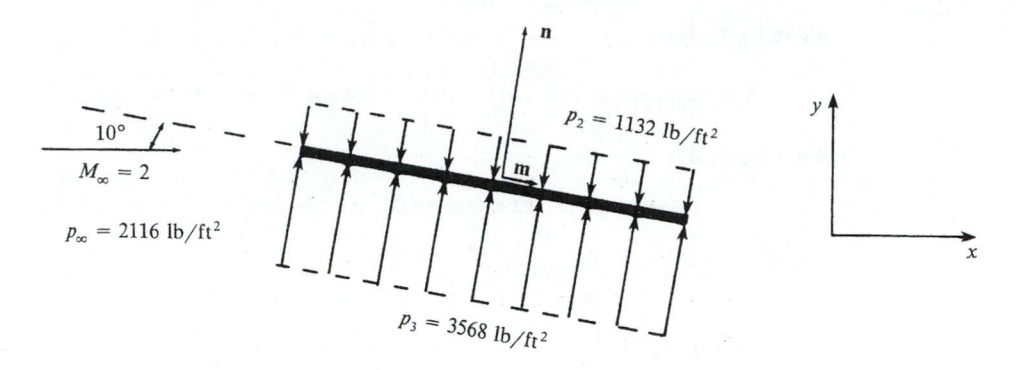

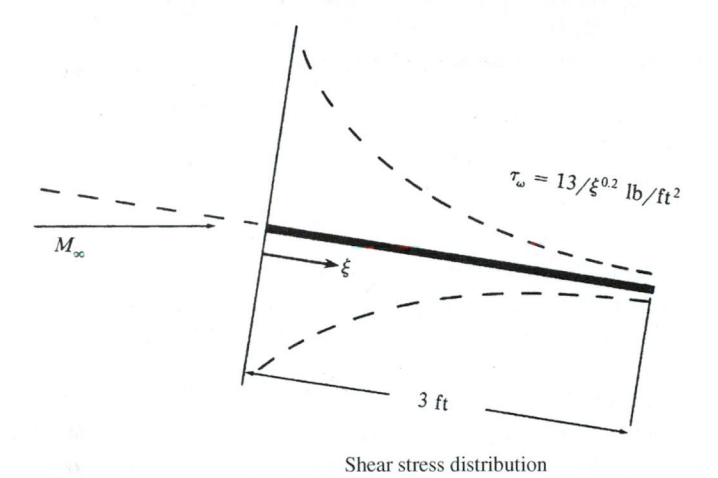

Figure 1.12 | Geometry for Example 1.8.

$\tau_w = 13/\xi^{0.2}$, where τ_w is in pounds per square feet and ξ is the distance in feet along the plate from the leading edge. Assume that the distribution of τ_w over the top and bottom surfaces is the same. (We make this assumption for simplicity in this example. In reality, the shear stress distributions over the top and bottom surfaces will be different because the flow properties over these two surfaces are different.) Both the pressure and shear stress distributions are sketched qualitatively in Fig. 1.12. Calculate the lift and drag per unit span on the plate.

■ **Solution**

Considering a unit span,

$$-\oiint p\, d\mathbf{S} = \left[-\int_0^3 p_2\, d\xi + \int_0^3 p_3\, d\xi \right] \mathbf{n} = [-(1132)(3) + (3568)(3)]\mathbf{n} = 7308\mathbf{n}$$

From Eq. (1.46)

$$L = y \text{ component of } \left[-\oiint p \, d\mathbf{S} \right] = 7308 \cos 10° = \boxed{7197 \text{ lb}} \text{ per unit span}$$

From Eq. (1.47)

$$\text{Pressure drag} = \text{wave drag} \equiv D_w = x \text{ component of } \left[-\oiint p \, d\mathbf{S} \right]$$

Hence

$$D_w = 7308 \sin 10° = \boxed{1269 \text{ lb}} \text{ per unit span}$$

Also from Eq. (1.46)

$$\text{Skin-friction drag} \equiv D_f = x \text{ component of } \left[\oiint \tau \mathbf{m} \, dS \right]$$

$$\oiint \tau \mathbf{m} \, dS = \left[13 \int_0^3 \xi^{-0.2} d\xi \right] \mathbf{m} = 16.25 \xi^{4/5} \Big|_0^3 \mathbf{m} = 39.13 \mathbf{m}$$

Hence, recalling that shear stress acts on both sides,

$$D_f = 2(39.13) \cos 10° = \boxed{77.1 \text{ lb}} \text{ per unit span}$$

The total drag is

$$D = D_w + D_f$$
$$D = 1269 \text{ lb} + 77.1 \text{ lb} = \boxed{1346 \text{ lb}}$$

Note: For this example, the drag is mainly wave drag; skin-friction drag accounts for only 5.7 percent of the total drag. This illustrates an important point. For supersonic flow over slender bodies at a reasonable angle of attack, the wave drag is the primary drag contributor at sea level, far exceeding the skin-friction drag. For such applications, the inviscid methods discussed in this book suffice, because the wave drag (pressure drag) can be obtained from such methods. We see here also why so much attention is focused on the reduction of wave drag—because it is frequently the primary drag component. At smaller angles of attack, the relative proportion of D_f to D increases. Also, at higher altitudes, where viscous effects become stronger (the Reynolds number is lower), the relative proportion of D_f to D increases.

1.6 | MODERN COMPRESSIBLE FLOW

In Sec. 1.1, we saw how the convergent-divergent steam nozzles of de Laval helped to usher compressible flow into the world of practical engineering applications. However, compressible flow did not begin to receive major attention until the advent of jet propulsion and high-speed flight during World War II. Indeed, between 1945 and 1960, the fundamentals and applications of compressible flow became essentially "classic," generally characterized by

1.   Treatment of a calorically perfect gas, i.e., constant specific heats.
2.   Exact solutions of flows in one dimension, but usually approximate solutions (based on linearized equations) for two- and three-dimensional flows. These

solutions were closed form, yielding equations or formulas for the desired information. Exceptions were the method of characteristics, an exact numerical approach applicable to certain classes of compressible flows (see Chap. 11), and the exact Taylor–Maccoll solution to the flow over a sharp, right-circular cone at zero angle of attack (see Chap. 10). Both of these exceptions required numerical solutions, which were laborious endeavors before the advent of the modern high-speed digital computer.

Many good textbooks on classical compressible flow have been written since 1945. Some of them are listed as Refs. 3 through 17 at the end of this book. The reader is strongly encouraged to study these references, because a thorough understanding of classical compressible flow is essential to modern applications.

Since approximately 1960, compressible flow has entered a "modern" period, characterized by

1. The necessity of dealing with high-temperature, chemically reacting gases associated with hypersonic flight and rocket engines, hence requiring a major extension and modification of the classical literature based on a calorically perfect gas. (See, for example, Ref. 119.)

2. The rise of computational fluid dynamics, which is a new third dimension in fluid dynamics, complementing the previous existing dimensions of pure experiment and pure theory. With the advent of modern high-speed digital computers, and the subsequent development of computational fluid dynamics as a distinct discipline, the practical solution of the exact governing equations for a myriad of complex compressible flow problems is now at hand. In brief, computational fluid dynamics is the art of replacing the governing partial differential equations of fluid flow with numbers, and advancing these numbers in space and/or time to obtain a final numerical description of the complete flowfield of interest. The end product of computational fluid dynamics is indeed a collection of numbers, in contrast to a closed-form analytical solution. However, in the long run the objective of most engineering analyses, closed-form or otherwise, is a quantitative description of the problem, i.e., numbers. (See, for example, Ref. 18.)

The modern compressible flow of today is a mutually supportive mixture of classical analyses along with computational techniques, with the treatment of noncalorically perfect gases as almost routine. The purpose of this book is to provide an understanding of compressible flow from this point of view. Its intent is to blend the important aspects of classical compressible flow with the recent techniques of computational fluid dynamics. Moreover, the first part of the book will deal almost exclusively with a calorically perfect gas. In turn, the second part will contain a logical extension to realms of high-temperature gases, and the results will be contrasted with those from classical analyses. In addition, various historical aspects of the development of compressible flow, both classical and modern, will be included along with the technical material. In this fashion, it is hoped that the reader will gain an appreciation of the heritage of the discipline. The author feels strongly that a knowledge of such historical traditions and events is important for a truly fundamental understanding of the discipline.

1.7 | SUMMARY

The compressibility is generically defined as

$$\tau = -\frac{1}{v}\frac{dv}{dp} \tag{1.2}$$

hence

$$d\rho = \rho\tau\,dp \tag{1.6}$$

From Eq. (1.6), a flow must be treated as compressible when the pressure gradients in the flowfield are large enough such that, in combination with a large enough value of the compressibility, τ, the resulting density changes are too large to ignore. For gases, this occurs when the flow Mach number is greater than about 0.3. In short, for high-speed flows, the density becomes a variable; such variable-density flows are called *compressible* flows.

High-speed, compressible flow is also high-energy flow. Thermodynamics is the science of energy and entropy; hence a study and application of compressible flow involves a coupling of purely fluid dynamic fundamentals with the results of thermodynamics.

Compressible flow pertains to flows at Mach numbers from 0.3 to infinity. In turn, this range of Mach number is subdivided into four regimes, each with its own distinguishing physical characteristics and different analytical methods. These regimes are subsonic, transonic, supersonic, and hypersonic flow. Each of these regimes is discussed at length in this book.

PROBLEMS

1.1 At the nose of a missile in flight, the pressure and temperature are 5.6 atm and 850°R, respectively. Calculate the density and specific volume. (*Note:* 1 atm = 2116 lb/ft².)

1.2 In the reservoir of a supersonic wind tunnel, the pressure and temperature of air are 10 atm and 320 K, respectively. Calculate the density, the number density, and the mole-mass ratio. (*Note:* 1 atm = 1.01×10^5 N/m².)

1.3 For a calorically perfect gas, derive the relation $c_p - c_v = R$. Repeat the derivation for a thermally perfect gas.

1.4 The pressure and temperature ratios across a given portion of a shock wave in air are $p_2/p_1 = 4.5$ and $T_2/T_1 = 1.687$, where 1 and 2 denote conditions ahead of and behind the shock wave, respectively. Calculate the change in entropy in units of (*a*) (ft · lb)/(slug · °R) and (*b*) J/(kg · K).

1.5 Assume that the flow of air through a given duct is isentropic. At one point in the duct, the pressure and temperature are $p_1 = 1800$ lb/ft² and $T_1 = 500°$R, respectively. At a second point, the temperature is 400°R. Calculate the pressure and density at this second point.

1.6 Consider a room that is 20 ft long, 15 ft wide, and 8 ft high. For standard sea level conditions, calculate the mass of air in the room in slugs. Calculate the weight in pounds. (*Note:* If you do not know what standard sea level conditions are, consult any aerodynamics text, such as Refs. 1 and 104, for these values. Also, they can be obtained from any standard atmosphere table.)

1.7 In the infinitesimal neighborhood surrounding a point in an inviscid flow, the small change in pressure, dp, that corresponds to a small change in velocity, dV, is given by the differential relation $dp = -\rho V \, dV$. (This equation is called Euler's Equation; it is derived in chapter 6.)

    a.   Using this relation, derive a differential relation for the fractional change in density, $d\rho/\rho$, as a function of the fractional change in velocity, dV/V, with the compressibility τ as a coefficient.

    b.   The velocity at a point in an isentropic flow of air is 10 m/s (a low speed flow), and the density and pressure are 1.23 kg/m³ and 1.01×10^5 N/m² respectively (corresponding to standard sea level conditions). The fractional change in velocity at the point is 0.01. Calculate the fractional change in density.

    c.   Repeat part (b), except for a local velocity at the point of 1000 m/s (a high-speed flow). Compare this result with that from part (b), and comment on the differences.

CHAPTER 2

Integral Forms of the Conservation Equations for Inviscid Flows

Mathematics up to the present day have been quite useless to us in regard to flying.
From the 14th Annual Report of the Aeronautical Society of Great Britain, 1879

Mathematical theories from the happy hunting grounds of pure mathematicians are found suitable to describe the airflow produced by aircraft with such excellent accuracy that they can be applied directly to airplane design.
Theodore von Karman, 1954

PREVIEW BOX

The common phrase "you can not get something for nothing," besides holding in everyday life, and besides representing a colloquial statement of the second law of thermodynamics, also is relevant to the present chapter. Most students of engineering are anxious to get to the exciting practical applications that form the core of their profession; this is usually the reason for their interest in engineering in the first place. A study of compressible flow is no different—we would love to jump right in and design a supersonic airplane, or learn about rocket engines. But at this early stage in our studies we have no theoretical tools to design anything or to gain an understanding of any exciting application. We first have to acquire the necessary theoretical tools—the fundamental equations that govern the flow of a compressible fluid. Such tool gathering is the main purpose of this chapter. Here we will convert three fundamental physical principles into equations that will be the first tools to go into our toolbox for the study of compressible flow. As we proceed through this book, other theoretical tools progressively will be added to our toolbox. In the process, these tools will enable us to understand and quantify

progressively more exciting and challenging applications. This chapter is all about fundamental equations. The deviation of these equations is an intellectual exercise that is interesting in and of itself. So sit back and let yourself enjoy the intellectual gems to be found here.

To further orient ourselves, return to Fig. 1.7, which is the general roadmap for this book. The present chapter deals with boxes 2 and 3 in Fig. 1.7. The roadmap for the present chapter is given in Fig. 2.1. This chapter deals exclusively with three familiar fundamental physical principles, and how they are applied to a compressible flow: (1) mass can be neither created nor destroyed; (2) Newton's second law; and (3) the first law of thermodynamics. This chapter is all about converting these word statements into corresponding equations labeled, respectively, the integral forms of the continuity, momentum, and energy equations. However, this chapter is not devoid of applications. We end the chapter with a detailed derivation of the thrust equation for a jet propulsion device. This is a beautiful application of the integral form of the momentum equation in order to obtain a very practical result.

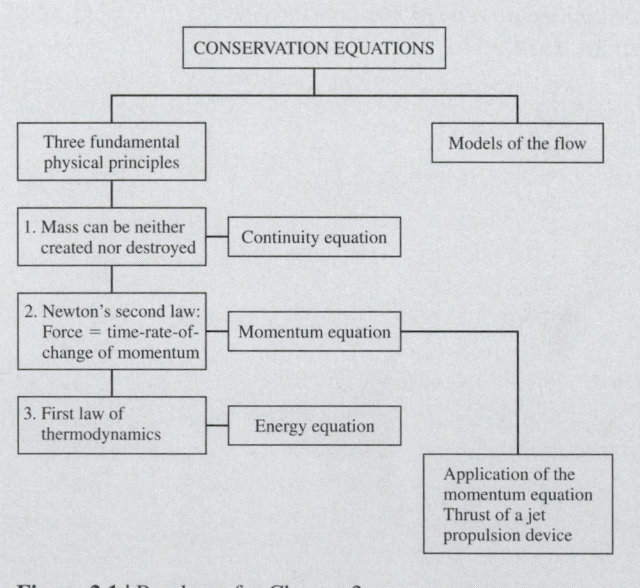

Figure 2.1 | Roadmap for Chapter 2.

2.1 | PHILOSOPHY

Consider the flowfield over an arbitrary aerodynamic body. We are interested in calculating the properties (p, ρ, T, $\mathbf{V}$, etc.) of the flowfield at all points within the flow. Why? Because, if we can calculate the flow properties throughout the flow, then we can certainly compute them on the *surface* of the body. In turn, from the surface distributions of p, T, ρ, $\mathbf{V}$, etc., we can compute the aerodynamic forces (lift and drag), moments, and heat transfer on the body. Indeed, the calculation of such practical information is one of the main functions of theoretical fluid mechanics, whether the body be a supersonic missile in flight, a submarine under water, or a high-rise apartment building in a hurricane. The essential point here is that in order to obtain practical information on engineering devices involving fluid flows, it is frequently necessary to approach the theoretical solution of the complete flowfield.

How do we calculate the flowfield properties? The answer is from equations, algebraic, differential, or integral, which relate p, ρ, T, $\mathbf{V}$, etc., to each other, along with suitable boundary conditions for the problem. The equations are obtained from the fundamental laws of nature applied to fluid flows. These laws and equations are a necessary prerequisite for an understanding of compressible flow. Therefore, let us proceed to establish these fundamental results.

2.2 | APPROACH

In obtaining the basic equations of fluid motion, the following approach is always taken:

1. Choose the appropriate fundamental physical principles from the laws of nature, such as
    a. Mass is conserved.
    b. Force = mass × acceleration.
    c. Energy is conserved.
2. Apply these physical principles to a suitable model of the flow.
3. From this application, extract the mathematical equations which embody such physical principles.

We first consider step 2, namely, what constitutes a suitable model of the flow? This is a somewhat subtle question. In contrast to the dynamics of well-defined solid bodies, on which it is usually apparent where to apply forces and moments, the dynamics of a fluid are complicated by the "squishy" nature of a rather elusive continuous medium that generally extends over large regions in space. Consequently, fluid dynamicists have to focus on specific regions of the flow, and apply the fundamental laws to a subscale model of the fluid motion. Three such models can be employed.

2.2.1   Finite Control Volume Approach

Consider a general flowfield, as represented by the streamlines in Fig. 2.2. Let us imagine a closed volume drawn within a finite region of the flow. This is defined as a *control volume* with volume $\mathcal{V}$ and surface area S. The control volume may be

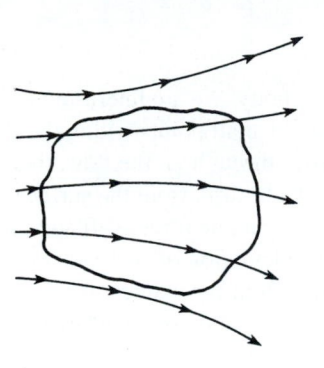
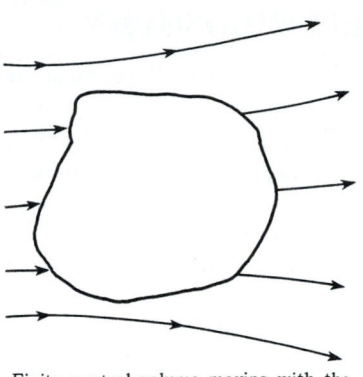

Finite control volume fixed in space with the fluid moving through it.

Finite control volume moving with the fluid such that the same fluid particles are always in the same control volume

Figure 2.2 | Finite control volume approach.

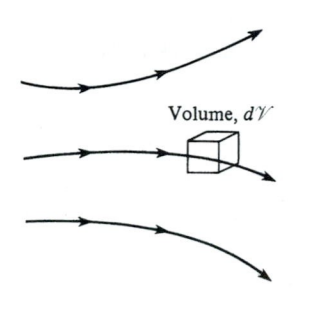
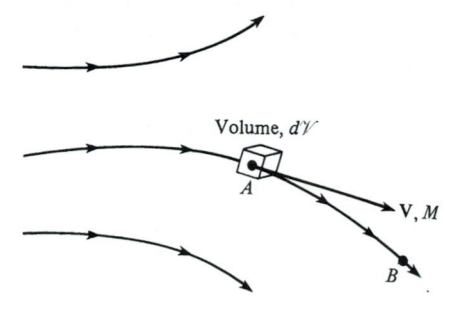

Infinitesimal fluid element fixed in space with the fluid moving through it

Infinitesimal fluid element moving along a streamline with the velocity **V** equal to the flow velocity at each point

Figure 2.3 | Infinitesimal fluid element approach.

either *fixed* in space with the fluid moving through it, or *moving* with the fluid such that the same fluid particles are always inside it.

With the application of the already mentioned fundamental physical principles to these finite control volumes, fixed or moving, integral equations for the fluid properties can be directly obtained. With some further manipulation, differential equations for the fluid properties can be indirectly extracted.

2.2.2 Infinitesimal Fluid Element Approach

Consider a general flowfield as represented by the streamlines in Fig. 2.3. Let us imagine an infinitesimally small *fluid element* in the flow, with volume $d\mathcal{V}$. The fluid element is infinitesimal in the same sense as differential calculus; however, it is large enough to contain a huge number of molecules so that it can be viewed as a continuous medium (see the discussion of a continuum in Sec. 1.3). The fluid element may be fixed in space with the fluid moving through it, or it may be moving along a streamline with velocity **V** equal to the flow velocity at each point. With the application of the fundamental physical principles to these fluid elements, fixed or moving, differential equations for the fluid properties can be directly obtained.

2.2.3   Molecular Approach

In actuality, of course, the motion of a fluid is a ramification of the mean molecular motion of its particles. Therefore, a third model of the flow can be a microscopic approach wherein the fundamental laws of nature are applied directly to the molecules, with suitable statistical averaging. This leads to the Boltzmann equation from kinetic theory, from which the governing differential equations for the fluid properties can be extracted. This is an elegant approach, with many advantages in the long run. However, it is beyond the scope of the present book. The reader should consult the authoritative book by Hirchfelder, Curtis, and Bird (Ref. 19) for more details.

In summary, although many variations on the theme can be found in different texts for the derivation of the general equations of fluid flow, the flow model can usually be categorized as one of the approaches described above. For the sake of consistency, the model of a fixed finite control volume will be employed for the remainder of this chapter.

2.3 | CONTINUITY EQUATION

2.3.1   Physical Principle

Mass Can Be Neither Created Nor Destroyed.   Let us apply this principle to the model of a fixed control volume in a flow, as illustrated in Fig. 2.4. The volume is $\mathcal{V}$, and the area of the closed surface is S. First, consider point B on the control surface and an elemental area around B, dS. Let **n** be a unit vector normal to the surface at B. Define $d\mathbf{S} = \mathbf{n}\, dS$. Also, let **V** and ρ be the local velocity and density at B.

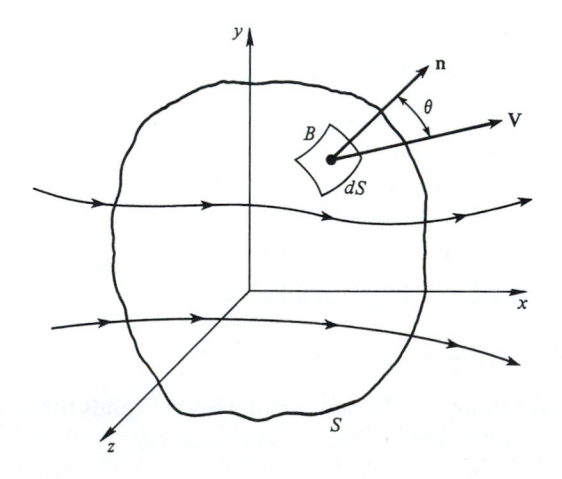

Figure 2.4 | Fixed control volume for derivation of the governing equations.

The mass flow (slug/s or kg/s) through any elemental surface arbitrarily oriented in a flowing fluid is equal to the product of density, the component of velocity normal to the surface, and the area. (Prove this to yourself.) Letting $\dot{m}$ denote the mass flow through dS, and referring to Fig. 2.4,

$$\dot{m} = \rho(V\cos\theta)\,dS = \rho V_n\,dS = \rho\mathbf{V}\cdot d\mathbf{S} \tag{2.1}$$

[*Note:* The product ρV_n is called the *mass flux,* i.e., the flow of mass per unit area per unit time. Whenever you see a product of (density $\times$ velocity) in fluid mechanics, it can always be interpreted as mass flow per second per unit area perpendicular to the velocity vector.] The net mass flow *into* the control volume through the entire control surface S is the sum of the elemental mass flows from Eq. (2.1), namely,

$$-\oiint_S \rho\mathbf{V}\cdot d\mathbf{S}$$

where the minus sign denotes inflow (in the opposite direction of $\mathbf{V}$ and $d\mathbf{S}$ in Fig. 2.4). Consider now an infinitesimal volume $d\mathcal{V}$ inside the control volume. The mass of this infinitesimal volume is $\rho\,d\mathcal{V}$. Hence, the total mass inside the control volume is the sum of these elemental masses, namely,

$$\oiiint_\mathcal{V} \rho\,d\mathcal{V}$$

The time rate of change of this mass inside the control volume is therefore

$$\frac{\partial}{\partial t}\oiiint_\mathcal{V} \rho\,d\mathcal{V}$$

Finally, the physical principle that mass is conserved (given at the beginning of this section) states that the net mass flow into the control volume must equal the rate of increase of mass inside the control volume. In terms of the integrals just given, a mathematical representation of this statement is simply

$$-\oiint_\mathbf{S} \rho\mathbf{V}\cdot d\mathbf{S} = \frac{\partial}{\partial t}\oiiint_\mathcal{V} \rho\,d\mathcal{V} \tag{2.2}$$

This equation is called the *continuity equation;* it is the integral formulation of the conservation of mass principle as applied to a fluid flow. Equation (2.2) is quite general; it applies to all flows, compressible or incompressible, viscous or inviscid.

2.4 | MOMENTUM EQUATION

2.4.1   Physical Principle

The Time Rate of Change of Momentum of a Body Equals the Net Force Exerted on It.   Written in vector form, this statement becomes

$$\frac{d}{dt}(m\mathbf{V}) = \mathbf{F} \tag{2.3}$$

For constant mass, Eq. (2.3) yields

$$\mathbf{F} = m\frac{d\mathbf{V}}{dt} = m\mathbf{a} \tag{2.4}$$

which is the more familiar form of Newton's second law, namely, that force = mass × acceleration. However, the physical principle with Eq. (2.3) is a more general statement of Newton's second law than Eq. (2.4). In this section, we wish to put Newton's second law [Eq. (2.3)] in fluid mechanic terms by employing the same control volume utilized in Sec. 2.3 and sketched in Fig. 2.4.

First, consider the forces on the control volume. Using some intuitive physical sense, we can visualize these forces as two types:

1. *Body forces* acting on the fluid inside $\mathscr{V}$. These forces stem from "action at a distance," such as gravitational and electromagnetic forces that may be exerted on the fluid inside $\mathscr{V}$ due to force fields acting through space. Let $\mathbf{f}$ represent the body force per unit mass of fluid. Considering an elemental volume, $d\mathscr{V}$, inside $\mathscr{V}$, the elemental body force on $d\mathscr{V}$ is equal to the product of its mass and the force per unit mass, namely, $(\rho\, d\mathscr{V})\mathbf{f}$. Hence, summing over the complete control volume,

$$\text{Total body force} = \iiint\limits_{\mathscr{V}} \rho\mathbf{f}\, d\mathscr{V} \tag{2.5}$$

2. *Surface forces* acting on the boundary of the control volume. As discussed in Sec. 1.5, surface forces in a fluid stem from two sources: pressure and shear stress distributions over the surface. Since we are dealing with inviscid flows here, the only surface force is therefore due to pressure. Consider the elemental area $d\mathbf{S}$ sketched in Fig. 2.4. The elemental surface force acting on this area is $-p\, d\mathbf{S}$, where the minus sign signifies that pressure acts inward, opposite to the outward direction of the vector $d\mathbf{S}$. Hence, summing over the complete control surface,

$$\text{Total surface force due to pressure} = -\iint\limits_{S} p\, d\mathbf{S} \tag{2.6}$$

Note that the sum of Eqs. (2.5) and (2.6) represent $\mathbf{F}$ in Eq. (2.3). That is, at any given instant in time, the total force $\mathbf{F}$ acting on the control volume is

$$\mathbf{F} = \iiint\limits_{\mathscr{V}} \rho\mathbf{f}\, d\mathscr{V} - \iint\limits_{S} p\, d\mathbf{S} \tag{2.7}$$

[Please note that, if an aerodynamic body were inserted inside the control volume, there would be an additional force on the fluid—the equal and opposite reaction to the force on the body. However, in dealing with control volumes, it is always possible to wrap the control surface around the body in such a fashion that the body is always *outside* the control volume, and the body force then shows up as part of the pressure distribution on the control surface. This is already taken into account by the last term in Eq. (2.7).]

Now consider the left-hand side of Eq. (2.3). In terms of our fluid dynamic model, how is the time rate of change of momentum, $m(d\mathbf{V}/dt)$, expressed? To answer this question, again use some physical intuition. Look at the control volume in Fig. 2.4. Because it is fixed in space, mass flows into the control volume from the left at the same time that other mass is streaming out toward the right. The mass flowing in brings with it a certain momentum. At the same time, the mass flowing out also has momentum. With this picture in mind, let $\mathbf{A}_1$ represent the *net* rate of flow of momentum across the surface S. The elemental mass flow across dS is given by Eq. (2.1) as $\rho\mathbf{V} \cdot d\mathbf{S}$. With this elemental mass flow is associated a momentum flow (or flux) $(\rho\mathbf{V} \cdot d\mathbf{S})\mathbf{V}$. Note from Fig. 2.4 that, when the direction of $\mathbf{V}$ is *away* from the control volume, this physically represents an *outflow* of momentum and mathematically represents a positive value of $\mathbf{V} \cdot d\mathbf{S}$. Conversely, when the direction of $\mathbf{V}$ is *toward* the control volume, this physically represents an *inflow* of momentum and mathematically represents a negative value of $\mathbf{V} \cdot d\mathbf{S}$. The *net* rate of flow of momentum, summed over the complete surface S, is

$$\mathbf{A}_1 = \oiint_S (\rho\mathbf{V} \cdot d\mathbf{S})\mathbf{V} \tag{2.8}$$

At this stage, it would be tempting to claim that $\mathbf{A}_1$ represents the left-hand side of Eq. (2.3). However, consider an *unsteady* flow, where, by definition, the flow properties at any given point in the flowfield are functions of time. Examples would be the flow over a body that is oscillating back and forth with time, and the flow through a nozzle where the supply valves are being twisted off and on. If our control volume in Fig. 2.4 were drawn in such an unsteady flow, then the momentum of the fluid *inside* the control volume would be fluctuating with time simply due to the time variations in ρ and $\mathbf{V}$. Therefore, $\mathbf{A}_1$ does *not* represent the whole contribution to the left-hand side of Eq. (2.3). There is, in addition, a time rate of change of momentum due to unsteady, transient effects in the flowfield inside $\mathcal{V}$. Let $\mathbf{A}_2$ represent this fluctuation in momentum. Also consider an elemental mass of fluid, $\rho\, d\mathcal{V}$. This mass has momentum $(\rho\, d\mathcal{V})\mathbf{V}$. Summing over the complete control volume $\mathcal{V}$, we have

$$\text{Total momentum inside } \mathcal{V} = \oiiint_{\mathcal{V}} \rho\mathbf{V}\, d\mathcal{V}$$

Hence, the *change* in momentum in $\mathcal{V}$ due to unsteady fluctuations in the local flow properties is

$$\mathbf{A}_2 = \frac{\partial}{\partial t} \oiiint_{\mathcal{V}} \rho\mathbf{V}\, d\mathcal{V} = \oiiint_{\mathcal{V}} \frac{\partial(\rho\mathbf{V})}{\partial t}\, d\mathcal{V} \tag{2.9}$$

[Note that in Eq. (2.9) the partial derivative can be taken inside the integral because we are considering a volume of integration that is fixed in space. If the limits of integration were not fixed, then Leibnitz's rule from calculus would yield a different form for the right-hand term of Eq. (2.9).]

Finally, the sum $\mathbf{A}_1 + \mathbf{A}_2$ represents the total instantaneous time rate of change of momentum of the fluid as it flows through the control volume. This is the fluid mechanical counterpart of the left-hand side of Eq. (2.3), i.e.,

$$\frac{d}{dt}(m\mathbf{V}) = \mathbf{A}_1 + \mathbf{A}_2 = \oiint_S (\rho\mathbf{V} \cdot d\mathbf{S})\mathbf{V} + \iiint_{\mathcal{V}} \frac{\partial(\rho\mathbf{V})}{\partial t} \, d\mathcal{V} \qquad (2.10)$$

Therefore, to repeat the physical principle stated at the beginning of this section, the time rate of change of momentum of the fluid that is flowing through the control volume at any instant is equal to the net force exerted on the fluid inside the volume. In turn, these words can be directly translated into an equation by combining Eqs. (2.3), (2.7), and (2.10):

$$\boxed{\oiint_S (\rho\mathbf{V} \cdot d\mathbf{S})\mathbf{V} + \iiint_{\mathcal{V}} \frac{\partial(\rho\mathbf{V})}{\partial t} \, d\mathcal{V} = \iiint_{\mathcal{V}} \rho\mathbf{f}\, d\mathcal{V} - \oiint_S p\, d\mathbf{S}} \qquad (2.11)$$

Equation (2.11) is called the *momentum equation;* it is the integral formulation of Newton's second law applied to inviscid fluid flows. Note that Eq. (2.11) does *not* include the effects of friction. If friction were to be included, it would appear as an additional surface force, namely, shear and normal viscous stresses integrated over the control surface. If $\mathbf{F}_{\text{viscous}}$ represents this surface integral, then Eq. (2.11), modified for the inclusion of friction, becomes:

$$\oiint_S (\rho\mathbf{V} \cdot d\mathbf{S})\mathbf{V} + \iiint_{\mathcal{V}} \frac{\partial(\rho\mathbf{V})}{\partial t} \, d\mathcal{V} = \iiint_{\mathcal{V}} \rho\mathbf{f}\, d\mathcal{V} - \oiint_S p\, d\mathbf{S} + \mathbf{F}_{\text{viscous}} \qquad (2.11a)$$

Since this book mainly treats inviscid flows, Eq. (2.11) is of primary interest here, rather than Eq. (2.11*a*).

2.5 | A COMMENT

The continuity equation, Eq. (2.2), and the momentum equation, Eq. (2.11), despite their complicated-looking integral forms, are powerful tools in the analysis and understanding of fluid flows. Although it may not be apparent at this stage in our discussion, these conservation equations will find definite practical applications in subsequent chapters. It is important to become familiar with these equations and with the energy equation to be discussed next, and to understand fully the physical fundamentals they embody.

For a study of incompressible flow, the continuity and momentum equations are sufficient tools to do the job. These equations govern the mechanical aspects of such flows. However, for a compressible flow, the principle of the conservation of energy must be considered in addition to the continuity and momentum equations, for the reasons discussed in Sec. 1.4. The energy equation is where thermodynamics enters the game of compressible flow, and this is our next item of business.

2.6 | ENERGY EQUATION

2.6.1 Physical Principle

Energy Can Be Neither Created Nor Destroyed; It Can Only Change in Form.
This fundamental principle is contained in the first law of thermodynamics, Eq. (1.24).
Let us apply the first law to the fluid flowing through the fixed control volume in
Fig. 2.4. Let

B_1 = rate of heat added to the fluid inside the control volume from the
      surroundings
B_2 = rate of work done on the fluid inside the control volume
B_3 = rate of change of the energy of the fluid as it flows through the
      control volume

From the first law,

$$B_1 + B_2 = B_3 \tag{2.12}$$

First, consider the rate of heat transferred to or from the fluid. This can be visualized as volumetric heating of the fluid inside the control volume due to the absorption of radiation orginating outside the system, or the local emission of radiation by
the fluid itself, if the temperature inside the control volume is high enough. Also, if
the flow were viscous, there could be heat transferred across the boundary by thermal
conduction and diffusion; however, these effects are not considered here. Finally, if
the flow were chemically reacting, it might be tempting to consider energy released
or absorbed by such reactions as a volumetric heating term. This is done in many
treatments of reacting flows. However, the energy exchange due to chemical reactions is more fundamentally treated as part of the overall internal energy of the gas
mixture and not as a separate heating term in the energy equation. This matter will be
discussed at length in Chaps. 16 and 17. In any event, we can simply handle the rate
of heat added to the control volume by first defining $\dot{q}$ to be the rate of heat added
per unit mass, and then writing the rate of heat added to an elemental volume as
$\dot{q}(\rho \, d\mathcal{V})$. Summing over the complete control volume,

$$B_1 = \iiint\limits_{\mathcal{V}} \dot{q}\rho \, d\mathcal{V} \tag{2.13}$$

Before considering the rate of work done on the fluid inside the control volume,
consider a simpler case of a solid object in motion, with a force **F** being exerted on
the object, as sketched in Fig. 2.5. The position of the object is measured from a fixed
origin by the radius vector **r**. In moving from position $\mathbf{r}_1$ to $\mathbf{r}_2$ over an interval of time
dt, the object is displaced through $d\mathbf{r}$. By definition, the work done on the object
in time dt is $\mathbf{F} \cdot d\mathbf{r}$. Hence, the time rate of doing work is simply $\mathbf{F} \cdot d\mathbf{r}/dt$. But
$d\mathbf{r}/dt = \mathbf{V}$, the velocity of the moving object. Hence, we can state that

$$\left[\begin{array}{l} \text{The rate of doing work} \\ \text{on a moving body} \end{array} \right] = \mathbf{F} \cdot \mathbf{V}$$

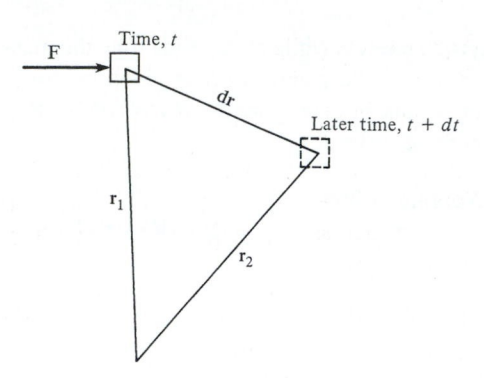

Figure 2.5 | Rate of doing work.

In words, the rate of work done on a moving body is equal to the product of its velocity and the component of force in the direction of the velocity.

This result leads to an expression for B_2, as follows. Consider the elemental area $d\mathbf{S}$ of the control surface in Fig. 2.4. The pressure force on this elemental area is $-p\,d\mathbf{S}$, as explained in Sec. 2.4. From the result just reached, the rate of work done on the fluid passing through $d\mathbf{S}$ with velocity $\mathbf{V}$ is $(-p\,d\mathbf{S}) \cdot \mathbf{V}$. Hence, summing over the complete control surface,

$$\left[\begin{array}{l}\text{Rate of work done on}\\\text{the fluid inside } \mathcal{V} \text{ due}\\\text{to pressure forces on } S\end{array}\right] = -\oiint_S (p\,d\mathbf{S}) \cdot \mathbf{V} \qquad (2.14)$$

In addition, consider an elemental volume inside the control volume. Recalling that $\mathbf{f}$ is the body force per unit mass, the rate of work done on the elemental volume due to body force is $(\rho\mathbf{f}\,d\mathcal{V}) \cdot \mathbf{V}$. Summing over the complete control volume,

$$\left[\begin{array}{l}\text{Rate of work done on}\\\text{the fluid inside } \mathcal{V} \text{ due}\\\text{to body forces}\end{array}\right] = \iiint_{\mathcal{V}} (\rho\mathbf{f}\,d\mathcal{V}) \cdot \mathbf{V} \qquad (2.15)$$

Thus, the total work done on the fluid inside the control volume is the sum of Eqs. (2.14) and (2.15),

$$B_2 = -\oiint_S p\mathbf{V} \cdot d\mathbf{S} + \iiint_{\mathcal{V}} \rho(\mathbf{f} \cdot \mathbf{V})\,d\mathcal{V} \qquad (2.16)$$

To visualize the energy inside the control volume, recall that in Sec. 1.4 the system was stationary and the energy inside the system was the *internal* energy e (per unit mass). However, the fluid inside the control volume in Fig. 2.4 is not stationary; it is moving at the local velocity $\mathbf{V}$ with a consequent *kinetic* energy per unit mass of $V^2/2$. Hence, the energy per unit mass of the moving fluid is the sum of both internal and kinetic energies, $e + V^2/2$.

Keep in mind that mass flows into the control volume of Fig. 2.4 from the left at the same time that other mass is streaming out towards the right. The mass flowing in brings with it a certain energy, while at the same time the mass flowing out also has energy. The elemental mass flow across dS is given by Eq. (2.1) as $\rho \mathbf{V} \cdot d\mathbf{S}$ and therefore the elemental flux of energy across dS is $(\rho \mathbf{V} \cdot d\mathbf{S})(e + V^2/2)$. Summing over the complete control surface,

$$\begin{bmatrix} \text{Net rate of flow} \\ \text{of energy across} \\ \text{the control surface} \end{bmatrix} = \oiint_S (\rho \mathbf{V} \cdot d\mathbf{S}) \left(e + \frac{V^2}{2} \right) \tag{2.17}$$

However, this is not necessarily the total energy change inside the control volume. Analogous to the discussion surrounding Eq. (2.9), if the flow is unsteady there is also a rate of change of energy due the local transient fluctuations of the flow-field variables inside the control volume. The energy of an elemental volume is $\rho(e + V^2/2)\, d\mathcal{V}$, and hence the energy inside the complete control volume at any instant in time is

$$\iiint_{\mathcal{V}} \rho \left(e + \frac{V^2}{2} \right) d\mathcal{V}$$

Therefore,

$$\begin{bmatrix} \text{Time rate of change} \\ \text{of energy inside } \mathcal{V} \text{ due} \\ \text{to transient variations} \\ \text{of the flowfield variables} \end{bmatrix} = \frac{\partial}{\partial t} \iiint_{\mathcal{V}} \rho \left(e + \frac{V^2}{2} \right) d\mathcal{V} \tag{2.18}$$

In turn, B_3 is the sum of Eqs. (2.17) and (2.18):

$$B_3 = \frac{\partial}{\partial t} \iiint_{\mathcal{V}} \rho \left(e + \frac{V^2}{2} \right) d\mathcal{V} + \oiint_S (\rho \mathbf{V} \cdot d\mathbf{S}) \left(e + \frac{V^2}{2} \right) \tag{2.19}$$

Repeating the physical principle stated at the beginning of this section, the rate of heat added to the fluid plus the rate of work done on the fluid is equal to the rate of change of energy of the fluid as it flows through the control volume, i.e., *energy is conserved.* In turn, these words can be directly translated into an equation by combining Eqs. (2.12), (2.13), (2.16), and (2.19):

$$\boxed{\begin{aligned} \iiint_{\mathcal{V}} \dot{q}\rho\, d\mathcal{V} &- \oiint_S p\mathbf{V} \cdot d\mathbf{S} + \iiint_{\mathcal{V}} \rho(\mathbf{f} \cdot \mathbf{V})\, d\mathcal{V} \\ &= \iiint_{\mathcal{V}} \frac{\partial}{\partial t} \left[\rho \left(e + \frac{V^2}{2} \right) \right] d\mathcal{V} + \oiint_S \rho \left(e + \frac{V^2}{2} \right) \mathbf{V} \cdot d\mathbf{S} \end{aligned}} \tag{2.20}$$

Equation (2.20) is called the *energy equation;* it is the integral formulation of the first law of thermodynamics applied to an inviscid fluid flow.

Note that Eq. (2.20) does *not* include these phenomena:

1. The rate of work done on the fluid inside the control volume by a rotating shaft that crosses the control surface, $\dot{W}_{shaft}$.
2. The rate of work done by viscous stresses on the control surface, $\dot{W}_{viscous}$.
3. The heat added across the control surface due to thermal conduction and diffusion. In combination with radiation, denote the total rate of heat addition from all these effects as $\dot{Q}$.

If all of these phenomena were included, then Eq. (2.20) would be modified as

$$\dot{Q} + \dot{W}_{shaft} + \dot{W}_{viscous} - \oiint_S p\mathbf{V} \cdot d\mathbf{S} + \iiint_\mathcal{V} \rho(\mathbf{f} \cdot \mathbf{V})\, d\mathcal{V}$$

$$= \iiint_\mathcal{V} \frac{\partial}{\partial t}\left[\rho\left(e + \frac{V^2}{2}\right)\right] d\mathcal{V} + \oiint_S \rho\left(e + \frac{V^2}{2}\right)\mathbf{V} \cdot d\mathbf{S} \qquad (2.20a)$$

For the inviscid flows treated in this book, there is no thermal conduction or diffusion and there is no work done by viscous stresses. Moreover, for the basic flow problems discussed in later chapters, there is no shaft work. Therefore, Eq. (2.20) is of primary interest here, rather than Eq. (2.20a).

2.7 | FINAL COMMENT

The three conservation equations derived, Eqs. (2.2), (2.11), and (2.20), in conjunction with the equation of state

$$p = \rho RT$$

and the thermodynamic relation

$$e = e(T, v)$$

(which simplifies to $e = c_v T$ for a calorically perfect gas) are sufficient tools to analyze inviscid compressible flows of an equilibrium gas—including equilibrium chemically reacting gases. The more complex case of a nonequilibrium gas will be treated in Chaps. 16 and 17. The conservation equations have been derived in integral form in this chapter; however, in Chap. 6 we will extract partial differential equations of continuity, momentum, and energy from these integral forms. In the meantime, we will do something even simpler: In the applications treated in Chaps. 3 through 5, the integral forms presented here will be applied to important, practical problems where algebraic equations fortunately can be extracted for the conservation principles.

Finally, note that Eqs. (2.2), (2.11), and (2.20) are written in vector notation, and therefore have the advantage of not being limited to any one particular coordinate system: cartesian, cylindrical, spherical, etc. These equations describe the motion of an inviscid fluid in three dimensions. They speak words—mass is conserved, force = mass × acceleration, and energy is conserved. Never let the mathematical formulation cause you to lose sight of the physical meaning of these equations. In their integral formulation they are particularly powerful equations from which all of our subsequent analyses will follow.

2.8 | AN APPLICATION OF THE MOMENTUM EQUATION: JET PROPULSION ENGINE THRUST

The integral form of the conservation equations is immediately useful for many practical applications. We discuss one such important application here—the calculation of the thrust of a jet propulsion device, such as a gas turbine jet engine, or a rocket engine. Our purpose here simply is to illustrate the power of the equations derived in this chapter. However, our choice of application to jet propulsion is not entirely arbitrary, because a study of flight propulsion is a fertile field for the principles of compressible flow, as discussed in the preview box for Chap. 1.

This section highlights two important principles that we have already discussed:

1. The force exerted on a body by the fluid flow over or through the body is due only to the pressure distribution and the shear stress distribution exerted over the entire exposed surface of the body [see Sec. 1.5 and Eq. (1.45)].

2. The integral form of the momentum equation [see Sec. 2.4 and Eq. (2.11)].

All jet propulsion engines—turbojet engines, turbofans, ramjets, rockets, etc.—depend on the flow of a gas through and around the engines. In turn, this gas flow creates a pressure and shear stress distribution that are exerted over all the exposed surface areas of the engine, and it is the net integrated result of these two local distributions that is the source of the thrust from the engine. The pressure and shear stress distributions can be very complex, such as those exerted over the compressor blades, combustor cans, turbine blades, and the nozzle of a turbojet engine, or more simple such as those exerted over the walls of the combustion chamber and exhaust nozzle of a rocket engine. In each case, however, it is these two hands of nature—the pressure and shear stress distributions—that reach out, grab hold of the engine, and create the thrust.

It would seem, therefore, that the calculation of the thrust of a jet propulsion device would require detailed theoretical or experimental measurements of pressure and shear stress distributions exerted over every component of the engine. Obtaining such complex data is most formidable to say the least. Fortunately, it is not necessary, because the integral form of the momentum equation leads to a much simpler means to calculate the thrust of a jet propulsion device. The purpose of this section is to show how this is done, and to obtain a straightforward equation for the thrust of a jet propulsion device. In the process, we will highlight the tremendous advantage that sometimes comes from the use of the integral forms of the conservation equations derived in this chapter.

The pressure distribution is by far the dominant contributor to the thrust; the shear stress distribution has only a very small effect. Therefore, in what follows we

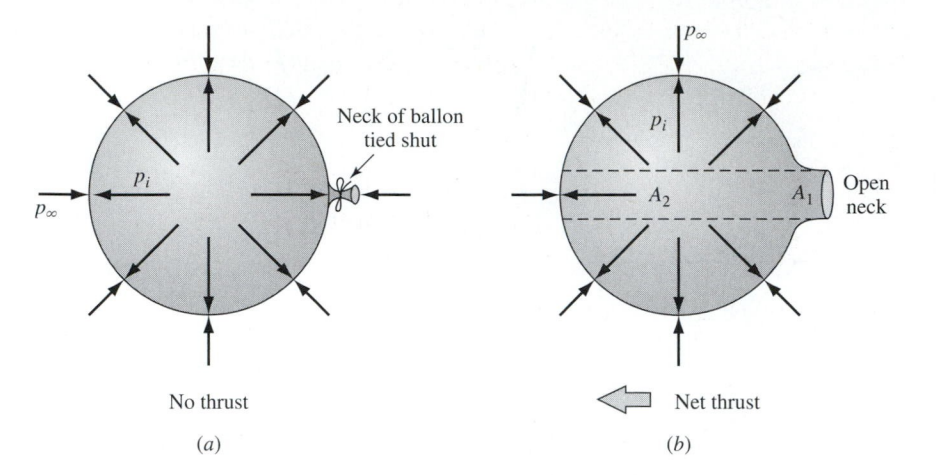

Figure 2.6 | Illustration of thrust on a balloon.

will neglect shear stress and consider the pressure distribution only. Also, the simplest example of how pressure creates thrust is to consider a toy rubber balloon, sketched in Fig. 2.6. Imagine that you inflate the balloon with air, tie the neck of the balloon shut, and let go. The balloon will gradually sink to the ground under its own weight, but it will not surge forward because there is no net thrust exerted on the balloon. This is because the pressure distribution over the inside and outside surfaces of the balloon integrates to a zero net force. This is sketched in Fig. 2.6a, where the external atmospheric pressure is p_∞ and the slightly higher internal pressure is p_i. The external pressure p_∞ is equal on all parts of the closed external surface, and hence integrates to a zero net force. Similarly, the internal pressure p_i is equal on all parts of the closed internal surface, and hence also integrates to a zero net force. As a result, there is no net pressure force on the balloon, i.e., no thrust. However, after you inflate the balloon, imagine that you do not tie the neck shut, but rather pinch it shut with your fingers for a moment, and then let go. The balloon will scoot forward and propel itself through the air for a few moments. This case is illustrated in Fig. 2.6b. Here, the neck of the balloon is open with area A_1. The equal projected area on the opposite side of the balloon is A_2. The internal pressure p_i acts on the rubber surface A_2, tending to push the balloon to the left. However, there is no corresponding rubber surface area at A_1 for p_i to push the balloon to the right, as is the case in Fig. 2.6a. As a result, there is an imbalance of forces on the balloon in Fig. 2.6b, resulting in a net thrust propelling the balloon to the left. The thrust is essentially equal to $(p_i - p_\infty)A_2$. This is the simplest example of how pressure distribution is the source of thrust for a jet propulsion device, the device in this case being an inflated balloon scooting through the air, with a jet of air exhausting in the opposite direction through its open neck. The fundamental idea is the same for all jet propulsion devices.

Let us now consider the generic jet propulsion device sketched in Fig. 2.7a. The device is represented by a duct through which air flows into the inlet at the left, is pressurized, is burned with fuel inside the duct, and is exhausted out the exit with an exit

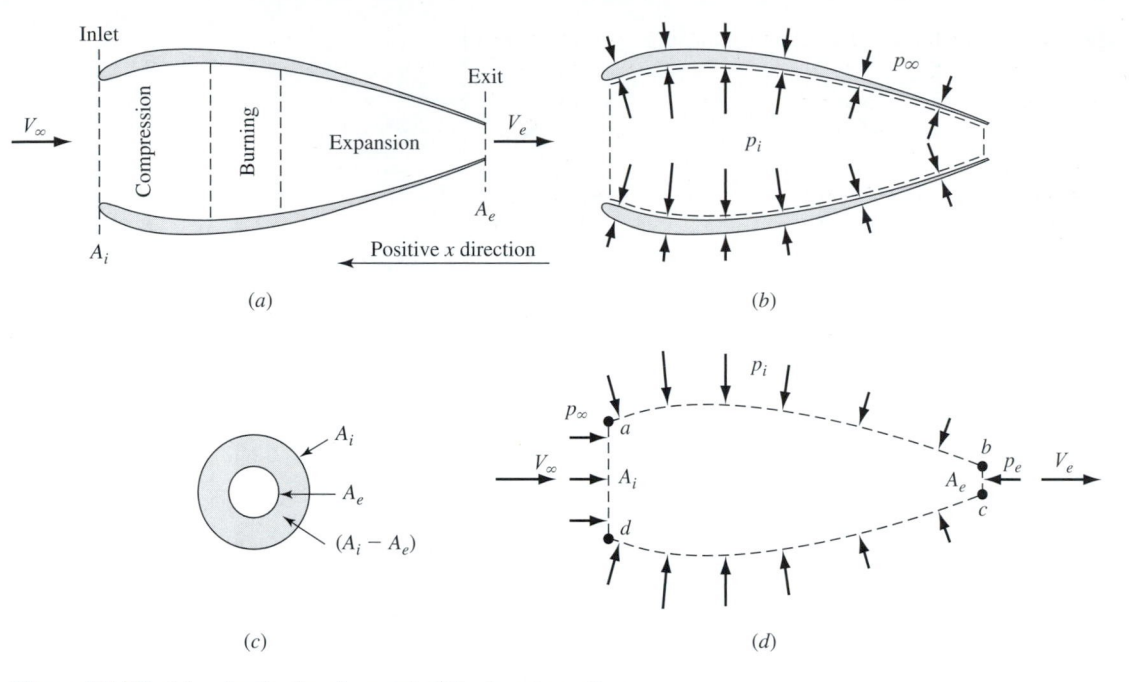

Figure 2.7 | Sketches for the development of the thrust equation.

jet velocity, V_e. The internal pressure acting on the inside surface of the engine is p_i, which varies with location inside the engine, as sketched in Fig. 2.7b. The external pressure acting on the outside surface of the engine is assumed to be the free-stream ambient pressure p_∞, constant over the outside surface. (This, of course, is not correct because the pressure will vary as the air flows over the curved outside surface. However, for an actual engine, the duct shown in Figs. 2.7a and b will be installed in some type of housing, or nacelle, on a flight vehicle, which will certainly affect the external air pressure. The assumption of constant p_∞ on the outer surface as sketched in Fig. 2.7b yields a thrust value that is *defined* as the *uninstalled* engine thrust. Hence, in this section we are deriving an equation for the uninstalled engine thrust.)

The net force on the engine due to the pressure distribution is given by Eq. (1.45). With the shear stress neglected, this yields

$$\mathbf{F} = -\oiint p \, d\mathbf{S} \tag{2.21}$$

Recall that the minus sign in Eq. (2.21) is due to $d\mathbf{S}$ being directed away from the surface, whereas the pressure exerts a force into the surface. The net force $\mathbf{F}$ is the thrust of the engine. Because of the symmetry of the flow and the engine shown in Fig. 2.7, $\mathbf{F}$ acts in the horizontal direction, which we will denote as the x direction. Hence, Eq. (2.21) can be written in scalar form as

$$T = -\int (p_i \, dS)_x - \int (p_\infty \, dS)_x \tag{2.22}$$

where the vector force $\mathbf{F}$ has been replaced by the scalar thrust T acting in the x direction. The subscript x denotes the x component of the vector $p\,d\mathbf{S}$, and the first and second terms on the right-hand side represent the integrated force due to the internal and external pressure distributions respectively. Let us take the positive x direction as that acting toward the *left*, as shown in Fig. 2.7*b*.

Consider the last term in Eq. (2.22). Since p_∞ is a constant value, the integral can be written as

$$\int (p_\infty\,dS)_x = p_\infty \int (dS)_x \qquad (2.23)$$

Recall from Fig. 2.7*b* that the integral is taken over the outer surface, and that the vector $d\mathbf{S}$ is directed away from the surface. For those vectors $d\mathbf{S}$ that are inclined towards the positive x direction (toward the left in Fig. 2.7*b*), $(dS)_x$ is positive, and for those that are inclined towards the negative x direction (toward the right in Fig. 2.7*b*), $(dS)_x$ is negative. Since $(dS)_x$ is the x component of the vector $d\mathbf{S}$, its absolute value is simply the projection of the elemental area as seen by looking along the x axis. Hence $|\int (dS)_x|$ is simply the net projected area of the solid surface as seen by looking along the x axis, which is the inlet area minus the exit area, $A_i - A_e$. This projected area is sketched in Fig. 2.7*c*. However, the *sign* of the integral $\int (dS)_x$ is determined by the net sum of the positive and negative components $(dS)_x$. When A_e is less than A_i, as is the case here, the sum of the negative components is greater than the sum of the positive components (more of the surface area has rearward sloping vectors $d\mathbf{S}$ than it has forward sloping). Hence, the sign of $\int (dS)_x$ is negative, and we must rewrite

$$\int (dS)_x = -|(dS)_x| = -(A_i - A_e)$$

Hence, Eq. (2.23) becomes

$$\int (p_\infty\,dS)_x = p_\infty \int (dS)_x = p_\infty(A_e - A_i) \qquad (2.24)$$

Substituting Eq. (2.24) into Eq. (2.22), we have

$$T = -\int (p_i\,dS)_x - p_\infty(A_e - A_i)$$

or,

$$T = -\int (p_i\,dS)_x + p_\infty(A_i - A_e) \qquad (2.25)$$

Recall that physically the last term in Eq. (2.25) is the force on the engine due to the constant p_∞ acting on the external surface. Since A_e is smaller than A_i, the force due to p_∞ acting on the rearward part of the surface pushing the engine toward the left in Fig. 2.7*b* is larger than the force due to p_∞ acting on the forward part of the surface, pushing the engine toward the right. Hence, physically the effect of p_∞ distributed over the external surface must be a force toward the left in Fig. 2.7*b*, i.e., *adding* to the thrust. The last term in Eq. (2.25), $p_\infty(A_i - A_e)$, is indeed a positive value, consistent with the physics discussed here.

Now consider the first term on the right-hand side of Eq. (2.25). Recall that it physically represents the force exerted *by the gas on the internal solid surface*. To make this explicit in the upcoming steps, we write Eq. (2.25) as

$$T = \underbrace{\left[- \int (p_i \, dS)_x \right.}_{\text{force on solid surface due to the gas}} \left. \vphantom{\int} \right] + p_\infty(A_i - A_e) \qquad (2.26)$$

To evaluate the integral in Eq. (2.26), we turn to the integral form of the momentum equation, Eq. (2.11). We apply this equation to the control volume defined by the dashed lines in Fig. 2.7*b*, where the upper and lower boundaries of the control volume are adjacent to the internal solid surface, and the left and right sides of the control volume are drawn perpendicular across the inlet and exit, respectively. The control volume is drawn in Fig. 2.7*d*. The dashed lines in Fig. 2.7*d* are not solid surfaces, but are simply the boundaries of the control volume that contains the gas that flows through the jet engine. We make the assumption that the gas flowing into the control volume through the inlet area A_i at the left enters at the free-stream velocity and pressure V_∞ and p_∞, respectively. The gas flowing out of the control volume through the exit area A_e at the right leaves at the exit velocity and pressure V_e and p_e, respectively. Along the upper and lower surfaces of the control volume, the *surroundings* (in this case the surroundings are the solid internal surfaces of the engine) exert a distributed pressure p_i directed into the control volume. This distributed pressure acting on the gas is *equal and opposite to* the distributed pressure acting on the solid surface as sketched in Fig. 2.7*b*. This is Newton's third law—for every action there is an equal and opposite reaction. For example, if you press your hand down on a desk with a force of 20 newtons, the desk presses back on your hand with an equal and opposite force of 20 newtons. By analogy, your hand is the gas exerting a pressure distribution on the internal surface of the engine (Fig. 2.7*b*), and the desk pressing back on your hand is the internal engine surface exerting an equal and opposite pressure distribution on the gas (Fig. 2.7*d*).

The flow through the control volume in Fig. 2.7*d* is steady with no body forces acting on it. Hence, for this case the momentum equation, Eq. (2.11), can be written as

$$\oiint (\rho \mathbf{V} \cdot d\mathbf{S}) \, \mathbf{V} = - \oiint p \, d\mathbf{S} \qquad (2.27)$$

Taking the *x* component of Eq. (2.27), we have

$$\int (\rho \mathbf{V} \cdot d\mathbf{S}) \, V_x = - \int (p \, d\mathbf{S})_x \qquad (2.28)$$

where V_x is the *x* component of the flow velocity, and the integrals are taken along the entire boundary of the control volume denoted by *abcda* in Fig. 2.7*d*. To evaluate the left side of Eq. (2.28), note that there is no flow across the upper and lower boundaries of the control volume, denoted by surfaces *ab* and *cd*, respectively, in

Fig. 2.7d, i.e., $\mathbf{V}$ and $d\mathbf{S}$ are everywhere mutually perpendicular along ab and cd, and hence the dot product $\rho \mathbf{V} \cdot d\mathbf{S} = 0$ along these boundaries. Thus,

$$\int_{ab} (\rho\,\mathbf{V} \cdot d\mathbf{S}) V_x = \int_{cd} (\rho\mathbf{V} \cdot d\mathbf{S}) V_x = 0 \tag{2.29}$$

Along the inlet boundary ad, $\mathbf{V}$ and $d\mathbf{S}$ are in opposite directions ($d\mathbf{S}$ always acts away from the control surface, in this case toward the left, whereas $\mathbf{V}$ is toward the right). Hence, the dot product $\rho\mathbf{V} \cdot d\mathbf{S}$ is negative. Also, along ad, V_x and ρ are uniform and equal to $-V_\infty$ and ρ_∞, respectively. (Note that the positive x direction is toward the left, as shown in Fig. 2.7b, and V_∞ is toward the right, hence along ad $V_x = -V_\infty$.) Thus,

$$\int_{ad} (\rho\,\mathbf{V} \cdot d\mathbf{S}) V_x = (-\rho_\infty V_\infty A_i)(-V_\infty)$$

Since $\rho_\infty V_\infty A_i$ is the mass flow across the inlet, denoted by $\dot{m}_i$, the last equation can be written as

$$\int_{ad} (\rho\,\mathbf{V} \cdot d\mathbf{S}) V_x = \dot{m}_i\, V_\infty \tag{2.30}$$

Along the exit boundary bc, $\mathbf{V}$ and $d\mathbf{S}$ are in the same direction, and V_x and ρ are uniform, equal to $-V_e$ and ρ_e, respectively. Hence,

$$\int_{bc} (\rho\,\mathbf{V} \cdot d\mathbf{S}) V_x = (\rho_e\, V_e\, A_e)(-V_e) = -\dot{m}_e\, V_e \tag{2.31}$$

where $\dot{m}_e$ is the mass flow across the exit boundary. Returning to Eq. (2.28), the left hand side can be written as

$$\int (\rho\,\mathbf{V} \cdot d\mathbf{S}) V_x = \int_{ab} (\rho\,\mathbf{V} \cdot d\mathbf{S}) V_x + \int_{cd} (\rho\,\mathbf{V} \cdot d\mathbf{S}) V_x$$

$$+ \int_{ad} (\rho\,\mathbf{V} \cdot d\mathbf{S}) V_x + \int_{bc} (\rho\,\mathbf{V} \cdot d\mathbf{S}) V_x$$

Substituting Eqs. (2.29), (2.30), and (2.31) into this, we have

$$\int (\rho\,\mathbf{V} \cdot d\mathbf{S}) V_x = 0 + 0 + \dot{m}_i\, V_\infty - \dot{m}_e\, V_e \tag{2.32}$$

Hence, Eq. (2.28) becomes

$$\dot{m}_i\, V_\infty - \dot{m}_e\, V_e = -\int (p\,dS)_x \tag{2.33}$$

Finally, the integral on the right side of Eq. (2.33) is also taken over the entire boundary of the control surface in Fig. 2.7d. Hence, in Eq. (2.33),

$$-\int (p\,dS)_x = -\int_{ad} (p\,dS)_x - \int_{bc} (p\,dS)_x - \int_{ab} (p\,dS)_x - \int_{cd} (p\,dS)_x \tag{2.34}$$

From Fig. 2.7d, note that along ad, $d\mathbf{S}$ acts to the left (the positive direction), and along bc, $d\mathbf{S}$ acts to the right (the negative direction),

$$\int_{ad} (p\,dS)_x = p_\infty A_i \tag{2.35}$$

$$\int_{bc} (p\,dS)_x = -p_e A_e \tag{2.36}$$

Along the boundaries ab and cd, p_i is the distributed pressure acting on the gas due to the equal and opposite reaction on the solid interior surface of the engine. Hence, we can write

$$\int_{ab} (p\,dS)_x + \int_{cd} (p\,dS)_x = \int_{abcd} (p_i\,dS)_x \tag{2.37}$$

Substituting Eqs. (2.35)–(2.37) into (2.34), we have

$$-\int (p\,dS)_x = -p_\infty A_i + p_e A_e - \int_{abcd} (p_i\,dS)_x \tag{2.38}$$

Substituting Eq. (2.38) into Eq. (2.33), we have

$$\dot{m}_i V_\infty - \dot{m}_e V_e = -p_\infty A_i + p_e A_e - \int_{abcd} (p_i\,dS)_x \tag{2.39}$$

The last term in Eq. (2.39) is physically the force on the gas due to the reaction from the solid interior surface of the engine, i.e.,

$$-\int_{abcd} (p_i\,dS)_x \equiv \left[-\int (p_i\,dS)_x\right]_{\text{force on the gas due to the solid surface}} \tag{2.40}$$

Hence, Eq. (2.39) can be written as

$$\dot{m}_i V_\infty - \dot{m}_e V_e = -p_\infty A_i + p_e A_e$$
$$+ \left[-\int (p_i\,dS)_x\right]_{\text{force on the gas due to the solid surface}} \tag{2.41}$$

or,

$$\left[-\int (p_i\,dS)_x\right]_{\text{force on the gas due to the solid surface}}$$
$$= \dot{m}_i V_\infty - \dot{m}_e V_e + p_\infty A_i - p_e A_e \tag{2.42}$$

Return to Eq. (2.26) for the engine thrust; here the bracketed term is the force on the solid surface due to the gas, which from Newton's third law is equal and opposite to the force on the gas due to the solid surface. That is,

$$\left[-\int (p_i \, dS)_x \right]_{\text{force on the solid surface due to the gas}}$$

$$= -\left[-\int (p_i \, dS)_x \right]_{\text{force on the gas due to the solid surface}} \tag{2.43}$$

Replacing the bracketed term on the right side of Eq. (2.43) with Eq. (2.42), we have

$$\left[-\int (p_i \, dS)_x \right]_{\text{force on the solid surface due to the gas}}$$

$$= \dot{m}_e \, V_e - \dot{m}_i \, V_\infty + p_e \, A_e - p_\infty \, A_i \tag{2.44}$$

Substituting Eq. (2.44) into Eq. (2.26), yields

$$T = \dot{m}_e \, V_e - \dot{m}_i \, V_\infty + p_e \, A_e - p_\infty \, A_i + p_\infty (A_i - A_e)$$

or,

$$\boxed{T = \dot{m}_e \, V_e - \dot{m}_i \, V_\infty + (p_e - p_\infty) A_e} \tag{2.45}$$

Equation (2.45) is the desired equation for the uninstalled engine thrust of a jet propulsion device.

The derivation of the thrust equation in this section has been quite lengthy, but our purpose was to illustrate an application of the integral form of the momentum equation with all its details. Notice what happened. We started with the concept that the thrust of the engine is due to the net integrated pressure distribution over all the exposed solid surfaces of the engine, which is the fundamental source of the thrust. However, for practical cases, the calculation or measurement of this detailed pressure distribution is usually so complex and costly in terms of personpower and money that is not done. On the other hand, we do not need the detailed pressure distribution to calculate the thrust. Through the beauty of the integral form of the momentum equation, where the details of the pressure distribution inside the engine are buried inside the control volume and hence do not explicitly appear in the integral form of the equation, the thrust of the engine can be calculated just by knowing the net time rate of change of the momentum of the gas exhausting out the exit compared to that entering through the inlet, which is the physical meaning of the term $(\dot{m}_e \, V_e - \dot{m}_i \, V_\infty)$ in Eq. (2.45), and by knowing the exit pressure p_e, which appears in the term $(p_e - p_\infty) A_e$ in Eq. (2.45). All of this simplification occurs with no loss of generality or accuracy. The derivation of the straightforward thrust equation is one of the triumphs of the integral form of the momentum equation.

Students of propulsion will recognize that the physical model sketched in Fig. 2.7 making the assumption that the streamtube of air entering the inlet is at free-stream conditions of V_∞, p_∞, and ρ_∞, is only a special "on-design" case. In actual flight, the conditions at the inlet can be slightly different than free-stream conditions.

For the derivation of the thrust equation in this case, the streamtube is extended far enough into the airflow ahead of the engine so that free-stream conditions do exist at the inlet to the streamtube. For such an extended streamtube, its inlet area will be different from the inlet area of the engine. However, in this case the resulting equation for the uninstalled engine thrust turns out to be the same as Eq. (2.45). See, for example, the definitive book by Mattingly, *Elements of Gas Turbine Propulsion,* McGraw-Hill, 1996, page 215, for more details.

EXAMPLE 2.1

Consider a turbojet-powered airplane flying at a velocity of 300 m/s at an altitude of 10 km, where the free-stream pressure and density are 2.65×10^4 N/m^2 and 0.414 kg/m^3, respectively. The turbojet engine has inlet and exit areas of 2 m^2 and 1 m^2, respectively. The velocity and pressure of the exhaust gas are 500 m/s and 2.3×10^4 N/m^2 respectively. The fuel-to-air mass ratio is 0.05. Calculate the thrust of the engine.

■ Solution

The mass flow of air through the inlet is

$$\dot{m}_i = \rho_\infty \, V_\infty \, A_i = (0.414)(300)(2) = 248.4 \, \text{kg/s}$$

Fuel is added and burned inside the engine at the ratio of 0.05 kg of fuel for every kg of air. Hence, the mass flow at the exit, $\dot{m}_e$, is

$$\dot{m}_e = 1.05 \, \dot{m}_i = 1.05(248.4) = 260.8 \, \text{kg/s}$$

From Eq. (2.45)

$$
\begin{aligned}
T &= \dot{m}_e \, V_e - \dot{m}_i \, V_\infty + (p_e - p_\infty) A_e \\
&= (260.8)(500) - (248.4)(300) + [(2.3 - 2.65) \times 10^4](1) \\
&= 1.304 \times 10^5 - 0.7452 \times 10^5 - 0.35 \times 10^4 \\
&= \boxed{5.238 \times 10^4 \, \text{N}}
\end{aligned}
$$

Since 4.45 N = 1 lb, the thrust in pounds is

$$T = \boxed{11,771 \, \text{lb}}$$

EXAMPLE 2.2

Consider a liquid-fueled rocket engine burning liquid hydrogen as the fuel and liquid oxygen as the oxidizer. The hydrogen and oxygen are pumped into the combustion chamber at rates of 11 kg/s and 89 kg/s, respectively. The flow velocity and pressure at the exit of the engine are 4000 m/s and 1.2×10^3 N/m^2, respectively. The exit area is 12 m^2. The engine is part of a rocket booster that is sending a payload into space. Calculate the thrust of the rocket engine as it passes through an altitude of 35 km, where the ambient pressure is 0.584×10^3 N/m^2.

■ **Solution**

For the case of a rocket engine, there is no mass flow of air through an inlet; the propellants are injected directly into the combustion chamber. Hence, for a rocket engine, Eq. (2.45) becomes, with $\dot{m}_i = 0$,

$$T = \dot{m}_e V_e + (p_e - p_\infty) A_e$$

Since the total mass flow of propellants pumped into the combustion chamber is $11 + 89 = 100$ kg/s, this is also the mass flow of the burned gases that exhausts through the rocket engine nozzle. That is, $\dot{m}_e = 100$ kg/s. Thus,

$$T = \dot{m}_e V_e + (p_e - p_\infty) A_e$$

$$= (100)(4000) + [(1.2 - 0.584) \times 10^3](12)$$

$$= 4 \times 10^5 + 7.392 \times 10^3 = \boxed{4.074 \times 10^5 \text{ N}}$$

In pounds,

$$T = \frac{4.074 \times 10^5}{4.45} = \boxed{91{,}549 \text{ lb}}$$

2.9 | SUMMARY

The analysis of compressible flow is based on three fundamental physical principles; in turn, these principles are expressed in terms of the basic flow equations. They are:

1. *Principle:* Mass can be neither created nor destroyed.
   Continuity equation:

$$\frac{\partial}{\partial t} \iiint_{\mathcal{V}} \rho \, d\mathcal{V} + \iint_{S} \rho \mathbf{V} \cdot d\mathbf{S} = 0 \tag{2.2}$$

2. *Principle:* Time rate of change of momentum of a body equals the net force exerted on it. (Newton's second law.)
   Momentum equation:

$$\frac{\partial}{\partial t} \iiint_{\mathcal{V}} \rho \mathbf{V} \, d\mathcal{V} + \iint_{S} (\rho \mathbf{V} \cdot d\mathbf{S}) \mathbf{V}$$

$$= \iiint_{\mathcal{V}} \rho \mathbf{f} \, d\mathcal{V} - \iint_{S} p \, d\mathbf{S} \tag{2.11}$$

3. *Principle:* Energy can be neither created nor destroyed, it can only change in form.

Energy equation:

$$\frac{\partial}{\partial t} \iiint_{\mathscr{V}} \rho \left(e + \frac{V^2}{2} \right) d\mathscr{V} + \iint_{\mathbf{S}} \rho \left(e + \frac{V^2}{2} \right) \mathbf{V} \cdot d\mathbf{S}$$

$$= \iiint_{\mathscr{V}} \dot{q} \rho \, d\mathscr{V} - \iint_{\mathbf{S}} p \mathbf{V} \cdot d\mathbf{S} + \iiint_{\mathscr{V}} \rho (\mathbf{f} \cdot \mathbf{V}) \, d\mathscr{V} \qquad (2.20)$$

These equations are expressed in integral form; such a form is particularly useful for the topics to be discussed in Chapters 3–5. In Chapter 6, the preceding integral forms will be reexpressed as partial differential equations.

PROBLEMS

2.1 When the National Advisory Committee for Aeronautics (NACA) measured the lift and drag on airfoil models in the 1930s and 40s in their specially designed airfoil wind tunnel at the Langley Aeronautical Laboratory, they made wings that spanned the entire test section, with the wing tips butted against the two side-walls of the tunnel. This was done to ensure that the flow over each airfoil section of the wing was essentially two-dimensional (no wing-tip effects). Such an arrangement prevented measuring the lift and drag with a force balance. Instead, using a Pitot tube, the NACA obtained the drag by measuring the velocity distribution behind the wing in a plane perpendicular to the plane of the wing, i.e., the Pitot tube, located a fixed distance downstream of the wing, traversed the height from the top to the bottom of the test section. Using a control volume approach, derive a formula for the drag per unit span on the model as a function of the integral of the measured velocity distribution. For simplicity, assume incompressible flow.

2.2 In the same tests described in problem 2.1, the NACA measured the lift per unit span by measuring the pressure distribution in the flow direction on the top and bottom walls of the wind tunnel. Using a control volume approach, derive a formula for the lift per unit span as a function of the integral of these pressure distributions.

CHAPTER 3

One-Dimensional Flow

The Aeronautical engineer is pounding hard on the closed door leading into the field of supersonic motion.

Theodore von Karman, 1941

PREVIEW BOX

With this chapter we begin to deal with supersonic shock waves—but a special class of shock waves, namely, shocks that are oriented perpendicular to the flow. Such shock waves are called *normal* shock waves. Normal shock waves are not as specialized as they may sound. A large number of supersonic flow problems involve normal shocks. For example, the gas properties at the nose of a blunt-nosed body moving at supersonic speeds are dictated by the flow through a normal shock wave. Also, the inlets of some jet aircraft are designed to have a normal shock across their entrance—so-called normal shock inlets. By the end of this chapter, you will have the tools to analyze such applications. The material in this chapter puts us immediately into the mainstream of basic shock behavior. It does more; it treats one-dimensional flow in general, i.e., flow in which the properties vary in only one direction. Imagine a flow

constrained to vary in only one direction where the stream tube of such a flow has constant area, such as flow through a constant-area duct. What can cause the flow variables to change if the cross-sectional area of the flow remains constant? A normal shock wave is one such case, where the flow properties can change dramatically across a region thinner than the paper on which this page is printed. Also, if heat is added or taken away from the flow—one-dimensional flow with heat addition—the flow properties will change. In addition, if there is a frictional shear stress tugging on the outside boundary of the flow, such as friction at the wall of the duct—one-dimensional flow with friction—the flow properties will change. In this chapter, we will deal separately with one-dimensional flow with heat addition and one-dimensional flow with friction. Although such flows are just approximate models of real flows with

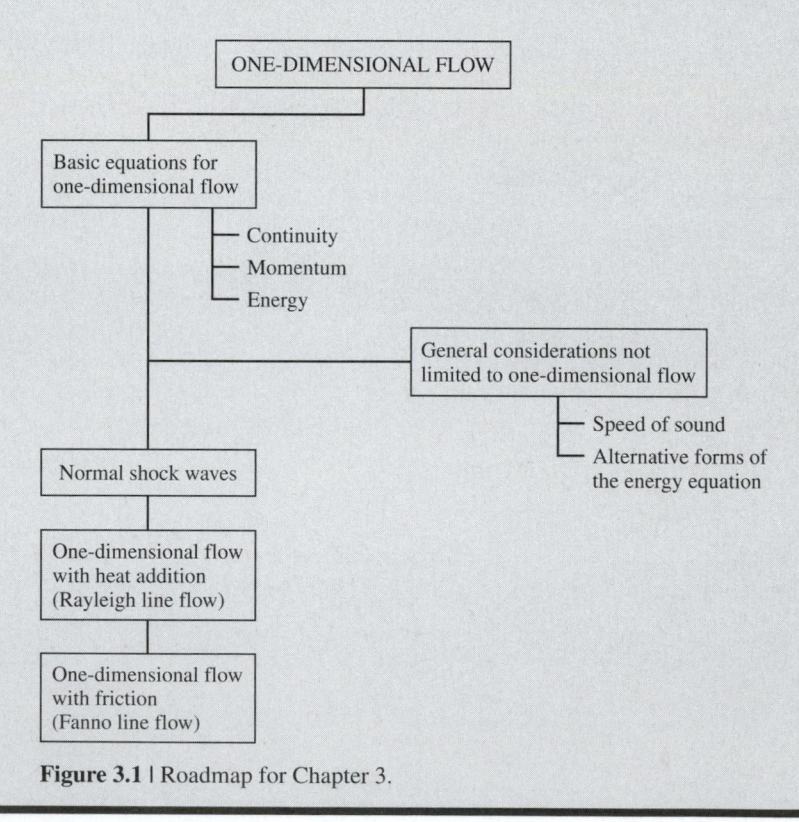

Figure 3.1 | Roadmap for Chapter 3.

heat addition and friction, one-dimensional flow with heat addition is very useful for estimating the effect of burning in a jet engine combustor can, and one-dimensional flow with friction provides an excellent method for the analysis of the gas flow through long pipes, to name just two important applications.

The roadmap for this chapter is given in Fig. 3.1. To begin our trip through this material, we first specialize the governing equations derived in Chap. 2 to the case of one-dimensional flow, obtaining the one-dimensional continuity, momentum and energy equation. These equations are algebraic equations, hence the mathematics in this chapter is simply algebra. Before moving on to the three types of one-dimensional flow that are high-lighted in this chapter, we take a side excursion to discuss some necessary general considerations that are not limited to just one-dimensional flow. This side excursion is shown at the right of the roadmap in Fig. 3.1, dealing with the speed of sound and some vital alternative forms of the energy equation. Then we return to the left side of the roadmap, and deal sequentially with normal shock waves, one-dimensional flow with heat addition, and one-dimensional flow with friction. As we cover the material highlighted in Fig. 3.1, we will be plunging into some of the most important physical and mathematical behavior that constitutes basic compressible flow. This is important stuff, so take a deep plunge, and make yourself very comfortable with this material.

3.1 | INTRODUCTION

On October 14, 1947, when Chuck Yeager nudged the Bell XS-1 to a speed slightly over Mach 1 (see Sec. 1.1), he entered a new flight regime where shock waves dominate the flowfield. At Mach 1.06, the bullet-shaped rocket-powered research airplane created a bow shock wave that was detached from the body, slightly upstream of the nose, as sketched in Fig. 3.2a. During a later flight, on March 26, 1948, Yeager pushed the XS-1 to Mach 1.45 in a dive. For this flight, the Mach number was high

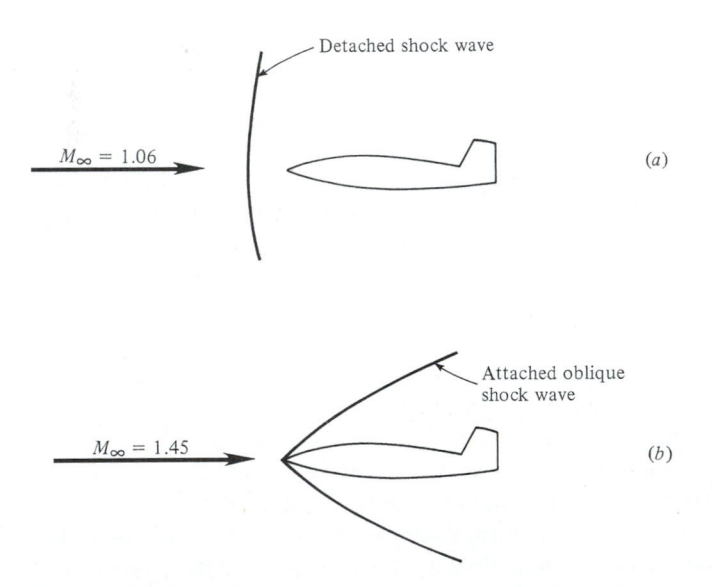

Figure 3.2 | Attached and detached shock waves on a supersonic vehicle.

Figure 3.3a | Shock wave on the Apollo command module. Wind tunnel model at $\alpha = 33°$ in the NASA Langley Mach 8 variable-density wind tunnel ion air. (*Courtesy of the NASA Langley Research Center.*)

enough that the shock wave attached itself to the pointed nose of the aircraft, as sketched in Fig. 3.2b. The difference between the two flows sketched in Fig. 3.2 is that the bow shock is nearly normal to the free-stream direction as in Fig. 3.2a, whereas the attached shock wave is oblique to the free-stream direction in Fig. 3.2b. For a blunt-nosed body in a supersonic flow, as shown in Fig. 3.3a, the bow shock wave is always detached from the body. Moreover, near the nose, the shock is nearly normal to the free stream; away from the nose, the shock gradually becomes oblique. For further illustration, photographs taken in supersonic wind tunnels of shock waves on various aerodynamic shapes are shown in Fig. 3.3.

The portions of the shock waves in Figs. 3.2 and 3.3 that are perpendicular to the free stream are called *normal shocks*. A normal shock wave is illustrated in Fig. 3.4, and it is an excellent example of a class of flowfields that is called *one-dimensional flow*. By definition, a one-dimensional flow is one in which the flow-field properties vary only with one coordinate direction—i.e., in Fig. 3.4, p, ρ, T, and the velocity u are functions of x only. In this chapter, we will examine the

Figure 3.3*b* | Shock waves on a sharp-nosed slender cone at angle of attack. (*Courtesy of the Naval Surface Weapons Center, White Oak, MD.*)

properties of such one-dimensional flows, with normal shock waves as one important example. As indicated in Figs. 3.2 and 3.3, normal shock waves play an important role in many supersonic flows.

Oblique shock waves are two-dimensional phenomena, and will be discussed in Chap. 4. Also, consider the two streamtubes in Fig. 3.5. In Fig. 3.5*a*, a truly one-dimensional flow is illustrated, where the flowfield variables are a function of x only, and as a consequence the streamtube area must be *constant* (as we shall prove later). On the other hand, there are many flow problems wherein the streamtube area varies with x, as sketched in Fig. 3.5*b*. For such a variable area streamtube, nature dictates that the flowfield is three-dimensional flow, where the flow properties in general are functions of x, y, and z. However, if the variation of area $A = A(x)$ is gradual, it is often convenient and sufficiently accurate to neglect the y and z flow variations, and to *assume* that the flow properties are functions of x only, as noted in Fig. 3.5*b*. This is tantamount to assuming uniform properties across the flow at every x station. Such a flow, where the area varies as $A = A(x)$ but where it is assumed that p, ρ, T, and u are still functions of x only, is defined as *quasi-one-dimensional flow*. This will be the subject of Chap. 5.

In summary, the present chapter will treat one-dimensional, hence constant-area, flows. The general integral conservation equations derived in Chap. 2 will be applied to one-dimensional flow, yielding straightforward algebraic relations which allow us to study the properties and characteristics of such flows.

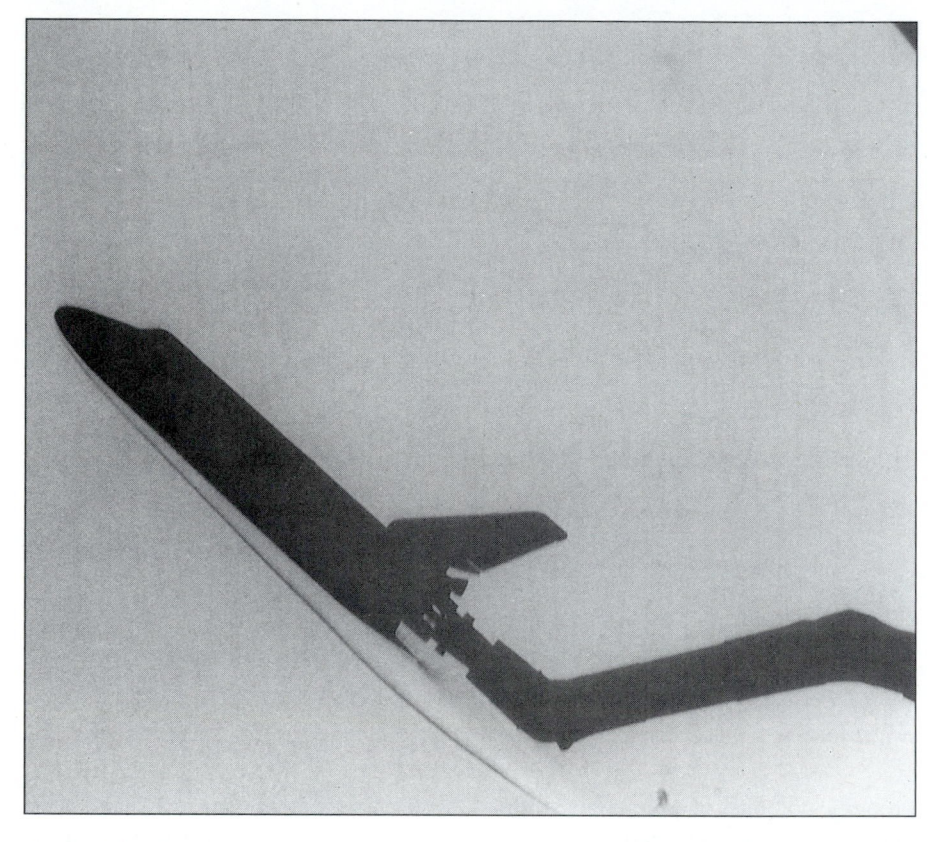

Figure 3.3c | Shock wave on a wind tunnel model of the space shuttle. (*Courtesy of the NASA Langley Research Center.*)

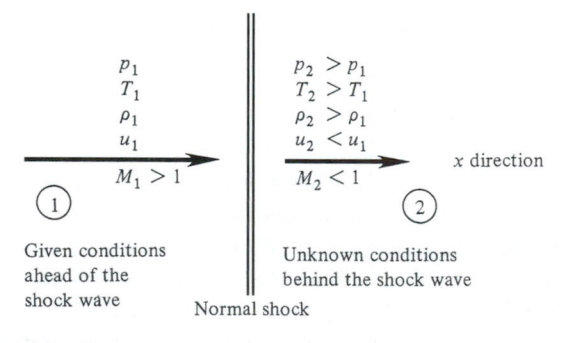

Figure 3.4 | Diagram of a normal shock.

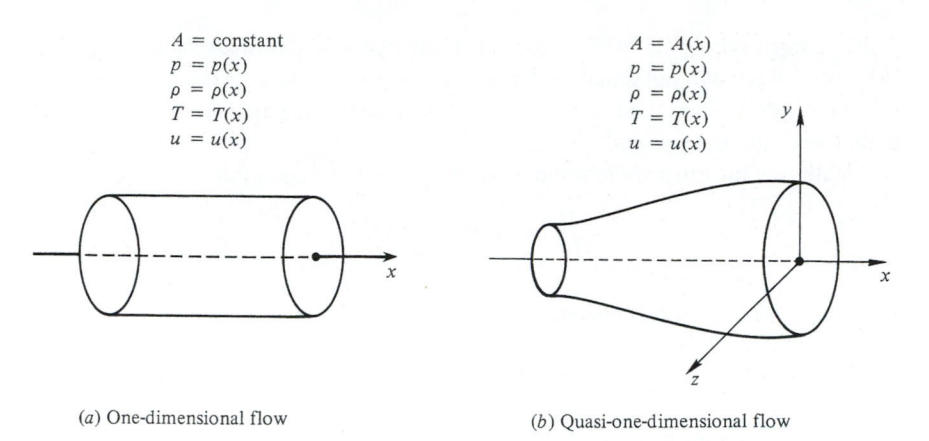

(a) One-dimensional flow   (b) Quasi-one-dimensional flow

Figure 3.5 | Comparison between one-dimensional and quasi-one-dimensional flows.

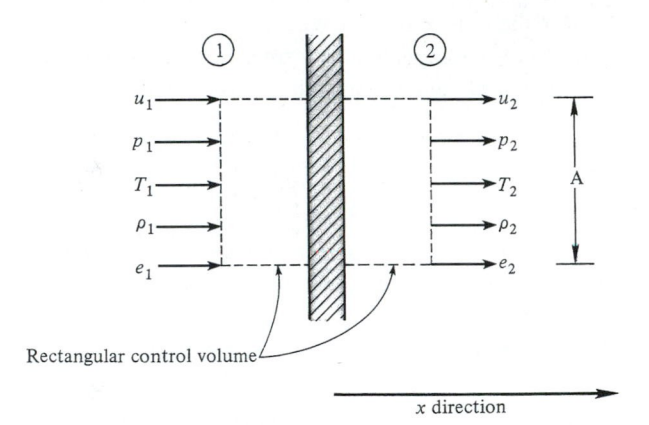

Figure 3.6 | Rectangular control volume for one-dimensional flow.

3.2 | ONE-DIMENSIONAL FLOW EQUATIONS

Consider the flow through a one-dimensional region, as represented by the shaded area in Fig. 3.6. This region may be a normal shock wave, or it may be a region with heat addition; in either case, the flow properties change as a function of x as the gas flows through the region. To the left of this region, the flowfield velocity, pressure, temperature, density, and internal energy are u_1, p_1, T_1, ρ_1, and e_1, respectively. To the right of this region, the properties have changed, and are given by u_2, p_2, T_2, ρ_2, and e_2. (Since we are now dealing with one-dimensional flow, we are using u to denote velocity. Later on, in dealing with multidimensional flows, u is the x component of velocity.) To calculate the changes, apply the integral conservation equations from Chap. 2 to the rectangular control volume shown by the dashed lines in Fig. 3.6. Since the flow is one-dimensional, u_1, p_1, T_1, ρ_1, and e_1 are uniform over the left-hand side of the control volume, and similarly u_2, p_2, T_2, ρ_2, and e_2 are uniform over the right-hand side of the control volume. Assume that the left- and right-hand sides each have an area equal to A perpendicular to the flow. Also, assume that the flow is *steady,* such that all derivatives with respect to time are zero, and assume that body forces are not present.

With this information in mind, write the continuity equation (2.2):

$$-\oiint_S \rho \mathbf{V} \cdot d\mathbf{S} = \frac{\partial}{\partial t} \iiint_{\mathscr{V}} \rho \, d\mathscr{V}$$

For steady flow, Eq. (2.2) becomes

$$\oiint_S \rho \mathbf{V} \cdot d\mathbf{S} = 0 \tag{3.1}$$

Evaluating the surface integral over the left-hand side, where $\mathbf{V}$ and $d\mathbf{S}$ are parallel but in opposite directions, we obtain $-\rho_1 u_1 A$; over the right-hand side, where $\mathbf{V}$ and $d\mathbf{S}$ are parallel and in the same direction, we obtain $\rho_2 u_2 A$. The upper and lower horizontal faces of the control volume both contribute nothing to the surface integral because $\mathbf{V}$ and $d\mathbf{S}$ are perpendicular to each other on these faces. Hence, from Eq. (3.1),

$$-\rho_1 u_1 A + \rho_2 u_2 A = 0$$

or

$$\boxed{\rho_1 u_1 = \rho_2 u_2} \tag{3.2}$$

Equation (3.2) is the *continuity equation* for steady one-dimensional flow.

The momentum equation (2.11) is repeated here for convenience:

$$\oiint_S (\rho \mathbf{V} \cdot d\mathbf{S}) \mathbf{V} + \iiint_{\mathscr{V}} \frac{\partial(\rho \mathbf{V})}{\partial t} \, d\mathscr{V} = \iiint_{\mathscr{V}} \rho \mathbf{f} \, d\mathscr{V} - \oiint_S p \, d\mathbf{S}$$

The second term is zero because we are considering steady flow. Also, because there are no body forces, the third term is zero. Hence, Eq. (2.11) becomes

$$\oiint_S (\rho \mathbf{V} \cdot d\mathbf{S}) \mathbf{V} = -\oiint_S p \, d\mathbf{S} \tag{3.3}$$

Equation (3.3) is a vector equation. However, since we are dealing with one-dimensional flow, we need to consider only the scalar x component of Eq. (3.3), which is

$$\oiint_S (\rho \mathbf{V} \cdot d\mathbf{S}) u = -\oiint_S (p \, d\mathbf{S})_x \tag{3.4}$$

In Eq. (3.4), the expression $(p \, d\mathbf{S})_x$ is the x component of the vector $p \, d\mathbf{S}$. Evaluating the surface integrals in Eq. (3.4) over the left- and right-hand sides of the dashed control volume in Fig. 3.6, we obtain

$$\rho_1(-u_1 A)u_1 + \rho_2(u_2 A)u_2 = -(-p_1 A + p_2 A)$$

or

$$\boxed{p_1 + \rho_1 u_1^2 = p_2 + \rho_2 u_2^2} \tag{3.5}$$

Equation (3.5) is the *momentum equation* for steady one-dimensional flow.

The energy equation (2.20) is written here for convenience:

$$\iiint_{\mathcal{V}} \dot{q}\rho\,d\mathcal{V} - \oiint_{S} p\mathbf{V}\cdot d\mathbf{S} + \iiint_{\mathcal{V}} \rho(\mathbf{f}\cdot\mathbf{V})\,d\mathcal{V}$$

$$= \iiint_{\mathcal{V}} \frac{\partial}{\partial t}\left[\rho\left(e+\frac{V^2}{2}\right)\right]d\mathcal{V} + \oiint_{S} \rho\left(e+\frac{V^2}{2}\right)\mathbf{V}\cdot d\mathbf{S}$$

The first term on the left physically represents the total rate of heat added to the gas inside the control volume. For simplicity, let us denote this volume integral by $\dot{Q}$. The third and fourth terms are zero because of zero body forces and steady flow, respectively. Hence, Eq. (2.20) becomes

$$\dot{Q} - \oiint_{S} p\mathbf{V}\cdot d\mathbf{S} = \oiint_{S} \rho\left(e+\frac{V^2}{2}\right)\mathbf{V}\cdot d\mathbf{S} \tag{3.6}$$

Evaluating the surface integrals over the left- and right-hand faces of the control volume in Fig. 3.6, we obtain

$$\dot{Q} - (-p_1 u_1 A + p_2 u_2 A) = -\rho_1\left(e_1+\frac{u_1^2}{2}\right)u_1 A + \rho_2\left(e_2+\frac{u_2^2}{2}\right)u_2 A$$

Rearranging,

$$\frac{\dot{Q}}{A} + p_1 u_1 + \rho_1\left(e_1+\frac{u_1^2}{2}\right)u_1 = p_2 u_2 + \rho_2\left(e_2+\frac{u_2^2}{2}\right)u_2 \tag{3.7}$$

Dividing by Eq. (3.2), i.e., dividing the left-hand side of Eq. (3.7) by $\rho_1 u_1$ and the right-hand side by $\rho_2 u_2$,

$$\frac{\dot{Q}}{\rho_1 u_1 A} + \frac{p_1}{\rho_1} + e_1 + \frac{u_1^2}{2} = \frac{p_2}{\rho_2} + e_2 + \frac{u_2^2}{2} \tag{3.8}$$

Considering the first term in Eq. (3.8), $\dot{Q}$ is the net rate of heat (energy/s) added to the control volume, and $\rho_1 u_1 A$ is the mass flow (mass/s) through the control volume. Hence, the ratio $\dot{Q}/\rho_1 u_1 A$ is simply the heat added per unit mass, q. Also, in Eq. (3.8) recall the definition of enthalpy, $h = e + pv$. Hence, Eq. (3.8) becomes

$$\boxed{h_1 + \frac{u_1^2}{2} + q = h_2 + \frac{u_2^2}{2}} \tag{3.9}$$

Equation (3.9) is the *energy equation* for steady one-dimensional flow.

In summary, Eqs. (3.2), (3.5), and (3.9) are the governing fundamental equations for steady one-dimensional flow. Look closely at these equations. They are algebraic equations that relate properties at two different locations, 1 and 2, along a one-dimensional, constant-area flow. The assumption of one-dimensionality has afforded us the luxury of a great simplification over the integral equations from Chap. 2. However, within the assumption of steady one-dimensional flow, the algebraic equations (3.2), (3.5), and (3.9) still represent the full authority and power of the integral equations from whence they came—i.e., they still say that mass is conserved [Eq. (3.2)], force equals time rate of change of momentum [Eq. (3.5)], and energy is conserved [Eq. (3.9)]. Also, keep in mind that Eq. (3.5) neglects body forces and viscous stresses, and that Eq. (3.9) does not include shaft work, work done by viscous stresses, heat transfer due to thermal conduction or diffusion, and changes in potential energy.

Returning to our roadmap in Fig. 3.1, we have finished the first box on the left-hand side. Before proceeding down the left-hand column, in Secs. 3.3–3.5 we will take the side excursion shown on the right-hand side of Fig. 3.1. Here we will deal with some important general aspects of compressible flow that are not limited to one-dimensional flow. It is necessary for us to define and discuss the speed of sound and to obtain some alternative forms of the energy equation before we can move on to address the remaining boxes in Fig. 3.1.

3.3 | SPEED OF SOUND AND MACH NUMBER

As you read this page, look up for a moment and consider the air around you. The air is composed of molecules that are moving about in a random motion with different instantaneous velocities and energies at different times. However, over a period of time, the average (mean) molecular velocity and energy can be defined, and for a perfect gas are functions of the temperature only. Now assume that a small firecracker detonates nearby. The energy released by the firecracker is absorbed by the surrounding air molecules, which results in an increase in their mean velocity. These faster molecules collide with their neighbors, transferring some of their newly acquired energy. In turn, these neighbors eventually collide with others, resulting in a net transfer or propagation of the firecracker energy through space. This wave of energy travels through the air at a velocity that must be somewhat related to the mean molecular velocity, because molecular collisions are propagating the wave. Through the wave, the energy increase also causes the pressure (as well as density, temperature, etc.) to change slightly. As the wave passes by you, this small pressure variation is picked up by your eardrum, and is transmitted to your brain as the sense of sound. Therefore, such a weak wave is defined as a *sound wave,* and the purpose of this section is to calculate how fast it is propagating through the air. As we will soon appreciate, the speed of sound through a gas is one of the most important quantities in a study of compressible flow.

Consider that the sound wave is moving with velocity a through the gas. Let us hop on the wave and move with it. As we ride along with the wave, we see that the air ahead of the wave moves toward the wave at the velocity a, as shown in Fig. 3.7. Because there are changes in the flow properties through the wave, the flow behind the wave moves away at a different velocity. However, these changes are slight. A sound wave, by definition, is a weak wave. (If the changes through the wave are strong, it is identified as a shock wave, which propagates at a higher velocity than a, as we will soon see.) Therefore, consider the change in velocity through the sound wave to be an infinitesimal quantity, da. Consequently, from our vantage point riding

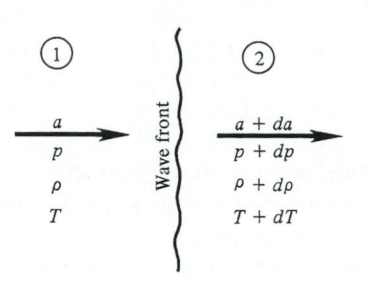

Figure 3.7 | Schematic of a sound wave.

along with the wave, we see the picture shown in Fig. 3.7 where the wave appears to be stationary, the flow ahead of it moves toward the wave at velocity a with pressure, density, and temperature p, ρ, and T, respectively, and the flow behind it moves away from the wave at velocity $a + da$ with pressure $p + dp$, density $\rho + d\rho$, and temperature $T + dT$.

The flow through the sound wave is one-dimensional and hence we can apply the equations from Sec. 3.2 to the picture in Fig. 3.7. If regions 1 and 2 are in front of and behind the wave, respectively, Eq. (3.2) yields

$$\rho a = (\rho + d\rho)(a + da)$$
$$\rho a = \rho a + a\,d\rho + \rho\,da + d\rho\,da \tag{3.10}$$

The product of two infinitesimal quantities $d\rho\,da$ is very small (of second order) in comparison to the other terms in Eq. (3.10), and hence can be ignored. Thus, from Eq. (3.10),

$$a = -\rho\,\frac{da}{d\rho} \tag{3.11}$$

Next, Eq. (3.5) yields

$$p + \rho a^2 = (p + dp) + (\rho + d\rho)(a + da)^2 \tag{3.12}$$

Ignoring products of differentials as before, Eq. (3.12) becomes

$$dp = -2a\rho\,da - a^2\,d\rho \tag{3.13}$$

Solve Eq. (3.13) for da:

$$da = \frac{dp + a^2\,d\rho}{-2a\rho} \tag{3.14}$$

Substitute Eq. (3.14) into Eq. (3.11):

$$a = -\rho\left[\frac{dp/d\rho + a^2}{-2a\rho}\right] \tag{3.15}$$

Solving Eq. (3.15) for a^2,

$$a^2 = \frac{dp}{d\rho} \tag{3.16}$$

Pause for a moment and consider the physical process occurring through a sound wave. First, the changes within the wave are slight, i.e., the flow gradients are small. This implies that the irreversible, dissipative effects of friction and thermal conduction are negligible. Moreover, there is no heat addition to the flow inside the wave (the gas is *not* being irradiated by a laser, for example). Hence, from Sec. 1.4, the process inside the sound wave must be *isentropic*. In turn, the rate of change of pressure with respect to density, $dp/d\rho$, which appears in Eq. (3.16) is an isentropic change, and Eq. (3.16) can be written as

$$\boxed{a^2 = \left(\frac{\partial p}{\partial \rho}\right)_s} \tag{3.17}$$

Equation (3.17) is a fundamental expression for the speed of sound. It shows that the speed of sound is a direct measure of the compressibility of a gas, as defined in Sec. 1.2. To see this more clearly, recall that $\rho = 1/v$, hence $d\rho = -dv/v^2$. Thus, Eq. (3.17) can be written as

$$a^2 = \left(\frac{\partial p}{\partial \rho}\right)_s = -\left(\frac{\partial p}{\partial v}\right)_s v^2 = -\frac{v}{(1/v)(\partial v/\partial p)_s}$$

Recalling the definition of isentropic compressibility, τ_s, given by Eq. (1.4), we find

$$\boxed{a = \sqrt{\left(\frac{\partial p}{\partial \rho}\right)_s} = \sqrt{\frac{v}{\tau_s}}} \tag{3.18}$$

This confirms the statement in Sec. 1.3 that incompressible flow ($\tau_s = 0$) implies an infinite speed of sound.

For a calorically perfect gas, Eq. (3.18) becomes more tractable. In this case, the isentropic relation [see Eq. (1.43)] becomes

$$pv^\gamma = c$$

where c is a constant. Differentiating, and recalling that $v = 1/\rho$, we find

$$\left(\frac{\partial p}{\partial \rho}\right)_s = \frac{\gamma p}{\rho}$$

Hence, Eq. (3.18) becomes

$$\boxed{a = \sqrt{\frac{\gamma p}{\rho}}} \tag{3.19}$$

Going one step further, from the equation of state, $p/\rho = RT$. Hence, Eq. (3.19) becomes

$$a = \sqrt{\gamma RT} \qquad (3.20)$$

In summary, Eq. (3.18) gives a general relation for the speed of sound in a gas; this reduces to Eqs. (3.19) and (3.20) for a perfect gas. Indeed, we will demonstrate in Chap. 17 that Eqs. (3.19) and (3.20) hold for thermally perfect as well as calorically perfect gases, but are invalid for chemically reacting gases or real gases. However, the general relation, Eq. (3.18), is valid for all gases.

Note that, for a perfect gas, Eq. (3.20) gives the speed of sound as a function of temperature only; indeed, it is proportional to the square root of the temperature. This is consistent with our previous discussion linking the speed of sound to the average molecular velocity, which from kinetic theory is given by $\sqrt{8RT/\pi}$. Note that the speed of sound is about three-quarters of the average molecular velocity.

The speed of sound in air at standard sea level conditions is a useful value to remember. It is

$$a_s = 340.9 \text{ m/s} = 1117 \text{ ft/s}$$

Finally, recall that the Mach number was defined in Sec. 1.3 as $M = V/a$, which leads to the following classifications of different flow regimes:

$$M < 1 \quad \text{(subsonic flow)}$$

$$M = 1 \quad \text{(sonic flow)}$$

$$M > 1 \quad \text{(supersonic flow)}$$

Also, it is interesting to attach some additional physical meaning to the Mach number at this stage of our discussion. Consider a fluid element moving along a streamline. The kinetic and internal energies per unit mass of this fluid element are $V^2/2$ and e, respectively. Forming their ratio, and recalling Eqs. (1.23) and (3.20), we have

$$\frac{V^2/2}{e} = \frac{V^2/2}{c_v T} = \frac{V^2/2}{RT/(\gamma - 1)} = \frac{(\gamma/2)V^2}{a^2/(\gamma - 1)} = \frac{\gamma(\gamma - 1)}{2} M^2$$

Thus, we see that, for a calorically perfect gas (where $e = c_v T$), the square of the Mach number is proportional to the ratio of kinetic to internal energy. It is a measure of the directed motion of the gas compared to the random thermal motion of the molecules.

3.4 | SOME CONVENIENTLY DEFINED FLOW PARAMETERS

In this chapter the fundamentals of one-dimensional compressible flow will be applied to the practical problems of normal shock waves, flow with heat addition, and flow with wall friction. However, before making these applications an inventory of useful definitions and supporting equations must be established. This is the purpose of Secs. 3.4 and 3.5.

To begin with, consider point *A* in an arbitrary flowfield, as sketched in Fig. 2.3. At this point a fluid element is traveling at some Mach number *M*, velocity *V*, with a static pressure and temperature *p* and *T*, respectively. Let us now *imagine* that we take this fluid element and *adiabatically* slow it down (if *M* > 1) or speed it up (if *M* < 1) until its Mach number at point *A* is 1. As we do this, common sense tells us that the temperature will change. When the fluid element arrives at *M* = 1 (in our imagination) from its initial state at *M* and *T* (its real properties at point *A*), the new temperature (that it has in our imagination at Mach 1) is *defined* as *T**. Furthermore, we now define the speed of sound at this hypothetical Mach 1 condition as *a**, where

$$a^* = \sqrt{\gamma R T^*}$$

Therefore, for any given flow with a given *M* and *T* at some point *A*, we can *associate* with it values of *T** and *a** at the same point, as already defined. Means of calculating *T** (and hence *a**) will be discussed in Sec. 3.5.

In the same spirit, consider again our fluid element at point *A* with velocity, temperature, and pressure equal to *V*, *T*, and *p*, respectively. Let us now *imagine* that we *isentropically* slow this fluid element to zero velocity, i.e., let us stagnate the fluid element. The pressure and temperature which the fluid element achieves when *V* = 0 are defined as *total pressure* p_o and *total temperature* T_o, respectively. (They are frequently called *stagnation pressure* and *temperature*; the adjectives "stagnation" and "total" are synonymous.) Both p_o and T_o are properties *associated* with the fluid element while it is in actuality moving at velocity *V* with an actual pressure and temperature equal to *p* and *T*, respectively. The actual *p* and *T* are called *static pressure* and *static temperature*, respectively, and are ramifications of the random molecular motion at point *A*.

Using these definitions, we can introduce other parameters:

Characteristic Mach number $M^* = V/a^*$. (Note that the real Mach number is $M = V/a$.)

Stagnation speed of sound $a_o = \sqrt{\gamma R T_o}$.

Total (or stagnation) density $\rho_o = p_o/R T_o$.

3.5 | ALTERNATIVE FORMS OF THE ENERGY EQUATION

Consider again Eq. (3.9). Assuming no heat addition, this becomes

$$\boxed{h_1 + \frac{u_1^2}{2} = h_2 + \frac{u_2^2}{2}} \tag{3.21}$$

where points 1 and 2 correspond to the regions 1 and 2 identified in Fig. 3.6. Specializing further to a calorically perfect gas, where $h = c_p T$, Eq. (3.21) becomes

$$\boxed{c_p T_1 + \frac{u_1^2}{2} = c_p T_2 + \frac{u_2^2}{2}} \tag{3.22}$$

Using Eq. (1.22), this becomes

$$\frac{\gamma R T_1}{\gamma - 1} + \frac{u_1^2}{2} = \frac{\gamma R T_2}{\gamma - 1} + \frac{u_2^2}{2} \tag{3.23}$$

Since $a = \sqrt{\gamma R T}$, Eq. (3.23) becomes

$$\boxed{\frac{a_1^2}{\gamma - 1} + \frac{u_1^2}{2} = \frac{a_2^2}{\gamma - 1} + \frac{u_2^2}{2}} \tag{3.24}$$

From Eq. (3.19), this can also be written as

$$\boxed{\frac{\gamma}{\gamma - 1}\left(\frac{p_1}{\rho_1}\right) + \frac{u_1^2}{2} = \frac{\gamma}{\gamma - 1}\left(\frac{p_2}{\rho_2}\right) + \frac{u_2^2}{2}} \tag{3.25}$$

Since Eq. (3.21) was written for no heat addition, it, as well as the corollary Eqs. (3.22) through (3.25), holds for an *adiabatic* flow. With this in mind, let us return to the definitions presented in Sec. 3.4. Let point 1 in these equations correspond to point A in Fig. 2.3, and let point 2 in these equations correspond to our *imagined* conditions where the fluid element is brought adiabatically to Mach 1 at point A. The actual speed of sound and velocity at point A are a and u, respectively. At the imagined condition of Mach 1 (point 2 in the above equations), the speed of sound is a^* and the flow velocity is sonic, hence $u_2 = a^*$. Thus, Eq. (3.24) yields

$$\frac{a^2}{\gamma - 1} + \frac{u^2}{2} = \frac{a^{*2}}{\gamma - 1} + \frac{a^{*2}}{2}$$

or

$$\boxed{\frac{a^2}{\gamma - 1} + \frac{u^2}{2} = \frac{\gamma + 1}{2(\gamma - 1)} a^{*2}} \tag{3.26}$$

Equation (3.26) provides a formula from which the defined quantity a^* can be calculated for the given actual conditions of a and u at any given point in a general flowfield. Remember, the actual flowfield itself does *not* have to be adiabatic from one point to the next, say from point A to point B in Fig. 2.3. In Eq. (3.26), the adiabatic process is just in our minds as part of the *definition* of a^* (see again Sec. 3.4). Applied at point A in Fig. 2.3, Eq. (3.26) gives us the value of a^* that is *associated* with point A. Denote this value as a_A^*. Similarly, applied at point B, Eq. (3.26) gives us the value of a^* that is *associated* with point B, namely, a_B^*. If the actual flowfield is *nonadiabatic* from A to B, then $a_A^* \neq a_B^*$. On the other hand, if the general flowfield in Fig. 2.3 is *adiabatic* throughout, then a^* is a *constant value* at every point in the flow. Since many practical aerodynamic flows are reasonably adiabatic, this is an important point to remember.

Now return to our definition of *total* conditions in Sec. 3.4. Let point 1 in Eq. (3.22) correspond to point A in Fig. 2.3, and let point 2 in Eq. (3.22) correspond to our *imagined* conditions where the fluid element is brought to rest isentropically at point A. If T and u are the actual values of static temperature and velocity, respectively, at point A, then $T_1 = T$ and $u_1 = u$. Also, by definition of total conditions, $u_2 = 0$ and $T_2 = T_o$. Hence, Eq. (3.22) becomes

$$c_p T + \frac{u^2}{2} = c_p T_o \tag{3.27}$$

Equation (3.27) provides a formula from which the defined total temperature, T_o, can be calculated for the given actual conditions of T and u at any point in a general flowfield. Remember that total conditions are defined in Sec. 3.4 as those where the fluid element is *isentropically* brought to rest. However, in the derivation of Eq. (3.27), only the energy equation for an adiabatic flow [Eq. (3.21)] is used. Isentropic conditions have not been imposed so far. Hence, the definition of T_o such as expressed in Eq. (3.27) is *less* restrictive than the definition of total conditions given in Sec. 3.4. From Sec. 1.4, isentropic flow implies reversible and adiabatic conditions; Eq. (3.27) tells us that, for the definition of T_o, only the "adiabatic" portion of the isentropic definition is required. That is, we can now redefine T_o as that temperature that would exist if the fluid element were brought to rest *adiabatically*. However, for the definition of total pressure, p_o, and total density, ρ_o, the imagined *isentropic* process is still necessary, as defined in Sec. 3.4.

Several very useful equations for total conditions are obtained as shown next. From Eqs. (3.27) and (1.22),

$$\frac{T_o}{T} = 1 + \frac{u^2}{2c_p T} = 1 + \frac{u^2}{2\gamma RT/(\gamma - 1)} = 1 + \frac{u^2}{2a^2/(\gamma - 1)} = 1 + \frac{\gamma - 1}{2}\left(\frac{u}{a}\right)^2$$

Hence,

$$\frac{T_o}{T} = 1 + \frac{\gamma - 1}{2}M^2 \tag{3.28}$$

Equation (3.28) gives the ratio of total to static temperature at a point in a flow as a function of the Mach number M at that point. Furthermore, for an isentropic process, Eq. (1.43) holds, such that

$$\frac{p_o}{p} = \left(\frac{\rho_o}{\rho}\right)^\gamma = \left(\frac{T_o}{T}\right)^{\gamma/(\gamma-1)} \tag{3.29}$$

Combining Eqs. (3.28) and (3.29), we find

$$\frac{p_o}{p} = \left(1 + \frac{\gamma - 1}{2}M^2\right)^{\gamma/(\gamma-1)} \tag{3.30}$$

$$\frac{\rho_o}{\rho} = \left(1 + \frac{\gamma - 1}{2}M^2\right)^{1/(\gamma-1)} \tag{3.31}$$

Equations (3.30) and (3.31) give the ratios of total to static pressure and density, respectively, at a point in the flow as a function of Mach number M at that point. Along with Eq. (3.28), they represent important relations for total properties—so important that their values are tabulated in Table A.1 (see Appendix A) as a function of M for $\gamma = 1.4$ (which corresponds to air at standard conditions).

It should be emphasized again that Eqs. (3.27), (3.28), (3.30), and (3.31) provide formulas from which the defined quantities T_o, p_o, and ρ_o can be calculated from the actual conditions of M, u, T, p, and ρ at a given point in a general flowfield, as sketched in Fig. 2.3. Again, the actual flowfield itself does *not* have to be adiabatic or isentropic from one point to the next. In these equations, the isentropic process is just in our minds as part of the *definition* of total conditions at a point. Applied at point A in Fig. 2.3, the above equations give us the values of T_o, p_o, and ρ_o *associated* with point A. Similarly, applied at point B, the earlier equations give us the values of T_o, p_o, and ρ_o associated with point B. If the actual flow between A and B is nonadiabatic and irreversible, then $T_{o_A} \neq T_{o_B}$, $p_{o_A} \neq p_{o_B}$, and $\rho_{o_A} \neq \rho_{o_B}$. On the other hand, if the general flowfield is *isentropic* throughout, then T_o, p_o, and ρ_o are *constant values* at every point in the flow. The idea of constant total (stagnation) conditions in an isentropic flow will be very useful in our later discussions of various practical applications in compressible flow—keep it in mind!

A few additional equations will be useful in subsequent sections. For example, from Eq. (3.24),

$$\boxed{\frac{a^2}{\gamma - 1} + \frac{u^2}{2} = \frac{a_o^2}{\gamma - 1}} \tag{3.32}$$

where a_o is the stagnation speed of sound defined in Sec. 3.4. From Eqs. (3.26) and (3.32),

$$\frac{\gamma + 1}{2(\gamma - 1)} a^{*2} = \frac{a_o^2}{\gamma - 1} \tag{3.33}$$

Solving Eq. (3.33) for a^*/a_o, and invoking Eq. (3.20),

$$\boxed{\left(\frac{a^*}{a_o}\right)^2 = \frac{T^*}{T_o} = \frac{2}{\gamma + 1}} \tag{3.34}$$

Recall that p^* and ρ^* are defined for conditions at Mach 1; hence, Eqs. (3.30) and (3.31) with $M = 1$ lead to

$$\boxed{\frac{p^*}{p_o} = \left(\frac{2}{\gamma + 1}\right)^{\gamma/(\gamma-1)}} \tag{3.35}$$

$$\boxed{\frac{\rho^*}{\rho_o} = \left(\frac{2}{\gamma + 1}\right)^{1/(\gamma-1)}} \tag{3.36}$$

For air at standard conditions, where $\gamma = 1.4$, these ratios are

$$\frac{T^*}{T_o} = 0.833$$

$$\frac{p^*}{p_o} = 0.528$$

$$\frac{\rho^*}{\rho_o} = 0.634$$

which will be useful numbers to keep in mind for subsequent discussions. Finally, dividing Eq. (3.26) by u^2, we have

$$\frac{(a/u)^2}{\gamma - 1} + \frac{1}{2} = \frac{\gamma + 1}{2(\gamma - 1)} \left(\frac{a^*}{u}\right)^2$$

$$\frac{(1/M)^2}{\gamma - 1} = \frac{\gamma + 1}{2(\gamma - 1)} \left(\frac{1}{M^*}\right)^2 - \frac{1}{2}$$

$$\boxed{M^2 = \frac{2}{[(\gamma + 1)/M^{*2}] - (\gamma - 1)}} \tag{3.37}$$

Equation (3.37) provides a direct relation between the actual Mach number M and the characteristic Mach number M^*, defined in Sec. 3.4. Note from Eq. (3.37) that

$$M^* = 1 \qquad \text{if } M = 1$$
$$M^* < 1 \qquad \text{if } M < 1$$
$$M^* > 1 \qquad \text{if } M > 1$$

$$M^* \to \sqrt{\frac{\gamma + 1}{\gamma - 1}} \qquad \text{if } M \to \infty$$

Hence, qualitatively, M^* acts in the same fashion as M, except when M goes to infinity. In future discussions involving shock and expansion waves, M^* will be a useful parameter because it approaches a finite number as M approaches infinity.

All the equations in this section, either directly or indirectly, are alternative forms of the original, fundamental energy equation for one-dimensional, adiabatic flow, Eq. (3.21). Make certain that you examine these equations and their derivations closely. It is important at this stage that you feel comfortable with these equations, especially those with a box around them for emphasis.

3.5.1   A Comment on Generality

This section began with Eq. (3.21), which was obtained from the one-dimensional energy equation, Eq. (3.9), specialized to adiabatic flow. The use of the x component of the flow velocity, u, in Eq. (3.21) clearly identifies it with one-dimensional flow. For one-dimensional flow, the velocity u *is* the velocity of the flow, and the use of the symbol u is simply consistent with the geometry of the flow. However, Eq. (3.21) is a general statement of the energy equation for any steady, adiabatic flow, whether in one, two, or three dimensions. For a general three-dimensional flow, the velocity at any point in the flow is denoted by V. For a three-dimensional, steady, adiabatic flow, Eq. (3.21) becomes

$$h_1 + \frac{V_1^2}{2} = h_2 + \frac{V_2^2}{2}$$

Similarly, for every form of the energy equation obtained in this section, u_1 and u_2 can be replaced by V_1 and V_2. So Eqs. (3.21)–(3.37) hold with u replaced by V everywhere. This general application of Eq. (3.21) to a three-dimensional case will be rigorously derived in Chap. 6.

EXAMPLE 3.1

At a point in the flow over an F-15 high-performance fighter airplane, the pressure, temperature, and Mach number are 1890 lb/ft^2, 450°R, and 1.5, respectively. At this point, calculate T_o, p_o, T^*, p^*, and the flow velocity.

■ **Solution**
From Table A.1, for $M = 1.5$: $p_o/p = 3.671$ and $T_o/T = 1.45$. Thus

$$p_o = 3.671p = 3.671(1890) = \boxed{6938 \text{ lb/ft}^2}$$

$$T_o = 1.45T = 1.45(450) = \boxed{652.5°\text{R}}$$

From Table A.1, for $M = 1.0$: $p_o/p^* = 1.893$ and $T_o/T^* = 1.2$. Keeping in mind that, for our imaginary process where the flow is slowed down isentropically to Mach 1, hence defining p^*, the total pressure is constant during this process; also, where the flow is slowed down adiabatically to Mach 1, hence defining T^*, the total temperature is constant. Thus

$$p^* = \frac{p^*}{p_o}\frac{p_o}{p}p = \frac{1}{1.893}(3.671)(1890) = \boxed{3665 \text{ lb/ft}^2}$$

$$T^* = \frac{T^*}{T_o}\frac{T_o}{T}T = \frac{1}{1.2}(1.45)(450) = \boxed{543.8°\text{R}}$$

Note: These answers exemplify the definitions of $p_o, T_o, p^*,$ and T^*. In the actual flow at Mach 1.5, the actual static pressure and static temperature are 1890 lb/ft^2 and 450°R, respectively. However, the *defined* values that are *associated* with the flow at this point (but not actually in existence at this point) are $p^* = 3665$ lb/ft^2, $p_o = 6938$ lb/ft^2, $T^* = 543.8$°R, and $T_o = 652.5$°R. Finally, the actual flow velocity is obtained from

$$V = Ma$$

where

$$a = \sqrt{\gamma RT} = \sqrt{(1.4)(1716)(450)} = 1040 \text{ ft/s}$$

$$V = (1.5)(1040) = \boxed{1560 \text{ ft/s}}$$

EXAMPLE 3.2

Return to Example 1.6. Calculate the Mach number and velocity at the exit of the rocket nozzle.

■ **Solution**

In the combustion chamber the flow velocity is very low; hence we can assume that the pressure and temperature in the combustion chamber are essentially p_o and T_o, respectively. Moreover, since the flow expansion through the nozzle is isentropic, then p_o and T_o are constant values throughout the nozzle flow. From Eq. (3.30), we have at the nozzle exit (denoted by the subscript 2)

$$\left(\frac{p_o}{p}\right)_2 = \left(1 + \frac{\gamma - 1}{2}M_2^2\right)^{\gamma/(\gamma - 1)}$$

Solving for M_2, we have

$$M_2 = \left\{\frac{2}{\gamma - 1}\left[\left(\frac{p_o}{p_2}\right)^{(\gamma-1)/\gamma} - 1\right]\right\}^{1/2}$$

$$= \left\{\frac{2}{0.2}\left[\left(\frac{15}{0.372}\right)^{0.167} - 1\right]\right\}^{1/2} = \boxed{2.919}$$

$$a_2 = \sqrt{\gamma R T_2} = \sqrt{(1.2)(692.8)(1350)} = 1059.4 \text{ m/s}$$

$$V_2 = M_2 a_2 = (2.919)(1059.4) = \boxed{3092 \text{ m/s}}$$

Note: An alternative solution to this problem, which constitutes a check on these results, is as shown next. From Eq. (3.22)

$$c_p T_2 + \frac{V_2^2}{2} = c_p T_o$$

[Recall from Sec. 3.5.1 that the various forms of the energy equation obtained in this section hold for flow of any dimensions—two or three dimensions as well as one dimension; this is because Eq. (3.21) is simply a statement that the *total* enthalpy, $h_o = h + V^2/2$, is constant for *any* adiabatic flow, no matter what the dimension. This will become clear repeatedly as we progress through the following chapters. Hence, Eqs. (3.21) through (3.37) are general, and are *not* in any way restricted to one-dimensional flow. Therefore, we can use Eq. (3.22) in the form given here to solve our rocket nozzle flow, even though such flow is not constant-area flow, i.e., it is *not* truly one-dimensional flow. Rather, this nozzle flow must be analyzed as either a quasi-one-dimensional flow as discussed in Chap. 5, or more precisely as a two-dimensional or axisymmetric flow as discussed in Chap. 11, because the flow through a nozzle encounters a *changing,* variable cross-sectional area as it expands through the nozzle.]

From Eq. (3.22) written above, solving for V_2,

$$V_2 = \sqrt{2c_p(T_o - T_2)} = \sqrt{2(4157)(2500 - 1350)} = 3092 \text{ m/s}$$

This agrees with the value already obtained. Of course, since $a_2 = 1059.4$ m/s as obtained, then

$$M_2 = \frac{V_2}{a_2} = \frac{3092}{1059.4} = 2.919$$

which also agrees with the earlier results.

EXAMPLE 3.3

Return to Example 1.1. Calculate the percentage density change between the given point on the wing and the free stream, *assuming compressible flow*.

■ **Solution**

The standard sea level values of density and temperature are 0.002377 slug/ft^3 and 519°R, respectively. Also, for air,

$$c_p = \frac{\gamma R}{\gamma - 1} = \frac{(1.4)(1716)}{0.4} = 6006 \text{ ft} \cdot \text{lb/slug} \cdot \text{°R}$$

Let points 1 and 2 in Eq. (3.22) denote the free stream and the wing points, respectively. *Note:* The flow over the wing is adiabatic and frictionless; hence it is *isentropic*. From Eq. (3.22)

$$c_p T_1 + \frac{V_1^2}{2} = c_p T_2 + \frac{V_2^2}{2}$$

$$T_2 = T_1 + \frac{V_1^2 - V_2^2}{2c_p} = 519 + \frac{(147)^2 - (220)^2}{2(6006)}$$

$$= 519 - 2.23 = 516.77\text{°R}$$

From Eq. (1.43)

$$\frac{\rho_2}{\rho_1} = \left(\frac{T_2}{T_1}\right)^{1/(\gamma-1)} = \left(\frac{516.77}{519}\right)^{2.5} = 0.9893$$

$$\rho_2 = 0.9893(0.002377) = 0.002352 \text{ slug/ft}^3$$

Thus

$$\frac{\rho_1 - \rho_2}{\rho_1} = \frac{0.000025}{0.002377} = 0.011$$

That is, the density changes by $\boxed{1.1 \text{ percent}}$. This is a *very* small change and clearly justifies the assumption of incompressible flow in the solution of Example 1.1. Moreover, note from this material that the temperature change is only 2.23°R, which represents a 0.43 percent change in temperature. This illustrates that low-speed flows are virtually constant temperature flows, and this is why, in the analysis of inviscid incompressible flow, the energy equation is never needed.

EXAMPLE 3.4

Consider again the rocket engine discussed in Examples 1.6 and 3.2. If the thrust of the engine is 4.5×10^5 N at an altitude where the ambient pressure is 0.372 atm, calculate the mass flow through the engine and the area of the exit.

■ Solution

From Example 1.6, the pressure at the exit is $p_2 = 0.372$ atm. From Example 3.2, the velocity at the exit is $V_2 = 3092$ m/s. From the thrust equation, Eq. (2.45), applied to a rocket engine, using the subscript 2 to denote exit conditions, we have

$$T = \dot{m}V_2 + (p_2 - p_\infty)A_e$$

Since $p_2 = p_\infty = 0.372$ atm, the pressure term on the right-hand side of this equation is zero, and we have

$$\dot{m} = \frac{T}{V_2} = \frac{4.5 \times 10^5}{3092} = \boxed{145.5 \text{ kg/s}}$$

From Example 1.6, we have for the specific gas constant of the gas expanding through the engine, $R = 692.8$ J/kg · K, and the temperature at the exit $T_2 = 1350$ K. Hence, from the equation of state the density at the exit is (recalling that 1 atm $= 1.01 \times 10^5$ N/m^2)

$$\rho_2 = \frac{p_2}{RT_2} = \frac{(0.372)(1.01 \times 10^5)}{(692.8)(1350)} = 0.04 \text{ kg/m}^3$$

The mass flow is given by

$$\dot{m} = \rho_2 A_2 V_2$$

or,

$$A_2 = \frac{\dot{m}}{\rho_2 V_2} = \frac{145.5}{(0.04)(3092)} = \boxed{1.18 \text{ m}^2}$$

3.6 | NORMAL SHOCK RELATIONS

Let us now apply the previous information to the practical problem of a normal shock wave. With this, we travel back to the left-hand side of our roadmap in Fig. 3.1, and start discussing the physical phenomena that can cause a change in properties of a one-dimensional (constant area) flow. Our first consideration is the case of a normal shock wave. As discussed in Sec. 3.1, normal shocks occur frequently as part of many supersonic flowfields. By definition, a normal shock wave is perpendicular to the flow, as sketched in Fig. 3.4. The shock is a very thin region (the shock thickness is usually on the order of a few molecular mean free paths, typically 10^{-5} cm for air at standard conditions). The flow is supersonic ahead of the wave, and subsonic behind it, as noted in Fig. 3.4. Furthermore, the static pressure, temperature, and density increase across the shock, whereas the velocity decreases, all of which we will demonstrate shortly.

Nature establishes shock waves in a supersonic flow as a solution to a perplexing problem having to do with the propagation of disturbances in the flow. To obtain some preliminary physical feel for the creation of such shock waves, consider a flat-faced cylinder mounted in a flow, as sketched in Fig. 3.8. Recall that the flow consists of individual molecules, some of which impact on the face of the cylinder. There is in general a change in molecular energy and momentum due to impact with the

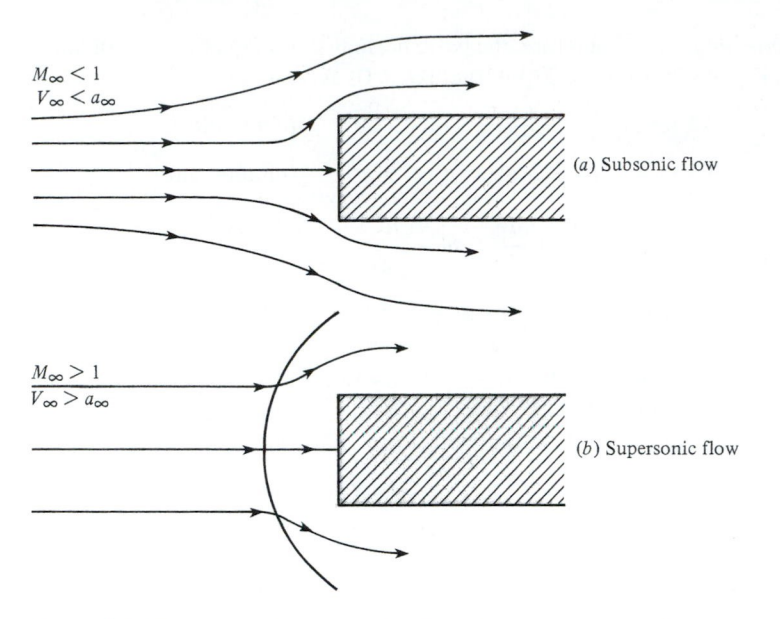

Figure 3.8 | Comparison between subsonic and supersonic streamlines for flow over a flat-faced cylinder or slab.

cylinder, which is seen as an obstruction by the molecules. Therefore, just as in our example of the creation of a sound wave in Sec. 3.3, the random motion of the molecules communicates this change in energy and momentum to other regions of the flow. The presence of the body tries to be propagated everywhere, including directly upstream, by sound waves. In Fig. 3.8a, the incoming stream is subsonic, $V_\infty < a_\infty$, and the sound waves can work their way upstream and forewarn the flow about the presence of the body. In this fashion, as shown in Fig. 3.8a, the flow streamlines begin to change and the flow properties begin to compensate for the body *far* upstream (theoretically, an infinite distance upstream). In contrast, if the flow is supersonic, then $V_\infty > a_\infty$, and the sound waves can no longer propagate upstream. Instead, they tend to coalesce a short distance ahead of the body. In so doing, their coalescence forms a thin shock wave, as shown in Fig. 3.8b. Ahead of the shock wave, the flow has no idea of the presence of the body. Immediately behind the normal shock, however, the flow is subsonic, and hence the streamlines quickly compensate for the obstruction. Although the picture shown in Fig. 3.8b is only one of many situations in which nature creates shock waves, the physical mechanism just discussed is quite general.

To begin a quantitative analysis of changes across a normal shock wave, consider again Fig. 3.4. Here, the normal shock is assumed to be a discontinuity across which the flow properties suddenly change. For purposes of discussion, assume that all conditions are known ahead of the shock (region 1), and that we want to solve for all conditions behind the shock (region 2). There is no heat added or taken away from the flow as it traverses the shock wave (for example, we are not putting the shock in a refrigerator, nor are we irradiating it with a laser); hence the flow across the shock wave is *adiabatic*. Therefore, the basic normal shock equations are obtained directly from Eqs. (3.2), (3.5), and (3.9) (with $q = 0$) as

$$\rho_1 u_1 = \rho_2 u_2 \qquad \text{(continuity)} \qquad (3.38)$$

$$p_1 + \rho_1 u_1^2 = p_2 + \rho_2 u_2^2 \qquad \text{(momentum)} \qquad (3.39)$$

$$h_1 + \frac{u_1^2}{2} = h_2 + \frac{u_2^2}{2} \qquad \text{(energy)} \qquad (3.40)$$

Equations (3.38) through (3.40) are general—they apply no matter what type of gas is being considered. Also, in general they must be solved numerically for the properties behind the shock wave, as will be discussed in Chap. 17 for the cases of thermally perfect and chemically reacting gases. However, for a calorically perfect gas, we can immediately add the thermodynamic relations

$$p = \rho R T \qquad (3.41)$$

and
$$h = c_p T \qquad (3.42)$$

Equations (3.38) through (3.42) constitute five equations with five unknowns: $\rho_2, u_2, p_2, h_2,$ and T_2. Hence, they can be solved algebraically, as follows.

First, divide Eq. (3.39) by (3.38):

$$\frac{p_1}{\rho_1 u_1} - \frac{p_2}{\rho_2 u_2} = u_2 - u_1 \qquad (3.43)$$

Recalling that $a = \sqrt{\gamma p / \rho}$, Eq. (3.43) becomes

$$\frac{a_1^2}{\gamma u_1} - \frac{a_2^2}{\gamma u_2} = u_2 - u_1 \qquad (3.44)$$

Equation (3.44) is a combination of the continuity and momentum equations. The energy equation (3.40) can be utilized in one of its alternative forms, namely, Eq. (3.26), which yields

$$a_1^2 = \frac{\gamma + 1}{2} a^{*2} - \frac{\gamma - 1}{2} u_1^2 \qquad (3.45)$$

and
$$a_2^2 = \frac{\gamma + 1}{2} a^{*2} - \frac{\gamma - 1}{2} u_2^2 \qquad (3.46)$$

Since the flow is adiabatic across the shock wave, a^* in Eqs. (3.45) and (3.46) is the same constant value (see Sec. 3.5). Substituting Eqs. (3.45) and (3.46) into (3.44), we obtain

$$\frac{\gamma + 1}{2} \frac{a^{*2}}{\gamma u_1} - \frac{\gamma - 1}{2\gamma} u_1 - \frac{\gamma + 1}{2} \frac{a^{*2}}{\gamma u_2} + \frac{\gamma - 1}{2\gamma} u_2 = u_2 - u_1$$

or
$$\frac{\gamma + 1}{2\gamma u_1 u_2} (u_2 - u_1) a^{*2} + \frac{\gamma - 1}{2\gamma} (u_2 - u_1) = u_2 - u_1$$

Dividing by $(u_2 - u_1)$,

$$\frac{\gamma + 1}{2\gamma u_1 u_2} a^{*2} + \frac{\gamma - 1}{2\gamma} = 1$$

Solving for a^*, this gives

$$\boxed{a^{*2} = u_1 u_2} \qquad (3.47)$$

Equation (3.47) is called the *Prandtl relation,* and is a useful intermediate relation for normal shocks. For example, from this simple equation we obtain directly

$$1 = \frac{u_1}{a^*} \frac{u_2}{a^*} = M_1^* M_2^*$$

or

$$\boxed{M_2^* = \frac{1}{M_1^*}} \qquad (3.48)$$

Based on our previous physical discussion, the flow ahead of a shock wave must be supersonic, i.e., $M_1 > 1$. From Sec. 3.5, this implies $M_1^* > 1$. Thus, from Eq. (3.48), $M_2^* < 1$ and thus $M_2 < 1$. Hence, the *Mach number behind the normal shock is always subsonic.* This is a general result, not just limited to a calorically perfect gas.

Recall Eq. (3.37), which, solved for M^*, gives

$$M^{*2} = \frac{(\gamma + 1)M^2}{2 + (\gamma - 1)M^2} \qquad (3.49)$$

Substitute Eq. (3.49) into (3.48):

$$\frac{(\gamma + 1)M_2^2}{2 + (\gamma - 1)M_2^2} = \left[\frac{(\gamma + 1)M_1^2}{2 + (\gamma - 1)M_1^2}\right]^{-1} \qquad (3.50)$$

Solving Eq. (3.50) for M_2^2:

$$\boxed{M_2^2 = \frac{1 + [(\gamma - 1)/2]M_1^2}{\gamma M_1^2 - (\gamma - 1)/2}} \qquad (3.51)$$

Equation (3.51) demonstrates that, for a calorically perfect gas with a constant value of γ, the Mach number behind the shock is a function of only the Mach number ahead of the shock. It also shows that when $M_1 = 1$, then $M_2 = 1$. This is the case of an infinitely weak normal shock, which is defined as a *Mach wave.* In contrast, as M_1 increases above 1, the normal shock becomes stronger and M_2 becomes progressively less than 1. However, in the limit, as $M_1 \to \infty$, M_2 approaches a finite minimum value, $M_2 \to \sqrt{(\gamma - 1)/2\gamma}$, which for air is 0.378.

The upstream Mach number M_1 is a powerful parameter which dictates shock wave properties. This is already seen in Eq. (3.51). Ratios of other properties across the shock can also be found in terms of M_1. For example, from Eq. (3.38) combined with (3.47),

$$\frac{\rho_2}{\rho_1} = \frac{u_1}{u_2} = \frac{u_1^2}{u_2 u_1} = \frac{u_1^2}{a^{*2}} = M_1^{*2} \tag{3.52}$$

Substituting Eq. (3.49) into (3.52),

$$\boxed{\frac{\rho_2}{\rho_1} = \frac{u_1}{u_2} = \frac{(\gamma + 1)M_1^2}{2 + (\gamma - 1)M_1^2}} \tag{3.53}$$

To obtain the pressure ratio, return to the momentum equation (3.39),

$$p_2 - p_1 = \rho_1 u_1^2 - \rho_2 u_2^2$$

which, combined with Eq. (3.38), yields

$$p_2 - p_1 = \rho_1 u_1 (u_1 - u_2) = \rho_1 u_1^2 \left(1 - \frac{u_2}{u_1}\right) \tag{3.54}$$

Dividing Eq. (3.54) by p_1, and recalling that $a_1^2 = \gamma p_1/\rho_1$, we obtain

$$\frac{p_2 - p_1}{p_1} = \gamma M_1^2 \left(1 - \frac{u_2}{u_1}\right) \tag{3.55}$$

Substitute Eq. (3.53) for u_1/u_2 into Eq. (3.55):

$$\frac{p_2 - p_1}{p_1} = \gamma M_1^2 \left[1 - \frac{2 + (\gamma - 1)M_1^2}{(\gamma + 1)M_1^2}\right] \tag{3.56}$$

Equation (3.56) simplifies to

$$\boxed{\frac{p_2}{p_1} = 1 + \frac{2\gamma}{\gamma + 1}\left(M_1^2 - 1\right)} \tag{3.57}$$

To obtain the temperature ratio, recall the equation of state, $p = \rho R T$. Hence

$$\frac{T_2}{T_1} = \left(\frac{p_2}{p_1}\right)\left(\frac{\rho_1}{\rho_2}\right) \tag{3.58}$$

Substituting Eqs. (3.57) and (3.53) into Eq. (3.58),

$$\boxed{\frac{T_2}{T_1} = \frac{h_2}{h_1} = \left[1 + \frac{2\gamma}{\gamma + 1}\left(M_1^2 - 1\right)\right]\left[\frac{2 + (\gamma - 1)M_1^2}{(\gamma + 1)M_1^2}\right]} \tag{3.59}$$

Examine Eqs. (3.51), (3.53), (3.57), and (3.59). For a calorically perfect gas with a given γ, they give M_2, ρ_2/ρ_1, p_2/p_1, and T_2/T_1 *as functions of* M_1 *only*. This is our first major demonstration of the importance of Mach number in the quantitative governance of compressible flowfields. In contrast, as will be shown in Chap. 17 for an equilibrium thermally perfect gas, the changes across a normal shock depend on both M_1 and T_1, whereas for an equilibrium chemically reacting gas they depend on

M_1, T_1, and p_1. Moreover, for such high-temperature cases, closed-form expressions such as Eqs. (3.51) through (3.59) are generally not possible, and the normal shock properties must be calculated numerically. Hence, the simplicity brought about by the calorically perfect gas assumption in this section is clearly evident. Fortunately, the results of this section hold reasonably accurately up to approximately $M_1 = 5$ in air at standard conditions. Beyond Mach 5, the temperature behind the normal shock becomes high enough that γ is no longer constant. However, the flow regime $M_1 < 5$ contains a large number of everyday practical problems, and therefore the results of this section are extremely useful.

The limiting case of $M_1 \to \infty$ can be visualized as $u_1 \to \infty$, where the calorically perfect gas assumption is invalidated by high temperatures, or as $a_1 \to 0$, where the perfect gas equation of state is invalidated by extremely low temperatures. Nevertheless, it is interesting to examine the variation of properties across the normal shock as $M_1 \to \infty$ in Eqs. (3.51), (3.53), (3.57), and (3.59). We find, for $\gamma = 1.4$,

$$\lim_{M_1 \to \infty} M_2 = \sqrt{\frac{\gamma - 1}{2\gamma}} = 0.378 \qquad \text{(as discussed previously)}$$

$$\lim_{M_1 \to \infty} \frac{\rho_2}{\rho_1} = \frac{\gamma + 1}{\gamma - 1} = 6$$

$$\lim_{M_1 \to \infty} \frac{p_2}{p_1} = \infty$$

$$\lim_{M_1 \to \infty} \frac{T_2}{T_1} = \infty$$

At the other extreme, for $M_1 = 1$, Eqs. (3.51), (3.53), (3.57), and (3.59) yield $M_1 = \rho_2/\rho_1 = p_2/p_1 = T_2/T_1 = 1$. This is the case of an infinitely weak normal shock degenerating into a Mach wave, where no finite changes occur across the wave. This is the same as the sound wave discussed in Sec. 3.3.

Earlier in this section, it was stated that the flow ahead of the normal shock wave must be supersonic. This is clear from our previous physical discussion on the formation of shocks. However, it is interesting to note that Eqs. (3.51), (3.53), (3.57), and (3.59) *mathematically* hold for $M_1 < 1$ as well as $M_1 > 1$. Therefore, to prove that these equations have *physical* meaning only when $M_1 > 1$, we must appeal to the second law of thermodynamics (see Sec. 1.4). From Eq. (1.36), repeated here,

$$s_2 - s_1 = c_p \ln \frac{T_2}{T_1} - R \ln \frac{p_2}{p_1}$$

with Eqs. (3.57) and (3.59), we have

$$s_2 - s_1 = c_p \ln \left\{ \left[1 + \frac{2\gamma}{\gamma + 1} \left(M_1^2 - 1 \right) \right] \left[\frac{2 + (\gamma - 1)M_1^2}{(\gamma + 1)M_1^2} \right] \right\}$$

$$- R \ln \left[1 + \frac{2\gamma}{\gamma + 1} \left(M_1^2 - 1 \right) \right] \tag{3.60}$$

Equation (3.60) demonstrates that the entropy change across the normal shock is also a function of M_1 only. Moreover, it shows that, if $M_1 = 1$ then $s_2 - s_1 = 0$, if $M_1 < 1$ then $s_2 - s_1 < 0$, and if $M_1 > 1$ then $s_2 - s_1 > 0$. Therefore, since it is necessary that $s_2 - s_1 \geq 0$ from the second law, the upstream Mach number M_1 must be greater than or equal to 1. Here is another example of how the second law tells us the direction in which a physical process will proceed. If M_1 is subsonic, then Eq. (3.60) says that the entropy decreases across the normal shock—an impossible situation. The only physically possible case is $M_1 \geq 1$, which in turn dictates from Eqs. (3.51), (3.53), (3.57), and (3.59) that $M_2 \leq 1$, $\rho_2/\rho_1 \geq 1$, $p_2/p_1 \geq 1$, and $T_2/T_1 \geq 1$. Thus, we have now established the phenomena sketched in Fig. 3.4, namely, that across a normal shock wave the pressure, density, and temperature increase, whereas the velocity decreases and the Mach number decreases to a subsonic value.

What really causes the entropy increase across a shock wave? To answer this, recall that the changes across the shock occur over a very short distance, on the order of 10^{-5} cm. Hence, the velocity and temperature gradients *inside* the shock structure itself are very large. In regions of large gradients, the viscous effects of viscosity and thermal conduction become important. In turn, these are dissipative, irreversible phenomena that generate entropy. Therefore, the net entropy increase predicted by the normal shock relations in conjunction with the second law of thermodynamics is appropriately provided by nature in the form of friction and thermal conduction *inside* the shock wave structure itself.

Finally, in this section we need to resolve one more question, namely, how do the total (stagnation) conditions vary across a normal shock wave? Consider Fig. 3.9, which illustrates the definition of total conditions before and after the shock. In region 1 ahead of the shock, a fluid element is moving with actual conditions of

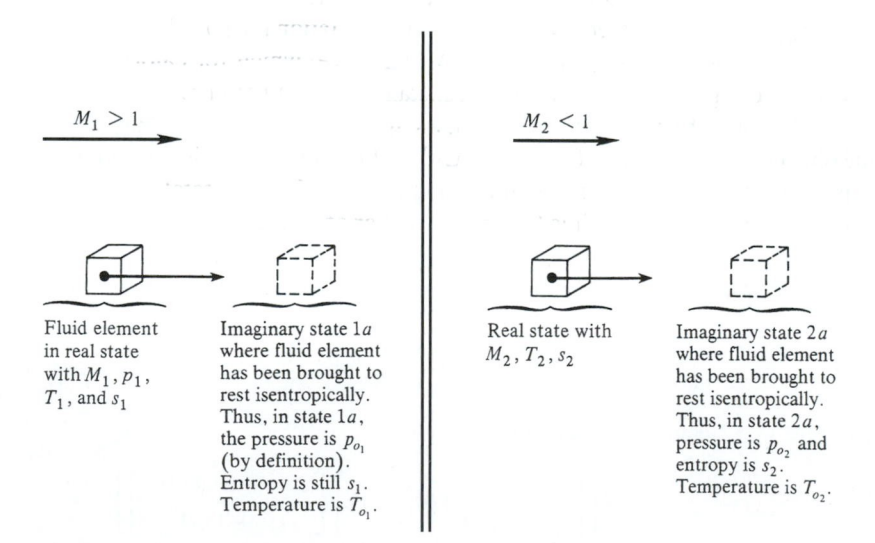

Figure 3.9 | Illustration of total (stagnation) conditions ahead of and behind a normal shock wave.

M_1, p_1, T_1, and s_1. Consider in this region the imaginary state $1a$ where the fluid element has been brought to rest isentropically. Thus, by definition, the pressure and temperature in state $1a$ are the total values p_{o_1}, and T_{o_1}, respectively. The entropy at state $1a$ is still s_1 because the stagnating of the fluid element has been done isentropically. In region 2 behind the shock, a fluid element is moving with actual conditions of M_2, p_2, T_2, and s_2. Consider in this region the imaginary state $2a$ where the fluid element has been brought to rest isentropically. Here, by definition, the pressure and temperature in state $2a$ are the total values of p_{o_2} and T_{o_2}, respectively. The entropy at state $2a$ is still s_2, by definition. The question is now raised how p_{o_2} and T_{o_2} behind the shock compare with p_{o_1} and T_{o_1}, respectively, ahead of the shock. To answer this question, consider Eq. (3.22), repeated here:

$$c_p T_1 + \frac{u_1^2}{2} = c_p T_2 + \frac{u_2^2}{2}$$

From Eq. (3.27), the total temperature is given by

$$c_p T_o = c_p T + \frac{u^2}{2}$$

Hence,

$$c_p T_{o_1} = c_p T_{o_2}$$

and thus

$$\boxed{T_{o_1} = T_{o_2}} \tag{3.61}$$

From Eq. (3.61), we see that the *total temperature is constant across a stationary normal shock wave*. [Note that Eq. (3.61), which holds for a calorically perfect gas, is a special case of the more general result that the total enthalpy is constant across the shock, as demonstrated by Eq. (3.40). For a stationary normal shock, the total enthalpy is always constant across the shock wave, which for calorically or thermally perfect gases translates into a constant total temperature across the shock. However, for a chemically reacting gas, the total temperature is *not* constant across the shock, as described in Chap. 17. Also, if the shock wave is not stationary—if it is moving through space—neither the total enthalpy nor total temperature are constant across the wave. This becomes a matter of reference systems, as discussed in Chap. 7.]

Considering Fig. 3.9 again, write Eq. (1.36) between the imaginary states $1a$ and $2a$:

$$s_{2a} - s_{1a} = c_p \ln \frac{T_{2a}}{T_{1a}} - R \ln \frac{p_{2a}}{p_{1a}} \tag{3.62}$$

However, $s_{2a} = s_2$, $s_{1a} = s_1$, $T_{2a} = T_o = T_{1a}$, $p_{2a} = p_{o_2}$, and $p_{1a} = p_{o_1}$. Hence, Eq. (3.62) becomes

$$\boxed{s_2 - s_1 = -R \ln \frac{p_{o_2}}{p_{o_1}}} \tag{3.63}$$

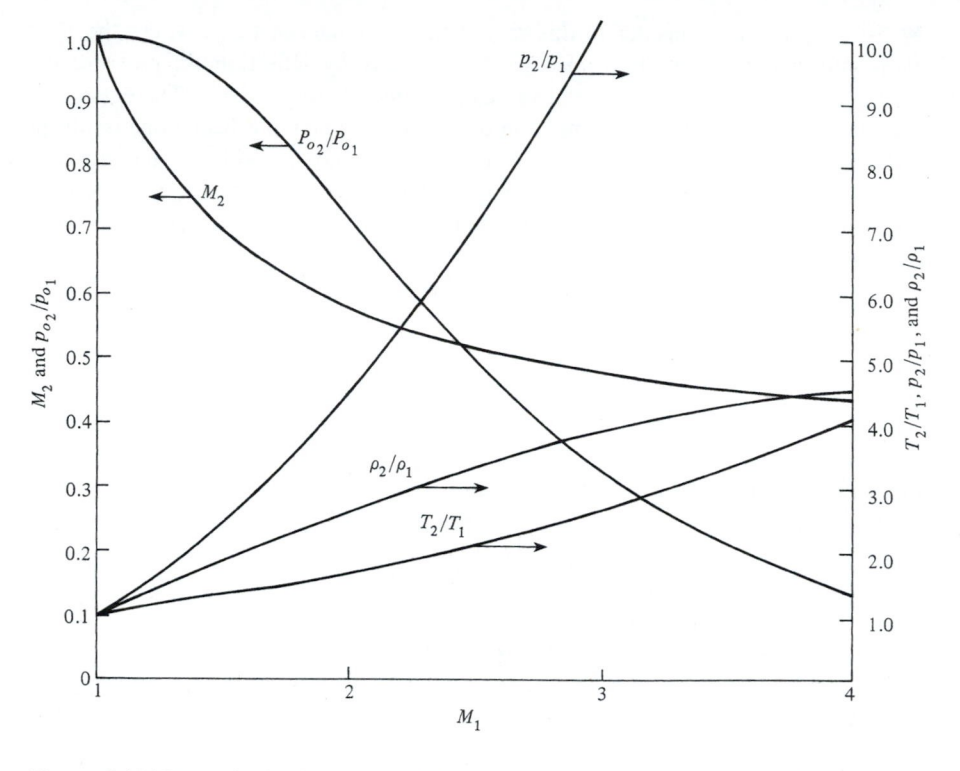

Figure 3.10 | Properties behind a normal shock wave as a function of upstream Mach number.

or
$$\frac{p_{o_2}}{p_{o_1}} = e^{-(s_2 - s_1)/R} \tag{3.64}$$

From Eqs. (3.64) and (3.60) we see that the ratio of total pressures across the normal shock depends on M_1 only. Also, because $s_2 > s_1$, Eqs. (3.63) and (3.64) show that $p_{o_2} < p_{o_1}$. *The total pressure decreases across a shock wave.*

The variations of p_2/p_1, ρ_2/ρ_1, T_2/T_1, p_{o_2}/p_{o_1}, and M_2 with M_1 as obtained from the above equations are tabulated in Table A.2 (in the Appendix A at the back of this book) for $\gamma = 1.4$. In addition, to provide more physical feel, these variations are also plotted in Fig. 3.10. Note that (as stated earlier) these curves show how, as M_1 becomes very large, T_2/T_1 and p_2/p_1 also become very large, whereas ρ_2/ρ_1 and M_2 approach finite limits.

EXAMPLE 3.5

A normal shock wave is standing in the test section of a supersonic wind tunnel. Upstream of the wave, $M_1 = 3$, $p_1 = 0.5$ atm, and $T_1 = 200$ K. Find M_2, p_2, T_2, and u_2 downstream of the wave.

■ **Solution**

From Table A.2, for $M_1 = 3$: $p_2/p_1 = 10.33$, $T_2/T_1 = 2.679$, and $M_2 = \boxed{0.4752}$. Hence

$$p_2 = \frac{p_2}{p_1}p_1 = 10.33(0.5) = \boxed{5.165 \text{ atm}}$$

$$T_2 = \frac{T_2}{T_1}T_1 = 2.679(200) = \boxed{535.8 \text{ K}}$$

$$a_2 = \sqrt{\gamma R T_2} = \sqrt{(1.4)(287)(535.8)} = 464 \text{ m/s}$$

$$u_2 = M_2 a_2 = (0.4752)(464) = \boxed{220 \text{ m/s}}$$

EXAMPLE 3.6

A blunt-nosed missile is flying at Mach 2 at standard sea level. Calculate the temperature and pressure at the nose of the missile.

■ **Solution**

The nose of the missile is a stagnation point, and the streamline through the stagnation point has also passed through the normal portion of the bow shock wave. Hence, the temperature and pressure at the nose are equal to the total temperature and pressure behind a normal shock. Also, at standard sea level, $T_1 = 519°\text{R}$ and $p_1 = 2116 \text{ lb/ft}^2$.

From Table A.1, for $M_1 = 2$: $T_{o_1}/T_1 = 1.8$ and $p_{o_1}/p_1 = 7.824$. Also, for adiabatic flow through a normal shock, $T_{o_2} = T_{o_1}$. Hence

$$T_{o_2} = T_{o_1} = \frac{T_{o_1}}{T_1}T_1 = 1.8(519) = \boxed{934.2°\text{R}}$$

From Table A.2, for $M_1 = 2$: $p_{o_2}/p_{o_1} = 0.7209$. Hence

$$p_{o_2} = \frac{p_{o_2}}{p_{o_1}}\frac{p_{o_1}}{p_1}p_1 = (0.7209)(7.824)(2116) = \boxed{11{,}935 \text{ lb/ft}^2}$$

EXAMPLE 3.7

Consider a point in a supersonic flow where the static pressure is 0.4 atm. When a Pitot tube is inserted in the flow at this point, the pressure measured by the Pitot tube is 3 atm. Calculate the Mach number at this point.

■ **Solution**

(We assume that the reader is familiar with the concept of a Pitot tube; see Sec. 8.7 of Ref. 104 for a discussion of the Pitot tube.) The pressure measured by a Pitot tube is the total pressure. However, when the tube is inserted into a supersonic flow, a normal shock is formed a short distance ahead of the mouth of the tube. In this case, the Pitot tube is sensing the total pressure *behind* the normal shock. Hence

$$\frac{p_{o_2}}{p_1} = \frac{3}{0.4} = 7.5$$

From Table A.2, for $p_{o_2}/p_1 = 7.5$: $M_1 = \boxed{2.35}$.

Note: As usual, in using the tables in Appendix A, we use the nearest entry for simplicity and efficiency; for improved accuracy, interpolation between the nearest entries should be used.

EXAMPLE 3.8

For the normal shock that occurs in front of the Pitot tube in Example 3.7, calculate the entropy change across the shock.

■ Solution
From Table A.2, for $M_1 = 2.35$: $p_{o_2}/p_{o_1} = 0.5615$. From Eq. (3.63)

$$\frac{s_2 - s_1}{R} = -\ln\frac{p_{o_2}}{p_{o_1}} = -\ln(0.5615) = 0.577$$

$$s_2 - s_1 = 0.577R = 0.577(1716) = \boxed{990.4 \text{ ft} \cdot \text{lb/slug} \cdot {}^\circ\text{R}}$$

EXAMPLE 3.9

Transonic flow is a mixed subsonic-supersonic flow where the local Mach number is near one. Such flows are discussed at length in Chap. 14, and are briefly described in Sec. 1.3. A typical example is the flow over the wing of a high-speed subsonic transport, such as the Boeing 777 shown in Fig. 1.4. When the airplane is flying at a free-stream Mach number on the order of 0.85, there will be a pocket of locally supersonic flow over the wing, as sketched in Fig. 1.10b. This pocket is terminated by a weak shock wave, also shown in Fig. 1.10b. Early numerical calculations of such transonic flows over an airfoil assumed the flow to be isentropic, hence ignoring the entropy increase and total pressure loss across the shock wave. Making the assumption that the shock wave in Fig. 1.10b is locally a normal shock, calculate the total pressure ratio and entropy increase across the shock for $M_1 = 1.04, 1.08, 1.12, 1.16$, and 1.2. Comment on the appropriateness of the isentropic flow assumption for the solution of transonic flows involving shocks of this nature.

■ Solution
From Table A.2, for $M_1 = 1.04$

$$\frac{p_{o_2}}{p_{o_1}} = 0.9999$$

From Eq. (3.63),

$$s_2 - s_1 = -R\ln\frac{p_{o_2}}{p_{o_1}} = -(287)\ln(0.9999) = \boxed{0.0287 \frac{\text{joule}}{\text{kg} \cdot \text{K}}}$$

Forming a table for the remaining calculations, we have

M_1	1.04	1.08	1.12	1.16	1.2
$\dfrac{p_{o_2}}{p_{o_1}}$	0.9999	0.9994	0.9982	0.9961	0.9928
$s_2 - s_1\left(\dfrac{\text{joule}}{\text{kg} \cdot \text{K}}\right)$	0.0287	0.172	0.517	1.12	2.07

From this table, the entropy increase across a normal shock with $M_1 = 1.04$ is very small; the shock is extremely weak. By comparison, the entropy increase for $M_1 = 1.12$ is 72 times larger than the case for $M_1 = 1.04$. The shock strength increases rapidly as M_1 increases above one. From these numbers, we might feel comfortable with the approximation of isentropic flow for transonic flows where the local Mach number in front of the shock is on the order of 1.08 or less. On the other hand, if the local Mach number is on the order of 1.2, the isentropic assumption is clearly suspect.

EXAMPLE 3.10

Consider two flows, one of helium and one of air, at the same Mach number of 5. Denoting the strength of a normal shock by the pressure ratio across the shock, p_2/p_1, which gas will result in the stronger shock? For a monatomic gas such as helium, $\gamma = 1.67$, and for a diatomic gas such as air, $\gamma = 1.4$.

■ **Solution**

For air, from Table A.2, for $M_1 = 5$

$$\frac{p_2}{p_1} = 29 \text{ (air)}$$

For helium, we cannot use Table A.2, which is for $\gamma = 1.4$ only. Returning to Eq. (3.57) for the pressure ratio across a normal shock,

$$\frac{p_2}{p_1} = 1 + \frac{2\gamma}{\gamma + 1}\left(M_1^2 - 1\right) = 1 + \frac{2(1.67)}{167 + 1}[(5)^2 - 1]$$

Hence,

$$\frac{p_2}{p_1} = 31 \text{ (helium)}$$

From this, we conclude that for equal upstream Mach numbers, the shock strength is greater in helium as compared to air.

EXAMPLE 3.11

Repeat Example 3.10, except assuming equal velocities of 1700 m/s and temperatures of 288 K for both gas flows.

■ **Solution**

For air, with $\gamma = 1.4$ and $R = 287$ joule/kg · K, the speed of sound at $T_1 = 288$ K is, from Eq. (3.20),

$$a_1 = \sqrt{\gamma R T_1} = \sqrt{(1.4)(287)(288)} = 340 \text{ m/s}$$

Hence,

$$M_1 = \frac{V_1}{a_1} = \frac{1700}{340} = 5$$

From Table A.2, we have

$$\frac{p_2}{p_1} = 29 \text{ (air)}$$

For helium, the molecular weight is 4. As given in Sec. 1.4,

$$R = \frac{\mathcal{R}}{M} = \frac{8314}{4} = 2078.5 \frac{\text{joule}}{\text{kg} \cdot \text{K}}$$

Hence,

$$a_1 = \sqrt{\gamma R T_1} = \sqrt{(1.67)(2078.5)(288)} = 999.8 \text{ m/s}$$

$$M_1 = \frac{V_1}{a_1} = \frac{1700}{3999.8} = 1.7$$

From Eq. (3.57)

$$\frac{p_2}{p_1} = 1 + \frac{2\gamma}{\gamma + 1}\left(M_1^2 - 1\right) = 1 + \frac{2(1.67)}{1.67 + 1}[(1.7)^2 - 1]$$

Hence

$$\frac{p_2}{p_1} = 3.36 \text{ (helium)}$$

From this, we conclude that, for equal upstream velocities and temperatures, the shock strength in helium is much weaker than in air. This is because the speed of sound in helium is much larger than air at the same temperature, due to the smaller molecular weight for helium. Since shock strength is dictated by Mach number, not velocity, the shock is much weaker in helium because of the much lower upstream Mach number.

3.7 | HUGONIOT EQUATION

The results obtained in Sec. 3.6 for the normal shock wave were couched in terms of velocities and Mach numbers—quantities which quite properly emphasize the fluid dynamic nature of shock waves. However, because the static pressure always increases across a shock wave, the wave itself can also be visualized as a thermodynamic device which compresses the gas. Indeed, the changes across a normal shock wave can be expressed in terms of purely thermodynamic variables without explicit reference to a velocity or Mach number, as follows.

From the continuity equation (3.38),

$$u_2 = u_1 \left(\frac{\rho_1}{\rho_2}\right) \tag{3.65}$$

Substitute Eq. (3.65) into the momentum equation (3.39):

$$p_1 + \rho_1 u_1^2 = p_2 + \rho_2 \left(\frac{\rho_1}{\rho_2} u_1\right)^2 \tag{3.66}$$

Solve Eq. (3.66) for u_1^2:

$$u_1^2 = \frac{p_2 - p_1}{\rho_2 - \rho_1} \left(\frac{\rho_2}{\rho_1} \right) \qquad (3.67)$$

Alternatively, writing Eq. (3.38) as

$$u_1 = u_2 \left(\frac{\rho_2}{\rho_1} \right)$$

and again substituting into Eq. (3.39), this time solving for u_2, we obtain

$$u_2^2 = \frac{p_2 - p_1}{\rho_2 - \rho_1} \left(\frac{\rho_1}{\rho_2} \right) \qquad (3.68)$$

From the energy equation (3.40),

$$h_1 + \frac{u_1^2}{2} = h_2 + \frac{u_2^2}{2}$$

and recalling that by definition $h = e + p/\rho$, we have

$$e_1 + \frac{p_1}{\rho_1} + \frac{u_1^2}{2} = e_2 + \frac{p_2}{\rho_2} + \frac{u_2^2}{2} \qquad (3.69)$$

Substituting Eqs. (3.67) and (3.68) into (3.69), the velocities are eliminated, yielding

$$e_1 + \frac{p_1}{\rho_1} + \frac{1}{2} \left[\frac{p_2 - p_1}{\rho_2 - \rho_1} \left(\frac{\rho_2}{\rho_1} \right) \right] = e_2 + \frac{p_2}{\rho_2} + \frac{1}{2} \left[\frac{p_2 - p_1}{\rho_2 - \rho_1} \left(\frac{\rho_1}{\rho_2} \right) \right] \qquad (3.70)$$

This simplifies to

$$e_2 - e_1 = \frac{(p_1 + p_2)}{2} \left(\frac{1}{\rho_1} - \frac{1}{\rho_2} \right) \qquad (3.71)$$

or

$$\boxed{e_2 - e_1 = \frac{p_1 + p_2}{2} (v_1 - v_2)} \qquad (3.72)$$

Equation (3.72) is called the *Hugoniot equation*. It has certain advantages because it relates only thermodynamic quantities across the shock. Also, we have made no assumption about the type of gas—Eq. (3.72) is a general relation that holds for a perfect gas, chemically reacting gas, real gas, etc. In addition, note that Eq. (3.72) has the form of $\Delta e = -p_{ave} \Delta v$, i.e., the change in internal energy equals the mean pressure across the shock times the change in specific volume. This strongly reminds us of the first law of thermodynamics in the form of Eq. (1.25), with $\delta q = 0$ for the adiabatic process across the shock.

In general, in equilibrium thermodynamics any state variable can be expressed as a function of any other two state variables, for example $e = e(p, v)$. This relation

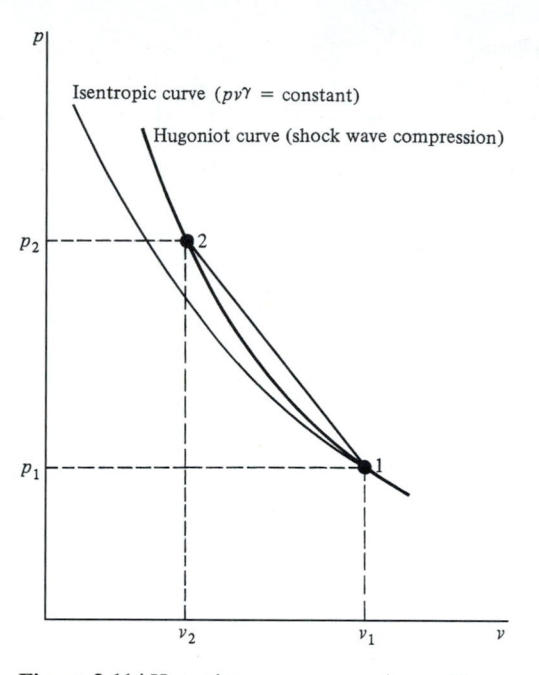

Figure 3.11 | Hugoniot curve; comparison with isentropic compression.

could be substituted into Eq. (3.72), resulting in a functional relation

$$p_2 = f(p_1, v_1, v_2) \tag{3.73}$$

For given conditions of p_1 and v_1 upstream of the normal shock, Eq. (3.73) represents p_2 as a function of v_2. A plot of this relation on a pv graph is called the *Hugoniot curve,* which is sketched in Fig. 3.11. This curve is the locus of all possible pressure-volume conditions behind normal shocks of various strengths for one specific set of upstream values for p_1 and v_1 (point 1 in Fig. 3.11). Each point on the Hugoniot curve in Fig. 3.11 therefore represents a different shock with a different upstream velocity u_1.

Now consider a specific shock with a specific value of upstream velocity u_1. How can we locate the specific point on the Hugoniot curve, point 2, which corresponds to this particular shock? To answer this question, return to Eq. (3.67), substituting $v = 1/\rho$:

$$u_1^2 = \frac{p_2 - p_1}{1/v_2 - 1/v_1} \left(\frac{v_1}{v_2} \right) \tag{3.74}$$

Rearranging Eq. (3.74), we obtain

$$\frac{p_2 - p_1}{v_2 - v_1} = -\left(\frac{u_1}{v_1} \right)^2 \tag{3.75}$$

Examining Eq. (3.75), the left-hand side is geometrically the slope of the straight line through points 1 and 2 in Fig. 3.11. The right-hand side is a known value, fixed by the upstream velocity and specific volume. Hence, by calculating $-(u_1/v_1)^2$ from the known upstream conditions, and by drawing a straight line through point 1 with this slope, the line will intersect the Hugoniot curve at point 2, as sketched in Fig. 3.11. Consequently, point 2 represents conditions behind the particular normal shock which has velocity u_1 with upstream pressure and specific volume p_1 and v_1, respectively.

Shock wave compression is a very effective (not necessarily efficient, but effective) process. For example, compare the isentropic and Hugoniot curves drawn through the same initial point (p_1, v_1) as sketched in Fig. 3.11. At this point, both curves have the same slope (prove this yourself, recalling that point 1 on the Hugoniot curve corresponds to an infinitely weak shock, i.e., a Mach wave). However, as v decreases, the Hugoniot curve climbs above the isentropic curve. Therefore, for a given decrease in specific volume, a shock wave creates a higher pressure increase than an isentropic compression. However, the shock wave costs more because of the entropy increase and consequent total pressure loss, i.e., the shock compression is less efficient than the isentropic compression.

Finally, noting that for a calorically perfect gas $e = c_v T$ and $T = pv/R$, Eq. (3.72) takes the form

$$\frac{p_2}{p_1} = \frac{\left(\dfrac{\gamma+1}{\gamma-1}\right)\dfrac{v_1}{v_2} - 1}{\left(\dfrac{\gamma+1}{\gamma-1}\right) - \dfrac{v_1}{v_2}}$$

Prove this to yourself.

EXAMPLE 3.12

Consider the normal shock wave properties calculated in Example 3.5. Show that these properties satisfy the Hugoniot equation for a calorically perfect gas.

■ Solution

The Hugoniot equation for a calorically perfect gas is given by the last equation in this section, namely,

$$\frac{p_2}{p_1} = \frac{\left(\dfrac{\gamma+1}{\gamma-1}\right)\dfrac{v_1}{v_2} - 1}{\left(\dfrac{\gamma+1}{\gamma-1}\right) - \dfrac{v_1}{v_2}}$$

Let us calculate v_1/v_2 from the information given in Example 3.5, substitute the value of v_1/v_2 into the last equation, and see if the resulting value of p_2/p_1 agrees with that obtained in Example 3.5.

From Example 3.5, $p_1 = 0.5$ atm, $T_1 = 200$ K, $p_2 = 5.165$ atm, and $T_2 = 535.8$ K. From the equation of state

$$\rho_1 = \frac{p_1}{RT_1} = \frac{0.5(1.01 \times 10^5)}{(287)(200)} = 0.8798 \text{ kg/m}^3$$

$$\rho_2 = \frac{p_2}{RT_2} = \frac{5.165(1.01 \times 10^5)}{(287)(535.8)} = 3.392 \text{ kg/m}^3$$

Hence,

$$\frac{\rho_2}{\rho_1} = \frac{v_1}{v_2} = \frac{3.392}{0.8798} = 3.855$$

From the Hugoniot equation,

$$\frac{p_2}{p_1} = \frac{\left(\dfrac{\gamma+1}{\gamma-1}\right)\dfrac{v_1}{v_2} - 1}{\left(\dfrac{\gamma+1}{\gamma-1}\right) - \dfrac{v_1}{v_2}} = \frac{\left(\dfrac{2.4}{0.4}\right)(3.855) - 1}{\left(\dfrac{2.4}{0.4}\right) - 3.855} = 10.32$$

From Example 3.5, the calculated pressure ratio was $p_2/p_1 = 10.33$, which agrees within round-off error with the result computed above from the Hugoniot equation. (*Please note:* All of the worked examples in this book were computed by the author using a hand calculator, hence the answers are subject to round-off errors that accumulate during the calculation.)

3.8 | ONE-DIMENSIONAL FLOW WITH HEAT ADDITION

Consider again Fig. 3.6, which illustrates a control volume for one-dimensional flow. Inside this control volume some action is occurring which causes the flow properties in region 2 to be different than in region 1. In the previous sections, this action has been due to a normal shock wave, where the large gradients inside the shock structure ultimately result in an increase in entropy via the effects of viscosity and thermal conduction. However, these effects are taking place *inside* the control volume in Fig. 3.6 and therefore the governing normal shock equations relating conditions in regions 1 and 2 did not require explicit terms accounting for friction and thermal conduction.

The action occurring inside the control volume in Fig. 3.6 can be caused by effects other than a shock wave. For example, if the flow is through a duct, friction between the moving fluid and the stationary walls of the duct causes changes between regions 1 and 2. This can be particularly important in long pipelines transferring gases over miles of land, for example. Another source of change in a one-dimensional flow is heat addition. If heat is added to or taken away from the gas inside the control volume in Fig. 3.6, the properties in region 2 will be different than those in region 1. This is a governing phenomenon in turbojet and ramjet engine burners, where heat is added in the form of fuel-air combustion. It also has an important effect on the supersonic flow in the cavities of modern gasdynamic and chemical lasers, where heat is effectively added by chemical reactions and molecular vibrational energy deactivation. Another example would be the heat added to an absorbing gas by an intense beam of radiation; such an idea has been suggested for laser-heated wind tunnels. In general, therefore, changes in a one-dimensional flow can be created by both friction and heat addition *without* the presence of a shock wave. One-dimensional flow with heat addition will be discussed in this section. Flow with friction, a somewhat analogous phenomenon, is the subject of Sec. 3.9.

Consider the one-dimensional flow in Fig. 3.6, with heat addition (or extraction) taking place between regions 1 and 2. The governing equations are Eqs. (3.2), (3.5), and (3.9), repeated here for convenience:

$$\rho_1 u_1 = \rho_2 u_2 \tag{3.2}$$

$$p_1 + \rho_1 u_1^2 = p_2 + \rho_2 u_2^2 \tag{3.5}$$

$$h_1 + \frac{u_1^2}{2} + q = h_2 + \frac{u_2^2}{2} \tag{3.9}$$

If conditions in region 1 are known, then for a specified amount of heat added per unit mass, q, these equations along with the appropriate equations of state can be solved for conditions in region 2. In general, a numerical solution is required. However, for the specific case of a calorically perfect gas, closed-form analytical expressions can be obtained—just as in the normal shock problem. Therefore, the remainder of this section will deal with a calorically perfect gas.

Solving Eq. (3.9) for q, with $h = c_p T$,

$$q = \left(c_p T_2 + \frac{u_2^2}{2} \right) - \left(c_p T_1 + \frac{u_1^2}{2} \right) \tag{3.76}$$

From the definition of total temperature, Eq. (3.27), the terms on the right-hand side of Eq. (3.76) simply result in

$$\boxed{q = c_p T_{o_2} - c_p T_{o_1} = c_p \left(T_{o_2} - T_{o_1} \right)} \tag{3.77}$$

Equation (3.77) clearly indicates that *the effect of heat addition is to directly change the total temperature of the flow.* If heat is added, T_o increases; if heat is extracted, T_o decreases.

Let us proceed to find the ratios of properties between regions 1 and 2 in terms of the Mach numbers M_1 and M_2. From Eq. (3.5), and noting that

$$\rho u^2 = \rho a^2 M^2 = \rho \frac{\gamma p}{\rho} M^2 = \gamma p M^2$$

we obtain

$$p_2 - p_1 = \rho_1 u_1^2 - \rho_2 u_2^2 = \gamma p_1 M_1^2 - \gamma p_2 M_2^2$$

Hence,

$$\boxed{\frac{p_2}{p_1} = \frac{1 + \gamma M_1^2}{1 + \gamma M_2^2}} \tag{3.78}$$

Also, from the perfect gas equation of state and Eq. (3.2),

$$\frac{T_2}{T_1} = \frac{p_2}{p_1} \frac{\rho_1}{\rho_2} = \frac{p_2}{p_1} \frac{u_2}{u_1} \tag{3.79}$$

From Eq. (3.20) and the definition of Mach number,

$$\frac{u_2}{u_1} = \frac{M_2 \, a_2}{M_1 \, a_1} = \frac{M_2}{M_1} \left(\frac{T_2}{T_1}\right)^{1/2} \tag{3.80}$$

Substituting Eqs. (3.78) and (3.80) into (3.79),

$$\boxed{\frac{T_2}{T_1} = \left(\frac{1 + \gamma M_1^2}{1 + \gamma M_2^2}\right)^2 \left(\frac{M_2}{M_1}\right)^2} \tag{3.81}$$

Since $\rho_2/\rho_1 = (p_2/p_1)(T_1/T_2)$, Eqs. (3.78) and (3.81) yield

$$\boxed{\frac{\rho_2}{\rho_1} = \left(\frac{1 + \gamma M_2^2}{1 + \gamma M_1^2}\right) \left(\frac{M_1}{M_2}\right)^2} \tag{3.82}$$

The ratio of total pressures is obtained directly from Eqs. (3.30) and (3.78),

$$\boxed{\frac{p_{o_2}}{p_{o_1}} = \frac{1 + \gamma M_1^2}{1 + \gamma M_2^2} \left(\frac{1 + \dfrac{\gamma - 1}{2} M_2^2}{1 + \dfrac{\gamma - 1}{2} M_1^2}\right)^{\gamma/(\gamma - 1)}} \tag{3.83}$$

The ratio of total temperatures is obtained directly from Eqs. (3.28) and (3.81),

$$\boxed{\frac{T_{o_2}}{T_{o_1}} = \left(\frac{1 + \gamma M_1^2}{1 + \gamma M_2^2}\right)^2 \left(\frac{M_2}{M_1}\right)^2 \left(\frac{1 + \dfrac{\gamma - 1}{2} M_2^2}{1 + \dfrac{\gamma - 1}{2} M_1^2}\right)} \tag{3.84}$$

Finally, the entropy change can be found from Eq. (1.36) with T_2/T_1 and p_2/p_1 given by Eqs. (3.81) and (3.78), respectively.

A scheme for the solution of one-dimensional flow with heat addition can now be outlined as follows. All conditions in region 1 are given. Therefore, for a given q, T_{o_2} can be obtained from Eq. (3.77). With this value of T_{o_2}, Eq. (3.84) can be solved for M_2. Once M_2 is known, then p_2/p_1, T_2/T_1, and ρ_2/ρ_1 are directly obtained from Eqs. (3.78), (3.81), and (3.82), respectively. This is a straightforward procedure; however, the solution of Eq. (3.84) for M_2 must be found by trial and error. Therefore, a more direct method of solving the problem of one-dimensional flow with heat addition is given below.

For convenience of calculation, we use sonic flow as a reference condition. Let $M_1 = 1$; the corresponding flow properties are denoted by $p_1 = p^*$, $T_1 = T^*$, $\rho_1 = \rho^*$, $p_{o_1} = p_o^*$, and $T_{o_1} = T_o^*$. The flow properties at any other value of M are then obtained by inserting $M_1 = 1$ and $M_2 = M$ into Eq. (3.78) and Eqs. (3.81) to

(3.84), yielding

$$\frac{p}{p^*} = \frac{1+\gamma}{1+\gamma M^2} \tag{3.85}$$

$$\frac{T}{T^*} = M^2 \left(\frac{1+\gamma}{1+\gamma M^2} \right)^2 \tag{3.86}$$

$$\frac{\rho}{\rho^*} = \frac{1}{M^2} \left(\frac{1+\gamma M^2}{1+\gamma} \right) \tag{3.87}$$

$$\frac{p_o}{p_o^*} = \frac{1+\gamma}{1+\gamma M^2} \left[\frac{2+(\gamma-1)M^2}{\gamma+1} \right]^{\gamma/(\gamma-1)} \tag{3.88}$$

$$\frac{T_o}{T_o^*} = \frac{(\gamma+1)M^2}{(1+\gamma M^2)^2} [2+(\gamma-1)M^2] \tag{3.89}$$

Equations (3.85) through (3.89) are tabulated as a function of M for $\gamma = 1.4$ in Table A.3. Note that, for a given flow, no matter what the local flow properties are, the reference sonic conditions (the starred quantities) are constant values. These starred values, although defined as conditions that exist at Mach 1, are fundamentally different than T^*, p^*, and ρ^* defined in Sec. 3.4. There, T^* was defined as the temperature that would exist at a point in the flow if the flow at that point were *imagined* to be locally slowed down (for a supersonic case) or speeded up (for a subsonic case) to Mach 1 *adiabatically*. In the present section we are dealing with a one-dimensional flow with heat addition—definitely a *nonadiabatic* process. Here, T^*, p^*, and ρ^* are those conditions in a one-dimensional flow that would exist if enough heat is added to achieve Mach 1. To see this more clearly, consider two different locations in a one-dimensional flow with heat addition, denoted by stations 1 and 2 as sketched in Fig. 3.12a. The flow at station 1 is given by M_1, p_1, and T_1. For the sake of discussion, let $M_1 = 3$. Now, let an amount of heat q_1 be added to this flow between stations 1 and 2. As a result, the flow properties at location 2 are M_2, p_2, and T_2 as shown in Fig. 3.12a. Assume that q_1 was a sufficient amount to result in $M_2 = 1.5$. (We will soon demonstrate that adding heat to a supersonic flow reduces the Mach number of the flow.) Now, return to station 1, where the local Mach number is $M_1 = 3$. Imagine that we add enough heat downstream of this station to cause the flow to slow down to Mach 1 as shown in Fig. 3.12b; denote this amount of heat by q_1^*. Clearly, $q_1^* > q_1$. The conditions in the duct where $M = 1$ after q_1^* is added are denoted by T^*, p^*, ρ^*, p_o^*, and T_o^*. Now, return to station 2, where $M_2 = 1.5$. Imagine that we add enough heat downstream of this station to cause the flow to slow down to Mach 1 as sketched in Fig. 3.12c; denote this amount of heat by q_2^*. The conditions in the duct where $M = 1$ after q_2^* is added are denoted by T^*, p^*, ρ^*, p_o^*, and T_o^*. These are precisely the same values that were obtained by adding q_1^* downstream of station 1. In other words, for a given one-dimensional flow, the values of T^*, p^*, ρ^*, etc., achieved when enough heat is added to bring the flow to Mach 1 are the same values, no matter whether the heat is added as q_1^* downstream of station 1

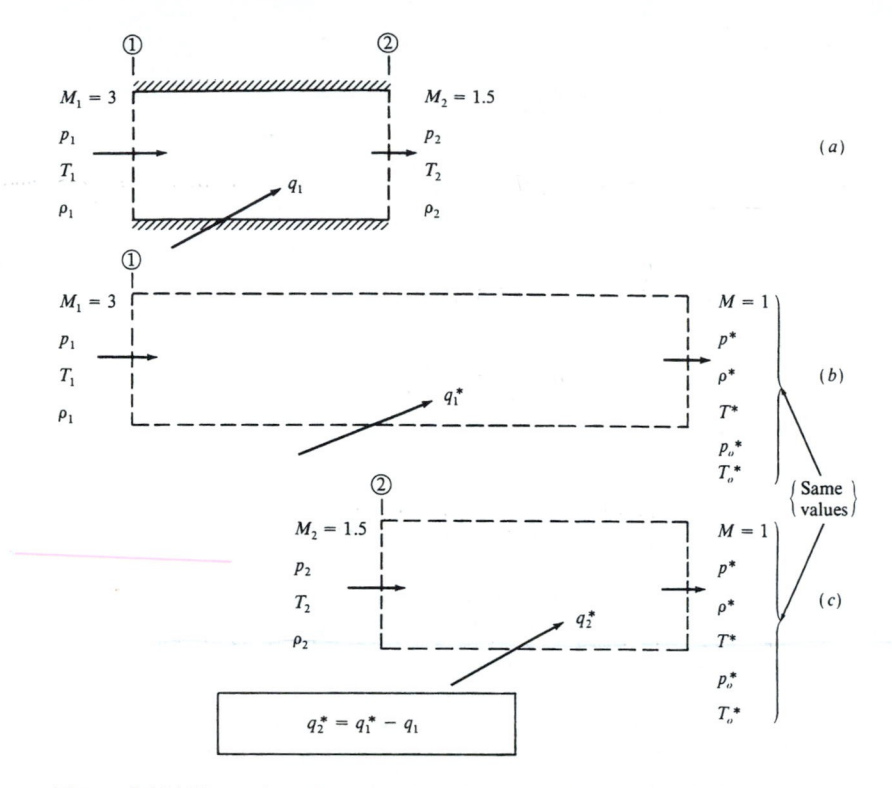

Figure 3.12 | Illustration of the meaning of the starred quantities at Mach 1 for one-dimensional flow with heat addition.

or as q_2^* downstream of station 2. This is why, in Eqs. (3.85) through (3.89), the starred quantities are simply reference quantities that are fixed values for a given flow entering a one-dimensional duct with heat addition. With this concept, Eqs. (3.85) through (3.89), or rather the tabulated values in Table A.3 obtained from these equations, simplify the calculation of problems involving one-dimensional flow with heat addition.

EXAMPLE 3.13

Air enters a constant-area duct at $M_1 = 0.2$, $p_1 = 1$ atm, and $T_1 = 273$ K. Inside the duct, the heat added per unit mass is $q = 1.0 \times 10^6$ J/kg. Calculate the flow properties M_2, p_2, T_2, ρ_2, T_{o_2}, and p_{o_2} at the exit of the duct.

■ **Solution**

From Table A.1, for $M_1 = 0.2$: $T_{o_1}/T_1 = 1.008$ and $p_{o_1}/p_1 = 1.028$. Hence

$$T_{o_1} = 1.008 T_1 = 1.008(273) = 275.2 \text{ K}$$

$$p_{o_1} = 1.028 p_1 = 1.028(1 \text{ atm}) = 1.028 \text{ atm}$$

$$c_p = \frac{\gamma R}{\gamma - 1} = \frac{(1.4)(287)}{0.4} = 1005 \text{ J/kg} \cdot \text{K}$$

From Eq. (3.77)

$$T_{o_2} = \frac{q}{c_p} + T_{o_1} = \frac{1.0 \times 10^6}{1005} + 275.2 = \boxed{1270\,\text{K}}$$

From Table A.3, for $M_1 = 0.2$: $T_1/T^* = 0.2066$, $p_1/p^* = 2.273$, $p_{o_1}/p_o^* = 1.235$, and $T_{o_1}/T_o^* = 0.1736$. Hence

$$\frac{T_{o_2}}{T_o^*} = \frac{T_{o_2}}{T_{o_1}}\frac{T_{o_1}}{T_o^*} = \frac{1270}{275.2}(0.1736) = 0.8013$$

From Table A.3, this corresponds to $\boxed{M_2 = 0.58}$.

Also from Table A3, for $M_2 = 0.58$: $T_2/T^* = 0.8955$, $p_2/p^* = 1.632$, $p_{o_2}/p_o^* = 1.083$. Hence

$$T_2 = \frac{T_2}{T^*}\frac{T^*}{T_1}T_1 = (0.8955)\left(\frac{1}{0.2066}\right)(273) = \boxed{1183\,\text{K}}$$

$$p_2 = \frac{p_2}{p^*}\frac{p^*}{p_1}p_1 = 1.632\frac{1}{2.273}1\,\text{atm} = \boxed{0.718\,\text{atm}}$$

$$p_{o_2} = \frac{p_{o_2}}{p_o^*}\frac{p_o^*}{p_{o_1}}p_{o_1} = 1.083\frac{1}{1.235}1.028 = \boxed{0.902\,\text{atm}}$$

Since $1\,\text{atm} = 1.01 \times 10^5\,\text{N/m}^2$,

$$\rho_2 = \frac{p_2}{RT_2} = \frac{(0.718)(1.01 \times 10^5)}{(278)(1183)} = \boxed{0.214\,\text{kg/m}^3}$$

EXAMPLE 3.14

Air enters a constant-area duct at $M_1 = 3$, $p_1 = 1$ atm, and $T_1 = 300$ K. Inside the duct, the heat added per unit mass is $q = 3 \times 10^5$ J/kg. Calculate the flow properties $M_2, p_2, T_2, \rho_2, T_{o_2}$, and p_{o_2} at the exit of the duct.

■ **Solution**

From Table A.1, for $M_1 = 3$: $T_{o_1}/T_1 = 2.8$. Hence

$$T_{o_1} = 2.8(300) = 840\,\text{K}$$

$$c_p = \frac{\gamma R}{\gamma - 1} = \frac{(1.4)(287)}{0.4} = 1004.5\,\text{J/kg} \cdot \text{K}$$

From Eq. (3.77)

$$q = c_p\left(T_{o_2} - T_{o_1}\right)$$

Thus

$$T_{o_2} = \frac{q}{c_p} + T_{o_1} = \frac{3 \times 10^5}{1004.5} + 840 = \boxed{1139\,\text{K}}$$

From Table A.3, for $M_1 = 3$: $p_1/p^* = 0.1765$, $T_1/T^* = 0.2803$, and $T_{o_1}/T_o^* = 0.6540$. Hence

$$\frac{T_{o_2}}{T_o^*} = \frac{T_{o_2}}{T_{o_1}}\frac{T_{o_1}}{T_o^*} = \frac{1139}{840}(0.6540) = 0.8868$$

From Table A.3, for $T_{o_2}/T_o^* = 0.8868$: $M_2 = \boxed{1.58}$. Also from Table A.3, $p_2/p^* = 0.5339$ and $T_2/T^* = 0.7117$. Thus

$$p_2 = \frac{p_2}{p^*}\frac{p^*}{p_1}p_1 = 0.5339\left(\frac{1}{0.1765}\right)(1\text{ atm}) = \boxed{3.025\text{ atm}}$$

$$T_2 = \frac{T_2}{T^*}\frac{T^*}{T_1}T_1 = 0.7117\left(\frac{1}{0.2803}\right)(300) = \boxed{761.7\text{ K}}$$

$$\rho_2 = \frac{p_2}{RT_2} = \frac{(3.025)(1.01 \times 10^5)}{(287)(761.7)} = \boxed{1.398\text{ kg/m}^3}$$

From Table A.3, for $M_1 = 3$: $p_{o_1}/p_o^* = 3.424$. For $M_2 = 1.58$: $p_{o_2}/p_o^* = 1.164$. Thus

$$\frac{p_{o_2}}{p_{o_1}} = \frac{p_{o_2}/p_o^*}{p_{o_1}/p_o^*} = \frac{1.164}{3.424} = 0.340$$

From Table A.1, For $M_1 = 3$: $p_{o_1}/p_1 = 36.73$. Hence

$$p_{o_2} = \frac{p_{o_2}}{p_{o_1}}\frac{p_{o_1}}{p_1}p_1 = (0.340)(36.73)(1\text{ atm}) = \boxed{12.49\text{ atm}}$$

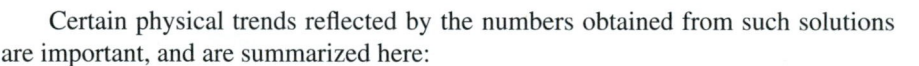

Certain physical trends reflected by the numbers obtained from such solutions are important, and are summarized here:

1. For *supersonic flow* in region 1, i.e., $M_1 > 1$, when heat is added
   a. Mach number decreases, $M_2 < M_1$
   b. Pressure increases, $p_2 > p_1$
   c. Temperature increases, $T_2 > T_1$
   d. Total temperature increases, $T_{o_2} > T_{o_1}$
   e. Total pressure decreases, $p_{o_2} < p_{o_1}$
   f. Velocity decreases, $u_2 < u_1$

2. For *subsonic flow* in region 1, i.e., $M_1 < 1$, when heat is added
   a. Mach number increases, $M_2 > M_1$
   b. Pressure decreases, $p_2 < p_1$
   c. Temperature increases for $M_1 < \gamma^{-1/2}$ and decreases for $M_1 > \gamma^{-1/2}$
   d. Total temperature increases, $T_{o_2} > T_{o_1}$
   e. Total pressure decreases, $p_{o_2} < p_{o_1}$
   f. Velocity increases, $u_2 > u_1$

For heat extraction (cooling of the flow), all of the above trends are opposite.

From the development here, it is important to note that heat addition always drives the Mach numbers toward 1, decelerating a supersonic flow and accelerating

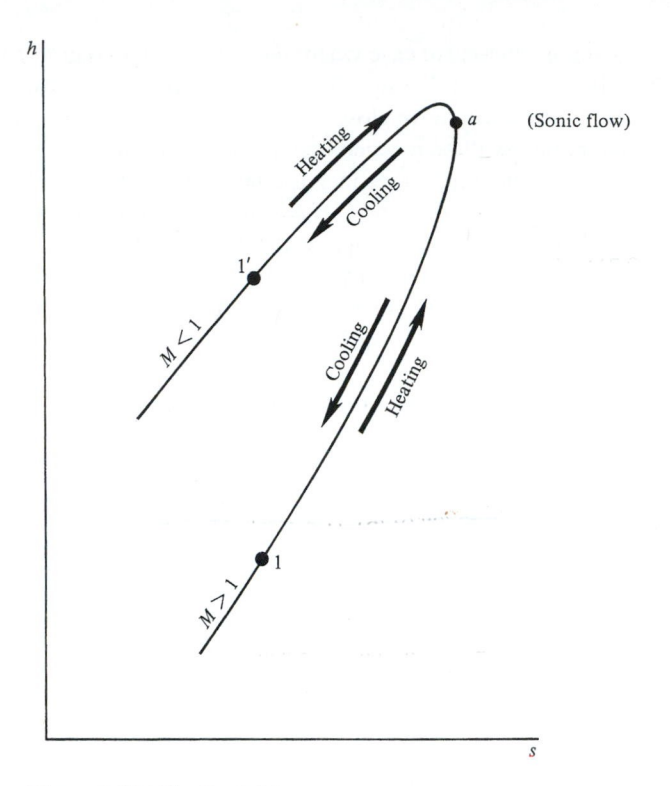

Figure 3.13 | The Rayleigh curve.

a subsonic flow. This is emphasized in Fig. 3.13, which is a Mollier diagram (enthalpy versus entropy) of the one-dimensional heat-addition process. The curve in Fig. 3.13 is called the *Rayleigh curve,* and is drawn for a set of given initial conditions. If the conditions in region 1 are given by point 1 in Fig. 3.13, then the particular Rayleigh curve through point 1 is the locus of all possible states in region 2. Each point on the curve corresponds to a different value of q added or taken away. Point a corresponds to maximum entropy; also at point a the flow is sonic. The lower branch of the Rayleigh curve below point a corresponds to supersonic flow; the upper branch above point a corresponds to subsonic flow. If the flow in region 1 of Fig. 3.6 is supersonic and corresponds to point 1 in Fig. 3.13, then heat addition will cause conditions in region 2 to move closer to point a, with a consequent decrease of Mach number towards unity. As q is made larger, conditions in region 2 get closer and closer to point a. Finally, for a certain value of q, the flow will become sonic in region 2. For this condition, the flow is said to be *choked,* because any further increase in q is not possible without a drastic revision of the upstream conditions in region 1. For example, if the initial supersonic conditions in region 1 were obtained by expansion through a supersonic nozzle, and if a value of q is added to the flow above that allowed for attaining Mach 1 in region 2, then a normal shock will form inside the nozzle and conditions in region 1 will suddenly become subsonic.

Now consider an alternative case where the initial flow in region 1 in Fig. 3.6 is subsonic, say given by point $1'$ in Fig. 3.13. If heat is added to the flow, conditions in the downstream region 2 will move closer to point a. If q is increased to a sufficiently high value, then point a will be reached and the flow in region 2 will be sonic. The flow is again choked, and any further increase in q is impossible without an adjustment of the initial conditions in region 1. If q is increased above this value, then a series of pressure waves will propagate upstream, and nature will adjust the conditions in region 1 to a lower subsonic Mach number, to the left of point $1'$ in Fig. 3.13.

Note from the Rayleigh curve in Fig. 3.13 that it is theoretically possible to decelerate a supersonic flow to a subsonic value by first heating it until sonic flow (point a) is reached, and then cooling it thereafter. Similarly, an initially subsonic flow can be made supersonic by first heating it until sonic flow (point a) is reached, and then cooling it thereafter.

Finally, just as in the case of a normal shock wave, heat addition to a flow— subsonic or supersonic—always *decreases* the total pressure. This effect is of prime importance in the design of jet engines and in the pressure recovery attainable in gas-dynamic and chemical lasers.

EXAMPLE 3.15

In Example 3.14, how much heat per unit mass must be added to choke the flow?

■ Solution

From Example 3.14, $T_{o_1} = 840$ K. Also from Table A.3, for $M_1 = 3$: $T_{o_1}/T_o^* = 0.6540$. Thus

$$T_o^* = \frac{T_{o_1}}{0.6540} = \frac{840}{0.6540} = 1284 \text{ K}$$

When the flow is choked, the Mach number at the end of the duct is $M_2 = 1$. Thus

$$T_{o_2} = T_o^* = 1284 \text{ K}$$

$$q = c_p \left(T_{o_2} - T_{o_1} \right) = (1004.5)(1284 - 840) = \boxed{4.46 \times 10^5 \text{ J/kg}}$$

EXAMPLE 3.16

Consider the supersonic inflow conditions given in Example 3.14. If an amount of heat equal to 6×10^5 J/kg is added to this flow, what will happen to it qualitatively and quantitatively?

■ Solution

From the result given in Example 3.15, we see that $q = 6 \times 10^5$ J/kg is *more* than that required to choke the flow. In this case, the flow mechanism that is producing the incoming flow at $M_1 = 3$ will be completely changed by strong pressure waves propagating upstream so that *new* inflow conditions will prevail that will accommodate this increased amount of heat addition, still choking the flow at the exit of the duct. Nature will change the originally supersonic inflow to a subsonic inflow with just the right value of $M_1 < 1$ such that the heat added will just choke the subsonic flow.

To calculate the new inflow Mach number, we assume that whatever mechanism that nature uses to change the supersonic inflow to a subsonic inflow will not change the total temperature of the inflow. For example, if the mechanism is that of a normal shock wave, the total temperature is not changed across the shock. Hence, T_{o_1} remains the same; $T_{o_1} = 840$ K. To calculate $T_{o_2} = T_o^*$, we have

$$q = c_p \left(T_{o_2} - T_{o_1} \right)$$

or

$$T_o^* = T_{o_2} = \frac{q}{c_p} + T_{o_1} = \frac{6 \times 10^5}{1004.5} + 840 = 1437 \, \text{K}$$

$$\frac{T_{o_1}}{T_o^*} = \frac{840}{1437} = 0.5846$$

From Table A.3, we find for $T_{o_2}/T_o^* = 0.5846$, $M_1 = 0.43$.

Hence, when $q = 6 \times 10^5$ J/kg is added to the flow, the initial supersonic inflow at $M_1 = 3$ will be modified through a complex transient process to become a subsonic inflow with $M_1 = 0.43$.

3.9 | ONE-DIMENSIONAL FLOW WITH FRICTION

With this section we arrive at the last box at the bottom of our roadmap in Fig. 3.1. Consider the one-dimensional flow of a compressible inviscid fluid in a constant-area duct. If the flow is steady, adiabatic, and shockless, Eqs. (3.2), (3.5), and (3.9) yield the trivial solution of constant property flow everywhere along the duct. However, in reality, all fluids are viscous, and the friction between the moving fluid and the stationary walls of the duct causes the flow properties to change along the duct. Although viscous flows are not the subject of this book, if the frictional effect is modeled as a shear stress at the wall acting on a fluid with uniform properties over any cross section, as illustrated in Fig. 3.14, then the equations developed in Sec. 3.2, with one modification, describe the mean properties of frictional flow in constant-area

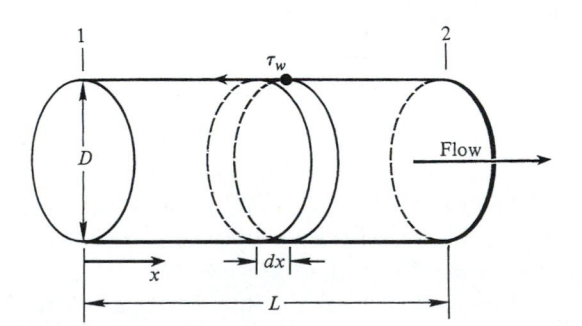

Figure 3.14 | Model of one-dimensional flow with friction.

ducts. The analysis and results are analogous to one-dimensional flow with heat addition, treated in Sec. 3.8.

The aforementioned modification applies to the momentum equation. As seen in Fig. 3.14, the frictional shear stress τ_w acts on the surface of the cylindrical control volume, thus contributing an additional surface force in the integral formulation of the momentum equation. Equation (3.4) is the x component of the momentum equation for an inviscid gas; with the shear stress included, this equation becomes

$$\oiint_S (\rho \mathbf{V} \cdot d\mathbf{S})u = -\oiint_S (p\, d\mathbf{S})_x - \oiint_S \tau_w\, dS \tag{3.90}$$

Applied to the cylindrical control volume of diameter D and length L sketched in Fig. 3.14, Eq. (3.90) becomes

$$-\rho_1 u_1^2 A + \rho_2 u_2^2 A = p_1 A - p_2 A - \int_0^L \pi D \tau_w\, dx \tag{3.91}$$

Since $A = \pi D^2/4$, Eq. (3.91) becomes

$$(p_2 - p_1) + \left(\rho_2 u_2^2 - \rho_1 u_1^2\right) = -\frac{4}{D} \int_0^L \tau_w\, dx \tag{3.92}$$

The shear stress τ_w varies with distance x along the duct, thus complicating the integration on the right-hand side of Eq. (3.92). This can be circumvented by taking the limit of Eq. (3.92) as L shrinks to dx, as shown in Fig. 3.14, resulting in the differential relation

$$dp + d(\rho u^2) = -\frac{4}{D}\tau_w\, dx \tag{3.93}$$

From Eq. (3.2), $\rho u = $ const. Hence, $d(\rho u^2) = \rho u\, du + u\, d(\rho u) = \rho u\, du + u(0) = \rho u\, du$. Thus Eq. (3.93) becomes

$$dp + \rho u\, du = -\frac{4}{D}\tau_w\, dx \tag{3.94}$$

The shear stress can be expressed in terms of a friction coefficient f, defined as $\tau_w = \frac{1}{2}\rho u^2 f$. Hence, Eq. (3.94) becomes

$$dp + \rho u\, du = -\frac{1}{2}\rho u^2 \frac{4f\, dx}{D} \tag{3.95}$$

Returning to Fig. 3.14, the driving force causing the mean cross-sectional flow properties to vary as a function of x is friction at the wall of the duct, and this variation is governed by Eq. (3.95). For practical calculations dealing with a calorically perfect gas, Eq. (3.95) is recast completely in terms of the Mach number M. This can be accomplished by recalling that, $a^2 = \gamma p/\rho$, $M^2 = u^2/a^2$, $p = \rho RT$, $\rho u = $ const, and $c_p T + u^2/2 = $ const. The derivation is left as an exercise for the reader; the result is

$$\frac{4f\, dx}{D} = \frac{2}{\gamma M^2}(1 - M^2)\left[1 + \tfrac{1}{2}(\gamma - 1)M^2\right]^{-1}\frac{dM}{M} \tag{3.96}$$

Integrating Eq. (3.96) between $x = x_1$ (where $M = M_1$) and $x = x_2$ (where $M = M_2$),

$$\int_{x_1}^{x_2} \frac{4f \, dx}{D} = \left[-\frac{1}{\gamma M^2} - \frac{\gamma + 1}{2\gamma} \ln \left(\frac{M^2}{1 + \frac{\gamma - 1}{2} M^2} \right) \right]_{M_1}^{M_2} \tag{3.97}$$

Equation (3.97) relates the Mach numbers at two different sections to the integrated effect of friction between the sections.

The ratios of static temperature, pressure, density, and total pressure between the two sections are readily obtained. The flow is adiabatic, hence $T_o = \text{const.}$ Thus, from Eq. (3.28), we have

$$\frac{T_2}{T_1} = \frac{T_o/T_1}{T_o/T_2} = \frac{2 + (\gamma - 1)M_1^2}{2 + (\gamma - 1)M_2^2} \tag{3.98}$$

Also, since $\rho_1 u_1 = \rho_2 u_2$, and $a^2 = \gamma p/\rho$, then

$$\frac{\gamma p_1 u_1}{a_1^2} = \frac{\gamma p_2 u_2}{a_2^2}$$

or

$$\frac{p_2}{p_1} = \frac{M_1 \, a_2}{M_2 \, a_1} = \frac{M_1}{M_2} \sqrt{\frac{T_2}{T_1}} \tag{3.99}$$

Substituting Eq. (3.98) into (3.99), we have

$$\frac{p_2}{p_1} = \frac{M_1}{M_2} \left[\frac{2 + (\gamma - 1)M_1^2}{2 + (\gamma - 1)M_2^2} \right]^{1/2} \tag{3.100}$$

From the equation of state, $\rho_2/\rho_1 = (p_2/p_1)(T_1/T_2)$. Substituting Eqs. (3.98) and (3.100) into this result, we obtain

$$\frac{\rho_2}{\rho_1} = \frac{M_1}{M_2} \left[\frac{2 + (\gamma - 1)M_1^2}{2 + (\gamma - 1)M_2^2} \right]^{-1/2} \tag{3.101}$$

Finally, from Eqs. (3.30) and (3.100), the ratio of total pressures is

$$\frac{p_{o_2}}{p_{o_1}} = \frac{p_{o_2}/p_2}{p_{o_1}/p_1} \frac{p_2}{p_2} = \left[\frac{2 + (\gamma - 1)M_2^2}{2 + (\gamma - 1)M_1^2} \right]^{\gamma/(\gamma - 1)} \frac{M_1}{M_2} \left[\frac{2 + (\gamma - 1)M_1^2}{2 + (\gamma - 1)M_2^2} \right]^{1/2}$$

$$\frac{p_{o_2}}{p_{o_1}} = \frac{M_1}{M_2} \left[\frac{2 + (\gamma - 1)M_2^2}{2 + (\gamma - 1)M_1^2} \right]^{(\gamma + 1)/[2(\gamma - 1)]} \tag{3.102}$$

Analogous to our previous discussion of one-dimensional flow with heat addition, calculations of flow with friction are expedited by using sonic flow reference conditions, where the flow properties are denoted by p^*, ρ^*, T^*, and p_o^*. From Eqs. (3.98) and (3.100) through (3.102),

$$\frac{T}{T^*} = \frac{\gamma + 1}{2 + (\gamma - 1)M^2} \tag{3.103}$$

$$\frac{p}{p^*} = \frac{1}{M}\left[\frac{\gamma + 1}{2 + (\gamma - 1)M^2}\right]^{1/2} \tag{3.104}$$

$$\frac{\rho}{\rho^*} = \frac{1}{M}\left[\frac{2 + (\gamma - 1)M^2}{\gamma + 1}\right]^{1/2} \tag{3.105}$$

$$\frac{p_o}{p_o^*} = \frac{1}{M}\left[\frac{2 + (\gamma - 1)M^2}{\gamma + 1}\right]^{(\gamma+1)/[2(\gamma-1)]} \tag{3.106}$$

Also, if we define $x = L^*$ as the station where $M = 1$, then Eq. (3.97) becomes

$$\int_0^{L^*} \frac{4f\,dx}{D} = \left[-\frac{1}{\gamma M^2} - \frac{\gamma + 1}{2\gamma}\ln\left(\frac{M^2}{1 + \frac{\gamma - 1}{2}M^2}\right)\right]_M^1$$

or

$$\frac{4\bar{f}L^*}{D} = \frac{1 - M^2}{\gamma M^2} + \frac{\gamma + 1}{2\gamma}\ln\left[\frac{(\gamma + 1)M^2}{2 + (\gamma - 1)M^2}\right] \tag{3.107}$$

where $\bar{f}$ is an average friction coefficient defined as

$$\bar{f} = \frac{1}{L^*}\int_0^{L^*} f\,dx$$

Equations (3.103) through (3.107) are tabulated versus Mach number in Table A.4 for $\gamma = 1.4$.

The local friction coefficient f depends on whether the flow is laminar or turbulent, and is a function of Mach number, Reynolds number, and surface roughness, among other variables. In almost all practical cases, the flow is turbulent, and the variation of f must be obtained empirically. Extensive friction coefficient data can be obtained from Schlicting's classical book (Ref. 20) among others; hence, no further elaboration will be given here. For our purposes, it is reasonable to assume an approximate constant value of $f = 0.005$, which holds for $R_e > 10^5$ and a surface roughness of $0.001D$.

EXAMPLE 3.17

Consider the flow of air through a pipe of inside diameter $= 0.15$ m and length $= 30$ m. The inlet flow conditions are $M_1 = 0.3$, $p_1 = 1$ atm, and $T_1 = 273$ K. Assuming $f = \text{const} = 0.005$, calculate the flow conditions at the exit, M_2, p_2, T_2, and p_{o_2}.

■ Solution
From Table A.1, for $M_1 = 0.3$: $p_{o_1}/p_1 = 1.064$. Thus

$$p_{o_1} = 1.064(1\text{ atm}) = 1.064\text{ atm}$$

From Table A.4, for $M_1 = 0.3$: $4\bar{f}L_1^*/D = 5.299$, $p_1/p^* = 3.619$, $T_1/T^* = 1.179$, and $p_{o_1}/p^* = 2.035$. Since $L = 30\,\text{m} = L_1^* - L_2^*$, then $L_2^* = L_1^* - L$ and

$$\frac{4\bar{f}L_2^*}{D} = \frac{4\bar{f}L_1^*}{D} - \frac{4\bar{f}L}{D} = 5.2993 - \frac{(4)(0.005)(30)}{0.15} = 1.2993$$

From Table A.4, for $4\bar{f}L^*/D = 1.2993$: $\boxed{M_2 = 0.475}$, $T_2/T^* = 1.148$, $p_2/p^* = 2.258$, and $p_{o_2}/p_o^* = 1.392$. Hence

$$p_2 = \frac{p_2}{p^*}\frac{p^*}{p_1}p_1 = 2.258\frac{1}{3.169}(1\,\text{atm}) = \boxed{0.713\,\text{atm}}$$

$$T_2 = \frac{T_2}{T^*}\frac{T^*}{T_1}T_1 = 1.148\frac{1}{1.179}273 = \boxed{265.8\,\text{K}}$$

$$p_{o_2} = \frac{p_{o_2}}{p_o^*}\frac{p_o^*}{p_{o_1}}p_{o_1} = 1.392\frac{1}{2.035}1.064 = \boxed{0.728\,\text{atm}}$$

EXAMPLE 3.18

Consider the flow of air through a pipe of inside diameter $= 0.4\,\text{ft}$ and length $= 5\,\text{ft}$. The inlet flow conditions are $M_1 = 3$, $p_1 = 1\,\text{atm}$, and $T_1 = 300\,\text{K}$. Assuming $f = \text{const} = 0.005$, calculate the flow conditions at the exit, M_2, p_2, T_2, and p_{o_2}.

■ Solution

$L_1^* - L_2^* = L$. Hence

$$\frac{4\bar{f}L_2^*}{D} = \frac{4\bar{f}L_1^*}{D} - \frac{4\bar{f}L}{D}$$

From Table A.4, for $M_1 = 3$: $4\bar{f}L_1^*/D = 0.5222$, $T_1/T^* = 0.4286$, and $p_1/p^* = 0.2182$. Thus

$$\frac{4\bar{f}L_2^*}{D} = 0.5222 - \frac{4(0.005)(5)}{0.4} = 0.2722$$

From Table A.4, for $4\bar{f}L_2^*/D = 0.2722$: $\boxed{M_2 = 1.9}$. Also from Table A.4: $T_2/T^* = 0.6969$ and $p_2/p^* = 0.4394$. Thus

$$T_2 = \frac{T_2}{T^*}\frac{T^*}{T_1}T_1 = (0.6969)\left(\frac{1}{0.4286}\right)(300) = \boxed{487.8\,\text{K}}$$

$$p_2 = \frac{p_2}{p^*}\frac{p^*}{p_1}p_1 = (0.4394)\left(\frac{1}{0.2182}\right)(1\,\text{atm}) = \boxed{2.014\,\text{atm}}$$

From Table A.4, for $M_1 = 3$: $p_{o_1}/p_o^* = 4.235$. Also for $M_2 = 1.9$: $p_{o_2}/p_o^* = 1.555$. Thus

$$\frac{p_{o_2}}{p_{o_1}} = \frac{p_{o_2}/p_o^*}{p_{o_1}/p_o^*} = \frac{1.555}{4.235} = 0.367$$

From Table A.1, for $M_1 = 3$: $p_{o_1}/p_1 = 36.73$. Thus

$$p_{o_2} = \frac{p_{o_2}}{p_{o_1}}\frac{p_{o_1}}{p_1}p_1 = (0.367)(36.73)(1\,\text{atm}) = \boxed{13.49\,\text{atm}}$$

Certain physical trends reflected by the numbers obtained from such solutions are summarized here:

1. For *supersonic* inlet flow, i.e., $M_1 > 1$, the effect of friction on the downstream flow is such that
    a. Mach number decreases, $M_2 < M_1$
    b. Pressure increases, $p_2 > p_1$
    c. Temperature increases, $T_2 > T_1$
    d. Total pressure decreases, $p_{o_2} < p_{o_1}$
    e. Velocity decreases, $u_2 < u_1$
2. For *subsonic* inlet flow, i.e., $M_1 < 1$, the effect of friction on the downstream flow is such that
    a. Mach number increases, $M_2 > M_1$
    b. Pressure decreases, $p_2 < p_1$
    c. Temperature decreases, $T_2 < T_1$
    d. Total pressure decreases, $p_{o_2} < p_{o_1}$
    e. Velocity increases, $u_2 > u_1$

From this, note that friction always drives the Mach number toward 1, decelerating a supersonic flow and accelerating a subsonic flow. This is emphasized in Fig. 3.15, which is a Mollier diagram of one-dimensional flow with friction. The curve in Fig. 3.15 is called the *Fanno curve,* and is drawn for a set of given initial conditions. Point *a* corresponds to maximum entropy, where the flow is sonic. This point splits the Fanno curve into subsonic (upper) and supersonic (lower) portions. If

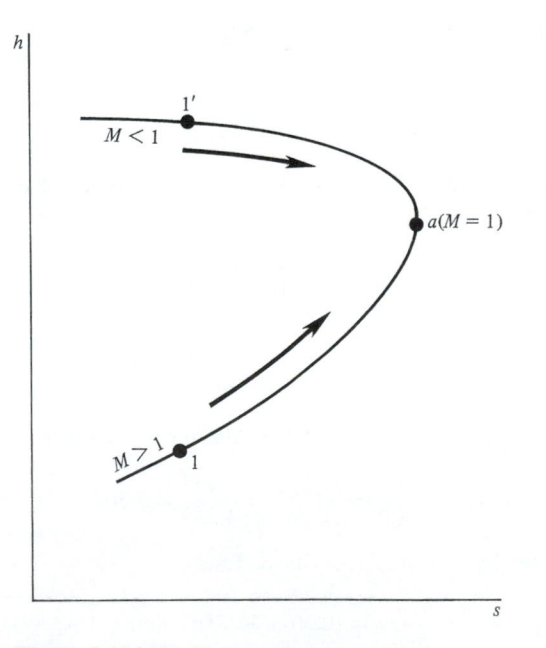

Figure 3.15 | The Fanno curve.

the inlet flow is supersonic and corresponds to point 1 in Fig. 3.15, then friction causes the downstream flow to move closer to point a, with a consequent decrease of Mach number toward unity. Each point on the curve between points 1 and a corresponds to a certain duct length L. As L is made larger, the conditions at the exit move closer to point a. Finally, for a certain value of L, the flow becomes sonic. For this condition, the flow is *choked*, because any further increase in L is not possible without a drastic revision of the inlet conditions. For example, if the inlet conditions at point 1 were obtained by expansion through a supersonic nozzle, and if L were larger than that allowed for attaining Mach 1 at the exit, then a normal shock would form inside the nozzle, and the duct inlet conditions would suddenly become subsonic.

Consider the alternative case where the inlet flow is subsonic, say given by point $1'$ in Fig. 3.15. As L increases, the exit conditions move closer to point a. If L is increased to a sufficiently large value, then point a is reached and the flow at the exit becomes sonic. The flow is again choked, and any further increase in L is impossible without an adjustment of the inlet conditions to a lower inlet Mach number, i.e., without moving the inlet conditions to the left of point $1'$ in Fig. 3.15.

Finally, note that friction always causes the total pressure to decrease whether the inlet flow is subsonic or supersonic. Also, unlike the Rayleigh curve for flow with heating and cooling, the upper and lower portions of the Fanno curve cannot be traversed by the same one-dimensional flow. That is, within the framework of one-dimensional theory, it is not possible to first slow a supersonic flow to sonic conditions by friction, and then further slow it to subsonic speeds also by friction. Such a subsonic deceleration would violate the second law of thermodynamics.

EXAMPLE 3.19

In Example 3.18, what is the length of the duct required to choke the flow?

■ Solution

From Table A.4, for $M_1 = 3$: $4\bar{f}L_1^*/D = 0.5222$. The length of the duct required to achieve Mach 1 at the exit of the duct is, by definition, L_1^*. Thus

$$L_1^* = 0.5222\frac{D}{4\bar{f}} = (0.5222)\frac{0.4}{(4)(0.005)} = \boxed{10.44\text{ ft}}$$

3.10 | HISTORICAL NOTE: SOUND WAVES AND SHOCK WAVES

Picking up the thread of history from Sec. 1.1, the following questions are posed: When was the speed of sound first calculated and properly understood? What is the origin of normal shock theory? Who developed the principal equations discussed in this chapter? Let us examine these questions further.

By the seventeenth century, it was clearly appreciated that sound propagates through the air at some finite velocity. Indeed, by the time Isaac Newton published the first edition of his *Principia* in 1687, artillery tests had already indicated that the speed of sound was approximately 1140 ft/s. These tests were performed by standing a known large distance away from a cannon, and noting the time delay between the light flash from the muzzle and the sound of the discharge. In Proposition 50, Book II, of his *Principia,* Newton correctly theorized that the speed of sound was related to the "elasticity" of the air (the reciprocal of the compressibility defined in Sec. 1.2). However, he made the erroneous assumption that a sound wave is an isothermal process, and consequently proposed the following incorrect expression for the speed of sound:

$$a = \sqrt{\frac{1}{\rho \tau_T}}$$

where τ_T is the isothermal compressibility defined in Sec. 1.1. Much to his dismay, Newton calculated a value of 979 ft/s from this expression—15 percent lower than the existing gunshot data. Undaunted, however, he followed a now familiar ploy of theoreticians; he proceeded to explain away the difference by the existence of solid dust particles and water vapor in the atmosphere. This misconception was corrected a century later by the famous French mathematician, Pierre Simon Marquis de Laplace, who in a paper entitled "Sur la vitesse du son dans l'aire et dan l'eau" from the *Annales de Chimie et de Physique* (1816) properly assumed that a sound wave was adiabatic, not isothermal. Laplace went on to derive the proper expression

$$a = \sqrt{\frac{1}{\rho \tau_s}}$$

where τ_s is the isentropic compressibility defined in Sec. 1.1. This equation is the same as Eq. (3.18) derived in Sec. 3.3. Therefore, by the time of the demise of Napoleon, the process and relationship for the propagation of sound in a gas was fully understood.

The existence of shock waves was also recognized by this time, and following the successful approach of Laplace to the calculation of the speed of sound, it was natural for the German mathematician G. F. Bernhard Riemann in 1858 to first attempt to calculate shock properties by also assuming isentropic conditions. Of course, this was doomed to failure. However, 12 years later, the first major break-through in shock wave theory was made by the Scottish engineer, William John Macquorn Rankine (1820–1872). (See Fig. 3.16.) Born in Edinburgh, Scotland, on July 5, 1820, Rankine was one of the founders of the science of thermodynamics. At the age of 25, he was offered the Queen Victoria Chair of Civil Engineering and Mechanics at the University of Glasgow, a post he occupied until his death on December 24, 1872. During this period, Rankine worked in the true sense as an engineer, applying scientific principles to the fatigue in metals of railroad-car axles, to new methods of mechanical construction, and to soil mechanics dealing with earth pressures and the stability of retaining walls. Perhaps his best-known contributions were in the field of steam engines and the development of a particular thermody-namic cycle bearing his name. Also, an engineering unit of absolute temperature was named in his honor.

Figure 3.16 | W. J. M. Rankine (1820–1872).

Rankine's contribution to shock wave theory came late in life—2 years before his death. In a paper published in 1870 in the *Philosophical Transactions of the Royal Society* entitled "On the Thermodynamic Theory of Waves of Finite Longitudinal Disturbance," Rankine clearly presented the proper normal shock equations for continuity, momentum, and energy in much the same form as our Eqs. (3.38) through (3.40). (It is interesting that in these equations Rankine defined a quantity he called "bulkiness," which is identical to what we now define as "specific volume." Apparently the usage of the term "bulkiness" later died out of its own cumbersomeness.) Moreover, Rankine properly assumed that the internal structure of the shock wave was not isentropic, but rather that it was a region of dissipation. He was thinking about thermal conduction, not the companion effect of viscosity within the shock. However, Rankine was able to successfully derive relationships for the thermodynamic changes across a shock wave analogous to the equations we have derived in Sec. 3.7. (It is also interesting to note that Rankine's paper coined the symbol γ for the ratio of specific heats, c_p/c_v; we are still following this notation a century later. He also recognized that the value of γ was "nearly 1.41 for air, oxygen, nitrogen, and hydrogen, and for steam-gas nearly 1.3.")

The equations obtained by Rankine were subsequently rediscovered by the French ballistician Pierre Henry Hugoniot. Not cognizant of Rankine's work, Hugoniot in 1887 published a paper in the *Journal de l'Ecole Polytechnique* entitled "Mémoire sur la propagation du Mouvement dans les Corps et Spécialement dans les Gases Parfaits" in which the equations for normal shock thermodynamic properties were presented, essentially the equations we have derived in Sec. 3.7. As a result of this pioneering work by Hugoniot and by Rankine before him, a rather modern

Figure 3.17 | Lord Rayleigh (1842–1919).

generic term has come into use for all equations dealing with changes across shock waves, namely, the *Rankine-Hugoniot relations*. This label appears frequently in modern gasdynamic literature.

However, the work of both Rankine and Hugoniot did not establish the *direction* of changes across a shock wave. Noted in both works is the mathematical possibility of either compression (pressure increases) or rarefaction (pressure decreases) shocks. This same possibility is discussed in Sec. 3.6. It was not until 1910 that this ambiguity was resolved. In two almost simultaneous and independent papers, first Lord Rayleigh (see Fig. 3.17) and then G. I. Taylor invoked the second law of thermodynamics to show that only compression shocks are physically possible—i.e., the Rankine-Hugoniot relations apply physically only to the case where the pressure behind the shock is greater than the pressure in front of the shock, Rayleigh's paper was published in Volume 84 of the *Proceedings of the Royal Society,* September 15, 1910, and was entitled "Aerial Plane Waves of Finite Amplitude." Here, Lord Rayleigh summarizes his results as follows:

> But here a question arises which Rankine does not seem to have considered. In order to secure the necessary transfers of heat by means of conduction it is an indispensable condition that the heat should pass from the hotter to the colder body. If maintenance of type be possible in a particular wave as a result of conduction, a reversal of the motion will

give a wave whose type cannot be so maintained. We have seen reason already for the conclusion that a dissipative agency can serve to maintain the type only when the gas passes from a less to a more condensed state.

In addition to applying the second law of thermodynamics, Rayleigh also showed that viscosity played as essential a role in the structure of a shock as conduction. (Recall that Rankine considered conduction, only; also, Hugoniot obtained his results without reference to any dissipative mechanism.)

One month later, in the same journal, a young G. I. Taylor (who was to become one of the leading fluid dynamicists of the twentieth century) published a short paper entitled "The Conditions Necessary for Discontinuous Motion in Gases," which supported Rayleigh's conclusions. Finally, over a course of 40 years, culminating in the second decade of this century, the theory of shock waves as presented in this chapter was fully established.

It should be noted that the shock wave studies by Rankine, Hugoniot, Rayleigh, and Taylor were viewed at the time as interesting basic mechanics research on a relatively academic problem. The on-rush of the application of this theory did not begin until 30 years later with blooming of interest in supersonic vehicles during World War II. However, this is a classic example of the benefits of basic research, even when such work appears obscure at the moment. Rapid advances in supersonic flight during the 1940s were clearly expedited because shock wave theory was sitting there, fully developed and ready for application.

3.11 | SUMMARY

This chapter has dealt with one-dimensional flow, i.e., where all flow properties are functions of one space dimension, say x, only. This implies flow with constant cross-sectional area. Three physical mechanisms that cause the flow properties to change with x even though the area is constant are: (1) a normal shock wave, (2) heat addition, and (3) friction. Return to the roadmap in Fig. 3.1, and review the flow of ideas that highlight this chapter.

The basic normal shock equations are:

Continuity:
$$\rho_1 u_1 = \rho_2 u_2 \tag{3.38}$$

Momentum:
$$p_1 + \rho_1 u_1^2 = p_2 + \rho_2 u_2^2 \tag{3.39}$$

Energy:
$$h_1 + \frac{u_1^2}{2} = h_2 + \frac{u_2^2}{2} \tag{3.40}$$

A combination of these equations, along with the equation of state leads to the Prandtl relation

$$a^{*2} = u_1 u_2 \tag{3.47}$$

which in turn leads to an expression for the Mach number behind a normal shock:

$$M_2^2 = \frac{1 + [(\gamma - 1)/2]M_1^2}{\gamma M_1^2 - (\gamma - 1)/2} \tag{3.51}$$

Further combinations of the basic normal shock equations give

$$\frac{p_2}{p_1} = 1 + \frac{2\gamma}{\gamma + 1}\left(M_1^2 - 1\right) \tag{3.57}$$

and

$$\frac{\rho_2}{\rho_1} = \frac{u_1}{u_2} = \frac{(\gamma + 1)M_1^2}{2 + (\gamma - 1)M_1^2} \tag{3.53}$$

Important: Note that the changes across a normal shock wave in a calorically perfect gas are functions of just M_1 and γ. For normal shock waves, the upstream Mach number is a pivotal quantity. Also, across a normal shock wave, T_o is constant, s increases, and p_o decreases. (However, if the gas is *not* calorically or thermally perfect, T_o is *not* constant across the shock.) A purely thermodynamic relation across a normal shock wave is the Hugoniot equation,

$$e_2 - e_1 = \frac{p_1 + p_2}{2}(v_1 - v_2) \tag{3.72}$$

a graph of which, on the $p - v$ plane, is called the Hugoniot curve.

The governing equations for one-dimensional flow with heat addition are:

Continuity: $$\rho_1 u_1 = \rho_2 u_2 \tag{3.2}$$

Momentum: $$p_1 + \rho_1 u_1^2 = p_2 + \rho_2 u_2^2 \tag{3.5}$$

Energy: $$h_1 + \frac{u_1^2}{2} + q = h_2 + \frac{u_2^2}{2} \tag{3.9}$$

The heat addition causes an increase in total temperature, given by

$$q = c_p\left(T_{o_2} - T_{o_1}\right) \tag{3.77}$$

for a calorically perfect gas. Also for this case, the governing equations lead to relationships for the flow properties before and after heat addition in terms of the Mach numbers M_1 and M_2 before and after heat addition, respectively. Note that heat added to an initially supersonic flow slows the flow. If enough heat is added, the flow after heat addition can be slowed to Mach 1; this is the case of thermal choking. Heat added to an initially subsonic flow increases the flow speed. If enough heat is added, the flow after heat addition can reach Mach 1, again becoming thermally choked. In both cases of choked flow, if additional heat is added, nature adjusts the *upstream* quantities to allow for the extra heat. An initially supersonic flow that becomes thermally choked will become totally subsonic when additional heat is added, i.e., the inlet Mach number is changed to a subsonic value. An initially subsonic flow that becomes thermally choked will have its inlet Mach number reduced when additional heat is added. A plot of the thermodynamic properties for one-dimensional flow with heat addition on a Mollier diagram is called a *Rayleigh curve;* hence, such flow with heat addition is called *Rayleigh-line flow.*

The governing equations for one-dimensional flow with friction are:

Continuity:
$$\rho_1 u_1 = \rho_2 u_2 \tag{3.2}$$

Momentum:
$$p_1 + \rho_1 u_1^2 - \frac{1}{A} \int_o^L \pi D \tau_w \, dx$$
$$= p_2 + \rho_2 u_2^2 \tag{3.91}$$

Energy:
$$h_1 + \frac{u_1^2}{2} = h_2 + \frac{u_2^2}{2} \tag{3.40}$$

This flow is adiabatic, hence T_o is constant. The entropy is increased due to the presence of friction. The governing equations lead to relationships for the flow properties at the inlet and exit in terms of M_1 and M_2 at the inlet and exit, respectively. M_2 is related to M_1 through Eq. (3.97). The same type of choking phenomena occurs here as the case of flow with heat addition. An initially supersonic flow slows due to the influence of friction; if the constant-area duct is long enough, the exit Mach number becomes unity, and the flow is said to be choked. If the duct is made longer after the flow is choked, nature readjusts the flow in the duct so as to become *subsonic* at the inlet. An initially subsonic flow experiences an increase in velocity due to friction— a seemingly incongruous result because intuition tells us that friction would always reduce the flow velocity. However, the *pressure gradient* along the duct in this case is one of decreasing pressure in the x direction; this is in order to obey the governing equations. This favorable pressure gradient tends to increase the flow velocity. Indeed, the effect of decreasing pressure in the flow direction dominates over the retarding effect of friction at the walls of the duct, and hence one-dimensional subsonic flow with friction results in an increase in velocity through the duct. Another way to look at this situation is to recognize that, in order to set up subsonic one-dimensional flow with friction, a high pressure must be exerted at the inlet and a lower pressure at the exit. A plot of the thermodynamic properties of flow with friction on a Mollier diagram is called a Fanno curve, and such flow is called *Fanno-line flow*.

In this chapter, a number of conveniently defined flow quantities are introduced: (1) total temperature, which is the temperature that would exist if the flow were reduced to zero velocity *adiabatically;* (2) total pressure, which is the pressure that would exist if the flow were reduced to zero velocity *isentropically;* (3) T^* (and hence $a^* = \sqrt{\gamma R T^*}$), which is the temperature that would exist if the flow were slowed down or speeded up (as the case may be) to Mach 1; (4) characteristic Mach number, $M^* = V/a^*$. Section 3.5 gives many alternative forms of the energy equation in terms of these quantities. Study this section carefully. Of particular importance are the following relations which hold for a calorically perfect gas:

$$\frac{T_o}{T} = 1 + \frac{\gamma - 1}{2} M^2 \tag{3.28}$$

$$\frac{p_o}{p} = \left(1 + \frac{\gamma - 1}{2} M^2\right)^{\gamma/(\gamma-1)} \tag{3.30}$$

$$\frac{\rho_o}{\rho} = \left(1 + \frac{\gamma - 1}{2} M^2\right)^{1/(\gamma - 1)}$$

(3.31)

PROBLEMS

(*Note:* Use the tables at the end of this book as extensively as you wish to solve the following problems. Also, when the words "pressure" and "temperature" are used without additional modification, they refer to the *static* pressure and temperature.)

3.1 At a given point in the high-speed flow over an airplane wing, the local Mach number, pressure and temperature are 0.7, 0.9 atm, and 250 K, respectively. Calculate the values of p_o, T_o, p^*, T^*, and a^* at this point.

3.2 At a given point in a supersonic wind tunnel, the pressure and temperature are 5×10^4 N/m^2 and 200 K, respectively. The total pressure at this point is 1.5×10^6 N/m^2. Calculate the local Mach number and total temperature.

3.3 At a point in the flow over a high-speed missile, the local velocity and temperature are 3000 ft/s and 500°R, respectively. Calculate the Mach number M and the characteristic Mach number M^* at this point.

3.4 Consider a normal shock wave in air. The upstream conditions are given by $M_1 = 3$, $p_1 = 1$ atm, and $\rho_1 = 1.23$ kg/m^3. Calculate the downstream values of p_2, T_2, ρ_2, M_2, u_2, p_{o_2}, and T_{o_2}.

3.5 Consider a Pitot static tube mounted on the nose of an experimental airplane. A Pitot tube measures the total pressure at the tip of the probe (hence sometimes called the *Pitot pressure*), and a Pitot static tube combines this with a simultaneous measurement of the free-stream static pressure. The Pitot and free-stream static measurements are given below for three different flight conditions. Calculate the free-stream Mach number at which the airplane is flying for each of the three different conditions:

a. Pitot pressure $= 1.22 \times 10^5$ N/m^2, static pressure $= 1.01 \times 10^5$ N/m^2

b. Pitot pressure $= 7222$ lb/ft^2, static pressure $= 2116$ lb/ft^2

c. Pitot pressure $= 13107$ lb/ft^2, static pressure $= 1020$ lb/ft^2

3.6 Consider the compression of air by means of (*a*) shock compression and (*b*) isentropic compression. Starting from the same initial conditions of p_1 and v_1, plot to scale the pv diagrams for both compression processes on the same graph. From the comparison, what can you say about the effectiveness of shock versus isentropic compression?

3.7 During the entry of the Apollo space vehicle into the Earth's atmosphere, the Mach number at a given point on the trajectory was $M = 38$ and the atmosphere temperature was 270 K. Calculate the temperature at the stagnation point of the vehicle, assuming a calorically perfect gas with $\gamma = 1.4$. Do you think this is an accurate calculation? If not, why? If not, is your answer an overestimate or underestimate?

3.8 Consider air entering a heated duct at $p_1 = 1$ atm and $T_1 = 288$ K. Ignore the effect of friction. Calculate the amount of heat per unit mass (in joules per kilogram) necessary to choke the flow at the exit of the duct, as well as the pressure and temperature at the duct exit, for an inlet Mach number of (*a*) $M_1 = 2.0$ (*b*) $M_1 = 0.2$.

3.9 Air enters the combustor of a jet engine at $p_1 = 10$ atm, $T_1 = 1000°R$, and $M_1 = 0.2$. Fuel is injected and burned, with a fuel-air ratio (by mass) of 0.06. The heat released during the combustion is 4.5×10^8 ft-lb per slug of fuel. Assuming one-dimensional frictionless flow with $\gamma = 1.4$ for the fuel-air mixture, calculate M_2, p_2, and T_2 at the exit of the combustor.

3.10 For the inlet conditions of Prob. 3.9, calculate the maximum fuel-air ratio beyond which the flow will be choked at the exit.

3.11 At the inlet to the combustor of a supersonic combustion ramjet (SCRAMjet), the flow Mach number is supersonic. For a fuel-air ratio (by mass) of 0.03 and a combustor exit temperature of $4800°R$, calculate the inlet Mach number above which the flow will be unchoked. Assume one-dimensional frictionless flow with $\gamma = 1.4$, with the heat release per slug of fuel equal to 4.5×10^8 ft · lb.

3.12 Air is flowing through a pipe of 0.02-m inside diameter and 40-m length. The conditions at the exit of the pipe are $M_2 = 0.5$, $p_2 = 1$ atm, and $T_2 = 270$ K. Assuming adiabatic, one-dimensional flow, with a local friction coefficient of 0.005, calculate M_1, p_1, and T_1 at the entrance to the pipe.

3.13 Consider the adiabatic flow of air through a pipe of 0.2-ft inside diameter and 3-ft length. The inlet flow conditions are $M_1 = 2.5$, $p_1 = 0.5$ atm, and $T_1 = 520°R$. Assuming the local friction coefficient equals a constant of 0.005, calculate the following flow conditions at the exit: M_2, p_2, T_2, and p_{o_2}.

3.14 The stagnation chamber of a wind tunnel is connected to a high-pressure air bottle farm which is outside the laboratory building. The two are connected by a long pipe of 4-in inside diameter. If the static pressure ratio between the bottle farm and the stagnation chamber is 10, and the bottle-farm static pressure is 100 atm, how long can the pipe be without choking? Assume adiabatic, subsonic, one-dimensional flow with a friction coefficient of 0.005.

3.15 Starting with Eq. (3.95), derive in detail Eq. (3.96).

3.16 Consider a Mach 2.5 flow of air entering a constant-area duct. Heat is added to this flow in the duct; the amount of heat added is equal to 30 percent of the total enthalpy at the entrance to the duct. Calculate the Mach number at the exit of the duct. Comment on the fluid dynamic significance of this problem, where the exit Mach number does not depend on a number for the actual heat added, but rather only on the dimensionless ratio of heat added to the total enthalpy of the inflowing gas.

CHAPTER 4

Oblique Shock and Expansion Waves

I believe we have now arrived at the stage where knowledge of supersonic aerodynamics should be considered by the aeronautical engineer as a necessary pre-requisite to his art.

Theodore von Karman, 1947

PREVIEW BOX

Figure 4.1 shows the computed shock wave and expansion wave pattern in the flow field over a hypersonic test vehicle at the moment of its separation from a booster rocket at Mach 7. This is NASA's Hyper-X supersonic-combustion ramjet (scramjet) powered unmanned test aircraft also designated the X-43, which should make its first flight in 2003. The flow field is a complex mixture of oblique shock and expansion waves. Figure 4.2 shows the computed detailed shock wave and expansion wave pattern in the internal flow through a scramjet engine. Again, the supersonic flow is dominated by a complex pattern of interacting oblique shock and expansion waves.

Oblique shock and expansion waves, and their various interactions, are the subject of this chapter. For the study of supersonic and hypersonic flow, this is a "bread-and-butter" chapter—it contains what is perhaps some

of the most important physical aspects of compressible flow. So get ready for a whirlwind and hopefully enjoyable ride through the ins and outs of the basic physics and mathematics of oblique shock and expansion waves.

The roadmap for this chapter is given in Fig. 4.3. After a discussion of the physical source of oblique waves, we will next discuss oblique shock waves and related items, as shown down the left side of Fig. 4.3. Then we move to the right side of the roadmap to study oblique expansion waves, concentrating on the special type labeled Prandtl–Meyer expansions. Finally, as shown at the bottom of Fig. 4.3, we combine these two types of oblique waves into a method of analysis called shock-expansion theory, which allows the direct and exact calculation of the lift and drag on a number of two-dimensional supersonic body shapes.

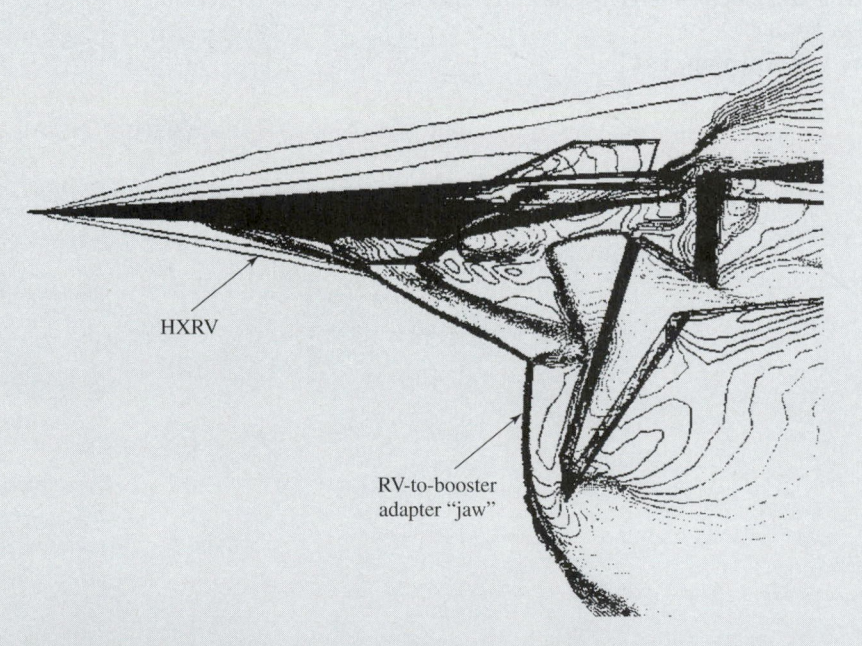

HXRV

RV-to-booster
adapter "jaw"

Figure 4.1 | Computational fluid dynamic solution for the shock wave pattern on NASA's Hyper-X hypersonic research vehicle at the instant of its separation from the boost vehicle at Mach 7. (Griffin Anderson, Charles McClinton, and John Weidner, "Scramjet Performance," in *Scramjet Propulsion,* edited by E. T. Curran and S. N. B. Murthy, AIAA Progress in Astronautics and Aeronautics, Vol. 189, Reston, Virginia, p. 431.)

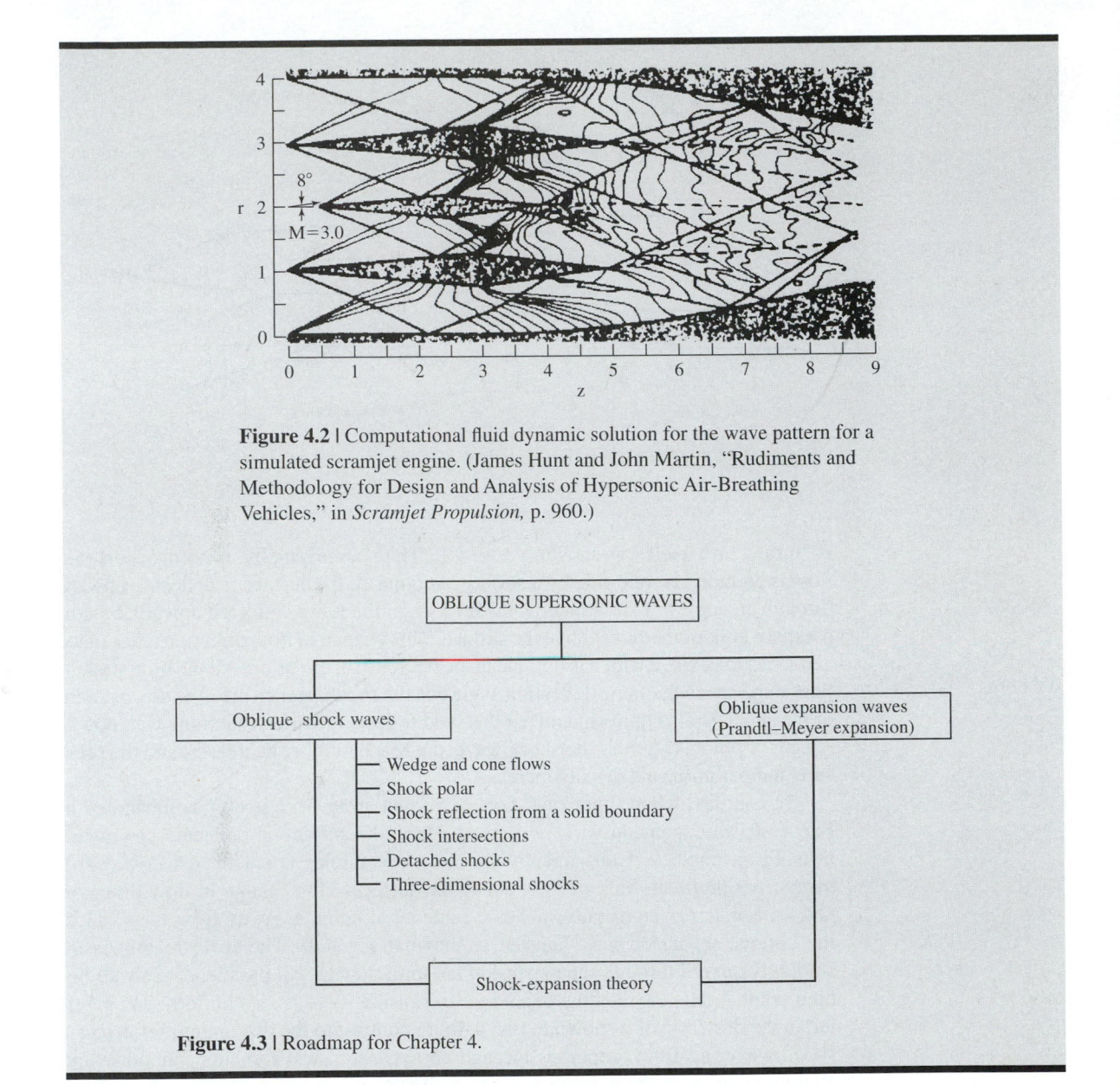

Figure 4.2 | Computational fluid dynamic solution for the wave pattern for a simulated scramjet engine. (James Hunt and John Martin, "Rudiments and Methodology for Design and Analysis of Hypersonic Air-Breathing Vehicles," in *Scramjet Propulsion*, p. 960.)

Figure 4.3 | Roadmap for Chapter 4.

4.1 | INTRODUCTION

The normal shock wave, as considered in Chap. 3, is a special case of a more general family of oblique waves that occur in supersonic flow. Oblique shock waves are illustrated in Figs. 3.2 and 3.3. Such oblique shocks usually occur when supersonic flow

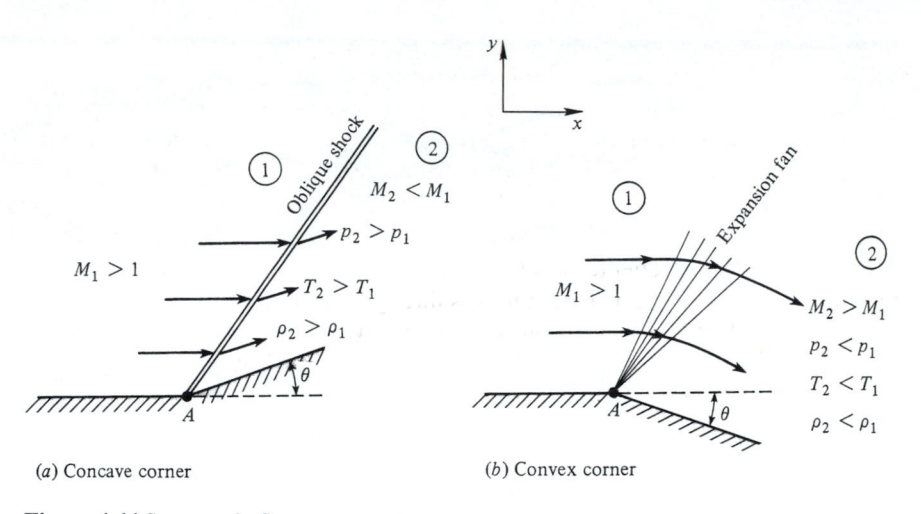

(a) Concave corner   (b) Convex corner

Figure 4.4 | Supersonic flow over a corner.

is "turned into itself," as shown in Fig. 4.4a. Here, an originally uniform supersonic flow is bounded on one side by a surface. At point A, the surface is deflected upward through an angle θ. Consequently, the flow streamlines are deflected upward, *toward* the main bulk of the flow above the surface. This change in flow direction takes place across a shock wave which is oblique to the free-stream direction. All the flow stream-lines experience the same deflection angle θ at the shock. Hence the flow downstream of the shock is also uniform and parallel, and follows the direction of the wall down-stream of point A. Across the shock wave, the Mach number decreases, and the pres-sure, temperature, and density increase.

In contrast, when supersonic flow is "turned away from itself" as illustrated in Fig. 4.4b, an expansion wave is formed. Here, the surface is deflected downward through an angle θ. Consequently the flow streamlines are deflected downward, *away from* the main bulk of flow above the surface. This change in flow direction takes place across an expansion wave, centered at point A. Away from the surface, this oblique expansion wave fans out, as shown in Fig. 4.4b. The flow streamlines are smoothly curved through the expansion fan until they are all parallel to the wall be-hind point A. Hence, the flow behind the expansion wave is also uniform and paral-lel, in the direction of θ shown in Fig. 4.4b. In contrast to the discontinuities across a shock wave, all flow properties through an expansion wave change smoothly and continuously, with the exception of the wall streamline which changes discontinu-ously at point A. Across the expansion wave, the Mach number increases and the pressure, temperature, and density decrease.

Oblique shock and expansion waves are prevalent in two- and three-dimensional supersonic flows. These waves are inherently two-dimensional in nature, in contrast to the one-dimensional normal shock waves in Chap. 3. That is, the flowfield proper-ties are functions of x and y in Fig. 4.4. The main thrust of this chapter is to present the properties of these two-dimensional waves.

4.2 | SOURCE OF OBLIQUE WAVES

Oblique waves are created by the same physical mechanism discussed at the beginning of Sec. 3.6—disturbances which propagate by molecular collisions at the speed of sound, some of which eventually coalesce into shocks and others of which spread out in the form of expansion waves. To more clearly see this process for an oblique wave, consider a moving point source of sound disturbances in a gas, as illustrated in Fig. 4.5. For lack of a better term, let us call this source a "beeper." The beeper is continually emitting sound waves as it moves through the stationary gas. Consider first the case when the beeper is moving at a velocity V, which is *less* than the speed of sound, as shown in Fig. 4.5a. When the beeper is at point A, it emits a sound disturbance which propagates in all directions at the speed of sound, a. After an interval of time t, this sound wave is represented by the circle of radius (at) in Fig. 4.5a. However, during this same time interval, the beeper has moved a distance Vt to point B. Moreover, during its transit from A to B, the beeper has emitted several other sound waves, which at time t are represented by the smaller circles in Fig. 4.5a. Note from this figure, which is a picture of the situation at time t, that the beeper always stays *inside* the family of circular sound waves, and that the waves continuously move ahead of the beeper. This is because the beeper is traveling at a subsonic speed, $V < a$. Now consider the case when the beeper is moving at supersonic speeds, $V > a$. This is illustrated in Fig. 4.5b. Again, when the beeper is at point A, it emits a sound wave. After an interval of time t, this wave is the circle with radius (at). During the same interval of time, the beeper has moved a distance Vt to point B. Moreover, during its transit from A to B, the beeper has emitted several other sound waves, which at time t are represented by the smaller circles in Fig. 4.5b. However, in contrast to the subsonic case, the beeper is now constantly *outside* the family of circular sound waves, i.e., it is moving ahead of the wave fronts because $V > a$. Moreover, something new is happening; these wave fronts form a disturbance envelope given by the straight line BC, which is tangent to the family of circles. This line of disturbances is defined as a *Mach wave*. In addition, the angle ABC which the Mach wave makes with respect to the direction of motion of the beeper is defined as the *Mach angle, μ*. The Mach angle is easily calculated from the geometry of Fig. 4.5b:

$$\sin \mu = \frac{at}{Vt} = \frac{a}{V} = \frac{1}{M}$$

Therefore, the Mach angle is simply determined by the local Mach number as

$$\boxed{\mu = \sin^{-1} \frac{1}{M}} \tag{4.1}$$

The propagation of weak disturbances and their coalescence into a Mach wave are clearly seen in Fig. 4.5c.

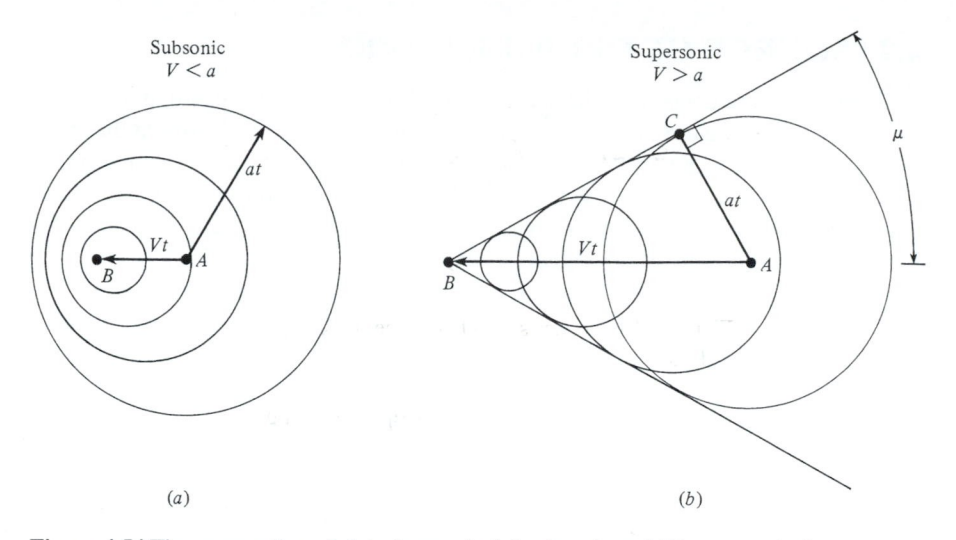

(a)   (b)

Figure 4.5 | The propagation of disturbances in (*a*) subsonic and (*b*) supersonic flow.

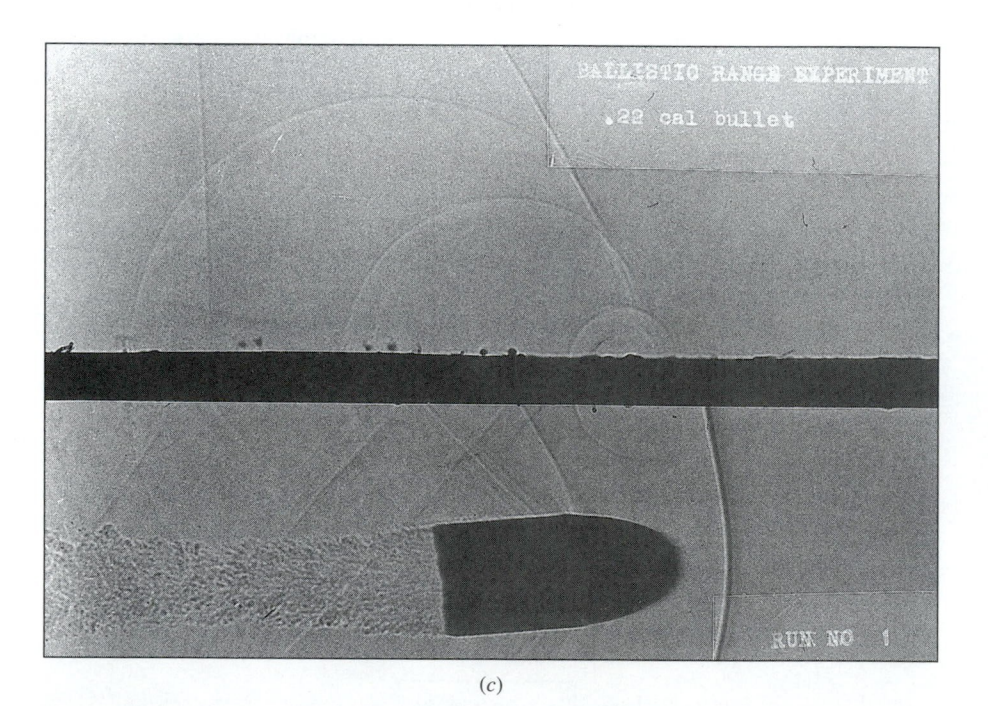

(c)

Figure 4.5 | Wave system established by a supersonic .22 caliber bullet passing under a perforated plate. The bow shock wave on the bullet, in passing over the holes in the plate, sends out weak disturbances above the plate which coalesce into a Mach wave above the plate. This is a photographic illustration of the schematic in Fig. 4.5*b*. (*Photo is courtesy of Daniel Bershader, Stanford University.*)

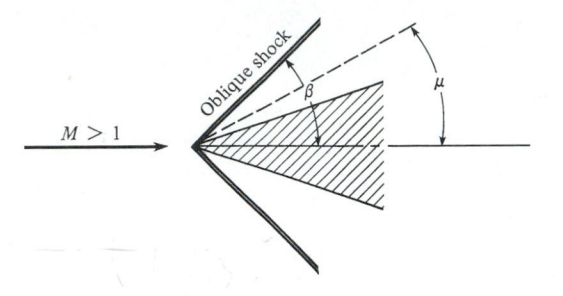

Figure 4.6 | Comparison between the wave angle
and the Mach angle.

If the disturbance is stronger than a small beeper emitting sound waves, such as
a wedge blasting its way through a gas at supersonic speeds as shown in Fig. 4.6, the
wave front becomes stronger than a Mach wave. The strong disturbances coalesce
into an oblique shock wave at an angle β to the free stream, where $\beta > \mu$. However,
the physical mechanism creating the oblique shock is essentially the same as that described above for the Mach wave. Indeed, a Mach wave is a limiting case for oblique
shocks, i.e., it is an infinitely weak oblique shock.

4.3 | OBLIQUE SHOCK RELATIONS

The geometry of flow through an oblique shock is given in Fig. 4.7. The velocity upstream of the shock is V_1, and is horizontal. The corresponding Mach number is M_1.
The oblique shock makes a wave angle β with respect to V_1. Behind the shock, the
flow is deflected toward the shock by the flow-deflection angle θ. The velocity and
Mach number behind the shock are V_2 and M_2, respectively. The components of V_1
perpendicular and parallel, respectively, to the shock are u_1 and w_1; the analogous
components of V_2 are u_2 and w_2, as shown in Fig. 4.7. Therefore, we can consider the
normal and tangential Mach numbers ahead of the shock to be M_{n_1} and M_{t_1}, respectively; similarly, we have M_{n_2} and M_{t_2} behind the shock.

The integral forms of the conservation equations from Chap. 2 were applied in
Sec. 3.2 to a specific control volume in one-dimensional flow, ultimately resulting
in the normal shock equations given in Sec. 3.6. Let us take a similar tack here. Consider the control volume drawn between two streamlines through an oblique shock, as
illustrated by the dashed lines at the top of Fig. 4.7. Faces a and d are parallel to the
shock wave. Apply the integral continuity equation (2.2) to this control volume for a
steady flow. The time derivative in Eq. (2.2) is zero. The surface integral evaluated over
faces a and d of the control volume in Fig. 4.7 yields $-\rho_1 u_1 A_1 + \rho_2 u_2 A_2$, where
$A_1 = A_2 =$ area of faces a and d. The faces b, c, e, and f of the control volume are
parallel to the velocity, and hence contribute nothing to the surface integral (i.e.,
$\mathbf{V} \cdot d\mathbf{S} = 0$ for these faces). Thus, the continuity equation for an oblique shock wave is

$$\rho_1 u_1 = \rho_2 u_2 \tag{4.2}$$

The integral form of the momentum equation (2.11) is a vector equation. Consider this equation resolved into two components, parallel and perpendicular to the

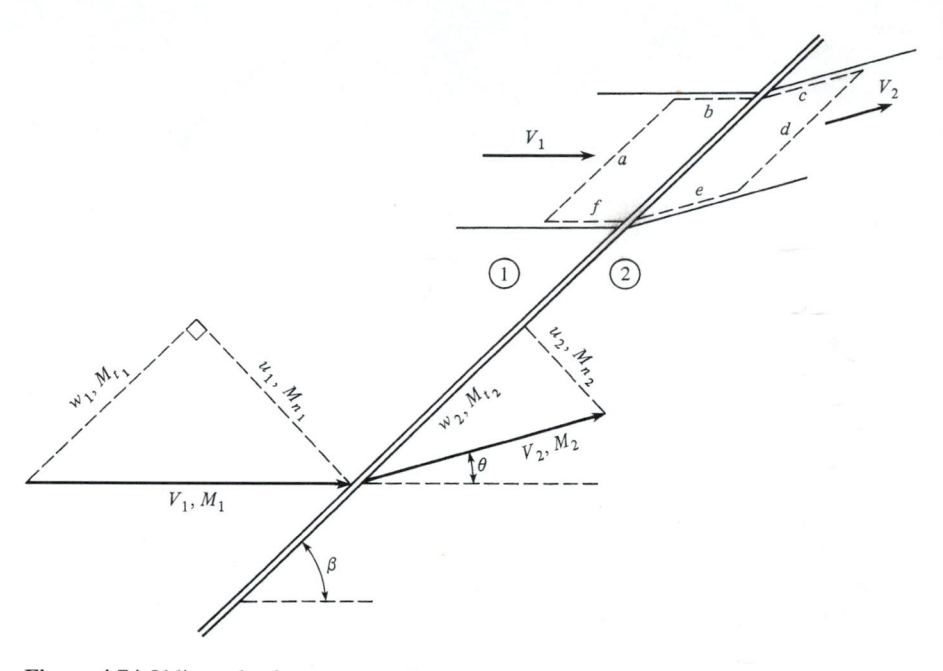

Figure 4.7 | Oblique shock wave geometry.

shock wave in Fig. 4.7. Again, considering steady flow with no body forces, the tangential component of Eq. (2.11) applied to the control surface in Fig. 4.7 yields (noting that the tangential component of $p\,d\mathbf{S}$ is zero on faces a and d, and that the components on b cancel those on f; similarly with faces c and e)

$$(-\rho_1 u_1)w_1 + (\rho_2 u_2)w_2 = 0 \tag{4.3}$$

Dividing Eq. (4.3) by (4.2), we find that

$$w_1 = w_2$$

This is a striking result—*the tangential component of the flow velocity is preserved across an oblique shock wave.*

Returning to Fig. 4.7, and applying the normal component of Eq. (2.11), we find

$$(-\rho_1 u_1)u_1 + (\rho_2 u_2)u_2 = -(-p_1 + p_2)$$

or

$$p_1 + \rho_1 u_1^2 = p_2 + \rho_2 u_2^2 \tag{4.3a}$$

The integral form of the energy equation is Eq. (2.20). Applied to the control volume in Fig. 4.7 for a steady adiabatic flow with no body forces, it yields

$$-(-p_1 u_1 + p_2 u_2) = -\rho_1\left(e_1 + \frac{V_1^2}{2}\right)u_1 + \rho_2\left(e_2 + \frac{V_2^2}{2}\right)u_2$$

or

$$\left(h_1 + \frac{V_1^2}{2}\right)\rho_1 u_1 = \left(h_2 + \frac{V_2^2}{2}\right)\rho_2 u_2 \tag{4.4}$$

Dividing Eq. (4.4) by (4.2),

$$h_1 + \frac{V_1^2}{2} = h_2 + \frac{V_2^2}{2} \tag{4.5}$$

However, recall from the geometry of Fig. 4.7 that $V^2 = u^2 + w^2$ and that $w_1 = w_2$. Hence,

$$V_1^2 - V_2^2 = \left(u_1^2 + w_1^2\right) - \left(u_2^2 + w_2^2\right) = u_1^2 - u_2^2$$

Therefore, Eq. (4.5) becomes

$$h_1 + \frac{u_1^2}{2} = h_2 + \frac{u_2^2}{2} \tag{4.6}$$

Look carefully at Eqs. (4.2), (4.3a), and (4.6). They are identical in form to the normal shock continuity, momentum, and energy equations (3.38) through (3.40). Moreover, in both sets of equations, the velocities are *normal* to the wave. Therefore, the changes across an oblique shock wave are governed by the normal component of the free-stream velocity. Furthermore, precisely the same algebra as applied to the normal shock equations in Sec. 3.6, when applied to Eqs. (4.2), (4.3a), and (4.6), will lead to identical expressions for changes across an oblique shock in terms of the normal component of the upstream Mach number M_{n_1}. That is, for an oblique shock wave with

$$M_{n_1} = M_1 \sin \beta \tag{4.7}$$

we have, for a calorically perfect gas,

$$\frac{\rho_2}{\rho_1} = \frac{(\gamma + 1)M_{n_1}^2}{(\gamma - 1)M_{n_1}^2 + 2} \tag{4.8}$$

$$\frac{p_2}{p_1} = 1 + \frac{2\gamma}{\gamma + 1}\left(M_{n_1}^2 - 1\right) \tag{4.9}$$

$$M_{n_2}^2 = \frac{M_{n_1}^2 + [2/(\gamma - 1)]}{[2\gamma/(\gamma - 1)]M_{n_1}^2 - 1} \tag{4.10}$$

and

$$\frac{T_2}{T_1} = \frac{p_2}{p_1}\frac{\rho_1}{\rho_2} \tag{4.11}$$

Note that the Mach number behind the oblique shock, M_2, can be found from M_{n_2} and the geometry of Fig. 4.7 as

$$M_2 = \frac{M_{n_2}}{\sin(\beta - \theta)} \tag{4.12}$$

In Sec. 3.6, we emphasized that changes across a normal shock were a function of one quantity only—the upstream Mach number. Now, from Eqs. (4.7) through (4.11), we see that changes across an oblique shock are a function of two quantities—both M_1 and β. We also see, in reality, normal shocks are just a special case of oblique shocks where $\beta = \pi/2$.

Equation (4.12) demonstrates that M_2 cannot be found until the flow deflection angle θ is obtained. However, θ is also a unique function of M_1 and β, as follows. From the geometry of Fig. 4.7,

$$\tan \beta = \frac{u_1}{w_1} \tag{4.13}$$

and

$$\tan(\beta - \theta) = \frac{u_2}{w_2} \tag{4.14}$$

Combining Eqs. (4.13) and (4.14), noting that $w_1 = w_2$, we have

$$\frac{\tan(\beta - \theta)}{\tan \beta} = \frac{u_2}{u_1} \tag{4.15}$$

Combining Eq. (4.15) with Eqs. (4.2), (4.7), and (4.8), we obtain

$$\frac{\tan(\beta - \theta)}{\tan \beta} = \frac{2 + (\gamma - 1)M_1^2 \sin^2 \beta}{(\gamma + 1)M_1^2 \sin^2 \beta} \tag{4.16}$$

With some trigonometric manipulation, this equation can be expressed as

$$\tan \theta = 2 \cot \beta \left[\frac{M_1^2 \sin^2 \beta - 1}{M_1^2(\gamma + \cos 2\beta) + 2} \right] \tag{4.17}$$

Equation (4.17) is called the θ-β-M *relation,* and specifies θ as a unique function of M_1 and β.

This relation is vital to an analysis of oblique shocks, and results obtained from it are plotted in Fig. 4.8 for $\gamma = 1.4$. Examine this figure closely. It is a plot of wave angle versus deflection angle, with the Mach number as a parameter. In particular, note that:

1. For any given M_1, there is a maximum deflection angle $\theta_{\max}$. If the physical geometry is such that $\theta > \theta_{\max}$, then no solution exists for a straight oblique shock wave. Instead, the shock will be curved and detached, as sketched in Fig. 4.9, which compares wedge and corner flow for situations where θ is less than or greater than $\theta_{\max}$.

2. For any given $\theta < \theta_{\max}$, there are *two* values of β predicted by the θ-β-M relation for a given Mach number, as sketched in Fig. 4.10. Because changes across the shock are more severe as β increases [see Eqs. (4.8) and (4.9), for example], the *large* value of β is called the *strong shock solution*; in turn, the *small* value of β is called the *weak shock solution*. In nature, the weak shock solution is favored, and usually occurs. For typical situations such as those

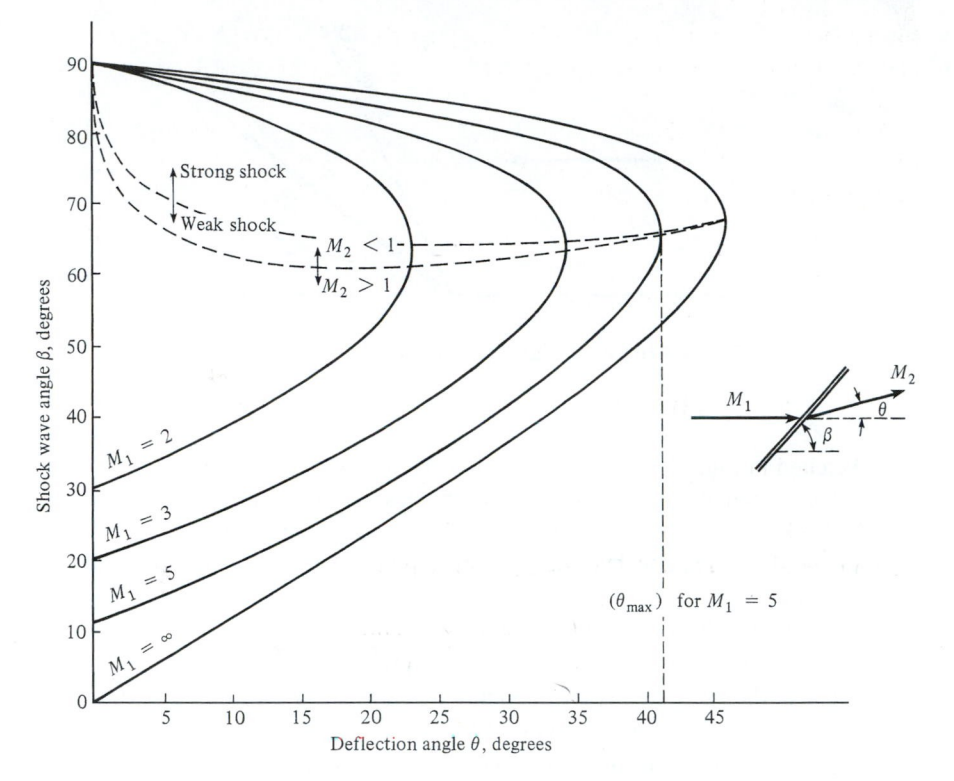

Figure 4.8 | θ-β-M curves. Oblique shock properties. *Important:* See front end pages for a more detailed chart.

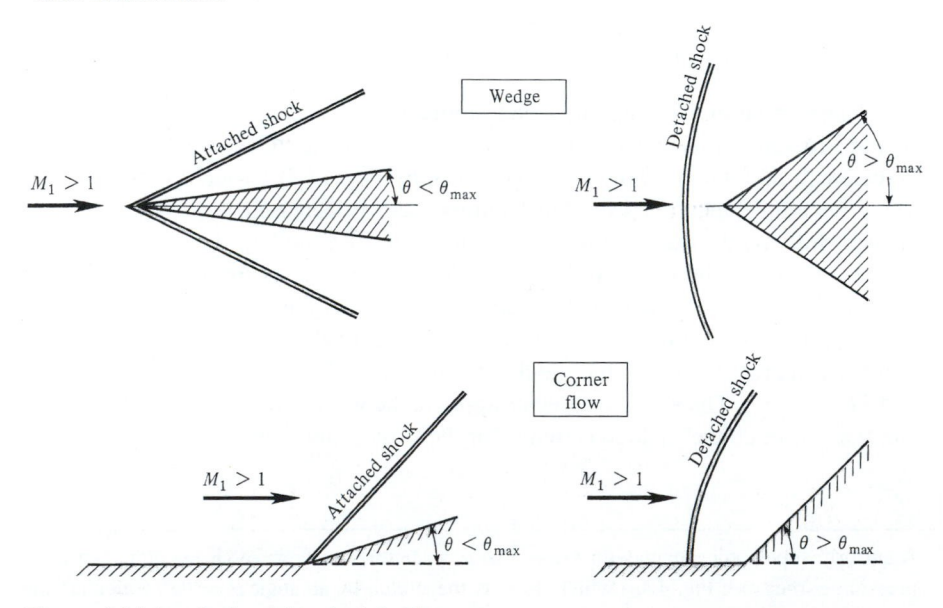

Figure 4.9 | Attached and detached shocks.

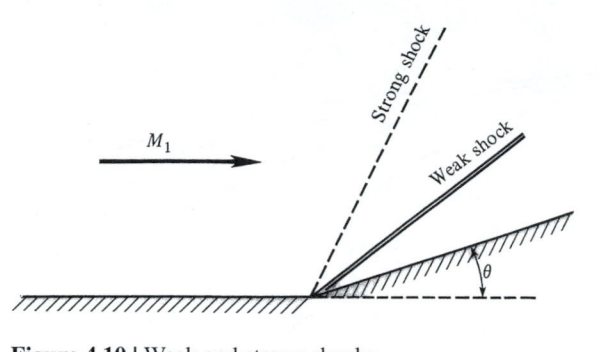

Figure 4.10 | Weak and strong shocks.

sketched in Fig. 4.10, the weak shock is the one we would normally see. However, whether the weak or strong shock solution occurs is determined by the backpressure; in Fig. 4.10, if the downstream pressure were increased by some independent mechanism, then the strong shock shown as the dashed line could be forced to occur. In the strong shock solution, M_2 is *subsonic*. In the *weak* shock solution, M_2 is *supersonic* except for a small region near θ_{max} (see Fig. 4.8).

3. If $\theta = 0$, then $\beta = \pi/2$ (corresponding to a normal shock) or $\beta = \mu$ (corresponding to a Mach wave).

4. For a fixed deflection angle θ, as the free-stream Mach number decreases from high to low supersonic values, the wave angle increases (for the weak shock solution). Finally, there is a Mach number below which no solutions are possible; at this Mach number, $\theta = \theta_{max}$. For lower Mach numbers the shock becomes detached, as sketched in Fig. 4.9.

These variations are important, and should be studied carefully. It is important to obtain a feeling for the physical behavior of oblique shocks. Considering Fig. 4.8 together with the oblique shock relations given by Eqs. (4.7) through (4.12), we can see, for example, that for a fixed Mach number, as θ is increased, β, p_2, T_2, and ρ_2 increase while M_2 decreases. However, if θ increases beyond θ_{max} the shock wave becomes detached. Alternatively, for a fixed θ, as M_1 increases from unity, the shock wave is first detached, then becomes attached when M_1 equals that value for which $\theta = \theta_{max}$. (See again Fig. 3.2 for the Bell XS-1 aircraft shock patterns.) As the Mach number is increased further, the shock remains attached, β decreases, and p_2, T_2, ρ_2, and M_2 increase. The above comments apply to the weak shock solutions; the reader can trace through the analogous trends for the strong shock case.

EXAMPLE 4.1

A uniform supersonic stream with $M_1 = 3.0$, $p_1 = 1$ atm, and $T_1 = 288$ K encounters a compression corner (see Fig. 4.4a) which deflects the stream by an angle $\theta = 20°$. Calculate the shock wave angle, and p_2, T_2, M_2, p_{o_2}, and T_{o_2} behind the shock wave.

■ **Solution**

For the geometrical picture, refer to Fig. 4.7. Also, from Fig. 4.8, for $M_1 = 3$ and $\theta = 20°$, $\boxed{\beta = 37.8°}$. Thus,

$$M_{n_1} = M_1 \sin\beta = 3 \sin 37.8° = 1.839$$

From Table A.2, for $M_{n_1} = 1.839$: $p_2/p_1 = 3.783$, $T_2/T_1 = 1.562$, $M_{n_2} = 0.6078$, and $p_{o_2}/p_{o_1} = 0.7948$. Hence,

$$p_2 = \frac{p_2}{p_1} p_1 = (3.783)(1) = \boxed{3.783 \text{ atm}}$$

$$T_2 = \frac{T_2}{T_1} T_1 = (1.562)(288) = \boxed{449.9 \text{ K}}$$

$$M_2 = \frac{M_{n_2}}{\sin(\beta - \theta)} = \frac{0.6078}{\sin 17.8°} = \boxed{1.988}$$

From Table A.1, for $M_1 = 3$: $p_{o_1}/p_1 = 36.73$ and $T_{o_1}/T_1 = 2.8$. Hence,

$$p_{o_2} = \frac{p_{o_2}}{p_{o_1}} \frac{p_{o_1}}{p_1} p_1 = (0.7948)(36.73)(1) = \boxed{29.19 \text{ atm}}$$

$$T_{o_2} = T_{o_1} = \frac{T_{o_1}}{T_1} T_1 = (2.8)(288) = \boxed{806.4 \text{ K}}$$

Note: In this example, we used the fact that the total pressure ratio across the oblique shock is dictated by the component of the upstream Mach number perpendicular to the shock, M_{n_1}. This is consistent with the fact that all thermodynamic properties across the shock are determined by M_{n_1}, including the entropy change $s_2 - s_1$. From Eq. (3.63), this determines the total pressure ratio, p_{o_2}/p_{o_1}. We can check the value of p_{o_2}/p_{o_1} obtained from Table A.2 by making an alternative calculation as

$$\frac{p_{o_2}}{p_{o_1}} = \frac{p_{o_2}}{p_2} \frac{p_2}{p_1} \frac{p_1}{p_{o_1}}$$

From Table A.1, for $M_2 = 1.988$, $p_{o_2}/p_2 = 7.681$ (obtained by interpolating between entries in the table). We have already obtained from the earlier calculations that $p_2/p_1 = 3.783$ and $p_{o_1}/p_1 = 36.73$. Hence,

$$\frac{p_{o_2}}{p_{o_1}} = \frac{p_{o_2}}{p_2} \frac{p_2}{p_1} \frac{p_1}{p_{o_1}} = (7.681)(3.783)\left(\frac{1}{36.73}\right) = 0.7911$$

This result compares within 0.46 percent with the value of 0.7948 read directly from Table A.2. The small inaccuracy is due to inaccuracy in reading β from the θ-β-M diagram, and in taking the nearest entries in Tables A.1 and A.2.

Comment on accuracy.   All the worked examples in this book that require the use of graphs and tabulated data will therefore have only graphical and tabulated accuracy. In many of our calculations using the tables, we will use the nearest entry in the table so as not to have to spend the time to interpolate between entries. Using the nearest entry is usually sufficient for our purposes.

EXAMPLE 4.2

In Example 4.1, the deflection angle is increased to $\theta = 30°$. Calculate the pressure and Mach number behind the wave, and compare these results with those of Example 4.1.

■ **Solution**

From the θ-β-M chart (see end pages), for $M_1 = 3$ and $\theta = 30°$: $\beta = 52°$. Hence

$$M_{n_1} = M_1 \sin \beta = 3 \sin 52° = 2.364$$

From Table A.2, for $M_{n_1} = 2.364$: $p_2/p_1 = 6.276$ (nearest entry) and $M_{n_2} = 0.5286$. Thus

$$p_2 = \frac{p_2}{p_1} p_1 = (6.276)(1) = \boxed{6.276 \text{ atm}}$$

$$M_2 = \frac{M_{n_2}}{\sin(\beta - \theta)} = \frac{0.5286}{\sin 22°} = \boxed{1.41}$$

Note: Compare the above results with those from Example 4.1. When θ is increased, the shock wave becomes *stronger,* as evidenced by the increased pressure behind the shock (6.276 atm compared to 3.783 atm). The Mach number behind the shock is reduced (1.41 compared to 1.988). Also, as θ is increased, β also increases (52° compared to 37.8°).

EXAMPLE 4.3

In Example 4.1, the free-stream Mach number is increased to 5. Calculate the pressure and Mach number behind the wave, and compare these results with those of Example 4.1.

■ **Solution**

From the θ-β-M chart, for $M_1 = 5$ and $\theta = 20°$: $\beta = 30°$. Hence,

$$M_{n_1} = M_1 \sin \beta = 5 \sin 30° = 2.5$$

From Table A.2, for $M_{n_1} = 2.5$: $p_2/p_1 = 7.125$ and $M_{n_2} = 0.513$. Thus,

$$p_2 = \frac{p_2}{p_1} p_1 = (7.125)(1) = \boxed{7.125 \text{ atm}}$$

$$M_2 = \frac{M_{n_2}}{\sin(\beta - \theta)} = \frac{0.513}{\sin 10°} = \boxed{2.95}$$

Note: Compare the above results with those from Example 4.1. When M_1 is increased, the shock wave becomes *stronger*, as evidenced by the increased pressure behind the shock (7.125 atm compared to 3.783 atm). The Mach number behind the shock is increased (2.95 compared to 1.988). Also, as M_1 is increased, β is decreased (30° compared to 37.8°).

The net results of Examples 4.1 through 4.3 are these basic variations.

1. Anything that increases the normal component of the Mach number ahead of the shock M_{n_1} increases the strength of the shock. In Example 4.2, M_{n_1} was increased by increasing the wave angle β; in turn, the increased β was brought about by increasing θ. In Example 4.3, M_{n_1} was increased by increasing M_1; although the wave angle β decreases in this case (which works to reduce M_{n_1}), the increased value of M_1 (which works to increase M_{n_1}) more than compensates, and the net result is a larger M_{n_1}.

2. It is a general rule that, as θ increases (holding M_1 constant), the shock wave becomes stronger, and β increases.

3. It is a general rule that, as M_1 increases (holding θ constant), the shock wave becomes stronger, and β decreases.

EXAMPLE 4.4

Consider a Mach 2.8 supersonic flow over a compression corner with a deflection angle of 15°. If the deflection angle is doubled to 30°, what is the increase in shock strength? Is it also doubled?

■ **Solution**

From the θ-β-M chart, for $\theta = 15°$, $\beta = 33.8°$, and for $\theta = 30°$, $\beta = 54.7°$

For $\theta = 15°$: $M_{n_1} = M_1 \sin \beta = 2.8 \sin 33.8° = 1.558$. From Table A.2, for $M_{n_1} = 1.56$ (nearest entry),

$$\frac{p_2}{p_1} = \boxed{2.673}$$

For $\theta = 30°$: $M_{n_1} = 2.8 \sin 54.7° = 2.285$. From Table A.2, for $M_{n_1} = 2.3$ (nearest entry)

$$\frac{p_2}{p_1} = \boxed{6.005}$$

Clearly, if the angle of the compression corner is doubled, the strength of the shock wave is more than doubled; in this case, the shock strength is increased by a factor of 2.3.

EXAMPLE 4.5

Consider a compression corner with a deflection angle of 28°. Calculate the shock strengths when $M_1 = 3$ and when M_1 is doubled to 6. Is the shock strength also doubled?

■ **Solution**

From the θ-β-M diagram for $M_1 = 3$, $\beta = 48.5°$. Hence,

$$M_{n_1} = M_1 \sin \beta = 3 \sin 48.5° = 2.247$$

From Table A.2, for $M_{n_1} = 2.25$ (nearest entry)

$$\frac{p_2}{p_1} = \boxed{5.74}$$

From the θ-β-M diagram for $M_1 = 6$, $\beta = 38.0°$. Hence,

$$M_{n_1} = M_1 \sin \beta = 6 \sin 38° = 3.69$$

From Table A.2, for $M_{n_1} = 3.7$ (nearest entry),

$$\frac{p_2}{p_1} = \boxed{15.8}$$

Clearly, if the Mach number is doubled, the strength of the shock is more than doubled; in this case, the shock strength is increased by a factor of 2.75.

The physical results in Examples 4.4 and 4.5 are reflective of the *nonlinear* behavior of shock waves. The nonlinearity of shock wave phenomena is mathematically reflected in the equations obtained in this section, such as Eqs. (4.7)–(4.12), where the Mach number appears as squared, and sometimes in an intricate fashion in the equations. This is especially true of the θ-β-M relation, Eq. (4.17). In Chap. 9 we will discuss an approximate theory for analyzing supersonic flows over bodies, where the theory involves linear equations. However, we will also see that such linearized theory deals with slender bodies at small angles of attack, where in reality the shock waves are weak. Indeed, linearized supersonic theory does not deal with shock waves explicitly—the theory pretends that they are not here. This will all make more sense when we discuss the material in Chap. 9. At present, we are just introducing a small precursor to the intellectual model contained in Chap. 9.

4.3.1   The β-θ-M Relation: An Alternative Form for the θ-β-M Relation

The θ-β-M relation expressed by Eq. (4.17) gives θ as an explicit function of β and M. In classical treatments of compressible flow, this is the equation used to relate deflection angle, wave angle, and Mach number. However, for many practical applications, we are given the deflection angle and upstream Mach number, because these are the parameters we can easily see and measure, and we want to find the corresponding wave angle, β. Equation (4.17) does not allow us to calculate β explicitly. Rather, we can plot the θ-β-M curves from Eq. (4.17) as shown in Fig. 4.8, and then find β from the graph as demonstrated in Examples 4.1–4.5. Alternatively, we can set up a short computer program to calculate β by iterating Eq. (4.17).

It is not commonly known that an alternative equation can be derived that relates β explicitly in terms of θ and M. There are at least four different derivations in the literature, found in Refs. 130–133. The key is to write Eq. (4.17) as a cubic equation, and then find the roots of this cubic equation. The earliest work along these lines appears to be that of Thompson (Ref. 130) who recognized that Eq. (4.17) can be expressed as a cubic in $\sin^2 \beta$:

$$\sin^6 \beta - \left(\frac{M^2 + 2}{M^2} + \gamma \sin^2 \theta \right) \sin^4 \beta$$

$$+ \left\{ \frac{2M^2 + 1}{M^4} + \left[\left(\frac{\gamma + 1}{2} \right)^2 + \frac{\gamma - 1}{M^2} \right] \sin^2 \beta \right\} \sin^2 \beta - \frac{\cos^2 \theta}{M^4} = 0$$

However, Emanuel found it more convenient analytically to express Eq. (4.17) as a cubic in $\tan \beta$:

$$\left(1 + \frac{\gamma - 1}{2} M^2 \right) \tan \theta \tan^3 \beta - (M^2 - 1) \tan^2 \beta$$

$$+ \left(1 + \frac{\gamma + 1}{2} M^2 \right) \tan \theta \tan \beta + 1 = 0 \qquad (4.18)$$

Emanuel observed that Eq. (4.18) has three real, unequal roots for an attached shock wave with a given θ and M. One root is negative, hence nonphysical. The other two positive roots correspond to the weak and strong shock solutions. These roots can be expressed as

$$\tan \beta = \frac{M^2 - 1 + 2\lambda \cos[(4\pi\delta + \cos^{-1} \chi)/3]}{3 \left(1 + \frac{\gamma - 1}{2} M^2 \right) \tan \theta} \qquad (4.19)$$

where $\delta = 0$ yields the strong shock solution, $\delta = 1$ yields the weak shock solution, and

$$\lambda = \left[(M^2 - 1)^2 - 3 \left(1 + \frac{\gamma - 1}{2} M^2 \right) \left(1 + \frac{\gamma + 1}{2} M^2 \right) \tan^2 \theta \right]^{1/2} \qquad (4.20)$$

and

$$\chi = \frac{(M^2 - 1)^3 - 9 \left(1 + \frac{\gamma - 1}{2} M^2 \right) \left(1 + \frac{\gamma - 1}{2} M^2 + \frac{\gamma + 1}{4} M^4 \right) \tan^2 \theta}{\lambda^3}$$

$$(4.21)$$

Equation (4.19) represents an alternative form of the relation between β, θ, and M; in analogy with Eq. (4.17), which is called the θ-β-M relation, we will label Eq. (4.19) as the β-θ-M relation. Eq. (4.19), along with Eqs. (4.20) and (4.21), allows an exact explicit calculation for β when θ and M are known, albeit a more lengthy calculation than that associated with Eq. (4.17). We emphasize that no simplifying mathematical assumptions go into the derivation of Eq. (4.19); it is an exact relationship.

EXAMPLE 4.6

Consider a Mach 4 flow over a compression corner with a deflection angle of 32°. Calculate the oblique shock wave angle for the weak shock case using (a) Fig. 4.8, and (b) the β-θ-M equation, Eq. (4.19). Compare the results from the two sets of calculations.

■ Solution

a.   From Fig. 4.8, we have for $M = 4$ and $\theta = 32°$, $\boxed{\beta = 48.2°}$.

b.   To use Eq. (4.19), we first calculate λ and χ from Eqs. (4.20) and (4.21), respectively.

In these equations, we have

$$(M^2 - 1)^2 = [(4)^2 - 1]^2 = (15)^2 = 225$$

$$(M^2 - 1)^3 = (15)^3 = 3375$$

$$\frac{\gamma - 1}{2} M^2 = \frac{1.4 - 1}{2}(4)^2 = 3.2$$

$$\frac{\gamma + 1}{2} M^2 = \frac{1.4 + 1}{2}(4)^2 = 19.2$$

$$\frac{\gamma + 1}{4} M^4 = \frac{1.4 + 1}{4}(4)^4 = 153.6$$

From Eq. (4.20),

$$\lambda = \left[(M^2 - 1)^2 - 3\left(1 + \frac{\gamma - 1}{2} M^2\right)\left(1 + \frac{\gamma + 1}{2} M^2\right)\tan^2 \theta \right]^{1/2}$$

$$= [225 - 3(4.2)(20.2)\tan^2 32°]^{1/2} = 11.208$$

From Eq. (4.21),

$$\chi = \frac{(M^2 - 1)^3 - 9\left(1 + \frac{\gamma - 1}{2} M^2\right)\left(1 + \frac{\gamma - 1}{2} M^2 + \frac{\gamma + 1}{4} M^4\right)\tan^2 \theta}{\lambda^3}$$

$$= \frac{(15)^3 - 9(4.2)(1 + 3.2 + 153.6)\tan^2 32°}{(11.208)^3}$$

$$= 0.7429$$

For Eq. (4.19), using $\delta = 1$ for the weak shock solution, we need

$$\cos^{-1} \chi = \cos^{-1}(0.7439) = 0.7334 \, \text{rad}$$

[*Note:* the factor in Eq. (4.19) involving $\cos^{-1} \chi$ is in radians]:

$$\frac{4\pi \delta + \cos^{-1} \chi}{3} = 4.433 \, \text{rad}$$

$$\cos\left(\frac{4\pi \delta + \cos^{-1} \chi}{3}\right) = \cos 4.433 = -0.2752$$

From Eq. (4.19),

$$\tan \beta = \frac{M^2 - 1 + 2\lambda \cos[4\pi\delta + \cos^{-1} \chi]/3}{3\left(1 + \frac{\gamma - 1}{2} M^2\right)\tan\theta}$$

$$= \frac{16 - 1 + 2(11.208)(-0.2752)}{3(4.2)\tan 32°}$$

$$= 1.1216$$

Hence,

$$\beta = \tan^{-1}(1.1216) = \boxed{48.28°}$$

This result agrees very well with the graphical solution obtained in part (a).

4.4 | SUPERSONIC FLOW OVER WEDGES AND CONES

The oblique shock properties discussed above represent the exact solution for the flow over a wedge or a two-dimensional compression corner, as sketched on the left-hand side of Fig. 4.9. The flow streamlines behind the shock are straight and parallel to the wedge surface. The pressure on the surface of the wedge is constant and equal to p_2, as further illustrated in Fig. 4.11a.

Straight oblique shocks are also attached to the tip of a sharp cone in supersonic flow, as sketched in Fig. 4.11b. The properties immediately behind this conical shock are given by the oblique shock relations. However, because the flow over a cone is inherently three-dimensional, the flowfield between the shock and cone surface is no longer uniform, as in the case of the wedge. As shown in Fig. 4.11b, the streamlines are curved, and the pressure at the cone surface p_s is not the same as p_2 immediately behind the shock. Moreover, the addition of a third dimension provides the flow with extra space to move through, hence relieving some of the obstructions set up by the presence of the body. This is called the "three-dimensional relieving effect," which is characteristic of all three-dimensional flows. For the flow over a cone, the three-dimensional relieving effect results in a weaker shock wave than for a wedge of the same angle. For example, Fig. 4.11 shows that a 20° half-angle wedge creates a 53° oblique shock for $M_1 = 2$; by comparison, the shock on a 20° half-angle cone is at a wave angle of 37°, with an attendant lower p_2, ρ_2, and T_2 immediately behind the shock. Because of these differences, the study in this book of supersonic flow over cones will be delayed until Chap. 10.

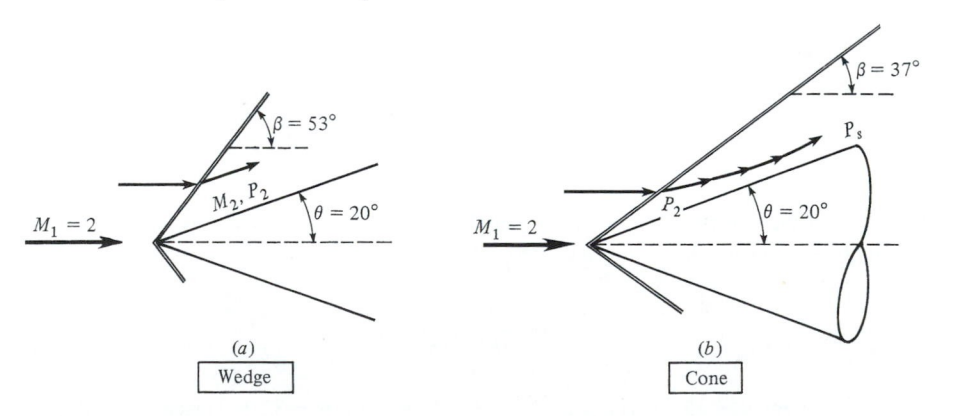

Figure 4.11 | Comparison between wedge and cone flow; illustration of the three-dimensional relieving effect.

EXAMPLE 4.7

A 10° half-angle wedge is placed in a "mystery flow" of unknown Mach number. Using a Schlieren system, the shock wave angle is measured as 44°. What is the free-stream Mach number?

■ **Solution**

From the θ-β-M chart, for $\theta = 10°$ and $\beta = 44°$, we have

$$\boxed{M_1 = 1.8}$$

Note: This technique has actually been used in some experiments for the measurement of Mach number. However, it is usually more accurate and efficient to use a Pitot tube to measure Mach number, as described in Example 3.7.

EXAMPLE 4.8

Consider a 15° half-angle wedge at zero angle of attack. Calculate the pressure coefficient on the wedge surface in a Mach 3 flow of air.

■ **Solution**

The pressure coefficient is defined as

$$C_p = \frac{p - p_\infty}{q_\infty}$$

where p_∞ is the free-stream pressure and q_∞ is the free-stream dynamic pressure, defined by $q_\infty = \frac{1}{2}\rho_\infty V_\infty^2$. For a calorically perfect gas, q_∞ can also be expressed in terms of p_∞ and M_∞ as

$$q_\infty \equiv \tfrac{1}{2}\rho_\infty V_\infty^2 = \frac{1}{2}\frac{\gamma p_\infty}{\gamma p_\infty}\rho_\infty V_\infty^2 = \frac{\gamma p_\infty}{2}\frac{V_\infty^2}{a_\infty^2} = \frac{\gamma}{2}p_\infty M_\infty^2$$

Thus, the pressure coefficient can be written as

$$C_p = \frac{p - p_\infty}{\dfrac{\gamma}{2}p_\infty M_\infty^2} = \frac{2}{\gamma M_\infty^2}\left(\frac{p}{p_\infty} - 1\right)$$

In terms of the nomenclature being used in this chapter, where the free-stream properties in front of the shock are denoted by a subscript 1, then C_p is written as

$$C_p = \frac{2}{\gamma M_1^2}\left(\frac{p_2}{p_1} - 1\right)$$

For $M_1 = 3$ and $\theta = 15°$, we have from the θ-β-M diagram $\beta = 32.2°$. Hence

$$M_{n_1} = M_1 \sin\beta = 3\sin 32.2 = 1.6$$

From Table A.2, for $M_{n_1} = 1.6$: $p_2/p_1 = 2.82$. Thus,

$$C_p = \frac{2}{(1.4)(3)^2}(2.82 - 1) = \boxed{0.289}$$

Note: For this example, we can deduce that C_p is strictly a function of γ and M_1.

EXAMPLE 4.9

Consider a 15° half-angle wedge at zero angle of attack in a Mach 3 flow of air. Calculate the drag coefficient. Assume that the pressure exerted over the base of the wedge, the base pressure, is equal to the free-stream pressure.

■ **Solution**

The physical picture is sketched in Fig. 4.12. The drag is the net force in the x direction; p_2 is exerted perpendicular to the top and bottom faces, and p_1 is exerted over the base. The chord length of the wedge is c. Consider a unit span of the wedge, i.e., a length of unity perpendicular to the xy plane. The drag per unit span, denoted by D', is

$$D' = 2\left[\frac{(c)(1)}{\cos 15°}p_2\right]\sin 15° - (2c\tan 15°)p_1$$

By definition, the drag coefficient is

$$c_d \equiv \frac{D'}{q_\infty S}$$

where S is the planform area (the projected area seen by viewing the wedge from the top). Thus, $S = (c)(1)$. Hence

$$c_d = \frac{D'}{q_\infty c}$$

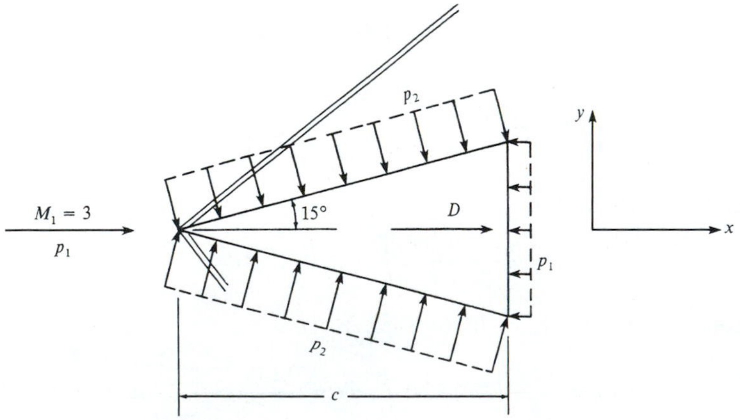

Figure 4.12 | Geometry for Example 4.9.

From Example 4.8, we saw that

$$q_\infty = \frac{\gamma}{2} p_\infty M_\infty^2 = \frac{\gamma}{2} p_1 M_1^2$$

Thus,

$$c_d = \frac{2D'}{\gamma p_1 M_1^2 (c)(1)}$$

or

$$c_d = \frac{2}{\gamma p_1 M_1^2 c} \left[\frac{(2)(c)(1)}{\cos 15°} p_2 \sin 15° - (2c \tan 15°) p_1 \right]$$

$$= \frac{4}{\gamma p_1 M_1^2} (p_2 - p_1) \tan 15° = \frac{4}{\gamma M_1^2} \left(\frac{p_2}{p_1} - 1 \right) \tan 15°$$

From Example 4.8, which deals with the same wedge at the same flow conditions, we have $p_2/p_1 = 2.82$. Thus

$$c_d = \frac{4}{(1.4)(3)^2} (2.82 - 1) \tan 15° = \boxed{0.155}$$

An *alternative* solution to this problem can be developed using the pressure coefficient given in Example 4.8. The drag coefficient for an aerodynamic body is given by the integral of the pressure coefficient over the surface, as shown in Sec. 1.5 of Ref. 104. To be specific, from Ref. 104 we have

$$c_d = \frac{1}{c} \int_{LE}^{TE} \left(C_{p_u} - C_{p_l} \right) dy$$

Here, the integral is taken over the surface from the leading edge (LE) to the trailing edge (TE), and C_{p_u} and C_{p_l} are the pressure coefficients over the upper and lower surfaces, respectively. In this problem, due to the symmetry, clearly $C_{p_u} = C_{p_l}$. On the upper surface,

$$dy = \frac{dy}{dx} dx = (\tan 15°) dx$$

On the lower surface (because y decreases as x increases),

$$dy = -\left(\frac{dy}{dx} \right) dx = -\tan 15° dx$$

Thus,

$$c_d = \frac{1}{c} \left[\int_0^c C_{p_u} (\tan 15°) dx - \int_0^c C_{p_l} (-\tan 15°) dx \right]$$

Since $\tan 15° = 0.2679$, then

$$c_d = \frac{0.268}{c} \int_0^c \left(C_{p_u} + C_{p_l} \right) dx$$

From Example 4.8, $C_{p_u} = C_{p_l} = 0.289$. Thus,

$$c_d = \frac{0.268}{c} (2)(0.289) \int_0^c dx = \frac{0.155}{c} c = \boxed{0.155}$$

This is the same answer as obtained from the first method described above.

Note: The only information given in this problem was the body shape, free-stream Mach number, and the fact that we are dealing with air (hence we know that $\gamma = 1.4$). To calculate the drag coefficient for a given body shape, we only need M_1 and γ. This is consistent with the results of dimensional analysis (see Chap. 1 of Ref. 104) that the drag coefficient for a compressible inviscid flow is a function of Mach number and γ only; c_d does *not* depend on the size of the body (denoted by c), the free-stream density, pressure, or velocity. It depends only on the Mach number and γ. Thus

$$c_d = f(M_1, \gamma)$$

This relation is verified by the results of this example. Also, the drag in this problem is due to the pressure distribution only; since we are dealing with an inviscid flow, shear stress due to friction is not included. The drag in this problem is therefore a type of "pressure drag"; it is frequently identified as *wave drag,* and hence c_d calculated here is the wave drag coefficient.

4.5 | SHOCK POLAR

Graphical explanations go a long way towards the understanding of supersonic flow with shock waves. One such graphical representation of oblique shock properties is given by the shock polar, described next.

Consider an oblique shock with a given upstream velocity V_1 and deflection angle θ_B, as sketched in Fig. 4.13. Also, consider an xy cartesian coordinate system with the x axis in the direction of V_1. Figure 4.13 is called the *physical plane.* Define V_{x_1}, V_{y_1}, V_{x_2}, and V_{y_2} as the x and y components of velocity ahead of and behind the shock, respectively. Now plot these velocities on a graph that uses V_x and V_y as axes, as shown in Fig. 4.14. This graph of velocity components is called the *hodograph plane.* The line OA represents V_1 ahead of the shock; the line OB represents V_2 behind the shock. In turn, *point A* in the hodograph plane of Fig. 4.14 represents the entire *flowfield* of region 1 in the physical plane of Fig. 4.13. Similarly, point B in the hodograph plane represents the entire flowfield of region 2 in the physical plane. If

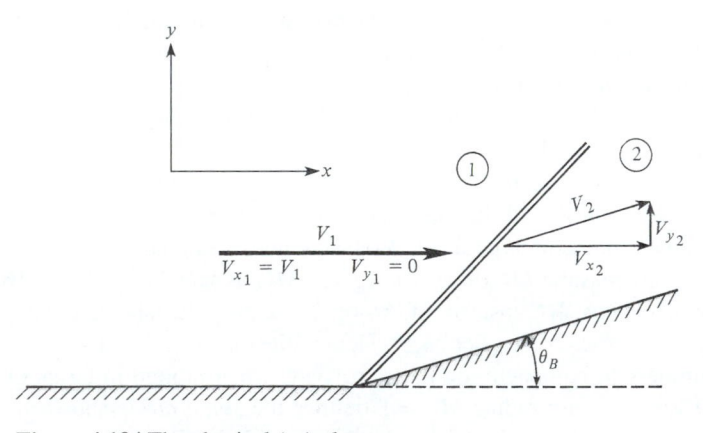

Figure 4.13 | The physical (xy) plane.

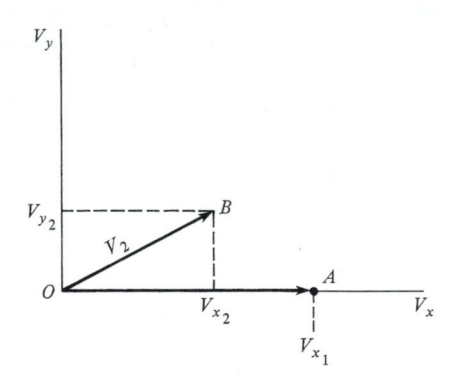

Figure 4.14 | The hodograph plane.

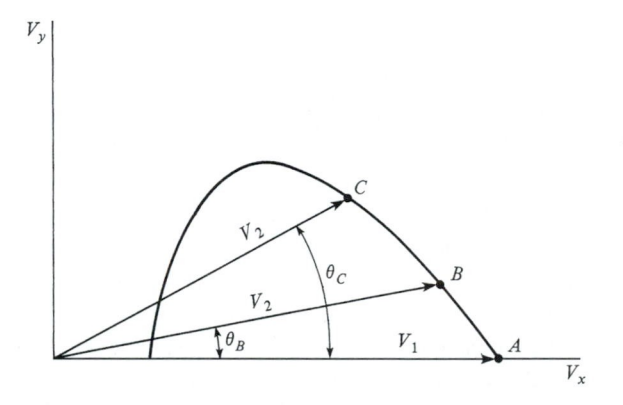

Figure 4.15 | Shock polar for a given V_1.

now the deflection angle in Fig. 4.13 is increased to a larger value, say θ_C, then the velocity V_2 is inclined further to angle θ_C, and its magnitude is decreased because the shock wave becomes stronger. This condition is shown as point C in the hodograph diagram of Fig. 4.15. Indeed, if the deflection angle θ in Fig. 4.12 is carried through all possible values for which there is an oblique shock solution ($\theta < \theta_{max}$), then the locus of all possible velocities behind the shock is given in Fig. 4.15. This locus is defined as a *shock polar*. Points A, B, and C in Figs. 4.14 and 4.15 are just three points on the shock polar for a given V_1.

For convenience, let us now nondimensionalize the velocities in Fig. 4.15 by a^*, defined in Sec. 3.4. Recall that the flow across a shock is adiabatic, hence a^* is the same ahead of and behind the shock. Consequently, we obtain a shock polar which is the locus of all possible M_2^* values for a given M_1^*, as sketched in Fig. 4.16. The convenience of using M^* instead of M or V to plot the shock polar is that, as $M \to \infty$, $M^* \to 2.45$ (see Sec. 3.5). Hence, the shock polars for a wide range of Mach numbers fit compactly on the same page when plotted in terms of M^*. Also note that a circle with radius $M^* = 1$ defines the *sonic circle* shown in Fig. 4.16. Inside this circle, all velocities are subsonic; outside it, all velocities are supersonic.

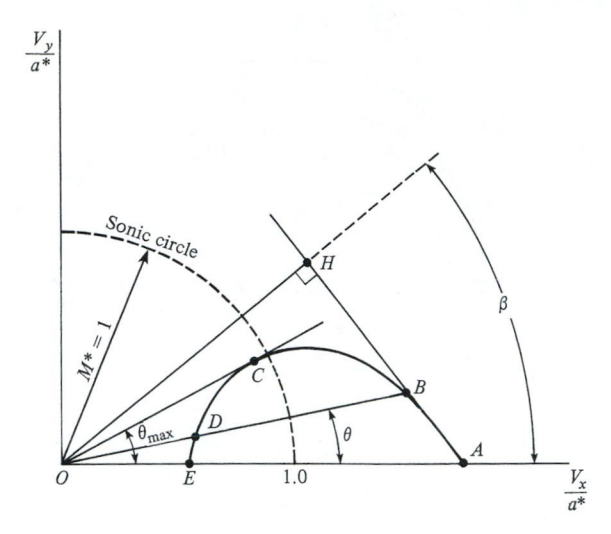

Figure 4.16 | Geometric constructions using the shock polar.

Several important properties of the shock polar are illustrated in Fig. 4.16:

1. For a given deflection angle θ, the shock polar is cut at two points B and D. Points B and D represent the weak and strong shock solutions, respectively. Note that D is inside the sonic circle, as would be expected.

2. The line OC drawn tangent to the shock polar represents the maximum deflection angle θ_{max} for the given M_1^* (hence also for the given M_1). For $\theta > \theta_{max}$, there is no oblique shock solution.

3. Points E and A represent flow with no deflection. Point E is the normal shock solution; point A corresponds to a Mach line.

4. If a line is drawn through A and B, and line OH is drawn perpendicular to AB, then the angle HOA is the wave angle β corresponding to the shock solution at point B. This can be proved by simple geometric argument, recalling that the tangential component of velocity is preserved across the shock wave. Try it yourself.

5. The shock polars for different Mach numbers form a family of curves, as drawn in Fig. 4.17. Note that the shock polar for $M_1^* = 2.45(M_1 \to \infty)$ is a circle.

The analytic equation for the shock polar (V_y/a^* versus V_x/a^*) can be obtained from the oblique shock equations given in Sec. 4.3. The derivation is given in such classic texts as those by Ferri (Ref. 5) or Shapiro (Ref. 16). The result is given here for reference:

$$\left(\frac{V_y}{a^*}\right)^2 = \frac{(M_1^* - V_x/a^*)^2[(V_x/a^*)M_1^* - 1]}{\dfrac{2}{\gamma + 1}(M_1^*)^2 - \left(\dfrac{V_x}{a^*}\right)M_1^* + 1} \tag{4.22}$$

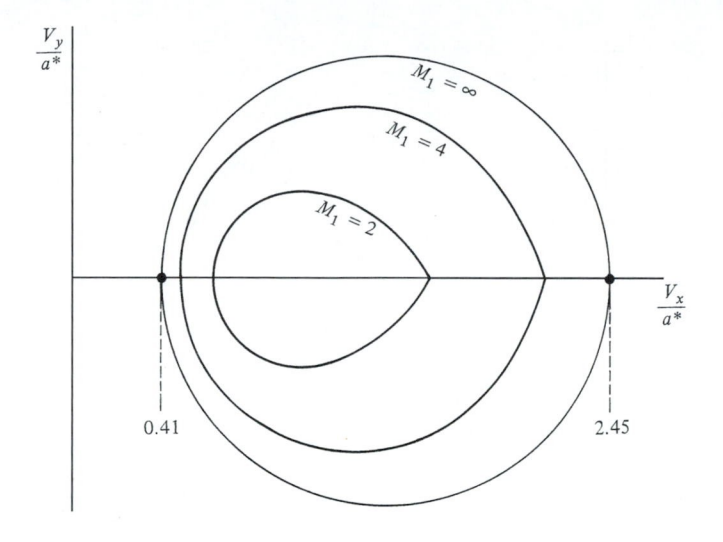

Figure 4.17 | Shock polars for different Mach numbers.

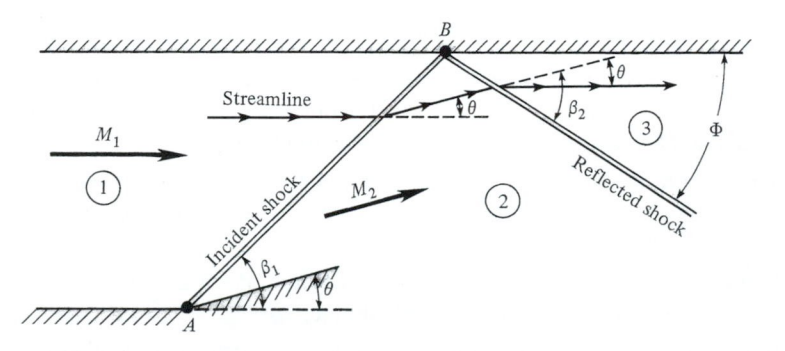

Figure 4.18 | Regular reflection from a solid boundary.

4.6 | REGULAR REFLECTION FROM A SOLID BOUNDARY

Consider an oblique shock wave incident on a solid wall, as sketched in Fig. 4.18. Question: Does the shock wave disappear at the wall, or is it reflected downstream? If it is reflected, at what angle and what strength? The answer lies in the physical boundary condition at the wall, where the flow immediately adjacent to the wall must be parallel to the wall. In Fig. 4.18, the flow in region 1 with Mach number M_1 is deflected through an angle θ at point A. This creates an oblique shock wave that impinges on the upper wall at point B. In region 2 behind this incident shock, the streamlines are inclined at an angle θ to the upper wall. All flow conditions in region 2 are uniquely defined by M_1 and θ through the oblique shock relations discussed in Sec. 4.5. At point B, in order for the flow to remain tangent to the upper wall, the streamlines in region 2 must be deflected downward through the angle θ. This can only be done by a second shock wave, originating at B, with sufficient strength to turn the flow through an angle θ, with an upstream Mach number of M_2. This second shock is called a *reflected shock;* its strength is uniquely defined by M_2 and θ, yielding the consequent properties in region 3. Because $M_2 < M_1$, the reflected shock wave is weaker than the incident shock, and the angle Φ it makes with the upper wall is *not* equal to β_1 (i.e., the reflected shock wave is not specularly reflected).

EXAMPLE 4.10

Consider a horizontal supersonic flow at Mach 2.8 with a static pressure and temperature of 1 atm and 519°R, respectively. This flow passes over a compression corner with a deflection angle of 16°. The oblique shock generated at the corner propagates into the flow, and is incident on a horizontal wall, as shown in Fig. 4.18. Calculate the angle Φ made by the reflected shock wave with respect to the wall, and the Mach number, pressure, and temperature behind the reflected shock.

■ **Solution**

The flowfield is as shown in Fig. 4.18. From the θ-β-M diagram, $\beta_1 = 35°$,

$$M_{n_1} = M_1 \sin \beta_1 = 2.8 \sin 35° = 1.606$$

From Table A.2, for $M_{n_1} = 1.606$: $p_2/p_1 = 2.82$, $T_2/T_1 = 1.388$, and $M_{n_2} = 0.6684$. Hence

$$M_2 = \frac{M_{n_2}}{\sin(\beta_1 - \theta)} = \frac{0.6684}{\sin(35 - 16)} = 2.053$$

From the θ-β-M diagram, for $M = 2.053$ and $\theta = 16°$: $\beta_2 = 45.5°$. The component of the Mach number ahead of the reflected shock normal to the shock is M_{n_2}, given by

$$M_{n_2} = M_2 \sin \beta_2 = 2.053 \sin 45.5° = 1.46$$

From Table A.2, for $M_{n_2} = 1.46$: $p_3/p_2 = 2.32$, $T_3/T_2 = 1.294$, and $M_{n_3} = 0.7157$, where M_{n_3} is the component of the Mach number behind the reflected shock normal to the shock. The Mach number in region 3 behind the reflected shock is given by

$$M_3 = \frac{M_{n_3}}{\sin(\beta_2 - \theta)} = \frac{0.7157}{\sin(45.5 - 16)} = \boxed{1.45}$$

Also

$$p_3 = \frac{p_3}{p_2} \frac{p_2}{p_1} p_1 = (2.32)(2.82)(1 \text{ atm}) = \boxed{6.54 \text{ atm}}$$

$$T_3 = \frac{T_3}{T_2} \frac{T_2}{T_1} T_1 = (1.294)(1.388)(519) = \boxed{932°\text{R}}$$

$$\Phi = \beta_2 - \theta = 45.5 - 16 = \boxed{29.5°}$$

Note: The incident shock makes the angle 35° with respect to the upper wall; the reflected shock wave lies closer to the wall, at an angle of 29.5°. Clearly, the shock wave is *not* specularly reflected.

EXAMPLE 4.11

Consider the geometry shown in Fig. 4.19. Here a supersonic flow with Mach number, pressure, and temperature M_1, p_1, and T_1, respectively, is deflected through an angle θ_1 by a compression corner at point A on the lower wall, creating an oblique shock wave emanating from point A. This shock impinges on the upper wall at point B. Also precisely at point B the upper wall is bent downward through the angle θ_2. The incident shock is reflected at point B, creating a reflected shock which propagates downward and to the right in Fig. 4.19. Consider a flow where $M_1 = 3$, $p_1 = 1$ atm, and $T_1 = 300$ K. Consider the geometry as sketched in Fig. 4.19 where $\theta_1 = 14°$ and $\theta_2 = 10°$. Calculate the Mach number, pressure, and temperature in region 3 behind the reflected shock wave.

■ **Solution**

From the θ-β-M diagram, $\beta_1 = 31.2°$,

$$M_{n_1} = M_1 \sin \beta_1 = 3 \sin 31.2° = 1.554$$

From Table A.2, for $M_{n_1} = 1.56$ (nearest entry),

$$\frac{p_2}{p_1} = 2.673, \quad \frac{T_2}{T_1} = 1.361, \quad M_{n_2} = 0.6809$$

$$M_2 = \frac{M_{n_2}}{\sin(\beta_1 - \theta_1)} = \frac{0.6809}{\sin(31.2 - 14)} = 2.30$$

The flow in region 2, at $M_2 = 2.3$, is deflected downward through the combined angle $\theta_1 + \theta_2 = 14° + 10° = 24°$. From the θ-β-M diagram for $M = 2.3$ and $\theta = 24°$, $\beta_2 = 52.5°$,

$$M_{n_2} = M_2 \sin \beta_2 = 2.3 \sin 52.5° = 1.82$$

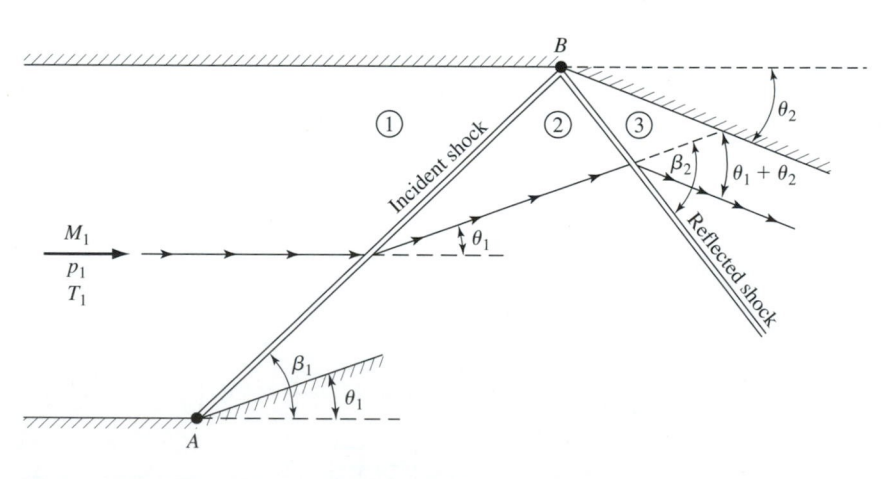

Figure 4.19 | Reflected shock geometry for Example 4.11.

From Table A.2, for $M = 1.82$,

$$\frac{p_3}{p_2} = 3.698, \quad \frac{T_3}{T_2} = 1.547, \quad M_{n_3} = 0.6121$$

$$M_3 = \frac{M_{n_3}}{\sin(\beta_2 - \theta_1 - \theta_2)} = \frac{0.6121}{\sin(52.5 - 24)} = \boxed{1.28}$$

$$p_3 = \frac{p_3}{p_2}\frac{p_2}{p_1}p_1 = (3.698)(2.673)(1) = \boxed{9.88 \text{ atm}}$$

$$T_3 = \frac{T_3}{T_2}\frac{T_2}{T_1}T_1 = (1.547)(1.361)(300) = \boxed{631.6 \text{ K}}$$

EXAMPLE 4.12

a. Consider the supersonic flow described in Example 4.10, where $M_1 = 2.8$, $p_1 = 1$ atm, and $M_3 = 1.45$. This flow is shown in Fig. 4.20a. Calculate the total pressure in region 3 where $M_3 = 1.45$.

b. Consider the supersonic flow shown in Fig. 4.20b, where the upstream Mach number and pressure are the same as in part (a), i.e., $M_1 = 2.8$ and $p_1 = 1$ atm. This flow is deflected through the angle θ such that the Mach number behind the single oblique shock in Fig. 4.20b is the same as that behind the reflected shock in Fig. 4.20a, i.e., $M_2 = 1.45$ in Fig. 4.20b. For the flow in Fig. 4.20b, calculate θ and the total pressure in region 2, p_{o_2}.

Comment on the relative values of the total pressure obtained in parts (a) and (b).

■ **Solution**

a. From Example 4.10, $M_{n_1} = 1.606$, and $M_{n_2} = 1.46$. From Table A.2, for $M_{n_1} = 1.606$,

$$\frac{p_{o_2}}{p_{o_1}} = 0.8952$$

From Table A.2, for $M_{n_2} = 1.46$,

$$\frac{p_{o_3}}{p_{o_2}} = 0.9420$$

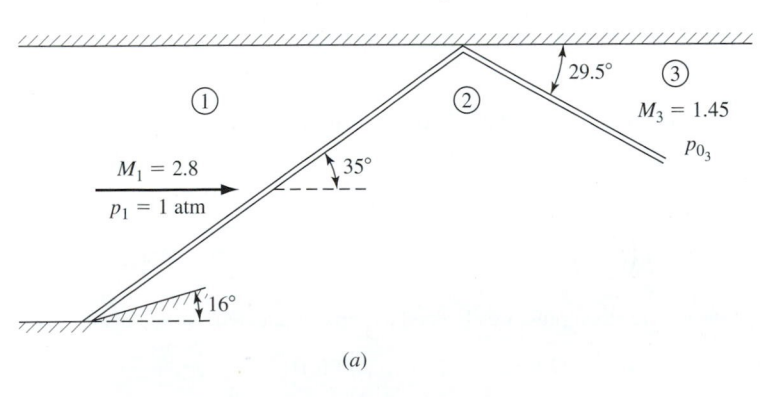

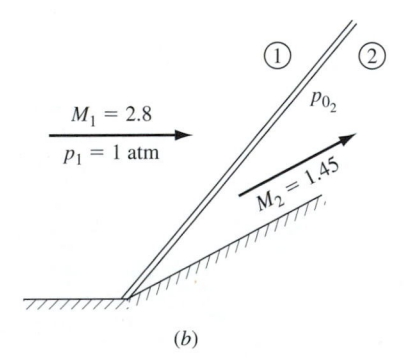

(a)   (b)

Figure 4.20 | Shock waves for Example 4.12.

From Table A.1, for $M_1 = 2.8$,

$$\frac{p_{o_1}}{p_1} = 27.14$$

Hence,

$$p_{o_3} = \left(\frac{p_{o_3}}{p_{o_2}}\right)\left(\frac{p_{o_2}}{p_{o_1}}\right)\left(\frac{p_{o_1}}{p_1}\right) p_1 = (0.9420)(0.8952)(27.14)(1) = \boxed{22.9 \, \text{atm}}$$

b.   For the single shock wave shown in Fig. 4.20b, to find θ such that $M_2 = 1.45$ when $M_1 = 2.8$, we have to carry out an iterative (trial-and-error) solution where we assume various values of θ, calculate M_2 for each value, and finally obtain the specific value of θ, which will yield $M_2 = 1.45$. To begin, we arbitrarily assume $\theta = 20°$. Using the θ-β-M diagram and Table A.2, we find

For $\theta = 20°$: $\beta = 39.4°$, $M_{n_1} = 1.777$, $M_{n_2} = 0.621$, $M_2 = 1.87$

Here, M_2 is too high. We need to assume a larger θ so that the shock is stronger. Assume $\theta = 30°$.

For $\theta = 30°$: $\beta = 54.7°$, $M_{n_1} = 2.27$, $M_{n_2} = 0.541$, $M_2 = 1.29$

Here, M_2 is too low. We need to assume a slightly smaller θ so that the shock is slightly weaker. Assume $\theta = 28°$.

For $\theta = 28°$: $\beta = 50.8°$, $M_{n_1} = 2.17$, $M_{n_2} = 0.554$, $M_2 = 1.43$

Here, M_2 is slightly too low. Assume $\theta = 27°$ so that the shock wave is marginally weaker.

For $\theta = 27°$: $\beta = 49°$, $M_{n_1} = 2.11$, $M_{n_2} = 0.5613$, $M_2 = 1.50$

Here, M_2 is slightly too high. The correct value of θ is somewhere between $27°$ and $28°$. Since this example is subject to graphical accuracy only, as well as the level of accuracy obtained by taking the nearest entry in Table A.2, let us simply interpolate between $\theta = 27°$ where $M_2 = 1.50$, and $\theta = 28°$ where $M_2 = 1.43$, to obtain θ where $M_2 = 1.45$:

$$\theta = 27° + (1°)\left(\frac{1.5 - 1.45}{1.5 - 1.43}\right) = 27° + 0.7° = \boxed{27.7°}$$

The total pressure in region 2 in Fig. 4.20b is obtained from Table A.2, using the nearest entry for $M_{n_1} = 2.15$, where $p_{o_2}/p_{o_1} = 0.6511$. Also, from Table A.1 for $M_1 = 2.8$, $p_{o_1}/p_1 = 27.14$. Hence,

$$p_{o_2} = \left(\frac{p_{o_2}}{p_{o_1}}\right)\left(\frac{p_{o_1}}{p_1}\right)(p_1) = (0.6511)(27.14)(1) = \boxed{17.67 \, \text{atm}}$$

Comparing the two values for total pressure obtained in parts (a) and (b), we see that

$$p_{o_3} = 22.9 \, \text{atm} \qquad \text{(from part (a))}$$

$$p_{o_2} = 17.67 \, \text{atm} \qquad \text{(from part (b))}$$

Clearly, the case of the flow through the single shock wave shown in Fig. 4.20*b* results in a lower total pressure than the case of the flow through the double shock system shown in Fig. 4.20*a*.

4.7 | COMMENT ON FLOW THROUGH MULTIPLE SHOCK SYSTEMS

The results of Example 4.12 illustrate an important physical phenomena associated with flow through shock waves. Here we have a flow with an initial Mach number of 2.8, which in both cases shown in Fig. 4.20 is slowed to a lower Mach number of 1.45. In Fig. 4.20*a*, this is accomplished by passing the flow through two weaker shocks, and in Fig. 4.20*b* this is accomplished by passing the flow through a single stronger shock. The process of slowing the flow to the same Mach number by means of two shocks compared to that of a single shock results in a higher total pressure. That is, the system shown in Fig. 4.20 results in a *smaller loss* of total pressure, hence it is an aerodynamically more efficient system. This phenomena has a major practical impact on engine inlet design for supersonic airplanes, and for the diffuser design in supersonic wind tunnels, where it is always preferable to slow the incoming supersonic flow by passing it through a multiple system of weaker shocks than through a single stronger shock. Problem 4.8 at the end of this chapter reinforces this fact. Also, the geometry for a simulated scramjet engine shown in Fig. 4.2 is designed specifically to initiate the multiple shock pattern in the flow seen in Fig. 4.2 in order to decrease the total pressure losses in the engine and therefore achieve better propulsion efficiency.

It is interesting to compare the sum of the two turning angles of the flow in Fig. 4.20*a* with the single turning angle in Fig. 4.20*b*. In Fig. 4.20*a*, the flow is first turned into itself through a deflection of 16° across the incident shock, and then turned again into itself through a deflection of 16° across the reflected shock, the sum of the turning angles being 32°. In contrast, the turning angle for the single shock in Fig. 4.20*b* is calculated (in Example 4.12) to be a smaller value, namely, 27.7°. Hence, the flow through the multiple shock system experiences a net turning angle that is actually larger than that for the single shock system. In spite of this, the multiple shock system is more efficient, resulting in a smaller loss of total pressure (hence a smaller increase in entropy). The reason for this is the highly nonlinear increase in entropy and decrease in total pressure as the Mach number ahead of a shock wave increases. Examine again Fig. 3.10, where the changes in physical properties across a normal shock are plotted versus upstream Mach number. Note the rapid and highly nonlinear decrease in the total pressure ratio, p_{o_2}/p_{o_1}, as M_1 increases. For example, doubling the upstream Mach number results in a much larger than proportional decrease in total pressure. Returning to the double shock system in Fig. 4.20*a*, the key to its better efficiency is that the Mach number ahead of the second shock has been reduced by first flowing across the first shock. Even though the flow is going through twice as many shocks with a net turning angle larger than the single shock case, the smaller local Mach number ahead of the second shock more than compensates by causing a sufficiently smaller increase in entropy across the second shock. Hence, the net total pressure loss across the multiple shock system is less than that across the single shock. The progressive slowing down of the flow through a multiple system of progressively weaker shocks is always more efficient than achieving the same decrease in Mach number across a single shock.

4.8 | PRESSURE-DEFLECTION DIAGRAMS

The shock wave reflection discussed in Sec. 4.6 is just one example of a wave interaction process—in the above case it was an interaction between the wave and a solid boundary. There are other types of interaction processes involving shock and expansion waves, and solid and free boundaries. To understand some of these interactions, it is convenient to introduce the pressure-deflection diagram, which is nothing more than the locus of all possible static pressures behind an oblique shock wave as a function of deflection angle for given upstream conditions. Consider Fig. 4.21, which at the top shows oblique shock waves of two different orientations. The top left shows a left-running wave—so called because, when standing at a point on the wave and looking downstream, you see the wave running off toward your left. The flow deflection angle θ_2 is upward, and is considered positive. In contrast, the top right shows a right-running wave; since an oblique shock wave always deflects the flow toward the wave, the deflection angle θ_2' is downward and is considered negative. The static pressure ahead of the wave, where $\theta = 0$, is p_1; the static pressure behind the left-running wave, where $\theta = \theta_2$, is p_2. These two conditions are illustrated by points 1 and 2, respectively, on a plot of pressure versus deflection at the bottom of

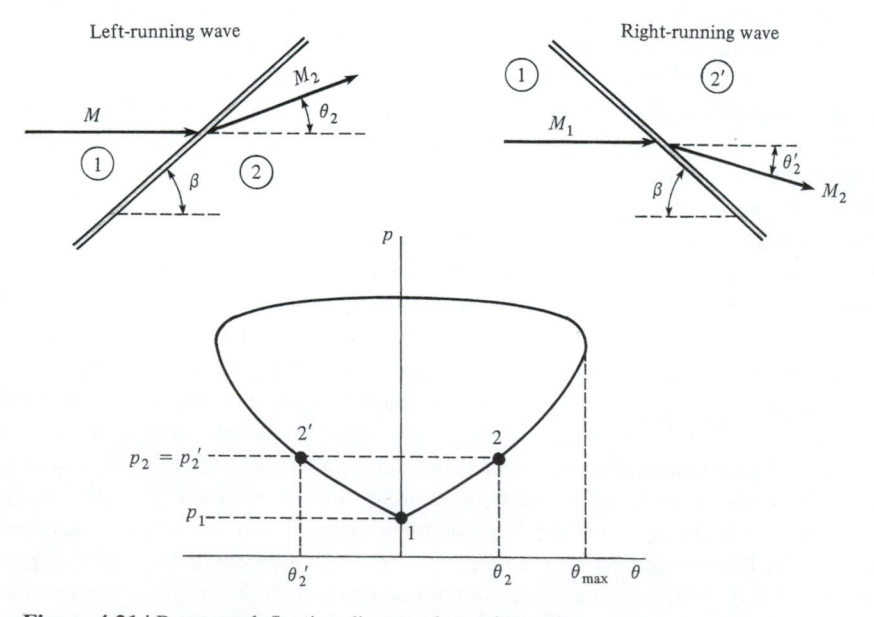

Figure 4.21 | Pressure-deflection diagram for a given M_1.

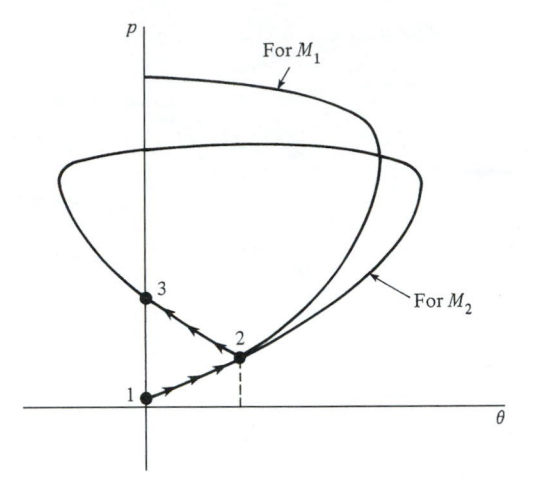

Figure 4.22 | The reflected shock process on a pressure-deflection diagram.

Fig. 4.21. For the right-running wave, if θ_2 and θ_2' are equal in absolute magnitude (but different in sign), the pressure in region $2'$ will also be p_2. This condition is given by point $2'$ on Fig. 4.21. When θ ranges over all possible values $|\theta| < \theta_{max}$ for an oblique shock solution, the locus of all possible pressures (for the given M_1 and p_1) is given by the pressure-deflection diagram, sketched in Fig. 4.21. The right-hand lobe of this figure corresponds to positive θ, the left-hand lobe to negative θ.

The shock reflection process of Sec. 4.6 is sketched in terms of pressure-deflection ($p\theta$) diagrams in Fig. 4.22. A $p\theta$ diagram is first drawn for M_1, where point 1 corresponds to the pressure in region 1 of Fig. 4.18. Conditions in region 2 are given by point 2 on the $p\theta$ diagram. At this point, a new pressure-deflection diagram is drawn for a free-stream Mach number equal to M_2. The vertex of this $p\theta$ diagram is at point 2 because the "free stream" of region 2 is already bent upward by the angle θ. Since the flow in region 3 must have $\theta = 0$, then we move along the left-hand lobe of this second $p\theta$ diagram until $\theta = 0$. This defines point 3 in Fig. 4.22, which yields the conditions behind the reflected shock. Hence, in Fig. 4.22, we move from point 1 to point 2 across the incident shock, and then from point 2 to point 3 across the reflected shock.

4.9 | INTERSECTION OF SHOCKS OF OPPOSITE FAMILIES

Consider the intersection of left- and right-running shocks as sketched in Fig. 4.23. The left- and right-running shocks are labeled A and B, respectively. Both are incident shocks, and correspond to deflections θ_2 and θ_3, respectively. These shocks continue as the *refracted* shocks C and D downstream of the intersection at point E. Assume $\theta_2 > \theta_3$. Then shock A is stronger than B, and a streamline going through the shock system A and C experiences a different entropy change than the streamline

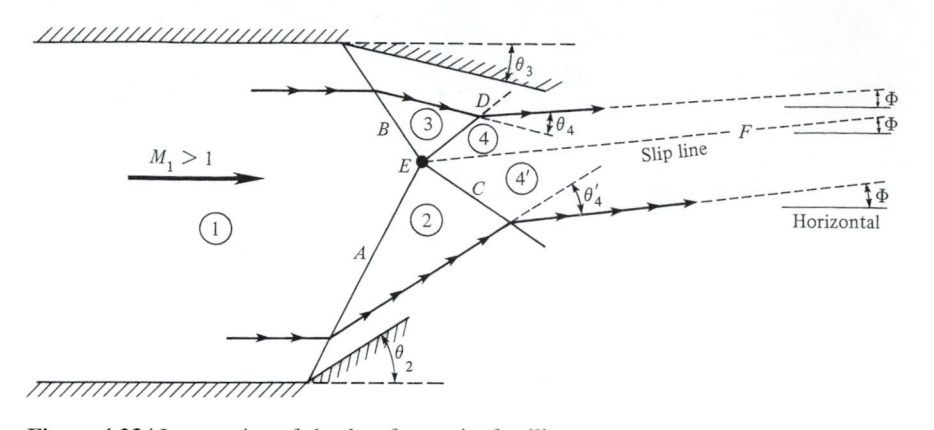

Figure 4.23 | Intersection of shocks of opposite families.

going through the shock system B and D. Therefore, the entropy in regions 4 and 4′ is different. Consequently, the dividing streamline EF between these two regions is a line across which the entropy changes discontinuously. Such a line is defined as a *slip line*. However, on a physical basis, these conditions must hold across the slip line in Fig. 4.23:

1. The pressure must be the same, $p_4 = p_{4'}$. Otherwise, the slip line would be curved, inconsistent with the geometry of Fig. 4.23.

2. The velocities in regions 4 and 4′ must be in the *same direction,* although they in general differ in magnitude. If the velocities were in different directions, there would be the chance of a complete void in the flowfield in the vicinity of the slip line—an untenable physical situation.

These two conditions, along with the known properties in region 1 as well as the known θ_2 and θ_3, completely determine the shock interaction in Fig. 4.23. Also, note that the temperature and density, as well as the entropy and velocity magnitude, are different in regions 4 and 4′.

Pressure-deflection diagrams are particularly useful in visualizing the solution of this shock interaction process. The $p\theta$ diagram corresponding to M_1 is drawn as the solid curve in Fig. 4.24. Point 1 denotes conditions in region 1, ahead of the shocks. In region 2 of Fig. 4.23, the flow is deflected through the angle θ_2. Therefore, point 2 on the $p\theta$ diagram is located by moving along the curve until $\theta = \theta_2$. At point 2, a new $p\theta$ diagram corresponding to M_2 is drawn, as shown by the dashed curve to the right in Fig. 4.24. Note that the pressure in region 4′ must lie on this curve. Similarly, point 3 is located by moving along the solid curve until θ_3 is reached; remember that this deflection is downward, hence we must move in the negative θ direction. Point 3 corresponds to region 3 in Fig. 4.23. At point 3, a new $p\theta$ diagram corresponding to M_3 is drawn, as shown by the dashed curve to the left in Fig. 4.24. The pressure in region 4 must lie on this curve. Because $p_4 = p_{4'}$, the point corresponding to regions 4 and 4′ in Fig. 4.24 is the intersection of the two dashed $p\theta$ diagrams. This point defines the flow direction (hence slip line direction) in regions 4 and 4′, namely the angle Φ in Figs. 4.23 and 4.24. In turn, the flow deflections across the refracted shocks D and C

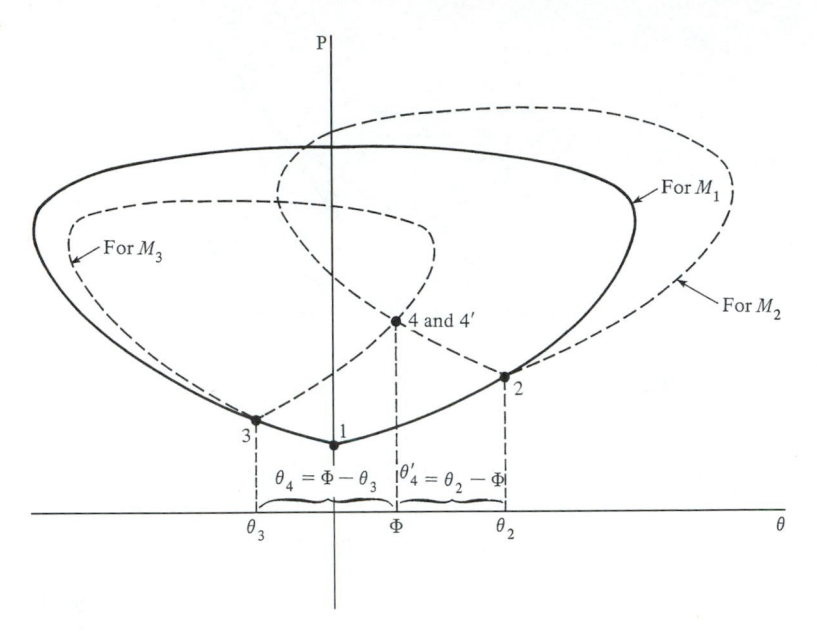

Figure 4.24 | Pressure-deflection diagrams for the shock intersection picture given in Fig. 4.23.

are determined: $\theta_4 = \Phi - \theta_3$ and $\theta_{4'} = \theta_2 - \Phi$. With these deflections, and with the Mach numbers in regions 3 and 2, respectively, the strengths of the refracted shocks D and C are now determined.

Note from Fig. 4.23 that, if $\theta_2 = \theta_3$, the intersecting shocks would be of equal strength, the flow pattern would be completely symmetrical, and there would be no slip line.

4.10 | INTERSECTION OF SHOCKS OF THE SAME FAMILY

Consider the compression corner sketched in Fig. 4.25, where the supersonic flow in region 1 is deflected through an angle θ, with the consequent oblique shock wave emanating from point B. Now consider a Mach wave generated at point A ahead of the shock. Will this Mach wave intersect the shock, or will it simply diverge, i.e., is μ_1 greater than or less than β? To find out, consider Eq. (4.7), which written in terms of velocities is

$$u_1 = V_1 \sin \beta$$

Hence,
$$\sin \beta = \frac{u_1}{V_1} \tag{4.23}$$

In addition, from Eq. (4.1),

$$\sin \mu_1 = \frac{a_1}{V_1} \tag{4.24}$$

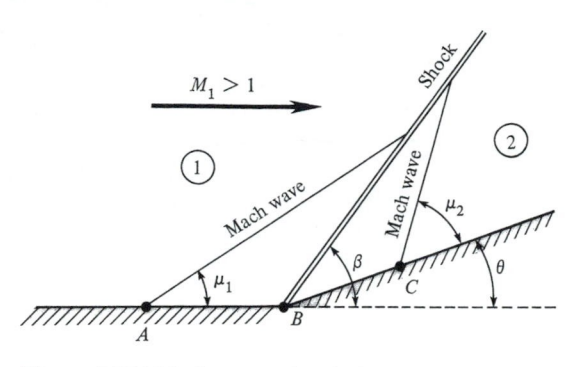

Figure 4.25 | Mach waves ahead of and behind a shock wave.

We have already proven that, for a shock to exist, the normal component of the flow velocity ahead of the shock wave must be supersonic. Thus, $u_1 > a_1$; consequently, from Eqs. (4.23) and (4.24), $\beta > \mu_1$. Therefore, referring to Fig. 4.25, the Mach wave at A must intersect the shock wave, as shown.

Now consider a Mach wave generated at point C behind the shock. From Eq. (4.12)

$$u_2 = V_2 \sin(\beta - \theta)$$

Hence,

$$\sin(\beta - \theta) = \frac{u_2}{V_2} \tag{4.25}$$

In addition, from Eq. (4.1),

$$\sin \mu_2 = \frac{a_2}{V_2} \tag{4.26}$$

We have already proven that the normal component of the flow velocity behind a shock wave is subsonic. Thus, $u_2 < a_2$; consequently, from Eqs. (4.25) and (4.26), $\beta - \theta < \mu_2$. Therefore, referring to Fig. 4.25, the Mach wave at C must intersect the shock wave, as shown.

It is now not difficult to extrapolate to the case of two left-running oblique shock waves generated at corners A and B in Fig. 4.26. Because shock wave BC must be inclined at a steeper angle than a Mach wave in region 2, and we have already shown that a left-running Mach wave will intersect a left-running shock, then it is obvious that shock waves AC and BC will intersect as shown in Fig. 4.26. Above the point of intersection C, a single shock CD will propagate.

Now consider a streamline passing through regions 1, 2, and 3 as sketched in Fig. 4.26. The pressure and flow direction in region 3 are p_3 and θ_3, respectively, and are determined by the upstream conditions in region 1, as well as the deflection angles θ_2 and θ_3. Properties in region 3 are processed by the dual shocks AC and BC. On the other hand, consider a streamline passing through regions 1 and 5. The pressure and flow direction in region 5 are p_5 and θ_5, respectively. Properties in region 5 are processed by the single shock CD. Therefore, the entropy change across this

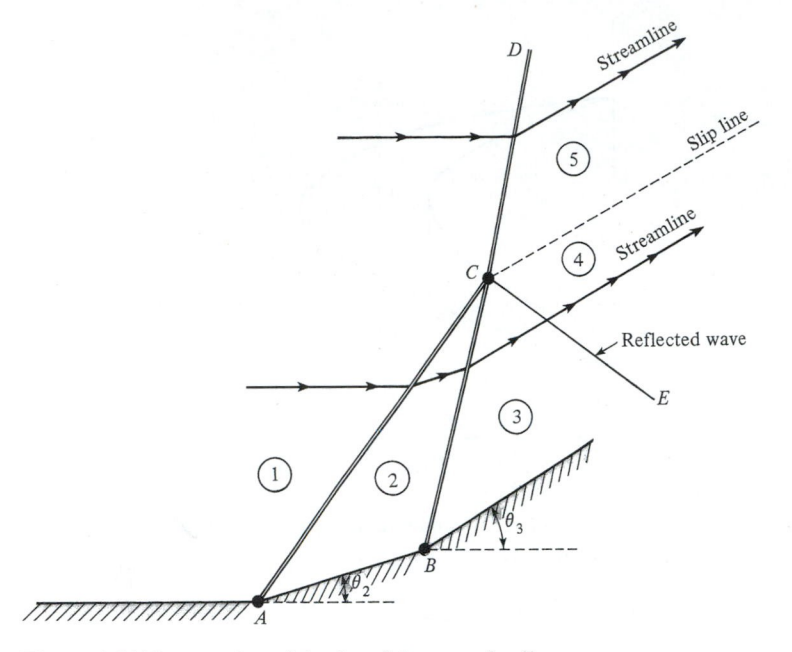

Figure 4.26 | Intersection of shocks of the same family.

single shock will be different than across the two shocks, and hence a slip line must exist downstream, originating at the intersection point C. As discussed in Sec. 4.9, the pressures and flow directions across the slip line must be the same. If no other wave existed in the system, this would require $p_5 = p_3$ and $\theta_5 = \theta_3$ simultaneously. However, it is generally not possible to find a single shock CD that will give simultaneously the same pressure and flow deflection as two intermediate shocks AC and BC, with both systems starting from the same upstream conditions in region 1. Therefore, nature removes this problem by creating a weak reflected wave from the intersection point C. Depending on the upstream conditions and θ_1 and θ_2, this reflected wave CE may be a weak shock or expansion wave. Its purpose is to process the flow in region 4 such that $p_4 = p_5$ and $\theta_4 = \theta_5$ simultaneously, thus satisfying the necessary physical conditions across a slip line. The flowfield can be solved numerically by iteratively adjusting waves CD and CE such that the above conditions between regions 4 and 5 are obtained.

4.11 | MACH REFLECTION

Return again to the shock wave reflection from a solid wall as discussed in Sec. 4.6 and as sketched in Fig. 4.18. The governing condition is that the flow must be deflected through the angle θ from regions 2 to 3 by the reflected shock so that the streamlines are parallel to the upper wall. In the discussion of Sec. 4.6, this value of θ was assumed to be less than θ_{max} for M_2, and hence a solution was allowed for a straight, attached reflected shock. Consider the θ-β-M curves for both M_1 and M_2, as

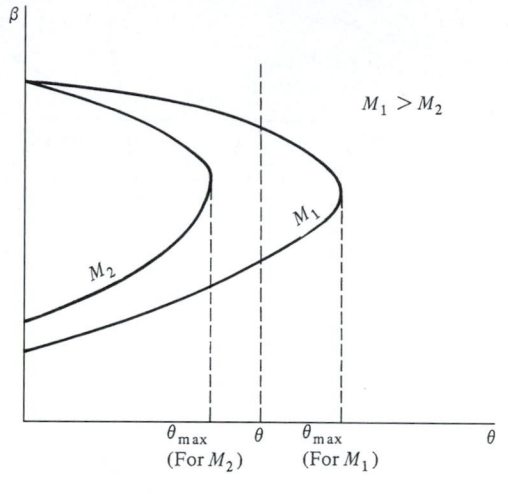

Figure 4.27 | Maximum deflection angle for two different Mach numbers.

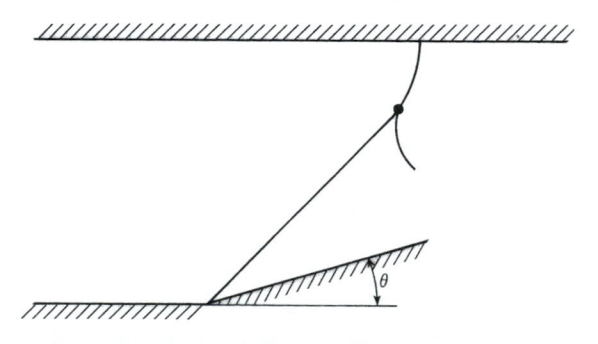

Figure 4.28 | Mach reflection.

sketched in Fig. 4.27. In Sec. 4.6, it was assumed that θ was to the left of θ_{max} for M_2 in Fig. 4.27. However, what happens when $(\theta_{max}$ for $M_2) < \theta < (\theta_{max}$ for $M_1)$? This situation is illustrated in Fig. 4.27. For the incident shock with an upstream Mach number of M_1, $\theta < \theta_{max}$, and hence the incident shock is an allowable straight oblique shock solution. This straight incident shock is sketched in Fig. 4.28. On the other hand, when the flow in region 2 at Mach number M_2 wants to again deflect through the angle θ via the reflected shock, it finds that $\theta > \theta_{max}$ for M_2, and a regular reflection is not possible. Instead, a normal shock is formed at the upper wall to allow the streamlines to continue parallel to the wall. Away from the wall, this normal shock transits into a curved shock which intersects the incident shock, with a curved reflected shock propagating downstream. This shock pattern is sketched in Fig. 4.28 and is labeled a *Mach reflection* in contrast to the regular reflection discussed in Sec. 4.6. The Mach reflection is characterized by large regions of subsonic flow behind the normal or near normal shocks, and its analysis must be carried out by the more sophisticated numerical techniques to be discussed in Chaps. 11 and 12.

4.12 | DETACHED SHOCK WAVE IN FRONT OF A BLUNT BODY

Consider the supersonic flow over a blunt-nosed body as illustrated in Fig. 4.29. A strong curved bow shock wave is created in front of this body, with the shock detached from the nose by a distance δ. At point a, the upstream flow is normal to the wave; hence point a corresponds to a normal shock wave. Away from the centerline, the shock wave becomes curved and weaker, eventually evolving into a Mach wave at large distances from the body (illustrated by point e in Fig. 4.29).

Moreover, between points a and e, the curved shock goes through all possible conditions allowed for oblique shocks for an upstream Mach number of M_1. To see this more clearly, consider the θ-β-M_1 curve sketched in Fig. 4.30. At point a, a normal shock exists. Slightly above the centerline at point b in Fig. 4.29, the shock is oblique but pertains to the strong-shock solution in Fig. 4.30. Further along the shock, point c is the dividing point between strong and weak solutions; the streamline through point c experiences the maximum deflection, $\theta_{\max}$. Slightly above point c in Fig. 4.29, at point c', the flow becomes sonic behind the shock. From points a to c', the flow behind the shock is subsonic. Above point c' the flow is supersonic behind the shock. Hence, the flowfield between the blunt body and its curved bow shock is

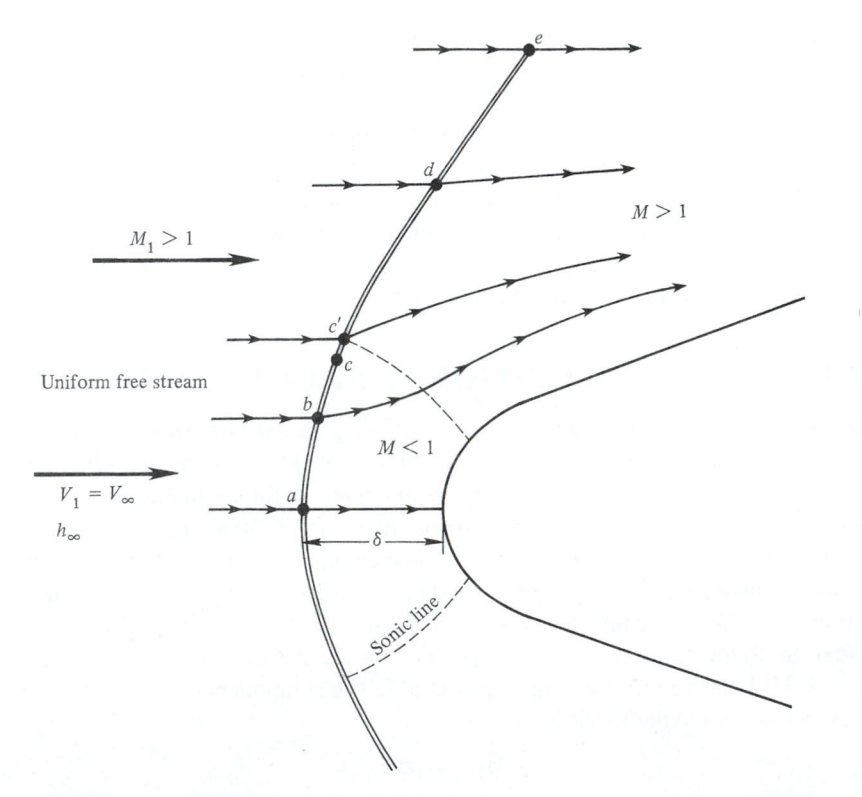

Figure 4.29 | Flow over a supersonic blunt body.

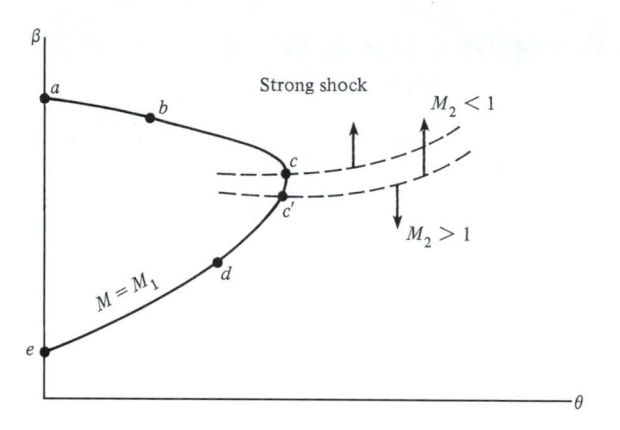

Figure 4.30 | θ-β-M diagram for the sketch in Fig. 4.23.

a mixed subsonic–supersonic flow, and the imaginary dividing curve between these two regions (where $M = 1$) is denoted as the *sonic line,* as shown in Fig. 4.29.

The shape of the detached shock wave, its detachment distance δ, and the complete flowfield (with curved streamlines) between the shock and the body depend on M_1 and the size and shape of the body. The solution of this flowfield is not trivial. Indeed, the supersonic blunt body problem was a major focus for supersonic aerodynamicists during the 1950s and 1960s spurred by the need to understand the high-speed flow over blunt-nosed missiles and reentry bodies. The situation in 1957 was precisely described in the classic text by Liepmann and Roshko (Ref. 9), where, in their discussion of blunt body flows, they categorically state that "the shock shape and detachment distance cannot, at present, be theoretically predicted." Indeed, it was not until a decade later that truly sufficient numerical techniques became available for satisfactory engineering solutions of supersonic blunt body flows. These modern techniques are discussed at length in Chap. 12.

4.13 | THREE-DIMENSIONAL SHOCK WAVES

In treating oblique shock waves in this chapter, two-dimensional (plane) flow has been assumed. However, many practical supersonic flow problems are three-dimensional, with correspondingly curved shock waves extending in three-dimensional space. The shock wave around a supersonic axisymmetric blunt body at angle of attack is one such example, as sketched in Fig. 4.31. For such three-dimensional shock waves, the two-dimensional theory of the present chapter is still appropriate for calculating properties *immediately* behind the shock surface at some local point. For example, consider an elemental area dS around point A on the curved shock surface shown in Fig. 4.31. Let $\mathbf{n}$ be the unit normal vector at A. The component of the upstream Mach number normal to the shock is then

$$M_{n_1} = (M_1\mathbf{i}) \cdot \mathbf{n} \tag{4.27}$$

With the Mach number component normal to the three-dimensional shock wave obtained from Eq. (4.27), values of p_2, ρ_2, T_2, h_2, and M_{n_2} can be calculated

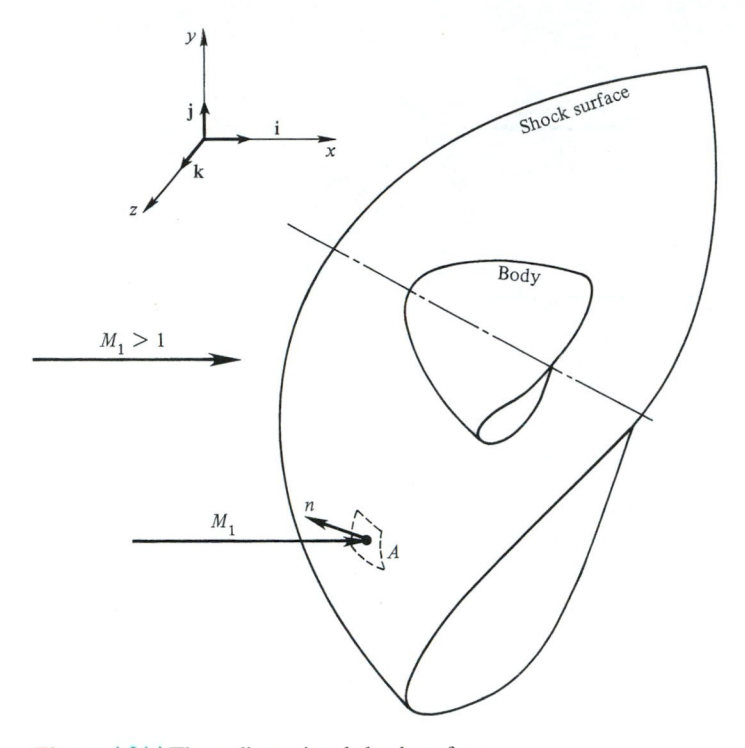

Figure 4.31 | Three-dimensional shock surface.

immediately behind the shock at point A from the shock wave relations given in Eqs. (4.8) through (4.11). We again emphasize that these results hold just immediately behind the shock surface at the local point A. Further downstream, the flowfield experiences a complex nonuniform variation which must be analyzed by appropriate three-dimensional techniques beyond the scope of this chapter. Such matters are discussed in Chap. 13.

4.14 | PRANDTL–MEYER EXPANSION WAVES

We have now finished our discussion of oblique shock waves as itemized in the left column of the roadmap in Fig. 4.3. We now move to the right side of the roadmap, which deals with expansion waves. When a supersonic flow is turned away from itself as discussed in Sec. 4.1, an expansion wave is formed as sketched in Fig. 4.4*b*. This is directly opposite to the situation when the flow is turned into itself, with the consequent shock wave as sketched in Fig. 4.4*a*. Expansion waves are the antithesis of shock waves. To appreciate this more fully, some qualitative aspects of flow through an expansion wave are itemized as follows (referring to Fig. 4.4*b*):

1. $M_2 > M_1$. An expansion corner is a means to *increase* the flow Mach number.
2. $p_2/p_1 < 1$, $\rho_2/\rho_1 < 1$, $T_2/T_1 < 1$. The pressure, density, and temperature *decrease* through an expansion wave.

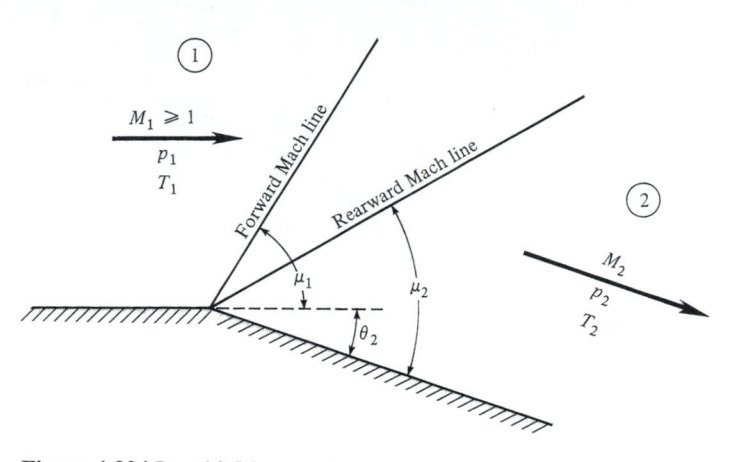

Figure 4.32 | Prandtl–Meyer expansion.

3. The expansion fan itself is a *continuous* expansion region, composed of an infinite number of Mach waves, bounded upstream by μ_1 and downstream by μ_2 (see Fig. 4.32), where $\mu_1 = \arcsin(1/M_1)$ and $\mu_2 = \arcsin(1/M_2)$.

4. Streamlines through an expansion wave are smooth curved lines.

5. Since the expansion takes place through a continuous succession of Mach waves, and $ds = 0$ for each Mach wave, the expansion is *isentropic*.

An expansion wave emanating from a sharp convex corner such as sketched in Figs. 4.4*b* and 4.32 is called a *centered* expansion fan. Moreover, because Prandtl in 1907, followed by Meyer in 1908, first worked out the theory for such a supersonic flow, it is denoted as a *Prandtl–Meyer expansion wave*.

The quantitative problem of a Prandtl–Meyer expansion wave can be stated as follows (referring to Fig. 4.32): For a given M_1, p_1, T_1, and θ_2, calculate M_2, p_2, and T_2. The analysis can be started by considering the infinitesimal changes across a very weak wave (essentially a Mach wave) produced by an infinitesimally small flow deflection, $d\theta$, as illustrated in Fig. 4.33. From the law of sines,

$$\frac{V + dV}{V} = \frac{\sin(\pi/2 + \mu)}{\sin(\pi/2 - \mu - d\theta)} \tag{4.28}$$

However, from trigonometric identities,

$$\sin\left(\frac{\pi}{2} + \mu\right) = \sin\left(\frac{\pi}{2} - \mu\right) = \cos\mu \tag{4.29}$$

$$\sin\left(\frac{\pi}{2} - \mu - d\theta\right) = \cos(\mu + d\theta) = \cos\mu\cos d\theta - \sin\mu\sin d\theta \tag{4.30}$$

Substitute Eqs. (4.29) and (4.30) into (4.28):

$$1 + \frac{dV}{V} = \frac{\cos\mu}{\cos\mu\cos d\theta - \sin\mu\sin d\theta} \tag{4.31}$$

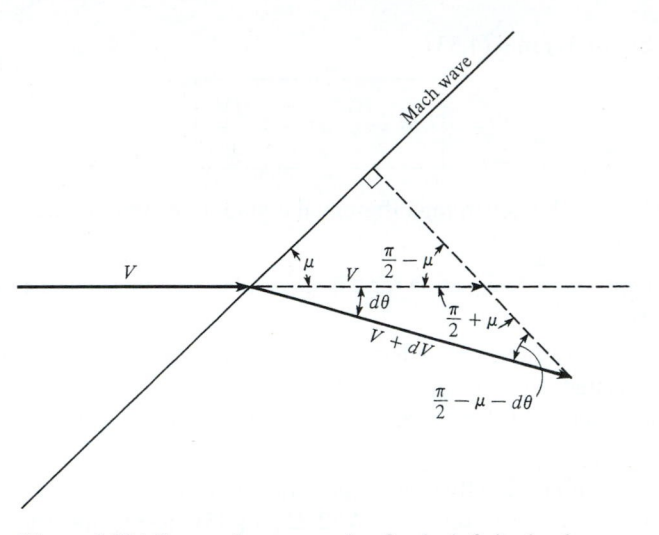

Figure 4.33 | Geometric construction for the infinitesimal changes across a Mach wave; for use in the derivation of the Prandtl–Meyer function. Note that the change in velocity across the wave is normal to the wave.

For small $d\theta$, we can make the small-angle assumptions $\sin d\theta \approx d\theta$ and $\cos d\theta \approx 1$. Then, Eq. (4.31) becomes

$$1 + \frac{dV}{V} = \frac{\cos\mu}{\cos\mu - d\theta\sin\mu} = \frac{1}{1 - d\theta\tan\mu} \tag{4.32}$$

Recalling the series expansion (for $x < 1$),

$$\frac{1}{1-x} = 1 + x + x^2 + x^3 + \cdots$$

Eq. (4.32) can be expanded as (ignoring terms of second and higher order)

$$1 + \frac{dV}{V} = 1 + d\theta\tan\mu + \cdots \tag{4.32a}$$

Thus, from Eq. (4.32a),

$$d\theta = \frac{dV/V}{\tan\mu} \tag{4.33}$$

However, from Eq. (4.1),

$$\mu = \sin^{-1}\frac{1}{M}$$

which can be written as

$$\tan\mu = \frac{1}{\sqrt{M^2 - 1}} \tag{4.34}$$

Substitute Eq. (4.34) into (4.33)

$$d\theta = \sqrt{M^2 - 1}\,\frac{dV}{V} \tag{4.35}$$

Equation (4.35) is the governing differential equation for Prandtl–Meyer flow. Note these aspects of it:

1. It is an approximate equation for a finite $d\theta$, but becomes a true equality as $d\theta \to 0$.

2. It was derived strictly on the basis of geometry, where the only real physics is that associated with the definition of a Mach wave. Hence, it is a general relation which holds for perfect gases, chemically reacting gases, and real gases.

3. It treats an infinitesimally small expansion angle, $d\theta$. To analyze the entire Prandtl–Meyer expansion in Fig. 4.32, Eq. (4.35) must be integrated over the complete angle θ_2. Integrating Eq. (4.35) from regions 1 to 2,

$$\int_{\theta_1}^{\theta_2} d\theta = \int_{M_1}^{M_2} \sqrt{M^2 - 1}\,\frac{dV}{V} \tag{4.36}$$

The integral on the right-hand side can be evaluated after dV/V is obtained in terms of M, as follows. From the definition of Mach number,

$$V = Ma$$

Hence, $$\ln V = \ln M + \ln a \tag{4.37}$$

Differentiating Eq. (4.37),

$$\frac{dV}{V} = \frac{dM}{M} + \frac{da}{a} \tag{4.38}$$

Specializing to a calorically perfect gas, the adiabatic energy equation can be written from Eq. (3.28) as

$$\left(\frac{a_o}{a}\right)^2 = \frac{T_o}{T} = 1 + \frac{\gamma - 1}{2}M^2$$

or, solving for a,

$$a = a_o\left(1 + \frac{\gamma - 1}{2}M^2\right)^{-1/2} \tag{4.39}$$

Differentiating Eq. (4.39),

$$\frac{da}{a} = -\left(\frac{\gamma - 1}{2}\right)M\left(1 + \frac{\gamma - 1}{2}M^2\right)^{-1}dM \tag{4.40}$$

Substituting Eq. (4.40) into (4.38), we obtain

$$\frac{dV}{V} = \frac{1}{1 + \dfrac{\gamma - 1}{2}M^2} \frac{dM}{M} \tag{4.41}$$

Equation (4.41) is the desired relation for dV/V in terms of M; substitute it into Eq. (4.36):

$$\int_{\theta_1}^{\theta_2} d\theta = \theta_2 - 0 = \int_{M_1}^{M_2} \frac{\sqrt{M^2 - 1}}{1 + \dfrac{\gamma - 1}{2}M^2} \frac{dM}{M} \tag{4.42}$$

In Eq. (4.42), the integral

$$\nu(M) = \int \frac{\sqrt{M^2 - 1}}{1 + \dfrac{\gamma - 1}{2}M^2} \frac{dM}{M} \tag{4.43}$$

is called the *Prandtl–Meyer function,* and is given the symbol ν. Performing the integration, Eq. (4.43) becomes

$$\boxed{\nu(M) = \sqrt{\frac{\gamma + 1}{\gamma - 1}} \tan^{-1} \sqrt{\frac{\gamma - 1}{\gamma + 1}(M^2 - 1)} - \tan^{-1} \sqrt{M^2 - 1}} \tag{4.44}$$

The constant of integration that would ordinarily appear in Eq. (4.44) is not important, because it drops out when Eq. (4.44) is substituted into (4.42). For convenience, it is chosen as zero such that $\nu(M) = 0$ when $M = 1$. Finally, we can now write Eq. (4.42), combined with (4.43), as

$$\boxed{\theta_2 = \nu(M_2) - \nu(M_1)} \tag{4.45}$$

where $\nu(M)$ is given by Eq. (4.44) for a calorically perfect gas. The Prandtl–Meyer function [Eq. (4.44)] is tabulated as a function of M in Table A.5 for $\gamma = 1.4$, along with values of the Mach angle μ, for convenience.

Returning again to Fig. 4.32, Eqs. (4.45) and (4.44) allow the calculation of a Prandtl–Meyer expansion wave, as follows:

1. Obtain $\nu(M_1)$ from Table A.5 for the given M_1.
2. Calculate $\nu(M_2)$ from Eq. (4.45) using the given θ_2 and $\nu(M_1)$ obtained in step 1.
3. Obtain M_2 from Table A.5 corresponding to the value of $\nu(M_2)$ from step 2.
4. Recognizing that the expansion is isentropic, and hence that T_o and p_o are constant through the wave, Eqs. (3.28) and (3.30) yield

$$\frac{T_1}{T_2} = \frac{1 + \dfrac{\gamma - 1}{2}M_2^2}{1 + \dfrac{\gamma - 1}{2}M_1^2}$$

and
$$\frac{p_1}{p_2} = \left[\frac{1 + \frac{\gamma - 1}{2}M_2^2}{1 + \frac{\gamma - 1}{2}M_1^2}\right]^{\gamma/(\gamma-1)}$$

EXAMPLE 4.13

A uniform supersonic stream with $M_1 = 1.5$, $p_1 = 1700\,\text{lb/ft}^2$, and $T_1 = 460°\text{R}$ encounters an expansion corner (see Fig. 4.32) which deflects the stream by an angle $\theta_2 = 20°$. Calculate M_2, p_2, T_2, p_{o_2}, T_{o_2}, and the angles the forward and rearward Mach lines make with respect to the upstream flow direction.

■ Solution

From Table A.5, for $M_1 = 1.5$: $\nu_1 = 11.91°$ and $\mu_1 = 41.81°$. So

$$\nu_2 = \nu_1 + \theta_1 = 11.91 + 20 = 31.91°$$

From Table A.5, for $\nu_2 = 31.91°$:

$$\boxed{M_2 = 2.207} \quad \text{and} \quad \mu_2 = 26.95°$$

From Table A.1, for $M_1 = 1.5$:

$$\frac{p_{o_1}}{p_1} = 3.671 \quad \text{and} \quad \frac{T_{o_1}}{T_1} = 1.45$$

From Table A.1, for $M_2 = 2.207$:

$$\frac{p_{o_2}}{p_2} = 10.81 \quad \text{and} \quad \frac{T_{o_2}}{T_2} = 1.974$$

The flow through an expansion wave is isentropic; hence $p_{o_2} = p_{o_1}$ and $T_{o_2} = T_{o_1}$. Thus,

$$p_2 = \frac{p_2}{p_{o_2}}\frac{p_{o_2}}{p_{o_1}}\frac{p_{o_1}}{p_1}p_1 = (10.81)^{-1}(1)(3.671)(1700) = \boxed{577.3\,\text{lb/ft}^2}$$

$$T_2 = \frac{T_2}{T_{o_2}}\frac{T_{o_2}}{T_{o_1}}\frac{T_{o_1}}{T_1}T_1 = (1.975)^{-1}(1)(1.45)(460) = \boxed{337.9°\text{R}}$$

$$p_{o_2} = p_{o_1} = \frac{p_{o_1}}{p_1}p_1 = (3.671)(1700) = \boxed{6241\,\text{lb/ft}^2}$$

$$T_{o_2} = T_{o_1} = \frac{T_{o_1}}{T_1}T_1 = (1.45)(460) = \boxed{667°\text{R}}$$

Returning to Fig. 4.32:

$$\text{Angle of forward Mach line} = \mu_1 = \boxed{41.81°}$$

$$\text{Angle of rearward Mach line} = \mu_2 - \theta_2 = 26.95 - 20 = \boxed{6.95°}$$

EXAMPLE 4.14

Consider the arrangement shown in Fig. 4.34. A 15° half-angle diamond wedge airfoil is in a supersonic flow at zero angle of attack. A Pitot tube is inserted into the flow at the location shown in Fig. 4.34. The pressure measured by the Pitot tube is 2.596 atm. At point a on the backface, the pressure is 0.1 atm. Calculate the free-stream Mach number M_1.

■ Solution

There will be a normal shock wave in front of the face of the Pitot tube immersed in region 3 in Fig. 4.34. Let the region immediately behind this normal shock be denoted as region 4. The Pitot tube senses the total pressure in region 4, i.e., p_{o_4}. The pressure at point a is the static pressure in region 3. Thus

$$\frac{p_{o_4}}{p_3} = \frac{2.596}{0.1} = 25.96$$

From Table A.2, for $p_{o_4}/p_3 = 25.96$: $M_3 = 4.45$. From Table A.5, for $M_3 = 4.45$, we have $v_3 = 71.27°$. From Eq. (4.45)

$$v_2 = v_3 - \theta = 71.27 - 30 = 41.27°$$

From Table A.5, for $v_2 = 41.27°$: $M_2 = 2.6$. In region 2, we have

$$M_{n_2} = M_2 \sin(\beta - \theta) = 2.6 \sin(\beta - 15°) \tag{E.1}$$

In this equation, both M_{n_2} and β are unknown. We must solve by trial and error, as follows.

Assume $M_1 = 4$. Then $\beta = 27°$, $M_{n_1} = M_1 \sin\beta = 4 \sin 27° = 1.816$. Hence, from Table A.2, $M_{n_2} = 0.612$. Putting these results into Eq. (E.1) above,

$$0.612 \overset{?}{=} 2.6 \sin 12° = 0.54$$

This does *not* check.

Assume $M_1 = 4.5$. Then $\beta = 25.5°$, $M_{n_1} = 4.5 \sin 25.5° = 1.937$. Hence, from Table A.2, $M_{n_2} = 0.588$. Putting these results into Eq. (E.1),

$$0.588 \overset{?}{=} 2.6 \sin 10.5° = 0.47$$

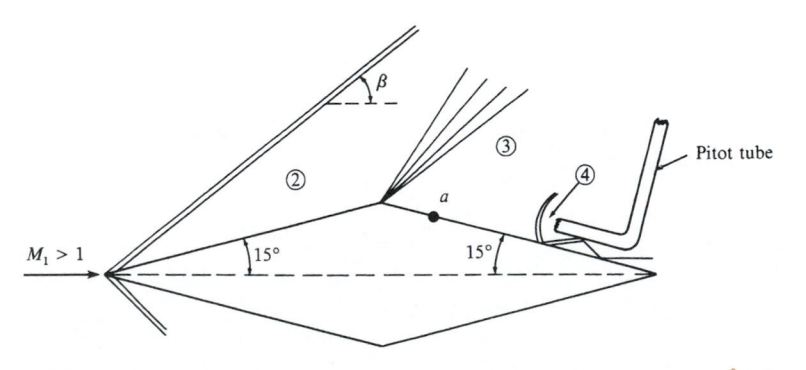

Figure 4.34 | Geometry for Example 4.14.

This does *not* check. We are going in the wrong direction.

  Assume $M_1 = 3.5$. Then $\beta = 29.2°$, $M_{n_1} = 3.5 \sin 29.2° = 1.71$. Hence, from Table A.2, $M_{n_2} = 0.638$. Putting these results into Eq. (E.1),

$$0.638 \stackrel{?}{=} 2.6 \sin 14.2° = 0.638$$

This *checks*. Thus

$$\boxed{M_1 = 3.5}$$

4.15 | SHOCK-EXPANSION THEORY

In this section we move to the bottom of our roadmap in Fig. 4.3 and discuss shock-expansion theory, which is a logical and natural combination of the items in both the left and right columns of the roadmap. The shock and expansion waves discussed in this chapter allow the exact calculation of the aerodynamic force on many types of two-dimensional supersonic airfoils made up of straight-line segments. For example, consider the symmetrical diamond-shaped airfoil at zero angle of attack in Fig. 4.35. The supersonic flow is first compressed and deflected through the angle ε by an oblique shock wave at the leading edge. At midchord, the flow is expanded through an angle 2ε by the expansion wave. At the trailing edge, the flow is again deflected through the angle ε by another oblique shock; this deflection is necessary to make the flow downstream of the airfoil parallel to the free-stream direction due to symmetry conditions. Hence, the surface pressure on segments a and c are found from oblique shock theory, and on segments b and d from Prandtl–Meyer expansion theory.

  At zero angle of attack, the only aerodynamic force on the diamond airfoil will be drag; the lift is zero because the pressure distributions on the top and bottom

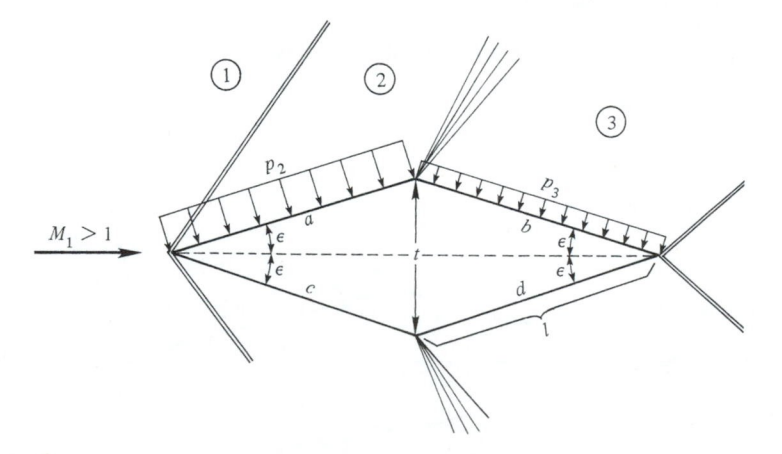

Figure 4.35 | Symmetrical diamond-wedge airfoil.

surfaces are the same. From Eq. (1.47), the pressure drag is

$$D = x \text{ component of } \left[-\oiint p \, d\mathbf{S} \right]$$

In terms of scalar quantities, and referring to Fig. 4.35, the surface integral yields for the drag per unit span

$$D = 2(p_2 l \sin \varepsilon - p_3 l \sin \varepsilon) = 2(p_2 - p_3)\frac{t}{2}$$

Hence,

$$D = (p_2 - p_3)t \tag{4.46}$$

It is a well-known aerodynamic result that two-dimensional inviscid flow over a wing of infinite span at subsonic velocity gives zero drag—a theoretical result given the name *d'Alembert's paradox*. (The paradox is removed by accounting for the effects of friction). In contrast, for supersonic inviscid flow over an infinite wing, Eq. (4.46) clearly demonstrates that the drag per unit span is *finite*. This new source of drag encountered when the flow is supersonic is called *wave drag,* and is inherently related to the loss of total pressure and increase of entropy across the oblique shock waves created by the airfoil.

EXAMPLE 4.15

Consider an infinitely thin flat plate at a 5° angle of attack in a Mach 2.6 free stream. Calculate the lift and drag coefficients.

■ Solution

From Table A.5, for $M_1 = 2.6$: $v_1 = 41.41°$. Thus, from Eq. (4.45)

$$v_2 = v_1 + \alpha = 41.41 + 5 = 46.41°$$

From Table A.5, for $v_2 = 46.41°$: $M_2 = 2.85$. From Table A.1, for $M_1 = 2.6$: $p_{o_1}/p_1 = 19.95$. From Table A.1, for $M_2 = 2.85$: $p_{o_2}/p_2 = 29.29$. Hence

$$\frac{p_2}{p_1} = \frac{p_2}{p_{o_2}} \frac{p_{o_2}}{p_{o_1}} \frac{p_{o_1}}{p_1} = \frac{1}{29.29}(1)(19.95) = 0.681$$

From the θ-β-M diagram, for $M_1 = 2.6$ and $\theta = \alpha = 5°$: $\beta = 26.5°$. Thus

$$M_{n_1} = M_1 \sin \beta = 2.6 \sin 26.5° = 1.16$$

From Table A.2, for $M_{n_1} = 1.16$: $p_3/p_1 = 1.403$. From Fig. 4.36, the lift per unit span L' is

$$L' = (p_3 - p_2)c \cos \alpha$$

The drag per unit span D' is

$$D' = (p_3 - p_2)c \sin \alpha$$

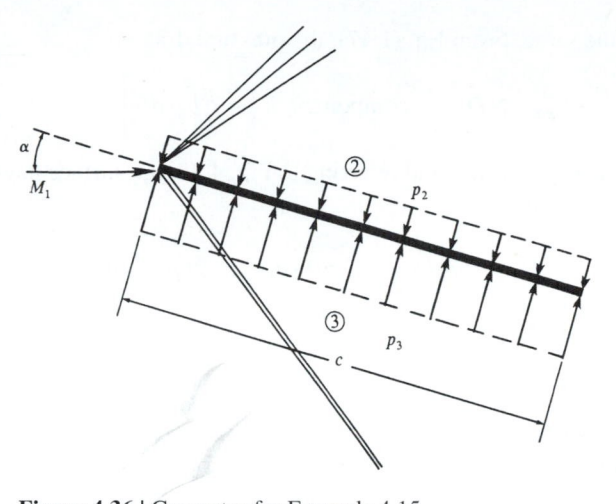

Figure 4.36 | Geometry for Example 4.15.

Recalling that $q_1 = (\gamma/2)p_1 M_1^2$, we have

$$c_l = \frac{L'}{q_1 c} = \frac{2}{\gamma M_1^2}\left(\frac{p_3}{p_1} - \frac{p_2}{p_1}\right)\cos\alpha$$

$$= \frac{2}{(1.4)(2.6)^2}(1.403 - 0.681)\cos 5° = \boxed{0.152}$$

$$c_d = \frac{D'}{q_1 c} = \frac{2}{\gamma M_1^2}\left(\frac{p_3}{p_1} - \frac{p_2}{p_1}\right)\sin\alpha$$

$$= \frac{2}{(1.4)(2.6)^2}(1.403 - 0.681)\sin 5° = \boxed{0.0133}$$

Figure 4.36 shows only part of the wave system associated with the supersonic flow over a flat plate at angle of attack. After the flow passes over the flat plate, it will move downstream of the trailing edge in approximately, but not exactly, the freestream direction. As shown in Fig. 4.37, the supersonic flow over the top surface is turned into itself at the trailing edge, hence generating a left-running shock wave emanating from the trailing edge. The supersonic flow over the bottom surface is turned away from itself at the trailing edge, hence generating a right-running expansion wave. The streamline ab trailing downstream from the trailing edge makes the angle Φ with respect to the free-stream direction. The flow in region 4, above ab, has passed through both the leading edge expansion wave and the trailing edge shock wave, and similarly the flow in region 5, below ab, has passed through both the leading edge shock wave and the trailing edge expansion wave. Because the strengths of both shock waves are different, the entropy in region 4 is different than that in region 5, $s_4 \neq s_5$. Therefore, ab is a slip line dividing the two regions of different entropy. As discussed in Section 4.9, the pressure is the same across the slip line,

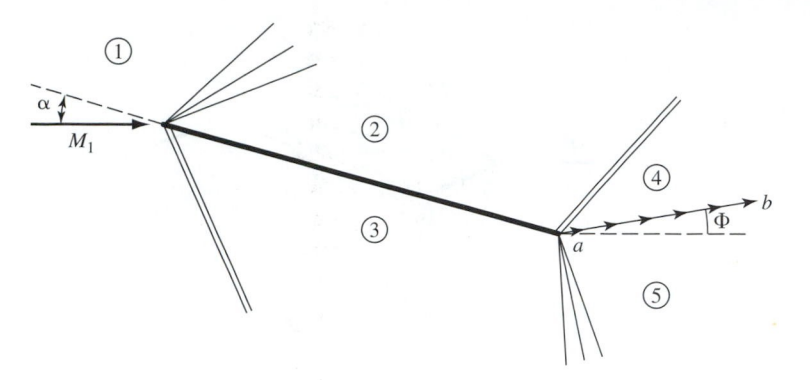

Figure 4.37 | Illustration of the tailing edge streamline for a flat plate at an angle of attack in a supersonic flow.

$p_4 = p_5$, and the flow velocities in regions 4 and 5 are in the same direction, but have different magnitudes. These two conditions dictate the properties of the flow downstream of the leading edge, including the flow direction angle Φ. Indeed, the ultimate physical reason why the flow downstream of the trailing edge does not return to exactly the free-stream conditions and direction is because the entropy of the downstream flow is increased by the shock waves, and hence the conditions downstream of the trailing edge can never be exactly the same as those in the free stream.

However, interestingly enough the downstream flow angle Φ is usually quite small, on the order of a degree or less. The precise value of Φ is a function of M_1 and angle of attack, as will be illustrated in Example 4.16. For values of M_1 above about 1.3, the downstream flow is canted upward, above the free-stream direction. This is the case shown in Fig. 4.37. This result may at first appear to be against our intuition, because the production of lift on an aerodynamic body creates a downward canting of the downstream flow (downwash). Indeed, Newton's third law dictates that if lift is generated on the body by the flow, the equal and opposite reaction pushes the airflow in the general downward direction downstream of the body. This is a general result for any flow, subsonic or supersonic. However, the flow sketched in Fig. 4.37 appears to violate physics. This paradox is resolved when the wave pattern over a much larger extent of the flow is examined, such as the wave interaction pattern in the far wake of the flat plate shown in Fig. 4.38. The overall effect of the flow through this much larger region results in an overall downwash when viewed over the whole domain. For example, the upwash (upward deflection of Φ) shown in Fig. 4.37 is compensated by a net downwash over other parts of the flowfield.

We note that the downstream flow shown in Figs. 4.37 and 4.38 does not affect the lift and drag on the plate. For an inviscid flow, the aerodynamic force on the plate is due only to the integrated pressure distribution on the surface of the plate, as sketched in Fig. 4.36. In steady supersonic flow, disturbances do not propagate upstream, and hence the flow downstream of the trailing edge does not affect the pressure distribution over the plate. This is a basic physical property of steady supersonic

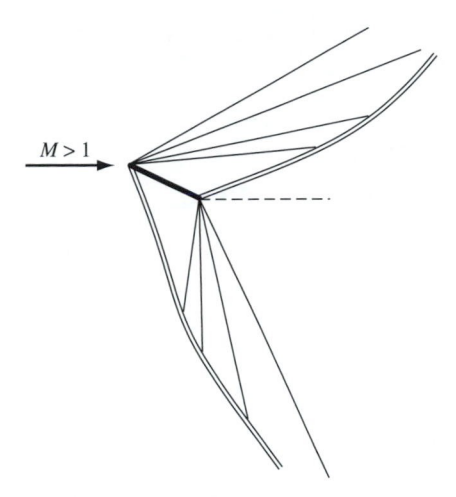

$M > 1$

Figure 4.38 | Schematic of the far-field
wave pattern downstream of a flat plate at
an angle of attack in a supersonic flow.

flow—disturbances can not feed upstream. In contrast, for a completely subsonic
flow, a disturbance initiated somewhere in the flow will eventually propagate
throughout the entire flowfield. These different physical phenomena for subsonic and
supersonic flow are ingrained in the sketches shown in Figs. 4.5*a* and *b*, respectively.

EXAMPLE 4.16

Consider an infinitely thin flat plate at an angle of attack of 20° in a Mach 3 free stream. Calculate the magnitude of the flow direction angle Φ downstream of the trailing edge, as
sketched in Fig. 4.37.

■ **Solution**
Figure 4.37 illustrates the nature of the flow over the flat plate. The flow properties in each
region shown in Fig. 4.37 are calculated as shown next.
   Region 2: This flow has passed through the leading edge expansion wave, where the
deflection angle $\theta = \alpha = 20°$ and $M_1 = 3$. From Table A.5, $\nu_1 = 49.76°$. Hence,

$$\nu_2 = \nu_1 + \theta = 49.76 + 20 = 69.76°$$

From Table A.5, for $\nu_2 = 69.76°$, $M_2 = 4.319$.

Note: Because Φ is generally a very small angle in this example, rather than using the nearest
entry, we will interpolate between entries in the table in order to obtain more accuracy.
   From Table A.1, for $M_1 = 3$, $p_{o_1}/p_1 = 36.73$. For $M_2 = 4.319$, $p_{o_2}/p_2 = 230.4$. Hence,

$$\frac{p_2}{p_1} = \left(\frac{p_2}{p_{o_2}}\right)\left(\frac{p_{o_2}}{p_{o_1}}\right)\left(\frac{p_{o_1}}{p_1}\right) = \left(\frac{1}{230.4}\right)(1)(36.73) = 0.1594$$

Region 3: This flow has passed through the leading edge shock wave, where $M_1 = 3$ and $\theta = 20°$. From the θ-β-M diagram, $\beta = 37.8°$.

$$M_{n_1} = M_1 \sin \beta = 3 \sin 37.8° = 1.839$$

From Table A.2 for $M_{n_1} = 1.839$,

$$\frac{p_3}{p_1} = 3.781, \quad \frac{p_{o3}}{p_{o1}} = 0.795, \quad M_{n_3} = 0.6079$$

$$M_3 = \frac{M_{n_3}}{\sin(\beta - \theta)} = \frac{0.6079}{\sin(37.8 - 20)} = 1.989$$

Regions 4 and 5: Here we have to set up an iterative solution in order to simultaneously match the pressures in regions 4 and 5. The steps are:

1. Assume a value for Φ.
2. Calculate the strength of the trailing edge shock for the local compression angle, $\alpha + \Phi$. From this, we can obtain p_4, or alternatively, p_4/p_1.
3. Calculate the strength of the trailing edge expansion wave for a local expansion angle, $\alpha + \Phi$. From this, we can obtain p_5, or alternatively, p_5/p_1.
4. Compare p_4/p_1, and p_5/p_1 from the steps 3 and 4. If they are different, assume a new value of Φ.
5. Repeat steps 2–4 until $p_4/p_1 = p_5/p_1$. When this condition is satisfied, the iteration has converged, and the flow downstream of the trailing edge is now determined.

Assume $\Phi = 0$: We know that this is not the answer, but the calculated wave strengths for this assumption provide a convenient base to start the iterations. For region 4, the oblique shock angle for $M_2 = 4.319$ and $\theta = 20°$ is $\beta = 31.5°$.

$$M_{n_2} = M_2 \sin \beta = 4.319 \sin 31.5° = 2.257$$
$$\frac{p_4}{p_2} = 5.777$$
$$\frac{p_4}{p_1} = \frac{p_4}{p_2}\frac{p_2}{p_1} = (5.777)(0.1594) = 0.921$$

For region 5, the expansion angle is $\theta = 20°$. Since $M_3 = 1.989$, $\nu_3 = 26.08°$. Then $\nu_5 = 26.08 + 20 = 46.08°$. Hence, $M_5 = 2.815$. From Table A.1, for $M_5 = 2.815$, $p_{o5}/p_5 = 27.79$.

$$\frac{p_5}{p_1} = \frac{p_5}{p_{o5}}\frac{p_{o5}}{p_{o3}}\frac{p_{o3}}{p_{o1}}\frac{p_{o1}}{p_1} = \left(\frac{1}{27.79}\right)(1)(0.795)(36.73) = 1.05$$

Comparing the values of $p_4/p_1 = 0.921$ and $p_5/p_1 = 1.05$, we need to assume Φ such as to strengthen both the trailing edge shock and expansion waves. This is done by choosing Φ such that line ab in Fig. 4.37 is canted upward slightly. Already we can see that the result will be an upwash, as discussed earlier.

Assume $\Phi = 1°$: The deflection angle for both waves will be $\alpha + \Phi = 20° + 1° = 21°$. Hence, $\beta = 33.6$, $M_{n_2} = 2.39$, and $p_4/p_2 = 6.498$.

$$\frac{p_4}{p_1} = \frac{p_4}{p_2}\frac{p_2}{p_1} = (6.498)(0.1594) = 1.036$$

For region 5, $\theta = 21°$, $\nu_5 = \nu_3 + \theta = 26.08 + 21 + 47.08°$. Hence, $M_5 = 2.865$. Thus $p_{o_5}/p_5 = 29.98$.

$$\frac{p_5}{p_1} = \frac{p_5}{p_{o_5}}\frac{p_{o_5}}{p_{o_3}}\frac{p_{o_3}}{p_{o_1}}\frac{p_{o_1}}{p_1} = \left(\frac{1}{29.98}\right)(1)(0.795)(36.73) = 0.974$$

Comparing $p_4/p_1 = 1.036$ and $p_5/p_1 = 0.974$, we see that $\Phi = 1°$ is slightly too large.

Since the two iterations carried out here clearly illustrate the technique, rather than carry out any more iterations, we can interpolate between the cases for $\Phi = 0°$ and $\Phi = 1°$. For the first iteration with $\Phi = 0°$, the difference between the two pressure ratios is $1.050 - 0.921 = 0.129$. For the second iteration with $\Phi = 1°$, the difference is $0.974 - 1.036 = -0.062$. Interpolating between these differences, where the correct value of Φ would give a zero pressure difference, we have

$$\Phi = 0° + \frac{0.129}{0.129 - (-0.062)} = 0.675°$$

Rounding off, we can state that, approximately,

$$\boxed{\Phi \approx 0.7°}$$

It is important to note that an expansion wave is a strong mechanism for turning a supersonic flow through large deflection angles. For example, return to the Prandtl–Meyer function given by Eq. (4.44). In the limit of $M \to \infty$, the terms in Eq. (4.44) involving the inverse tangent become $90°$ because the $\tan 90° \to \infty$. Hence, from Eq. (4.44)

$$\nu(\infty) = \left[\sqrt{\frac{\gamma + 1}{\gamma - 1}} - 1\right]90° = 130.45°$$

This means that an initially sonic flow over a flat surface theoretically can be expanded through a maximum deflection angle of $130.45°$, as sketched in Fig. 4.39. The corresponding pressure and temperature downstream of this expansion are both zero—a physically impossible situation. For upstream Mach numbers larger than one, the maximum deflection angle is correspondingly smaller. However, the case shown in Fig. 4.39 clearly demonstrates that large deflection angles can occur through expansion waves.

In this light, return to Example 4.9 and Fig. 4.12. There, we did not account for the expansion waves that trail downstream from the upper and lower corners of the base, and in Example 4.9 we simply assumed that a constant pressure was exerted over the base of the wedge, equal to freestream pressure. In reality, the flow downstream of the base, and the variation of pressure over the base, is much more complicated than the picture shown in Fig. 4.12. Base flow and the corresponding base pressure distribution are influenced by flow separation in the base region, which in turn is governed in part by viscous flow effects that are beyond the scope of this book. However, in Example 4.17 we make some arbitrary assumptions about the effect of the corner expansion waves on the base pressure, and recalculate the drag coefficient for the wedge. In this fashion, we wish to demonstrate the effect that base pressure can have on the overall drag coefficient.

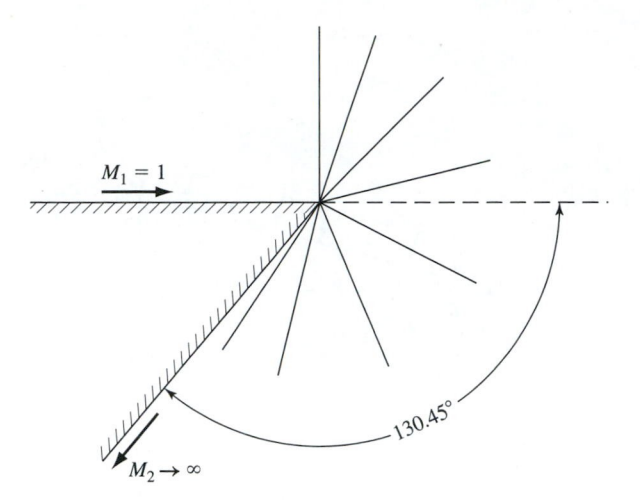

Figure 4.39 | Maximum expansion angle for a
Prandtl–Meyer centered expansion wave.

EXAMPLE 4.17

Consider the $15°$ half-angle wedge shown in Fig. 4.40. This is the same flow problem sketched
in Fig. 4.12, with the added feature of the expansion waves at the corners of the base. We make
the assumptions that (1) the flow separates at the corners, with the streamlines trailing
downstream of the corners deflected toward the base at an angle of $15°$ from the horizontal,
as shown in Fig. 4.40, and (2) the base pressure p_B is the arithmetic average between the
pressure downstream of the expansion waves, p_3, and the freestream pressure, p_1, i.e.,
$p_B = 1/2(p_3 + p_1)$. We emphasize that both of these assumptions are purely arbitrary;
they represent a qualitative model of the flow with arbitrary numbers, and do not necessarily
reflect the actual quantitative flowfield values that actually exist in the base flow region. On the
basis of the model flow sketched in Fig. 4.40, calculate the drag coefficient of the wedge, and
compare with the result obtained in Example 4.9 where the base pressure was assumed to
equal p_1.

■ **Solution**

From Example 4.8, we have these results for the leading edge shock wave and properties in
region 2 behind the shock: $\theta = 15°$, $\beta = 32.2°$, $M_{n_1} = 1.6$, $p_2/p_1 = 2.82$. From Table A.2,
we obtain $M_{n_2} = 0.6684$. Hence,

$$M_2 = \frac{M_{n_2}}{\sin(\beta - \theta)} = \frac{0.6684}{\sin(32.2 - 15)} = 2.26$$

From Table A.1, for $M_2 = 2.26$, $p_{o_2}/p_2 = 11.75$. From Table A.5, for $M_2 = 2.26$, $\nu_2 = 33.27°$. Examining Fig. 4.40, the flow expands from region 2 to region 3 through a total
deflection angle of $15° + 15° = 30°$. Hence,

$$\nu_3 = 33.27 + 30 = 63.27°$$

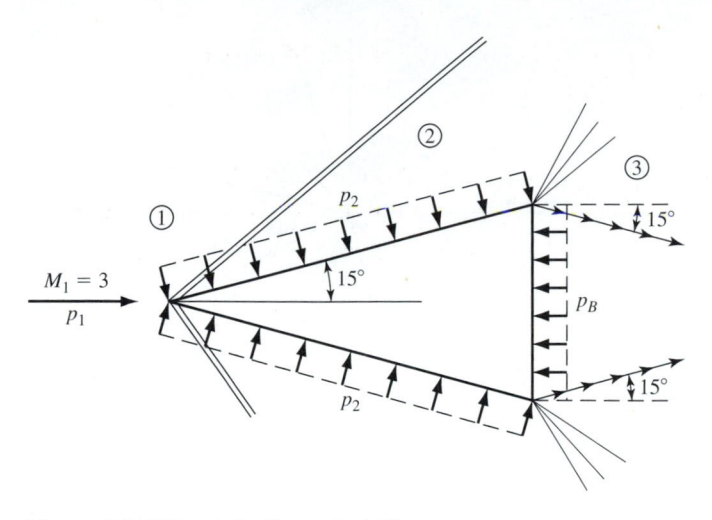

Figure 4.40 | Sketch for Example 4.17.

From Table A.5, for $\nu_3 = 63.27°$ we obtain $M_3 = 3.82$. From Table A.1, for $M_3 = 3.82$, $p_{o_3}/p_3 = 119.1$. Hence,

$$\frac{p_3}{p_1} = \frac{p_3}{p_{o_3}} \frac{p_{o_3}}{p_{o_2}} \frac{p_{o_2}}{p_2} \frac{p_2}{p_1} = \left(\frac{1}{119.1}\right)(1)(11.75)(2.82) = 0.278$$

Assume $p_B = 1/2(p_1 + p_3)$. Hence

$$\frac{p_B}{p_1} = \frac{1}{2}\left(1 + \frac{p_3}{p_1}\right) = \frac{1}{2}(1 + 0.278) = 0.639$$

From Example 4.9, the drag coefficient for the wedge, with the base pressure now denoted by p_B, is given by

$$c_d = \frac{4}{\gamma p_1 M_1^2}(p_2 - p_B)\tan 15°$$

$$= \frac{4}{\gamma M_1^2}\left(\frac{p_2}{p_1} - \frac{p_B}{p_1}\right)\tan 15°$$

$$= \frac{4}{(1.4)(3)^2}(2.82 - 0.639)\tan 15° = \boxed{0.186}$$

The value of c_d obtained from Example 4.9 was the lower value of 0.155. The present example indicates that a 36 percent reduction in base pressure results in a 20 percent increase in drag coefficient.

The result of Example 4.17 illustrates the important effect that base pressure has on the drag coefficient on the wedge shown in Fig. 4.40. The accurate calculation of base pressure for real flow situations involving any aerodynamic body shape with a blunt base is difficult to achieve, even with modern techniques in computational fluid dynamics. The accurate determination of base pressure remains today a state-of-the-art research problem.

4.16 | HISTORICAL NOTE: PRANDTL'S EARLY RESEARCH ON SUPERSONIC FLOWS AND THE ORIGIN OF THE PRANDTL–MEYER THEORY

The small German city of Göttingen nestles on the Leine River, which winds its way through lush countryside once part of the great Saxon empire. Göttingen was chartered in 1211, and quickly became a powerful member of the mercantilistic Hanseatic League in the fourteenth century. The wall around the town, many narrow cobblestone streets, and numerous medieval half-timbered houses survive to this day as reminders of Göttingen's early origin. However, this quaint appearance belies the fact that Göttingen is the home of one of the most famous universities in Europe— the Georgia Augusta University founded in 1737 by King George II of England (the Hanover family that ruled England during the eighteenth century was of German origin). The university, simply known as "Göttingen" throughout the world, has been the home of many giants of science and mathematics—Gauss, Weber, Riemann, Planck, Hilbert, Born, Lorentz, Runge, Nernst, and Heisenberg, among others.

One such man, equal in stature to those above, was Ludwig Prandtl. Born in Friesing, Germany, on February 4, 1875, Prandtl became a professor of applied mechanics at Göttingen in 1904. In that same year, at the Congress of Mathematicians in Heidelberg, Prandtl introduced his concept of the boundary layer—an approach that was to revolutionize theoretical fluid mechanics in the twentieth century. Later, during the period from 1912 to 1919, he evolved a theoretical approach for calculating lift and induced drag on finite wings—Prandtl's lifting line and lifting surface theories. This work established Prandtl as the leading fluid dynamicist of modern times; he has clearly been accepted as the father of aerodynamics. Although no Nobel Prize has ever been awarded to a fluid dynamicist, Prandtl probably came closest to deserving such an accolade. (See Sec. 9.10 for a more complete biographical sketch of Prandtl.)

It is not recognized by many students that Prandtl also made major contributions to the theory and understanding of compressible flow. However, in 1905, he built a small Mach 1.5 supersonic nozzle for the purpose of studying steam turbine flows and (of all things) the movement of sawdust in sawmills. For the next 3 years, he was curious about the flow patterns associated with such supersonic nozzles; Fig. 4.41 shows some stunning photographs made in Prandtl's laboratory during this period which clearly illustrate a progression of expansion and oblique shock waves emanating from the exit of a supersonic nozzle. (Using nomenclature to be introduced in Chap. 5, the flow progresses from an "underexpanded" nozzle at the top of Fig. 4.41 to an "overexpanded" nozzle at the bottom of the figure. At the top of the figure, we see expansion waves; at the bottom are shock waves followed by expansion waves.) The dramatic aspect of these photographs is that Prandtl was learning about *supersonic flow* at the same time that the Wright brothers were just introducing practical powered airplane flight to the world, with maximum velocities no larger than 40 mi/h!

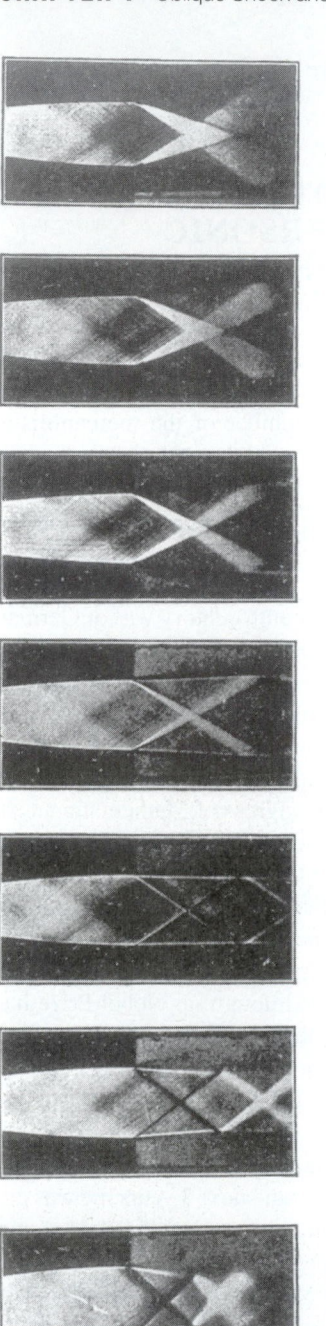

Figure 4.41 | Schlieren photographs of wave patterns downstream of the exit of a supersonic nozzle. The photographs were obtained by Prandtl and Meyer during 1907–1908.

The observation of such shock and expansion waves naturally prompted Prandtl to explore their theoretical properties. Consequently, Theodor Meyer, one of Prandtl's students at Göttingen, presented his doctoral dissertation in 1908 entitled "Ueber Zweidimensionale Bewegungsvorgänge in einem Gas, das mit Ueberschallgeschwindigkeit Stromt" ("On the Two-Dimensional Flow Processes in a Gas Flowing at Supersonic Velocities"). In this dissertation, Meyer presents the first practical theoretical development of the relations for both expansion waves and oblique shock waves—essentially the same theory as developed in this chapter. He begins by first defining a Mach wave and Mach angle as given by Eq. (4.1). Then, starting with geometry similar to that shown in Fig. 4.32, he derives the Prandtl–Meyer function [see Eq. (4.44) in Sec. 4.14] and tabulates it, not versus Mach number, but rather as a function of p/p_o. (It is interesting to note that the term "Mach number" had not yet been coined; it was introduced by Jakob Ackeret 20 years later in honor of Ernst Mach, an Austrian scientist and philosopher who studied high-speed flow for a brief period in the 1870s. So Mach number is of fairly recent use.) In the same dissertation, Meyer follows these fundamental results with a companion study of oblique shock waves, deriving relations similar to those discussed in this chapter, and presenting limited shock wave tables of wave angle, deflection angle, and pressure ratio. Almost without fanfare, Meyer ends his paper with a spectacular photograph of internal flow within a supersonic nozzle, reproduced here as Fig. 4.42. The walls of the nozzle have been intentionally roughened so that weak waves—essentially Mach waves—will be

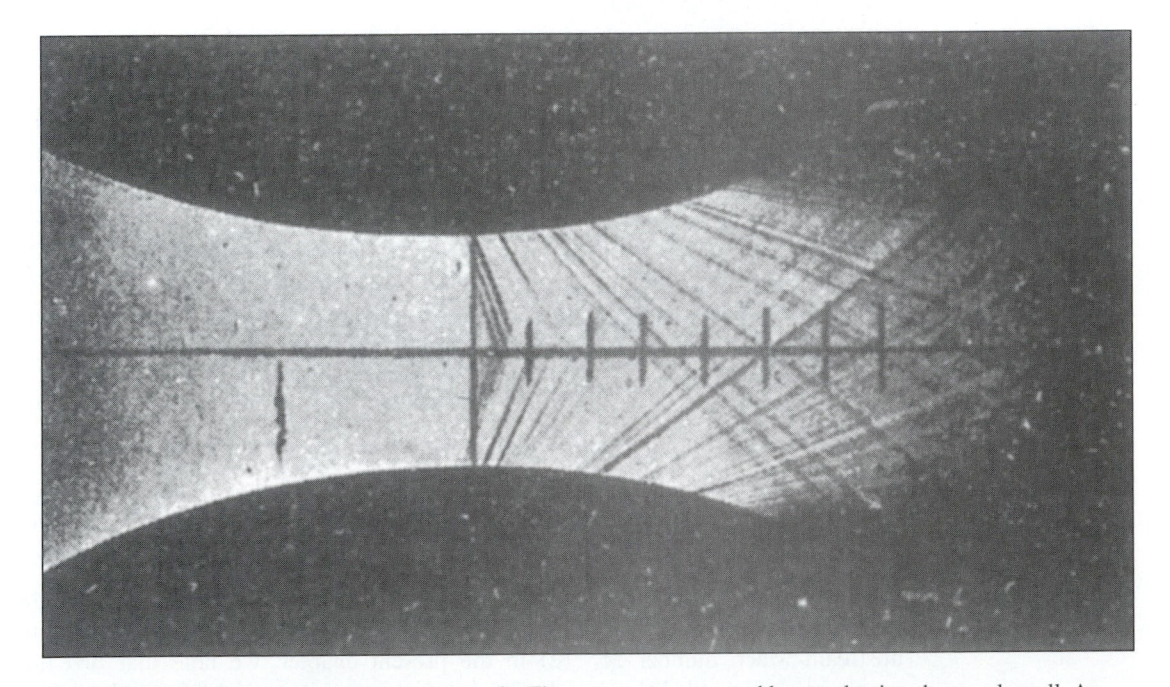

Figure 4.42 | Mach waves in a supersonic nozzle. The waves are generated by roughening the nozzle wall. An original photograph from Meyer's Ph.D. dissertation, 1908.

visible in the schlieren photograph. The reader should marvel over such a picture being taken in 1908; it has the appearance of coming from a modern supersonic laboratory in the 2000s.

We emphasize that Prandtl's and Meyer's work on expansion and oblique shock waves was contemporary with the normal shock studies of Rayleigh and Taylor in 1910 (see Sec. 3.10). So once again we are reminded of the value of basic research on problems that appear purely academic at the time. The true practical value of Meyer's dissertation did not come to fruition until the advent of supersonic flight in the 1940s.

Throughout subsequent decades, Prandtl maintained his interest in high-speed compressible flow; for example, his work on compressibility corrections for subsonic flow in the 1920s will be discussed in Sec. 9.9. Moreover, many of his students went on to distinguish themselves in high-speed flow research, most notably Theodore von Karman and Adolf Busemann. But this is the essence of other stories, to be told in later sections.

4.17 | SUMMARY

Whenever a supersonic flow is turned into itself, shock waves can occur; when the flow is turned away from itself, expansion waves can occur. In either case, if the wave is infinitely weak, it becomes a Mach wave, which makes an angle μ with respect to the upstream flow direction; μ is called the Mach angle, defined as

$$\mu = \sin^{-1} \frac{1}{M} \tag{4.1}$$

Across an oblique shock wave, the tangential components of velocity in front of and behind the wave are equal. (However, the tangential components of Mach number are *not* the same.) The thermodynamic properties across the oblique shock are dictated by the normal component of the upstream Mach number M_{n_1}. The values of p_2/p_1, ρ_2/ρ_1, T_2/T_1, $s_2 - s_1$, and p_{o_2}/p_{o_1} across the oblique shock are the same as for a normal shock wave with an upstream Mach number of M_{n_1}. In this fashion, the normal shock tables in Appendix A.2 can be used for oblique shocks. The value of M_{n_1} depends on both M_1 and the wave angle, β, via

$$M_{n_1} = M_1 \sin \beta \tag{4.7}$$

In turn, β is related to M_1 and the flow deflection angle θ through the θ-β-M relation

$$\tan \theta = 2 \cot \beta \left[\frac{M_1^2 \sin^2 \beta - 1}{M_1^2 (\gamma + \cos 2\beta) + 2} \right] \tag{4.17}$$

In light of this, we can make the following comparison: (1) In Chap. 3, we noted that the changes across a normal shock depended *only* on one flow parameter, namely the upstream Mach number M_1. (2) In the present chapter, we note that *two* flow parameters are needed to uniquely define the changes across an oblique shock. Any combination of two parameters will do. For example, an oblique shock is uniquely defined by any one of the following pairs of parameters: M_1 and β, M_1 and θ, θ and β, M_1 and p_2/p_1, β and p_2/p_1 etc.

For the solution of shock wave problems, especially cases involving shock intersections and reflections, the graphical constructions associated with the shock polar and the pressure-deflection diagrams are instructional.

For the curved, detached bow shock wave in front of a supersonic blunt body, the properties at any point immediately behind the shock are given by the oblique shock relations studied in this chapter, for the values of M_1 and the local β. Indeed, the oblique shock relations studied here apply in general to points immediately behind *any* curved, three-dimensional shock wave, so long as the component of the upstream Mach number *normal* to the shock at a given point is used to obtain the shock properties.

The properties through and behind a Prandtl–Meyer expansion fan are dictated by the differential relation

$$d\theta = \sqrt{M^2 - 1}\, \frac{dV}{V} \tag{4.35}$$

When integrated across the wave, this equation becomes

$$\theta_2 = \nu(M_2) - \nu(M_1) \tag{4.45}$$

where θ_1 is assumed to be zero and ν is the Prandtl–Meyer function given by

$$\nu(M) = \sqrt{\frac{\gamma + 1}{\gamma - 1}}\, \tan^{-1} \sqrt{\frac{\gamma - 1}{\gamma + 1}(M^2 - 1)} - \tan^{-1} \sqrt{M^2 - 1} \tag{4.44}$$

The flow through an expansion wave is isentropic; from the local Mach numbers obtained from the above relations, all other flow properties are given by the isentropic flow relations discussed in Section 3.5.

PROBLEMS

4.1 Consider an oblique shock wave with a wave angle equal to 35°. Upstream of the wave, $p_1 = 2000\ \text{lb/ft}^2$, $T_1 = 520°\text{R}$, and $V_1 = 3355$ ft/s. Calculate p_2, T_2, V_2, and the flow deflection angle.

4.2 Consider a wedge with a half-angle of 10° flying at Mach 2. Calculate the ratio of total pressures across the shock wave emanating from the leading edge of the wedge.

4.3 Calculate the maximum surface pressure (in newtons per square meter) that can be achieved on the forward face of a wedge flying at Mach 3 at standard sea level conditions ($p_1 = 1.01 \times 10^5\ \text{N/m}^2$) with an attached shock wave.

4.4 In the flow past a compression corner, the upstream Mach number and pressure are 3.5 and 1 atm, respectively. Downstream of the corner, the pressure is 5.48 atm. Calculate the deflection angle of the corner.

4.5 Consider a 20° half-angle wedge in a supersonic flow at Mach 3 at standard sea level conditions ($p_1 = 2116\ \text{lb/ft}^2$ and $T_1 = 519°\text{R}$). Calculate the wave angle, and the surface pressure, temperature, and Mach number.

4.6   A supersonic stream at $M_1 = 3.6$ flows past a compression corner with a deflection angle of $20°$. The incident shock wave is reflected from an opposite wall which is parallel to the upstream supersonic flow, as sketched in Fig. 4.18. Calculate the angle of the reflected shock relative to the straight wall.

4.7   An incident shock wave with wave angle $= 30°$ impinges on a straight wall. If the upstream flow properties are $M_1 = 2.8$, $p_1 = 1$ atm, and $T_1 = 300\,\text{K}$, calculate the pressure, temperature, Mach number, and total pressure downstream of the reflected wave.

4.8   Consider a streamline with the properties $M_1 = 4.0$ and $p_1 = 1$ atm. Consider also the following two different shock structures encountered by such a streamline: (*a*) a single normal shock wave, and (*b*) an oblique shock with $\beta = 40°$, followed by a normal shock. Calculate and compare the total pressure behind the shock structure of each (*a*) and (*b*) above. From this comparison, can you deduce a general principle concerning the efficiency of a single normal shock in relation to an oblique shock plus normal shock in decelerating a supersonic flow to subsonic speeds (which, for example, is the purpose of an inlet of a conventional jet engine)?

4.9   Consider the intersection of two shocks of opposite families, as sketched in Fig. 4.23. For $M_1 = 3$, $p_1 = 1$ atm, $\theta_2 = 20°$, and $\theta_3 = 15°$, calculate the pressure in regions 4 and 4′, and the flow direction Φ, behind the refracted shocks.

4.10 Consider the flow past a $30°$ expansion corner, as sketched in Fig. 4.32. The upstream conditions are $M_1 = 2$, $p_1 = 3$ atm, and $T_1 = 400\,\text{K}$. Calculate the following downstream conditions: M_2, p_2, T_2, T_{o_2}, and p_{o_2}.

4.11 For a given Prandtl–Meyer expansion, the upstream Mach number is 3 and the pressure ratio across the wave is $p_2/p_1 = 0.4$. Calculate the angles of the forward and rearward Mach lines of the expansion fan relative to the free-stream direction.

4.12 Consider a supersonic flow with an upstream Mach number of 4 and pressure of 1 atm. This flow is first expanded around an expansion corner with $\theta = 15°$, and then compressed through a compression corner with equal angle $\theta = 15°$ so that it is returned to its original upstream direction. Calculate the Mach number and pressure downstream of the compression corner.

4.13 Consider the incident and reflected shock waves as sketched in Fig. 4.17. Show by means of sketches how you would use shock polars to solve for the reflected wave properties.

4.14 Consider a supersonic flow past a compression corner with $\theta = 20°$. The upstream properties are $M_1 = 3$ and $p_1 = 2116\,\text{lb/ft}^2$. A Pitot tube is inserted in the flow downstream of the corner. Calculate the value of pressure measured by the Pitot tube.

4.15 Can shock polars be used to solve the intersection of shocks of opposite families, as sketched in Fig. 4.23? Explain.

4.16 Using shock-expansion theory, calculate the lift and drag (in pounds) on a symmetrical diamond airfoil of semiangle $\varepsilon = 15°$ (see Fig. 4.35) at an angle of attack to the free stream of 5° when the upstream Mach number and pressure are 2.0 and 2116 lb/ft^2, respectively. The maximum thickness of the airfoil is $t = 0.5$ ft. Assume a unit length of 1 ft in the span direction (perpendicular to the page in Fig. 4.35).

4.17 Consider a flat plate with a chord length (from leading to trailing edge) of 1 m. The free-stream flow properties are $M_1 = 3$, $p_1 = 1$ atm, and $T_1 = 270$ K. Using shock-expansion theory, tabulate and plot on graph paper these properties as functions of angle of attack from 0 to 30° (use increments of 5°):

   a. Pressure on the top surface

   b. Pressure on the bottom surface

   c. Temperature on the top surface

   d. Temperature on the bottom surface

   e. Lift per unit span

   f. Drag per unit span

   g. Lift/drag ratio

(*Note:* The results from this problem will be used for comparison with linear supersonic theory in Chap. 9.)

4.18 A flat plate is immersed in a Mach 2 flow at standard sea level conditions at an angle of attack of 2°. Assuming the same shear stress distribution given in Example 1.8, calculate, per unit span: (*a*) lift, (*b*) wave drag, and (*c*) skin friction drag. What percentage of the total drag is skin-friction drag? Compare this percentage with the 10° angle of attack case discussed in Example 1.8.

4.19 Calculate the drag coefficient for a wedge with a 20° half-angle at Mach 4. Assume the base pressure is free-stream pressure.

4.20 The flow of a chemically reacting gas is sometimes approximated by the use of relations obtained assuming a calorically perfect gas, such as in this chapter, but using an "effective gamma", a ratio of specific heats less than 1.4. Consider the Mach 3 flow of chemically reacting air, where the flow is approximated by a ratio of specific heats equal to 1.2. If this gas flows over a compression corner with a deflection angle of 20 degrees, calculate the wave angle of the oblique shock. Compare this result with that for ordinary air with a ratio of specific heats equal to 1.4. What conclusion can you make about the general effect of a chemically reacting gas on wave angle?

4.21 For the two cases treated in Problem 4.20, calculate and compare the pressure ratio (shock strength) across the oblique shock wave. What can you conclude about the effect of a chemically reacting gas on shock strength?

CHAPTER 5

Quasi-One-Dimensional Flow

The whole problem of aerodynamics, both subsonic and supersonic, may be summed up in one sentence: Aerodynamics is the science of slowing-down the air without loss, after it has once been accelerated by any device, such as a wing or a wind tunnel. It is thus good aerodynamic practice to avoid accelerating the air more than is necessary.

W. F. Hilton, 1951

PREVIEW BOX

This chapter has to do with nozzles and diffusers—vital elements of high-speed wind tunnels. A typical large supersonic wind tunnel is shown in Fig. 5.1, so large that it fills a building all by itself. To the left (upstream) of the test section (labeled 2 in Fig. 5.1) is a convergent-divergent nozzle designed to smoothly accelerate an airflow to supersonic speeds. To the right (downstream) of the test section is a convergent-divergent diffuser designed to slow the supersonic stream to subsonic velocity via a pattern of multiple shock waves with a minimum of total pressure loss. In Fig. 5.2, a model is shown mounted in the supersonic test section of the same wind tunnel. In this chapter, you will learn how nozzles and diffusers work, and how to calculate the overall properties of the flows through these and other ducts. The applications of this chapter do not stop there.

The same principles govern the performance of rocket engines, such as that shown in Fig. 5.3. Indeed, much of the material in this chapter has direct application to propulsion devices in general.

In this chapter, an important assumption is made regarding the physical nature of the flow through a variable-area duct. We will assume that the flow properties vary only in one direction along the duct, for example, the x direction shown in Fig. 5.4. However, changes of the flow properties in the x direction are brought about by *area change* of the duct. This is in contrast to the purely one-dimensional flows through a constant area duct studied in Chap. 4. To distinguish between these flows, the flow sketched in Fig. 5.4 is called *quasi-one-dimensional flow*. To assume that the flow properties vary only in the x direction through a duct that has a

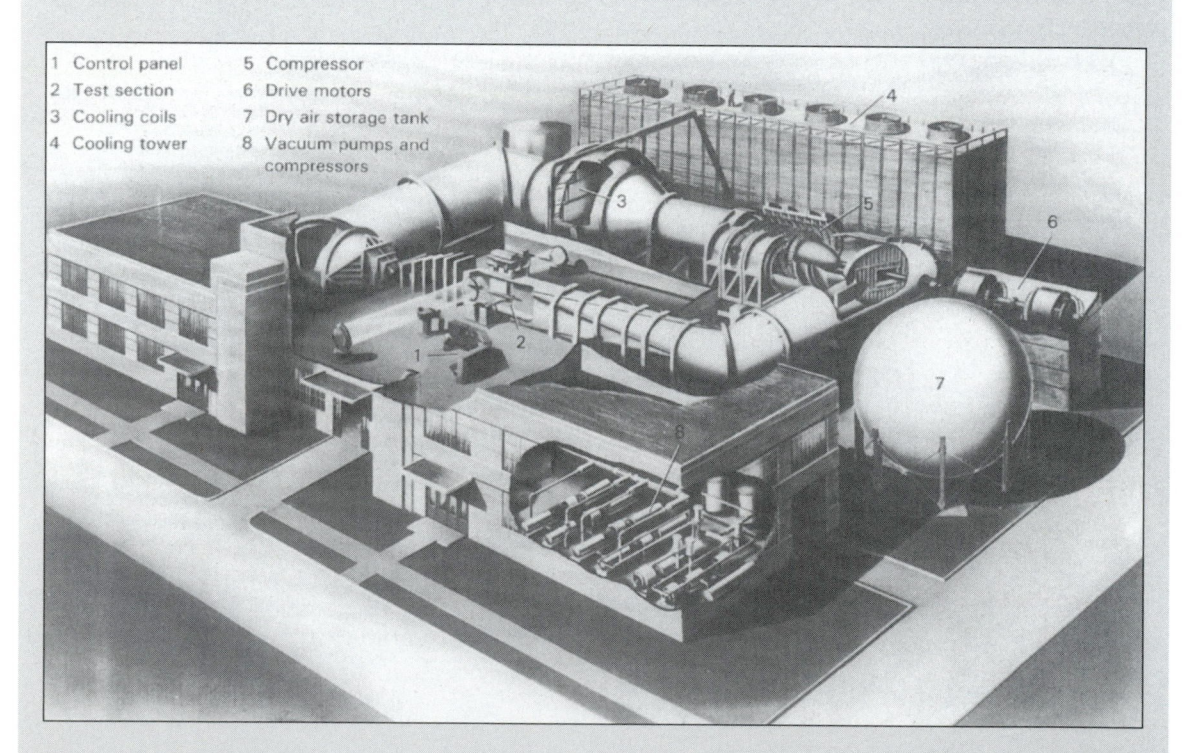

1	Control panel	5	Compressor
2	Test section	6	Drive motors
3	Cooling coils	7	Dry air storage tank
4	Cooling tower	8	Vacuum pumps and compressors

Figure 5.1 | The NASA Ames 6 × 6-foot supersonic wind tunnel with supporting facilities. The 6 × 6-foot label applies to the test section with a square cross-section six feet on each side.

two- or three-dimensional shape, such as sketched in Fig. 5.4, seems contradictory. We will discuss and resolve this apparent contradiction in the present chapter.

The roadmap for the present chapter is given in Fig. 5.5. Under the banner of quasi-one-dimensional flow, we first move to the left side of the roadmap and obtain the fundamental equations that govern such flows. Then we move to the right side to obtain a very special and important relation for quasi-one-dimensional flow called the area-velocity relation, which will tell us a lot about the physics of such flows. With these equations and relations, we go to the main features of this chapter, the study of flows through nozzles and diffusers. The material of this chapter is pivotal to many applications in compressible flow—please pay close attention to it.

Figure 5.2 | An aircraft model mounted in the test section of the Ames 6 × 6-foot supersonic wind tunnel. The test section is labeled as item 2 in Fig. 5.1.

(*continued on next page*)

(*continued from page 193*)

Figure 5.3 | Space Shuttle main engine.

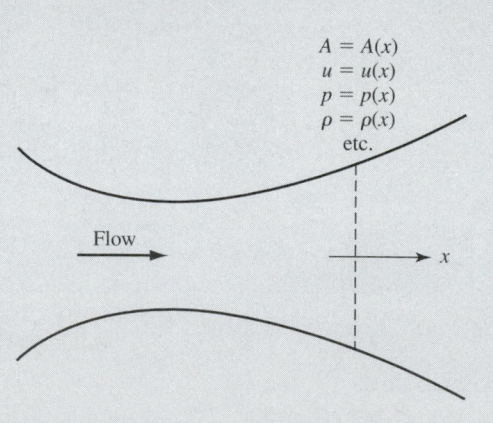

$$A = A(x)$$
$$u = u(x)$$
$$p = p(x)$$
$$\rho = \rho(x)$$
etc.

Flow

x

Figure 5.4 | Quasi-one-dimensional flow.

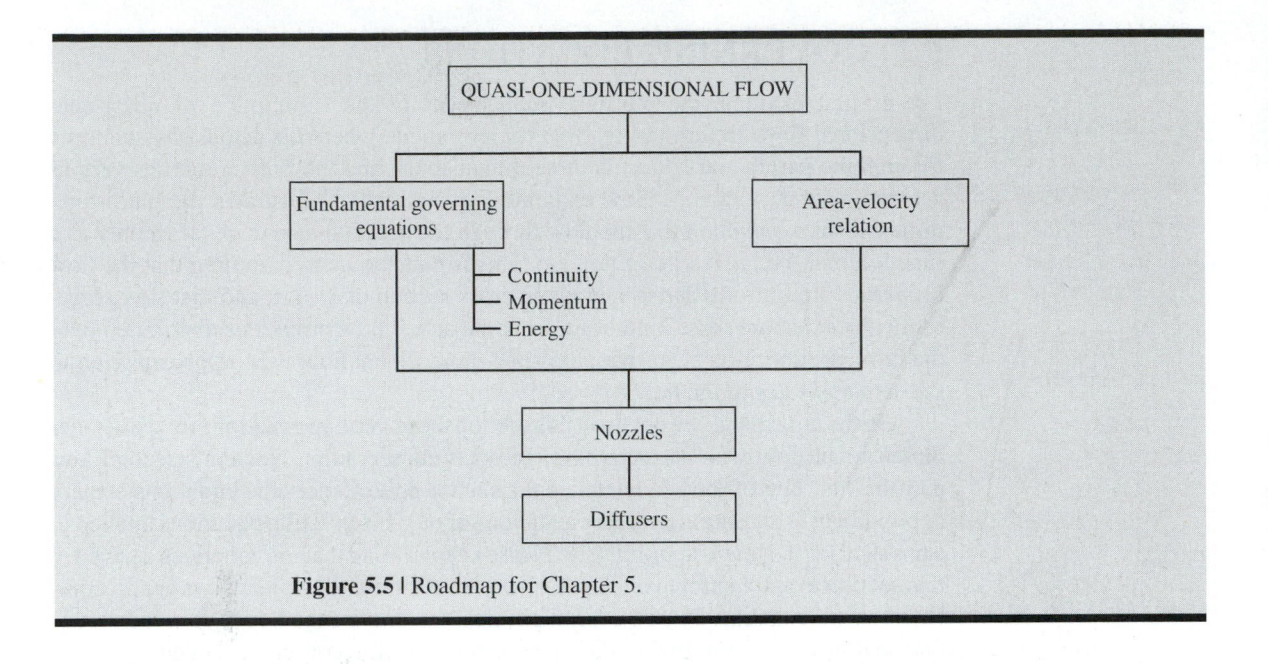

Figure 5.5 | Roadmap for Chapter 5.

5.1 | INTRODUCTION

The distinction between one-dimensional flow and quasi-one-dimensional flow was discussed in Sec. 3.1, which should be reviewed by the reader before proceeding further. In Sec. 3.1, as throughout all of Chap. 3, one-dimensional flow was treated as strictly constant-area flow. In the present chapter, this restriction will be relaxed by allowing the streamtube area A to vary with distance x, as shown in Figs. 3.5b and 5.4. At the same time, we will continue to assume that all flow properties are uniform across any given cross section of the flow, and hence are functions of x only (and time t if the flow is unsteady). Such a flow, where $A = A(x)$, $p = p(x)$, $\rho = \rho(x)$, and $V = u = u(x)$ for steady flow, is defined as *quasi-one-dimensional flow*. For this flow, it is the *area change* that causes the flow properties to vary as a function of x; in contrast, for the purely one-dimensional constant area flow treated in Chap. 3, it is a normal shock, heat addition and/or friction that causes the flow properties to vary as a function of x. In Sec. 5.2, the governing equations for steady quasi-one-dimensional flow will be derived by applying our conservation principles to a control volume of variable area. In the process, the reader is cautioned that quasi-one-dimensional flow is an approximation—the flow in the variable-area streamtube shown in Figs. 3.5b and 5.4 is (strictly speaking) three-dimensional, and its exact solution must be carried out by methods such as those discussed in Chaps. 11 and 12. However, for a wide variety of engineering problems, such as the study of flow through wind tunnels and rocket engines, quasi-one-dimensional results are frequently sufficient. Indeed, the material developed in this chapter is used virtually daily by practicing gas dynamicists and aerodynamicists, and is indispensable toward a full understanding of compressible flow.

5.2 | GOVERNING EQUATIONS

Let us first examine the physical implications of the assumption of quasi-one-dimensional flow. Return to Fig. 3.5*b* for a moment, where the actual physical flow through the variable-area duct is three-dimensional, and the flow properties vary as a function of x, y, and z. Now examine Fig. 5.4, which illustrates the quasi-one-dimensional *assumption* that the flow through the variable-area duct varies only as a function of x, i.e., $u = u(x)$, $p(x)$, etc. This is tantamount to assuming that the flow properties are uniform across any given cross section of area A, and that they represent values that are some kind of mean of the actual flow properties distributed over the cross section. It is clear that quasi-one-dimensional flow is an *approximation* to the actual *physics* of the flow.

On the other hand, we obtain in this section the governing equations for quasi-one-dimensional flow which *exactly* enforce mass conservation, Newton's second law, and the first law of thermodynamics for such a flow. Hence, the equations are *not* approximate—they are exact representations of our conservation equations applied to a physical model that is approximate. Please keep in mind that the equations derived in this section exactly enforce our basic flow conservation principles; there are no compromises here in regard to the overall physical integrity of the flow. We preserve this physical integrity by utilizing the integral forms of the conservation equations obtained in Chap. 2, applied in a mathematically exact manner to the model of the flow shown in Fig. 5.4, which is *physically* approximate. Let us see how this is done.

Algebraic equations for steady quasi-one-dimensional flow can be obtained by applying the integral form of the conservation equations to the variable-area control volume sketched in Fig. 5.6. For example, the continuity equation, Eq. (2.2), repeated here for convenience,

$$- \oiint_S \rho \mathbf{V} \cdot d\mathbf{S} = \frac{\partial}{\partial t} \iiint_{\mathscr{V}} \rho \, d\mathscr{V}$$

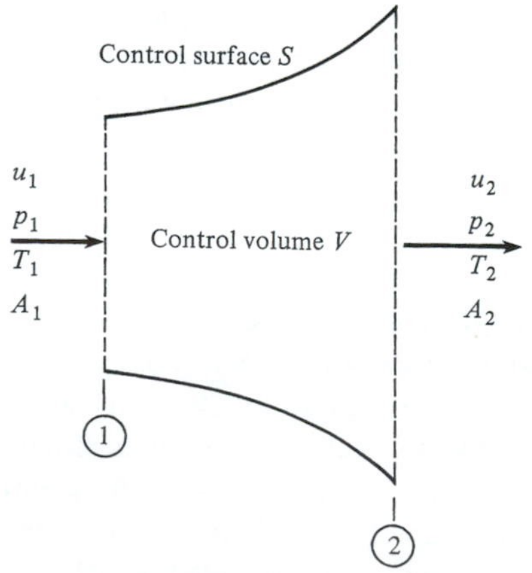

Control surface S

u_1
p_1 Control volume V
T_1
A_1

u_2
p_2
T_2
A_2

① ②

Figure 5.6 | Finite control volume for quasi-one-dimensional flow.

when integrated over the control volume in Fig. 5.6 leads, for steady flow, directly to

$$\boxed{\rho_1 u_1 A_1 = \rho_2 u_2 A_2}$$

(5.1)

This is the continuity equation for steady quasi-one-dimensional flow. Note that in Eq. (5.1) the term $\rho_1 u_1 A_1$ is the surface integral over the cross section at location 1, and $\rho_2 u_2 A_2$ is the surface integral over the cross section at location 2. The surface integral taken over the side of the control surface between locations 1 and 2 is zero, because the control surface is a streamtube; hence $\mathbf{V}$ is assumed oriented along the surface, and hence $\mathbf{V} \cdot d\mathbf{S} = 0$ along the side.

The integral form of the momentum equation, repeated from Eq. (2.11), is

$$\oiint_S (\rho \mathbf{V} \cdot d\mathbf{S}) \mathbf{V} + \iiint_{\mathcal{V}} \frac{\partial (\rho \mathbf{V})}{\partial t} \, d\mathcal{V} = \iiint_{\mathcal{V}} \rho \mathbf{f} \, d\mathcal{V} - \oiint_S p \, d\mathbf{S}$$

Applied to Fig. 5.6, assuming steady flow and no body forces, it directly becomes

$$\boxed{p_1 A_1 + \rho_1 u_1^2 A_1 + \int_{A_1}^{A_2} p \, dA = p_2 A_2 + \rho_2 u_2^2 A_2}$$

(5.2)

This is the momentum equation for steady quasi-one-dimensional flow. Note that it is not strictly an algebraic equation because of the integral term which represents the pressure force on the sides of the control surface between locations 1 and 2.

The integral form of the energy equation, repeated from Eq. (2.20), is

$$\iiint_{\mathcal{V}} \dot{q} \rho \, d\mathcal{V} - \oiint_S p \mathbf{V} \cdot d\mathbf{S} + \iiint_{\mathcal{V}} \rho (\mathbf{f} \cdot \mathbf{V}) \, d\mathcal{V}$$

$$= \iiint_{\mathcal{V}} \frac{\partial}{\partial t} \left[\rho \left(e + \frac{V^2}{2} \right) \right] d\mathcal{V} + \oiint_S \rho \left(e + \frac{V^2}{2} \right) \mathbf{V} \cdot d\mathbf{S}$$

Applied to Fig. 5.6, and assuming steady adiabatic flow with no body forces, it directly yields

$$-(-p_1 u_1 A_1 + p_2 u_2 A_2) = \rho_1 \left(e_1 + \frac{u_1^2}{2} \right) (-u_1 A_1) + \rho_2 \left(e_2 + \frac{u_2^2}{2} \right) u_2 A_2$$

Rearranging,

$$p_1 u_1 A_1 + \rho_1 u_1 A_1 \left(e_1 + \frac{u_1^2}{2} \right) = p_2 u_2 A_2 + \rho_2 u_2 A_2 \left(e_2 + \frac{u_2^2}{2} \right)$$

(5.3)

Divide Eq. (5.3) by (5.1):

$$\frac{p_1}{\rho_1} + e_1 + \frac{u_1^2}{2} = \frac{p_2}{\rho_2} + e_2 + \frac{u_2^2}{2}$$

(5.4)

Noting that $h = e + p/\rho$, Eq. (5.4) becomes

$$h_1 + \frac{u_1^2}{2} = h_2 + \frac{u_2^2}{2} \qquad (5.5)$$

This is the energy equation for steady adiabatic quasi-one-dimensional flow—it states that the total enthalpy is constant along the flow:

$$h_o = \text{const} \qquad (5.6)$$

Note that Eqs. (5.5) and (5.6) are identical to the adiabatic one-dimensional energy equation derived in Chap. 3 [see Eq. (3.40)]. Indeed, this is a general result; in any adiabatic steady flow, the total enthalpy is constant along a streamline—a result that will be proven in Chap. 6. Also note that Eqs. (5.1) and (5.2), when applied to the special case where $A_1 = A_2$, reduce to the corresponding one-dimensional results expressed in Eqs. (3.2) and (3.5).

In Chap. 6, the general conservation laws will be expressed in differential rather than integral or algebraic forms, as done so far. As a precursor to this, differential expressions for the steady quasi-one-dimensional continuity, momentum, and energy equations will be of use to us now. For example, from Eq. (5.1),

$$\rho u A = \text{const}$$

Hence,

$$d(\rho u A) = 0 \qquad (5.7)$$

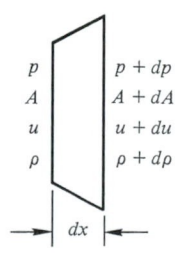

To obtain a differential form of the momentum equation, apply Eq. (5.2) to the infinitesimal control volume sketched in Fig. 5.7, where the length in the x direction is dx:

$$pA + \rho u^2 A + p\,dA = (p + dp)(A + dA) + (\rho + d\rho)(u + du)^2(A + dA)$$

Dropping all second-order terms involving products of differentials, this becomes

$$A\,dp + Au^2\,d\rho + \rho u^2\,dA + 2\rho u A\,du = 0 \qquad (5.8)$$

Expanding Eq. (5.7), and multiplying by u,

$$\rho u^2\,dA + \rho u A\,du + Au^2\,d\rho = 0$$

Subtracting this equation from Eq. (5.8), we obtain

$$dp = -\rho u\,du \qquad (5.9)$$

Equation (5.9) is called *Euler's equation*, to be discussed in Sec. 6.4. Finally, a differential form of the energy equation is obtained from Eq. (5.5), which states that

$$h + \frac{u^2}{2} = \text{const}$$

Figure 5.7 |
Incremental
volume.

Hence,

$$\boxed{dh + u\,du = 0} \tag{5.10}$$

To reinforce the comments made at the beginning of this section, we emphasize that Eqs. (5.1), (5.2), (5.5), (5.7), (5.9), and (5.10) are *exact* representations of physics as applied to the *approximate* model of quasi-one-dimensional flow. So the basic fundamental physical principles stated in Chap. 2 are not compromised here. The only compromise with the true nature of the flow is the use of the simplified *model* of quasi-one-dimensional flow.

Return to the roadmap in Fig. 5.5. We have completed the left column, and we are now ready to use the fundamental governing equations for quasi-one-dimensional flow to study the properties of nozzle and diffuser flows. However, before going to these applications, we move to the right side of the roadmap and obtain the area-velocity relation. This relation is vital to understanding the *physics of the flow*, and we need this understanding before we go to the applications.

5.3 | AREA-VELOCITY RELATION

A wealth of physical information regarding quasi-one-dimensional flow can be obtained from a particular combination of the differential forms of the conservation equations presented at the end of Sec. 5.2 as shown next. From Eq. (5.7),

$$\frac{d\rho}{\rho} + \frac{du}{u} + \frac{dA}{A} = 0 \tag{5.11}$$

To eliminate $d\rho/\rho$ from Eq. (5.11), consider Eq. (5.9):

$$\frac{dp}{\rho} = \frac{dp}{d\rho}\frac{d\rho}{\rho} = -u\,du \tag{5.12}$$

Recall that we are considering adiabatic, inviscid flow, i.e., there are no dissipative mechanisms such as friction, thermal conduction, or diffusion acting on the flow. Thus, the flow is isentropic. Hence, any change in pressure, dp, in the flow is accompanied by a corresponding isentropic change in density, $d\rho$. Therefore, we can write

$$\frac{dp}{d\rho} = \left(\frac{\partial p}{\partial \rho}\right)_s = a^2 \tag{5.13}$$

Combining Eqs. (5.12) and (5.13),

$$a^2 \frac{d\rho}{\rho} = -u\,du$$

or

$$\frac{d\rho}{\rho} = -\frac{u\,du}{a^2} = -\frac{u^2\,du}{a^2 u} = -M^2 \frac{du}{u} \tag{5.14}$$

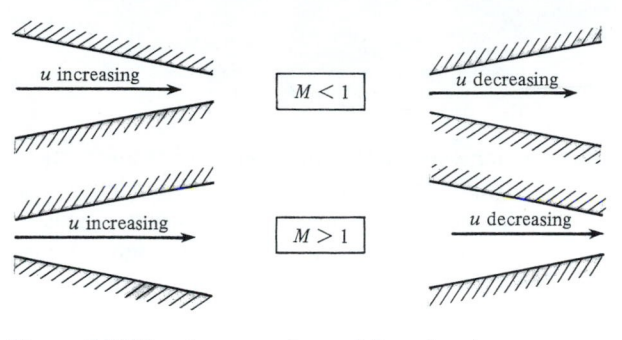

Figure 5.8 | Flow in converging and diverging ducts.

Substituting Eq. (5.14) into Eq. (5.11),

$$\frac{dA}{A} = (M^2 - 1)\frac{du}{u} \tag{5.15}$$

Equation (5.15) is an important result. It is called the *area-velocity relation,* and it tells us this information:

1. For $M \to 0$, which in the limit corresponds to incompressible flow, Eq. (5.15) shows that $Au = $ const. This is the familiar continuity equation for incompressible flow.

2. For $0 \le M < 1$ (subsonic flow), an *increase* in velocity (positive du) is associated with a *decrease* in area (negative dA), and vice versa. Therefore, the familiar result from incompressible flow that the velocity increases in a converging duct and decreases in a diverging duct still holds true for subsonic compressible flow (see top of Fig. 5.8).

3. For $M > 1$ (supersonic flow), an *increase* in velocity is associated with an *increase* in area, and vice versa. Hence, we have a striking difference in comparison to subsonic flow. For supersonic flow, the velocity increases in a diverging duct and decreases in a converging duct (see bottom of Fig. 5.8).

4. For $M = 1$ (sonic flow), Eq. (5.15) yields $dA/A = 0$, which mathematically corresponds to a minimum or maximum in the area distribution. The minimum in area is the only physically realistic solution, as described next.

These results clearly show that for a gas to expand isentropically from subsonic to supersonic speeds, it must flow through a convergent-divergent duct (or stream-tube), as sketched at the top of Fig. 5.9. Moreover, at the minimum area that divides the convergent and divergent sections of the duct, we know from item 4 above that the flow must be sonic. This minimum area is called a *throat.* Conversely, for a gas to compress isentropically from supersonic to subsonic speeds, it must also flow through a convergent-divergent duct, with a throat where sonic flow occurs, as sketched at the bottom of Fig. 5.9.

From this discussion, we recognize why rocket engines have large, bell-like nozzle shapes as sketched in Fig. 5.10—to expand the exhaust gases to high-velocity,

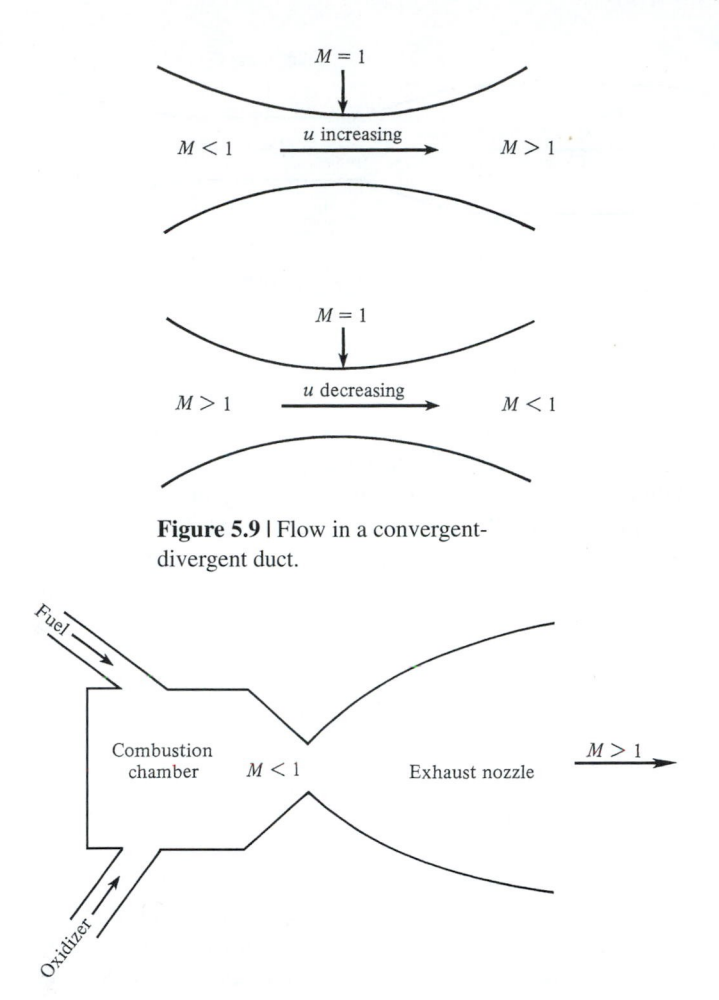

Figure 5.9 | Flow in a convergent-divergent duct.

Figure 5.10 | Schematic of a rocket engine.

supersonic speeds. This bell-like shape is clearly evident in the photograph of the space shuttle main engine shown in Fig 5.3. Moreover, we can infer the configuration of a supersonic wind tunnel, which is designed to first expand a stagnant gas to supersonic speeds for aerodynamic testing, and then compress the supersonic stream back to a low-speed subsonic flow before exhausting it to the atmosphere. This general configuration is illustrated in Fig. 5.11. Stagnant gas is taken from a reservoir and expanded to high subsonic velocities in the convergent portion of the nozzle. At the minimum area (the first throat), sonic flow is achieved. Downstream of the throat, the flow goes supersonic in the divergent portion of the nozzle. At the end of the nozzle, designed to achieve a specified Mach number, the supersonic flow enters the test section, where a test model or other experimental device is usually situated. Downstream of the test section, the supersonic flow enters a diffuser, where it is slowed down in a convergent duct to sonic flow at the second throat, and then further slowed to low subsonic speeds in a divergent duct, finally being exhausted to the atmosphere.

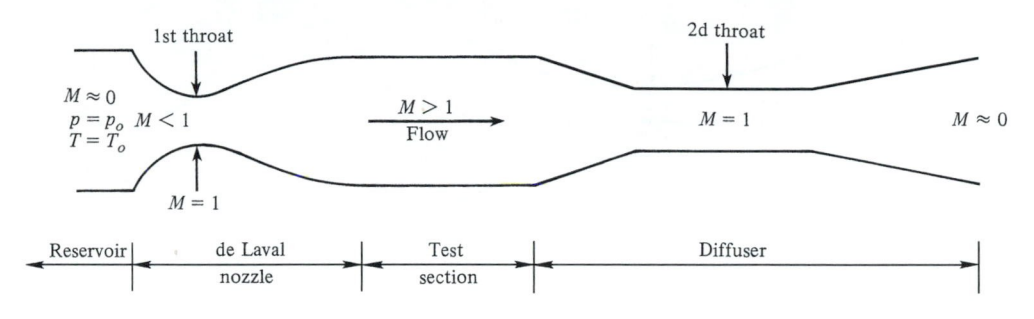

Figure 5.11 | Schematic of a supersonic wind tunnel.

This discussion, along with Fig. 5.11, is a simplistic view of real supersonic wind tunnels, but it serves to illustrate the basic phenomena as revealed by the area-velocity relation, Eq. (5.15). Also note that a convergent-divergent nozzle is sometimes called a *de Laval* (or *Laval*) *nozzle,* after Carl G. P. de Laval, who first used such a configuration in his steam turbines in the late nineteenth century, as described in Secs. 1.1 and 5.8.

The derivation of Eq. (5.15) utilized only the basic conservation equations—no assumption as to the type of gas was made. Hence, Eq. (5.15) is a general relation which holds for real gases and chemically reacting gases, as well as for a perfect gas—as long as the flow is isentropic. We will visit this matter again in Chap. 17.

The area-velocity relation is a differential relation, and in order to make quantitative use of it, we need to integrate Eq. (5.15). However, there is a more direct way of obtaining quantitative relations for quasi-one-dimensional flow, which we will see in the next section. The primary importance of the area-velocity relation is the invaluable *physical* information it provides, as we have already discussed.

We now move to the bottom of our roadmap in Fig. 5.5. Using the fundamental governing equations as well as the physical information provided by the area-velocity relation, we examine the first of the two central applications in this chapter—flows through nozzles.

5.4 | NOZZLES

The analysis of flows through variable-area ducts in a general sense requires numerical solutions such as those to be discussed in Chap. 17. However, based on our experience obtained in Chaps. 3 and 4, we suspect (correctly) that we can obtain closed-form results for the case of a calorically perfect gas. We will divide our discussion into two parts: (1) purely isentropic subsonic-supersonic flow through nozzles and (2) the effect of different pressure ratios across nozzles.

5.4.1   Isentropic Subsonic-Supersonic Flow of a Perfect Gas through Nozzles

Consider the duct shown in Fig. 5.12. At the throat, the flow is sonic. Hence, denoting conditions at sonic speed by an asterisk, we have, at the throat, $M^* = 1$ and

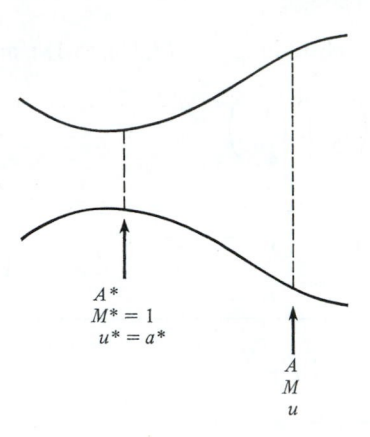

Figure 5.12 | Geometry for derivation of the area Mach number relation.

$u^* = a^*$. The area of the throat is A^*. At any other section of the duct, the local area, Mach number, and velocity are A, M, and u, respectively. Apply Eq. (5.1) between these two locations:

$$\rho^* u^* A^* = \rho u A \tag{5.16}$$

Since $u^* = a^*$, Eq. (5.16) becomes

$$\frac{A}{A^*} = \frac{\rho^*}{\rho} \frac{a^*}{u} = \frac{\rho^*}{\rho_o} \frac{\rho_o}{\rho} \frac{a^*}{u} \tag{5.17}$$

where ρ_o is the stagnation density defined in Sec. 3.4, and is constant throughout the isentropic flow. Repeating Eq. (3.31),

$$\frac{\rho_o}{\rho} = \left(1 + \frac{\gamma - 1}{2} M^2\right)^{1/(\gamma - 1)}$$

and apply this to sonic conditions, we have

$$\frac{\rho_o}{\rho^*} = \left(\frac{\gamma + 1}{2}\right)^{1/(\gamma - 1)} \tag{5.18}$$

Also, by definition, and from Eq. (3.37),

$$\left(\frac{u}{a^*}\right)^2 = M^{*2} = \frac{\frac{\gamma + 1}{2} M^2}{1 + \frac{\gamma - 1}{2} M^2} \tag{5.19}$$

Squaring Eq. (5.17), and substituting Eqs. (3.31), (5.18), and (5.19), we have

$$\left(\frac{A}{A^*}\right)^2 = \left(\frac{\rho^*}{\rho_o}\right)^2 \left(\frac{\rho_o}{\rho}\right)^2 \left(\frac{a^*}{u}\right)^2$$

$$\left(\frac{A}{A^*}\right)^2 = \left(\frac{2}{\gamma+1}\right)^{2/(\gamma-1)} \left(1 + \frac{\gamma-1}{2}M^2\right)^{2/(\gamma-1)} \left(\frac{1 + \frac{\gamma-1}{2}M^2}{\frac{\gamma+1}{2}M^2}\right)$$

$$\left(\frac{A}{A^*}\right)^2 = \frac{1}{M^2}\left[\frac{2}{\gamma+1}\left(1 + \frac{\gamma-1}{2}M^2\right)\right]^{(\gamma+1)/(\gamma-1)} \tag{5.20}$$

Equation (5.20) is called the *area–Mach number relation,* and it contains a striking result. Turned inside out, Eq. (5.20) tells us that $M = f(A/A^*)$, i.e., the Mach number at any location in the duct is a function of the *ratio* of the local duct area to the sonic throat area. As seen from Eq. (5.15), A must be greater than or at least equal to A^*; the case where $A < A^*$ is physically not possible in an isentropic flow. Also, from Eq. (5.20) there are two values of M that correspond to a given $A/A^* > 1$, a subsonic and a supersonic value. The solution of Eq. (5.20) is plotted in Fig. 5.13,

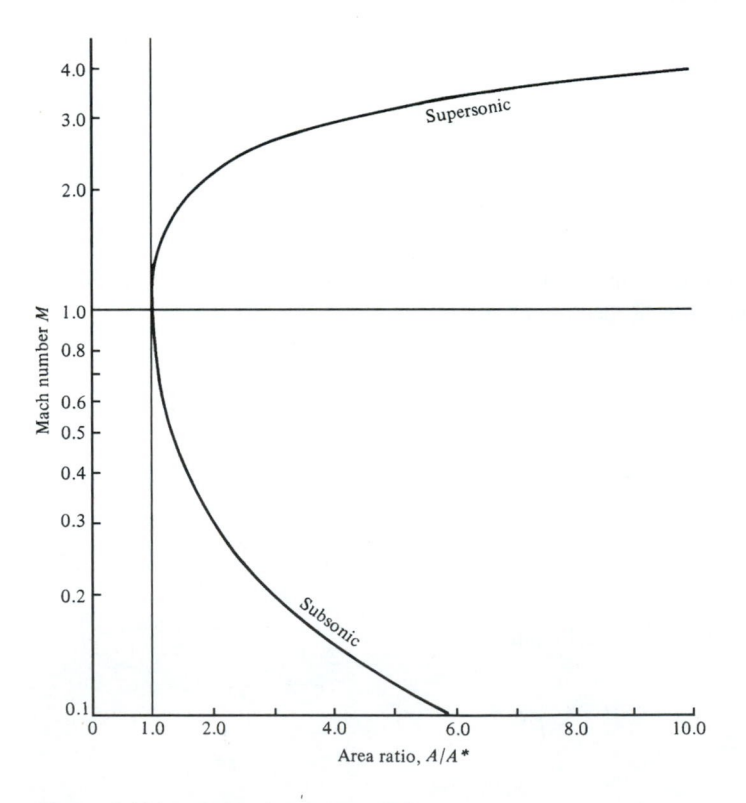

Figure 5.13 | Area-Mach number relation.

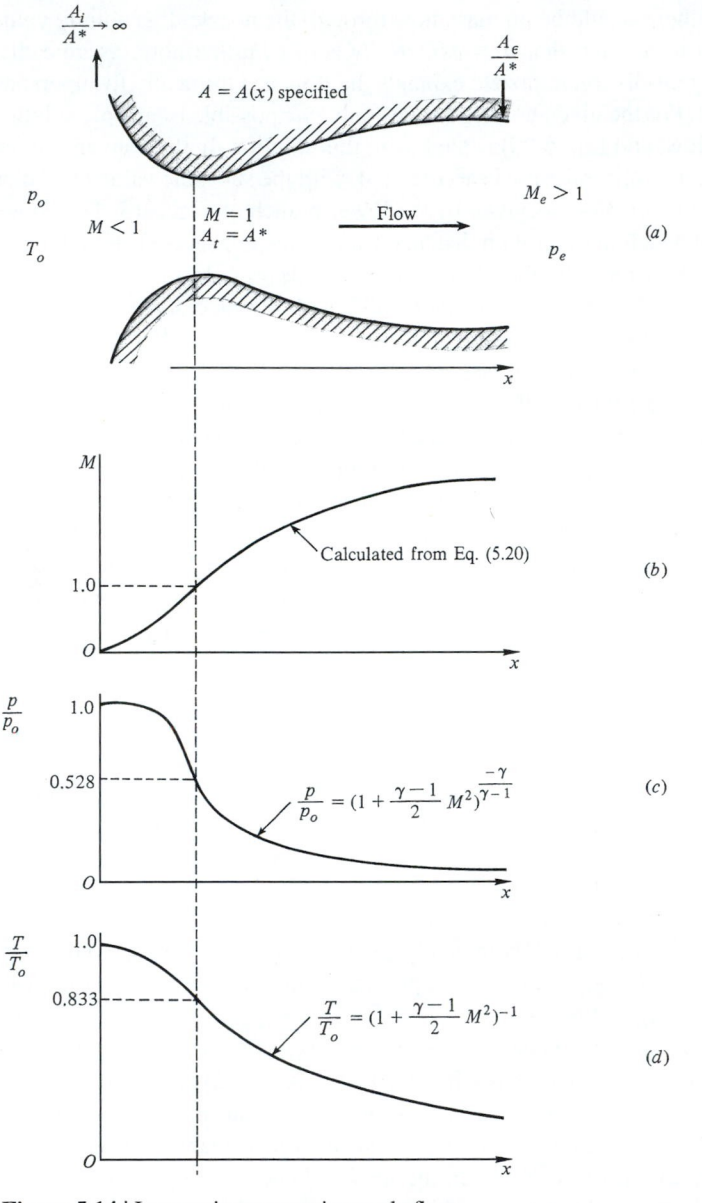

Figure 5.14 | Isentropic supersonic nozzle flow.

which clearly delineates the subsonic and supersonic branches. Values of A/A^* as a function of M are tabulated in Table A.1 for both subsonic and supersonic flow.

Consider a given convergent-divergent nozzle, as sketched in Fig. 5.14a. Assume that the area ratio at the inlet A_i/A^* is very large, $A_i/A^* \to \infty$, and that the inlet is fed with gas from a large reservoir at pressure and temperature p_o and T_o, respectively. Because of the large inlet area ratio, $M \approx 0$; hence p_o and T_o are essentially stagnation (or total) values. (The Mach number cannot be precisely zero in the reservoir, or else there would be no mass flow through the nozzle. It is a finite value, but small enough to assume that it is *essentially* zero.) Furthermore, assume that the given convergent-divergent nozzle expands the flow isentropically to supersonic speeds at the exit. For the given nozzle, there is only *one* possible isentropic solution for supersonic flow, and Eq. (5.20) is the key to this solution. In the convergent portion of the nozzle, the subsonic flow is accelerated, with the subsonic value of M dictated by the local value of A/A^* as given by the lower branch of Fig. 5.13. The consequent variation of Mach number with distance x along the nozzle is sketched in Fig. 5.14b. At the throat, where the throat area $A_t = A^*$, $M = 1$. In the divergent portion of the nozzle, the flow expands supersonically, with the supersonic value of M dictated by the local value of A/A^* as given by the upper branch of Fig. 5.13. This variation of M with x in the divergent nozzle is also sketched in Fig. 5.14b. Once the variation of Mach number through the nozzle is known, the variations of static temperature, pressure, and density follow from Eqs. (3.28), (3.30), and (3.31), respectively. The resulting variations of p and T are shown in Figs. 5.14c and d, respectively. Note that the pressure, density, and temperature decrease continuously throughout the nozzle. Also note that the exit pressure, density, and temperature ratios, p_e/p_o, ρ_e/ρ_o, and T_e/T_o depend only on the exit area ratio, A_e/A^* via Eq. (5.20). If the nozzle is part of a supersonic wind tunnel, then the test section conditions are completely determined by A_e/A^* (a geometrical design condition) and p_o and T_o (gas properties in the reservoir).

5.4.2   The Effect of Different Pressure Ratios Across a Given Nozzle

If a convergent-divergent nozzle is simply placed on a table, and nothing else is done, obviously nothing is going to happen; the air is not going to start rushing through the nozzle of its own accord. To accelerate a gas, a pressure difference must be exerted, as clearly stated by Euler's equation, Eq. (5.9). Therefore, in order to establish a flow through any duct, the exit pressure must be lower than the inlet pressure, i.e., $p_e/p_o < 1$. Indeed, for completely shockfree isentropic supersonic flow to exist in the nozzle of Fig. 5.14a, the exit pressure ratio must be precisely the value of p_e/p_o shown in Fig. 5.14c.

What happens when p_e/p_o is *not* the precise value as dictated by Fig. 5.14c? In other words, what happens when the backpressure downstream of the nozzle exit is independently governed (say by exhausting into an infinite reservoir with controllable pressure)? Consider a convergent-divergent nozzle as sketched in Fig. 5.15a. Assume that no flow exists in the nozzle, hence $p_e = p_o$. Now assume that p_e is minutely reduced below p_o. This small pressure difference will cause a small wind to blow through the duct at low subsonic speeds. The local Mach number will increase slightly through the convergent portion of the nozzle, reaching a maximum at the throat, as shown by curve 1 of Fig. 5.15b. This maximum will *not* be sonic; indeed it will be a low subsonic value. Keep in mind that the value A^* defined earlier is the *sonic* throat area, i.e., that area where $M = 1$. In the case we are now considering, where $M < 1$ at the minimum-area section of the duct, the real throat area of the duct, A_t, is larger than A^*, which for completely subsonic flow takes on the

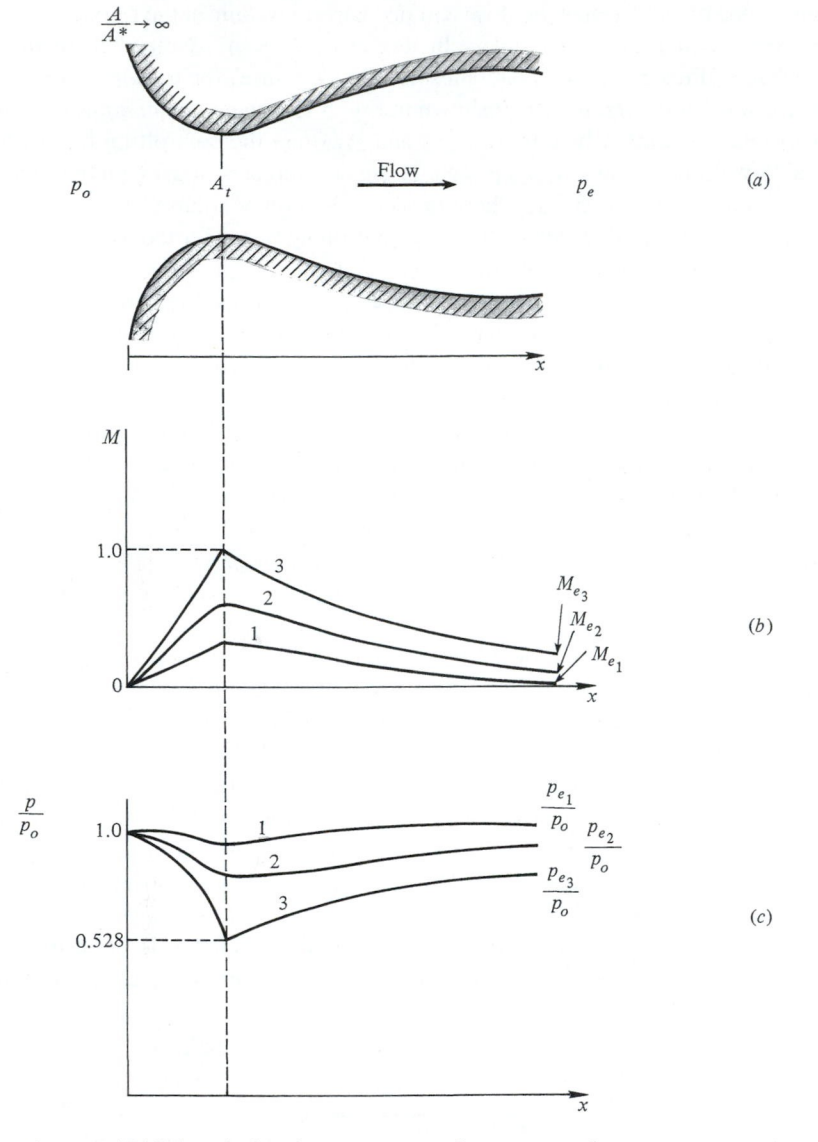

Figure 5.15 | Subsonic flow in a convergent-divergent nozzle.

character of a reference quantity different from the actual geometric throat area. Downstream of the throat, the subsonic flow encounters a diverging duct, and hence M decreases as shown in Fig. 5.15b. The corresponding variation of static pressure is given by curve 1 in Fig. 5.15c. Now assume p_e is further reduced. This stronger pressure ratio between the inlet and exit will now accelerate the flow more, and the variations of subsonic Mach number and static pressure through the duct will be larger, as indicated by curve 2 in Figs. 5.15b and c. If p_e is further reduced, there will be some value of p_e at which the flow will just barely go sonic at the throat, as given by the curve 3 in Figs. 5.15b and c. In this case, $A_t = A^*$. Note that all the cases sketched in Figs 5.15b and c are subsonic flows. Hence, for *subsonic* flow through the convergent-divergent nozzle shown in Fig. 5.15a, there are an *infinite number of isentropic solutions,* where both p_e/p_o and A/A_t are the controlling factors for the local flow properties at any given section. This is a direct contrast with the supersonic case discussed in Sec. 5.4.1, where only *one* isentropic solution exists for a given duct, and where A/A^* becomes the only controlling factor for the local flow properties (relative to reservoir properties).

For the cases shown in Figs. 5.15a, b, and c, the mass flow through the duct increases as p_e decreases. This mass flow can be calculated by evaluating Eq. (5.1) at the throat, $\dot{m} = \rho_t A_t u_t$. When p_e is reduced to p_{e_3}, where sonic flow is attained at the throat, then $\dot{m} = \rho^* A^* a^*$. If p_e is now reduced further, $p_e < p_{e_3}$, the Mach number at the throat cannot increase beyond $M = 1$; this is dictated by Eq. (5.15). Hence, the flow properties at the throat, and indeed throughout the entire subsonic section of the duct, become "frozen" when $p_e < p_{e_3}$, i.e., the subsonic flow becomes unaffected and the mass flow remains constant for $p_e < p_{e_3}$. This condition, after sonic flow is attained at the throat, is called *choked flow*. No matter how low p_e is made, after the flow becomes choked, the mass flow remains constant. This phenomenon is illustrated in Fig. 5.16. Note from Eq. (3.35) that sonic flow at the throat corresponds to a pressure ratio $p^*/p_o = 0.528$ for $\gamma = 1.4$; however, because of the divergent duct downstream of the throat, the value of p_{e_3}/p_o required to attain sonic flow at the throat is larger than 0.528, as shown in Figs. 5.15c and 5.16.

What happens in the duct when p_e is reduced below p_{e_3}? In the convergent portion, as we stated, nothing happens. The flow properties remain as given by the subsonic portion of curve 3 in Fig. 5.15b and c. However, a lot happens in the divergent portion of the duct. No isentropic solution is allowed in the divergent duct until p_e is adequately reduced to the specified low value dictated by Fig. 5.14c. For values of exit pressure above this, but below p_{e_3}, a normal shock wave exists inside the divergent duct. This situation is sketched in Fig. 5.17. Let the exit pressure be given by p_{e_4}. There is a region of supersonic flow ahead of the shock. Behind the

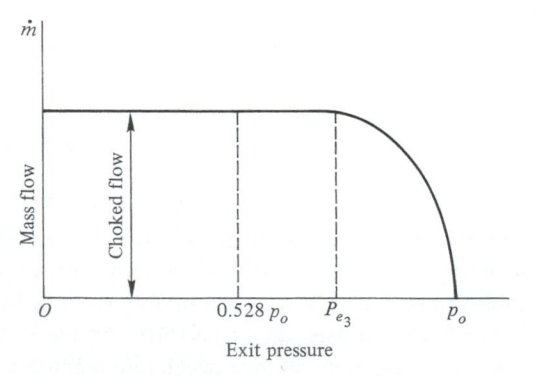

Figure 5.16 | Variation of mass flow with exit pressure; illustration of choked flow.

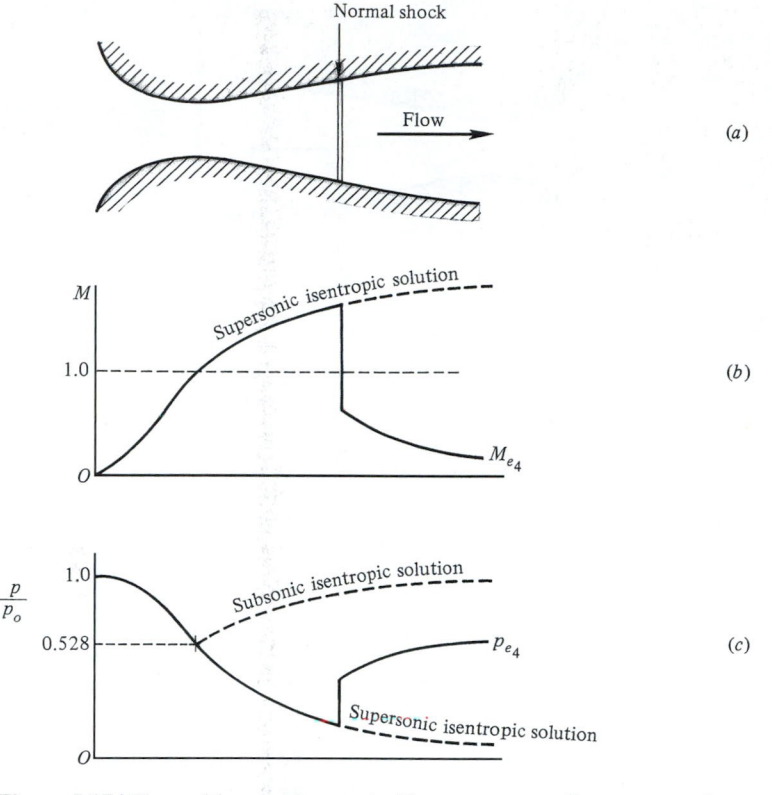

Figure 5.17 | Flow with a shock wave inside a convergent-divergent nozzle.

shock, the flow is subsonic, hence the Mach number decreases towards the exit and the static pressure increases to p_{e_4} at the exit. The location of the normal shock wave in the duct is determined by the requirement that the increase of static pressure across the wave plus that in the divergent portion of the subsonic flow behind the shock be just right to achieve p_{e_4} at the exit. As the exit pressure is reduced further, the normal shock wave will move downstream, closer to the nozzle exit. It will stand precisely at the exit when $p_e = p_{e_5}$, where p_{e_5} is the static pressure behind a normal shock at the design Mach number of the nozzle. This is illustrated in Figs. 5.18a, b, and c. In Fig. 5.18c, p_{e_6} represents the proper isentropic value for the design exit Mach number, which exists immediately upstream of the normal shock wave standing at the exit. When the downstream backpressure p_B is further decreased such that $p_{e_6} < p_B < p_{e_5}$, the flow inside the nozzle is fully supersonic and isentropic, with the behavior the same as given earlier in Figs. 5.14 a, b, c, and d. The increase to the backpressure takes place across an oblique shock attached to the nozzle exit, but outside the duct itself. This is sketched in Fig. 5.18d. If the backpressure is further reduced below p_{e_6}, equilibration of the flow takes place across expansion waves outside the duct, as shown in Fig. 5.18e.

When the situation in Fig. 5.18d exists, the nozzle is said to be *overexpanded,* because the pressure at the exit has expanded below the back pressure, $p_{e_6} < p_B$.

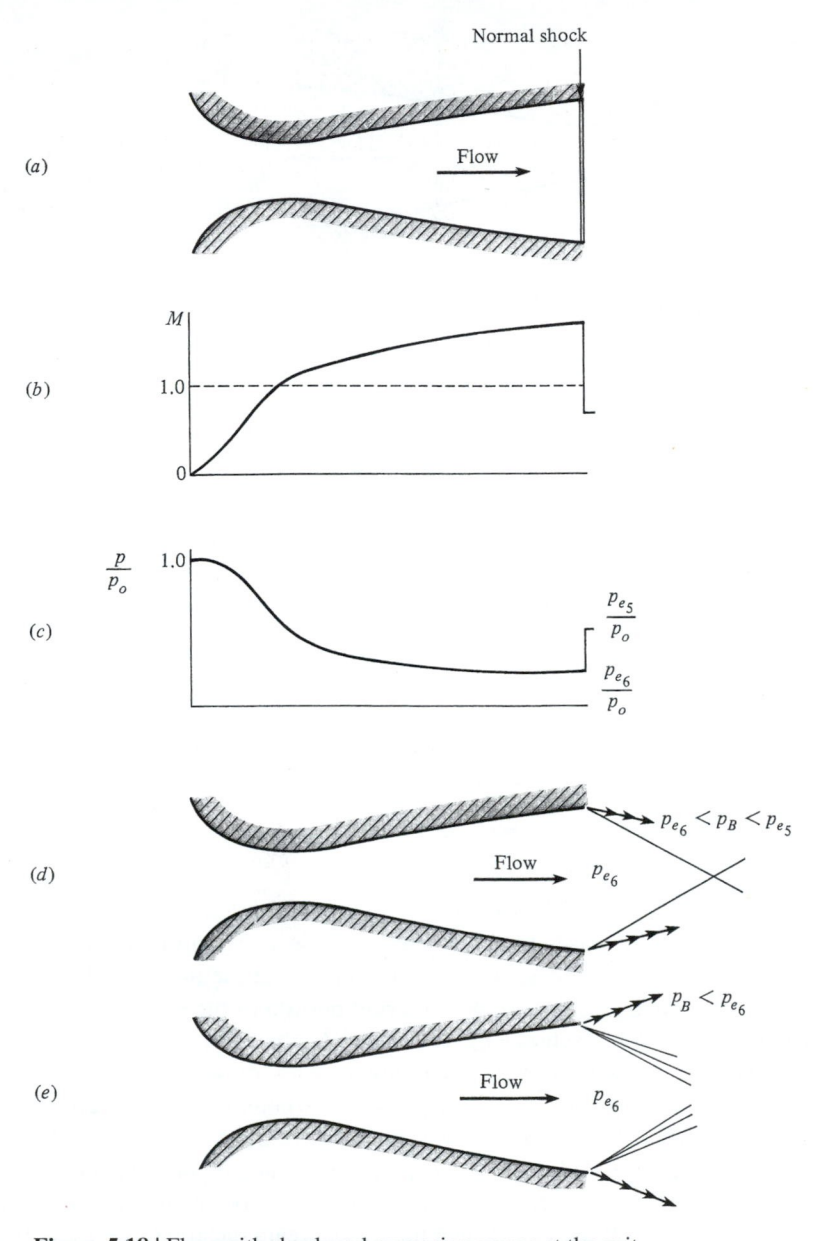

Figure 5.18 | Flow with shock and expansion waves at the exit of a convergent-divergent nozzle.

Conversely, when the situation in Fig. 5.18e exists, the nozzle is said to be *underexpanded,* because the exit pressure is higher than the back pressure, $p_{e_6} > p_B$, and hence the flow is capable of additional expansion after leaving the nozzle.

The results of this section are particularly important and useful. The reader should make certain to reread this section until he or she feels comfortable with the concepts and results before proceeding further. Also, keep in mind that these quasi-one-dimensional considerations allow the analysis of cross-sectional averaged properties inside a nozzle of given shape. They do not tell us much about how to design the *contour* of a nozzle—especially that for a supersonic nozzle in order to ensure shockfree, isentropic flow. If the shape of the walls of a supersonic nozzle is not just right, oblique shock waves can occur inside the nozzle. The proper contour for a supersonic nozzle can be determined from the method of characteristics, to be discussed in Chap. 11.

EXAMPLE 5.1

Consider the isentropic subsonic-supersonic flow through a convergent-divergent nozzle. The reservoir pressure and temperature are 10 atm and 300 K, respectively. There are two locations in the nozzle where $A/A^* = 6$: one in the convergent section and the other in the divergent section. At each location, calculate M, p, T, and u.

■ **Solution**

In the *convergent* section, the flow is subsonic. From the front of Table A.1, for subsonic flow with $A/A^* = 6$: $\boxed{M = 0.097}$, $p_o/p = 1.006$, and $T_o/T = 1.002$. Hence

$$p = \frac{p}{p_o}p_o = (1.006)^{-1}(10) = \boxed{9.94\ \text{atm}}$$

$$T = \frac{T}{T_o}T_o = (1.002)^{-1}(300) = \boxed{299.4\ \text{K}}$$

$$a = \sqrt{\gamma RT} = \sqrt{(1.4)(287)(299.4)} = 346.8\ \text{m/s}$$

$$u = Ma = (0.097)(346.8) = \boxed{33.6\ \text{m/s}}$$

In the *divergent* section, the flow is supersonic. From the supersonic section of Table A.1, for $A/A^* = 6$: $\boxed{M = 3.368}$, $p_o/p = 63.13$, and $T_o/T = 3.269$. Hence

$$p = \frac{p}{p_o}p_o = (63.13)^{-1}(10) = \boxed{0.1584\ \text{atm}}$$

$$T = \frac{T}{T_o}T_o = (3.269)^{-1}(300) = \boxed{91.77\ \text{K}}$$

$$a = \sqrt{\gamma RT} = \sqrt{(1.4)(287)(91.77)} = 192.0\ \text{m/s}$$

$$u = Ma = (3.368)(192.0) = \boxed{646.7\ \text{m/s}}$$

EXAMPLE 5.2

A supersonic wind tunnel is designed to produce Mach 2.5 flow in the test section with standard sea level conditions. Calculate the exit area ratio and reservoir conditions necessary to achieve these design conditions.

■ **Solution**

From Table A.1, for $M_e = 2.5$:

$$\boxed{A_e/A^* = 2.637} \quad p_o/p_e = 17.09 \quad T_o/T_e = 2.25$$

Also, at standard sea level conditions, $p_e = 1$ atm and $T_e = 288$ K. Hence,

$$p_o = \frac{p_o}{p_e} p_e = (17.09)(1) = \boxed{17.09 \text{ atm}}$$

$$T_o = \frac{T_o}{T_e} T_e = (2.25)(288) = \boxed{648 \text{ K}}$$

EXAMPLE 5.3

Consider a rocket engine burning hydrogen and oxygen; the combustion chamber temperature and pressure are 3517 K and 25 atm, respectively. The molecular weight of the chemically reacting gas in the combustion chamber is 16, and $\gamma = 1.22$. The pressure at the exit of the convergent-divergent rocket nozzle is 1.174×10^{-2} atm. The area of the throat is 0.4 m^2. Assuming a calorically perfect gas and isentropic flow, calculate: (a) the exit Mach number, (b) the exit velocity, (c) the mass flow through the nozzle, and (d) the area of the exit.

■ **Solution**

Note that for this problem, where $\gamma = 1.22$, the compressible flow tables in the appendix *cannot* be used since the tables are calculated for $\gamma = 1.4$. Thus, to solve this problem, we have to use the governing equations directly.

a.   To obtain the exit Mach number, use the isentropic relation given by Eq. (3.30):

$$\frac{p_o}{p_e} = \left(1 + \frac{\gamma - 1}{2} M_e^2\right)^{\gamma/(\gamma-1)}$$

or

$$M_e^2 = \frac{2}{\gamma - 1}\left[\left(\frac{p_o}{p_e}\right)^{(\gamma-1)/\gamma} - 1\right] = \frac{2}{0.22}\left[\left(\frac{25}{1.174 \times 10^{-2}}\right)^{0.22/1.22} - 1\right] = 27.116$$

$$\boxed{M_e = 5.21}$$

To obtain the exit velocity:

b.

$$\frac{T_e}{T_o} = \left(\frac{p_e}{p_o}\right)^{(\gamma-1)/\gamma} = \left(\frac{1.174 \times 10^{-2}}{25}\right)^{0.180} = 0.2517$$

$$T_e = 0.2517 T_o = 0.2517(3517) = 885.3 \text{ K}$$

From Sec. 1.4, we know that

$$R = \frac{\mathscr{R}}{\mathscr{M}} = \frac{8314}{16} = 519.6 \text{ J/kg} \cdot \text{K}$$

$$a_e = \sqrt{\gamma R T_e} = \sqrt{(1.22)(519.6)(885.3)} = 749.1 \text{ m/s}$$

$$V_e = M_e a_e = (5.21)(749.1) = \boxed{3903 \text{ m/s}}$$

c.   Since we are given $A^* = 0.4 \text{ m}^2$, let us calculate the mass flow at the throat. First, obtain ρ_o from the equation of state:

$$\rho_o = \frac{p_o}{R T_o} = \frac{(25)(1.01 \times 10^5)}{(519.6)(3517)} = 1.382 \text{ kg/m}^3$$

From Eq. (3.36)

$$\frac{\rho^*}{\rho_o} = \left(\frac{2}{\gamma+1}\right)^{1/(\gamma-1)} = \left(\frac{2}{2.22}\right)^{4.545} = 0.622$$

$$\rho^* = 0.622\rho_o = (0.622)(1.382) = 0.860 \text{ kg/m}^3$$

From Eq. (3.34)

$$\frac{T^*}{T_o} = \frac{2}{\gamma+1} = \frac{2}{2.22} = 0.9$$

$$T^* = 0.9T_o = (0.9)(3517) = 3168 \text{ K}$$

$$a^* = \sqrt{\gamma R T^*} = \sqrt{(1.22)(519.6)(3168)} = 1417 \text{ m/s}$$

$$\dot{m} = \rho A V = \rho^* A^* a^* = (0.860)(0.4)(1417) = \boxed{487.4 \text{ kg/s}}$$

d. At the exit, since $\dot{m} = \text{const}$,

$$\dot{m} = \rho_e A_e V_e = 487.4 \text{ kg/s}$$

$$\rho_e = \frac{p_e}{RT_e} = \frac{(1.174 \times 10^{-2})(1.01 \times 10^5)}{(519.6)(885.3)} = 0.00258 \text{ kg/m}^3$$

$$A_e = \frac{\dot{m}}{\rho_e V_e} = \frac{487.4}{(0.00258)(3903)} = \boxed{48.4 \text{ m}^2}$$

EXAMPLE 5.4

Consider the flow through a convergent-divergent duct with an exit-to-throat area ratio of 2. The reservoir pressure is 1 atm, and the exit pressure is 0.95 atm. Calculate the Mach numbers at the throat and at the exit.

■ Solution

First, let us analyze this problem. If the flow were supersonic in the divergent portion, then from Table A.1, for an area ratio of $A_e/A^* = 2$, $p_o/p_e = 10.69$; thus p_e would have to be $p_e = p_o/10.69 = (1 \text{ atm})/10.69 = 0.0935 \text{ atm}$. This is considerably less than the given $p_e = 0.95 \text{ atm}$. Therefore, we do not have a subsonic-supersonic isentropic flow as was the case in Examples 5.1 through 5.3. *Question:* Is the flow completely subsonic? If this were the case, the throat area A_t is *not* equal to A^*, and $A_t > A^*$. Let us examine A_t and A^*. From Table A.1, for $p_o/p_e = 1/0.95 = 1.053$, $A_e/A^* = 2.17$ (nearest entry). However, for the given problem, $A_e/A_t = 2$. Thus, $A_t > A^*$, and the flow is completely subsonic. From Table A.1, since $p_o/p_e = 1.053$, we have

$$\boxed{M_e = 0.28}$$

At the throat,

$$\frac{A_t}{A^*} = \frac{A_t}{A_e}\frac{A_e}{A^*} = \tfrac{1}{2}(2.17) = 1.085$$

From Table A.1, for $A_t/A^* = 1.085$, we have

$$\boxed{M_t = 0.72}$$

EXAMPLE 5.5

Consider a convergent-divergent duct with an exit-to-throat area ratio of 1.6. Calculate the exit-to-reservoir pressure ratio required to achieve sonic flow at the throat, but subsonic flow everywhere else.

■ Solution

Since $M = 1$ at the throat, $A_t = A^*$. Thus

$$\frac{A_e}{A_t} = \frac{A_e}{A^*} = 1.6$$

From Table A.1, the subsonic entry that corresponds to $A_e/A^* = 1.6$ is $p_o/p_e = 1.1117$. Hence

$$\frac{p_e}{p_o} = \frac{1}{1.1117} = \boxed{0.9}$$

For this area ratio of $A_e/A_t = 1.6$, if the exit-to-reservoir pressure ratio is greater than 0.9, the flow through the duct is completely subsonic. If this pressure ratio is less than 0.9, then the flow will expand to supersonic speed downstream of the throat. However, unless $p_e/p_o = 1/7.128 = 0.1403$, which corresponds to an isentropic expansion to the exit, there will be shock waves either at the lip of the nozzle (overexpanded case) or a normal shock somewhere inside the duct. Which of these cases hold depends upon the prescribed value of p_e/p_o.

EXAMPLE 5.6

Consider a convergent-divergent nozzle with an exit-to-throat area ratio of 3. A normal shock wave is inside the divergent portion at a location where the local area ratio is $A/A_t = 2$. Calculate the exit-to-reservoir pressure ratio.

■ Solution

For this case, we have an isentropic subsonic-supersonic expansion through the part of the nozzle upstream of the normal shock. Let the subscripts 1 and 2 denote conditions immediately upstream and downstream of the shock, respectively. The local Mach number M_1 just ahead of the shock is obtained from Table A.1 for $A_1/A_1^* = 2$, namely $M_1 = 2.2$. From Table A.2, for $M_1 = 2.2$, $M_2 = 0.5471$ and $p_{o_2}/p_{o_1} = 0.6281$. From Table A.1, for $M_2 = 0.5471$, we have $A_2/A_2^* = 1.27$. Note an important fact at this stage of our calculation. The normal shock is assumed to be infinitely thin, hence $A_1 = A_2$. However, we have previously shown that $A_1/A_1^* = 2$ and $A_2/A_2^* = 1.27$. *Clearly, the value of A^* changes across the shock wave.* This is due to the entropy increase across the shock. A_1^* is the flow area necessary to achieve Mach 1 isentropically in the flow upstream of the shock, and A_2^* is the flow area necessary to achieve Mach 1 isentropically in the flow downstream of the shock. Since the entropy is different for these two flows, then A^* is different for the two flows. Proceeding with the calculation,

$$\frac{A_e}{A_2^*} = \frac{A_e}{A_2}\frac{A_2}{A_2^*} = \frac{A_e}{A_t}\frac{A_t}{A_2}\frac{A_2}{A_2^*} = (3)\left(\tfrac{1}{2}\right)(1.27) = 1.905$$

The flow is *subsonic* behind the normal shock wave, and hence is subsonic throughout the remainder of the divergent portion downstream of the shock. Thus, from the subsonic entries in Table A.1, we have for $A_e/A_2^* = 1.905$, $M_e = 0.32$ and $p_{o_e}/p_e = 1.074$. Thus, since $p_o = p_{o_1}$ and $p_{o_e} = p_{o_2}$, we have

$$\frac{p_e}{p_o} = \frac{p_e}{p_{o_e}} \frac{p_{o_e}}{p_{o_2}} \frac{p_{o_2}}{p_{o_1}} \frac{p_{o_1}}{p_o} = \left(\frac{1}{1.074}\right)(1)(0.6281)(1) = \boxed{0.585}$$

Example 5.6 treated the case of a normal shock standing inside a nozzle. In this example, the location of the normal shock inside the nozzle was given, and the exit-to-reservoir pressure ratio, p_e/p_o, was calculated. This is a straightforward calculation, as demonstrated in Example 5.6. However, in most applications we are not given the location of the shock, but rather we know the pressure ratio p_e/p_o across the nozzle, and we want to find the location of the shock (i.e., the value of A/A_t, where the shock is standing). In this situation, we can take either of two approaches.

The first approach is an iterative solution. Assume the location of the shock in the nozzle, i.e., assume the value of A/A_t for the shock. Then calculate the pressure ratio p_e/p_o that would correspond to the shock in this assumed location, using the approach taken in Example 5.6. Check to see if p_e/p_o from this calculation agrees with the specified value of p_e/p_o. If not, assume another location of the shock, and calculate the new value of p_e/p_o corresponding to this new shock location. Repeat this iterative process until the proper shock location is found that will yield a calculated p_e/p_o that agrees with the specified value.

The second approach is direct, but more elaborate. Consider a normal shock standing inside a nozzle, as sketched in Fig. 5.19. The reservoir pressure is p_o and the static pressure at the exit is p_e; the pressure ratio across the nozzle is therefore p_e/p_o. Immediately upstream of the shock (condition 1), the total pressure is p_{o_1}. Because the flow is isentropic between the reservoir and location 1, $p_{o_1} = p_o$. Recall that A^* is a constant value everywhere upstream of the shock, and is equal to the throat area, A_t. Denote this value of A^* by A_1^*. Immediately downstream of the shock (condition 2), the total pressure is p_{o_2}. Also, recall that the value of A^* changes across the shock. Denote the value of A^* downstream of the shock by A_2^*, which is a constant value

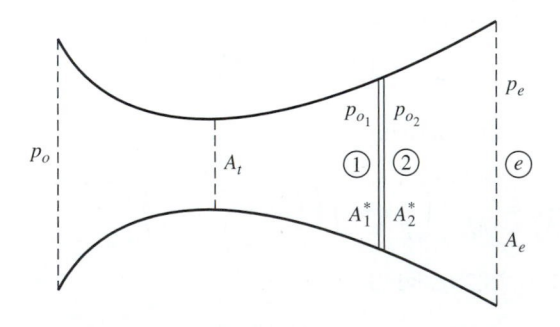

Figure 5.19 | Conditions associated with a normal shock standing inside a nozzle.

everywhere downstream of the shock. The mass flow at any location in the nozzle is $\dot{m} = \rho u A$. In Problem 5.6 at the end of this chapter, you are asked to derive this equation for the mass flow through a choked nozzle:

$$\dot{m} = \frac{p_o A^*}{\sqrt{T_o}} \sqrt{\frac{\gamma}{R} \left(\frac{2}{\gamma+1}\right)^{(\gamma+1)/(\gamma-1)}} \tag{5.21}$$

where A^* is equal to the throat area, and p_o and T_o are the reservoir pressure and temperature, respectively. Since Eq. (5.21) is of the form

$$\dot{m} = \frac{p_o A^*}{\sqrt{T_o}} f(\gamma, R)$$

we see that mass flow is directly proportional to $p_o A^*/(T_o)^{1/2}$. Since both the mass flow and T_o are constant across the shock wave in Fig. 5.19, we have from Eq. (5.21):

$$p_o A^* = \text{constant across a shock wave}$$

or

$$p_{o_1} A_1^* = p_{o_2} A_2^* \tag{5.22}$$

Referring to Fig. 5.19, since the flow is isentropic from location 2 to the exit, $p_{o_e} = p_{o_2}$ and $A_e^* = A_2^*$. Thus, Eq. (5.22) becomes

$$p_{o_1} A_1^* = p_{o_e} A_e^* \tag{5.23}$$

Hence, from Eq. (5.23) we can write

$$\frac{p_e A_e}{p_{o_e} A_e^*} = \frac{p_e A_e}{p_{o_1} A_1^*} = \left(\frac{p_e}{p_{o_1}}\right)\left(\frac{A_e}{A_1^*}\right) \tag{5.24}$$

In Eq. (5.24), p_e/p_{o_1} is the specified pressure ratio across the given nozzle. Also, A_e/A_1^* is the known exit-to-throat area ratio for the given nozzle. Hence the right-hand side of Eq. (5.24) is a known number, and therefore the ratio $(p_e A_e)/(p_{o_e} A_e^*)$ is a known number. This ratio can be expressed in terms of the exit Mach number as shown next.

From Eq. (3.30), we can write

$$\frac{p_e}{p_{o_e}} = \left(1 + \frac{\gamma-1}{2} M_2^2\right)^{-\gamma/(\gamma-1)} \tag{5.25}$$

and from Eq. (5.20) we can write

$$\frac{A_e}{A_e^*} = \frac{1}{M_e}\left[\left(\frac{2}{\gamma+1}\right)\left(1 + \frac{\gamma-1}{2} M_e^2\right)\right]^{(\gamma+1)/[2(\gamma-1)]} \tag{5.26}$$

The product of Eqs. (5.25) and (5.26) is

$$\frac{p_e}{p_{o_e}} \frac{A_e}{A_e^*} = \frac{1}{M_e}\left(\frac{2}{\gamma+1}\right)^{(\gamma+1)/[2(\gamma-1)]}\left(1 + \frac{\gamma-1}{2} M_e^2\right)^{-1/2} \tag{5.27}$$

Solving Eq. (5.27) for M_e^2, we have

$$M_e^2 = -\frac{1}{\gamma - 1} + \sqrt{\frac{1}{(\gamma - 1)^2} + \left(\frac{2}{\gamma - 1}\right)\left(\frac{2}{\gamma + 1}\right)^{(\gamma+1)/(\gamma-1)}\left(\frac{p_{o_e}A_e^*}{p_e A_e}\right)^2}$$

(5.28)

Since $p_{o_e}A_e^*/p_e A_e$ is a known number from Eq. (5.24), Eq. (5.28) allows the direct calculation of the exit Mach number. Keep in mind that for the flow shown in Fig. 5.19, M_e will be a subsonic value.

The remaining steps required to solve for the location of the normal shock are

1. For the value of M_e obtained from Eq. (5.28), obtain p_{o_e}/p_e from Table A.1.
2. Calculate the ratio of the total pressure across the shock from

$$\frac{p_{o_2}}{p_{o_1}} = \frac{p_{o_e}}{p_{o_1}} = \frac{p_{o_e}}{p_e}\frac{p_e}{p_{o_1}}$$

(5.29)

where p_e/p_{o_1} is the specified pressure ratio across the nozzle.
3. For the value of p_{o_2}/p_{o_1} calculated from Eq. (5.29), obtain M_1 from Table A.2.
4. For the value of M_1, obtain A_1/A_1^* from Table A.1.

Since $A_1/A_1^* = A_1/A_t$, the value of A_1/A_1^* obtained from step 4 is the location of the normal shock wave inside the nozzle.

EXAMPLE 5.7

Consider a convergent-divergent nozzle with an exit-to-throat area ratio of 3. The inlet reservoir pressure is 1 atm and the exit static pressure is 0.5 atm. For this pressure ratio, a normal shock will stand somewhere inside the divergent portion of the nozzle. Calculate the location of the shock wave using (a) a trial-and-error solution and (b) the direct solution. Compare the results.

■ **Solution**

a. Assume $A/A_t = A/A_1^* = 2.3$. From Table A.1, $M_1 = 2.35$. From Table A.2, $M_2 = 0.5286$ and $p_{o_2}/p_{o_1} = 0.5615$. From Table A.1, for $M_2 = 0.5286$, $A/A_2^* = 1.303$. (Recall that we are using nearest entries in the table.) Hence,

$$\frac{A_e}{A_2^*} = \frac{A_e}{A_1^*}\frac{A_1^*}{A}\frac{A}{A_2^*} = (3)\left(\frac{1}{2.3}\right)(1.303) = 1.7$$

For $A_e/A_2^* = 1.7$, from Table A.1, $M_e = 0.36$, and $p_{o_e}/p_e = 1.094$. Hence,

$$p_e = \frac{p_e}{p_{o_e}}\frac{p_{o_2}}{p_{o_1}}p_{o_1} = \frac{1}{1.094}(0.5615)(1) = 0.513\ \text{atm}$$

Since p_e should be 0.5 atm, assume a new A/A_1^* (closer to the exit), and start over again. Assume $A/A_1^* = 2.4$. For this, $M_1 = 2.4$, $M_2 = 0.5231$, $p_{o_2}/p_{o_1} = 0.5401$, and

$A/A_2^* = 1.303$. (Again, recall that we are using nearest entries.) Hence, $A_e/A_2^* = (3)(1/2.4)(1.303) = 1.629$. With this, $M_e = 0.39$ and $p_{o_e}/p_e = 1.111$. Hence,

$$p_e = \frac{p_e}{p_{o_e}}\frac{p_{o_2}}{p_{o_1}}p_{o_1} = \frac{1}{1.111}(0.5401)(1) = 0.486 \text{ atm}$$

Since p_e should be 0.5 atm, the value of 0.486 atm is too low by about the same amount as the first iteration is too high. Splitting the difference, the correct location of the normal shock wave is approximately $\boxed{A/A_t = 2.35}$.

b. Using the direct method, from the specified conditions

$$\left(\frac{p_e}{p_{o_1}}\right)\left(\frac{A_e}{A_1^*}\right) = \left(\frac{0.5}{1.0}\right)(3) = 1.5$$

From Eq. (5.24),

$$\frac{p_e A_e}{p_{o_e} A_e^*} = 1.5$$

From Eq. (5.28)

$$M_e^2 = -\frac{1}{\gamma - 1} + \sqrt{\frac{1}{(\gamma - 1)^2} + \left(\frac{2}{\gamma - 1}\right)\left(\frac{2}{\gamma + 1}\right)^{(\gamma+1)/(\gamma-1)}\left(\frac{p_{o_e} A_e^*}{p_e A_e}\right)^2}$$

$$= -2.5 + \sqrt{(2.5)^2 + (5)(0.8333)^6 \left(\frac{1}{1.5}\right)^2}$$

$$= -2.5 + \sqrt{6.994} = 0.1447$$

Hence,

$$M_e = 0.38$$

From Table A.1 for $M_e = 0.38$, $p_{o_e}/p_e = 1.094$. From Eq. (5.29),

$$\frac{p_{o_2}}{p_{o_1}} = \frac{p_{o_e}}{p_e}\frac{p_e}{p_{o_1}} = (1.094)\left(\frac{0.5}{1}\right) = 0.547$$

From Table A.2, for $p_{o_2}/p_{o_1} = 0.547$, $M_1 = 2.38$. From Table A.1, for $M_1 = 2.38$, $A/A_1^* = A/A_t = \boxed{2.36}$. This direct answer compares to that obtained with the iteration in part (a) to within 0.4 percent.

5.5 | DIFFUSERS

Let us go through a small thought experiment. Assume that we want to design a supersonic wind tunnel with a test section Mach number of 3 (see Fig. 5.11). Some immediate information about the nozzle is obtained from Table A.1; at $M = 3$, $A_e/A^* = 4.23$ and $p_o/p_e = 36.7$. Assume the wind tunnel exhausts to the atmosphere. What value of total pressure p_o must be provided by the reservoir to drive the tunnel? There are several possible alternatives. The first is to simply exhaust the nozzle directly to the atmosphere, as sketched in Fig. 5.20. In order to avoid shock

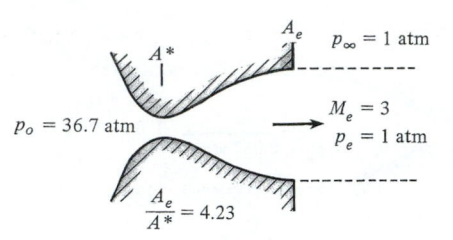

Figure 5.20 | Nozzle exhausting directly to the atmosphere.

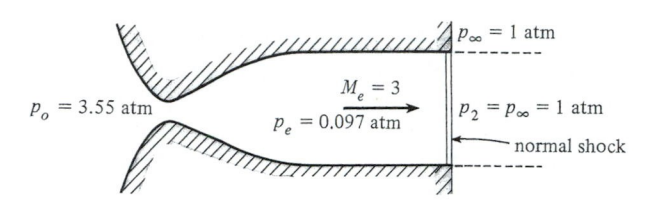

Figure 5.21 | Nozzle with a normal shock at the exit, exhausting to the atmosphere.

or expansion waves in the test region downstream of the exit, the exit pressure p_e must be equal to the surrounding atmospheric pressure, i.e., $p_e = 1$ atm. Since $p_o/p_e = 36.7$, the driving reservoir pressure for this case must be 36.7 atm. However, a second alternative is to exhaust the nozzle into a constant-area duct which serves as the test section, and to exhaust this duct into the atmosphere, as sketched in Fig. 5.21. In this case, because the testing area is inside the duct, shock waves from the duct exit will not affect the test section. Therefore, assume a normal shock stands at the duct exit. The static pressure behind the normal shock is p_2, and because the flow is subsonic behind the shock, $p_2 = p_\infty = 1$ atm. In this case, the reservoir pressure p_o is obtained from

$$p_o = \frac{p_o}{p_e} \frac{p_e}{p_2} p_\infty = 36.7 \frac{1}{10.33} 1 = 3.55 \text{ atm}$$

where p_2/p_e is the static pressure ratio across a normal shock at Mach 3, obtained from Table A.2. Note that, by the simple addition of a constant-area duct with a normal shock at the end, the reservoir pressure required to drive the wind tunnel has markedly dropped from 36.7 to 3.55 atm. Now, as a third alternative, add a divergent duct behind the normal shock in Fig. 5.21 in order to slow the already subsonic flow to a lower velocity before exhausting to the atmosphere. This is sketched in Fig. 5.22. At the duct exit, the Mach number is a very low subsonic value, and for all practical purposes the local total and static pressure are the same. Moreover, assuming an isentropic flow in the divergent duct behind the shock, the total pressure at the duct exit is equal to the total pressure behind the normal shock. Consequently,

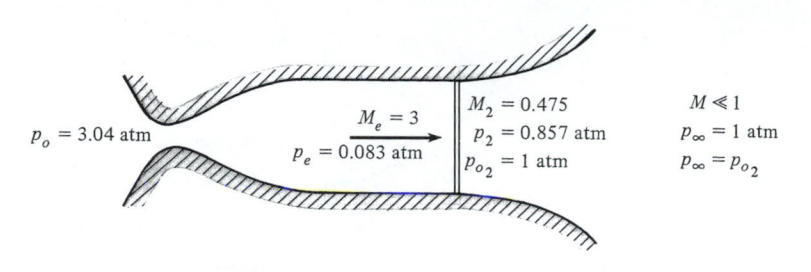

Figure 5.22 | Nozzle with a normal-shock diffuser. The normal shock is slightly upstream of the divergent duct.

$p_{o_2} \approx p_\infty = 1$ atm. From Table A.2, the Mach number immediately behind the shock is $M_2 = 0.475$, and the ratio of total to static pressure at this Mach number (from Table A.1) is $p_{o_2}/p_2 = 1.17$. Hence

$$p_o = \frac{p_o}{p_e}\frac{p_e}{p_2}\frac{p_2}{p_{o_2}}p_\infty = 36.7\frac{1}{10.33}\frac{1}{1.17}1 = 3.04 \text{ atm}$$

This is even better yet—the total pressure required to drive the wind tunnel has been further reduced to 3.04 atm.

Take a look at what has happened! From Table A.2, note the ratio of total pressures across a normal shock wave at Mach 3 is $p_{o_2}/p_{o_1} = 0.328$. Hence $p_{o_1}/p_{o_2} = 1/0.328 = 3.04$; this is precisely the pressure ratio required to drive the wind tunnel in Fig. 5.22! Thus, from this thought experiment, we infer that the reservoir pressure required to drive a supersonic wind tunnel (and hence the power required from the compressors) is considerably reduced by the creation of a normal shock and subsequent isentropic diffusion to $M \approx 0$ at the tunnel exit, and that this pressure is simply determined by the total pressure loss across a normal shock wave at the test section Mach number.

The normal shock and divergent exhaust duct in Fig. 5.22 are acting as a specific mechanism to slow the air to low subsonic speeds before exhausting to the atmosphere. Such mechanisms are called *diffusers,* and their function is to *slow the flow with as small a loss of total pressure as possible.* Of course, the ideal diffuser would compress the flow isentropically, hence with no loss of total pressure. For example, consider the wind tunnel sketched in Fig. 5.11. After isentropically expanding through the supersonic nozzle and passing through the test section, conceptually the supersonic flow could be isentropically compressed by the convergent part of the diffuser to sonic velocity at the second throat, and then further isentropically compressed to low velocity in the divergent section downstream of the throat. This would take place with no loss in total pressure, and hence the pressure ratio required to drive the tunnel would be unity—a perpetual motion machine! Obviously, something is wrong. The problem can be seen by reflecting on the results of Chap. 4. When the convergent part of the diffuser changes the direction of the supersonic flow at the wall, it is extremely difficult to prevent oblique shock waves from occurring inside the duct. Moreover, even without shocks, the real-life effects of friction between the flow and the diffuser surfaces cause a loss of total pressure. Therefore, the design of a perfect isentropic diffuser is physically impossible.

Accepting the fact that a perfect diffuser cannot be built, can we still hope to do better than the normal shock diffuser sketched in Fig. 5.22? The answer is yes, because it can easily be shown that the total pressure loss across a series of oblique shocks and a terminating weak normal shock is less than that across a single strong normal shock at the same upstream Mach number. (See Example 4.12 and Sec. 4.7.) Therefore, it would appear wise to replace the normal shock diffuser in Fig. 5.22 with an oblique shock diffuser as sketched in Fig. 5.23. Here, the test section flow at Mach number M_e and static pressure p_e is slowed down through a series of oblique shock waves initiated by a compression corner at the inlet of the diffuser, further slowed by a weak normal shock wave at the end of the constant-area section, and then subsonically compressed by a divergent section which exhausts to the atmosphere. At the diffuser exit, the static pressure is p_d, which for subsonic flow at the exit is equal to p_∞. In concept, this oblique shock diffuser should provide greater pressure recovery (smaller loss in total pressure) than a normal shock diffuser. However, in practice, the interaction of the shock waves in Fig. 5.23 with the viscous boundary layer on the diffuser walls creates an additional total pressure loss which tends to partially mitigate the advantages of an oblique shock diffuser. The real flow through an oblique shock diffuser is shown in the photograph of Fig. 5.24. The shock waves and boundary layers are made visible by a schlieren system—an optical technique sensitive to density gradients in the flow. Note the decay of the diamond-shaped oblique shock

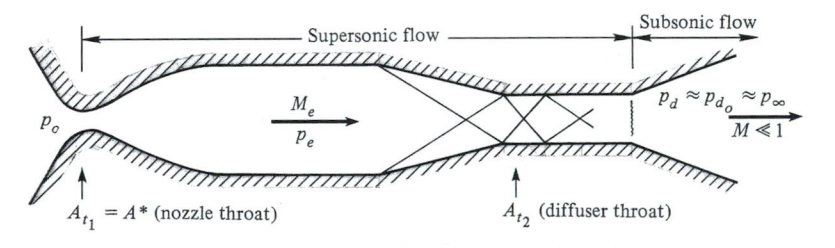

Figure 5.23 | Nozzle with a conventional supersonic diffuser.

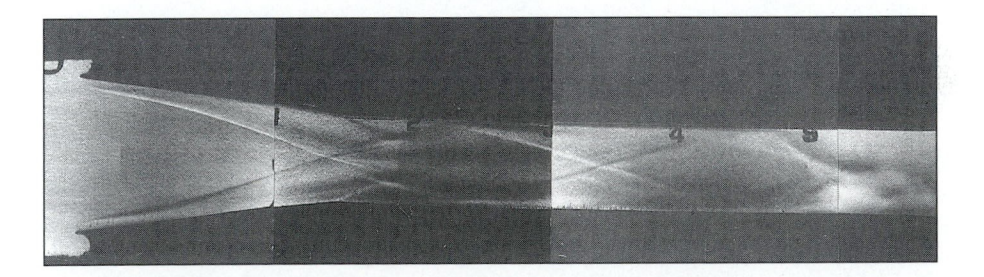

Figure 5.24 | Oblique shock pattern in a two-dimensional supersonic diffuser. The flow is from left to right, and the inlet Mach number is 5. (*Photo was taken by the author at the Aerospace Research Laboratory, Wright-Patterson Air Force Base, OH.*)

pattern due to viscous interaction downstream. The net result is that the full potential of an oblique shock diffuser is never fully achieved.

In the literature, there are several figures of merit used to denote the efficiency of diffusers. For wind tunnel work, the most common definition of diffuser efficiency is to compare the actual total pressure ratio across the diffuser, p_{d_o}/p_o, with the total pressure ratio across a hypothetical normal shock wave at the test section Mach number, p_{o_2}/p_{o_1} (using the nomenclature of Fig. 3.9). Let η_D denote diffuser efficiency. Then

$$\eta_D = \frac{\left(p_{d_o}/p_o\right)_{\text{actual}}}{\left(p_{o_2}/p_{o_1}\right)_{\text{normal shock at } M_e}} \tag{5.30}$$

If, $\eta_D = 1$, then the actual diffuser is performing as if it were a normal shock diffuser. For low supersonic test section Mach numbers, diffusers in practice usually perform slightly better than normal shock ($\eta_D > 1$); however, for hypersonic conditions, normal shock recovery is about the best to be expected, and usually $\eta_D < 1$.[†]

Note from Figs. 5.11 and 5.23 that oblique shock diffusers have a minimum-area section, i.e., a throat. In wind tunnel nomenclature, the nozzle throat is called the *first throat*, with cross-sectional area $A_{t_1} = A^*$; the diffuser throat is called the *second throat*, with area A_{t_2}. Due to the entropy increase in the diffuser, $A_{t_2} > A_{t_1}$. To prove this, assume that sonic flow exists at both the first and second throats. From Eq. (5.1) evaluated between the two throats,

$$\rho_1^* A_{t_1} a_1^* = \rho_2^* A_{t_2} a_2^* \tag{5.31}$$

or

$$\frac{A_{t_2}}{A_{t_1}} = \frac{\rho_1^* \, a_1^*}{\rho_2^* \, a_2^*} \tag{5.32}$$

From Secs. 3.4 and 3.5, a^* and hence T^* are constant throughout a given adiabatic flow. Thus, $a_1^*/a_2^* = 1$, and Eq. (5.32) becomes

$$\frac{A_{t_2}}{A_{t_1}} = \frac{\rho_1^*}{\rho_2^*} \tag{5.33}$$

However, from the equation of state,

$$\frac{\rho_1^*}{\rho_2^*} = \frac{p_1^*/RT_1^*}{p_2^*/RT_2^*} = \frac{p_1^*}{p_2^*} \tag{5.34}$$

Substituting Eq. (5.34) into (5.33),

$$\frac{A_{t_2}}{A_{t_1}} = \frac{p_1^*}{p_2^*} \tag{5.35}$$

[†] For a more extensive discussion of supersonic diffusers, as well as their application in a modern situation, see Chap. 12 of Ref. 21.

Since $M_1 = M_2 = 1$, and from Eq. (3.30) evaluated at locations 1 and 2,

$$\frac{p_{o_1}}{p_1^*} = \left(1 + \frac{\gamma - 1}{2} M_1^2\right)^{\gamma/(\gamma-1)} = \left(\frac{\gamma + 1}{2}\right)^{\gamma/(\gamma-1)}$$

$$\frac{p_{o_2}}{p_2^*} = \left(1 + \frac{\gamma - 1}{2} M_2^2\right)^{\gamma/(\gamma-1)} = \left(\frac{\gamma + 1}{2}\right)^{\gamma/(\gamma-1)}$$

Eq. (5.35) can be written as

$$\boxed{\frac{A_{t_2}}{A_{t_1}} = \frac{p_{o_1}}{p_{o_2}}} \tag{5.36}$$

Since the total pressure always decreases across shock waves and within boundary layers, p_{o_2} will always be less than p_{o_1}. Thus, from Eq. (5.36), the second throat must always be larger than the first throat. Indeed, if we know the values of total pressure at the two throats, then Eq. (5.36) tells us precisely how large to make the second throat. If A_{t_2} is made smaller than demanded by Eq. (5.36), the mass flow through the tunnel cannot be handled by the diffuser; the diffuser "chokes," and supersonic flow in the nozzle and test section is not possible. Note from Eq. (5.36) that only for a hypothetical perfect diffuser (with isentropic flow throughout) would the area of the second throat be equal to that of the first throat.

For typical supersonic diffusers, the efficiency η_D is very sensitive to A_{t_2}, as sketched in Fig. 5.25. Note that as A_{t_2} is decreased from a large value, η_D first increases, reaches a peak value, then rapidly decreases. The peak efficiency is obtained by a value of A_{t_2} slightly larger than given by Eq. (5.36). Keep in mind that the value of A_{t_2} obtained from Eq. (5.36) is the minimum allowed value that will pass the incoming mass flow from the nozzle. Below this value, the flow will be choked, and the diffuser efficiency plummets. The value of A_{t_2} from Eq. (5.36) is represented by the dashed vertical line in Fig. 5.25. At much higher values of A_{t_2}, there are no problems with passing the incoming mass flow; however, the diffuser efficiency is compromised because the supersonic flow from the inlet is not sufficiently compressed and hence remains supersonic in the second throat. In the downstream divergent portion, this supersonic flow first accelerates, and then passes through a normal shock near the diffuser exit. Since the Mach number is fairly high in front of the shock, the total pressure loss across the normal shock is large. This defeats the purpose of an oblique shock diffuser (namely, to have a weak normal shock occur at the second throat in a near sonic flow). As a result, for large A_{t_2}, the diffuser efficiency is low, as sketched in Fig. 5.25.

Up to this stage in our discussion, the most serious problem with diffusers has not yet been mentioned—the starting problem. Consider again the wind tunnel sketched in Fig. 5.11. When the flow through this tunnel is first started (say by rapidly opening a pressure valve from the reservoir), a complicated transient flow pattern is established, which after a certain time interval settles to the familiar steady flow which we have been discussing in this chapter. The starting process is complex

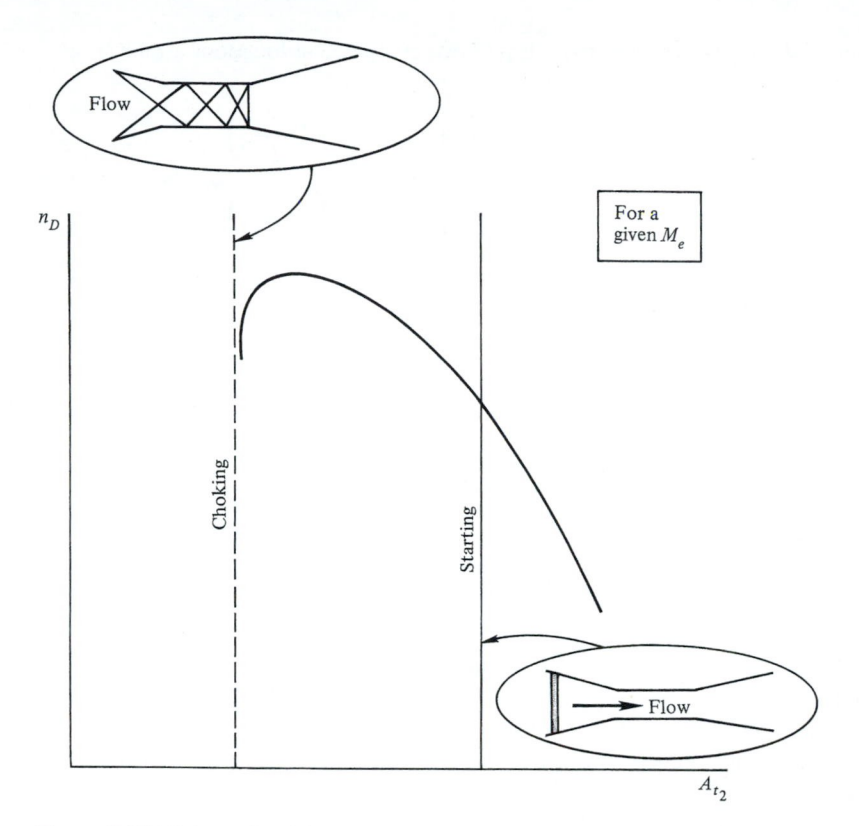

Figure 5.25 | Schematic of the variation of diffuser efficiency with second throat area.

and is still not perfectly understood. However, it is usually accompanied by a normal shock wave that sweeps through the complete duct from the nozzle to the diffuser. When this starting normal shock wave is momentarily at the inlet to the diffuser, the second throat area must be large enough to pass the mass flow behind a normal shock. This value of A_{t_2} is given by Eq. (5.36) where now p_{o_2}/p_{o_1} is the total pressure ratio across a normal shock at the test section Mach number. This starting value of A_{t_2} is represented by the solid vertical line in Fig. 5.25, and is always larger than the throat area for peak efficiency. If A_{t_2} is less than the starting value, the normal shock will remain upstream of the diffuser, and the tunnel flow will not start properly. If A_{t_2} is equal to or greater than the starting value, the normal shock will proceed through (be "swallowed" by) the diffuser, and the tunnel flow will start properly. Therefore, examining Fig. 5.25, we see that a fixed-geometry diffuser designed with a second throat area large enough to allow the flow to start will operate at an efficiency less than maximum. Herein lies the advantage of variable-geometry diffusers, where the throat area can be changed by some mechanical or fluid dynamic means. In such a diffuser, the throat area is made large enough to start the flow, and then later is decreased to obtain higher efficiency during running of the tunnel. However, the design and fabrication of variable-geometry diffusers is usually complex and expensive, and for this reason most operational wind tunnels use fixed-geometry diffusers.

Our discussion on diffusers has focused on a wind tunnel application for illustration of the general phenomena. However, the analysis of the flow through inlets and diffusers for air-breathing jet engines follows similar arguments. The reader is encouraged to read Shapiro (Ref. 16) or Zucrow and Hoffman (Ref. 17) for extensive discussions on such supersonic inlets.

The reader is cautioned not to take this discussion on diffusers too literally. The actual flow through diffusers is a complicated three-dimensional interaction of shock waves and boundary layers which is not well understood—even after a half-century of serious work on diffusers. Therefore, *diffuser design is more of an art than a science*. Diffuser efficiency is influenced by a myriad of parameters such as A_{t_2}/A_{t_1}, M_e, entrance angle, second throat length, etc. Therefore, the design of a diffuser for a given application must be based on empirical data and inspiration. Rarely is the first version of the new diffuser ever completely successful. In this context, the discussion of diffusers in this section is intended for general guidance only.

EXAMPLE 5.8

Consider the wind tunnel described in Example 5.2. Estimate the ratio of diffuser throat area to nozzle throat area required to allow the tunnel to start. Also, assuming that the diffuser efficiency is 1.2 after the tunnel has started, calculate the pressure ratio across the tunnel necessary for running, i.e., calculate the ratio of total pressure at the diffuser exit to the reservoir pressure.

■ **Solution**

From Table A.2, for $M = 2.5$: $p_{o_2}/p_{o_1} = 0.499$. From Eq. (5.36)

$$\frac{A_{t_2}}{A_{t_1}} = \frac{p_{o_1}}{p_{o_2}} = \frac{1}{0.499} = \boxed{2.00}$$

From Eq. (5.30)

$$\left(\frac{p_{d_o}}{p_o}\right)_{actual} = \eta_D \left(\frac{p_{o_2}}{p_{o_1}}\right)_{normal\ shock} = (1.2)(0.499) = \boxed{0.599}$$

Note: In Example 5.2, standard sea level conditions were stipulated in the test section. For this case, the pressure at the diffuser exit is far above atmospheric pressure. Specifically, from Example 5.2, $p_o = 17.09$ atm; hence $p_{d_o} = (0.599)(17.09) = 10.23$ atm. If the diffuser exhausted directly to the atmosphere, the flow would rapidly expand to supersonic velocity in the free jet downstream of the tunnel exit, with accompanying tremendous losses. Therefore, for this particular wind tunnel, a *closed circuit* design is by far the best. That is, the low subsonic flow at the exit of the diffuser is ducted right back to the entrance of the nozzle. The tunnel forms a closed loop, and the pressure loss in passing through the tunnel and the return loop is made up by a fan with a motor drive. Since the gas is also heated by the addition of power from this motor drive, a cooler must also be inserted in the return loop. See Chap. 5 of Ref. 9 for a more detailed discussion of the design of a closed-loop (or closed-return) supersonic wind tunnel.

5.6 | WAVE REFLECTION FROM A FREE BOUNDARY

Although they are not inherently quasi-one-dimensional flows, the wave patterns shown emanating from the nozzle exit in Figs. 5.18d and e are frequently encountered in the study of nozzle flows. Therefore, it is appropriate to discuss them at this stage.

The gas jet from a nozzle which exhausts into the atmosphere has a boundary surface which interfaces with the surrounding quiescent gas. As in the case of the slip lines discussed in Chap. 4, the pressure across this boundary must be preserved; hence the jet boundary pressure must equal p_∞ along its complete length. Therefore, the oblique shock waves shown in Fig. 5.18d and the expansion waves sketched in Fig. 5.18e must reflect from the jet boundary in such a fashion as to preserve the pressure at the boundary downstream of the nozzle exit. This jet boundary is not a *solid* surface as treated in Chap. 4; rather, it is a free boundary which can change in size and direction. For example, consider the incident shock wave impinging on a constant-pressure free boundary as shown in Fig. 5.26. In region 1, the pressure is p_∞, equal to the surrounding atmosphere. In region 2 behind the incident shock, $p_2 > p_\infty$. However, at the edge of the jet boundary (the dashed line in Fig. 5.26), the pressure must always be p_∞. Therefore, when the incident shock hits the boundary, it must be reflected in such a fashion as to obtain p_∞ in region 3 behind the reflected wave. Since $p_3 = p_\infty < p_2$, this reflected wave must be an expansion wave, as sketched in Fig. 5.26. In turn, the flow is deflected upward by both the incident shock and reflected expansion, causing the free boundary to deflect upward also. The strength of the reflected expansion wave is readily obtained from the theory presented in Chap. 4.

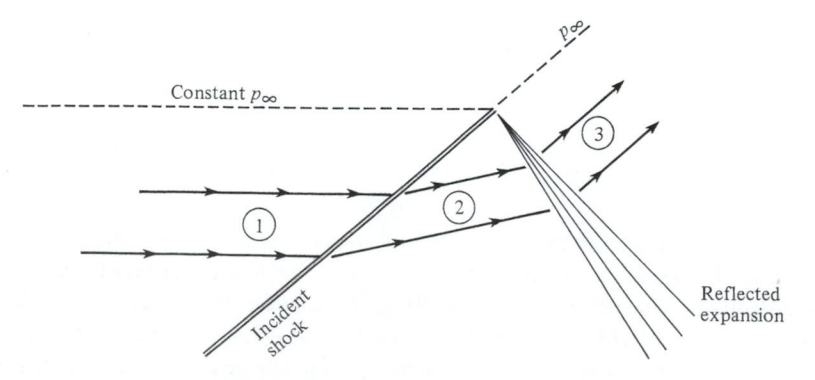

Figure 5.26 | Shock wave incident on a constant-pressure boundary.

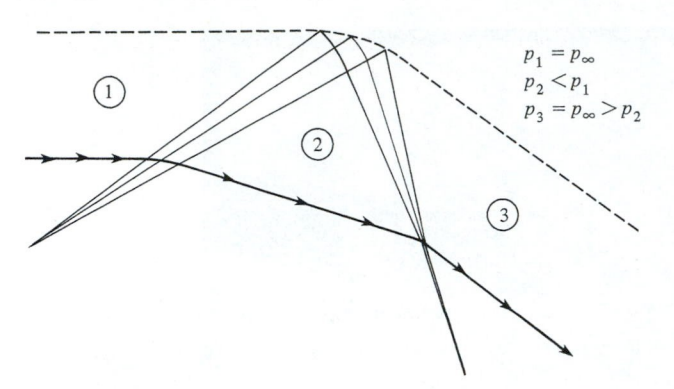

Figure 5.27 | Reflection of an expansion wave incident
on a constant-pressure boundary.

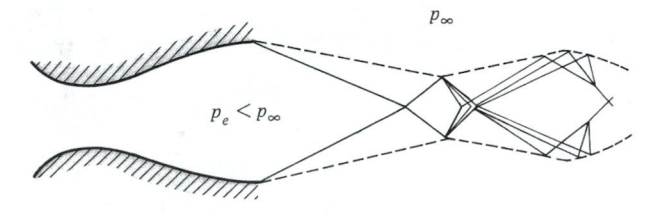

Figure 5.28 | Schematic of the diamond wave pattern in the
exhaust from a supersonic nozzle.

Analogously, the incident expansion wave shown in Fig. 5.27 is reflected from a
free boundary as a compression wave. This finite compression wave quickly coalesces into a shock wave, as shown. The wave interaction shown in Fig. 5.27 must be
analyzed by the method of characteristics, to be discussed in Chap. 11.

From this discussion combined with our results of Chap. 4, we conclude that

1. Waves incident on a solid boundary reflect in like manner, i.e., a compression
    wave reflects as a compression and an expansion wave reflects as an
    expansion.
2. Waves incident on a free boundary reflect in opposite manner, i.e., a
    compression wave reflects as an expansion and an expansion wave reflects
    as a compression.

Considering the overexpanded nozzle flow in Fig. 5.18d, the flow pattern downstream of the nozzle exit will appear as sketched in Fig. 5.28. The various reflected
waves form a diamond-like pattern throughout the exhaust jet. Such a diamond wave
pattern is visible in the exhaust from the free jet shown in Fig. 5.29. The reader is
left to sketch the analogous wave pattern for the underexpanded nozzle flow in
Fig. 5.18e.

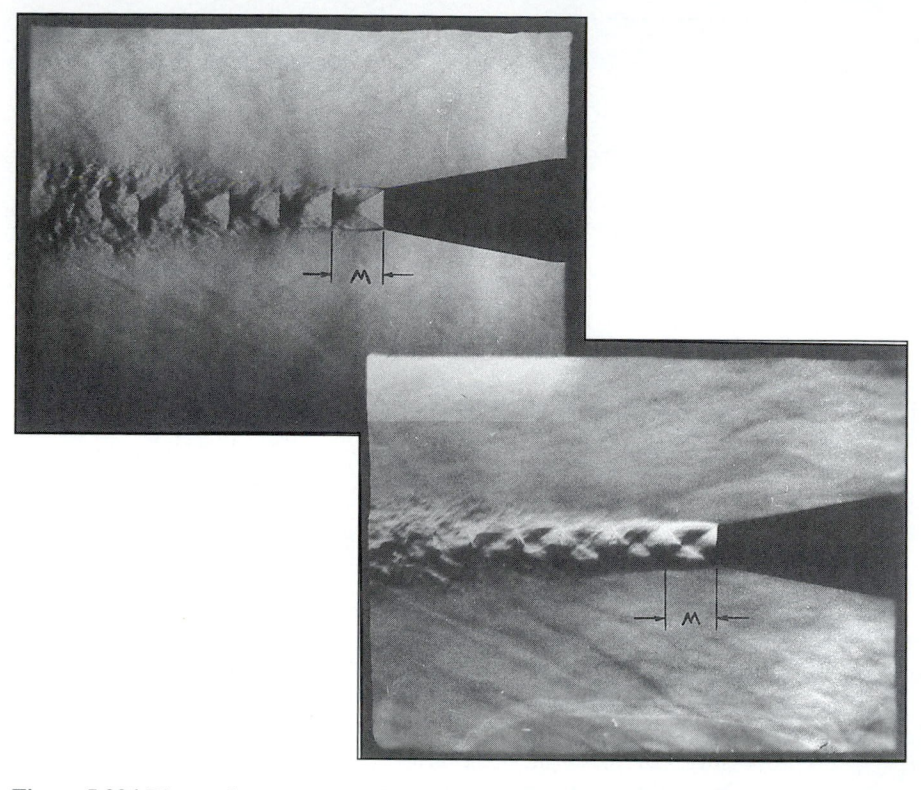

Figure 5.29 | Diamond wave patterns from an axisymmetric free jet (similar to the exhaust from a rocket engine). Taken from E. S. Love, C. E. Grigsby, L. P. Lee, and M. J. Woodling, "Experimental and Theoretical Studies of Axisymmetric Free Jets," NASA Tech. Report No. TR R-6, 1959. M is the wavelength of the first diamond.

5.7 | SUMMARY

This brings to an end the technical discussion of the present chapter. The quasi-one-dimensional duct flows discussed herein, in concert with the shock and expansion waves discussed in Chaps. 3 and 4, constitute a first tier in the overall structure of compressible flow. You should take this material very seriously, and should make certain that you feel comfortable with the major concepts and results. This will promote a smoother excursion into the remaining chapters.

5.8 | HISTORICAL NOTE: DE LAVAL— A BIOGRAPHICAL SKETCH

The first practical use of a convergent-divergent supersonic nozzle was made before the twentieth century. As related in Sec. 1.1, the Swedish engineer, Carl G. P. de Laval, designed a steam turbine in the late 1800s which incorporated supersonic expansion nozzles upstream of the turbine blades (see Fig. 1.8). For this reason, such convergent-divergent nozzles are frequently referred to as "Laval nozzles" in the literature. Who was de Laval? What prompted him to design a supersonic nozzle for steam turbines? What kind of man was he? Let us take a closer look.

Carl Gustaf Patrick de Laval was born at Blasenborg, Sweden, on May 9, 1845. The son of a Swedish army captain, de Laval showed an early interest in mechanical mechanisms, disassembling and then reassembling such devices as watches and gun locks. His parents encouraged his development along these lines, and at the age of 18 de Laval entered the University of Upsala, graduating in 1866 with high honors in engineering. He was then employed by a Swedish mining company, the Stora Kopparberg, where he quickly realized that he needed more education. (This is a phenomenon which has affected young engineers through the ages.) Therefore, he returned to Upsala, where he studied chemistry, physics, and mathematics, and graduated with a Ph.D. in 1872. From there, he returned to the Stora Company for 3 years, and then joined the Kloster Iron Works in Germany in 1875. By this time, his inventive genius was beginning to surface: he developed a sieve for improving the distribution of air in bessemer converters, and a new apparatus for galvanizing processes. Also, during his time with Kloster, de Laval was experimenting with centrifugal machines for the separation of cream in milk. Unable to convince Kloster to manufacture his cream separator, de Laval resigned in 1877, moved to Stockholm, and started his own company. Within 30 years, he had sold more than a million de Laval cream separators, and to the present day he is better known in Europe for cream separators then for steam turbines.

However, it was with his steam turbine designs that de Laval made a lasting contribution to the advancement of compressible flow. In 1882, he constructed his first steam turbine using rather conventional nozzles. Such nozzles were convergent shapes, indeed nothing more than orifices in some designs of that day. In turn, the kinetic energy of the steam entering the rotor blades was low, resulting in low rotational turbine speeds. The cause of this deficiency was recognized—the pressure ratio across such nozzles was never less than one-half. Today, as described in Secs. 5.3 and 5.4, we know that such nozzles were choked, and that the flow exhausted from the nozzle exit at a velocity that was not greater than sonic. However, in 1882, engineers did not fully understand such phenomena. Finally, in 1888, de Laval hit upon the system of further expanding the gas by adding a divergent section to the original convergent shape. Suddenly, his steam turbines began to operate at incredible rotational speeds—over 30,000 r/min. Overcoming the many mechanical problems introduced by such an improvement in rotational speed, de Laval developed his turbine business into a large corporation in Stockholm, and quickly obtained a number of international affiliates, in France, Germany, England, the Netherlands, Austria-Hungary, Russia, and the United States. Subsequently, his design was demonstrated at the World Columbian Exposition in Chicago in 1893, as related in Sec. 1.1.

In addition to his successes as an engineer and businessman, de Laval was also adroit in his social relations. He was respected and liked by his social peers and employees. He held national office—being elected to the Swedish Parliament during

1888 to 1890, and later becoming a member of the Senate. He was awarded numerous honors and decorations, and was a member of the Swedish Royal Academy of Science.

After a full and productive life, Carl G. P. de Laval died in Stockholm in 1912 at the age of 67. However, his influence and his company have lasted to the present day.

It is interesting to note that, on a technical basis, de Laval and other contemporary engineers in 1888 were not quite certain that supersonic flow actually existed in the "Laval nozzle." This was a point of contention that was not properly resolved until the experiments of Stodola in 1903. But Stodola's story is told in the next section.

5.9 | HISTORICAL NOTE: STODOLA, AND THE FIRST DEFINITIVE SUPERSONIC NOZZLE EXPERIMENTS

The innovative steam turbine nozzle design by de Laval (see Secs. 1.1 and 5.8) sparked interest in the fluid mechanics of flow through convergent-divergent nozzles at the turn of the century. Leading this interest was an Hungarian-born engineer by the name of Aurel Boleslav Stodola, who was to eventually become the leading expert in Europe on steam turbines. However, whereas de Laval was an idea and design man, Stodola was a scholarly professor who tied up the loose scientific and technical strings associated with Laval nozzles. Stodola is a major figure in the advancement of compressible flow, thermodynamics, and steam turbines. Let us see why, and at the same time take a look at the man himself.

Stodola was born on May 10, 1859, in Liptovsky Mikulas, Hungary, a small Slovakian town at the foot of the High Tatra mountains. The second son of a leather manufacturer, he attended the Budapest Technical University for 1 year in 1876. He was an exceptional student, and in 1877 he shifted to the University of Zurich in Switzerland, and then to the Eidgenossische Technische Hochschule in 1878, also in Zurich. Here, he graduated in 1880 with a mechanical engineering degree. Subsequently, he served a brief time with Ruston and Company in Prague, where he was responsible for the design of several different types of steam engines. However, his superb performance as a student soon earned him a "Chair for Thermal Machinery" back at the Eidgenossische Technische Hochschule in Zurich, a position he held until his retirement in 1929.

There, Stodola established a glowing academic career which included teaching, industrial consultation, and engineering design. However, his main contributions were in applied research. Stodola had a synergistic combination of high mathematical competence with an intense devotion to practical applications. Moreover, he understood the importance of engineering research at a time when it was virtually nonexistent throughout the world. In 1903 (the same year as the Wright brothers' first powered airplane flight), Stodola wrote:

> We engineers of course know that machine building, through widely extended practical experimenting, has solved problems, with the utmost ease, which baffled scientific investigation for years. But this "cut and try method," as engineers ironically term it, is often

extremely costly; and one of the most important questions of all technical activity, that of efficiency, should lead us not to underestimate the results of scientific technical work.

This commentary on the role of basic scientific research was aimed primarily at the design of steam turbines. But it was prophetic of the massive and varied research programs to come during the latter half of the twentieth century.

The importance of Stodola to our consideration in the present book lies in his pioneering work on the flow of steam through Laval nozzles. As mentioned in Sec. 5.8, the possibility of supersonic flow in such nozzles, although theoretically established, had not been experimentally verified, and therefore was a matter of controversy. To study this problem, Stodola constructed a convergent-divergent nozzle with the shape illustrated at the top of Fig. 5.30. He could vary the backpressure over

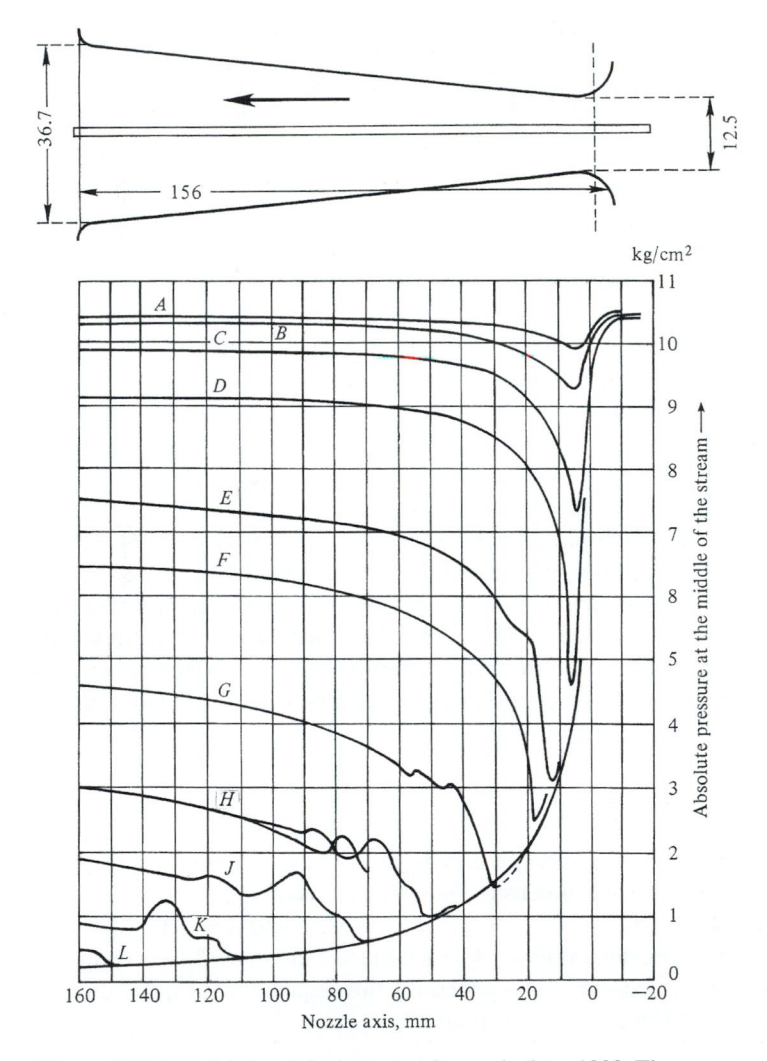

Figure 5.30 | Stodola's original supersonic nozzle data, 1903. The curves are pressure distributions for different backpressures.

any desired range by closing a valve downstream of the nozzle exit. With pressure taps in a long, thin tube extended through the nozzle along its centerline (also shown in Fig. 5.30), Stodola measured the axial pressure distributions associated with different backpressures. These data are shown below the nozzle configuration in Fig. 5.30. This figure is taken directly from Stodola's original publication, a book entitled *Steam Turbines,* first published in 1903. Here, for the first time in history, the characteristics of the flow through a supersonic nozzle were experimentally confirmed. In Fig. 5.30, the lowest curve corresponds to a complete isentropic expansion (as illustrated in Fig. 5.14*c*). The curves *D* through *L* in Fig. 5.30 correspond to a shock wave inside the nozzle, induced by higher backpressures (as illustrated in Fig. 5.17*c*). The curves *A*, *B*, and *C* in Fig. 5.30 correspond to completely subsonic flow induced by high backpressures (as illustrated in Fig. 5.15*c*). With regard to the large jumps in pressure shown by some of the data in Fig. 5.30, Stodola comments:

> I see in these extraordinary heavy increases of pressure a realization of the "compression shock" theoretically derived by von Riemann; because steam particles possessed of great velocity strike against a slower moving steam mass and are therefore compressed to a higher degree.

(In this quote, Stodola is referring to G. F. Bernhard Riemann mentioned in Sec. 3.10; however, he would be historically more correct to refer instead to Rankine and Hugoniot, as described in Sec. 3.10.) Stodola's nozzle experiments, as described, and his original data shown in Fig. 5.30, represented a quantum-jump in the understanding of supersonic nozzle flows. Taken in conjunction with de Laval's contributions, Stodola's work represents the original historical underpinning for the material given in this chapter. Furthermore, this work was quickly picked up by Ludwig Prandtl at Göttingen, who went on to make dramatic schlerien photographs of waves in supersonic nozzle flows, as described in Sec. 4.16.

Stodola died in Zurich on December 25, 1942, at the age of 83. During his lifetime, he became the leading world expert on steam turbines, and his students permeated the Swiss steam turbine manufacturing companies, making those companies into international leaders in this field. Moreover, he had exceptional personal charm. The loyalty of his friends was extraordinary, and he acquired an almost disciplelike group during his long life in Zurich. Even upon his death, the number and persuasiveness of his eulogies were exceptional. Clearly, Stodola has left a permanent mark in the history of compressible flow.

5.10 | SUMMARY

Quasi-one-dimensional flow is defined as flow wherein all the flow properties are functions of one space dimension only, say *x*, whereas the flow cross-sectional area is a variable, i.e., $u = u(x)$, $p = p(x)$, $T = T(x)$, and $A = A(x)$. This is in contrast to the purely one-dimensional flows discussed in Chap. 3, where the flow cross-sectional area is constant. The governing flow equations for quasi-one-dimensional flow, obtained from a control volume model, are

Continuity:
$$\rho_1 u_1 A_1 = \rho_2 u_2 A_2 \tag{5.1}$$

Momentum: $\qquad p_1 A_1 + \rho_1 u_1^2 A_1 + \displaystyle\int_{A_1}^{A_2} p\, dA = p_2 A_2 + \rho_2 u_2^2 A_2$ $\qquad\qquad$ (5.2)

Energy: $\qquad\qquad\qquad\qquad h_1 + \dfrac{u_1^2}{2} = h_2 + \dfrac{u_2^2}{2}$ $\qquad\qquad\qquad$ (5.5)

The differential forms of these equations are:

Continuity: $\qquad\qquad\qquad\qquad d(\rho u A) = 0$ $\qquad\qquad\qquad\qquad$ (5.7)

Momentum: $\qquad\qquad\qquad\qquad dp = -\rho u\, du$ $\qquad\qquad\qquad\qquad$ (5.9)

Energy: $\qquad\qquad\qquad\qquad dh + u\, du = 0$ $\qquad\qquad\qquad\qquad$ (5.10)

These equations hold for inviscid, adiabatic flow—hence isentropic flow. They can be combined to yield the *area-velocity relation*

$$\frac{dA}{A} = (M^2 - 1)\frac{du}{u} \qquad\qquad (5.15)$$

which states, among other aspects, that

1. If the flow is subsonic, an *increase* in velocity corresponds to a *decrease* in area.
2. If the flow is supersonic, an *increase* in velocity corresponds to an *increase* in area.
3. If the flow is sonic, the area is at a local minimum.

These results clearly state that, in order to expand an isentropic flow from subsonic to supersonic speeds, a convergent-divergent duct must be used, where Mach 1 will occur at the minimum area (the throat) of the duct.

Quasi-one-dimensional isentropic flow is dictated by the *area–Mach number relation,*

$$\left(\frac{A}{A^*}\right)^2 = \frac{1}{M^2}\left[\frac{2}{\gamma+1}\left(1 + \frac{\gamma-1}{2}M^2\right)\right]^{(\gamma+1)/(\gamma-1)} \qquad (5.20)$$

where A^* is the flow area at a local value of Mach 1. From Eq. (5.20) we note the pivotal result that local Mach number is a function of *only* A/A^* (and, of course, γ).

To understand the various flowfields possible in a quasi-one-dimensional, convergent-divergent duct, imagine that the reservoir pressure is held fixed and the backpressure downstream of the exit is progressively reduced. These cases are possible, as we progressively reduce the backpressure:

1. First, the flow is completely subsonic, including both the convergent and the divergent sections. The maximum value of the Mach number (still subsonic) occurs at the throat. The mass flow continually increases as the backpressure is reduced.

2. At some specific value of the backpressure, the flow at the throat becomes sonic. The Mach numbers both upstream and downstream of the throat are still subsonic. The mass flow reaches a maximum value; when the backpressure is further reduced, the mass flow remains constant. The flow is *choked*.

3. As the backpressure is further reduced, a region of supersonic flow occurs downstream of the throat, terminated by a normal shock wave standing inside the divergent region.

4. At some specific value of the backpressure, the normal shock will be located exactly at the exit. The fully isentropic, subsonic-supersonic flow pattern now exists throughout the entire duct, except right at the exit.

5. As the backpressure is further reduced, the normal shock is replaced by oblique shocks emanating from the edge of the nozzle exit. This is called an *overexpanded* nozzle flow.

6. At some specific value of the backpressure, corresponding to the isentropic flow value, no waves of any kind will exist in the flow; we will have the purely isentropic subsonic-supersonic expansion through the nozzle, with no waves at the exit.

7. Finally, for a lower backpressure, expansion waves will emanate from the edge of the nozzle exit. This is called an *underexpanded* nozzle flow.

The function of a diffuser is to slow a flow with the smallest possible loss of total pressure. For a supersonic or hypersonic wind tunnel, the diffuser must slow the flow to a low subsonic speed at the end of the tunnel. For a measure of how efficient the diffuser is, the normal shock diffuser efficiency is defined as

$$\eta_D = \frac{\left(p_{d_o}/p_o\right)_{\text{actual}}}{\left(p_{o_2}/p_{o_1}\right)_{\text{normal shock at } M_e}} \tag{5.30}$$

where p_{d_o}/p_o is the actual ratio of total pressure between the exit of the diffuser and the nozzle reservoir, and p_{o_2}/p_{o_1} is the usual total pressure ratio across a normal shock wave at the design Mach number at the nozzle exit. A supersonic diffuser has a local minimum of cross-sectional area called the second throat; the ratio of the second throat area (diffuser) to the first throat area (nozzle) is given by

$$\frac{A_{t_2}}{A_{t_1}} = \frac{p_{o_1}}{p_{o_2}} \tag{5.36}$$

PROBLEMS

5.1 A supersonic wind tunnel is designed to produce flow in the test section at Mach 2.4 at standard atmospheric conditions. Calculate:

  a. The exit-to-throat area ratio of the nozzle

  b. Reservoir pressure and temperature

5.2 The reservoir pressure of a supersonic wind tunnel is 10 atm. A Pitot tube inserted in the test section measures a pressure of 0.627 atm. Calculate the test section Mach number and area ratio.

5.3 The reservoir pressure of a supersonic wind tunnel is 5 atm. A static pressure probe is moved along the center-line of the nozzle, taking measurements at various stations. For these probe measurements, calculate the local Mach number and area ratio:

   a.   4 atm

   b.   2.64 atm

   c.   0.5 atm

5.4 Consider the purely subsonic flow in a convergent-divergent duct. The inlet, throat, and exit area are 1 m^2, 0.7 m^2, and 0.85 m^2, respectively. If the inlet Mach number and pressure are 0.3 and $0.8 \times 10^5 \text{ N/m}^2$, respectively, calculate:

   a. M and p at the throat

   b. M and p at the exit

5.5 Consider the subsonic flow through a divergent duct with area ratio $A_2/A_1 = 1.7$. If the inlet conditions are $T_1 = 300 \text{ K}$ and $u_1 = 250 \text{ m/s}$, and the pressure at the exit is $p_2 = 1$ atm, calculate:

   a.   Inlet pressure p_1

   b.   Exit velocity u_2.

5.6 The mass flow of a calorically perfect gas through a choked nozzle is given by

$$\dot{m} = \frac{p_o A^*}{\sqrt{T_o}} \sqrt{\frac{\gamma}{R} \left(\frac{2}{\gamma + 1} \right)^{(\gamma+1)/(\gamma-1)}}$$

Derive this relation.

5.7 When the reservoir pressure and temperature of a supersonic wind tunnel are 15 atm and 750 K, respectively, the mass flow is 1.5 kg/s. If the reservoir conditions are changed to $p_o = 20$ atm and $T_o = 600 \text{ K}$, calculate the mass flow.

5.8 A blunt-nosed aerodynamic model is mounted in the test section of a supersonic wind tunnel. If the tunnel reservoir pressure and temperature are 10 atm and 800°R, respectively, and the exit-to-throat area ratio is 25, calculate the pressure and temperature at the nose of the model.

5.9 Consider a flat plate mounted in the test section of a supersonic wind tunnel. The plate is at an angle of attack of 10° and the static pressure on the top surface of the plate is 1.0 atm. The nozzle throat area is 0.05 m^2 and the exit area is 0.0844 m^2. Calculate the reservoir pressure of the tunnel.

5.10 Consider a supersonic nozzle with a Pitot tube mounted at the exit. The reservoir pressure and temperature are 10 atm and 500 K, respectively. The

pressure measured by the Pitot tube is 0.6172 atm. The throat area is 0.3 m^2.
Calculate:

a.   Exit Mach number M_e

b.   Exit area A_e

c.   Exit pressure and temperature p_e and T_e

d.   mass flow through the nozzle

5.11 Consider a convergent-divergent duct with exit and throat areas of 0.5 m^2 and
0.25 m^2, respectively. The inlet reservoir pressure is 1 atm and the exit static
pressure is 0.6 atm. For this pressure ratio, the flow will be supersonic in a
portion of the nozzle, terminating with a normal shock inside the nozzle.
Calculate the local area ratio (A/A^*) at which the shock is located inside the
nozzle.

5.12 Consider a supersonic wind tunnel where the nozzle area ratio is $A_e/A_{t_1} =$
104.1. The throat area of the nozzle is $A_{t_1} = 1.0\,\text{cm}^2$. Calculate the minimum
area of the diffuser throat, A_{t_2}, which will allow the tunnel to start.

5.13 At the exit of the diffuser of a supersonic wind tunnel which exhausts directly
to the atmosphere, the Mach number is very low (≈ 0.1). The reservoir
pressure is 1.8 atm, and the test section Mach number is 2.6. Calculate the
diffuser efficiency η_D.

5.14 In a supersonic nozzle flow, the exit-to-throat area ratio is 10, $p_o = 10$ atm,
and the backpressure $p_B = 0.04$ atm. Calculate the angle θ through which the
flow is deflected immediately after leaving the edge (or lip) of the nozzle exit.

5.15 Consider an oblique shock wave with $M_1 = 4.0$ and $\beta = 50°$. This shock
wave is incident on a constant-pressure boundary, as sketched in Fig. 5.26.
For the flow downstream of the reflected expansion wave, calculate the Mach
number M_3 and the flow direction relative to the flow upstream of the shock.

5.16 Consider a rocket engine burning hydrogen and oxygen. The combustion
chamber temperature and pressure are 4000 K and 15 atm, respectively. The
exit pressure is 1.174×10^{-2} atm. Calculate the Mach number at the exit.
Assume that $\gamma = \text{constant} = 1.22$ and that $R = 519.6\,\text{J/kg K}$.

5.17 We wish to design a Mach 3 supersonic wind tunnel, with a static pressure
and temperature in the test section of 0.1 atm and 400°R, respectively.
Calculate:

a.   The exit-to-throat area ratio of the nozzle

b.   The ratio of diffuser throat area to nozzle throat area

c.   Reservoir pressure

d.   Reservoir temperature

5.18 Consider two hypersonic wind tunnels with the same reservoir temperature of
3000 K in air. (a) One tunnel has a test-section Mach number of 10. Calculate
the flow velocity in the test section. (b) The other tunnel has a test-section
Mach number of 20. Calculate the flow velocity in the test section.
(c) Compare the answers from (a) and (b), and discuss the physical
significance of this comparison.

5.19 Consider a hypersonic wind tunnel with a reservoir temperature of 3000 K in air. Calculate the theoretical maximum velocity obtainable in the test section. Compare this result with the results of Problem 5.18 (a) and (b).

5.20 As Problems 5.18 and 5.19 reflect, the air temperature in the test section of conventional hypersonic wind tunnels is low. In reality, air liquefies at a temperature of about 50 K (depending in part on the local pressure as well). In the practical operation of a hypersonic wind tunnel, liquefaction of the test stream gas should be avoided; when liquefaction occurs, the test stream is a two-phase flow, and the test data is compromised. For a Mach 20 tunnel using air, calculate the minimum reservoir temperature required to avoid liquefaction in the test section.

5.21 The reservoir temperature calculated in Problem 5.20 is beyond the capabilities of heaters in the reservoir of continuous-flow wind tunnels using air. This is why you do not see a Mach 20 continuous-flow tunnel using air. On the other hand, consider the flow of helium, which has a liquefaction temperature of 2.2 K at the low pressures in the test section. This temperature is much lower than that of air. For a Mach 20 wind tunnel using helium, calculate the minimum reservoir temperature required to avoid liquefaction in the test section. For helium, the ratio of specific heats is 1.67.

5.22 The result from Problem 5.21 shows that the reservoir temperature for a Mach 20 helium tunnel can be very reasonable. This is why several very high Mach number helium hypersonic wind tunnels exist. For the helium wind tunnel in Problem 5.21, calculate the nozzle exit-to-throat area ratio. Compare this with the exit-to-throat area ratio required for an air Mach 20 tunnel.

CHAPTER 6

Differential Conservation Equations for Inviscid Flows

The information needed by design engineers of either aircraft or flow machinery is the pressure, the shearing stress, the temperature, and the heat flux vector imposed by the moving fluid over the surface of a specified solid body or bodies in a fluid stream of specified conditions. To supply this information is the main purpose of the discipline of gasdynamics.

H. S. Tsien, 1953

PREVIEW BOX

In 1913 Gertrude Stein wrote the familiar phrase: "Rose is a rose is a rose is a rose." We borrow from this phrase for the present chapter and say: "A conservation equation is a conservation equation is a conservation equation is a conservation equation." However, like roses that come in a multitude of varieties, the conservation equations come in a variety of forms. We have derived and utilized one form of the conservation equations in the previous chapters, namely, the integral form. In the present chapter, we expand our horizons, and obtain the governing conservation equations in the form of partial differential equations, which will be necessary for the study of the multidimensional and/or unsteady flows highlighted in the remainder of this book.

The panoply of the various forms of the conservation equations are frequently confusing to new students of fluid dynamics because they look so different and yet they speak the same physics. For example, there are at least a dozen different forms of the energy equation, yet in the final analysis they are essentially the same and they all are a statement of the first law of thermodynamics. In the simplest sense, the origin of the different forms stem from the particular model of the flow used to

obtain them. For example, four different models of the flow are illustrated in Figs. 2.2 and 2.3. When we apply the fundamental principle of mass conservation to each of these four models, four different forms of the continuity equation are directly obtained. Although these four equations look completely different, they can be reworked into each other's form by mathematical manipulation, and all four forms represent the same physical principle, namely, that mass is conserved. A purpose of the present chapter is to somewhat demystify the different forms of the equations for you, and to emphasize those forms that will be useful for the remainder of this book. For a complete discussion of the fundamental governing equations, their different forms, and where they come from, see Chap. 2 of the author's book *Computational Fluid Dynamics: The Basics With Applications* (Ref. 18), where a major effort is made to demystify the equations in their many different forms.

By now you get the point. This chapter is all about equations; it is essentially a continuation of Chap. 2, which also was all about equations. The comments made in our preview of Chap. 2 also apply here. The present chapter is very important because it adds many useful

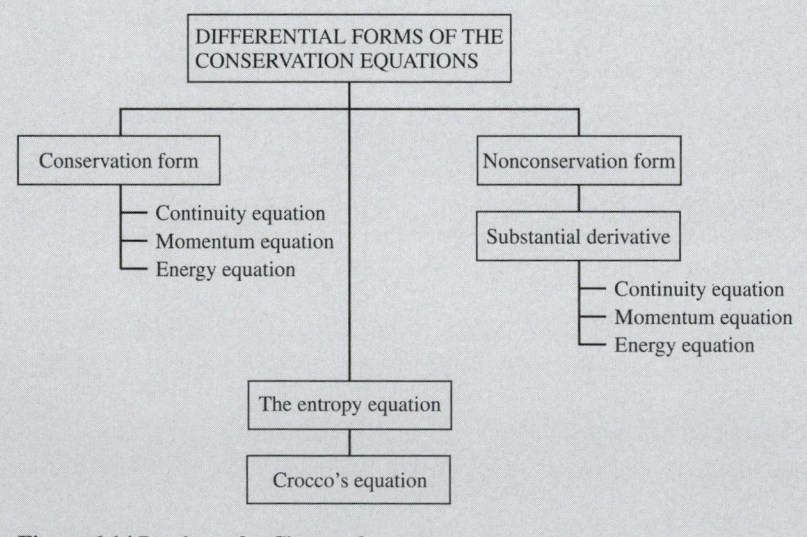

Figure 6.1 | Roadmap for Chapter 6.

tools to our toolbox. These tools will make it possible for us to examine a number of exciting applications later in the book.

The roadmap for this chapter is given in Fig. 6.1. We derive two forms of the differential conservation equations: the conservation form and the nonconservation form. (Do not be put off by the term "nonconservation form"—it is strictly nomenclature and does not imply any violation of the physics. The classification of the equations under the conservation and nonconservation forms is a fairly recent artifact that has come from the rise of computational fluid dynamics, and because this nomenclature is becoming more widespread, we use it here.) The chapter ends with two additional equations, the entropy equation and Crocco's theorem, which have certain special applications to our further studies.

6.1 | INTRODUCTION

The analysis of problems in fluid dynamics requires three primary steps:

1. Determine a model of the fluid.
2. Apply the basic principles of physics to this model in order to obtain appropriate mathematical equations embodying these principles.
3. Use the resulting equations to solve the specific problem of interest.

In Chap. 2, the model of the fluid chosen was a control volume. The basic principles of mass conservation, Newton's second law, and energy conservation were applied to a finite control volume to obtain integral forms of the conservation equations. In turn, these equations were applied to specific problems in Chaps. 3, 4, and 5. These applications were such that the integral conservation equations nicely reduced to algebraic equations describing properties at different cross sections of the flow. However, we are now climbing to a higher tier in our study of compressible flow, where most of the previous algebraic equations no longer hold. We will soon be dealing with problems of unsteady flow, as well as flows with two or three spatial dimensions. For such cases, the integral forms of the conservation equations from Chap. 2 must be applied to a small neighborhood surrounding a *point* in the flow, resulting in *differential equations,* which describe flow properties *at that point*. To expedite our analysis, we will make use of these vector identities:

$$\oiint_S \mathbf{A} \cdot d\mathbf{S} = \oiiint_{\mathcal{V}} (\nabla \cdot \mathbf{A}) \, d\mathcal{V} \tag{6.1}$$

$$\oiint_S \mathbf{\Phi} \, d\mathbf{S} = \oiiint_{\mathcal{V}} (\nabla \mathbf{\Phi}) \, d\mathcal{V} \tag{6.2}$$

where $\mathbf{A}$ and $\mathbf{\Phi}$ are vector and scalar functions, respectively, of time and space, and $\mathcal{V}$ is a control volume surrounded by a closed control surface S, as sketched in Fig. 2.4.

6.2 | DIFFERENTIAL EQUATIONS IN CONSERVATION FORM

6.2.1 Continuity Equation

Repeating for convenience the continuity equation, Eq. (2.2),

$$-\oiint_S \rho \mathbf{V} \cdot d\mathbf{S} = \iiint_{\mathscr{V}} \frac{\partial \rho}{\partial t} \, d\mathscr{V}$$

and using Eq. (6.1) in the form

$$\oiint_S (\rho \mathbf{V}) \cdot d\mathbf{S} = \iiint_{\mathscr{V}} \nabla \cdot (\rho \mathbf{V}) \, d\mathscr{V} \tag{6.3}$$

we combine Eqs. (2.2) and (6.3) to obtain

$$\iiint_{\mathscr{V}} \left[\frac{\partial \rho}{\partial t} + \nabla \cdot (\rho \mathbf{V}) \right] d\mathscr{V} = 0 \tag{6.4}$$

It might be argued that a control volume could be chosen such that, in some special case, integration of Eq. (6.4) over one part of the volume would exactly cancel the integration over the remaining part, giving zero for the right-hand side. However, the control volume is an arbitrary shape and size, and in general the only way Eq. (6.4) can be satisfied is for the integrand to be zero *at each point* within the volume. Hence,

$$\boxed{\frac{\partial \rho}{\partial t} + \nabla \cdot (\rho \mathbf{V}) = 0} \tag{6.5}$$

Equation (6.4) is the *differential form of the continuity equation.*

6.2.2 Momentum Equation

Repeating for convenience the momentum equation, Eq. (2.11),

$$\iiint_{\mathscr{V}} \rho \mathbf{f} \, d\mathscr{V} - \oiint_S p \, d\mathbf{S} = \iiint_{\mathscr{V}} \frac{(\partial \rho \mathbf{V})}{\partial t} \, d\mathscr{V} + \oiint_S (\rho \mathbf{V} \cdot d\mathbf{S})\mathbf{V}$$

and using Eq. (6.2) in the form

$$\oiint_S p \, d\mathbf{S} = \iiint_{\mathscr{V}} (\nabla p) \, d\mathscr{V} \tag{6.6}$$

we combine Eqs. (2.11) and (6.6) to obtain

$$\oiiint_{\mathscr{V}} \rho\mathbf{f}\, d\mathscr{V} - \oiiint_{\mathscr{V}} \nabla p\, d\mathscr{V} = \oiiint_{\mathscr{V}} \frac{\partial(\rho\mathbf{V})}{\partial t}\, d\mathscr{V} + \oiint_{S}(\rho\mathbf{V}\cdot d\mathbf{S})\mathbf{V} \qquad (6.7)$$

Equation (6.7) is a vector equation; for convenience, let us consider cartesian scalar components in the x, y, and z directions, respectively (see Fig. 2.4). The x component of Eq. (6.7) is

$$\oiiint_{\mathscr{V}} \rho f_x\, d\mathscr{V} - \oiiint_{\mathscr{V}} \frac{\partial p}{\partial x}\, d\mathscr{V} = \oiiint_{\mathscr{V}} \frac{\partial(\rho u)}{\partial t}\, d\mathscr{V} + \oiint_{S}(\rho\mathbf{V}\cdot d\mathbf{S})u \qquad (6.8)$$

However, from Eq. (6.1),

$$\oiint_{S}(\rho\mathbf{V}\cdot d\mathbf{S})u = \oiint_{S}(\rho u\mathbf{V})\cdot d\mathbf{S} = \oiiint_{\mathscr{V}} \nabla\cdot(\rho u\mathbf{V})\, d\mathscr{V} \qquad (6.9)$$

Substituting Eq. (6.9) into (6.8),

$$\oiiint_{\mathscr{V}} \left[\rho f_x - \frac{\partial p}{\partial x} - \frac{\partial(\rho u)}{\partial t} - \nabla\cdot(\rho u\mathbf{V})\right] d\mathscr{V} = 0 \qquad (6.10)$$

By the same reasoning used to obtain Eq. (6.5) from Eq. (6.4), Eq. (6.10) yields

$$\boxed{\frac{\partial(\rho u)}{\partial t} + \nabla\cdot(\rho u\mathbf{V}) = -\frac{\partial p}{\partial x} + \rho f_x} \qquad (6.11)$$

Equation (6.11) is the *differential form of the x component of the momentum equation*. The analogous y and z components are

$$\boxed{\frac{\partial(\rho v)}{\partial t} + \nabla\cdot(\rho v\mathbf{V}) = -\frac{\partial p}{\partial y} + \rho f_y} \qquad (6.12)$$

$$\boxed{\frac{\partial(\rho w)}{\partial t} + \nabla\cdot(\rho w\mathbf{V}) = -\frac{\partial p}{\partial z} + \rho f_z} \qquad (6.13)$$

6.2.3 Energy Equation

Repeating for convenience the energy equation, Eq. (2.20),

$$\oiiint_{\mathscr{V}} \rho\dot{q}\, d\mathscr{V} - \oiint_{S} p\mathbf{V}\cdot d\mathbf{S} + \oiiint_{\mathscr{V}} \rho(\mathbf{f}\cdot\mathbf{V})\, d\mathscr{V}$$

$$= \oiiint_{\mathscr{V}} \frac{\partial}{\partial t}\left[\rho\left(e + \frac{V^2}{2}\right)\right] d\mathscr{V} + \oiint_{S} \rho\left(e + \frac{V^2}{2}\right)\mathbf{V}\cdot d\mathbf{S}$$

and using Eq. (6.1) in the forms

$$\oiint_S \rho \left(e + \frac{V^2}{2} \right) \mathbf{V} \cdot d\mathbf{S} = \iiint_{\mathscr{V}} \nabla \cdot \left[\rho \left(e + \frac{V^2}{2} \right) \mathbf{V} \right] d\mathscr{V} \tag{6.14}$$

and

$$\oiint_S p\mathbf{V} \cdot d\mathbf{S} = \iiint_{\mathscr{V}} \nabla \cdot (p\mathbf{V}) \, d\mathscr{V} \tag{6.15}$$

we combine Eqs. (2.20), (6.14), and (6.15) to obtain

$$\iiint_{\mathscr{V}} \left\{ \rho\dot{q} - \nabla \cdot (p\mathbf{V}) + \rho(\mathbf{f} \cdot \mathbf{V}) - \frac{\partial}{\partial t} \left[\rho \left(e + \frac{V^2}{2} \right) \right] \right.$$

$$\left. - \nabla \cdot \left[\rho \left(e + \frac{V^2}{2} \right) \mathbf{V} \right] \right\} d\mathscr{V} = 0 \tag{6.16}$$

Setting the integrand equal to zero, we obtain

$$\boxed{ \frac{\partial}{\partial t} \left[\rho \left(e + \frac{V^2}{2} \right) \right] + \nabla \cdot \left[\rho \left(e + \frac{V^2}{2} \right) \mathbf{V} \right] = -\nabla \cdot (p\mathbf{V}) + \rho\dot{q} + \rho(\mathbf{f} \cdot \mathbf{V}) }$$

$$\tag{6.17}$$

Equation (6.17) is the *differential form of the energy equation*.

6.2.4   Summary

Equations (6.5), (6.11) through (6.13), and (6.17) are general equations that apply at any point in an unsteady, three-dimensional flow of a compressible inviscid fluid. They are nonlinear partial differential equations, and they contain all of the physical information and importance of the integral equations from which they were extracted. For virtually the remainder of this book, such differential forms of the basic conservation equations will be employed. Also, note that these equations contain divergence terms of the quantities $\rho\mathbf{V}$, $\rho u\mathbf{V}$, $\rho v\mathbf{V}$, $\rho w\mathbf{V}$, and $\rho(e + V^2/2)\mathbf{V}$. For this reason, these equations are said to be in *divergence form*. This form of the equations is also called the *conservation form* since they stem directly from the integral conservation equations applied to a fixed control volume. However, other forms of these equations are frequently used, as will be derived in Secs. 6.3 and 6.4. We have now finished the left-hand column of our roadmap in Fig. 6.1, and we move on to the right-hand column.

6.3 | THE SUBSTANTIAL DERIVATIVE

Consider a small fluid element moving through cartesian space as illustrated in Figs. 6.2a and b. The x, y, and z axes in these figures are fixed in space. Figure 6.2a shows the fluid element at point 1 at time $t = t_1$. Figure 6.2b shows the *same* fluid

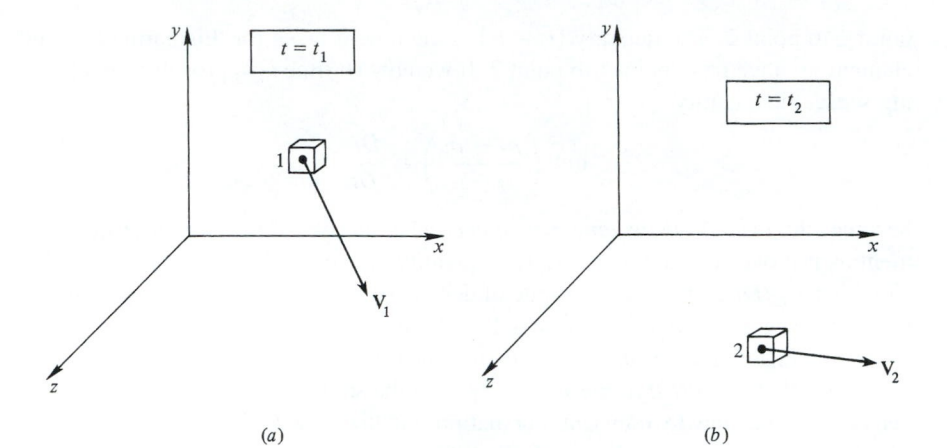

Figure 6.2 | Illustration of the substantial derivative (the xyz coordinate system above is fixed in space, and the fluid element is moving from point 1 to point 2).

element at point 2 in the flowfield at some later time, t_2. Throughout the (x, y, z) space, the velocity field is given by

$$V = u\mathbf{i} + v\mathbf{j} + w\mathbf{k}$$

where

$$u = u(x, y, z, t)$$
$$v = v(x, y, z, t)$$
$$w = w(x, y, z, t)$$

and $\mathbf{i}$, $\mathbf{j}$, and $\mathbf{k}$ are unit vectors in the x, y, and z directions, respectively. In addition, the density field is given by

$$\rho = \rho(x, y, z, t)$$

At time t_1, the density of the fluid element is $\rho_1 = \rho(x_1, y_1, z_1, t_1)$. At time t_2, the density of the *same* fluid element is $\rho_2 = \rho(x_2, y_2, z_2, t_2)$. Since $\rho = \rho(x, y, z, t)$, we can expand this function in a Taylor's series about point 1 as follows:

$$\rho_2 = \rho_1 + \left(\frac{\partial\rho}{\partial x}\right)_1 (x_2 - x_1) + \left(\frac{\partial\rho}{\partial y}\right)_1 (y_2 - y_1)$$

$$+ \left(\frac{\partial\rho}{\partial z}\right)_1 (z_2 - z_1) + \left(\frac{\partial\rho}{\partial t}\right)_1 (t_2 - t_1) + \text{higher-order term}$$

Dividing by $(t_2 - t_1)$, and ignoring higher-order terms,

$$\frac{\rho_2 - \rho_1}{t_2 - t_1} = \left(\frac{\partial\rho}{\partial x}\right)_1 \frac{(x_2 - x_1)}{(t_2 - t_1)} + \left(\frac{\partial\rho}{\partial y}\right)_1 \frac{(y_2 - y_1)}{(t_2 - t_1)} + \left(\frac{\partial\rho}{\partial z}\right)_1 \frac{(z_2 - z_1)}{(t_2 - t_1)} + \left(\frac{\partial\rho}{\partial t}\right)_1$$

$$(6.18)$$

Keep in mind the physical meaning of the left-hand side of Eq. (6.18). The quantity $(\rho_2 - \rho_1)$ is the change of density of the particular fluid element as it moves from point 1 to point 2. The quantity $(t_2 - t_1)$ is the time it takes for this particular fluid element to move from point 1 to point 2. If we now let time t_2 approach t_1 in a limiting sense, the quantity

$$\lim_{t_2 \to t_1} \left(\frac{\rho_2 - \rho_1}{t_2 - t_1} \right) \equiv \frac{D\rho}{Dt}$$

becomes the *instantaneous time rate of change* of density of the particular fluid element as it moves through point 1. This quantity is denoted by the symbol $D\rho/Dt$. Note that $D\rho/Dt$ is the rate of change of density of a *given fluid element* as it moves through space. Here, our eyes are fixed on the fluid element as it is moving. This is physically different than $(\partial \rho/\partial t)_1$, which is the time rate of change of density at the *fixed* point 1. For $(\partial \rho/\partial t)_1$, we fix our eyes on the stationary point 1 and watch the density change due to transient fluctuations in the flowfield. Thus, $D\rho/Dt$ and $(\partial \rho/\partial t)_1$ are physically and numerically different quantities.

Continuing with our limiting procedures, and again remembering that we are following a given fluid element,

$$\lim_{t_2 \to t_1} \frac{(x_2 - x_1)}{(t_2 - t_1)} \equiv u$$

$$\lim_{t_2 \to t_1} \frac{(y_2 - y_1)}{(t_2 - t_1)} \equiv v$$

$$\lim_{t_2 \to t_1} \frac{(z_2 - z_1)}{(t_2 - t_1)} \equiv w$$

Hence, returning to Eq. (6.18) and taking the limit as $t_2 \to t_1$, we obtain

$$\frac{D\rho}{Dt} = u\frac{\partial \rho}{\partial x} + v\frac{\partial \rho}{\partial y} + w\frac{\partial \rho}{\partial z} + \frac{\partial \rho}{\partial t}$$

From this, we can define the notation

$$\frac{D}{Dt} \equiv \frac{\partial}{\partial t} + u\frac{\partial}{\partial x} + v\frac{\partial}{\partial y} + w\frac{\partial}{\partial z} = \frac{\partial}{\partial t} + (\mathbf{V} \cdot \nabla) \tag{6.19}$$

as the *substantial derivative*. The time rate of change of *any* quantity associated with a particular moving fluid element is given by the substantial derivative. For example,

$$\frac{De}{Dt} = \frac{\partial e}{\partial t} + u\frac{\partial e}{\partial x} + v\frac{\partial e}{\partial y} + w\frac{\partial e}{\partial z} = \frac{\partial e}{\partial t} + (\mathbf{V} \cdot \nabla)e$$

where De/Dt is the time rate of change of internal energy per unit mass of the fluid element as it moves through a point in the flowfield, $\partial e/\partial t$ is the *local time derivative* at the point, and

$$u\frac{\partial e}{\partial x} + v\frac{\partial e}{\partial y} + w\frac{\partial e}{\partial z}$$

is the *convective derivative*. Again, physically, the properties of the fluid element are changing as it moves past a point in a flow because the flowfield itself may be fluctuating with time (the local derivative) and because the fluid element is simply on its way to another point in the flowfield where the properties are different (the convective derivative).

This example will help to reinforce the physical meaning of the substantial derivative. Consider the substantial derivative of the temperature, which from Eq. (6.19) is written as

$$\frac{DT}{Dt} = \frac{\partial T}{\partial t} + (\mathbf{V} \cdot \nabla)T \tag{6.19a}$$

Imagine that you are hiking in the mountains on a summer day, and you are about to enter a cave. The air temperature inside the cave is cooler than outside. Thus, as you walk through the mouth of the cave, you feel a temperature decrease—this is analogous to the convective derivative, $(\mathbf{V} \cdot \nabla)T$, in Eq. (6.19a). Moreover, being in the mountains, assume that some patches of snow remain from the previous winter. Imagine that you are with a friend who scoops up some of this snow and makes a snowball. Consider a point at the entrance to the cave. If the snowball were thrown through this point, there would be a momentary fluctuation in local temperature at the point due to the cold snowball. This temperature fluctuation is the local time derivative, $\partial T/\partial t$, in Eq. (6.19a). Imagine now that your friend throws the snowball past the entrance of the cave at the same instant you are walking through the entrance, hitting you with the snowball. You will feel an additional, but momentary, temperature drop when the snowball hits you—analogous to the local time derivative in Eq. (6.19a). The *net* temperature drop you feel as you walk through the mouth of the cave is therefore a combination of both the act of moving into the cave, where it is cooler, and being struck by the snowball at the same instant—this net temperature drop is analogous to the substantial derivative, DT/Dt, in Eq. (6.19a).

6.4 | DIFFERENTIAL EQUATIONS IN NONCONSERVATION FORM

6.4.1   Continuity Equation

Returning to Eq. (6.5) and expanding the divergence term (recalling the vector identity that $\nabla \cdot (a\mathbf{B}) = a\nabla \cdot \mathbf{B} + \mathbf{B} \cdot \nabla a$, where a is a scalar and $\mathbf{B}$ is a vector), we have

$$\frac{\partial \rho}{\partial t} + \rho\nabla \cdot \mathbf{V} + \mathbf{V} \cdot \nabla\rho = 0 \tag{6.20}$$

Slightly rearranging Eq. (6.20),

$$\frac{\partial \rho}{\partial t} + (\mathbf{V} \cdot \nabla)\rho + \rho\nabla \cdot \mathbf{V} = 0 \tag{6.21}$$

Incorporating the nomenclature of Eq. (6.19) into (6.21),

$$\boxed{\frac{D\rho}{Dt} + \rho \nabla \cdot \mathbf{V} = 0} \tag{6.22}$$

Equation (6.22) is an alternative form of the continuity equation given by Eq. (6.5). Physically, Eq. (6.22) says that the mass of a fluid element made up of a fixed set of particles (molecules and atoms) is constant as the fluid element moves through space. [For a chemically reacting flow, we have to think in terms of a fluid element made up of a fixed set of electrons and nuclei because the molecules and atoms inside the fluid element may increase or decrease due to chemical reaction; nevertheless, Eq. (6.22) is still valid for a chemically reacting flow.]

6.4.2   Momentum Equation

Returning to Eq. (6.11) and again expanding the divergence term as well as the time derivative,

$$\rho \frac{\partial u}{\partial t} + u \frac{\partial \rho}{\partial t} + u \nabla \cdot (\rho \mathbf{V}) + \rho \mathbf{V} \cdot \nabla u = -\frac{\partial p}{\partial x} + \rho f_x \tag{6.23}$$

Multiply Eq. (6.5) by u:

$$u \frac{\partial \rho}{\partial t} + u \nabla \cdot (\rho \mathbf{V}) = 0 \tag{6.24}$$

Subtract Eq. (6.24) from (6.23):

$$\rho \frac{\partial u}{\partial t} + \rho \mathbf{V} \cdot \nabla u = -\frac{\partial p}{\partial x} + \rho f_x \tag{6.25}$$

Using the substantial derivative given in Eq. (6.19),

$$\boxed{\rho \frac{Du}{Dt} = -\frac{\partial p}{\partial x} + \rho f_x} \tag{6.26}$$

By similar manipulation of Eqs. (6.12) and (6.13), we have

$$\boxed{\rho \frac{Dv}{Dt} = -\frac{\partial p}{\partial y} + \rho f_y} \tag{6.27}$$

$$\boxed{\rho \frac{Dw}{Dt} = -\frac{\partial p}{\partial z} + \rho f_z} \tag{6.28}$$

In vector form, Eqs. (6.26) through (6.28) can be written as

$$\boxed{\rho \frac{D\mathbf{V}}{Dt} = -\nabla p + \rho \mathbf{f}} \tag{6.29}$$

Equations (6.26) through (6.29) are different forms of *Euler's equation,* which is an alternative form of the momentum equation given in Eqs. (6.11) through (6.13). Euler's equation physically is a statement of Newton's second law, $\mathbf{F} = m\mathbf{a}$, applied to a moving fluid element of fixed identity.

6.4.3 Energy Equation

Returning to Eq. (6.17) and expanding,

$$\rho \frac{\partial(e + V^2/2)}{\partial t} + \left(e + \frac{V^2}{2}\right)\frac{\partial \rho}{\partial t} + \left(e + \frac{V^2}{2}\right)\nabla \cdot (\rho \mathbf{V})$$

$$+ \rho \mathbf{V} \cdot \nabla \left(e + \frac{V^2}{2}\right) = -\nabla \cdot (p\mathbf{V}) + \rho \dot{q} + \rho(\mathbf{f} \cdot \mathbf{V}) \qquad (6.30)$$

The second and third terms of Eq. (6.30), from the continuity equation, Eq. (6.5), give

$$\left(e + \frac{V^2}{2}\right)\left[\frac{\partial \rho}{\partial t} + \nabla \cdot (\rho \mathbf{V})\right] = \left(e + \frac{V^2}{2}\right)(0) = 0$$

Hence, along with the substantial derivative nomenclature, Eq. (6.30) becomes

$$\rho \frac{D(e + V^2/2)}{Dt} = -\nabla \cdot (p\mathbf{V}) + \rho \dot{q} + \rho(\mathbf{f} \cdot \mathbf{V}) \qquad (6.31)$$

Equation (6.31) is an alternative form of the energy equation given in Eq. (6.17). Equation (6.31) is a physical statement of the first law of thermodynamics applied to a moving fluid element of fixed identity; however, note that for a moving fluid, the energy is the *total* energy, $e + V^2/2$, i.e., the sum of both internal and kinetic energies per unit mass.

The energy equation is multifaceted—it can be written in many different forms, all of which you will sooner or later encounter in the literature. Therefore, it is important to sort out these different forms now. For example, let us obtain a form of Eq. (6.31) in terms of internal energy e only. Consider the left-hand side of Eq. (6.31),

$$\rho \frac{D(e + V^2/2)}{Dt} = \rho \frac{De}{Dt} + \rho \frac{D(V^2/2)}{Dt} = \rho \frac{De}{Dt} + \frac{\rho}{2}\frac{D(\mathbf{V} \cdot \mathbf{V})}{Dt}$$

$$= \rho \frac{De}{Dt} + \rho \mathbf{V} \cdot \frac{D\mathbf{V}}{Dt} \qquad (6.32)$$

Considering the first term of the right-hand side of Eq. (6.31),

$$\nabla \cdot (p\mathbf{V}) = p\nabla \cdot \mathbf{V} + \mathbf{V} \cdot \nabla p \qquad (6.33)$$

Substitute Eqs. (6.32) and (6.33) into Eq. (6.31):

$$\rho \frac{De}{Dt} + \rho \mathbf{V} \cdot \frac{D\mathbf{V}}{Dt} = -p\nabla \cdot \mathbf{V} - \mathbf{V} \cdot \nabla p + \rho \dot{q} + \rho(\mathbf{f} \cdot \mathbf{V}) \qquad (6.34)$$

Form the scalar product of **V** with the vector form of Euler's equation, Eq. (6.29):

$$\rho \mathbf{V} \cdot \frac{D\mathbf{V}}{Dt} = -\mathbf{V} \cdot \nabla p + \rho(\mathbf{f} \cdot \mathbf{V}) \tag{6.35}$$

Subtracting Eq. (6.35) from (6.34),

$$\boxed{\rho \frac{De}{Dt} = -p\nabla \cdot \mathbf{V} + \rho\dot{q}} \tag{6.36}$$

Equation (6.36) is an alternative form of the energy equation dealing with the rate of change of the internal energy of a moving fluid element.

Let us now obtain a form of the energy equation in terms of enthalpy h only. By definition of enthalpy,

$$h = e + pv = e + p/\rho$$

Thus,

$$\frac{Dh}{Dt} = \frac{De}{Dt} + \frac{D(p/\rho)}{Dt}$$

Rearranging,

$$\frac{De}{Dt} = \frac{Dh}{Dt} - \frac{D(p/\rho)}{Dt} = \frac{Dh}{Dt} - \left[\frac{\rho(Dp/Dt) - p(D\rho/Dt)}{\rho^2}\right]$$

Hence,

$$\frac{De}{Dt} = \frac{Dh}{Dt} - \frac{1}{\rho}\frac{Dp}{Dt} + \frac{p}{\rho^2}\frac{D\rho}{Dt} \tag{6.37}$$

However, recall Eq. (6.22), where

$$\frac{D\rho}{Dt} = -\rho\nabla \cdot \mathbf{V} \tag{6.38}$$

Combining Eqs. (6.37) and (6.38),

$$\frac{De}{Dt} = \frac{Dh}{Dt} - \frac{1}{\rho}\frac{Dp}{Dt} - \frac{p}{\rho}\nabla \cdot \mathbf{V} \tag{6.39}$$

and substituting Eq. (6.39) into (6.36), we have

$$\boxed{\rho\frac{Dh}{Dt} = \frac{Dp}{Dt} + \rho\dot{q}} \tag{6.40}$$

Equation (6.40) is an alternative form of the energy equation dealing with the rate of change of static enthalpy of a moving fluid element.

Let us now obtain a form of the energy equation in terms of total enthalpy $h_o = h + V^2/2$. Add Eqs. (6.31) and (6.40):

$$\rho \frac{D(e + V^2/2 + h)}{Dt} = -\nabla \cdot (p\mathbf{V}) + \frac{Dp}{Dt} + 2\rho\dot{q} + \rho(\mathbf{f} \cdot \mathbf{V}) \qquad (6.41)$$

Recalling that $Dp/Dt = \partial p/\partial t + \mathbf{V} \cdot \nabla p$, and subtracting Eq. (6.36) from (6.41),

$$\rho \frac{D(h + V^2/2)}{Dt} = -\nabla \cdot (p\mathbf{V}) + \frac{\partial p}{\partial t} + \mathbf{V} \cdot \nabla p + p\nabla \cdot \mathbf{V} + \rho\dot{q} + \rho(\mathbf{f} \cdot \mathbf{V})$$

$$= -p\nabla \cdot \mathbf{V} - \mathbf{V} \cdot \nabla p + \frac{\partial p}{\partial t} + \mathbf{V} \cdot \nabla p$$

$$+ p\nabla \cdot \mathbf{V} + \rho\dot{q} + \rho(\mathbf{f} \cdot \mathbf{V}) \qquad (6.42)$$

Cancelling terms in Eq. (6.42), and writing $h_o \equiv h + V^2/2$, we have

$$\boxed{\rho \frac{Dh_o}{Dt} = \frac{\partial p}{\partial t} + \rho\dot{q} + \rho(\mathbf{f} \cdot \mathbf{V})} \qquad (6.43)$$

Of all the alternative forms of the energy equation obtained to this point, Eq. (6.43) is probably the most useful and revealing. It states physically that the total enthalpy of a moving fluid element in an inviscid flow can change due to

1. Unsteady flow, i.e., $\partial p/\partial t \neq 0$
2. Heat transfer, i.e., $\dot{q} \neq 0$
3. Body forces, i.e., $\mathbf{f} \cdot \mathbf{V} \neq 0$

As we have already seen, many inviscid problems in compressible flow are also *adiabatic* with *no body forces*. For this case, Eq. (6.43) becomes

$$\boxed{\rho \frac{Dh_o}{Dt} = \frac{\partial p}{\partial t}} \qquad (6.44)$$

Furthermore, for a *steady flow,* Eq. (6.44) reduces to

$$\rho \frac{Dh_o}{Dt} = 0$$

which when integrated, yields

$$\boxed{h_o = \text{const}} \qquad (6.45)$$

Equation (6.45) is an important result—for an inviscid, adiabatic steady flow with no body forces, the total enthalpy is constant along a given streamline. This is to be expected almost from intuition and common sense; it is presaged by the steady shock wave results of Chaps. 3 and 4, and by the steady adiabatic duct flows of Chap. 5, where the total enthalpy is constant throughout the flow. Equation (6.45) holds only along a streamline because in the previous equations we are following a moving fluid element as it makes its way along a streamline. However, if the particular flowfield under study originates from a reservoir of common total enthalpy, such as the free stream far ahead of a body moving in the atmosphere, then the total enthalpy is the same value for all streamlines, and hence Eq. (6.45) holds throughout the complete flowfield. Finally, note that Eq. (6.45) is a simple algebraic statement of a fundamental physical result which holds no matter how complex the geometry of the flow may be. Although the continuity and momentum equations have to be dealt with as partial differential equations, the energy equation can be utilized as Eq. (6.45), subject of course to the stated restrictions. This will prove to be extremely useful in our subsequent discussions.

Let us obtain yet another alternative form of the energy equation. Solve Eq. (6.22) for $\nabla \cdot \mathbf{V}$,

$$\nabla \cdot \mathbf{V} = -\frac{1}{\rho} \frac{D\rho}{Dt} \tag{6.46}$$

Substitute Eq. (6.46) into (6.36):

$$\rho \frac{De}{Dt} = \frac{p}{\rho} \frac{D\rho}{Dt} + \rho \dot{q} \tag{6.47}$$

Recalling that $1/\rho = v$, hence

$$\frac{D\rho}{Dt} = -\frac{1}{v^2} \frac{Dv}{Dt}$$

then Eq. (6.47) becomes

$$\rho \frac{De}{Dt} = -\rho p \frac{Dv}{Dt} + \rho \dot{q}$$

$$\rho \left[\frac{De}{Dt} + p \frac{Dv}{Dt} - \dot{q} \right] = 0$$

$$\boxed{\frac{De}{Dt} + p \frac{Dv}{Dt} - \dot{q} = 0} \tag{6.48}$$

Compare Eq. (6.48) with the first law of thermodynamics as given by Eq. (1.25)—the two are *identical*. However, in Eq. (6.48), the changes in internal energy and specific volume are those taking place in a moving fluid element, and hence the differentials de and dv in Eq. (1.25) are physically replaced by the substantial derivatives De/Dt and Dv/Dt. Indeed, in hindsight, Eq. (6.48) could have been derived directly by applying Eq. (1.25) to a moving fluid element. Instead, we chose to derive Eq. (6.48) from a consistent evolution of our general energy equation for a moving fluid, Eq. (6.31), where we recognized that the energy of the fluid is both internal energy and kinetic energy. In the process, we have obtained a rather striking physical result—the internal and kinetic energies of a moving fluid can be separated such that the first law written strictly in terms of internal energy only does indeed apply to a moving fluid element, as clearly proven by Eq. (6.48).

6.4.4 Comment

All the forms of the equations derived in the present section are labeled the *nonconservation* form of the governing equations. They involve changes of fluid properties of a given fluid element as it *moves through the flowfield,* and hence they all involve substantial derivatives. This is in contrast to the conservation form derived in Sec. 6.2, which was obtained from the point of view of a control volume *fixed in space*. The label "nonconservation" is perhaps misleading. This does not mean that the physics of the flow is being violated and that something physically in the flow is not being conserved that should be conserved. Indeed, either form of the governing equations—conservation or nonconservation—are equally valid theoretical descriptions of the flowfield variables as a function of space and time. The label "nonconservation" is an artifact from computational fluid dynamics, where it has some numerical implications. Indeed, if you were to pick up a standard classical fluid dynamics text book and look for the words "conservation form" or "nonconservation form" in the index, you would most likely not find them. This nomenclature is a recent artifact from the discipline of computational fluid dynamics (CFD). Prior to the advent of CFD, the form of the governing equations used was purely arbitrary. To carry out an aerodynamic analysis, the choice of the form of the equations was, and still is, purely a matter of personal preference. The theoretical results are the same, no matter which form is used. So there is no need to make any real distinction between the different forms except when dealing with CFD.

However, CFD is an emerging discipline that plays a strong role in the study and applications of fluid dynamics. Indeed, I am of the opinion that CFD today takes on a role equal to those of pure experiment and pure theory in the practice of fluid dynamics. Therefore, it is appropriate in this book to at least identify the various forms of the governing equations as to conservation or nonconservation form, because you will encounter those labels with increasing frequency in your future work in fluid dynamics. Moreover, this matter will be addressed again in the discussions of CFD applications in Chaps. 11, 12, and 16. Therefore, you should examine these equations carefully enough such that you feel comfortable with them in both forms.

6.5 | THE ENTROPY EQUATION

Consider the combined form of the first and second laws of thermodynamics, as given by Eq. (1.30). From Sec. 6.4, we are justified in applying Eq. (1.30) directly to a moving fluid element, where it takes the form

$$T\frac{Ds}{Dt} = \frac{De}{Dt} + p\frac{Dv}{Dt} \qquad (6.49)$$

Equation (6.49) is labeled simply the *entropy equation,* and it holds in general for a nonadiabatic viscous flow. However, for an inviscid adiabatic flow, Eq. (6.48) says that

$$\frac{De}{Dt} + p\frac{Dv}{Dt} = 0 \tag{6.50}$$

Combining Eqs. (6.49) and (6.50), we have

$$\boxed{\frac{Ds}{Dt} = 0} \tag{6.51}$$

or

$$\boxed{s = \text{const}} \tag{6.52}$$

Equations (6.51) and (6.52) say that the entropy of a moving fluid element is constant. If the flow is steady, the entropy is constant along a streamline in an adiabatic, inviscid flow. Moreover, if the flow originates in a constant entropy reservoir, such as the free stream far ahead of a moving body, each streamline has the same value of entropy, and hence Eq. (6.52) holds throughout the complete flowfield. (In some literature, this is denoted as "homentropic" flow.) Note that Eqs. (6.51) and (6.52) are valid for both steady and unsteady flows.

For the solution of most problems in compressible flow, the continuity, momentum, and energy equations are sufficient; the entropy equation is not needed except to calculate the direction in which a given process may be occurring. However, for isentropic flows, Eqs. (6.51) or (6.52) are frequently a convenience, and may be used to *substitute* for either the energy or momentum equations. This advantage will be demonstrated in subsequent discussions.

6.6 | CROCCO'S THEOREM: A RELATION BETWEEN THE THERMODYNAMICS AND FLUID KINEMATICS OF A COMPRESSIBLE FLOW

Consider again an element of fluid as it moves through a flowfield. The movement of this fluid element is both translational and rotational. The translational motion is denoted by the velocity $\mathbf{V}$. The rotational motion is denoted by the angular velocity, ω. In any basic fluid mechanic text, it is readily shown that $\omega = \frac{1}{2}\nabla \times \mathbf{V}$; hence the curl of the velocity field at any point is a measure of the rotation of a fluid element at that point. The quantity $\nabla \times \mathbf{V}$ is itself denoted as the *vorticity* of the fluid; the vorticity is equal to twice the angular velocity.

In this section, we will derive a relationship between the fluid vorticity (a kinematic property of the flow) and the pertinent thermodynamic properties. To begin, consider Euler's equation, Eq. (6.29), without body forces,

$$\rho\frac{D\mathbf{V}}{Dt} = -\nabla p \tag{6.53}$$

Writing out the substantial derivative, Eq. (6.53) is

$$\rho \frac{\partial \mathbf{V}}{\partial t} + \rho (\mathbf{V} \cdot \nabla) \mathbf{V} = -\nabla p \tag{6.54}$$

Recall the combined first and second laws of thermodynamics in the form of Eq. (1.32). In terms of changes in three-dimensional space, the differentials in Eq. (1.32) can be replaced by the gradient operator,

$$T \nabla s = \nabla h - v \nabla p = \nabla h - \frac{\nabla p}{\rho} \tag{6.55}$$

Combining Eqs. (6.54) and (6.55),

$$T \nabla s = \nabla h - \frac{1}{\rho} \left[-\rho \frac{\partial \mathbf{V}}{\partial t} - \rho (\mathbf{V} \cdot \nabla) \mathbf{V} \right]$$

or
$$T \nabla s = \nabla h + \frac{\partial \mathbf{V}}{\partial t} + (\mathbf{V} \cdot \nabla) \mathbf{V} \tag{6.56}$$

However, from the definition of total enthalpy,

$$h \equiv h_o - \frac{V^2}{2}$$

Hence,
$$\nabla h = \nabla h_o - \nabla \left(\frac{V^2}{2} \right) \tag{6.57}$$

Substitute Eq. (6.57) into (6.56):

$$T \nabla s = \nabla h_o - \nabla \left(\frac{V^2}{2} \right) + \frac{\partial \mathbf{V}}{\partial t} + (\mathbf{V} \cdot \nabla) \mathbf{V} \tag{6.58}$$

Using the vector identity

$$\nabla \left(\frac{V^2}{2} \right) - (\mathbf{V} \cdot \nabla) \mathbf{V} = \mathbf{V} \times (\nabla \times \mathbf{V})$$

Eq. (6.58) becomes

$$\boxed{T \nabla s = \nabla h_o - \mathbf{V} \times (\nabla \times \mathbf{V}) + \frac{\partial \mathbf{V}}{\partial t}} \tag{6.59}$$

Equation (6.59) is called *Crocco's theorem*, because it was first obtained by L. Crocco in 1937 in a paper entitled "Eine neue Stromfunktion fur die Erforschung der Bewegung der Gase mit Rotation," *Z. Angew. Math. Mech.* vol. 17, 1937, pp. 1–7.

For steady flow, Crocco's theorem becomes

$$\boxed{T \nabla s = \nabla h_o - \mathbf{V} \times (\nabla \times \mathbf{V})} \tag{6.60}$$

Keep in mind that Eqs. (6.59) and (6.60) hold for an inviscid flow with no body forces.

Rearranging Eq. (6.60),

$$\underbrace{\mathbf{V} \times (\nabla \times \mathbf{V})}_{\text{vorticity}} = \underbrace{\nabla h_o}_{\substack{\text{total enthalpy} \\ \text{gradient}}} - \underbrace{T \nabla s}_{\substack{\text{gradient of} \\ \text{entropy}}} \tag{6.61}$$

Equation (6.61) has an important physical interpretation. When a steady flow-field has gradients of total enthalpy and/or entropy, Eq. (6.61) dramatically shows that it is *rotational*. This has definite practical consequences in the flow behind a curved shock wave, as sketched in Fig. 4.29. In region 1 ahead of the curved shock, all streamlines in the uniform free stream have the same total enthalpy, $h_{o_1} = h_\infty + V_\infty^2/2$. Across the stationary shock wave, the total enthalpy does not change; hence, in region 2 behind the shock, $h_{o_2} = h_{o_1}$. Hence, all streamlines in the flow behind the shock have the same total enthalpy; thus, behind the shock, $\nabla h_o = 0$. However, in Fig. 4.29 streamline (*b*) goes through a strong portion of the curved shock and hence experiences a higher entropy increase than streamline (*d*), which crosses a weaker portion of the shock. Therefore, in region 2, $\nabla s \neq 0$. Consequently, from Crocco's theorem as given in Eq. (6.61), $\mathbf{V} \times (\nabla \times \mathbf{V}) \neq 0$ behind the shock. Thus,

$$\nabla \times \mathbf{V} \neq 0 \quad \text{behind the shock}$$

Hence, Crocco's theorem shows that the *flowfield behind a curved shock is rotational*. This is unfortunate, because rotational flowfields are inherently more difficult to analyze than flows without rotation (irrotational flows). We will soon come to appreciate the full impact of this statement.

6.7 | HISTORICAL NOTE: EARLY DEVELOPMENT OF THE CONSERVATION EQUATIONS

In his *Principia* of 1687, Isaac Newton devoted the entire second book to the study of fluid mechanics. To some extent, there was a practical reason for Newton's interest in the flow of fluids—England had become a major sea power under Queen Elizabeth, and its growing economic influence was extended through the world by means of its merchant marine. Consequently, by the time Newton was laying the foundations for rational mechanics, there was intense practical interest in the calculation of the resistance of ship hulls as they move through water, with the ultimate objective of improving ship design. However, the analysis of fluid flow is conceptually more difficult than the dynamics of solid bodies; a solid body is usually geometrically well-defined, and its motion is therefore relatively easy to describe. On the other hand, a fluid is a "squishy" substance, and in Newton's time, it was difficult to decide even how to qualitatively model its motion, let alone obtain quantitative relationships. As will be described in more detail in Sec. 12.4, Newton considered a fluid flow as a uniform, rectilinear stream of particles, much like a cloud of pellets from a shotgun blast. Newton assumed that, upon striking a surface inclined at an angle θ to the stream, the particles would transfer their normal momentum to the surface, but their tangential momentum would be preserved. Hence, after collision with the surface, the particles would then move along the surface. As derived in Sec. 12.4, this leads to an expression for the hydrodynamic force on the surface which varies as $\sin^2 \theta$. This is Newton's famous "sine-squared" law; however, its accuracy left much to be desired, and of course the physical model was not appropriate. Indeed, it was not until the advent of hypersonic aerodynamics in the 1950s that Newton's sine-squared law could be used in an environment that actually reasonably approached Newton's physical model. This is described in more detail in Secs. 12.4 and 12.9. Nevertheless, Newton's efforts at the end of the seventeenth century represent the first meaningful fluid dynamic analysis, and they stimulated the interest of other scientists.

The discipline of fluid dynamics first bloomed under the influence of Daniel and Johann Bernoulli, and especially through the work of Leonhard Euler, during the period 1730 to 1760. Euler had great physical insight that allowed him to visualize a fluid as a collection of moving fluid elements. Moreover, he recognized that pressure was a point property that varied throughout a flow, and that differences in this pressure provided a mechanism to accelerate the fluid elements. He put these ideas in terms of an equation, obtaining for the first time in history those relations we have derived as Eqs. (6.26) through (6.29) in this chapter. Therefore, the momentum equation in the form we frequently use in modern compressible flow dates back to 1748, as derived by Euler during his residence in St. Petersburg, Russia. Euler went further to explain that the force on an object moving in a fluid is due to the *pressure distribution* over the object's surface. Although he completely ignored the influence of friction, Euler had established the modern idea for one important source of the aerodynamic force on a body (see Sec. 1.5).

The origin of the continuity equation in the form of Eq. (6.5) also stems back to the mideighteenth century. Although Newton had postulated the obvious fact that the mass of a specified object was constant, this principle was not appropriately applied to fluid mechanics until 1749. In this year, the famous French scientist, Jean le Rond d'Alembert gave a paper in Paris entitled "Essai d'une nouvelle theorie de la resitance des fluides" in which he formulated differential equations for the conservation of mass in special applications to plane and axisymmetric flows. However, the general equation in the form of Eq. (6.5) was first expressed 8 years later by Euler in a series of three basic papers on fluid mechanics that appeared in 1757.

It is therefore interesting to observe that two of the three basic conservation equations used today in modern compressible flow were well-established long before the American Revolutionary War, and that such equations were contemporary with the time of George Washington and Thomas Jefferson!

The origin of the energy equation in the form of Eqs. (6.17) or (6.31) has its roots in the development of thermodynamics in the nineteenth century. It is known that as early as 1839 B. de Saint Venant used a one-dimensional form of the energy equation to derive an expression for the exit velocity from a nozzle in terms of the pressure ratio across the nozzle. But the precise first use of Eq. (6.17) or its derivatives is obscure and is buried somewhere in the rapid development of physical science in the nineteenth century.

The reader who is interested in a concise and interesting history of fluid mechanics in general is referred to the excellent discussion by R. Giacomelli and E. Pistolesi in Volume I of the series *Aerodynamic Theory,* edited by W. F. Durand in 1934. (See Ref. 22.) Here, the evolution of fluid mechanics from antiquity to 1930 is presented in a very cohesive fashion. You are also referred to the author's recent book *A History of Aerodynamics* (Ref. 134) for a presentation on the evolution of our intellectual understanding of aerodynamics starting with ancient Greek science.

6.8 | HISTORICAL NOTE: LEONHARD EULER—THE MAN

Euler was a giant among eighteenth-century mathematicians and scientists. As a result of his contributions, his name is associated with numerous equations and techniques, e.g. the Euler numerical solution of ordinary differential equations, Eulerian angles in geometry, and the momentum equations for inviscid fluid flow [Eqs. (6.26) through (6.29) in this book]. As indicated in Sec. 6.7, Euler played the primary role in establishing fluid mechanics as a rational science. Who was this man whose philosophy and results still pervade modern fluid mechanics? Let us take a closer look.

Leonhard Euler was born on April 15, 1707, in Basel, Switzerland. His father was a Protestant minister who enjoyed mathematics as a pastime. Therefore, Euler grew up in a family atmosphere that encouraged intellectual activity. At the age of 13, Euler entered the University of Basel, which at that time had about 100 students and 19 professors. One of those professors was Johann Bernoulli, who tutored Euler in mathematics. Three years later, Euler received his master's degree in philosophy.

It is interesting that three of the people most responsible for the early development of theoretical fluid dynamics—Johann Bernoulli, his son Daniel, and Euler—lived in the same town of Basel, were associated with the same University, and were contemporaries. Indeed, Euler and the Bernoullis were close and respected friends—so much so that, when Daniel Bernoulli moved to teach and study at the St. Petersburg Academy in 1725, he was able to convince the Academy to hire Euler as well. At this invitation, Euler left Basel for Russia; he never returned to Switzerland, although he remained a Swiss citizen throughout his life.

Euler's interaction with the Bernoullis in the development of fluid mechanics grew strong during these early years at St. Petersburg. There, Daniel Bernoulli formulated most of the concepts that were eventually published in his book *Hydrodynamica* in 1738. The book's contents ranged over such topics as jet propulsion, monometers, and flow in pipes. Bernoulli also attempted to obtain a relation between pressure and velocity in a fluid, but his derivation was obscure. In fact, even though the familiar Bernoulli's equation [Eq. (1.1) in this book] is usually ascribed to Daniel via his *Hydrodynamica,* the precise equation is not to be found in the book! Some improvement was made by his father, Johann, who about the same time also published a book entitled *Hydraulica.* It is clear from this latter book that the father understood Bernoulli's equation better than the son—Daniel thought of pressure strictly in terms of the height of a monometer column, whereas Johann had the more fundamental understanding that pressure was a force acting on the fluid. However, it was Euler a few years later who conceived of pressure as a point property that can vary from point to point throughout a fluid, and obtained a differential equation relating pressure and velocity [Eq. (6.29) in this book]. In turn, Euler integrated the differential equation to obtain, for the first time in history, Bernoulli's equation [Eq. (1.1)]. Hence we see that Bernoulli's equation is really a historical misnomer; credit for it is legitimately shared by Euler.

Daniel Bernoulli returned to Basel in 1733, and Euler succeeded him at St. Petersburg as a professor of physics. Euler was a dynamic and prolific man; by 1741 he had prepared 90 papers for publication and written the two-volume book *Mechanica*. The atmosphere surrounding St. Petersburg was conducive to such achievement. Euler wrote in 1749: "I and all others who had the good fortune to be for some time with the Russian Imperial Academy cannot but acknowledge that we owe everything which we are and possess to the favorable conditions which we had there."

However, in 1740, political unrest in St. Petersburg caused Euler to leave for the Berlin Society of Sciences, at that time just formed by Frederick the Great. Euler lived in Berlin for the next 25 years, where he transformed the Society into a major Academy. In Berlin, Euler continued his dynamic mode of working, preparing at least 380 papers for publication. Here, as a competitor with d'Alembert and others, Euler formulated the basis for mathematical physics.

In 1766, after a major disagreement with Frederick the Great over some financial aspects of the Academy, Euler moved back to St. Petersburg. This second period of his life in Russia became one of physical suffering. In that same year, he became blind in one eye after a short illness. An operation in 1771 resulted in restoration of his sight, but only for a few days. He did not take proper precautions after the operation, and within a few days he was completely blind. However, with the help of others, he continued with his work. His mind was as sharp as ever, and his spirit did not diminish. His literary output even increased—about half his total papers were written after 1765!

On September 18, 1783, Euler conducted business as usual—giving a mathematics lesson, making calculations of the motion of balloons, and discussing with friends the planet of Uranus which had recently been discovered. About 5 P.M. he suffered a brain hemorrhage. His only words before losing consciousness were "I am dying." By 11 P.M., one of the greatest minds in history had ceased to exist.

Euler is considered to be the "great calculator" of the eighteenth century. He made lasting contributions to mathematical analysis, theory of numbers, mechanics, astronomy, and optics. He participated in the founding of the calculus of variations, theory of differential equations, complex variables, and special functions. He invented the concept of finite differences (to be used so extensively in modern fluid dynamics, as described in Chaps. 11 and 12). In retrospect, his work in fluid dynamics was just a small percentage of his total impact on mathematics and science.

Someday, when you have nothing better to do, count the number of times Euler's equations are used and referenced throughout this book. In so doing, you will enhance your appreciation of just how much that eighteenth century giant dominates the foundations of modern compressible flow today.

6.9 | SUMMARY

This chapter, though it may appear to be virtually wall-to-wall equations, is extremely important for our further discussions. Therefore you should become very familiar with, and feel at home with, all the equations in boxes—they are the primary results—as well as how they were obtained. Therefore, before proceeding to the next chapter, take the time to reread the present chapter until these equations become firmly fixed in your mind.

The equations in this chapter describe the general unsteady, three-dimensional flow of an inviscid compressible fluid. They are nonlinear partial differential equations. Moreover, the continuity, momentum, and energy equations are coupled, and must be solved simultaneously. There is no general solution to these equations. Their solution for given problems (hence given boundary conditions) constituted the principle effort of theoretical gasdynamicists and aerodynamicists over the past half-century. Their efforts are still going on.

Historically, because no general closed-form solution of these nonlinear equations has been found, they have been linearized by the imposition of simplifying assumptions. In turn, the linearized equations can be solved by existing analytical techniques, and although approximate, yield valuable information on some specialized problems of interest. This will be the subject of Chap. 9.

Also historically, there have been a few specific problems that have lent themselves to an exact solution of the governing nonlinear equations. The unsteady one-dimensional expansion waves to be discussed in Chap. 7, and the flow over a sharp right-circular cone at zero angle of attack to be discussed in Chap. 10, are two such examples. Even these solutions require some type of limited numerical technique for completion.

In recent years, the high-speed digital computer has provided a new dimension to the solution of compressible flow problems. With such computers, the method of characteristics, an exact numerical technique which was applied laboriously by hand in the 1930s, 1940s, and 1950s, is now routinely employed to solve many nonlinear compressible flow problems of interest. The method of characteristics for unsteady one-dimensional flow will be discussed in Chap. 7, and for two- and three-dimensional steady flows in Chap. 11. But the major impact of computers has been the growth of computational fluid dynamic solutions of the nonlinear governing equations for a whole host of important problems; some computational fluid dynamic techniques will be discussed in Chaps. 11, 12, and 17. Thus, the advent of computational fluid dynamics has recently opened new vistas for the solution of compressible flow problems, and one purpose of the present book is to incorporate these modern vistas into a general study of the discipline. (See also the discussion of computational fluid dynamics in Sec. 1.6.)

CHAPTER 7

Unsteady Wave Motion

A wave of sudden rarefaction, though mathematically possible, is an unstable condition of motion; any deviation from absolute suddenness tending to make the disturbance become more and more gradual. Hence the only wave of sudden disturbance whose permanency of type is physically possible, is one of sudden compression.

**W. J. M. Rankine, 1870, attributed by him to a comment
from Sir William Thomson**

PREVIEW BOX

This chapter is all about *traveling* waves—pressure waves that propagate with finite velocity relative to a fixed coordinate system. This is in contrast to our previous discussions, where we considered shock waves and expansion waves to be stationary relative to a fixed coordinate system, and the gas ahead of the wave moves with a finite velocity. As far as the waves are concerned, these two pictures are equivalent; as we will soon appreciate, the wave properties depend on the velocity of the gas ahead of the wave *relative to the wave,* no matter whether the wave is propagating into a stagnant gas or the gas is moving through a stationary wave. However, from our point of view, there is a big difference in the methods used to analyze such waves. A wave traveling through the laboratory creates an *unsteady* flow relative to the laboratory, whereas the flow through a stationary wave relative to the laboratory is steady. Steady flow is inherently easier to calculate, and that is why we treated the waves as stationary in the previous chapters. For example, the shock waves generated by a supersonic airplane flying overhead are moving past us on the ground at the same velocity as the airplane—this is an unsteady

flow relative to us. However, there is no need to study the shock waves from this unsteady point of view, because we can hop on the airplane and ride with it; in this case, the shocks appear stationary relative to us, and we can use the steady flow techniques of the previous chapters to study the waves. This is analogous to placing the airplane in a hypothetical very large supersonic wind tunnel and blowing air at supersonic speeds over the stationary airplane. Our perspective in the previous chapters was that of the airplane fixed in the wind tunnel with the air blowing over it. The physical properties of the shock waves are the same in either case, so we take the easier path and study the shock phenomena from a steady flow point of view.

On the other hand, there are some devices and applications that make direct use of the unsteady flows generated by traveling waves, and in these situations we have to study the actual unsteady flow problem. For example, the exhaust system on an internal combustion reciprocating engine powering a motor vehicle is full of unsteady pressure waves propagating along the exhaust pipes, and it is important to understand and calculate this

Figure 7.1a | Naval Ordnance Laboratory (NOL) shock tube. Extended view of the shock tube length. (Courtesy of Dr. John S. Vamos, Naval Surface Warfare Center.)

Figure 7.1b | Close-up view of the end-wall flange, rectangular test cavity, and dump tank of the NOL shock tube. With the test cavity and dump tank in this installation, the facility is operating as a shock tunnel. (Dr. John S. Vamos)

unsteady flow in order to properly tune the design of the exhaust system.

Another important application is a shock tube, which is a laboratory device for producing high-temperature, high-pressure gases for the purpose of studying the thermodynamic and chemical properties of such gases at temperatures and pressures higher than obtainable in other laboratory devices. A typical shock tube is shown in Figs. 7.1a and b. The shock tube is a very long pipe, as can be seen in Fig. 7.1a, in which a strong shock wave is generated inside the tube and propagates along the tube (from left to right in Fig. 7.1), producing a high-temperature, high-pressure gas behind it. In the particular shock tube shown in Figs. 7.1a and b, the shock runs into an end wall at the flange seen in the middle of Fig. 7.1b, and reflects back to the left, producing an even higher temperature gas behind the reflected shock wave. In the NOL shock tube in the configuration shown here, this slug of very high temperature, high-pressure gas then expands through a bank of small supersonic nozzles, creating a supersonic flow in the rectangular shaped test section seen in the middle of Fig. 7.1b, which subsequently exhausts into the large dump tank shown at the right. In this special configuration shown in Fig. 7.1, with CO_2 and N_2 as the test gas, the supersonic flow in the rectangular test section becomes a laser gas, and the shock tube is configured to be a gasdynamic laser, an exciting device for a high power laser that was studied extensively in the 1960s and 70s (see Ref. 21 for more details on gasdynamic lasers). To understand and calculate the operation of such a device as shown in Fig. 7.1, the material in this chapter on unsteady wave motion is essential. Indeed, a major focus of this chapter is the understanding and calculation of shock tube wave patterns and flowfields.

The roadmap for this chapter is given in Fig. 7.2. We begin our study of unsteady wave motion by considering moving normal shock waves, which is the left branch shown in Fig. 7.2. Then we move to the right branch to study moving expansion waves, which requires as preliminaries a discussion of linear sound wave propagation (acoustic theory) and of nonlinear finite wave motion. Finally, we combine both branches to examine the flowfield in shock tubes, which is a combination of shock and expansion wave motion. When you reach the end of this roadmap, you will have the essential tools to understand the unsteady wave motion in mechanical devices such as the shock tube shown in Fig. 7.1.

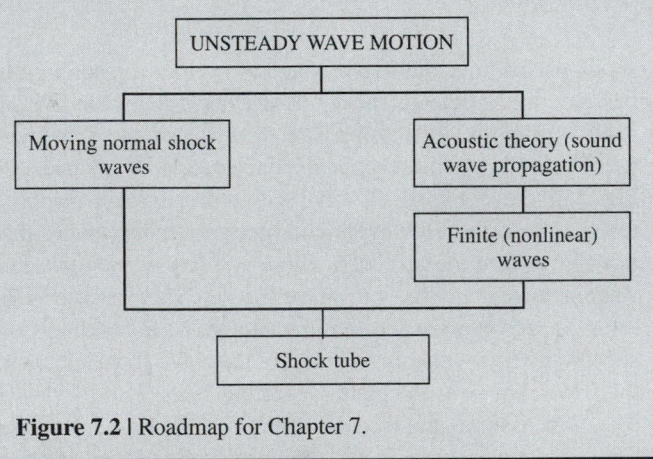

Figure 7.2 | Roadmap for Chapter 7.

7.1 | INTRODUCTION

Consider again the normal shock wave, as discussed in Chap. 3. In that discussion the shock is viewed as a stationary wave, fixed in space, as sketched in Fig. 7.3a. However, in Secs. 3.3 and 3.6, the wave is described as a physical disturbance in the flow, where the wave is propagated by molecular collisions. Hence, sound waves and

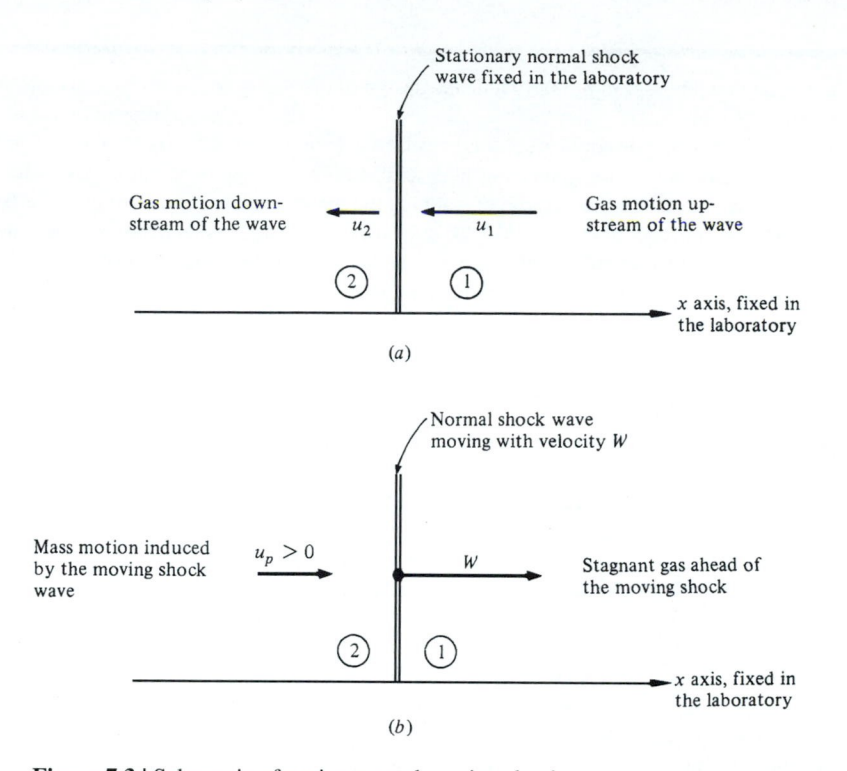

Stationary normal shock
wave fixed in the laboratory

Gas motion down-
stream of the wave u_2 u_1 Gas motion up-
                                                          stream of the wave

②   ① x axis, fixed in
                       the laboratory

(a)

Normal shock wave
moving with velocity W

Mass motion induced $u_p > 0$
by the moving shock W Stagnant gas ahead of
wave   the moving shock

②   ① x axis, fixed in
                       the laboratory

(b)

Figure 7.3 | Schematic of stationary and moving shock waves.

shock waves have definite propagation velocities, sonic in the case of sound and supersonic in the case of shocks. However, if the wave is propagating into a flow that
itself is moving in the opposite direction at the same velocity magnitude as the
wave velocity, then the wave appears stationary in space. This is the case shown in
Fig. 7.3a; here, the shock wave with a propagation velocity of u_1 is trying to move
toward the right. However, it is precisely balanced by the upstream gas which is
moving toward the left, also with a velocity of u_1. Consequently, the normal shock
wave appears *stationary* in space (i.e., the shock wave is *fixed* "relative to the laboratory"), and we see the familiar picture of a standing normal shock wave with a
supersonic flow velocity u_1 ahead of the wave and a subsonic flow velocity u_2 behind
the wave. This was the picture used in Chaps. 3 and 4.

Now assume that the flow velocity u_1 in Fig. 7.3a is turned off, i.e., let $u_1 = 0$.
Then the shock wave is no longer constrained, and it propagates through space to
the right. This picture is sketched in Fig. 7.3b; here we relabel the wave propagation
velocity as W to emphasize that the wave is now propagating through the laboratory. The magnitude of u_1 in Fig. 7.3a and W in Fig. 7.3b are the same. However,
in Fig. 7.3b we are now watching a normal shock wave propagate with velocity W
(relative to the laboratory) into a quiescent gas. In the process, the moving wave
induces the gas behind it to move in the same direction as the wave; this mass motion
is shown as u_p in Fig. 7.3b. Returning to the stationary wave in Fig. 7.3a, all

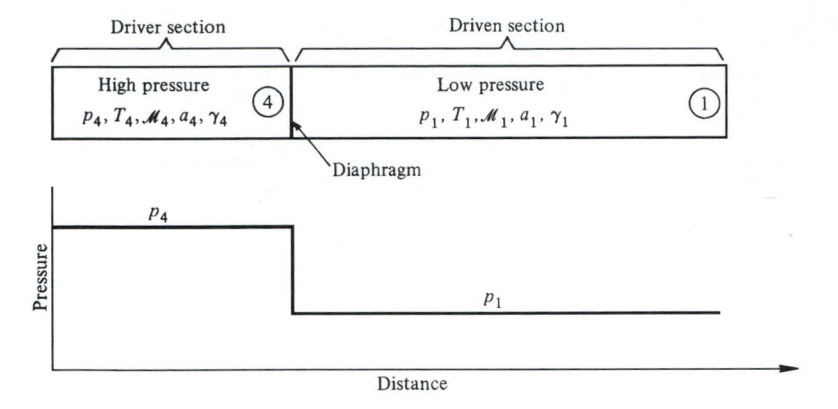

Figure 7.4 | Initial conditions in a pressure-driven shock tube.

properties of the flowfield depend on x only, i.e., $\rho = \rho(x)$, $T = T(x)$, $u = u(x)$, etc. This is a steady flow. In contrast, for the moving wave in Fig. 7.3b, all properties of the flowfield depend on both x and t, i.e., $\rho = \rho(x, t)$, $T = T(x, t)$, $u = u(x, t)$, etc. This is an *unsteady flow,* and hence the picture in Fig. 7.3b is that of *unsteady wave motion.* Such unsteady wave motion is the subject of this chapter.

An important application of unsteady wave motion is a shock tube, sketched in Fig. 7.4. This is a tube closed at both ends, with a diaphragm separating a region of high-pressure gas on the left (region 4) from a region of low-pressure gas on the right (region 1). The pressure distribution is also illustrated in Fig. 7.4. The gases in regions 1 and 4 can be at different temperatures and have different molecular weights, $\mathscr{M}_1$ and $\mathscr{M}_4$. In Fig. 7.4, region 4 is called the *driver section,* and region 1 is the *driven section.* When the diaphragm is broken (for example, by electrical current, or by mechanical means), a shock wave propagates into section 1 and an expansion wave propagates into section 4. This picture is sketched in Fig. 7.5. As the normal shock wave propagates to the right with velocity W, it increases the pressure of the gas behind it (region 2), and induces a mass motion with velocity u_p. The interface between the driver and driven gases is called the *contact surface,* which also moves with velocity u_p. This contact surface is somewhat like the slip lines discussed in Chap. 4; across it the entropy changes discontinuously. However, the pressure and velocity are preserved: $p_3 = p_2$ and $u_3 = u_2 = u_p$. The expansion wave propagates to the left, smoothly and continuously decreasing the pressure in region 4 to the lower value p_3 behind the expansion wave. The flowfield in the tube after the diaphragm is broken (Fig. 7.5) is completely determined by the given conditions in regions 1 and 4 before the diaphragm is broken (Fig. 7.4).

Shock tubes are valuable gasdynamic instruments; they have important applications in the study of high-temperature gases in physics and chemistry, in the testing of supersonic bodies and hypersonic entry vehicles, and more recently in the development of high-power gasdynamic and chemical lasers. Many of the high-temperature thermodynamic and chemical kinetic properties to be discussed in Chap. 16 were measured in shock tubes. They are basic tools in the understanding

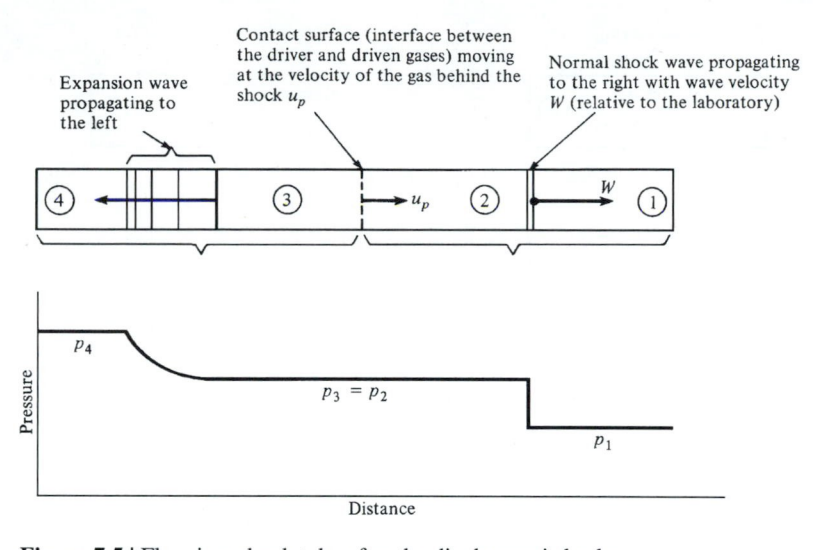

Figure 7.5 | Flow in a shock tube after the diaphragm is broken.

of high-speed compressible flow. Therefore, this chapter first discusses unsteady normal shock waves, followed by a treatment of unsteady one-dimensional finite wave motion, and then focuses the results on the important application of shock tubes.

7.2 | MOVING NORMAL SHOCK WAVES

Consider again the stationary normal shock wave sketched in Fig. 7.3a. For this picture, we know from Eqs. (3.38) through (3.40) that the continuity, momentum, and energy equations are, respectively,

$$\rho_1 u_1 = \rho_2 u_2 \tag{3.38}$$

$$p_1 + \rho_1 u_1^2 = p_2 + \rho_2 u_2^2 \tag{3.39}$$

$$h_1 + u_1^2/2 = h_2 + u_2^2/2 \tag{3.40}$$

Looking at Fig. 7.3a, a literal interpretation of u_1 and u_2 is easily seen as

$u_1 = $ velocity of the gas *ahead* of the shock wave, *relative* to the wave

$u_2 = $ velocity of gas *behind* the shock wave, *relative* to the wave

It just so happens in Fig. 7.3a that the shock wave is stationary, so therefore u_1 and u_2 are also the flow velocities we see relative to the laboratory. However, the interpretation of u_1 and u_2 as relative to the *shock wave* is more fundamental; Eqs. (3.38) through (3.40) always hold for gas velocities relative to the shock wave, no matter whether the shock is moving or stationary. Therefore, examining the moving shock in Fig. 7.3b, we immediately deduce from the geometry that

$W = $ velocity of the gas *ahead* of the shock wave, *relative* to the wave

$W - u_p = $ velocity of the gas *behind* the shock wave, *relative* to the wave

Hence, for the picture of the moving shock wave in Fig. 7.3b, the normal-shock continuity, momentum, and energy equations, Eqs. (3.38) through (3.40), become

$$\rho_1 W = \rho_2(W - u_p) \tag{7.1}$$

$$p_1 + \rho_1 W^2 = p_2 + \rho_2(W - u_p)^2 \tag{7.2}$$

$$h_1 + \frac{W^2}{2} = h_2 + \frac{(W - u_p)^2}{2} \tag{7.3}$$

Equations (7.1) through (7.3) are the governing normal-shock equations for a shock *moving* with velocity W into a stagnant gas.

Let us rearrange these equations into a more convenient form. From Eq. (7.1)

$$W - u_p = W \frac{\rho_1}{\rho_2} \tag{7.4}$$

Substitute Eq. (7.4) into (7.2):

$$p_1 + \rho_1 W^2 = p_2 + \rho_2 W^2 \left(\frac{\rho_1}{\rho_2}\right)^2$$

and rearranging,

$$p_2 - p_1 = \rho_1 W^2 \left(1 - \frac{\rho_1}{\rho_2}\right)$$

$$W^2 = \frac{p_2 - p_1}{\rho_1(1 - \rho_1/\rho_2)}$$

$$W^2 = \frac{p_2 - p_1}{\rho_2 - \rho_1} \left(\frac{\rho_2}{\rho_1}\right) \tag{7.5}$$

Returning to Eq. (7.1),

$$W = (W - u_p) \frac{\rho_2}{\rho_1} \tag{7.6}$$

Substitute Eq. (7.6) into (7.5):

$$(W - u_p)^2 \left(\frac{\rho_2}{\rho_1}\right)^2 = \frac{p_2 - p_1}{\rho_2 - \rho_1} \left(\frac{\rho_2}{\rho_1}\right)$$

or

$$(W - u_p)^2 = \frac{p_2 - p_1}{\rho_2 - \rho_1} \left(\frac{\rho_1}{\rho_2}\right) \tag{7.7}$$

Substitute Eqs. (7.5) and (7.7) into (7.3), and recall that $h = e + p/\rho$, to obtain

$$e_1 + \frac{p_1}{\rho_1} + \frac{1}{2} \left[\frac{p_2 - p_1}{\rho_2 - \rho_1} \left(\frac{\rho_2}{\rho_1}\right) \right] = e_2 + \frac{p_2}{\rho_2} + \frac{1}{2} \left[\frac{p_2 - p_1}{\rho_2 - \rho_1} \left(\frac{\rho_1}{\rho_2}\right) \right] \tag{7.8}$$

Equation (7.8) algebraically simplifies to

$$e_2 - e_1 = \frac{p_1 + p_2}{2} \left(\frac{1}{\rho_1} - \frac{1}{\rho_2} \right)$$

or
$$\boxed{e_2 - e_1 = \frac{p_1 + p_2}{2} (v_1 - v_2)} \qquad (7.9)$$

Equation (7.9) is the *Hugoniot equation,* and is identically the same form as Eq. (3.72) for a stationary shock. In hindsight, this is to be expected; the Hugoniot equation relates changes of *thermodynamic* variables across a normal shock wave, and these are physically independent of whether or not the shock is moving.

In general, Eqs. (7.1) through (7.3) must be solved numerically. However, let us specialize to the case of a calorically perfect gas. In this case, $e = c_v T$, and $v = RT/p$; hence Eq. (7.9) becomes

$$\boxed{\frac{T_2}{T_1} = \frac{p_2}{p_1} \left(\frac{\dfrac{\gamma + 1}{\gamma - 1} + \dfrac{p_2}{p_1}}{1 + \dfrac{\gamma + 1}{\gamma - 1} \dfrac{p_2}{p_1}} \right)} \qquad (7.10)$$

Similarly,

$$\boxed{\frac{\rho_2}{\rho_1} = \frac{1 + \dfrac{\gamma + 1}{\gamma - 1} \left(\dfrac{p_2}{p_1} \right)}{\dfrac{\gamma + 1}{\gamma - 1} + \dfrac{p_2}{p_1}}} \qquad (7.11)$$

Note that Eqs. (7.10) and (7.11) give the density and temperature ratios across the shock wave as a function of pressure ratio. Unlike a stationary shock wave, where it is convenient to think of Mach number M_1 as the governing parameter for changes across the wave, for a moving shock wave it now becomes convenient to think of p_2/p_1 as the major parameter governing changes across the wave. To reinforce this statement, define the moving shock Mach number as

$$M_s = \frac{W}{a_1}$$

Incorporating this definition along with the calorically perfect gas relations into Eqs. (7.1) through (7.3), and proceeding with a derivation identical to that used to obtain Eq. (3.57) for a stationary shock, we obtain

$$\frac{p_2}{p_1} = 1 + \frac{2\gamma}{\gamma + 1} \left(M_s^2 - 1 \right) \qquad (7.12)$$

Solving Eq. (7.12) for M_s,

$$M_s = \sqrt{ \frac{\gamma + 1}{2\gamma} \left(\frac{p_2}{p_1} - 1 \right) + 1 } \qquad (7.13)$$

However, since $M_s = W/a_1$, Eq. (7.13) yields

$$\boxed{W = a_1 \sqrt{\frac{\gamma + 1}{2\gamma}\left(\frac{p_2}{p_1} - 1\right) + 1}}$$

(7.14)

Equation (7.14) is important; it relates the wave velocity of the moving shock wave to the *pressure ratio* across the wave and the *speed of sound* of the gas into which the wave is propagating.

As mentioned earlier, a shock wave propagating into a stagnant gas induces a mass motion with velocity u_p behind the wave. From Eq. (7.1),

$$u_p = W\left(1 - \frac{\rho_1}{\rho_2}\right)$$

(7.15)

Substituting Eqs. (7.10) and (7.14) into Eq. (7.15), and simplifying, we obtain

$$\boxed{u_p = \frac{a_1}{\gamma}\left(\frac{p_2}{p_1} - 1\right)\left(\frac{\dfrac{2\gamma}{\gamma + 1}}{\dfrac{p_2}{p_1} + \dfrac{\gamma - 1}{\gamma + 1}}\right)^{1/2}}$$

(7.16)

Note from Eq. (7.16) that, as in the case of W, the mass-motion velocity u_p also depends on the pressure ratio across the wave and the speed of sound of the gas ahead of the wave.

In summary, for a given pressure ratio p_2/p_1 and speed of sound a_1, the corresponding values of ρ_2/ρ_1, T_2/T_1, W, and u_p are obtained from Eqs. (7.10), (7.11), (7.14), and (7.16), respectively.

Before leaving this section, let us further explore the characteristics of the induced mass motion behind the moving shock wave. The velocity of this mass motion, u_p, is relative to the laboratory, i.e., it is what we would observe if we were standing motionless in the laboratory and a shock wave swept by us with velocity W. After the wave passed by, we would feel a rush of air in the same direction as the wave motion, and the velocity of this rush of air is u_p. How large a value can u_p obtain? Can it ever be a supersonic velocity? To answer these questions, note that the Mach number of the induced motion (relative to the laboratory) is u_p/a_2, where

$$\frac{u_p}{a_2} = \frac{u_p}{a_1}\frac{a_1}{a_2} = \frac{u_p}{a_1}\sqrt{\frac{T_1}{T_2}}$$

(7.17)

Substitute Eqs. (7.11) and (7.16) into (7.17):

$$\frac{u_p}{a_2} = \frac{1}{\gamma}\left(\frac{p_2}{p_1} - 1\right)\left(\frac{\dfrac{2\gamma}{\gamma + 1}}{\dfrac{p_2}{p_1} + \dfrac{\gamma - 1}{\gamma + 1}}\right)^{1/2}\left[\frac{1 + \dfrac{\gamma + 1}{\gamma - 1}\left(\dfrac{p_2}{p_1}\right)}{\dfrac{\gamma + 1}{\gamma - 1}\left(\dfrac{p_2}{p_1}\right) + \left(\dfrac{p_2}{p_1}\right)^2}\right]^{1/2}$$

(7.18)

Consider an infinitely strong shock, where $p_2/p_1 \to \infty$. From Eq. (7.18),

$$\lim_{p_2/p_1 \to \infty} \left(\frac{u_p}{a_2} \right) = \sqrt{\frac{2}{\gamma(\gamma - 1)}} \tag{7.19}$$

For $\gamma = 1.4$, Eq. (7.19) shows that $u_p/a_2 \to 1.89$ as $p_2/p_1 \to \infty$. Hence, we see that u_p is not always a gentle wind—it can be a high-velocity flow, even supersonic. However, the Mach number cannot exceed a limiting value, which in general turns out to be moderately supersonic. As already calculated for a calorically perfect gas with $\gamma = 1.4$, the Mach number of the induced flow cannot exceed 1.89. Nevertheless, it is important to recognize that a strong moving shock wave can induce a supersonic mass motion behind it.

There is a fundamental distinction between steady and unsteady wave motion that must be appreciated—the stagnation properties of the two flows are different. For example, consider again the steady wave in Fig. 7.3a. In Chap. 3 we have shown that the total enthalpy (hence, for a calorically perfect gas, the total temperature) is constant across the stationary wave, i.e., $h_{o_2} = h_{o_1}$. In contrast, for the moving shock wave in Fig. 7.3b, the total enthalpy is *not* constant across the shock wave, i.e., $h_{o_2} \neq h_{o_1}$. This is easily seen by inspection. In front of the moving wave the gas is motionless, and hence $h_{o_1} = h_1$. However, behind the wave, $h_{o_2} = h_2 + u_p^2/2$; since $h_2 > h_1$ and because u_p is finite, obviously $h_{o_2} > h_{o_1}$. Similarly, the total pressure behind the moving shock wave, p_{o_2}, is *not* given by Eq. (3.63), which holds only for a stationary shock. Rather, p_{o_2} for a moving shock must be calculated from the known properties of the induced mass motion.

The above is a special example of a general result: "In an *unsteady* adiabatic inviscid flow, the total enthalpy is *not* constant." This is easily proven from an examination of the energy equation in the form of Eq. (6.44), repeated here:

$$\rho \frac{Dh_o}{Dt} = \frac{\partial p}{\partial t}$$

Clearly, if the flow is unsteady, $\partial p/\partial t \neq 0$, and hence h_o is not constant.

EXAMPLE 7.1

Consider a normal shock wave propagating into stagnant air where the ambient temperature is 300 K. The pressure ratio across the shock is 10. Calculate the shock wave velocity, the velocity of the induced mass motion behind the shock wave, and the temperature ratio across the wave, using (a) the equations of this section and (b) the tabulated numbers in Table A.2. Compare the two sets of results.

■ **Solution**

a. The speed of sound in the ambient air is

$$a_1 = \sqrt{\gamma R T_1} = \sqrt{(1.4)(287)(300)} = 347.2 \text{ m/s}$$

From Eq. (7.14),

$$W = a_1 \sqrt{\frac{\gamma+1}{2\gamma} \left(\frac{p_2}{p_1} - 1 \right) + 1} = 347.2 \sqrt{\frac{2.4}{2.8}(10-1)+1} = \boxed{1024.9 \text{ m/s}}$$

From Eq. (7.16),

$$u_p = \frac{a_1}{\gamma}\left(\frac{p_2}{p_1} - 1\right)\left[\frac{\dfrac{2\gamma}{\gamma+1}}{\dfrac{p_2}{p_1} + \dfrac{\gamma-1}{\gamma+1}}\right]^{1/2}$$

$$= \frac{347.2}{1.4}(10-1)\left[\frac{1.167}{10+0.167}\right]^{1/2} = \boxed{756.2 \text{ m/s}}$$

From Eq. (7.10),

$$\frac{T_2}{T_1} = \frac{p_2}{p_1}\left[\frac{\dfrac{\gamma+1}{\gamma-1} + \dfrac{p_2}{p_1}}{1 + \dfrac{\gamma+1}{\gamma-1}\left(\dfrac{p_2}{p_1}\right)}\right] = 10\left[\frac{6+10}{1+6(10)}\right] = \boxed{2.623}$$

b. From Table A.2, for $p_2/p_1 = 10$, the upstream Mach number is 2.95 (nearest entry). This is the Mach number of the gas ahead of the wave, relative to the wave. Since the gas ahead of the wave is motionless relative to the laboratory, then this is also the Mach number of the moving shock wave relative to the laboratory. Hence,

$$M_s = \frac{W}{a_1} = 2.95$$

Thus,

$$W = 2.95a_1 = 2.95(347.2) = \boxed{1024.2 \text{ m/s}}$$

This result obtained from the tables compares within 0.07 percent with that obtained from the exact equation in part (a).

Also, from Table A.2, $T_2/T_1 = \boxed{2.621}$. This compares within 0.08 percent of that obtained from the exact equation in part (a).

From Table A.2, $M_2 = 0.4782$. This is the Mach number of the gas behind the shock relative to the shock. The speed of sound in the gas behind the shock is

$$a_2 = \sqrt{\gamma R T_2} = \sqrt{\gamma R(T_2/T_1)T_1} = \sqrt{(1.4)(287)(2.621)(300)} = 562.1 \text{ m/s}$$

Hence, the velocity of the gas behind the shock relative to the shock is

$$u_2 = M_2 a_2 = 0.4782(562.1) = 268.8 \text{ m/s}$$

The velocity of the induced mass motion behind the shock, relative to the laboratory, is denoted by u_p, where

$$u_p = W - u_2 = 1024.2 - 268.8 = \boxed{755.4 \text{ m/s}}$$

This compares within 0.1 percent of that obtained from the exact equation in part (a).

EXAMPLE 7.2

Calculate the change in total enthalpy across the moving shock wave in Example 7.1.

■ **Solution**

From Eq. (1.22), for air

$$c_p = \frac{\gamma R}{\gamma - 1} = \frac{(1.4)(287)}{0.4} = 1004.5 \text{ J/kg} \cdot \text{K}$$

In region 1, in the stagnant gas ahead of the moving wave, the velocity is zero. Hence the total enthalpy is the same as the static enthalpy.

$$h_{o_1} = h_1 + \frac{u_1^2}{2} = h_1 + 0 = c_p T_1 = (1004.5)(300) = 3.014 \times 10^5 \text{ J/kg}$$

The temperature of the gas in region 2 behind the shock is $T_2 = (T_2/T_1)T_1 = 2.623(300) = 786.9$ K. The velocity of the gas behind the shock relative to the laboratory is u_p. Hence

$$h_{o_2} = h_2 + \frac{u_p^2}{2} = c_p T_2 + \frac{u_p^2}{2}$$

$$= (1004.5)(786.9) + \frac{(756.2)^2}{2}$$

$$= 10.76 \times 10^5 \text{ J/kg}$$

Thus,

$$h_{o_2} - h_{o_1} = (10.76 - 3.014) \times 10^5 = \boxed{7.746 \times 10^5 \text{ J/kg}}$$

The total enthalpy *increases* by the factor of 3.57 across the moving shock wave, clearly demonstrating that the total enthalpy is *not* constant across a moving shock wave.

EXAMPLE 7.3

Consider the same shock wave as in Example 7.1 propagating into air that is not stagnant, but rather is moving with a velocity of 200 m/s relative to the laboratory in a direction opposite to that of the wave motion. Calculate the velocity of the wave relative to the laboratory, and the velocity of the induced mass motion of the gas behind the wave relative to the laboratory.

■ Solution

Since the wave velocity $W = 1024.9$ m/s calculated from Eq. (7.14) is the same as the velocity of the gas ahead of the shock wave relative to the wave, then in the present example:

Velocity of wave relative to the laboratory $= 1024.9 - 200 = \boxed{824.9 \text{ m/s}}$

Since $W - u_p$ is the velocity of the gas behind the shock relative to the shock, and from Example 7.1, $W - u_p = 1024.9 - 756.2 = 268.7$ m/s, then the velocity of the gas behind the shock relative to the laboratory in the present example where the shock is moving at a velocity of 824.9 m/s relative to the laboratory is:

Velocity of gas behind the wave relative to the laboratory $= 824.9 - 268.7 = \boxed{556.2 \text{ m/s}}$

and it is the same direction in which the shock is moving.

Note: The two answers in this example could have been obtained more directly by subtracting the velocity of the air ahead of the wave relative to the laboratory, namely, 200 m/s, from both W and u_p obtained in Example 7.1. For example, the velocity of 556.2 m/s obtained here for the velocity of the gas behind the wave relative to the laboratory is simply $u_p - 200 = 756.2 - 200 = 556.2$ m/s. Hence, we have proven that when the gas in front of the shock is given some finite velocity relative to the laboratory, the other velocities relative to the laboratory are simply changed by the same amount.

EXAMPLE 7.4

For the case treated in Example 7.3, calculate the change in the total enthalpy across the shock wave.

■ Solution

Designate the velocity of the air ahead of the shock relative to the laboratory by V_1. In this case, $V_1 = 200$ m/s. Also, designate the velocity of the air behind the shock relative to the laboratory by V_2. In this case, $V_2 = 556.2$ m/s. For the gas ahead of the shock,

$$h_{o_1} = h_1 + \frac{V_1^2}{2} = c_p T_1 + \frac{V_1^2}{2} = (1004.5)(300) + \frac{(200)^2}{2}$$

$$= 3.214 \times 10^5 \text{ J/kg} \cdot \text{m}$$

For the gas behind the shock, the static temperature is still $T_2 = 786.9$ K (from Example 7.2). Hence,

$$h_{o_2} = c_p T_2 + \frac{V_2^2}{2} = (1004.5)(786.9) + \frac{(556.2)^2}{2} = 9.45 \times 10^5 \text{ J/kg} \cdot \text{m}$$

Thus,

$$h_{o_2} - h_{o_1} = (9.45 - 3.214) \times 10^5 = \boxed{6.236 \times 10^5 \text{ J/kg}}$$

Compare this result with that from Example 7.2. It is *different,* even though the strength of the shock is the same in both cases, namely with a pressure ratio $p_2/p_1 = 10$. This is a further demonstration that for unsteady wave motion, the total enthalpy changes across the shock, and this change depends not only on the strength of the shock but also on the velocity of the gas relative to the laboratory into which the shock is propagating.

7.3 | REFLECTED SHOCK WAVE

Consider a normal shock wave propagating to the right with velocity W, as shown in Fig. 7.6a. Assume this moving shock is incident on a flat endwall, as also sketched in Fig. 7.6a. In front of the incident shock, the mass motion $u_1 = 0$. Behind the incident shock, the mass velocity is u_p toward the endwall. At the instant the incident shock wave impinges on the endwall, it would appear that the flow velocity at the wall would be u_p, directed *into* the wall. However, this is physically impossible; the wall is solid, and the flow velocity normal to the surface must be zero. To avoid this ambiguity, nature immediately creates a *reflected normal shock wave* which travels to the left with velocity W_R (relative to the laboratory), as shown in Fig. 7.6b. The strength of this reflected shock (hence the value of W_R) is such that the originally induced mass motion with velocity u_p is stopped dead in its tracks. The mass motion behind the reflected shock wave must be zero, i.e., $u_5 = 0$ in Fig. 7.6b. Thus, the zero-velocity boundary condition is preserved by the reflected shock wave. (This is directly analogous to the steady reflected oblique shock wave discussed in Sec. 4.6, where the reflected shock is necessary to preserve, at the surface, flow tangent to

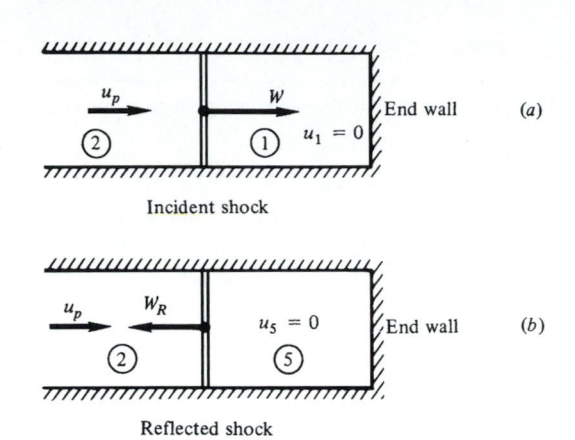

Incident shock

Reflected shock

Figure 7.6 | Incident and reflected shock waves.

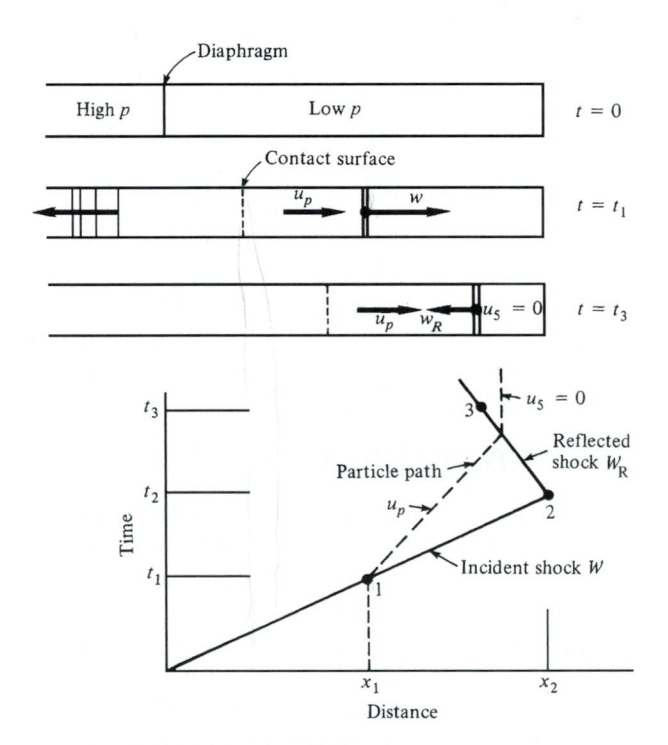

Figure 7.7 | Wave diagram (xt diagram).

the wall.) Indeed, for an incident normal shock of specified strength, the reflected normal shock strength is completely determined by imposing the boundary condition $u_5 = 0$.

In dealing with unsteady wave motion, it is convenient to construct wave diagrams (xt diagrams) such as sketched in Fig. 7.7. A wave diagram is a plot of the wave motion on a graph of t versus x. At time $t = 0$, the incident shock wave is just starting at the diaphragm location. Therefore, at $t = 0$, the incident shock is at location $x = 0$. At some instant later, say time $t = t_1$, the shock wave is traveling to the right, and is located at point $x = x_1$. This is labeled as point 1 in the xt diagram. Note that the path of the incident shock is a straight line in the wave diagram. When the incident shock hits the wall at $x = x_2$ (point 2 in Fig. 7.7), it reflects toward the left with velocity W_R. At some later instant $t = t_3$, the reflected shock is at location $x = x_3$ (point 3 in Fig. 7.7). The path of the reflected shock wave is also a straight line in the wave diagram. The slopes of the incident and reflected shock paths are $1/W$ and $1/W_R$, respectively. Also note as a general characteristic of reflected shocks that $W_R < W$; hence the reflected shock path is more steeply inclined than the incident shock path.

In addition to wave motion, particle motion can also be sketched on the xt diagram. For example, consider a fluid element originally located at $x = x_1$. During the time interval $0 \leq t \leq t_1$, the incident shock has not yet passed over the element, and hence the element simply stands still. This is indicated by the vertical dashed line through point 1 in Fig. 7.7. At time t_1, the incident shock passes over the fluid element located at x_1, and sets it into motion with velocity u_p. The path of the particle is then given by the inclined dashed line above point 1. The fluid element continues along this path until it encounters the reflected shock, which brings the element to a standstill again. The complete dashed curve in Fig. 7.7 represents a *particle path* in the xt diagram.

Return again to the picture of a reflected shock as sketched in Fig. 7.6b. By inspection, we note that

$W_R + u_p =$ velocity of the gas *ahead* of the shock wave *relative* to the wave

$W_R =$ velocity of the gas *behind* the shock wave *relative* to the wave

Hence, from Eqs. (3.48) through (3.50) and the literal interpretation of the velocities u_1 and u_2, we can write for the *reflected shock:*

$$\rho_2(W_R + u_p) = \rho_5 W_R \tag{7.20}$$

$$p_2 + \rho_2(W_R + u_p)^2 = p_5 + \rho_5 W_R^2 \tag{7.21}$$

$$h_2 + \frac{(W_R + u_p)^2}{2} = h_5 + \frac{W_R^2}{2} \tag{7.22}$$

These are the continuity, momentum, and energy equations, respectively, for a reflected shock wave.

Examine Figs. 7.6a and b. The incident shock propagates into the gas ahead of it with a Mach number $M_s = W/a_1$. The reflected shock propagates into the gas ahead of it with a Mach number $M_R = (W_R + u_p)/a_2$. From the incident shock equations, Eqs. (7.1) through (7.3), and the reflected shock equations, Eqs. (7.20) through (7.22), and specializing to a calorically perfect gas, a relation between M_R

and M_s can be obtained as

$$\frac{M_R}{M_R^2 - 1} = \frac{M_s}{M_s^2 - 1} \sqrt{1 + \frac{2(\gamma - 1)}{(\gamma + 1)^2} \left(M_s^2 - 1\right) \left(\gamma + \frac{1}{M_s^2}\right)} \qquad (7.23)$$

The derivation is left as an exercise for the reader. However, Eq. (7.23) explicitly dramatizes that the reflected shock properties are a unique function of the incident shock strength—a result that only makes common sense.

    With this we have finished our basic discussions of moving normal shock waves. Returning to our roadmap in Fig. 7.2, we have finished the left-hand branch. We now move on to the right-hand branch and prepare for the discussion of moving expansion waves.

EXAMPLE 7.5

Consider the normal shock in Example 7.1 to be an incident shock on an end wall. Calculate the reflected shock Mach number, the pressure ratio across the reflected shock, and the gas temperature behind the shock.

■ Solution

From Example 7.1, $M_s = 2.95$, $T_2/T_1 = 2.623$, and $T_1 = 300$ K. From Eq. (7.23),

$$\frac{M_R}{M_R^2 - 1} = \frac{M_s}{M_s^2 - 1} \sqrt{1 + \frac{2(\gamma - 1)}{(\gamma + 1)^2} \left(M_s^2 - 1\right) \left(\gamma + \frac{1}{M_s^2}\right)}$$

$$= \frac{2.95}{(2.95)^2 - 1} \sqrt{1 + \frac{0.8}{5.76}[(2.95)^2 - 1]\left[1.4 + \frac{1}{(2.95)^2}\right]}$$

$$= 0.62$$

Thus,

$$0.62 M_R^2 - M_R - 0.62 = 0$$

    Solving the quadratic, $M_R = \boxed{2.09}$ (we throw away the negative root). This is the Mach number of the reflected wave relative to the gas ahead of it. From Table A.2, for $M_R = 2.09$, we have for the pressure ratio across the reflected shock,

$$\frac{p_5}{p_2} = \boxed{4.978} \quad \text{(nearest entry)}$$

Also,

$$\frac{T_5}{T_2} = 1.77 \quad \text{(nearest entry).}$$

Hence,

$$T_5 = \left(\frac{T_5}{T_2}\right)\left(\frac{T_2}{T_1}\right) T_1 = (1.77)(2.623)(300) = \boxed{1393 \text{ K}}$$

Note: The temperature increase across the *incident* shock is $T_2 - T_1 = 786.9 - 300 = 486.9$ K. The temperature increase across the *reflected* shock is $T_5 - T_2 = 1393 - 786.9 = 606.1$ K, even larger than that across the incident shock. So the reflected shock is a useful mechanism for obtaining high temperatures in a gas, and many shock tubes are designed to use the very hot slug of gas behind the reflected shock at the end wall as the test gas.

7.4 | PHYSICAL PICTURE OF WAVE PROPAGATION

Refer again to the flow in a shock tube illustrated in Fig. 7.5. In Secs. 7.2 and 7.3, we have discussed the traveling shock waves that propagate into the driven gas. We now proceed to examine the expansion wave that propagates into the driver gas. This topic will be introduced in the present section by considering a physical definition of finite wave propagation, followed in Sec. 7.5 by a study of the special aspect of the propagation of a sound wave in one dimension. Then in Secs. 7.6–7.9, the quantitative aspects of finite compression and expansion waves will be developed.

Consider a long duct where properties vary only in the x direction, as sketched in Fig. 7.8a. At time $t = t_1$, let all properties be constant except in some small local region near $x = x_1$. For example, the density distribution is a constant value ρ_∞, except near $x = x_1$, where there is a change in density $\Delta\rho$, as sketched in Fig. 7.8b.

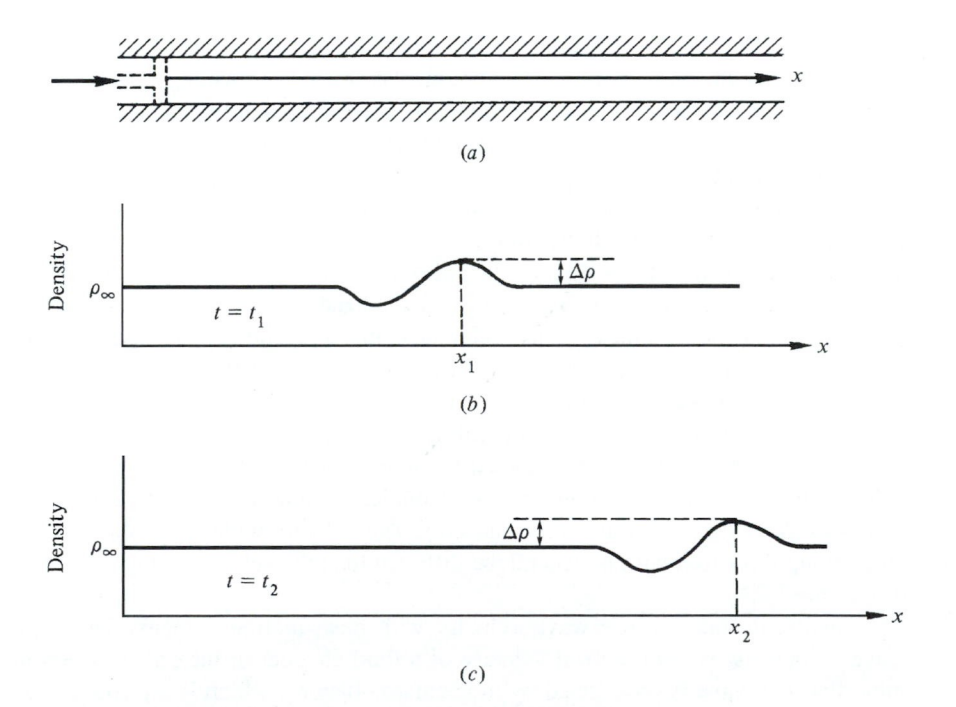

Figure 7.8 | Propagation of a pulse in a one-dimensional tube.

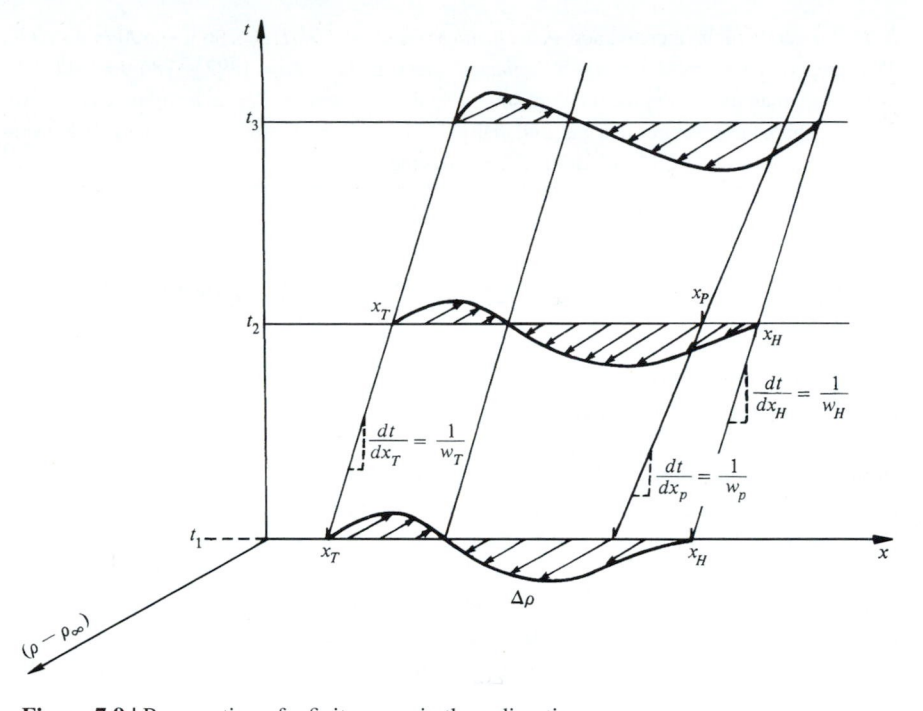

Figure 7.9 | Propagation of a finite wave in the x direction.

This little pulse in density, $\Delta\rho$, can be imagined as created by pushing a piston in the x direction for a moment and then stopping it, as illustrated at the left of Fig. 7.8a. The pulse $\Delta\rho$ moves to the right so that, at a later time $t = t_2$, it is located at $x = x_2$, as sketched in Fig. 7.8c.

The motion of this pulse on an xt diagram (with ρ added as a third axis for additional clarification) is illustrated in Fig. 7.9. Here, x_H denotes the location of the *head* of the pulse, x_T the location of the tail of the pulse, and x_p the location of the *peak* value of the pulse. As shown in Fig. 7.9, the head, tail, and peak are propagating relative to the laboratory with velocities w_H, w_T, and w_p, respectively. In the most general case, $w_H \neq w_T \neq w_P$; hence the *shape* of the pulse continually deforms as it propagates along the x axis. Because the disturbance $\Delta\rho$ moves along the x axis, the region where $\Delta\rho \neq 0$ is called a *finite wave*. The velocity with which an element of this wave moves is called the *local wave velocity w*. In general, the value of w varies through the wave. For example, consider two specific numerical values of $\Delta\rho$ within the wave, $\Delta\rho_1$ and $\Delta\rho_2$. The velocity with which $\Delta\rho_1$ propagates along the x axis will, in general, be different than the velocity with which $\Delta\rho_2$ propagates.

Finally, do not confuse wave velocity with mass-motion velocity. The local wave velocity w is *not* the local velocity of a fluid element of the gas, u. Keep in mind that the wave is propagated by molecular collisions, which is a phenomenon superimposed on top of the mass motion of the gas.

7.5 | ELEMENTS OF ACOUSTIC THEORY

In order to calculate the local value of such wave properties as $\Delta\rho$ and w we must apply the physical principles of conservation of mass, momentum, and energy as embodied in our general equations of motion for an inviscid adiabatic flow. For example, consider Eqs. (6.5), (6.29), and (6.51) repeated here:

$$\frac{\partial\rho}{\partial t} + \nabla \cdot (\rho\mathbf{V}) = 0 \tag{6.5}$$

$$\rho\frac{D\mathbf{V}}{Dt} = -\nabla p \tag{6.29}$$

$$\frac{Ds}{Dt} = 0 \tag{6.51}$$

Let us apply these equations to the flowfield in Figs. 7.8a, b, and c, keeping in mind that the local change in density, $\Delta\rho$, is accompanied by corresponding changes in the other flowfield variables, such as a change in the mass-motion velocity, Δu. Both $\Delta\rho$ and Δu are called *perturbations;* in general they are not necessarily small. Because the undisturbed density and velocity are ρ_∞ and $u_\infty = 0$, respectively, we can express the local density and velocity, ρ and u, respectively, as

$$\rho = \rho_\infty + \Delta\rho \tag{7.24}$$

$$u = u_\infty + \Delta u = 0 + \Delta u = \Delta u \tag{7.25}$$

Note that both $\Delta\rho$ and Δu are functions of x and t. From Eq. (6.5), written for one-dimensional flow,

$$\frac{\partial\rho}{\partial t} + \frac{\partial(\rho u)}{\partial x} = 0$$

or

$$\frac{\partial\rho}{\partial t} + \rho\frac{\partial u}{\partial x} + u\frac{\partial\rho}{\partial x} = 0 \tag{7.26}$$

Substituting Eqs. (7.24) and (7.25) into Eq. (7.26), we have

$$\frac{\partial(\rho_\infty + \Delta\rho)}{\partial t} + (\rho_\infty + \Delta\rho)\frac{\partial(\Delta u)}{\partial x} + \Delta u\frac{\partial(\rho_\infty + \Delta\rho)}{\partial x} = 0 \tag{7.27}$$

Because ρ_∞ is constant, Eq. (7.27) becomes

$$\frac{\partial\Delta\rho}{\partial t} + \rho_\infty\frac{\partial\Delta u}{\partial x} + \Delta\rho\frac{\partial\Delta u}{\partial x} + \Delta u\frac{\partial\Delta\rho}{\partial x} = 0 \tag{7.28}$$

Consider Eq. (6.29) for one-dimensional flow:

$$\rho\frac{\partial u}{\partial t} + \rho u\frac{\partial u}{\partial x} = -\frac{\partial p}{\partial x} \tag{7.29}$$

Consider also the discussion of thermodynamics in Chap. 1, where it was stated that, for a gas in equilibrium, any thermodynamic state variable is uniquely specified by any two other state variables. For example,

$$p = p(\rho, s)$$

Hence,
$$dp = \left(\frac{\partial p}{\partial \rho}\right)_s d\rho + \left(\frac{\partial p}{\partial s}\right)_\rho ds \tag{7.30}$$

However, for the physical picture as shown in Figs. 7.8a, b, and c, *before* the initiation of the wave the gas properties are constant throughout the one-dimensional space. This includes the entropy, which is the same for all fluid elements. Equation (6.51) states that the entropy of a given fluid element remains constant. Therefore, for the inviscid adiabatic wave motion considered here, $s =$ const in both time and space; i.e., the wave motion is *isentropic*. Thus in Eq. (7.30), $ds = 0$, and we have

$$dp = \left(\frac{\partial p}{\partial \rho}\right)_s d\rho \tag{7.31}$$

Considering changes of p and ρ in the x direction, Eq. (7.31) becomes

$$\frac{\partial p}{\partial x} = \left(\frac{\partial p}{\partial \rho}\right)_s \frac{\partial \rho}{\partial x} \tag{7.32}$$

Let $(\partial p/\partial \rho)_s = a^2$. A quick glance at Eq. (3.17) reveals that a is the local speed of sound. However, at this stage in our analysis, we do not as yet have to identify a as the speed of sound; indeed, it will be proven as part of the solution. Thus, for the time being, simply consider a^2 as an abbreviation for $(\partial p/\partial \rho)_s$, and assume we do not identify it with the speed of sound. Then, Eq. (7.32) becomes

$$\frac{\partial p}{\partial x} = a^2 \frac{\partial \rho}{\partial x} \tag{7.33}$$

Substitute Eq. (7.33) into (7.29):

$$\rho \frac{\partial u}{\partial t} + \rho u \frac{\partial u}{\partial x} = -a^2 \frac{\partial \rho}{\partial x} \tag{7.34}$$

Substitute Eqs. (7.24) and (7.25) into Eq. (7.34):

$$(\rho_\infty + \Delta\rho)\frac{\partial \Delta u}{\partial t} + (\rho_\infty + \Delta\rho)\Delta u \frac{\partial \Delta u}{\partial x} = -a^2 \frac{\partial (\rho_\infty + \Delta\rho)}{\partial x}$$

or
$$\rho_\infty \frac{\partial \Delta u}{\partial t} + \Delta\rho \frac{\partial \Delta u}{\partial t} + \rho_\infty \Delta u \frac{\partial \Delta u}{\partial x} + \Delta\rho \Delta u \frac{\partial \Delta u}{\partial x} = -a^2 \frac{\partial \Delta\rho}{\partial x} \tag{7.35}$$

Let us recapitulate at this stage. Equations (7.28) and (7.35) represent the continuity and combined momentum and energy equations, respectively. Although they are in terms of the perturbation quantities $\Delta\rho$ and Δu, they are still *exact* equations for one-dimensional isentropic flow. Also, keep in mind that they are *nonlinear* equations.

Now let us consider the wave in Fig. 7.8*b* and *c* to be *very weak,* i.e., consider $\Delta\rho$ and Δu as *very small perturbations*. In this case, the wave becomes, by definition, a *sound wave*. Here, $\Delta\rho \ll \rho_\infty$ and $\Delta u \ll a$. Also, since $a^2 = (\partial p/\partial\rho)_s$ is a thermodynamic state variable, we can consider it as a function of any two other state variables, say $a^2 = a^2(\rho, s)$. But $s = $ const, so $a^2 = a^2(\rho)$. Expand a^2 in a Taylor's series about the point ρ_∞:

$$a^2 = a_\infty^2 + \left(\frac{\partial a^2}{\partial\rho}\right)(\rho - \rho_\infty) + \cdots$$

$$= a_\infty^2 + \left(\frac{\partial a^2}{\partial\rho}\right)_\infty \Delta\rho + \cdots \tag{7.36}$$

In Eq. (7.36), a^2 is the value at any point in the wave, whereas a_∞^2 is the value of a^2 in the undisturbed gas. Substitute Eq. (7.36) into (7.35):

$$\rho_\infty \frac{\partial\Delta u}{\partial t} + \Delta\rho \frac{\partial\Delta u}{\partial t} + \rho_\infty \Delta u \frac{\partial\Delta u}{\partial x} + \Delta\rho\Delta u \frac{\partial\Delta u}{\partial x}$$

$$= -\left[a_\infty^2 + \left(\frac{\partial a^2}{\partial\rho}\right)_\infty \Delta\rho + \cdots\right]\frac{\partial\Delta\rho}{\partial x} \tag{7.37}$$

Since $\Delta\rho$ and Δu are very small quantities, products of these quantities and their derivatives are extremely small. That is, the *second-order terms* $(\Delta u)^2$, $(\Delta u)(\Delta\rho)$, $(\Delta u)(\partial\Delta\rho/\partial t)$, etc., are very small when compared with the *first-order terms* $\rho_\infty(\partial\Delta u/\partial t)$, $\rho_\infty(\partial\Delta u/\partial x)$, etc. In Eqs. (7.28) and (7.37), *ignore* the second-order terms as being inconsequentially small. The resulting equations are

$$\boxed{\frac{\partial\Delta\rho}{\partial t} + \rho_\infty \frac{\partial\Delta u}{\partial x} = 0} \tag{7.38}$$

$$\boxed{\rho_\infty \frac{\partial\Delta u}{\partial t} = -a_\infty^2 \frac{\partial\Delta\rho}{\partial x}} \tag{7.39}$$

Equations (7.38) and (7.39) are called the *acoustic equations* because they describe the motion of a gas induced by the passage of a sound wave. Due to our assumption of small perturbations, and ignoring higher-order terms, these equations are no longer exact—they are *approximate equations,* which become more and more accurate as the perturbations become smaller and smaller. However, they have one tremendous advantage—they are *linear equations,* and hence can be readily solved in closed form.

For future reference, it is important to note that the above analysis is a specific example of general *small perturbation theory,* leading to linearized equations of motion. Such *linearized theory* is discussed at length in Chap. 9.

Let us now solve Eqs. (7.38) and (7.39). Differentiate Eq. (7.38) with respect to t:

$$\frac{\partial^2 \Delta \rho}{\partial t^2} = -\rho_\infty \frac{\partial^2 \Delta u}{\partial x \, \partial t} \tag{7.40}$$

Differentiate Eq. (7.39) with respect to x:

$$\rho_\infty \frac{\partial^2 \Delta u}{\partial x \, \partial t} = -a_\infty^2 \frac{\partial^2 \Delta \rho}{\partial x^2} \tag{7.41}$$

Substitute Eq. (7.41) into (7.40):

$$\frac{\partial^2 \Delta \rho}{\partial t^2} = a_\infty^2 \frac{\partial^2 \Delta \rho}{\partial x^2} \tag{7.42}$$

The reader may note that Eq. (7.42) is the one-dimensional form of the classic *wave equation* from mathematical physics. Its solution is of the form

$$\boxed{\Delta \rho = F(x - a_\infty t) + G(x + a_\infty t)} \tag{7.43}$$

This is easily proven as follows. From Eq. (7.43),

$$\frac{\partial \Delta \rho}{\partial t} = \frac{\partial F}{\partial (x - a_\infty t)} \frac{\partial (x - a_\infty t)}{\partial t} + \frac{\partial G}{(x + a_\infty t)} \frac{\partial (x + a_\infty t)}{\partial t}$$

or

$$\frac{\partial \Delta \rho}{\partial t} = F'(-a_\infty) + G'(a_\infty)$$

Hence,

$$\frac{\partial^2 \Delta \rho}{\partial t^2} = a_\infty^2 F'' + a_\infty^2 G'' \tag{7.44}$$

where the primes denote differentiation with respect to the argument of F and G, respectively. Also from Eq. (7.43),

$$\frac{\partial \Delta \rho}{\partial x} = \frac{\partial F}{\partial (x - a_\infty t)} \frac{(x - a_\infty t)}{\partial x} + \frac{\partial G}{\partial (x + a_\infty t)} \frac{\partial (x + a_\infty t)}{\partial x}$$

or

$$\frac{\partial \Delta \rho}{\partial x} = F' + G'$$

Hence,

$$\frac{\partial^2 \Delta \rho}{\partial x^2} = F'' + G'' \tag{7.45}$$

Substituting Eqs. (7.44) and (7.45) into Eq. (7.42), we find the identity

$$a_\infty^2 F'' + a_\infty^2 G'' = a_\infty^2 (F'' + G'')$$

Hence, Eq. (7.43) is indeed a solution of Eq. (7.42). Moreover, the acoustic equations, Eqs. (7.38) and (7.39), can be manipulated in an analogous fashion to solve for u as

$$\boxed{\Delta u = f(x - a_\infty t) + g(x + a_\infty t)} \tag{7.46}$$

In both Eqs. (7.43) and (7.46), F, G, f, and g are *arbitrary* functions of their argument. Thus, it would appear that our solution for the flow induced by a sound wave is still not specific enough. However, a very powerful physical interpretation lurks behind Eqs. (7.43) and (7.46). For example, consider Eq. (7.43). For simplicity, since F and G are arbitrary, let $G = 0$. Then, from Eq. (7.43),

$$\Delta\rho = F(x - a_\infty t) \tag{7.47}$$

Consider a wave propagating along the x axis as sketched in Fig. 7.8. Let us watch the propagation of a given constant value of $\Delta\rho$, say $\Delta\rho_1$. Since $\Delta\rho_1$ is chosen as a constant magnitude, Eq. (7.47) becomes

$$\Delta\rho_1 = F(x - a_\infty t) = \text{const}$$

Hence, $(x - a_\infty t)$ must be constant, and thus

$$\boxed{x = a_\infty t + \text{const}} \tag{7.48}$$

Equation (7.48) dictates that the fixed value of the disturbance $\Delta\rho_1$ must move such that $(x - a_\infty t)$ remains constant. Thus, $\Delta\rho_1$ moves with a velocity $dx/dt = a_\infty$ in the positive x direction. Moreover, all other parts of the wave also move with velocity a_∞. Indeed, from this discussion, we can infer that in the wave equation

$$\frac{\partial^2 \Phi}{\partial t^2} = a_\infty^2 \frac{\partial^2 \Phi}{\partial x^2}$$

the constant coefficient a_∞^2 always represents the square of the speed of propagation of the general quantity Φ.

For the sound wave discussed in this section, a figure analogous to Fig. 7.9 can be drawn, as shown in Fig. 7.10. Here, all parts of the sound wave propagate with the same velocity a_∞. The shape of the wave stays the same for all time. This is a

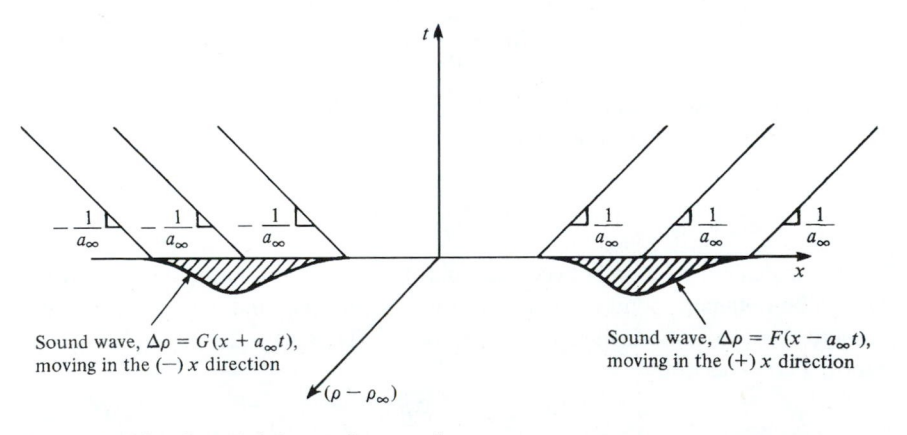

Sound wave, $\Delta\rho = G(x + a_\infty t)$, moving in the $(-)x$ direction

Sound wave, $\Delta\rho = F(x - a_\infty t)$, moving in the $(+)x$ direction

Figure 7.10 | Left- and right-running sound waves.

consequence of our linearized equations as obtained above. If in Eq. (7.43) we assume that $F = 0$, then

$$\Delta\rho = G(x + a_\infty t)$$

represents a sound wave moving to the left, as also illustrated in Fig. 7.10.

Look what has happened! As a direct result of the above analysis, we have proven that the quantity a_∞^2, defined as $[(\partial p/\partial\rho)_s]_\infty$, is indeed the velocity of propagation of the wave. Moreover, the wave we are considering is a sound wave. Therefore, we have just proven from acoustic theory that the velocity of sound is given by $(\partial p/\partial\rho)_s$ evaluated locally in the gas through which it is propagating. Note that a completely separate derivation led to the same result in Eq. (3.17).

Equations (7.43) and (7.46) give $\Delta\rho$ and Δu, respectively. However, we should have enough fluid dynamic intuition by now to suspect that $\Delta\rho$ and Δu are not independent. Indeed, for a given change in density, there is a corresponding change in mass-motion velocity. The relation between Δu and $\Delta\rho$ for a sound wave is obtained as follows. From Eq. (7.46), letting $g = 0$, we obtain $\Delta u = f(x - a_\infty t)$. Hence,

$$\frac{\partial\Delta u}{\partial x} = f'$$

and

$$\frac{\partial\Delta u}{\partial t} = -a_\infty f'$$

Hence,

$$\frac{\partial\Delta u}{\partial x} = -\frac{1}{a_\infty}\frac{\partial\Delta u}{\partial t} \tag{7.49}$$

Substitute Eq. (7.49) into the linearized continuity equation (7.38):

$$\frac{\partial\Delta\rho}{\partial t} - \frac{\rho_\infty}{a_\infty}\frac{\partial\Delta u}{\partial t} = 0$$

$$\frac{\partial}{\partial t}\left(\Delta\rho - \frac{\rho_\infty}{a_\infty}\Delta u\right) = 0$$

$$\Delta\rho - \frac{\rho_\infty}{a_\infty}\Delta u = \text{const} \tag{7.50}$$

The constant is easily evaluated by applying Eq. (7.50) in the undisturbed gas, where $\Delta\rho = \Delta u = 0$. Hence, the constant is zero, and Eq. (7.50) yields

$$\Delta u = \frac{a_\infty}{\rho_\infty}\Delta\rho \tag{7.51}$$

This is the desired relation between Δu and $\Delta\rho$. A similar relation between Δu and Δp can be obtained by noting that the flow is isentropic, and hence any change in pressure Δp causes an isentropic change in $\Delta\rho$. Thus, $\Delta p/\Delta\rho = (\partial p/\partial\rho)_s = a_\infty^2$, and Eq. (7.51) becomes

$$\Delta u = \frac{\Delta p}{\rho_\infty a_\infty} \tag{7.52}$$

Recall that Eqs. (7.51) and (7.52) were obtained by assuming $g = 0$ in Eq. (7.46); hence they apply to a wave moving to the right, as shown in Fig. 7.10. For a wave moving to the left, as also shown in Fig. 7.10, let $f = 0$ in Eq. (7.46). This results in expressions similar to Eqs. (7.51) and (7.52), except with a negative sign. The results are therefore generalized as

$$\Delta u = \pm \frac{a_\infty}{\rho_\infty} \Delta \rho = \pm \frac{\Delta p}{\rho_\infty a_\infty} \tag{7.53}$$

where the $+$ and $-$ signs pertain to right- and left-running waves, respectively. Also note that a positive Δu denotes mass motion in the positive x direction (to the right), and a negative Δu denotes mass motion in the negative x direction (to the left).

In acoustic terminology, that part of a sound wave where $\Delta \rho > 0$ is called a *condensation,* and that part where $\Delta \rho < 0$ is called a *rarefaction.* Note from Eq. (7.53) that for a condensation (where p and ρ increase above ambient conditions), the induced mass motion of the gas is always in the *same* direction as the wave motion, analogous to the effect of a traveling shock wave. For a rarefaction (where p and ρ decrease below ambient conditions), the induced mass motion is always in the *opposite* direction as the wave motion. As we shall find in the following sections, this is analogous to the effect of a traveling expansion wave.

7.6 | FINITE (NONLINEAR) WAVES

In Sec. 7.5 we studied the properties of a traveling wave where the perturbation from ambient conditions, say $\Delta \rho$, was small. This type of wave was defined as a weak wave, or a sound wave. In the present section, the previous constraint will be lifted, and $\Delta \rho$ will not necessarily be small. Such waves, where the perturbations can be large, are called *finite waves*.

Consider a finite wave propagating to the right, as shown in Fig. 7.11. Here, the density, temperature, local speed of sound, and mass motion are sketched as functions of x for some instant in time. At the leading portion of the wave, (around $x = x_2$), ρ is higher than ambient; at the trailing portion (around $x = x_1$), ρ is lower than ambient. Because the flow is isentropic, the temperature follows the density via Eq. (1.43). Since $a = \sqrt{\gamma R T}$, the local speed of sound also varies through the wave, in the same manner as T. With regard to the mass-motion velocity u, we can induce from the results of Sec. 7.5 that it will be positive (in the direction of wave motion) where the density is above ambient, and negative (opposite to the direction of wave motion) where the density is below ambient. In Fig. 7.11, the portions of the wave where the density is increasing (ahead of x_2 and behind x_1) are called *finite compression regions,* and the portion where the density is decreasing (between x_1 and x_2) is called an *expansion region.*

In contrast to the linearized sound wave discussed in Sec. 7.5, different parts of the finite wave in Fig. 7.11 propagate at different velocities relative to the laboratory.

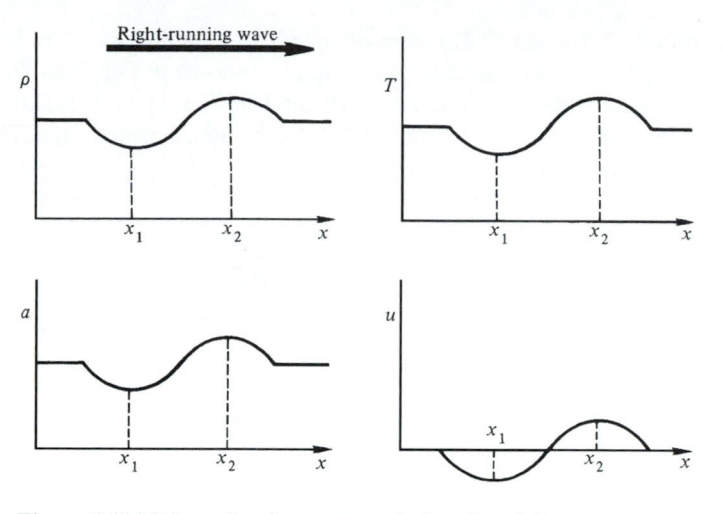

Figure 7.11 | Schematic of property variations in a finite wave.

Consider a fluid element located at x_2 in Fig. 7.11. At this point, it is moving to the right with velocity u_2. In addition, the wave is propagating through the gas due to molecular collisions. In fact, if we are riding along with the fluid element, we see the wave propagating by us at the local velocity of sound, a_2. Therefore, relative to the laboratory, the portion of the wave at location x_2 is propagating at the velocity $w_2 = u_2 + a_2$. Indeed, all portions of the wave are propagating at a velocity $u + a$ relative to the laboratory, where u and a are *local* values of mass velocity and speed of sound, respectively. Physically, the propagation of a local part of the finite wave is the local speed of sound superimposed on top of the local gas mass motion.

Again, reflecting on Fig. 7.11 at x_2 the mass velocity u_2 is toward the right, whereas at x_1 the mass velocity u_1 is toward the left. Moreover, at x_2 the speed of sound is larger than at x_1. Therefore $u_2 + a_2 > u_1 + a_1$, and the portion of the wave around x_2 is traveling *faster* to the right than the portion around x_1. Indeed, if u_1 is a large enough negative number, larger in magnitude than a_1, then the trailing portion of the wave will actually propagate to the left in such a case. So it is clearly evident that the wave shape will distort as it propagates through space. The compression wave will continually steepen until it coalesces into a shock wave, whereas the expansion wave will continually spread out and become more gradual. This distortion of the wave form is illustrated in Fig. 7.9.

Let us now contrast a sound wave with a finite wave. For an *acoustic wave:*

1. $\Delta\rho$, ΔT, Δu, etc., are very small.
2. All parts of the wave propagate with the same velocity relative to the laboratory, namely, at the velocity a_∞.
3. The wave shape stays the same.
4. The flow variables are governed by linear equations.
5. This is an ideal situation, which is closely approached by audible sound waves.

For a *finite wave:*

1. $\Delta\rho$, ΔT, Δu, etc., can be large.
2. Each local part of the wave propagates at the local velocity $u + a$ relative to the laboratory.
3. The wave shape *changes* with time.
4. The flow variables are governed by the full nonlinear equations.
5. This is the "real-life" situation, followed by nature for all real waves.

To develop the governing equations for a finite wave, first consider the continuity equation in the form of Eq. (6.22):

$$\frac{D\rho}{Dt} + \rho(\nabla \cdot \mathbf{V}) = 0 \qquad (6.22)$$

Recall that, from thermodynamics, $\rho = \rho(p, s)$. Hence,

$$d\rho = \left(\frac{\partial\rho}{\partial p}\right)_s dp + \left(\frac{\partial\rho}{\partial s}\right)_p ds \qquad (7.54)$$

For isentropic flow, $ds = 0$. Thus, Eq. (7.54), written in terms of the substantial derivative following a fluid element, becomes

$$\frac{D\rho}{Dt} = \frac{1}{a^2}\frac{Dp}{Dt} \qquad (7.55)$$

Substitute Eq. (7.55) into Eq. (6.22):

$$\frac{1}{a^2}\frac{Dp}{Dt} + \rho(\nabla \cdot \mathbf{V}) = 0 \qquad (7.56)$$

Write Eq. (7.56) for one-dimensional flow:

$$\frac{1}{a^2}\left(\frac{\partial p}{\partial t} + u\frac{\partial p}{\partial x}\right) + \rho\frac{\partial u}{\partial x} = 0 \qquad (7.57)$$

Now consider the momentum equation in the form of Eq. (6.29), without body forces:

$$\rho\frac{D\mathbf{V}}{Dt} = -\nabla p$$

For one-dimensional flow, this becomes

$$\rho\frac{\partial u}{\partial t} + \rho u\frac{\partial u}{\partial x} + \frac{\partial p}{\partial x} = 0$$

or

$$\frac{\partial u}{\partial t} + u\frac{\partial u}{\partial x} + \frac{1}{\rho}\frac{\partial p}{\partial x} = 0 \qquad (7.58)$$

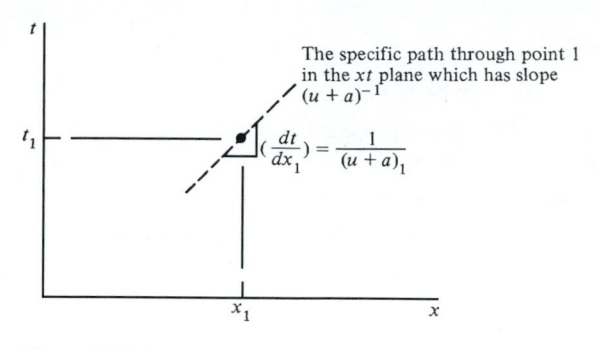

Figure 7.12 I A preferred path in the xt plane.

Adding Eqs. (7.57) and (7.58),

$$\left[\frac{\partial u}{\partial t} + (u + a)\frac{\partial u}{\partial x}\right] + \frac{1}{\rho a}\left[\frac{\partial p}{\partial t} + (u + a)\frac{\partial p}{\partial x}\right] = 0 \qquad (7.59)$$

Subtracting Eq. (7.57) from Eq. (7.58),

$$\left[\frac{\partial u}{\partial t} + (u - a)\frac{\partial u}{\partial x}\right] - \frac{1}{\rho a}\left[\frac{\partial p}{\partial t} + (u - a)\frac{\partial p}{\partial x}\right] = 0 \qquad (7.60)$$

Examine Eqs. (7.59) and (7.60). In principle, a solution of these equations gives $u = u(x, t)$ and $p = p(x, t)$, where (x, t) is any point in the xt plane, as sketched in Fig. 7.12. Moreover, from the definition of a differential,

$$du = \frac{\partial u}{\partial t}\, dt + \frac{\partial u}{\partial x}\, dx \qquad (7.61)$$

In general, we can consider arbitrary changes in t and x, say dt and dx, and calculate the corresponding change in u, given by du from Eq. (7.61). However, let us *not* consider arbitrary values of dt and dx; rather, let us consider a specific path through point 1 in Fig. 7.12. This specific path is chosen so that it satisfies the equation

$$dx = (u + a)\, dt \qquad (7.62)$$

That is, the path we are defining is the dashed line in Fig. 7.12 that goes through point 1 and has a slope $(dt/dx)_1 = 1/(u_1 + a_1)$. Hence, from Eq. (7.61) combined with (7.62), the value of du that corresponds to dt and dx *constrained* to move along the path in Fig. 7.12 is

$$du = \left[\frac{\partial u}{\partial t} + \frac{\partial u}{\partial x}(u + a)\right] dt \qquad (7.63)$$

Similarly for dp,

$$dp = \left[\frac{\partial p}{\partial t} + \frac{\partial p}{\partial x}(u + a)\right] dt \qquad (7.64)$$

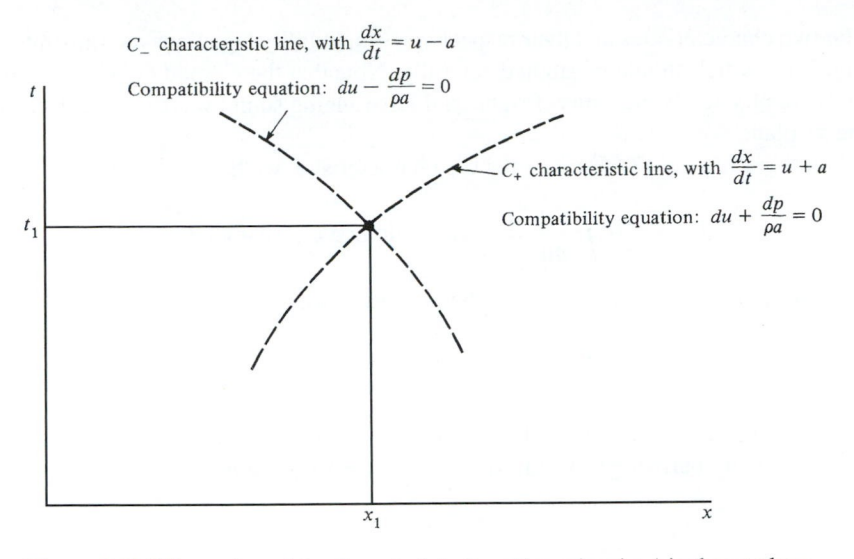

Figure 7.13 | Illustration of the characteristic lines through point 1 in the xt plane.

Substituting Eqs. (7.63) and (7.64) into Eq. (7.59),

$$du + \frac{dp}{\rho a} = 0 \qquad (7.65)$$

where du and dp are changes along a specific path defined by the slope $dx/dt = u + a$ in the xt plane. [Note the similarity between Eq. (7.65) for finite waves and Eq. (7.53) for sound waves.]

We now interject the fact that the above analysis is a specific example of a powerful technique in compressible flow—*the method of characteristics*. Consider any given point (x_1, t_1) in the xt plane as shown in Fig. 7.13. In this analysis, we have found a path through (x_1, t_1) along which the governing partial differential equation (7.59) reduces to an *ordinary* differential equation (7.65). The path is called a C_+ *characteristic line* in the xt plane, and Eq. (7.65) is called the *compatibility equation* along the C_+ characteristic. Equation (7.65) holds *only along the characteristic line*. The method of characteristics will be discussed at length in Chap. 11; the specific application to finite unsteady wave motion in this chapter serves as an illustrative introduction to some of the general concepts.

From Eq. (7.60) we can find another characteristic line, C_-, through the point (x_1, t_1) in Fig. 7.11, where the slope of the C_- characteristic is $dx/dt = u - a$, and along which the following compatibility equation holds:

$$du - \frac{dp}{\rho a} = 0 \qquad (7.66)$$

The two characteristics and their respective compatibility equations are illustrated in Fig. 7.13, which should be studied carefully. Note that the C_+ and C_- characteristic lines are physically the paths of right- and left-running sound waves, respectively, in the xt plane.

Integrating Eq. (7.65) along the C_+ characteristic, we have

$$J_+ = u + \int \frac{dp}{\rho a} = \text{const} \quad \text{(along a } C_+ \text{ characteristic)} \tag{7.67}$$

Integrating Eq. (7.66) along the C_- characteristic, we have

$$J_- = u - \int \frac{dp}{\rho a} = \text{const} \quad \text{(along a } C_- \text{ characteristic)} \tag{7.68}$$

In Eqs. (7.67) and (7.68) J_+ and J_- are called the *Riemann invariants*. Specializing to a calorically perfect gas, from Eq. (3.19), $a^2 = \gamma p/\rho$; thus

$$\rho = \gamma p/a^2 \tag{7.69}$$

Also, since the process is isentropic,

$$p = c_1 T^{\gamma/(\gamma-1)} = c_2 a^{2\gamma/(\gamma-1)} \tag{7.70}$$

where c_1 and c_2 are constants. Differentiating Eq. (7.70), we have

$$dp = c_2 \left(\frac{2\gamma}{\gamma - 1} \right) a^{[2\gamma/(\gamma-1)-1]} \, da \tag{7.71}$$

Substitute Eq. (7.70) into (7.69):

$$\rho = c_2 \gamma a^{[2\gamma/(\gamma-1)-2]} \tag{7.72}$$

Substitute Eqs. (7.71) and (7.72) into Eqs. (7.67) and (7.68):

$$\boxed{J_+ = u + \frac{2a}{\gamma - 1} = \text{const} \quad \text{(along a } C_+ \text{ characteristic)}} \tag{7.73}$$

$$\boxed{J_- = u - \frac{2a}{\gamma - 1} = \text{const} \quad \text{(along a } C_- \text{ characteristic)}} \tag{7.74}$$

Equations (7.73) and (7.74) give the Riemann invariants for a calorically perfect gas. The usefulness of the Riemann invariants is clearly seen by solving Eqs. (7.73) and (7.74) for u and a:

$$\boxed{a = \frac{\gamma - 1}{4}(J_+ - J_-)} \tag{7.75}$$

$$\boxed{u = \tfrac{1}{2}(J_+ + J_-)} \tag{7.76}$$

If the values of J_+ and J_- are known at a given point in the xt plane, then Eqs. (7.75) and (7.76) immediately give the local values of u and a at that point.

Considering again the shock tube in Fig. 7.5, with the above analysis we now have enough tools to solve the flowfield in a one-dimensional expansion wave. This is the subject of Sec. 7.7. Also, this brings us to the bottom of the right-hand column in our roadmap in Fig. 7.2.

7.7 | INCIDENT AND REFLECTED EXPANSION WAVES

Consider the high- and low-pressure regions separated by a diaphragm in a tube, as sketched in Fig. 7.14. When the diaphragm is removed, as discussed in Sec. 7.1, an expansion wave travels to the left, as also shown in Fig. 7.14. With the removal of the diaphragm, the gas in region 4 feels as if a piston is being withdrawn to the right with velocity u_3, as sketched in Fig. 7.14. The piston is purely imaginary in this picture; u_3 is really the mass-motion velocity of the gas (relative to the laboratory) behind the expansion wave. The expansion wave is shown on an xt diagram in Fig. 7.15, where $x = 0$ is the location of the diaphragm. The head of the expansion wave moves to the left into region 4. Recall from Sec. 7.6 that any part of a right-running finite wave moves with the local velocity $u + a$. The same reasoning shows that any part of a left-running wave moves with the local velocity $u - a$. The expansion in Figs. 7.14 and 7.15 is a left-running wave, and hence the local velocity of any part of the wave is $u - a$. In region 4, the mass-motion velocity is zero; hence the head of the wave propagates to the left with a velocity $u_4 - a_4 = 0 - a_4 = -a_4$. Therefore, the path of the head of the wave in the xt plane is a straight line with $dx/dt = u_4 - a_4 = -a_4$. In light of Sec. 7.6, this path must therefore also be a C_- characteristic, as shown in Fig. 7.15.

Within the expansion wave, the induced mass motion is u, and it is directed toward the right. Also, the temperature, and hence a, is reduced inside the wave. Therefore, although the head of the wave advances into region 4 at the speed of sound, other parts of the wave propagate at slower velocities (relative to the laboratory). Hence, the expansion wave spreads out as it propagates down the tube. This is clearly seen in Fig. 7.15, where several C_- characteristics have been sketched for internal

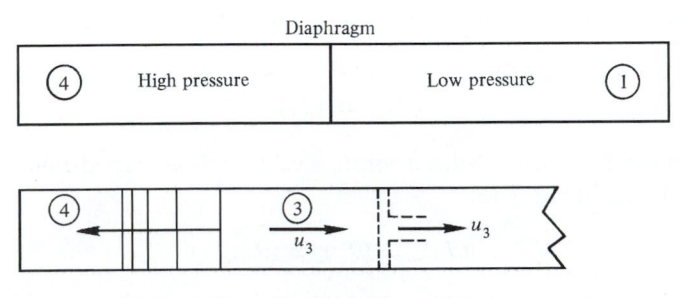

Figure 7.14 | Generation of an expansion wave.

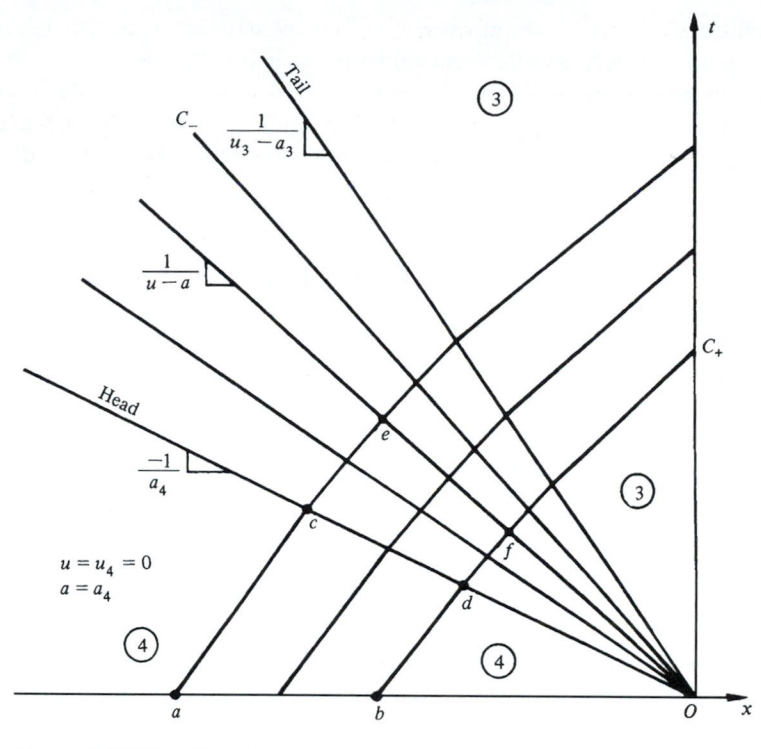

Figure 7.15 | The C_+ and C_- characteristics for a centered expansion wave (on an xt diagram).

portions of the wave. Note that the tail of the wave propagates at the velocity $dx/dt = u_3 - a_3$. Also note that, if u_3 is supersonic, i.e., larger than a_3, the *tail* of the wave will actually move toward the right relative to the laboratory, although the wave is a left-running wave.

In Fig. 7.15, the C_- characteristics have been drawn as straight lines. We need to prove that this is indeed the case. To do this, add the C_+ characteristics to the picture, as also shown in Fig. 7.15. In the constant-property region 4, $u_4 = 0$ and a_4 is a constant. Thus, in region 4, all the C_+ characteristics have the same slope. Moreover, J_+ is the same everywhere in region 4. Hence, considering the two points a and b in Fig. 7.15,

$$(J_+)_a = (J_+)_b \tag{7.77}$$

However, recall from Sec. 7.6 that a constant value of J_+ is carried along a C_+ characteristic. Hence, in Fig. 7.15,

$$(J_+)_a = (J_+)_c = (J_+)_e \tag{7.78}$$

and

$$(J_+)_b = (J_+)_d = (J_+)_f \tag{7.79}$$

Comparing Eqs. (7.78) and (7.79) with (7.77), we have

$$(J_+)_e = (J_+)_f \qquad (7.80)$$

Points e and f, by definition, are on the same C_- characteristic, and recalling that a constant value of J_- is carried along a C_- characteristic, we have

$$(J_-)_e = (J_-)_f \qquad (7.81)$$

Thus, substituting Eqs. (7.80) and (7.81) into Eqs. (7.75) and (7.76), we have $a_e = a_f$ and $u_e = u_f$. Therefore, at points e and f on the C_- characteristic, the value $dx/dt = u - a$ is the same; since points e and f are any arbitrary points on the same C_- characteristic, the slope is the same at all points; the C_- characteristic must therefore be a straight line in Fig. 7.15. Moreover, we have just shown that the values of u and a, and hence of p, ρ, T, etc., are constant along the given straight-line C_- characteristic.

The pictures shown in Figs. 7.14 and 7.15 are for a wave propagating into a constant-property region (region 4). Such a wave is defined as a *simple wave;* a left-running simple wave has straight C_- characteristics along which the flow properties are constant. Similarly, a right-running simple wave has straight C_+ characteristics along which the flow properties are constant. Moreover, because the wave in Figs. 7.14 and 7.15 originates at a given point (the origin in the xt plane), it is called a *centered wave.* Note the analogy between an unsteady one-dimensional centered expansion wave (Fig. 7.13) and the steady two-dimensional Prandtl–Meyer expansion wave in Fig. 4.32.

Repeating, a simple wave is one for which one family of characteristics is straight lines; this can only be the case when the wave is propagating into a uniform region. Note from Fig. 7.15 that the other family (in this case, the C_+ characteristics) can be curved through the wave. In contrast, a *nonsimple* wave has both families of characteristics as curved lines. This is the case, for example, of a reflected expansion wave during part of its reflection process. When the head of the expansion wave in Fig. 7.15 impinges on the endwall, the mass motion must remain zero at the wall. Therefore, the expansion wave must reflect toward the right. The head of the reflected expansion wave, now a right-running wave, propagates through the incident left-running wave. This region of mixed left- and right-running waves is called a *nonsimple region,* and is sketched in Fig. 7.16. The properties of the reflected expansion wave in both the nonsimple and simple regions can be calculated throughout the grid shown in Fig. 7.16 by applying the method of characteristics discussed in Sec. 7.6, and by using the boundary condition that $u = 0$ at the endwall. This becomes a numerical (or graphical) procedure, where the characteristic lines and the compatibility conditions (the Riemann invariants) are pieced together point by point.

In contrast, the solution for a simple centered expansion wave can be obtained in closed analytical form, as follows. Returning to Fig. 7.15 we have shown that J_+ at all the points a, b, c, d, e, f, etc., is the same value, i.e., J_+ *is constant through the expansion wave.* From Eq. (7.73), therefore,

$$u + \frac{2a}{\gamma - 1} = \text{const through the wave} \qquad (7.82)$$

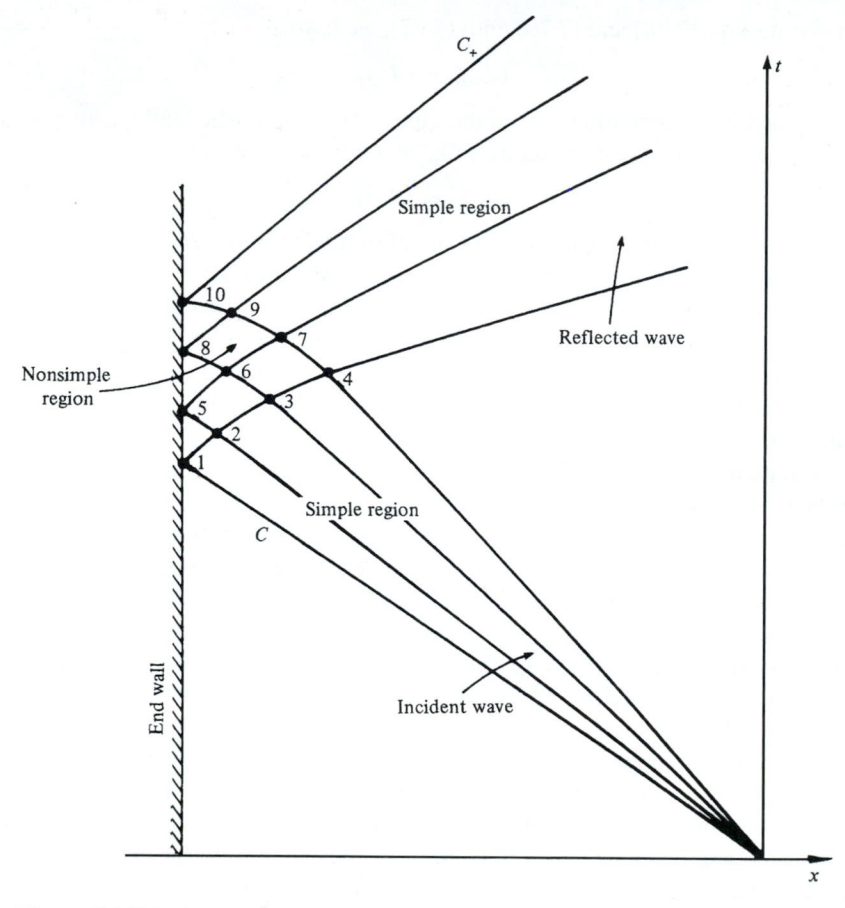

Figure 7.16 | Reflected expansion wave on an xt diagram.

Evaluate the constant by applying Eq. (7.82) in region 4:

$$u_4 + \frac{2a_4}{\gamma - 1} = 0 + \frac{2a_4}{\gamma - 1} = \text{const} \tag{7.83}$$

Combining Eqs. (7.82) and (7.83),

$$\boxed{\frac{a}{a_4} = 1 - \frac{\gamma - 1}{2}\left(\frac{u}{a_4}\right)} \tag{7.84}$$

Equation (7.84) relates a and u at any local point in a simple expansion wave. Because $a = \sqrt{\gamma RT}$, Eq. (7.84) also gives

$$\boxed{\frac{T}{T_4} = \left[1 - \frac{\gamma - 1}{2}\left(\frac{u}{a_4}\right)\right]^2} \tag{7.85}$$

Also, because the flow is isentropic, $p/p_4 = (\rho/\rho_4)^\gamma = (T/T_4)^{\gamma/(\gamma-1)}$. Hence, Eq. (7.85) yields

$$\frac{p}{p_4} = \left[1 - \frac{\gamma-1}{2}\left(\frac{u}{a_4}\right)\right]^{2\gamma/(\gamma-1)} \tag{7.86}$$

$$\frac{\rho}{\rho_4} = \left[1 - \frac{\gamma-1}{2}\left(\frac{u}{a_4}\right)\right]^{2/(\gamma-1)} \tag{7.87}$$

Equations (7.84) through (7.87) give the properties in a simple expansion wave as a function of the local gas velocity in the wave.

To obtain the variation of properties in a centered expansion wave as a function of x and t, consider the C_- characteristics in Fig. 7.15. The equation of any C_- characteristic is

$$\frac{dx}{dt} = u - a$$

or, because the characteristic is a straight line through the origin,

$$x = (u - a)t \tag{7.88}$$

Combining Eqs. (7.84) and (7.88), we have

$$x = \left(u - a_4 + \frac{\gamma-1}{2}u\right)t$$

or

$$u = \frac{2}{\gamma+1}\left(a_4 + \frac{x}{t}\right) \tag{7.89}$$

Equation (7.89) holds for the region between the head and tail of the centered expansion wave in Fig. 7.15, i.e., $-a_4 \le x/t \le u_3 - a_3$.

In summary, for a centered expansion wave moving toward the left as shown in Fig. 7.15, Eq. (7.89) gives u as a function of x and t. In turn, a, T, p, and ρ as functions of x and t are obtained by substituting $u = f(x, t)$ into Eqs. (7.84) through (7.87). The results are sketched in Fig. 7.17, which illustrates the spatial variations of u, ρ, T, and p through the wave at some instant in time. Note from Eq. (7.89) that u varies *linearly* with x through a centered expansion wave. For the left-running wave we have been considering, Eq. (7.89) also shows that u is positive, i.e., the mass motion is toward the right, opposite to the direction of propagation of the wave. Also note that the density, temperature, and pressure all decrease through the wave, with the strongest gradients at the head of the wave.

Analogous relations and results are obtained for a right-running expansion wave, except some of the signs in the equations are changed. The analog of Eqs. (7.82) through (7.89) for a right-running centered expansion wave is left for the reader to derive.

Referring again to Fig. 7.16, properties at the grid points defined by the intersection of C_+ and C_- characteristics in the nonsimple region are obtained from

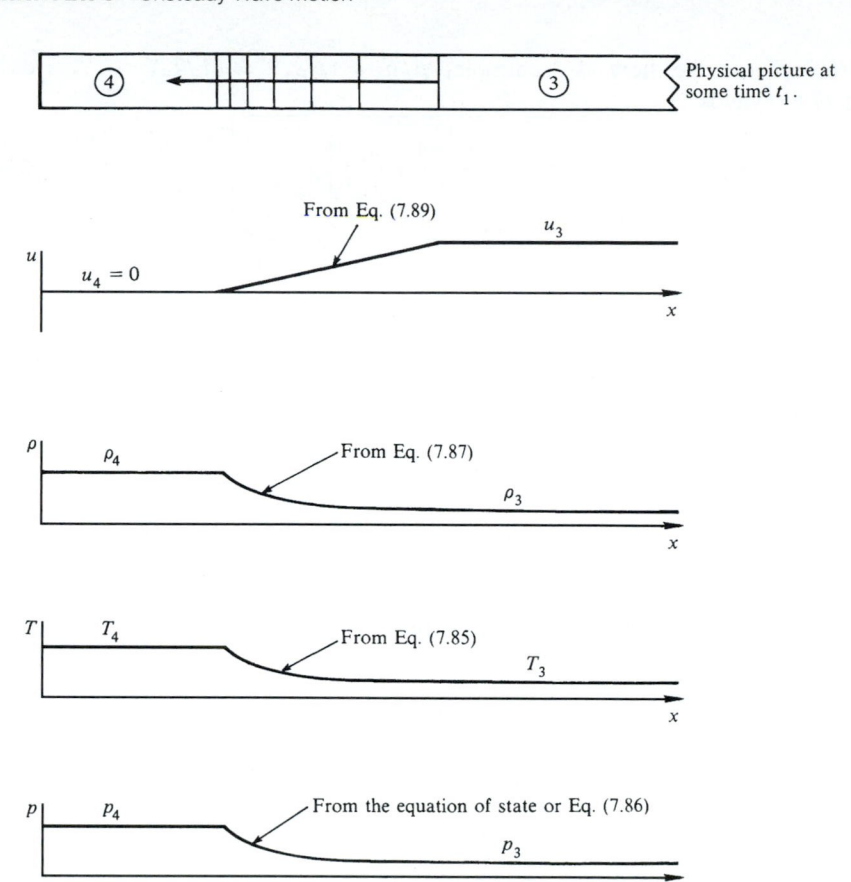

Figure 7.17 | Variation of physical properties within a centered expansion wave.

Eqs. (7.73) through (7.76). For example, J_+ and J_- at points 1, 2, 3, and 4 are known from the incident expansion wave. At point 5, a_5 is determined by $(J_-)_5 = (J_-)_2$ and by the boundary condition $u_5 = 0$. At point 6, both a_6 and u_6 are determined from Eqs. (7.75) and (7.76), knowing that $(J_-)_6 = (J_-)_3$ and $(J_+)_6 = (J_+)_5$. The location of point 6 in the xt space is found by the intersection of the C_- characteristic through point 3 and the C_+ characteristic through point 5. These characteristics are drawn as straight lines with slopes that are averages between the connecting points. For example, for line 3–6,

$$\tan^{-1}\left(\frac{dt}{dx}\right) = \frac{1}{2}\left[\tan^{-1}\left(\frac{1}{u-a}\right)_3 + \tan^{-1}\left(\frac{1}{u-a}\right)_6\right]$$

and for line 5–6,

$$\tan^{-1}\left(\frac{dt}{dx}\right) = \frac{1}{2}\left[\tan^{-1}\left(\frac{1}{u+a}\right)_5 + \tan^{-1}\left(\frac{1}{u+a}\right)_6\right]$$

In this fashion, the flow properties in the entire nonsimple region can be obtained.

Finally, the properties behind the reflected expansion wave after it completely leaves the interaction region are equal to the calculated properties at point 10.

With this we have completed the right-hand column of our roadmap in Fig. 7.2. We are now ready to combine our knowledge of moving shock waves (left column) and moving expansion waves (right column) in order to study the properties of shock tubes, the last box at the bottom of our roadmap. We have come full circle back to the type of application represented by the shock tube shown in Fig. 7.1.

7.8 | SHOCK TUBE RELATIONS

Consider again the shock tube sketched in Figs. 7.4 and 7.5. Initially, a high-pressure gas with molecular weight $\mathcal{M}_4$ and ratio of specific heats γ_4 is separated from a low-pressure gas with corresponding $\mathcal{M}_1$ and γ_1 by a diaphragm. The ratio p_4/p_1 is called the *diaphragm pressure ratio*. Along with the initial conditions of the driver and driven gas, p_4/p_1 determines uniquely the strengths of the incident shock and expansion waves that are set up after the diaphragm is removed. We are now in a position to calculate these waves from the given initial conditions.

As discussed in Sec. 7.1, $u_3 = u_2 = u_p$, and $p_2 = p_3$ across the contact surface. Repeating Eq. (7.16) for the mass motion induced by the incident shock,

$$u_p = u_2 = \frac{a_1}{\gamma_1} \left(\frac{p_2}{p_1} - 1 \right) \left(\frac{\dfrac{2\gamma_1}{\gamma_1 + 1}}{\dfrac{p_2}{p_1} + \dfrac{\gamma_1 - 1}{\gamma_1 + 1}} \right)^{1/2} \tag{7.16}$$

Also, applying Eq. (7.86) between the head and tail of the expansion wave,

$$\frac{p_3}{p_4} = \left[1 - \frac{\gamma_4 - 1}{2} \left(\frac{u_3}{a_4} \right) \right]^{2\gamma_4/(\gamma_4 - 1)} \tag{7.90}$$

Solving Eq. (7.90) for u_3, we have

$$u_3 = \frac{2a_4}{\gamma_4 - 1} \left[1 - \left(\frac{p_3}{p_4} \right)^{(\gamma_4 - 1)/2\gamma_4} \right] \tag{7.91}$$

However, since $p_3 = p_2$, Eq. (7.91) becomes

$$u_3 = \frac{2a_4}{\gamma_4 - 1} \left[1 - \left(\frac{p_2}{p_4} \right)^{(\gamma_4 - 1)/2\gamma_4} \right] \tag{7.92}$$

Recall that $u_2 = u_3$; Eqs. (7.16) and (7.92) can be equated as

$$\frac{a_1}{\gamma_1} \left(\frac{p_2}{p_1} - 1 \right) \left(\frac{\dfrac{2\gamma_1}{\gamma_1 + 1}}{\dfrac{p_2}{p_1} + \dfrac{\gamma_1 - 1}{\gamma_1 + 1}} \right)^{1/2} = \frac{2a_4}{\gamma_4 - 1} \left[1 - \left(\frac{p_2}{p_4} \right)^{(\gamma_4 - 1)/2\gamma_4} \right] \tag{7.93}$$

Equation (7.93) can be algebraically rearranged to give

$$\frac{p_4}{p_1} = \frac{p_2}{p_1}\left\{1 - \frac{(\gamma_4 - 1)(a_1/a_4)(p_2/p_1 - 1)}{\sqrt{2\gamma_1[2\gamma_1 + (\gamma_1 + 1)(p_2/p_1 - 1)]}}\right\}^{-2\gamma_4/(\gamma_4-1)} \qquad (7.94)$$

Equation (7.94) gives the incident shock strength p_2/p_1 as an implicit function of the diaphragm pressure ratio p_4/p_1. Although it is difficult to see from inspection of Eq. (7.94), an evaluation of this relation shows that, for a given diaphragm pressure ratio p_4/p_1, the incident shock strength p_2/p_1 will be made stronger as a_1/a_4 is made smaller. Because $a = \sqrt{\gamma RT} = \sqrt{\gamma(\mathcal{R}/\mathcal{M})T}$, the speed of sound in a light gas is faster than in a heavy gas. Thus, to maximize the incident shock strength for a given p_4/p_1, the driver gas should be a low-molecular-weight gas at high temperature (hence high a_4), and the driven gas should be a high-molecular-weight gas at low temperature (hence low a_1). For this reason, many shock tubes in practice use H_2 or He for the driver gas, and heat the driver gas by electrical means (arc-driven shock tubes) or by chemical combustion (combustion-driven shock tubes).

The analysis of the flow of a calorically perfect gas in a shock tube is now straightforward. For a given diaphragm pressure ratio p_4/p_1:

1. Calculate p_2/p_1 from Eq. (7.94). This defines the strength of the incident shock wave.

2. Calculate all other incident shock properties from Eqs. (7.10), (7.11), (7.14), and (7.16).

3. Calculate $p_3/p_4 = (p_3/p_1)/(p_4/p_1) = (p_2/p_1)/(p_4/p_1)$. This defines the strength of the incident expansion wave.

4. All other thermodynamic properties immediately behind the expansion wave can be found from the isentropic relations

$$\frac{p_3}{p_4} = \left(\frac{\rho_3}{\rho_4}\right)^{\gamma} = \left(\frac{T_3}{T_4}\right)^{\gamma/(\gamma-1)}$$

5. The local properties *inside* the expansion wave can be found from Eqs. (7.84) through (7.87) and (7.89).

7.9 | FINITE COMPRESSION WAVES

Consider the sketch shown in Fig. 7.18. Here, a piston is gradually accelerated from zero to some constant velocity to the right in a tube. The piston path is shown in the xt diagram. When the piston is first started at $t = 0$, a wave propagates to the right into the quiescent gas with the local speed of sound, $w_H = a_\infty$. This is the head of a *compression wave,* because the piston is moving in the same direction as the wave,

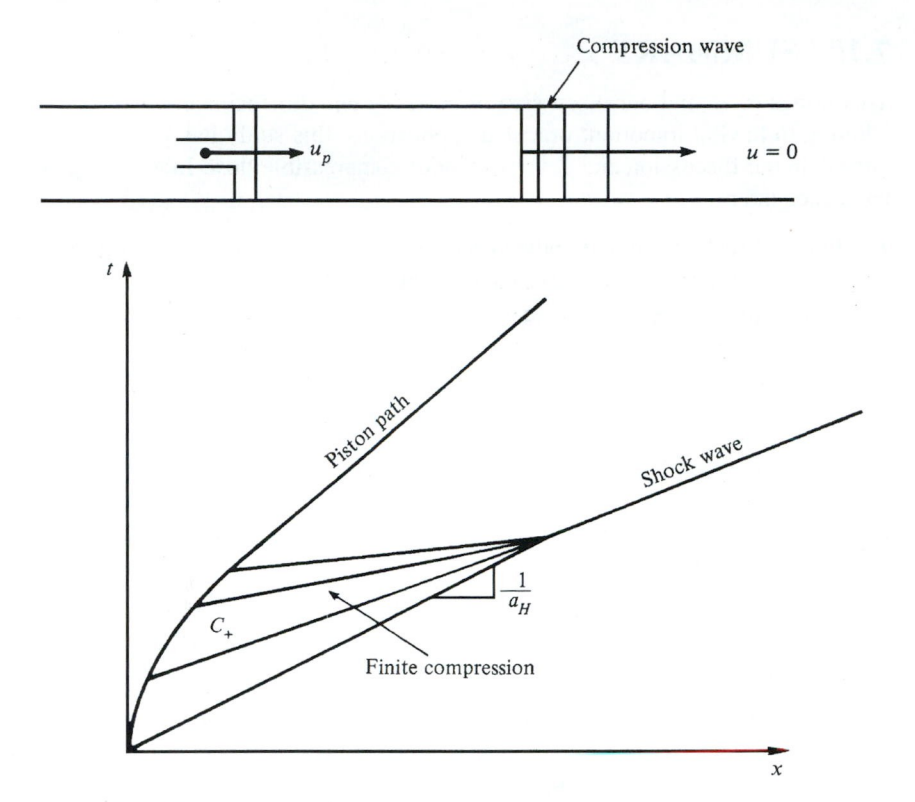

Figure 7.18 | Finite compression wave.

causing a local increase in pressure and temperature. Indeed, inside the wave, the local speed of sound increases, $a > a_\infty$, and there is an induced mass motion u toward the right. Hence, inside the wave, $u + a > w_H$. Since the characteristic lines are given by $dx/dt = u + a$, we see that the C_+ characteristics in Fig. 7.18 progressively approach each other, coalescing into a shock wave. The tail of the compression wave travels faster than the head, and therefore a finite compression wave will always ultimately become a discontinuous shock wave. This is in contrast to an expansion wave, which, as we have already seen, always spreads out as it propagates. These phenomena were recognized as early as 1870; witness the quotation at the beginning of this chapter.

In regard to our discussion of shock tubes, it is interesting to note that, after the breaking of the diaphragm, the incident shock is not formed instantly. Rather, in the immediate region downstream of the diaphragm location, a series of finite compression waves are first formed because the diaphragm breaking process is a complex three-dimensional picture requiring a finite amount of time. These compression waves quickly coalesce into the incident shock wave in a manner analogous to that shown in Fig. 7.18.

7.10 | SUMMARY

This brings to an end our discussion of unsteady one-dimensional wave motion. In addition to having important practical applications, this study has given us several "firsts" in our discussion and development of compressible flow. In this chapter, we have encountered

1. Our first real need to apply the general conservation equations in the form of partial differential equations as derived in Chap. 6.
2. Our first introduction to the idea and results of *linearized* flow—acoustic theory.
3. Our first introduction to the concept of the method of characteristics—finite wave motion.

In subsequent chapters, these philosophies and concepts will be greatly expanded.

PROBLEMS

7.1   Starting with Eq. (7.9), derive Eqs. (7.10) and (7.11).

7.2   Consider a normal shock wave moving with a velocity of 680 m/s into still air at standard atmospheric conditions ($p_1 = 1$ atm and $T_1 = 288$ K).

   a.   Using the equations of Sec. 7.2, calculate T_2, p_2, and u_p behind the shock wave.

   b.   The normal shock tables, Table A.2, can be used to solve moving shock wave problems simply by noting that the tables pertain to flow velocities (hence Mach numbers) *relative to the wave*. Use Table A.2 to obtain T_2, p_2, and u_p for this problem.

7.3   For the conditions of Prob. 7.2, calculate the total pressure and temperature of the gas behind the moving shock wave.

7.4   Consider motionless air with $p_1 = 0.1$ atm and $T_1 = 300$ K in a constant-area tube. We wish to accelerate this gas to Mach 1.5 by sending a normal shock wave through the tube. Calculate the necessary value of the wave velocity relative to the tube.

7.5   Consider an incident normal shock wave that reflects from the end wall of a shock tube. The air in the driven section of the shock tube (ahead of the incident wave) is at $p_1 = 0.01$ atm and $T_1 = 300$ K. The pressure ratio across the incident shock is 1050. With the use of Eq. (7.23), calculate

   a.   The reflected shock wave velocity relative to the tube

   b.   The pressure and temperature behind the reflected shock

7.6   The reflected shock wave associated with a given incident shock can be calculated strictly from the use of Table A.2, *without* using Eq. (7.23). However, the use of Table A.2 for this case requires a trial-and-error solution, converging on the proper boundary condition of zero mass motion behind the reflected shock wave. Repeat Prob. 7.5, using Table A.2 *only*.

7.7 Consider a blunt-nosed aerodynamic model mounted inside the driven section
of a shock tube. The axis of the model is aligned parallel to the axis of the
shock tube, and the nose of the model faces towards the on-coming incident
shock wave. The driven gas is air initially at a temperature and pressure of
300 K and 0.1 atm, respectively. After the diaphragm is broken, an incident
shock wave with a pressure ratio of $p_2/p_1 = 40.4$ propagates into the driven
section.

    a. Calculate the pressure and temperature at the nose of the model shortly
after the incident shock sweeps by the model.

    b. Calculate the pressure and temperature at the nose of the model after the
reflected shock sweeps by the model.

7.8 Consider a centered, one-dimensional, unsteady expansion wave propagating
into quiescent air with $p_4 = 10$ atm and $T_4 = 2500$ K. The strength of the
wave is given by $p_3/p_4 = 0.4$. Calculate the velocity and Mach number of
the induced mass motion behind the wave, relative to the laboratory.

7.9 The driver section of a shock tube contains He at $p_4 = 8$ atm and $T_4 = 300$ K.
$\gamma_4 = 1.67$. Calculate the maximum strength of the expansion wave formed
after removal of the diaphragm (minimum p_3/p_4) for which the incident
expansion wave will remain completely in the driver section.

7.10 The driver and driven gases of a pressure-driven shock tube are both air at
300 K. If the diaphragm pressure ratio is $p_4/p_1 = 5$, calculate:

    a. Strength of the incident shock (p_2/p_1)

    b. Strength of the reflected shock (p_5/p_2)

    c. Strength of the incident expansion wave (p_3/p_4)

7.11 For the shock tube in Prob. 7.10, the lengths of the driver and driven sections
are 3 and 9 m, respectively. On graph paper, plot the wave diagram
(*xt* diagram) showing the wave motion in the shock tube, including the
incident and reflected shock waves, the contact surface, and the incident and
reflected expansion waves. To construct the nonsimple region of the reflected
expansion wave, use the method of characteristics as outlined in Sec. 7.6. Use
at least four characteristic lines to define the incident expansion wave, as
shown in Fig. 7.16.

7.12 Let the uniform region behind the reflected expansion wave be denoted
by the number 6. For the shock tube in Probs. 7.10 and 7.11, calculate
the pressure ratio p_6/p_3 and the temperature T_6 behind the reflected
expansion wave.

7.13 In Probs. 5.20 and 5.21, we noted that the reservoir temperature required for a
continuous flow air Mach 20 hypersonic wind tunnel was beyond the
capabilities of heaters in the reservoir. On the other hand, as discussed in
regard to Fig. 7.1, the high temperature gas behind the reflected shock wave
at the end-wall of a shock tube can be expanded through a nozzle mounted
at the end of the tube. This device is called a shock tunnel, wherein very
large reservoir temperatures can be created. The flow duration through a

shock tunnel, however, is limited typically to a few milliseconds. This is the trade-off necessary to achieve a very high reservoir temperature. Consider a shock tunnel with a Mach 20 nozzle using air. The air temperature in the region behind the reflected shock (the reservoir temperature for the shock tunnel) is 4050 K. In the driven section of the shock tube, before the tube diaphram is broken, the air temperature is 288 K. Calculate the Mach number of the incident shock wave required to obtain a temperature of 4050 K behind the reflected shock.

CHAPTER 8

General Conservation Equations Revisited: Velocity Potential Equation

Dynamics of compressible fluids, like other subjects in which the nonlinear character of the basic equations plays a decisive role, is far from the perfection envisaged by Laplace as the goal of a mathematical theory.
   Richard Courant and K. O. Friedrichs, 1948

PREVIEW BOX

Here we go again—building another tool for our compressible flow toolbox. This new tool is more specialized—it works only for an irrotational flow. But this is not a problem because in many compressible flow applications the actual flow fields are so close to being irrotational that we can assume them as such and proceed to use the tool developed in this chapter. The tool is called the *velocity potential equation*. It is particularly powerful because it combines the continuity, momentum, and energy equations into one equation, and it has a single dependent variable, the velocity potential, from which all the other primitive flow variables, such as velocity, pressure, and temperature can ultimately be obtained. We will use this tool in subsequent chapters to great advantage.

This chapter is short and direct, therefore we have no need for a roadmap. On the other hand, it is important to return to Fig. 1.7, the overall roadmap for the book, and orient ourselves. Looking at Fig. 1.7, we have finished the left-hand column (boxes 3–6), where subjects such as one-dimensional flow, oblique waves, and quasi-one-dimensional flow were studied using the integral form of the governing conservation equations. The

mathematics for these studies was essentially algebra. Then we moved to the middle column in Fig. 1.7, climbing to a higher mathematical tier and dealing with the governing conservation equations in the form of partial differential equations (box 7). This allowed us to study unsteady shock and expansion wave motion (box 10). These same governing conservation equations in the form of partial differential equations will allow us to deal with many other important and interesting flows itemized in boxes 11–16. However, before continuing down the middle column in Fig. 1.7, we detour to boxes 8 and 9 at the right in Fig. 1.7. The material here, the velocity potential equation and its application to obtain linear solutions of subsonic and supersonic flows, is classical and elegant. After about 80 years since its first development, such linearized theory is still extremely powerful. It provides a quick and simple means to obtain pressure distributions over the surfaces of many types of aerodynamic shapes, and hence to obtain the lift, wave drag, and moments on these bodies. So enjoy this detour through boxes 8 and 9, keeping in mind that you will be seeing and learning some very classical compressible flow theory with very useful practical application.

8.1 | INTRODUCTION

In this chapter, the general conservation equations derived in Chap. 6 are simplified for the special case of irrotational flow, discussed below. This simplification is quite dramatic; it allows the separate continuity, momentum, and energy equations with the requisite dependent variables ρ, p, $\mathbf{V}$, T, etc., to cascade into one governing equation with one dependent variable—a new variable defined below as the *velocity potential*. In this chapter, the velocity potential equation will be derived; in turn, in Chap. 9 it will be employed for the approximate solution of several important problems in compressible flow.

8.2 | IRROTATIONAL FLOW

The concept of rotation in a moving fluid was introduced in Sec. 6.6. The vorticity is a point property of the flow, and is given by $\nabla \times \mathbf{V}$. Vorticity is twice the angular velocity of a fluid element, $\nabla \times \mathbf{V} = 2\boldsymbol{\omega}$. A flow where $\nabla \times \mathbf{V} \neq 0$ throughout is called a *rotational flow*. Some typical examples of rotational flows are illustrated in

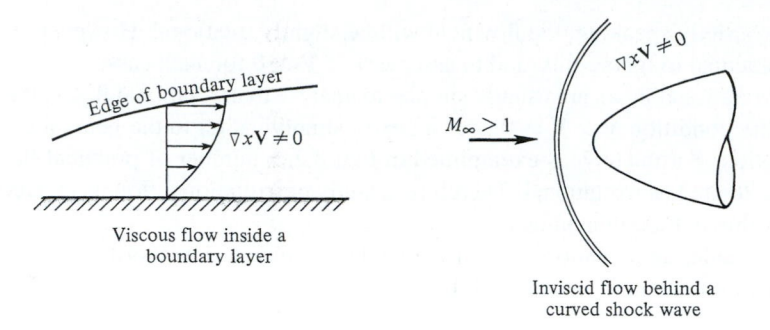

Figure 8.1 | Examples of rotational flows.

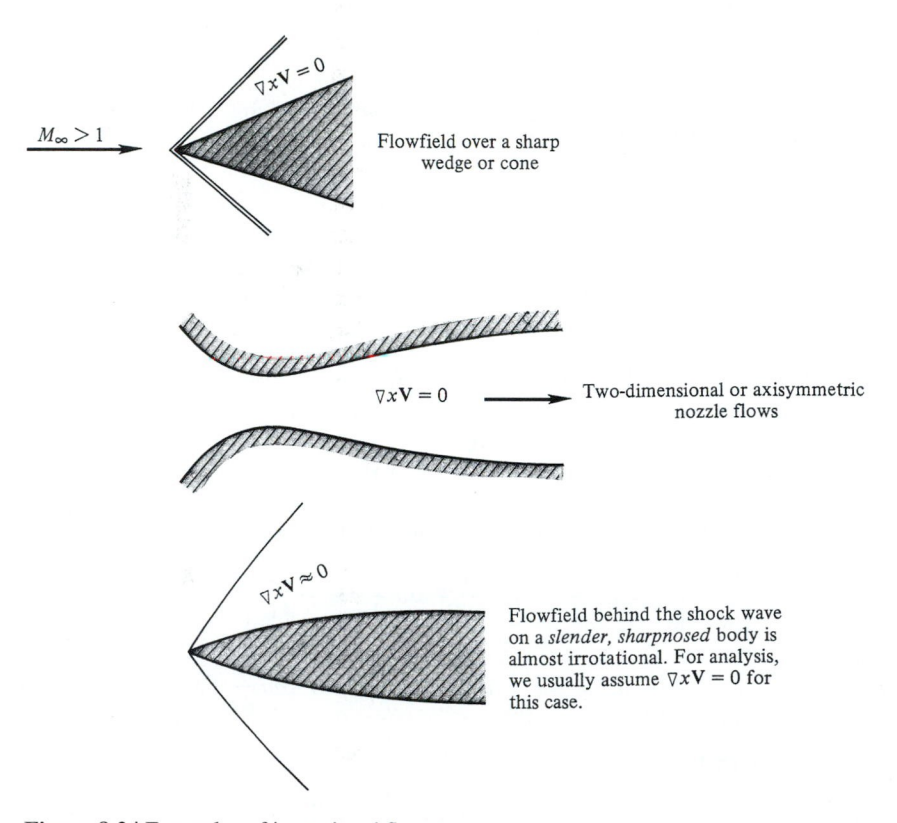

Figure 8.2 | Examples of irrotational flows.

Fig. 8.1 for the region inside a boundary layer and the inviscid flow behind a curved shock wave (see Sec. 6.6). In contrast, a flow where $\nabla \times \mathbf{V} = 0$ everywhere is called an *irrotational flow*. Some typical examples of irrotational flows are shown in Fig. 8.2 for the flowfield over a sharp wedge or cone, the two-dimensional or axisymmetric flow through a nozzle, and the flow over slender bodies. If the slender body is moving supersonically, the attendant shock wave will be slightly curved, and hence, strictly speaking, the flowfield will be slightly rotational. However, it is usually practical to ignore this, and to assume $\nabla \times \mathbf{V} \approx 0$ for such cases.

Irrotational flows are usually simpler to analyze than rotational flows; the irrotationality condition $\nabla \times \mathbf{V} = 0$ adds an extra simplification to the general equations of motion. Fortunately, as exemplified in Fig. 8.2, a number of practical flowfields can be treated as irrotational. Therefore, a study of irrotational flow is of great practical value in fluid dynamics.

Consider an irrotational flow in more detail. In cartesian coordinates, the mathematical statement of irrotational flow is

$$\nabla \times \mathbf{V} = \begin{vmatrix} \mathbf{i} & \mathbf{j} & \mathbf{k} \\ \dfrac{\partial}{\partial x} & \dfrac{\partial}{\partial y} & \dfrac{\partial}{\partial z} \\ u & v & w \end{vmatrix}$$

$$= \mathbf{i}\left(\frac{\partial w}{\partial y} - \frac{\partial v}{\partial z}\right) - \mathbf{j}\left(\frac{\partial w}{\partial x} - \frac{\partial u}{\partial z}\right) + \mathbf{k}\left(\frac{\partial v}{\partial x} - \frac{\partial u}{\partial y}\right) = 0$$

For this equality to hold at every point in the flow,

$$\boxed{\frac{\partial w}{\partial y} = \frac{\partial v}{\partial z} \qquad \frac{\partial w}{\partial x} = \frac{\partial u}{\partial z} \qquad \frac{\partial v}{\partial x} = \frac{\partial u}{\partial y}} \tag{8.1}$$

Equations (8.1) are called the *irrotationality conditions*. Now consider Euler's equation [Eq. (6.29)] without body forces.

$$\rho \frac{D\mathbf{V}}{Dt} = -\nabla p$$

For steady flow, the x component of this equation is

$$\rho u \frac{\partial u}{\partial x} + \rho v \frac{\partial u}{\partial y} + \rho w \frac{\partial u}{\partial z} = -\frac{\partial p}{\partial x}$$

or

$$-\frac{\partial p}{\partial x}\,dx = \rho u \frac{\partial u}{\partial x}\,dx + \rho v \frac{\partial u}{\partial y}\,dx + \rho w \frac{\partial u}{\partial z}\,dx \tag{8.2}$$

But from Eq. (8.1),

$$\frac{\partial u}{\partial y} = \frac{\partial v}{\partial x} \quad \text{and} \quad \frac{\partial u}{\partial z} = \frac{\partial w}{\partial x}$$

Substituting the above relations into Eq. (8.2), we have

$$-\frac{\partial p}{\partial x}\,dx = \rho u \frac{\partial u}{\partial x}\,dx + \rho v \frac{\partial v}{\partial x}\,dx + \rho w \frac{\partial w}{\partial x}\,dx$$

or

$$-\frac{\partial p}{\partial x}\,dx = \frac{1}{2}\rho \frac{\partial u^2}{\partial x}\,dx + \frac{1}{2}\rho \frac{\partial v^2}{\partial x}\,dx + \frac{1}{2}\rho \frac{\partial w^2}{\partial x}\,dx \tag{8.3}$$

Similarly, by considering the y and z components of Euler's equation,

$$-\frac{\partial p}{\partial y}\, dy = \frac{1}{2}\rho\frac{\partial u^2}{\partial y}\, dy + \frac{1}{2}\rho\frac{\partial v^2}{\partial y}\, dy + \frac{1}{2}\rho\frac{\partial w^2}{\partial y}\, dy \qquad (8.4)$$

$$-\frac{\partial p}{\partial z}\, dz = \frac{1}{2}\rho\frac{\partial u^2}{\partial z}\, dz + \frac{1}{2}\rho\frac{\partial v^2}{\partial z}\, dz + \frac{1}{2}\rho\frac{\partial w^2}{\partial z}\, dz \qquad (8.5)$$

Adding Eqs. (8.3) through (8.5), we obtain

$$-\left(\frac{\partial p}{\partial x}\, dx + \frac{\partial p}{\partial y}\, dy + \frac{\partial p}{\partial z}\, dz\right) = \frac{1}{2}\rho\frac{\partial V^2}{\partial x}\, dx + \frac{1}{2}\rho\frac{\partial V^2}{\partial y}\, dy + \frac{1}{2}\rho\frac{\partial V^2}{\partial z}\, dz$$

$$(8.6)$$

where $V^2 = u^2 + v^2 + w^2$.

Equation (8.6) is in the form of perfect differentials, and can be written as

$$-dp = \tfrac{1}{2}\rho\, d(V^2)$$

or

$$\boxed{dp = -\rho V\, dV} \qquad (8.7)$$

Equation (8.7) is a special form of Euler's equation which holds for any direction throughout an *irrotational* inviscid flow with no body forces. If the flow were rotational, Eq. (8.7) would hold only along a streamline. However, for an irrotational flow, the changes in pressure dp and velocity dV in Eq. (8.7) can be taken in any direction, not necessarily just along a streamline.

Euler's equation embodies one of the most fundamental physical characteristics of fluid flow—a physical characteristic that is easily seen in the form given by Eq. (8.7). Namely, in an inviscid flow if the pressure decreases along a given direction [dp is negative in Eq. (8.7)], the velocity must increase in the same direction [in Eq. (8.7), dV must be positive]; similarly, if the pressure increases along a given direction [dp is positive in Eq. (8.7)], the velocity must decrease in the same direction [in Eq. (8.7), dV must be negative]. In the popular literature this is sometimes called the "Bernoulli principle" because in the early eighteenth century Daniel Bernoulli observed this physical effect. Although he worked hard to properly quantify it, he was unsuccessful. His friend and colleague, Leonard Euler, was the first to obtain the proper quantitative relation, namely Eq. (8.7). This equation dates from 1753. (See Reference 134 for more historical details on Bernoulli and Euler, and their contribution to fluid dynamics.)

The Bernoulli principle is very easy to understand physically. Consider a fluid element moving with velocity V in the s direction as sketched in Fig. 8.3. If the pressure decreases in the s direction as shown in Fig. 8.3a (this is defined as a *favorable* pressure gradient), the pressure on the left face will be higher than that on the right face, exerting a net force on the fluid element acting toward the right, and hence

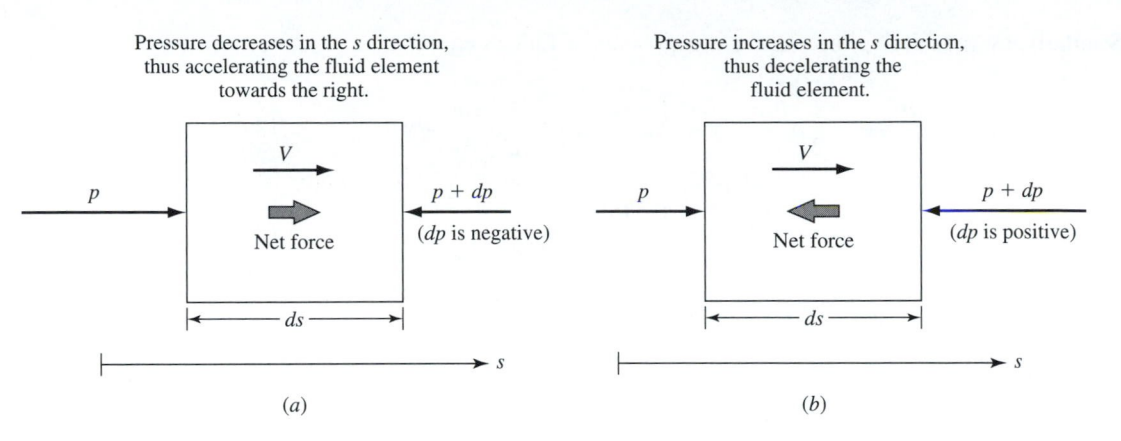

Figure 8.3 | Illustration of pressure gradient effect on the velocity of a fluid element. (a) Decreasing pressure in the flow direction increases the velocity. (b) Increasing pressure in the flow direction decreases the velocity.

accelerating it in the s direction. Clearly, in a region of decreasing pressure, the fluid element will increase its velocity. Conversely, if the pressure increases in the s direction as shown in Fig. 8.3b (this is defined as an *adverse* pressure gradient), the pressure on the right face will be higher than that on the left face, exerting a net force on the fluid element acting toward the left, and hence decelerating it in the s direction. Clearly, in a region of increasing pressure, the fluid element will decrease its velocity.

8.3 | THE VELOCITY POTENTIAL EQUATION

Consider a vector **A**. If $\nabla \times \mathbf{A} = 0$ everywhere, then **A** can always be expressed as $\nabla \zeta$, where ζ is a scalar function. This stems directly from the vector identity, curl (grad) $\equiv 0$. Hence,

$$\nabla \times \nabla \zeta = 0$$

where ζ is any scalar function. For *irrotational flow,* $\nabla \times \mathbf{V} = 0$. Hence, we can define a scalar function, $\mathbf{\Phi} = \mathbf{\Phi}(x, y, z)$, such that

$$\boxed{\mathbf{V} \equiv \nabla \mathbf{\Phi}} \tag{8.8}$$

where $\mathbf{\Phi}$ is called the *velocity potential.* In cartesian coordinates, since

$$\mathbf{V} = u\mathbf{i} + v\mathbf{j} + w\mathbf{k}$$

and

$$\nabla \mathbf{\Phi} = \frac{\partial \mathbf{\Phi}}{\partial x}\mathbf{i} + \frac{\partial \mathbf{\Phi}}{\partial y}\mathbf{j} + \frac{\partial \mathbf{\Phi}}{\partial z}\mathbf{k}$$

then, by comparison,

$$u = \frac{\partial \Phi}{\partial x} \quad v = \frac{\partial \Phi}{\partial y} \quad w = \frac{\partial \Phi}{\partial z} \tag{8.9}$$

Hence, if the velocity potential is known, the velocity can be obtained directly from Eq. (8.8) or (8.9).

As derived next, the velocity potential can be obtained from a single partial differential equation which physically describes an irrotational flow. In addition, we will assume steady, isentropic flow. For simplicity, we will adopt subscript notation for derivatives of Φ as follows: $\partial \Phi/\partial x \equiv \Phi_x$, $\partial \Phi/\partial y \equiv \Phi_y$, $\partial \Phi/\partial z \equiv \Phi_z$, etc. Thus, the continuity equation, Eq. (6.5), for steady flow becomes

$$\nabla \cdot (\rho \mathbf{V}) = 0$$

$$\frac{\partial (\rho u)}{\partial x} + \frac{\partial (\rho v)}{\partial y} + \frac{\partial (\rho w)}{\partial z} = 0$$

$$\frac{\partial}{\partial x} \rho \Phi_x + \frac{\partial}{\partial y} \rho \Phi_y + \frac{\partial}{\partial z} \rho \Phi_z = 0$$

$$\rho (\Phi_{xx} + \Phi_{yy} + \Phi_{zz}) + \Phi_x \frac{\partial \rho}{\partial x} + \Phi_y \frac{\partial \rho}{\partial y} + \Phi_z \frac{\partial \rho}{\partial z} = 0 \tag{8.10}$$

Since we are striving for an equation completely in terms of Φ, we eliminate ρ from Eq. (8.10) by using Euler's equation in the form of Eq. (8.7), which for an irrotational flow applies in any direction:

$$dp = -\rho V \, dV = -\frac{\rho}{2} d(V^2) = -\frac{\rho}{2} d(u^2 + v^2 + w^2)$$

$$dp = -\rho \, d \left(\frac{\Phi_x^2 + \Phi_y^2 + \Phi_z^2}{2} \right) \tag{8.11}$$

From the speed of sound, $a^2 = (\partial p/\partial \rho)_s$. Recalling that the flow is isentropic, any change in pressure dp in the flow is followed by a corresponding isentropic change in density, $d\rho$. Hence,

$$\frac{dp}{d\rho} = \left(\frac{\partial p}{\partial \rho} \right)_s = a^2$$

$$d\rho = \frac{dp}{a^2} \tag{8.12}$$

Combining Eqs. (8.11) and (8.12):

$$d\rho = -\frac{\rho}{a^2} d \left(\frac{\Phi_x^2 + \Phi_y^2 + \Phi_z^2}{2} \right) \tag{8.13}$$

Considering changes in the x direction, Eq. (8.13) directly yields

$$\frac{\partial \rho}{\partial x} = -\frac{\rho}{a^2} \frac{\partial}{\partial x} \left(\frac{\Phi_x^2 + \Phi_y^2 + \Phi_z^2}{2} \right)$$

or

$$\frac{\partial \rho}{\partial x} = -\frac{\rho}{a^2} \left(\Phi_x \Phi_{xx} + \Phi_y \Phi_{yx} + \Phi_z \Phi_{zx} \right) \tag{8.14}$$

Similarly,

$$\frac{\partial \rho}{\partial y} = -\frac{\rho}{a^2} \left(\Phi_x \Phi_{xy} + \Phi_y \Phi_{yy} + \Phi_z \Phi_{zy} \right) \tag{8.15}$$

$$\frac{\partial \rho}{\partial z} = -\frac{\rho}{a^2} \left(\Phi_x \Phi_{xz} + \Phi_y \Phi_{yz} + \Phi_z \Phi_{zz} \right) \tag{8.16}$$

Substituting Eqs. (8.14) through (8.16) into Eq. (8.10), canceling the ρ that appears in each term, and factoring out the second derivatives of Φ, we have

$$\boxed{\begin{aligned} &\left(1 - \frac{\Phi_x^2}{a^2} \right) \Phi_{xx} + \left(1 - \frac{\Phi_y^2}{a^2} \right) \Phi_{yy} + \left(1 - \frac{\Phi_z^2}{a^2} \right) \Phi_{zz} \\ &- \frac{2\Phi_x \Phi_y}{a^2} \Phi_{xy} - \frac{2\Phi_x \Phi_z}{a^2} \Phi_{xz} - \frac{2\Phi_y \Phi_z}{a^2} \Phi_{yz} = 0 \end{aligned}} \tag{8.17}$$

Equation (8.17) is called the *velocity potential equation.*

Equation (8.17) is not strictly in terms of Φ only; the variable speed of sound a still appears. We need to express a in terms of Φ. From the energy equation, Eq. (6.45),

$$h_o = \text{const}$$

Hence, for a calorically perfect gas, this equation can be expressed as

$$c_p T + \frac{V^2}{2} = c_p T_o$$

$$\frac{\gamma R T}{\gamma - 1} + \frac{V^2}{2} = \frac{\gamma R T_o}{\gamma - 1}$$

$$\frac{a^2}{\gamma - 1} + \frac{V^2}{2} = \frac{a_o^2}{\gamma - 1}$$

$$a^2 = a_o^2 - \frac{\gamma - 1}{2} V^2 = a_o^2 - \frac{\gamma - 1}{2} (u^2 + v^2 + w^2)$$

$$\boxed{a^2 = a_o^2 - \frac{\gamma - 1}{2} \left(\Phi_x^2 + \Phi_y^2 + \Phi_z^2 \right)} \tag{8.18}$$

Since a_o is a known constant of the flow, Eq. (8.18) gives the speed of sound a as a function of $\mathbf{\Phi}$.

In summary, Eq. (8.17) coupled with Eq. (8.18) represents a single equation for the unknown variable $\mathbf{\Phi}$. Equation (8.18) represents a combination of the continuity, momentum, and energy equations. This leads to a general procedure for the solution of irrotational, isentropic flowfields:

1. Solve for $\mathbf{\Phi}$ from Eqs. (8.17) and (8.18) for the specified boundary conditions of the given problem.
2. Calculate u, v, and w from Eq. (8.9). Hence, $V = \sqrt{u^2 + v^2 + w^2}$.
3. Calculate a from Eq. (8.18).
4. Calculate $M = V/a$.
5. Calculate T, p, and ρ from Eqs. (3.28), (3.30), and (3.31) respectively.

Hence, we see that once $\mathbf{\Phi} = \mathbf{\Phi}(x, y, z)$ is obtained, *the whole flowfield is known.* This demonstrates the importance of $\mathbf{\Phi}$.

Note that Eq. (8.17) combined with (8.18) is a *nonlinear* partial differential equation. It applies to any irrotational, isentropic flow: subsonic, transonic, supersonic, or hypersonic. It also applies to incompressible flow, where $a \to \infty$, hence yielding the familiar Laplace's equation,

$$\mathbf{\Phi}_{xx} + \mathbf{\Phi}_{yy} + \mathbf{\Phi}_{zz} = 0$$

Moreover, the combined Eqs. (8.17) and (8.18) is an *exact* equation within the framework of isentropic, irrotational flow. No mathematical assumptions (such as small perturbations) have been applied at this stage of our presentation. There is no general closed-form solution to the velocity potential equation, and hence its solution is usually approached in one of these ways:

1. *Exact numerical solutions.* This approach makes it difficult to formulate general trends and rules—the results are raw numbers which have to be analyzed, just like experimental data obtained in the laboratory. However, the techniques of modern computational fluid dynamics are rendering numerical solutions as everyday occurrences in compressible flow, allowing solutions to complicated applications where there would ordinarily be no solution at all. We will study aspects of computational fluid dynamics in Chaps. 11, 12, and 17, emphasizing methods of characteristic and finite-difference solutions.
2. *Transformation of variables* in order to make the velocity potential equation linear, but still exact. Examples of this approach are scarce. One such method is the hodograph solution for subsonic flow, as described by Shapiro (see Ref. 16). Due to its limited usefulness, this technique will not be considered here.
3. *Linearized solutions.* Here, we find linear equations that are *approximations* to the exact nonlinear equations, but which lend themselves to closed-form analytic solution. A large number of real engineering problems lend themselves

to reasonable approximations which linearize the velocity potential equation. Aerodynamic theory historically abounds in linearized theories. This will be the subject of Chap. 9.

8.4 | HISTORICAL NOTE: ORIGIN OF THE CONCEPTS OF FLUID ROTATION AND VELOCITY POTENTIAL

The French mathematician Augustin Cauchy, famous for his contributions to partial differential equations and complex variables, was also active in the theory of fluid flow. In a paper presented to the Paris Academy of Sciences in 1815, he introduced the *average* rotation at a point in the flow. The extension of this idea to the concept of instantaneous rotation of a fluid element was made by the Englishman George Stokes at Cambridge in 1847. (See Fig. 8.4.) In a paper dealing with the viscous flow of fluids. Stokes was the first person to visualize the motion of a fluid element as the resolution of three components: pure translation, pure rotation, and pure strain. The concept of rotation of a fluid element was then applied to inviscid flows about 15 years later by Hermann von Helmholtz.

Figure 8.4 | Sir George Stokes (1819–1903).

Figure 8.5 | Hermann von
Helmholtz (1821–1894).

Helmholtz (see Fig. 8.5) is generally known to fluid dynamicists as a towering
giant during the nineteenth century, with his accomplishments equivalent in stature
to those of Euler and d'Alembert. However, it is interesting to note that Helmholtz is
mainly recognized by the rest of civilization for his work in medicine, acoustics, optics, and electromagnetic theory. Born in Potsdam, Germany, on August 31, 1821,
Helmholtz studied medicine in Berlin, and became a noted physiologist, holding professional positions in medicine at Königsberg, Bonn, and Heidelberg between 1855
to 1871. After that, he became a professor of physics at the University of Berlin until
his death in 1894.

Helmholtz made substantial contributions to the theory of incompressible inviscid flow during the nineteenth century. We note here only one such contribution, relevant to this chapter. In 1858 he published a paper entitled "On the Integrals of the
Hydrodynamical Equations Corresponding to Vortex Motions," in which he observed that the velocity components along all three axes in a flow could be expressed
as a derivative of a single function. He called this function *potential of velocity,*
which is identical to Φ in Eq. (8.8). This was the first practical use of a velocity potential in fluid mechanics, although Louis de Lagrange (1736–1813), in his book
Mechanique Analytic published in 1788, had first introduced the basic concept of this
potential. Moreover, Helmholtz concluded "that in the cases in which a potential of
the velocity exists the smallest fluid particles do not possess rotatory motions,
whereas when no such potential exists, at least a portion of these particles is found in
rotary motion."

Therefore, the general concepts in this chapter dealing with irrotational and rotational flows, as well as the definition of the velocity potential, were established
more than a century ago.

CHAPTER 9

Linearized Flow

Geometry which should only obey physics, when united to the latter, sometimes commands it. If it happens that a question which we wish to examine is too complicated to permit all its elements to enter into the analytical relation which we wish to set up, we separate the more inconvenient elements, we substitute for them other elements less troublesome, but also less real, and then we are surprised to arrive, notwithstanding our painful labor, at a result contradicted by nature; as if after having disguised it, cut it short, or mutilated it, a purely mechanical combination would give it back to us.

Jean le Rond d'Alembert, 1752

PREVIEW BOX

The Boeing 777 jet transport, shown in Fig. 1.4, has swept wings, like many aircraft designed to fly near or beyond Mach one. Why swept wings? We answer this question in the present chapter. From a purely physical point of view, we will explain the advantages of a swept wing for high-speed flight. The discussion is centered around the concept of the critical Mach number, which is introduced toward the end of the chapter.

However, the main thrust of this chapter is to present a fast and simple means of estimating the pressure distribution over bodies moving at high subsonic or supersonic speeds. Recall from Eq. (1.45) that, for an inviscid flow, the aerodynamic force on a body is due only to the surface pressure distribution integrated over the complete body surface. If we had a simple means of calculating the pressure distributions over the body surface, then the calculation of the lift and drag on the body would be straightforward. This chapter presents such simple means of calculating pressure distributions. But as we have stated before, you cannot get something for nothing. The simple calculations of pressure distribution are only approximate results, and they can be made only for slender bodies (thin bodies, not thick ones) at small angles of attack in a flow that is either subsonic or supersonic, not transonic or hypersonic. However, there

are a lot of thin body shapes that fly at small angles of attack at subsonic or supersonic speeds, so the material in this chapter is much more useful than it seems at first thought.

The techniques presented here are based on a linearized form of the velocity potential equation; hence the label "linearized flows." This material is classical; it was developed in the 1920s, and remained the primary means for calculating pressure distributions over thin bodies at small angles of attack until the advent of computational fluid dynamics (CFD) in the 1960s and 70s. Today, linearized theory still has an important role to play in the study of compressible flow, mainly because of the simple formulas it provides for surface pressure distributions. The results of linearized theory give us fast, "back-of-the-envelope" calculations for surface pressure distributions that are tailor-made for preliminary design studies. So even in this modern world of computers and sophisticated CFD, classical linearized theory still has its place. This is why we devote a chapter to it in the present book on "modern" compressible flow.

The roadmap for this chapter is given in Fig. 9.1. We begin by discussing the appropriate physical assumptions that lead to the reduction of the nonlinear velocity

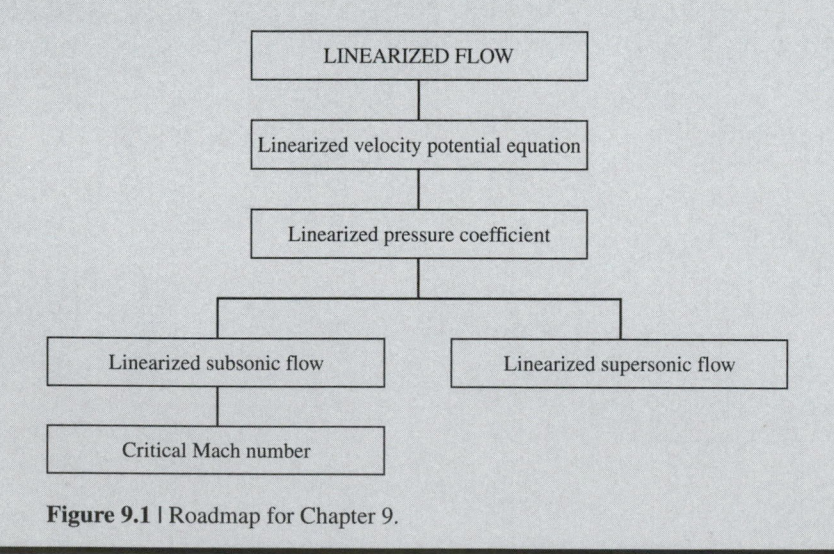

Figure 9.1 | Roadmap for Chapter 9.

potential equation to a linear partial differential equation, and then derive this linear equation in detail. Then we define the pressure coefficient, and proceed to obtain an approximate linear expression for the pressure coefficient that is consistent with the degree of accuracy represented by the linearized velocity potential equation. These tools are shown as the center column in our roadmap in Fig. 9.1. The tools apply equally well to subsonic and supersonic flows. Hence, we next move to the left column in Fig. 9.1, and study high-speed, compressible, subsonic flow. Then we move to the box at the right and study supersonic flow. Finally, we venture back to the left column and define the critical Mach number, discuss how it can be calculated, and examine its physical implications. It is here where we explain the aerodynamic functioning of swept wings.

9.1 | INTRODUCTION

Transport yourself back in time to the year 1940, and imagine that you are an aerodynamicist responsible for calculating the lift on the wing of a high-performance fighter plane. You recognize that the airspeed is high enough so that the well-established incompressible flow techniques of the day will give inaccurate results. Compressibility must be taken into account. However, you also recognize that the governing equations for compressible flow are nonlinear, and that no general solution exists for these equations. Numerical solutions are out of the question— high-speed digital computers are still 15 years in the future. So, what do you do? The only practical recourse is to seek assumptions regarding the physics of the flow, which will allow the governing equations to become linear, but which at the same time do not totally compromise the accuracy of the real problem. In turn, these linear equations can be attacked by conventional mathematical techniques.

In this context, it is easy to appreciate why linear solutions to flow problems dominated the history of aerodynamics and gasdynamics up to the middle 1950s. In modern compressible flow, with the advent of the high-speed computer, the importance of linearized flow has been relaxed. Linearized solutions now take their proper role as closed-form analytic solutions useful for explicitly identifying trends and governing parameters, for highlighting some important physical aspects of the flow, and for providing practical formulas for the rapid estimation of aerodynamic forces and pressure distributions. In modern practice, whenever accuracy is desired the full nonlinear equations are solved numerically on a computer, as described in subsequent chapters.

This chapter deals exclusively with linearized flow, but not to the extent that most earlier classical texts do. The reader is strongly urged to consult the classic texts listed as Refs. 3 through 17, especially those by Ferri, Hilton, Shapiro, and Liepmann and Roshko, for a more in-depth presentation. Our purpose here is to put linearized flow into proper perspective with modern techniques and to glean important physical trends from the linearized results.

Finally, there are a number of practical aerodynamic problems where, on a physical basis, a uniform flow is changed, or *perturbed,* only *slightly.* One such example is the flow over a thin airfoil illustrated in Fig. 9.2. The flow is characterized by only a small deviation of the flow from its original uniform state. The analyses of such

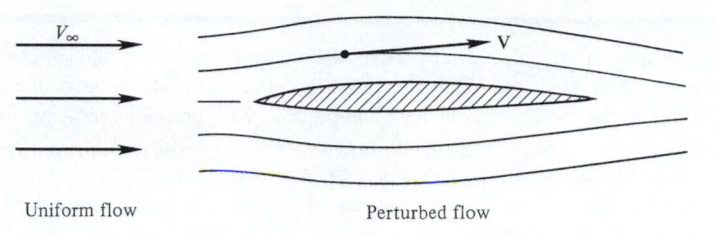

Figure 9.2 | Comparison between uniform and perturbed flows.

flows are usually called *small-perturbation theories*. Small-perturbation theory is frequently (but not always) linear theory, an example is the acoustic theory discussed in Sec. 7.5, where the assumption of small perturbations allowed a linearized solution. Linearized solutions in compressible flow always contain the assumption of small perturbations, but small perturbations do not always guarantee that the governing equations can be linearized, as we shall soon see.

9.2 | LINEARIZED VELOCITY POTENTIAL EQUATION

Consider a slender body immersed in a uniform flow, as sketched in Fig. 9.2. In the uniform flow, the velocity is V_∞ and is oriented in the x direction. In the perturbed flow, the local velocity is $\mathbf{V}$, where $\mathbf{V} = V_x \mathbf{i} + V_y \mathbf{j} + V_z \mathbf{k}$, and where V_x, V_y, and V_z are now used to denote the x, y, and z components of velocity, respectively. In this chapter, u', v', and w' denote *perturbations* from the uniform flow, such that

$$V_x = V_\infty + u'$$

$$V_y = v'$$

$$V_z = w'$$

Here, u', v', and w' are the *perturbation velocities* in the x, y, and z directions, respectively. Also in the perturbed flow, the pressure, density, and temperature are p, ρ, and T, respectively. In the uniform stream, $V_x = V_\infty$, $V_y = 0$, and $V_z = 0$. Also in the uniform stream, the pressure, density, and temperature are p_∞, ρ_∞, and T_∞, respectively.

In terms of the velocity potential,

$$\nabla \mathbf{\Phi} = \mathbf{V} = (V_\infty + u')\mathbf{i} + v'\mathbf{j} + w'\mathbf{k}$$

where $\mathbf{\Phi}$ is now denoted as the "total velocity potential" (introduced in Chap. 8). Let us now define a new velocity potential, the *perturbation velocity potential* ϕ, such that

$$\frac{\partial \phi}{\partial x} = u' \qquad \frac{\partial \phi}{\partial y} = v' \qquad \frac{\partial \phi}{\partial z} = w'$$

Then,

$$\mathbf{\Phi}(x, y, z) = V_\infty x + \phi(x, y, z)$$

where
$$V_x = V_\infty + u' = \frac{\partial \Phi}{\partial x} = V_\infty + \frac{\partial \phi}{\partial x}$$

$$V_y = v' = \frac{\partial \Phi}{\partial y} = \frac{\partial \phi}{\partial y}$$

$$V_z = w' = \frac{\partial \Phi}{\partial z} = \frac{\partial \phi}{\partial z}$$

Also,
$$\Phi_{xx} = \frac{\partial^2 \phi}{\partial x^2}$$

$$\Phi_{yy} = \frac{\partial^2 \phi}{\partial y^2}$$

$$\Phi_{zz} = \frac{\partial^2 \phi}{\partial z^2}$$

Consider again the velocity potential equation, Eq. (8.17). Multiplying this equation by a^2 and substituting $\Phi = V_\infty x + \phi$, we have

$$\left[a^2 - \left(V_\infty + \frac{\partial \phi}{\partial x}\right)^2\right]\frac{\partial^2 \phi}{\partial x^2} + \left[a^2 - \left(\frac{\partial \phi}{\partial y}\right)^2\right]\frac{\partial^2 \phi}{\partial y^2} + \left[a^2 - \left(\frac{\partial \phi}{\partial z}\right)^2\right]\frac{\partial^2 \phi}{\partial z^2}$$

$$- 2\left(V_\infty + \frac{\partial \phi}{\partial x}\right)\frac{\partial \phi}{\partial y}\frac{\partial^2 \phi}{\partial x \partial y} - 2\left(V_\infty + \frac{\partial \phi}{\partial x}\right)\frac{\partial \phi}{\partial z}\frac{\partial^2 \phi}{\partial x \partial z} - 2\frac{\partial \phi}{\partial y}\frac{\partial \phi}{\partial z}\frac{\partial^2 \phi}{\partial y \partial z} = 0$$

$$(9.1)$$

Equation (9.1) is called the *perturbation-velocity potential equation*. To obtain better physical insight, we recast Eq. (9.1) in terms of velocities:

$$[a^2 - (V_\infty + u')^2]\frac{\partial u'}{\partial x} + [a^2 - v'^2]\frac{\partial v'}{\partial y} + [a^2 - w'^2]\frac{\partial w'}{\partial z}$$

$$- 2(V_\infty + u')v'\frac{\partial u'}{\partial y} - 2(V_\infty + u')w'\frac{\partial u'}{\partial z} - 2v'w'\frac{\partial v'}{\partial z} = 0 \qquad (9.2)$$

Since the total enthalpy is constant throughout the flow,

$$h_\infty + \frac{V_\infty^2}{2} = h + \frac{V^2}{2} = h + \frac{(V_\infty + u')^2 + v'^2 + w'^2}{2}$$

or
$$\frac{a_\infty^2}{\gamma - 1} + \frac{V_\infty^2}{2} = \frac{a^2}{\gamma - 1} + \frac{(V_\infty + u')^2 + v'^2 + w'^2}{2}$$

$$a^2 = a_\infty^2 - \frac{\gamma - 1}{2}(2u'V_\infty + u'^2 + v'^2 + w'^2) \qquad (9.3)$$

Substituting Eq. (9.3) into (9.2), and algebraically rearranging,

$$
\left(1 - M_\infty^2\right) \frac{\partial u'}{\partial x} + \frac{\partial v'}{\partial y} + \frac{\partial w'}{\partial z}
$$

$$
= M_\infty^2 \left[(\gamma + 1)\frac{u'}{V_\infty} + \left(\frac{\gamma + 1}{2}\right) \frac{u'^2}{V_\infty^2} + \left(\frac{\gamma - 1}{2}\right) \left(\frac{v'^2 + w'^2}{V_\infty^2}\right) \right] \frac{\partial u'}{\partial x}
$$

$$
+ M_\infty^2 \left[(\gamma - 1)\frac{u'}{V_\infty} + \left(\frac{\gamma + 1}{2}\right) \frac{v'^2}{V_\infty^2} + \left(\frac{\gamma - 1}{2}\right) \left(\frac{w'^2 + u'^2}{V_\infty^2}\right) \right] \frac{\partial v'}{\partial y}
$$

$$
+ M_\infty^2 \left[(\gamma - 1)\frac{u'}{V_\infty} + \left(\frac{\gamma + 1}{2}\right) \frac{w'^2}{V_\infty^2} + \left(\frac{\gamma - 1}{2}\right) \left(\frac{u'^2 + v'^2}{V_\infty^2}\right) \right] \frac{\partial w'}{\partial z}
$$

$$
+ M_\infty^2 \left[\frac{v'}{V_\infty} \left(1 + \frac{u'}{V_\infty}\right) \left(\frac{\partial u'}{\partial y} + \frac{\partial v'}{\partial x}\right) + \frac{w'}{V_\infty} \left(1 + \frac{u'}{V_\infty}\right) \left(\frac{\partial u'}{\partial z} + \frac{\partial w'}{\partial x}\right) \right.
$$

$$
\left. + \frac{u'w'}{V_\infty^2} \left(\frac{\partial w'}{\partial y} + \frac{\partial v'}{\partial z}\right) \right]
\tag{9.4}
$$

Equation (9.4) is still an exact equation for irrotational, isentropic flow. It is simply an expanded form of the perturbation-velocity potential equation. Note that the left-hand side of Eq. (9.4) is linear, but the right-hand side is not. Also recall that we have not said anything about the *size* of the perturbation velocities u', v', and w'. They could be large or small. Equation (9.4) holds for both cases.

We now specialize to the case of *small perturbations*, i.e., we assume the u', v', and w' are small compared to V_∞:

$$
\frac{u'}{V_\infty}, \frac{v'}{V_\infty}, \text{ and } \frac{w'}{V_\infty} \ll 1 \qquad \left(\frac{u'}{V_\infty}\right)^2, \left(\frac{v'}{V_\infty}\right)^2, \text{ and } \left(\frac{w'}{V_\infty}\right)^2 \ll 1
$$

With this in mind, compare like terms (coefficients of like derivatives) on the left- and right-hand sides of Eq. (9.4):

1. For $0 \le M_\infty \le 0.8$ and for $M_\infty \ge 1.2$, the magnitude of

$$
M_\infty^2 \left[(\gamma + 1)\frac{u'}{V_\infty} + \cdots \right] \frac{\partial u'}{\partial x}
$$

is small in comparison to the magnitude of

$$
\left(1 - M_\infty^2\right) \frac{\partial u'}{\partial x}
$$

Thus, *ignore* the former term.

2. For $M_\infty \le 5$ (approximately),

$$
M_\infty^2 \left[(\gamma - 1)\frac{u'}{V_\infty} + \cdots \right] \frac{\partial v'}{\partial y}
$$

is small in comparison to $\partial v'/\partial y$,

$$M_\infty^2 \left[(\gamma - 1)\frac{u'}{V_\infty} + \cdots \right] \frac{\partial w'}{\partial z}$$

is small in comparison to $\partial w'/\partial z$, and

$$M_\infty^2 \left[\frac{v'}{V_\infty}\left(1 + \frac{u'}{V_\infty}\right)\left(\frac{\partial u'}{\partial y} + \frac{\partial v'}{\partial x}\right) + \cdots \right] \approx 0$$

Thus, ignore these terms in comparison to those on the left-hand side of Eq. (9.4).

With these order-of-magnitude comparisons, Eq. (9.4) reduces to

$$\left(1 - M_\infty^2\right)\frac{\partial u'}{\partial x} + \frac{\partial v'}{\partial y} + \frac{\partial w'}{\partial z} = 0 \qquad (9.5)$$

or, in terms of the perturbation velocity potential,

$$\boxed{\left(1 - M_\infty^2\right)\frac{\partial^2 \phi}{\partial x^2} + \frac{\partial^2 \phi}{\partial y^2} + \frac{\partial^2 \phi}{\partial z^2} = 0} \qquad (9.6)$$

Note that Eqs. (9.5) and (9.6) are *approximate* equations: they no longer represent the exact physics of the flow. However, look what has happened. The original nonlinear equations, Eqs. (9.1) through (9.4), have been reduced to linear equations, namely, Eqs. (9.5) and (9.6). Inasmuch as Eq. (9.1) is called the perturbation-velocity potential equation, Eq. (9.6) is called the *linearized* perturbation-velocity potential equation. However, a price has been paid for this linearization. The approximate equation (9.6) is much more restrictive than the exact equation (9.1), for these reasons:

1. The perturbations must be *small*.
2. From item 1 in the list above, we see that *transonic flow* ($0.8 \leq M_\infty \leq 1.2$) is excluded.
3. From item 2 in that same list we see that *hypersonic flow* ($M_\infty \geq 5$) is excluded.

Thus, Eq. (9.6) is valid for subsonic and supersonic flow only—an important point to remember. However, Eq. (9.6) has the striking advantage that it is *linear*.

In summary, we have demonstrated that subsonic and supersonic flows lend themselves to approximate, linearized theory for the case of irrotational, isentropic flow with small perturbations. In contrast, transonic and hypersonic flows cannot be linearized, even with small perturbations. This is another example of the consistency of nature. Note some of the *physical* problems associated with transonic flow (mixed subsonic-supersonic regions with possible shocks, and extreme sensitivity to geometry changes at sonic conditions) and with hypersonic flow (strong shock waves close to the geometric boundaries, i.e., thin shock layers, as well as high enthalpy, and hence high-temperature conditions in the flow). Just on an intuitive basis, we would expect such physically complicated flows to be inherently nonlinear. For the remainder of this chapter, we will consider linear flows only; thus, we will deal with subsonic and supersonic flows.

9.3 | LINEARIZED PRESSURE COEFFICIENT

The pressure coefficient C_p is defined as

$$C_p \equiv \frac{p - p_\infty}{\frac{1}{2}\rho_\infty V_\infty^2} \tag{9.7}$$

where p is the local pressure, and p_∞, ρ_∞, and V_∞ are the pressure, density, and velocity, respectively, in the uniform free stream. The pressure coefficient is simply a nondimensional pressure difference; it is extremely useful in fluid dynamics.

An alternative form of the pressure coefficient, convenient for compressible flow, can be obtained as follows:

$$\frac{1}{2}\rho_\infty V_\infty^2 = \frac{1}{2}\frac{\gamma p_\infty}{\gamma p_\infty}\rho_\infty V_\infty^2 = \frac{\gamma}{2}p_\infty\frac{V_\infty^2}{a_\infty^2} = \frac{\gamma}{2}p_\infty M_\infty^2 \tag{9.8}$$

Hence, Eq. (9.7) becomes

$$C_p = \frac{p - p_\infty}{(\gamma/2)p_\infty M_\infty^2} = \frac{p_\infty(p/p_\infty - 1)}{(\gamma/2)p_\infty M_\infty^2} \tag{9.9}$$

Hence,

$$\boxed{C_p = \frac{2}{\gamma M_\infty^2}\left(\frac{p}{p_\infty} - 1\right)} \tag{9.10}$$

Equation (9.10) is an alternative form of Eq. (9.7), expressed in terms of γ and M_∞ rather than ρ_∞ and V_∞. It is still an exact representation of the definition of C_p.

We now proceed to obtain an approximate expression for C_p that is consistent with linearized theory. Since the total enthalpy is constant,

$$h + \frac{V^2}{2} = h_\infty + \frac{V_\infty^2}{2}$$

For a calorically perfect gas, this becomes

$$T + \frac{V^2}{2c_p} = T_\infty + \frac{V_\infty^2}{2c_p}$$

$$T - T_\infty = \frac{V_\infty^2 - V^2}{2c_p} = \frac{V_\infty^2 - V^2}{2\gamma R/(\gamma - 1)}$$

$$\frac{T}{T_\infty} - 1 = \frac{\gamma - 1}{2}\frac{V_\infty^2 - V^2}{\gamma R T_\infty} = \frac{\gamma - 1}{2}\frac{V_\infty^2 - V^2}{a_\infty^2} \tag{9.11}$$

Since

$$V^2 = (V_\infty + u')^2 + v'^2 + w'^2$$

Eq. (9.11) becomes

$$\frac{T}{T_\infty} = 1 - \frac{\gamma - 1}{2a_\infty^2}(2u'V_\infty + u'^2 + v'^2 + w'^2) \tag{9.12}$$

Since the flow is isentropic, $p/p_\infty = (T/T_\infty)^{\gamma/(\gamma-1)}$, and Eq. (9.12) gives

$$\frac{p}{p_\infty} = \left[1 - \frac{\gamma - 1}{2a_\infty^2}(2u'V_\infty + u'^2 + v'^2 + w'^2)\right]^{\gamma/(\gamma-1)}$$

or
$$\frac{p}{p_\infty} = \left[1 - \frac{\gamma - 1}{2}M_\infty^2\left(\frac{2u'}{V_\infty} + \frac{u'^2 + v'^2 + w'^2}{V_\infty^2}\right)\right]^{\gamma/(\gamma-1)} \tag{9.13}$$

Equation (9.13) is still an exact expression. However, considering small perturbations: $u'/V_\infty \ll 1$: u'^2/V_∞, v'^2/V_∞^2, and $w'^2/V_\infty^2 \lll 1$. Hence, Eq. (9.13) is of the form

$$\frac{p}{p_\infty} = (1 - \varepsilon)^{\gamma/(\gamma-1)}$$

where ε is small. Hence, from the binomial expansion, neglecting higher-order terms,

$$\frac{p}{p_\infty} = 1 - \frac{\gamma}{\gamma - 1}\varepsilon + \cdots \tag{9.14}$$

Thus, Eq. (9.13) can be expressed in the form of Eq. (9.14) as seen next, neglecting higher-order terms:

$$\frac{p}{p_\infty} = 1 - \frac{\gamma}{2}M_\infty^2\left(\frac{2u'}{V_\infty} + \frac{u'^2 + v'^2 + w'^2}{V_\infty^2}\right) + \cdots \tag{9.15}$$

Substitute Eq. (9.15) into Eq. (9.10):

$$C_p = \frac{2}{\gamma M_\infty^2}\left[1 - \frac{\gamma}{2}M_\infty^2\left(\frac{2u'}{V_\infty} + \frac{u'^2 + v'^2 + w'^2}{V_\infty^2}\right) + \cdots - 1\right]$$

$$= -\frac{2u'}{V_\infty} - \frac{u'^2 + v'^2 + w'^2}{V_\infty^2} + \cdots \tag{9.16}$$

Since u'^2/V_∞^2, v'^2/V_∞^2, and $w'^2/V_\infty^2 \lll 1$, Eq. (9.16) becomes

$$\boxed{C_p = -\frac{2u'}{V_\infty}} \tag{9.17}$$

Equation (9.17) gives the *linearized pressure coefficient,* valid for *small perturbations*. Note its particularly simple form; the linearized pressure coefficient depends only on the *x* component of the perturbation velocity.

9.4 | LINEARIZED SUBSONIC FLOW

As mentioned in Sec. 9.1, historically a major impetus for the development of linearized theory for subsonic compressible flow grew out of the need to predict aerodynamic forces and moments on airfoils. Throughout the 1930s, this question became increasingly compelling: How can we take incompressible results (theory or experiment), and modify them to take compressibility into account? In this section we will develop an answer by utilizing the linearized equations developed in Secs. 9.2 and 9.3. The development will deal explicitly with the two-dimensional flow over an airfoil; however, it applies for any two-dimensional shape which satisfies the assumptions of small perturbations, e.g., the flow over a bumpy or wavy wall.

Consider the compressible subsonic flow over a thin airfoil at small angle of attack (hence small perturbations), as sketched in Fig. 9.3. The usual inviscid flow boundary condition must hold at the surface, i.e., the flow velocity must be tangent to the surface. Referring to Fig. 9.3, at the surface this boundary condition is

$$\frac{df}{dx} = \frac{v'}{V_\infty + u'} = \tan\theta \tag{9.18}$$

For small perturbations, $u' \ll V_\infty$, and $\tan\theta \approx \theta$; hence, Eq. (9.18) becomes

$$\frac{df}{dx} = \frac{v'}{V_\infty} = \theta \tag{9.19}$$

Since $v' = \partial\phi/\partial y$, Eq. (9.19) is written as

$$\frac{\partial\phi}{\partial y} = V_\infty \frac{df}{dx} \tag{9.20}$$

Equation (9.20) represents the appropriate boundary condition at the surface, consistent with linearized theory.

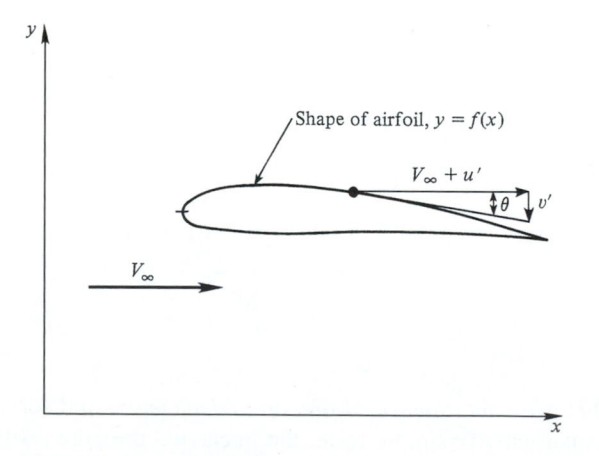

Figure 9.3 | Airfoil in physical space.

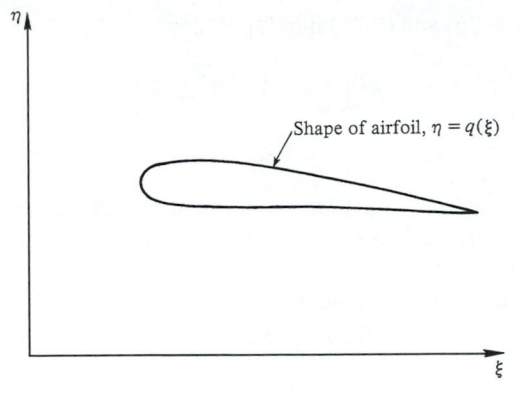

Figure 9.4 | Airfoil in transformed space.

The subsonic compressible flow over the airfoil in Fig. 9.3 is governed by the linearized perturbation-velocity potential equation (9.6). For two-dimensional flow, this becomes

$$\beta^2 \phi_{xx} + \phi_{yy} = 0 \tag{9.21}$$

where $\beta \equiv \sqrt{1 - M_\infty^2}$. Equation (9.21) can be transformed to a familiar incompressible form by considering a transformed coordinate system (ξ, η), such that

$$\xi = x \tag{9.22}$$

$$\eta = \beta y \tag{9.23}$$

In this transformed space, sketched in Fig. 9.4, a transformed perturbation velocity potential $\bar{\phi}(\xi, \eta)$ is defined such that

$$\bar{\phi}(\xi, \eta) = \beta \phi(x, y) \tag{9.24}$$

To couch Eq. (9.21) in terms of the transformed variables, note that

$$\frac{\partial \xi}{\partial x} = 1 \qquad \frac{\partial \xi}{\partial y} = 0 \qquad \frac{\partial \eta}{\partial x} = 0 \qquad \frac{\partial \eta}{\partial y} = \beta$$

Therefore, the derivatives of ϕ in (x, y) space are related to the derivatives of $\bar{\phi}$ in (ξ, η) space, according to

$$\phi_x = \frac{\partial \phi}{\partial x} = \frac{1}{\beta} \frac{\partial \bar{\phi}}{\partial x} = \frac{1}{\beta} \left[\frac{\partial \bar{\phi}}{\partial \xi} \frac{\partial \xi}{\partial x} + \frac{\partial \bar{\phi}}{\partial \eta} \frac{\partial \eta}{\partial x} \right] = \frac{1}{\beta} \frac{\partial \bar{\phi}}{\partial \xi} = \frac{\bar{\phi}_\xi}{\beta} \tag{9.25}$$

$$\phi_{xx} = \frac{1}{\beta} \bar{\phi}_{\xi\xi} \tag{9.26}$$

$$\phi_y = \frac{\partial \phi}{\partial y} = \frac{1}{\beta} \frac{\partial \bar{\phi}}{\partial y} = \frac{1}{\beta} \left[\frac{\partial \bar{\phi}}{\partial \xi} \frac{\partial \xi}{\partial y} + \frac{\partial \bar{\phi}}{\partial \eta} \frac{\partial \eta}{\partial y} \right] = \frac{\partial \bar{\phi}}{\partial \eta} = \bar{\phi}_\eta \tag{9.27}$$

$$\phi_{yy} = \bar{\phi}_{\eta\eta} \beta \tag{9.28}$$

Substituting Eqs. (9.26) and (9.28) into Eq. (9.21),

$$\beta^2 \left(\frac{1}{\beta} \bar{\phi}_{\xi\xi} \right) + \beta \bar{\phi}_{\eta\eta} = 0$$

or

$$\bar{\phi}_{\xi\xi} + \bar{\phi}_{\eta\eta} = 0 \tag{9.29}$$

Equation (9.29) is Laplace's equation, which governs incompressible flow. Hence, $\bar{\phi}$ represents an *incompressible* flow in (ξ, η) space, which is related to a *compressible* flow ϕ in (x, y) space.

The shape of the airfoil is given by $y = f(x)$ and $\eta = q(\xi)$ in (x, y) and (ξ, η) space, respectively. From Eq. (9.20) in (x, y) space, we have

$$V_\infty \frac{df}{dx} = \frac{\partial \phi}{\partial y} = \frac{1}{\beta} \frac{\partial \bar{\phi}}{\partial y} = \frac{\partial \bar{\phi}}{\partial \eta} \tag{9.30}$$

Applying Eq. (9.20) in (ξ, η) space,

$$V_\infty \frac{dq}{d\xi} = \frac{\partial \bar{\phi}}{\partial \eta} \tag{9.31}$$

The right-hand sides of Eqs. (9.30) and (9.31) are equal; hence, equating the left-hand sides,

$$\frac{df}{dx} = \frac{dq}{d\xi} \tag{9.32}$$

Equation (9.32) is an important result; it demonstrates that the *shape* of the airfoil in (x, y) and (ξ, η) space is the *same*. Hence, the above transformation relates the compressible flow over an airfoil in (x, y) space to the incompressible flow in (ξ, η) space over the *same* airfoil.

The practicality of the above development is in the pressure coefficient. For the compressible flow in Fig. 9.3, the pressure coefficient is, from Eq. (9.17),

$$C_p = -\frac{2u'}{V_\infty} = -\frac{2}{V_\infty} \frac{\partial \phi}{\partial x} = -\frac{2}{V_\infty} \frac{1}{\beta} \frac{\partial \bar{\phi}}{\partial x} = -\frac{2}{V_\infty} \frac{1}{\beta} \frac{\partial \bar{\phi}}{\partial \xi} \tag{9.33}$$

Denoting the incompressible perturbation velocity in the ξ direction by $\bar{u}$, where $\bar{u} = \partial \bar{\phi}/\partial \xi$, Eq. (9.33) becomes

$$C_p = \frac{1}{\beta} \left(-\frac{2\bar{u}}{V_\infty} \right) \tag{9.34}$$

Since (ξ, η) space corresponds to incompressible flow, Eq. (9.17) yields

$$-\frac{2\bar{u}}{V_\infty} = C_{p_o} \tag{9.35}$$

where C_{p_o} is the *incompressible* pressure coefficient. Combining Eqs. (9.34) and (9.35),

$$\boxed{C_p = \frac{C_{p_o}}{\sqrt{1 - M_\infty^2}}} \tag{9.36}$$

Equation (9.36) is called the *Prandtl–Glauert* rule; it is a similarity rule which relates *incompressible* flow over a given two-dimensional profile to *subsonic compressible* flow over the *same* profile. Moreover, consider the aerodynamic lift L and moment M on this airfoil. We define the lift and moment coefficients, C_L and C_M, respectively, as

$$C_L = \frac{L}{\frac{1}{2}\rho_\infty V_\infty^2 S}$$

$$C_M = \frac{M}{\frac{1}{2}\rho_\infty V_\infty^2 Sl}$$

where S is a reference area (for a wing, usually the platform area of the wing), and l is a reference length (for an airfoil, usually the chord length). In Sec. 1.5, the lift was defined as the component of aerodynamic force perpendicular to the free-stream velocity. As explained in Sec. 1.5, the sources of all aerodynamic forces and moments on a body are the pressure and shear stress distributions over the surface. Since we are dealing with an inviscid flow, the shear stress is zero. Moreover, Eq. (1.46) gives an equation for the lift in terms of the integral of the pressure distribution. Since both L and M are due to the pressure acting on the surface, and surface pressure for subsonic compressible flow is related to surface pressure for incompressible flow through Eq. (9.36), it can readily be shown that (see, for example, Ref. 1)

$$\boxed{C_L = \frac{C_{L_o}}{\sqrt{1 - M_\infty^2}}} \qquad (9.37a)$$

$$\boxed{C_M = \frac{C_{M_o}}{\sqrt{1 - M_\infty^2}}} \qquad (9.37b)$$

Equations 9.37a and 9.37b are also called the *Prandtl–Glauert rule*. They are exceptionally practical aerodynamic formulas for the approximate compressibility correction to low-speed lift and moments on slender two-dimensional aerodynamic shapes. Note that the effect of compressibility is to increase the magnitudes of C_L and C_M.

Equations (9.36) through (9.37) are results from linearized theory. They indicate that the aerodynamic forces go to infinity as M_∞ goes to unity—an impossible result. This quandary is resolved, of course, by recalling that linearized theory breaks down in the transonic regime (near $M_\infty = 1$). Indeed, the Prandtl–Glauert rule is reasonably valid only up to a Mach number of approximately 0.7. More accurate compressibility corrections will be discussed in Sec. 9.5.

An important effect of compressibility on subsonic flowfields can be seen by noting that

$$u' = \frac{\partial \phi}{\partial x} = \frac{1}{\beta}\frac{\partial \bar{\phi}}{\partial x} = \frac{1}{\beta}\frac{\partial \bar{\phi}}{\partial \xi} = \frac{\bar{u}}{\beta} = \frac{\bar{u}}{\sqrt{1 - M_\infty^2}} \qquad (9.38)$$

Comparing the extreme left- and right-hand sides of Eq. (9.38) at a given location in the flow, as M_∞ increases, the perturbation velocity u' increases. Compressibility strengthens the disturbance to the flow introduced by a solid body. From another perspective, in comparison to incompressible flow, a perturbation of given strength reaches further away from the surface in compressible flow. The spatial extent of the disturbed flow region is increased by compressibility. Also, the disturbance reaches out in all directions, both upstream and downstream.

In classical inviscid incompressible flow theory, a two-dimensional closed body experiences no aerodynamic drag. This is the well-known d'Alembert's paradox, and is due to the fact that, without the effects of friction and its associated separated flow, the pressure distributions over the forward and rearward portions of the body exactly cancel in the flow direction. Does the same result occur for inviscid subsonic compressible flow? The answer can be partly deduced from Eq. (9.36). The compressible pressure coefficient C_p differs from the incompressible value C_{p_o} by only a constant scale factor. Hence, if the distribution of C_{p_o} results in zero drag, the distribution of C_p will also cancel in the flow direction and result in zero drag. Similar results are obtained from nonlinear subsonic calculations (thick bodies at large angle of attack). Hence, d'Alembert's paradox can be generalized to include subsonic compressible flow as well as incompressible flow.

EXAMPLE 9.1

Consider a subsonic flow with an upstream Mach number of M_∞. This flow moves over a wavy wall with a contour given by $y_w = h\cos(2\pi x/l)$, where y_w is the ordinate of the wall, h is the amplitude, and l is the wavelength. Assume that h is small. Using the small perturbation theory of this chapter, derive an equation for the velocity potential and the surface pressure coefficient.

■ **Solution**

The wall shape is sketched in Fig. 9.5. Assume that h/l is small. Therefore, the flowfield above the wall is characterized by *small perturbations* from the uniform flow conditions. Hence, the perturbation-velocity potential equation, Eq. (9.6), applies. In two dimensions, this becomes

$$\left(1 - M_\infty^2\right)\frac{\partial^2 \phi}{\partial x^2} + \frac{\partial^2 \phi}{\partial y^2} = 0 \tag{E.1}$$

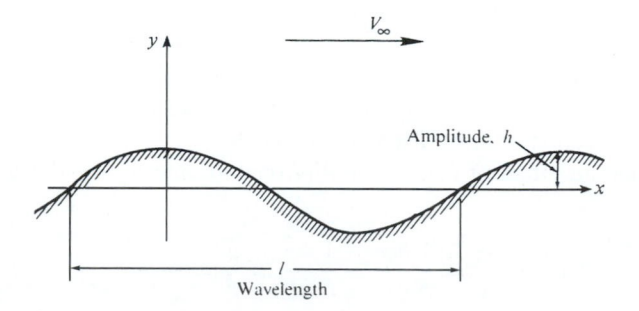

Figure 9.5 | Geometry of a wavy wall.

Recall that Eq. (E.1) is linear, and a standard approach to the solution of linear partial differential equations is separation of variables. Assume that ϕ, which is a function of x and y, can be expressed as a product of functions x only and y only, i.e.,

$$\phi(x, y) = F(x)G(y) \tag{E.2}$$

Substitute Eq. (E.2) into Eq. (E.1):

$$\left(1 - M_\infty^2\right) G \frac{d^2 F}{dx^2} + F \frac{d^2 G}{dy^2} = 0$$

or

$$\frac{1}{F} \frac{d^2 F}{dx^2} + \frac{1}{\left(1 - M_\infty^2\right)G} \frac{d^2 G}{dy^2} = 0 \tag{E.3}$$

Equation (E.3) must hold for any arbitrary values of x and y. In particular, if x is held constant but y is varied, $(1/F)(d^2 F/dx^2)$ is constant. However, Eq. (E.3) dictates that $[1/(1 - M_\infty^2)G](d^2 G/dy^2)$ must also be constant; indeed, it must be equal to the negative value of the former constant in order for the two terms in Eq. (E.3) to always add to zero. Let this constant be denoted by k^2. Hence, Eq. (E.3) yields

$$\frac{1}{\left(1 - M_\infty^2\right)G} \frac{d^2 G}{dy^2} = k^2 \tag{E.4}$$

and

$$\frac{1}{F} \frac{d^2 F}{dx^2} = -k^2 \tag{E.5}$$

From Eq. (E.4),

$$\frac{d^2 G}{dy^2} - k^2\left(1 - M_\infty^2\right)G = 0 \tag{E.6}$$

Equation (E.6) is a second-order linear ordinary differential equation with constant coefficients; its solution is (see any standard text on differential equations)

$$G(y) = A_1 e^{-k\sqrt{1-M_\infty^2}\,y} + A_2 e^{k\sqrt{1-M_\infty^2}\,y} \tag{E.7}$$

From Eq. (E.5),

$$\frac{d^2 F}{dx^2} + k^2 F = 0 \tag{E.8}$$

The standard solution of Eq. (E.8) is

$$F(x) = B_1 \sin kx + B_2 \cos kx \tag{E.9}$$

In Eqs. (E.7) and (E.9), the constants of integration, A_1, A_2, B_1, and B_2, and the parameter k are determined from the physical boundary conditions of the problem as

1. As $y \to \infty$, V and hence $\nabla\phi$ must remain *finite* (i.e., they cannot increase to an infinite value, because nature abhors infinities).
2. The flow at the wall must be *tangent* to the wall. Hence,

$$\frac{dy_w}{dx} = \frac{v'_w}{V_\infty + u'_w} \tag{E.10}$$

In Eq. (E.10), small perturbations dictate that $u'_w \ll V_\infty$; hence, Eq. (E.10) becomes

$$\frac{dy_w}{dx} = \frac{v'_w}{V_\infty} = \frac{1}{V_\infty} \left(\frac{\partial \phi}{\partial y} \right)_w \tag{E.11}$$

Combining Eqs. (E.11) and the wall equation, we have

$$\left(\frac{\partial \phi}{\partial y} \right)_w = -V_\infty h \left(\frac{2\pi}{l} \right) \sin \left(\frac{2\pi x}{l} \right) \tag{E.12}$$

Consistent with our assumption of small perturbations, y_w is small. Hence, Eq. (E.12), which strictly speaking is applied at the wall surface, can be evaluated at $y = 0$ without compromising the first-order accuracy of the solution. That is,

$$\left(\frac{\partial \phi}{\partial y} \right)_w \approx \left(\frac{\partial \phi}{\partial y} \right)_{y=0}$$

In turn, Eq. (E.12) becomes

$$\left(\frac{\partial \phi}{\partial y} \right)_{y=0} = -V_\infty h \left(\frac{2\pi}{l} \right) \sin \left(\frac{2\pi x}{l} \right) \tag{E.13}$$

Returning to Eq. (E.7), for the first boundary condition listed above to hold, $A_2 = 0$. This ensures that $\mathbf{V}$ remains finite at $y \to \infty$. Also, combining Eqs. (E.2) and (E.7) and (E.9), with $A_2 = 0$, we have

$$\phi(x, y) = (B_1 \sin kx + B_2 \cos kx) A_1 e^{-k\sqrt{1-M_\infty^2}\, y} \tag{E.14}$$

Hence,

$$\frac{\partial \phi}{\partial y} = (B_1 \sin kx + B_2 \cos kx) A_1 (-k) \sqrt{1 - M_\infty^2}\, e^{-k\sqrt{1-M_\infty^2}\, y} \tag{E.15}$$

Evaluating Eq. (E.15) at the wall ($y = 0$ as already described):

$$\left(\frac{\partial \phi}{\partial y} \right)_{y=0} = -A_1 k \sqrt{1 - M_\infty^2} (B_1 \sin kx + B_2 \cos kx) \tag{E.16}$$

Combining Eq. (E.16) with the second boundary condition, Eq. (E.13), we have

$$-A_1 k \sqrt{1 - M_\infty^2} (B_1 \sin kx + B_2 \cos kx) = -V_\infty h \left(\frac{2\pi}{l} \right) \sin \left(\frac{2\pi x}{l} \right) \tag{E.17}$$

By inspection, we see that Eq. (E.17) is satisfied if

$$B_2 = 0$$

$$k = \frac{2\pi}{l}$$

$$A_1 k \sqrt{1 - M_\infty^2}\, B_1 = V_\infty h \left(\frac{2\pi}{l} \right)$$

or
$$A_1 B_1 = \frac{V_\infty h}{\sqrt{1 - M_\infty^2}}$$

Hence, Eq. (E.14) becomes

$$\phi(x, y) = \frac{V_\infty h}{\sqrt{1 - M_\infty^2}} \exp\left(\frac{-2\pi\sqrt{1 - M_\infty^2}\, y}{l}\right) \sin\left(\frac{2\pi x}{l}\right) \qquad \text{(E.18)}$$

Equation (E.18) is the solution to the problem. From it all other physical properties can be found. For example,

$$u' = \frac{\partial \phi}{\partial x} = \frac{V_\infty h}{\sqrt{1 - M_\infty^2}} \left(\frac{2\pi}{l}\right) \left(\cos\frac{2\pi x}{l}\right) \exp\left(\frac{-2\pi\sqrt{1 - M_\infty^2}\, y}{l}\right) \qquad \text{(E.19)}$$

Also, from Eq. (9.17), combined with (E.19),

$$C_p = -\frac{2u'}{V_\infty} = -\frac{4\pi}{\sqrt{1 - M_\infty^2}} \left(\frac{h}{l}\right) \exp\left(\frac{-2\pi\sqrt{1 - M_\infty^2}\, y}{l}\right) \cos\left(\frac{2\pi x}{l}\right) \qquad \text{(E.20)}$$

Since $y = 0$ approximately corresponds to the wall, then the pressure coefficient at the wall C_{p_w} can be obtained from Eq. (E.20) as

$$C_{p_w} = -\frac{4\pi}{\sqrt{1 - M_\infty^2}} \left(\frac{h}{l}\right) \cos\left(\frac{2\pi x}{l}\right) \qquad \text{(E.21)}$$

Let us interpret the results as embodied in Eqs. (E.18) through (E.21). To begin with, a comparison of Eq. (E.21) with the wall equation shows that the pressure coefficient at the wall has the same cosine variation as the shape of the wall, but it is 180° out of phase [due to the negative sign in Eq. (E.21)]. This comparison is illustrated in Fig. 9.6 which shows a schematic of the C_{p_w} variation positioned above the wall shape. Clearly, the pressure variation is symmetrical with the wall shape. The pressure distribution is illustrated by the arrows normal to the surface. Due to the symmetry of this distribution, there is *no pressure force in the x direction on the wall. That is, there is no drag*. This is an example of a general result, namely: *For two-dimensional, inviscid, adiabatic, subsonic compressible flow, a body experiences no aerodynamic drag*. This is a generalization of the well-known d'Alembert's paradox which predicts zero drag for a two-dimensional body immersed in an incompressible potential flow.

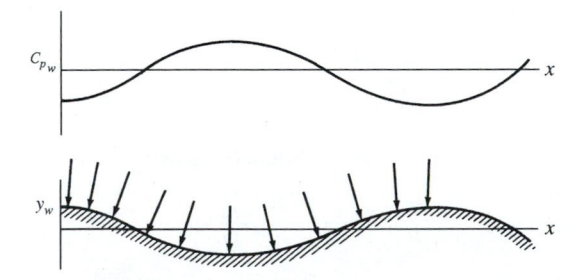

Figure 9.6 | Schematic of pressure variation on a wavy wall over which a subsonic flow is moving.

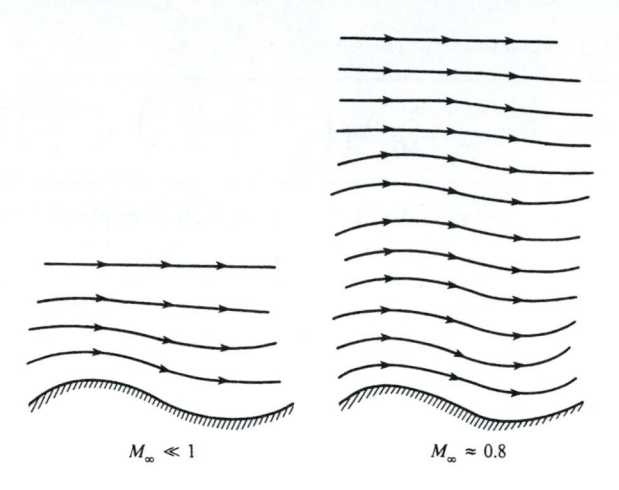

$M_\infty \ll 1$ $M_\infty \approx 0.8$

Figure 9.7 | Linearized subsonic flow over a wavy wall;
effects of compressibility on streamline shapes.

With regard to the Mach number effects on both the flowfield and C_{p_w}, first consider
Eq. (E.18), which shows that

$$\phi \propto \frac{1}{\sqrt{1 - M_\infty^2}} \exp\left(\frac{-2\pi \sqrt{1 - M_\infty^2}\, y}{l}\right)$$

Thus, for any fixed subsonic value of M_∞, $\phi \to 0$ as $y \to \infty$. That is, the disturbances introduced by the presence of the wall virtually disappear at large distances from the wall—they
attenuate with distance. However, the distance to which a disturbance of a given magnitude
reaches out, away from the wall, increases with increasing M_∞, as can be seen from the above
proportionality. Thus, in a subsonic flow, as M_∞ increases, the disturbances reach out further
from the wall. This is shown schematically in Fig. 9.7, which compares streamlines between
low and high subsonic Mach numbers.

The most important effect of Mach number in a subsonic flow is, by far, its influence on
surface pressure coefficient, as demonstrated by Eq. (E.21):

$$C_{p_w} \propto \frac{1}{\sqrt{1 - M_\infty^2}}$$

Let M_{∞_1} and M_{∞_2} be two different free-stream Mach numbers. Then, from Eq. (E.21),

$$\frac{\left(C_{p_w}\right)_{M_{\infty_1}}}{\left(C_{p_w}\right)_{M_{\infty_2}}} = \sqrt{\frac{1 - M_{\infty_2}^2}{1 - M_{\infty_1}^2}} \tag{E.22}$$

Furthermore, if $M_{\infty_2} \approx 0$, which corresponds to incompressible flow, then Eq. (E.22) yields

$$C_{p_w} = \frac{\left(C_{p_w}\right)_o}{\sqrt{1 - M_\infty^2}}$$

which is the Prandtl–Glauert rule derived earlier.

At the end of Example 9.1, the statement was made that $M_\infty \approx 0$ corresponds to incompressible flow. This provides a good opportunity to examine a physical (or should we say "metaphysical") implication of incompressible flow. Precisely speaking, for a purely incompressible flow, the Mach number is precisely zero, $M = 0$. At first thought, how can this be? Incompressible flows have a finite velocity, or else there would be no "flow." But a finite velocity does not necessarily mean a finite Mach number. An incompressible flow is a constant density flow, hence, from Eq. (1.5), where $d\rho = 0$, the compressibility $\tau = 0$. In turn, from Eq. (3.18), the speed of sound is infinite in an incompressible flow. Since $M = V/a$, the Mach number in a purely incompressible flow is always zero, even though V is finite. This result is consistent with the definitions of an incompressible flow. From time to time you will see results in the literature for flows labeled as $M = 0$. Just recognize that this is a label for incompressible flow results. In retrospect, the paradox discussed here is a consequence of the fact that purely incompressible flow is a myth—it does not exist in nature. It is simply an intellectual construct made by human beings to model a class of real flows in nature that closely resemble a defined incompressible flow.

9.5 | IMPROVED COMPRESSIBILITY CORRECTIONS

Linearized solutions are influenced predominantly by free-stream conditions; they do not fully recognize changes in local regions of the flow. Such local changes are basically nonlinear phenomena. For example, as shown in Sec. 7.5, the wave velocity of each portion of a linearized acoustic wave propagates at the free-stream speed of sound a_∞. Later in Chap. 7 we saw the true case where each element of a finite wave propagates at the local value of $u \pm a$, and therefore the wave shape distorts in the process—a nonlinear phenomena. Another example is contained in Sec. 9.4. Linearized subsonic flow is governed by M_∞, not the local Mach number M. Witness Eqs. (9.36) through (9.37), where M_∞ is the dominant parameter.

In an effort to obtain an improved compressibility correction, Laitone (see Ref. 23) applied Eq. (9.36) locally in the flow, i.e.,

$$C_p = \frac{C_{p_o}}{\sqrt{1 - M^2}}$$

where M is the local Mach number. In turn, M can be related to M_∞ and the pressure coefficient through the isentropic flow relations. The resulting compressibility correction is

$$C_p = \frac{C_{p_o}}{\sqrt{1 - M_\infty^2} + \left[M_\infty^2 \left(1 + \frac{\gamma - 1}{2} M_\infty^2 \right) \Big/ 2\sqrt{1 - M_\infty^2} \right] C_{p_o}} \tag{9.39}$$

Note that, as C_{p_o} becomes small, Eq. (9.39) approaches the Prandtl–Glauert rule.

Another compressibility correction that has been adopted widely is that due to von Karman and Tsien (see Refs. 24 and 25). Utilizing a hodograph solution of the nonlinear equations of motion along with a simplified "tangent gas" equation of state, this result was obtained:

$$C_p = \frac{C_{p_o}}{\sqrt{1 - M_\infty^2} + \left(\dfrac{M_\infty^2}{1 + \sqrt{1 - M_\infty^2}}\right)\dfrac{C_{p_o}}{2}}$$

(9.40)

Equation (9.40) is called the *Karman–Tsien rule*.

Figure 9.8 contains experimental measurements of the C_p variation with M_∞ at the 0.3 chord location on an NACA 4412 airfoil; these measurements are compared

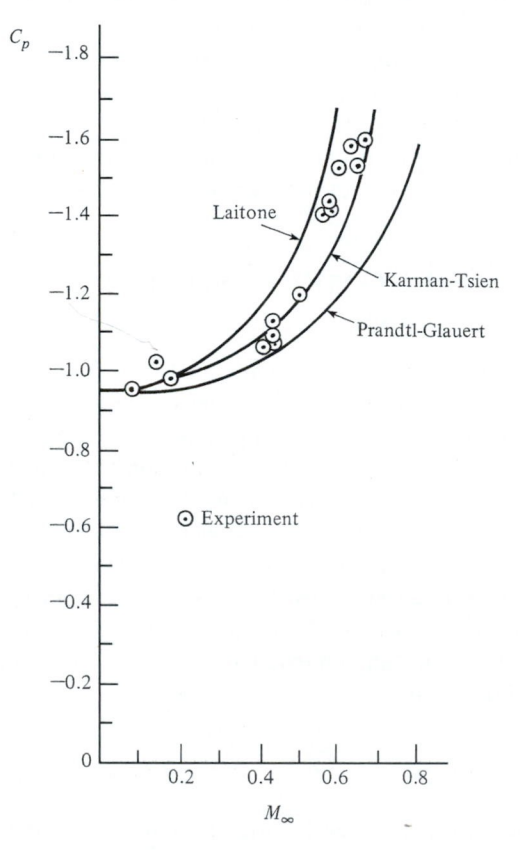

Figure 9.8 | Comparison of several compressibility corrections with experiment for an NACA 4412 airfoil at an angle of attack $\alpha = 1°53'$. The experimental data are chosen for their historical significance; they are from John Stack, W. F. Lindsey, and Robert E. Littell. "The compressibility Burble and the Effect of Compressibility on Pressures and Forces Acting on an Airfoil." NACA Report No. 646, 1938. This was the first major NACA publication to address the compressibility problem in a systematic fashion; it covered work performed in the 24-in-high speed tunnel at Langley Aeronautical Laboratory and was carried out during 1935–1936.

with the Prandtl–Glauert, Laitone, and Karman–Tsien rules. Note that the Prandtl–Glauert rule, although the simplest to apply, underpredicts the experimental values, whereas the improved compressibility corrections are clearly more accurate. This is because both the Laitone and Karman–Tsien rules bring in the nonlinear aspects of the flow.

9.6 | LINEARIZED SUPERSONIC FLOW

From Eq. (9.6) the linearized perturbation-velocity potential equation for two-dimensional flow takes the form of

$$\beta^2 \phi_{xx} + \phi_{yy} = 0 \tag{9.41}$$

for subsonic flow, where $\beta = \sqrt{1 - M_\infty^2}$, and the form of

$$\lambda^2 \phi_{xx} - \phi_{yy} = 0 \tag{9.42}$$

for supersonic flow, where $\lambda = \sqrt{M_\infty^2 - 1}$. The difference between Eqs. (9.41) and (9.42) is fundamental, for they are *elliptic* and *hyperbolic* partial differential equations, respectively. A discussion of the distinction between elliptic and hyperbolic equations is deferred until Chap. 11; suffice it to say here that the equations reflect fundamental physical differences between subsonic and supersonic flows—differences which will be highlighted in this and subsequent sections.

Consider the supersonic flow over a body or surface which introduces small changes in the flowfield, i.e., flow over a thin airfoil, over a mildly wavy wall, or over a small hump in a surface. The latter is sketched in Fig. 9.9. Equation (9.42), which

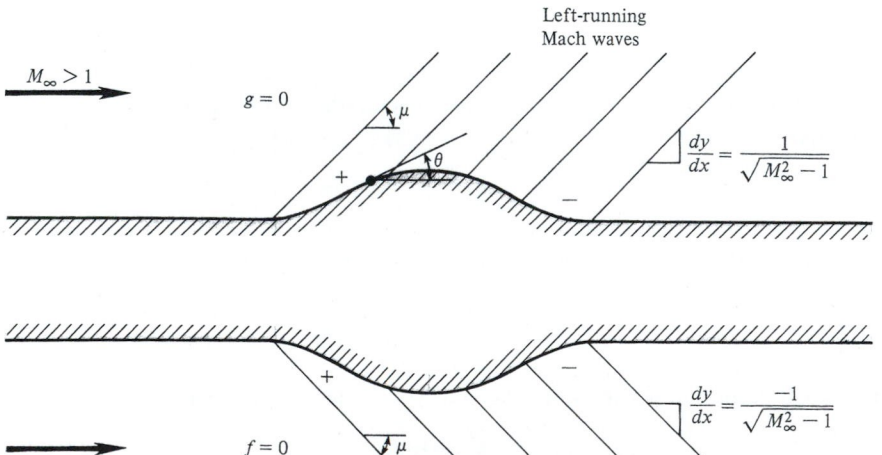

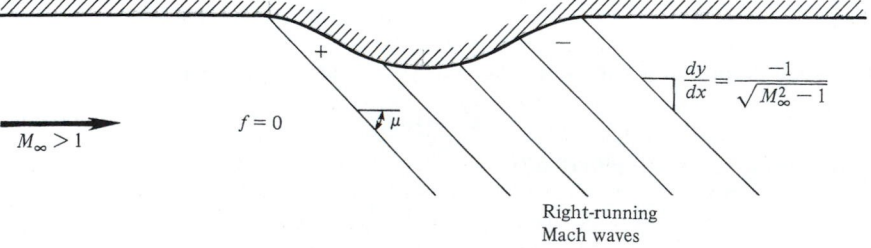

Figure 9.9 | Linearized supersonic flow over a bump.

governs this flow, is of the form of the classical wave equation first discussed in Sec. 7.5 in conjunction with acoustic theory. Its general solution is

$$\phi = f(x - \lambda y) + g(x + \lambda y) \tag{9.43}$$

which can be verified by direct substitution into Eq. (9.42). Examining the particular solution where $g = 0$, and hence $\phi = f(x - \lambda y)$, we see that lines of constant ϕ correspond to $x - \lambda y = \text{const}$, or

$$\frac{dy}{dx} = \frac{1}{\lambda} = \frac{1}{\sqrt{M_\infty^2 - 1}} \tag{9.44}$$

Recalling that the Mach angle $\mu = \arcsin(1/M_\infty) = \arctan(1/\sqrt{M_\infty^2 - 1})$. Eq. (9.44) states that lines of constant ϕ are the family of left-running Mach lines, as sketched in the upper half of Fig. 9.9. In turn, if $f = 0$ in Eq. (9.43), then lines of constant ϕ are the family of right-running Mach lines shown in the lower half of Fig. 9.9.

Hence, Fig. 9.9 illustrates a basic physical difference between subsonic and supersonic flow. When $M_\infty < 1$, it was shown in Sec. 9.4 that disturbances propagate everywhere in the flowfield, including upstream as well as downstream. In contrast, for $M_\infty > 1$. Fig. 9.9 illustrates that weak disturbances propagate along Mach lines, and hence the flowfield upstream of a disturbance does not feel the presence of the disturbance. In steady supersonic flows, disturbances do not propagate upstream; they are limited to a region downstream of the source of disturbance.

Returning to Eq. (9.43), letting $g = 0$, we have

$$\phi = f(x - \lambda y)$$

Hence,

$$u' = \frac{\partial \phi}{\partial x} = f' \tag{9.45}$$

and

$$v' = \frac{\partial \phi}{\partial y} = -\lambda f' \tag{9.46}$$

where f' represents the derivative with respect to the argument, $(x - \lambda y)$. Combining Eqs. (9.45) and (9.46),

$$u' = -\frac{v'}{\lambda} \tag{9.47}$$

Equation (9.18) gives the boundary condition on the surface as

$$\tan \theta = \frac{dy}{dx} = \frac{v'}{V_\infty + u'} \tag{9.48}$$

For small perturbations, $u' \ll V_\infty$ and $\tan \theta \approx \theta$. Hence, Eq. (9.48) becomes

$$v' = V_\infty \theta \tag{9.49}$$

Substituting Eq. (9.49) into (9.47),

$$u' = -\frac{V_\infty \theta}{\lambda} \tag{9.50}$$

Therefore, from Eqs. (9.17) and (9.50), the pressure coefficient on the surface is

$$C_p = -\frac{2u'}{V_\infty} = \frac{2\theta}{\lambda}$$

or

$$\boxed{C_p = \frac{2\theta}{\sqrt{M_\infty^2 - 1}}} \tag{9.51}$$

Equation (9.51) is an important result. It is the linearized supersonic surface pressure coefficient, and it states that C_p is directly proportional to the local surface inclination with respect to the free stream. It holds for any slender two-dimensional shape. For example, consider the biconvex airfoil shown in Fig. 9.10. At two arbitrary points A and B on the top surface,

$$C_{pA} = \frac{2\theta_A}{\sqrt{M_\infty^2 - 1}} \quad \text{and} \quad C_{pB} = \frac{2\theta_B}{\sqrt{M_\infty^2 - 1}}$$

respectively. Note in Fig. 9.10 that θ_A is positive and θ_B is negative, and hence C_p varies from positive on the forward surface to negative on the rearward surface. This is consistent with our earlier discussions in Chap. 4: We know from inspection of

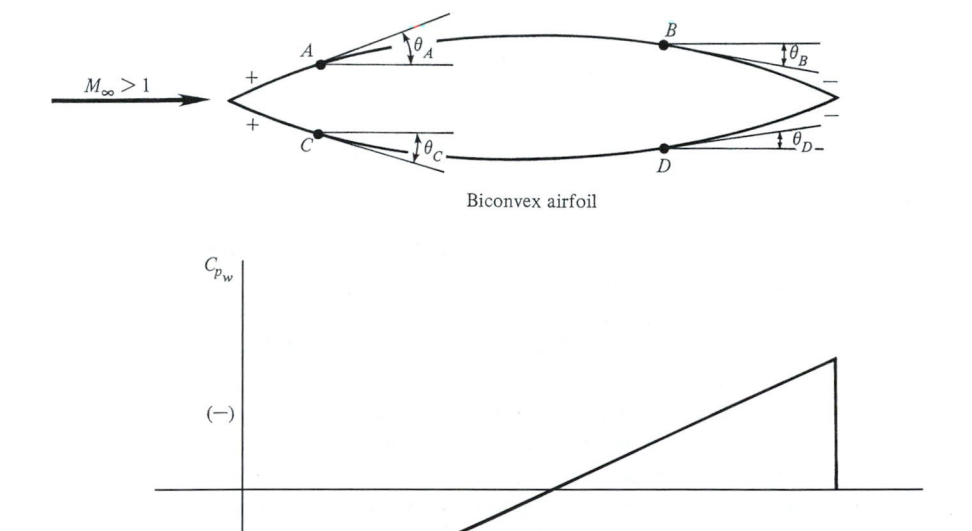

Figure 9.10 | Schematic of the linearized pressure coefficient over a biconvex airfoil.

Fig. 9.10 that the front and rear surfaces are compression and expansion surfaces, respectively.

Equation (9.51) was derived by setting $g = 0$ in Eq. (9.43). Thus it holds for a surface generating a family of left-running waves, i.e., the top surfaces in Figs. 9.9 and 9.10. If we set $f = 0$ in Eq. (9.43), the surface pressure coefficient becomes

$$C_p = \frac{-2\theta}{\sqrt{M_\infty^2 - 1}} \tag{9.52}$$

which holds for a surface generating right-running waves, i.e., the bottom surfaces in Figs. 9.9 and 9.10. In both Eqs. (9.51) and (9.52), θ is measured positive *above* the local flow direction and negative *below* the local flow direction. Hence, on the bottom surface of the biconvex airfoil in Fig. 9.10, θ_C is negative and θ_D is positive. In conjunction with Eq. (9.52), this still yields a positive C_p on the forward compression surface and a negative C_p on the rearward expansion surface.

There is no real need to worry about the formal sign conventions mentioned above. For any practical application, this author suggests the use of Eq. (9.51) along with common sense to single out the compression and expansion surfaces on a body. If the surface is a compression surface, C_p from Eq. (9.51) must be positive, no matter whether the surface is on the top or bottom of the body. Similarly, if the surface is an expansion surface, C_p from Eq. (9.51) must be negative.

This leads to another basic difference between subsonic and supersonic inviscid flows. Recall that, for $M_\infty < 1$, a two-dimensional body experiences no drag. For $M_\infty > 1$, however, as denoted by the $+$ and $-$ signs in Figs. 9.9 and 9.10, C_p is positive on the front surfaces and negative on the rear surface. Consequently, there is a net pressure imbalance which creates a drag force on the body. This force is the *wave drag,* first introduced in Sec. 4.15. Although shock waves do not appear explicitly within the framework of linearized theory, their consequence in terms of wave drag are reflected in the linearized results. Hence, d'Alembert's paradox does not apply to supersonic flows.

Further contrast between subsonic and supersonic flows is seen by comparing Eqs. (9.36) and (9.51). In subsonic flow, Eq. (9.36) shows that C_p increases when M_∞ increases. However, for supersonic flow, Eq. (9.51) shows that C_p decreases when M_∞ increases. These important trends are illustrated in Fig. 9.11.

Finally, to examine the accuracy of Eq. (9.51), Fig. 9.12 compares linearized theory with exact results for C_p on the surface of a wedge of semiangle θ. The exact results are obtained from oblique shock theory as described in Chap. 4. Note that the agreement between exact and linear theories is good at small θ, but deteriorates rapidly as θ increases. For $M_\infty = 2$ as shown in Fig. 9.11, linearized theory yields reasonably accurate results for C_p when $\theta < 4°$.

Although the linearized pressure distribution from Eq. (9.51) becomes inaccurate beyond a deflection angle of approximately $4°$, when it is integrated over the surface of an airfoil, these inaccuracies tend to compensate over the top and bottom surfaces. As a result the linearized values for C_L and C_D are more accurate at larger angles of attack than one would initially expect. Some of these trends are illustrated in the problems at the end of this chapter.

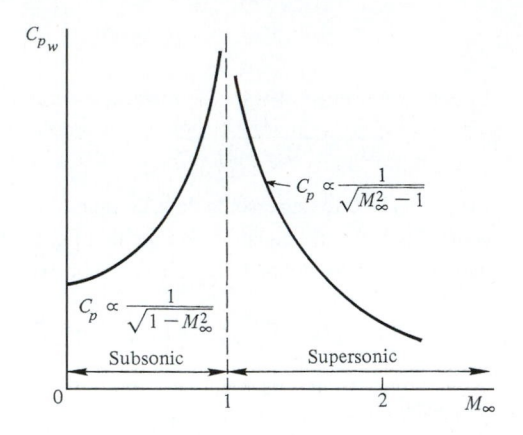

Figure 9.11 | Variation of the linearized pressure coefficient with Mach number.

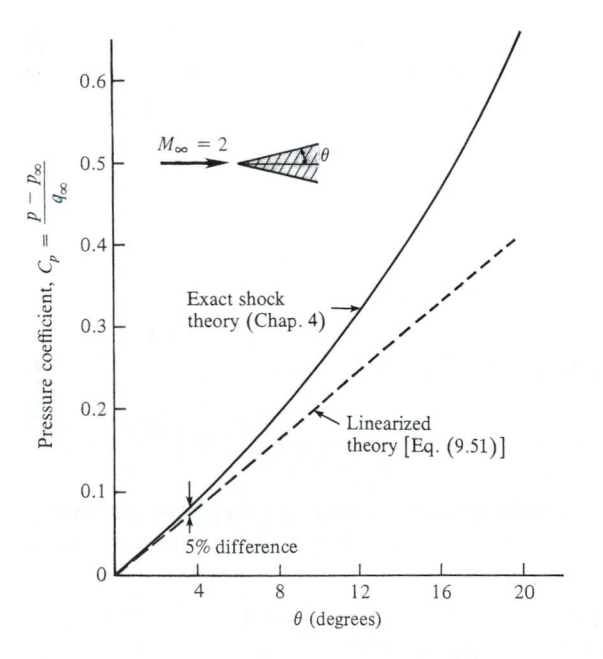

Figure 9.12 | Comparison between linearized theory and exact shock results for the pressure on a wedge in supersonic flow.

EXAMPLE 9.2

Consider a supersonic flow with an upstream Mach number of M_∞. This flow moves over the same wavy wall as first shown in Fig. 9.5, and as given in Example 9.1. For small h, use linear theory to derive an equation for the velocity potential and surface pressure coefficient.

■ Solution

From Eq. (9.42),

$$\frac{\partial^2 \phi}{\partial x} - \frac{1}{(M_\infty^2 - 1)} \frac{\partial^2 \phi}{\partial y^2} = 0 \qquad (G.1)$$

Keeping in mind that $(M_\infty^2 - 1) > 0$ for supersonic flow, compare Eq. (G.1) with Eq. (7.42), which was identified as the classical wave equation. We see that Eq. (G.1) is also of the form of the simple wave equation. Hence, a solution to Eq. (G.1) can be expressed as

$$\phi(x, y) = f\left(x - \sqrt{M_\infty^2 - 1}\,y\right) + g\left(x + \sqrt{M_\infty^2 - 1}\,y\right) \qquad (G.2)$$

Let $g = 0$. Then Eq. (G.2) becomes

$$\phi(x, y) = f\left(x - \sqrt{M_\infty^2 - 1}\,y\right) \qquad (G.3)$$

and

$$\frac{\partial \phi}{\partial y} = \left[f'\left(x - \sqrt{M_\infty^2 - 1}\,y\right)\right]\left(-\sqrt{M_\infty^2 - 1}\right) \qquad (G.4)$$

where f' denotes the derivative of f with respect to its argument, $(x - \sqrt{M_\infty^2 - 1}\,y)$. Recall the boundary conditions at the wall discussed in Sec. 9.4:

$$\frac{\partial \phi}{\partial y} = V_\infty \frac{dy_w}{dx}$$

Thus

$$\sqrt{M_\infty^2 - 1}\,f'(x) = V_\infty h\left(\frac{2\pi}{l}\right)\sin\left(\frac{2\pi x}{l}\right) \qquad (G.5)$$

where Eq. (G.5) holds at the wall. Thus, from Eq. (G.5),

$$f'(x) = \frac{V_\infty h}{\sqrt{M_\infty^2 - 1}}\left(\frac{2\pi}{l}\right)\sin\left(\frac{2\pi x}{l}\right) \qquad (G.6)$$

Integrating Eq. (G.6) with respect to its argument [note that the argument is $(x - \sqrt{M_\infty^2 - 1}\,y)$, but with $y = 0$], we have

$$f(x) = -\frac{V_\infty h}{\sqrt{M_\infty^2 - 1}}\cos\left(\frac{2\pi x}{l}\right) + \text{const} \qquad (G.7)$$

Since $f(x)$ is defined throughout the flow, not just at the wall, and because it has the form of Eq. (G.7), where x represents the argument of f, then Eq. (G.3) can be written as

$$\boxed{\begin{aligned} \phi(x, y) &= f\left(x - \sqrt{M_\infty^2 - 1}\,y\right) \\ &= -\frac{V_\infty h}{\sqrt{M_\infty^2 - 1}}\cos\left[\frac{2\pi}{l}\left(x - \sqrt{M_\infty^2 - 1}\,y\right)\right] + \text{const} \end{aligned}} \qquad (G.8)$$

Therefore, from Eqs. (9.17) and (G.8)

$$C_p = -\frac{2u'}{V_\infty} = -\frac{2}{V_\infty}\frac{\partial\phi}{\partial x} = -\frac{4\pi}{\sqrt{M_\infty^2-1}}\left(\frac{h}{l}\right)\sin\left[\frac{2\pi}{l}\left(x - \sqrt{M_\infty^2-1}y\right)\right] \qquad (G.9)$$

At the wall, Eq. (G.9) becomes

$$C_{p_w} = -\frac{4\pi}{\sqrt{M_\infty^2-1}}\left(\frac{h}{l}\right)\sin\left(\frac{2\pi x}{l}\right) \qquad (G.10)$$

Equations (G.8) through (G.10) represent the solution for the linearized supersonic flow over a wavy wall.

Let us examine these results closely. First, in contrast to the previous results for subsonic flow, no exponential attenuation factor occurs. For supersonic flow, the perturbations do *not* disappear at $y \to \infty$. Moreover, the *magnitude* of a disturbance (magnitude of ϕ or C_p, for example) is constant for $(x - \sqrt{M_\infty^2 - 1}y) = $ const. That is, the effect of the wall is propagated to infinity with constant strength along the lines $x - \sqrt{M_\infty^2 - 1}y = $ const. Hence, these lines have a slope

$$\frac{dy}{dx} = \frac{1}{\sqrt{M_\infty^2 - 1}}$$

and are therefore identical to *Mach lines,* with the angle μ to the free-stream direction,

$$\mu = \sin^{-1}\left(\frac{1}{M_\infty}\right)$$

These lines are sketched in Fig. 9.13 where they are also identified as *characteristic lines.* The proof that Mach lines are indeed the same as characteristic lines in the sense defined in Chap. 7 will be made in Chap. 11. We simply note the fact here. Also note, in contrast to

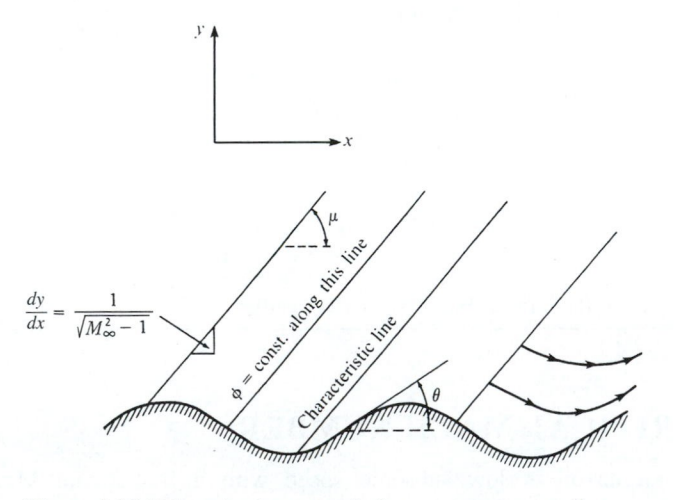

Figure 9.13 | Linearized supersonic flow over a wavy wall.

subsonic flow, that Eq. (G.8) yields streamlines that are *unsymmetrical* about a vertical line through a crest or trough of the wall. Instead, the streamlines remain geometrically similar between two inclined Mach waves, as sketched in Fig. 9.13.

Two additional physical results of great importance can be interpreted from Eq. (G.10). First, note that unlike subsonic flow, the surface pressure distribution is no longer symmetrical about the wall [Eq. (G.10) is a sine variation, whereas the wall is a cosine shape]. Hence, for supersonic flow, the surface pressure distributions do not cancel in the x direction; instead, there is a net force in the x direction, in the same direction as the free stream. This force is called *wave drag*.

Second, Eq. (G.10) for the pressure coefficient can be couched in a simpler form by noting that the equation of the wall is

$$y_w = h \cos\left(\frac{2\pi x}{l}\right)$$

Hence,

$$\frac{dy_w}{dx} = -2\pi \left(\frac{h}{l}\right) \sin\left(\frac{2\pi x}{l}\right) \tag{G.11}$$

Combining Eqs. (G.10) and (G.11),

$$C_p = \frac{2\left(\dfrac{dy_w}{dx}\right)}{\sqrt{M_\infty^2 - 1}} \tag{G.12}$$

However, letting θ denote the angle of the wall as sketched in the Fig. 9.13, at any point on the surface,

$$\tan\theta = \frac{dy_w}{dx} \tag{G.13}$$

Compatible with linearized theory, which assumes small perturbations, i.e., slender bodies, θ is assumed small. Hence, from Eq. (G.13)

$$\tan\theta = \frac{dy_w}{dx} \simeq \theta \tag{G.14}$$

Thus, combining Eqs. (G.12) and (G.14),

$$\boxed{C_{p_w} = \frac{2\theta}{\sqrt{M_\infty^2 - 1}}} \tag{G.15}$$

Equation (G.15) is the same as Eq. (9.51) derived earlier.

9.7 | CRITICAL MACH NUMBER

Consider an airfoil at low subsonic speed with a free-stream Mach number $M_\infty = 0.3$, as shown in Fig. 9.14a. The flow expands around the top surface of the airfoil, dropping to a minimum pressure at point A. At this point, the local Mach

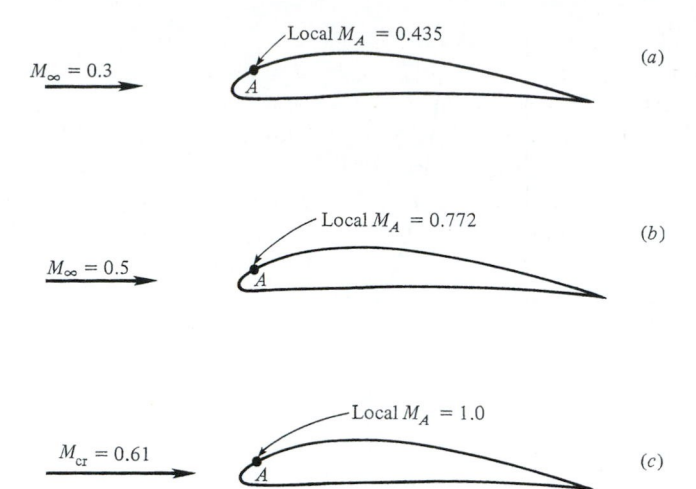

Figure 9.14 | Definition of critical Mach number. Point A is the location of minimum pressure on the top surface of the airfoil.

number on the surface will be a maximum, in this case $M_A = 0.435$. Now assume that we increase M_∞ to 0.5. The local Mach number at the minimum pressure point will correspondingly increase to 0.772, as shown in Fig. 9.14b. Now let us increase M_∞ to just the right value such that $M_A = 1.0$ at the minimum-pressure point. This value is $M_\infty = 0.61$, as shown in Fig. 9.14c. When this occurs, M_∞ is called the *critical Mach number, M_{cr}.* By definition, the critical Mach number is that *free-stream* Mach number at which sonic flow is first encountered on the airfoil.

The critical Mach number can be calculated as follows. Assuming isentropic flow throughout the flowfield, Eq. (3.30) gives

$$\frac{p_A}{p_\infty} = \left(\frac{1 + \dfrac{\gamma - 1}{2} M_\infty^2}{1 + \dfrac{\gamma - 1}{2} M_A^2} \right)^{\gamma/(\gamma-1)} \tag{9.53}$$

Combining Eqs. (9.10) and (9.53), the pressure coefficient at point A is

$$C_{pA} = \frac{2}{\gamma M_\infty^2} \left[\left(\frac{1 + \dfrac{\gamma - 1}{2} M_\infty^2}{1 + \dfrac{\gamma - 1}{2} M_A^2} \right)^{\gamma/(\gamma-1)} - 1 \right] \tag{9.54}$$

From Eq. (9.54), for a given M_∞ the values of local pressure coefficient and local Mach number are uniquely related at any given point A. Now assume as before that point A is the minimum-pressure (hence maximum-velocity) point on the airfoil. Furthermore, assume $M_A = 1$. Then, by definition, $M_\infty \equiv M_{\mathrm{cr}}$. Also, for this case the value of the pressure coefficient is defined as the critical pressure coefficient $C_{p_{\mathrm{cr}}}$.

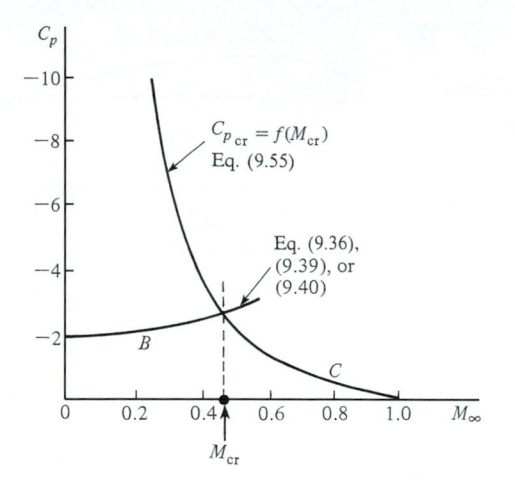

Figure 9.15 | Calculation of critical Mach number.

Setting $M_A = 1$, $M_\infty = M_{cr}$, and $C_p \equiv C_{p_{cr}}$ in Eq. (9.54), we obtain

$$
C_{p_{cr}} = \frac{2}{\gamma M_{cr}^2} \left[\left(\frac{1 + \dfrac{\gamma - 1}{2} M_{cr}^2}{1 + \dfrac{\gamma - 1}{2}} \right)^{\gamma/(\gamma - 1)} - 1 \right]
\tag{9.55}
$$

Note that $C_{p_{cr}}$ is a unique function of M_{cr}; this variation is plotted as curve C in Fig. 9.15.

Equation (9.55), along with one of the compressibility rules such as Eqs. (9.36), (9.39), or (9.40), provides enough tools to calculate the critical Mach number for a given airfoil:

1. Obtain as given data a measured or calculated value of the incompressible pressure coefficient at the minimum pressure point, C_{p_o}.
2. Using one of the compressibility corrections, plot C_p as a function of M_∞, shown as curve B in Fig. 9.15.
3. Using Eq. (9.55) plot $C_{p_{cr}}$ as a function of M_{cr}, shown as curve C in Fig. 9.15.
4. The intersection of curves B and C defines the critical Mach number for the given airfoil.

Note in Fig. 9.15 that curve C [from Eq. (9.55)] is a result of the fundamental gasdynamics of the flow; it is unique, and does not depend on the size or shape of the airfoil. In contrast, curve B is different for different airfoils. For example, consider two airfoils, one thin and one thick. For the thin airfoil, the flow experiences only a mild expansion over the top surface, and hence $|C_{p_o}|$ is small. Combined with the chosen compressibility correction, curve B in Fig. 9.15 is low on the graph, resulting in a high value of M_{cr}. For the thick airfoil, $|C_{p_o}|$ is naturally larger because the flow experiences a stronger expansion over the top surface. Curve B is higher on the graph, resulting in a lower value of M_{cr}. Hence, an airfoil designed for a high critical Mach number must have a thin profile.

When the free-stream Mach number exceeds M_{cr}, a finite region of supersonic flow exists on the top surface of the airfoil. At a high enough subsonic Mach number, this embedded supersonic region will be terminated by a weak shock wave. The total pressure loss associated with the shock will be small; however, the adverse pressure gradient induced by the shock tends to separate the boundary layer on the top surface, causing a large pressure drag. The net result is a dramatic increase in drag. The free-stream Mach number at which the large drag rise begins is defined as the *drag-divergence Mach number;* it is always slightly larger than M_{cr}. The massive increase in drag encountered at the drag-divergence Mach number is the technical base of the "sound barrier" which was viewed with much trepidation before 1947.

The relationship between the critical Mach number, the drag-divergence Mach number, and Mach one is sketched in Fig. 9.16, which shows the qualitative variation of the drag coefficient for a given shaped body (such as an airfoil, wing, or whole airplane) as a function of free-stream Mach number. At low subsonic speeds, the drag coefficient is relatively constant as M_∞ increases. Point a denotes the critical Mach number. As M_∞ is increased slightly above M_{cr}, C_D remains constant. Then, at some value of M_∞ slightly larger than M_{cr}, the value of C_D skyrockets. The free-stream Mach number at which this large drag increase occurs is the drag-divergence Mach number, denoted by point b in Fig. 9.16.

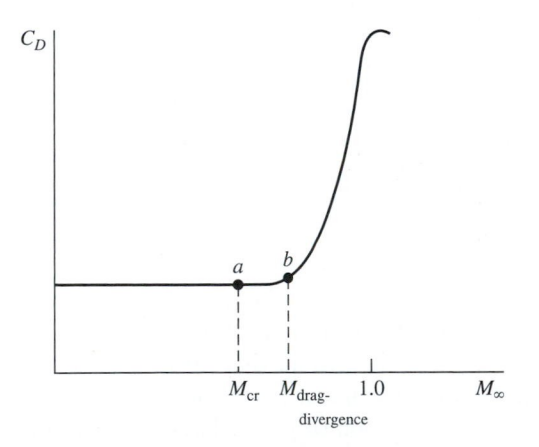

Figure 9.16 | Generic sketch of the variation of drag coefficient with freestream Mach number, showing the relative locations of the critical Mach number and the drag-divergence Mach number, both of which are less than Mach one.

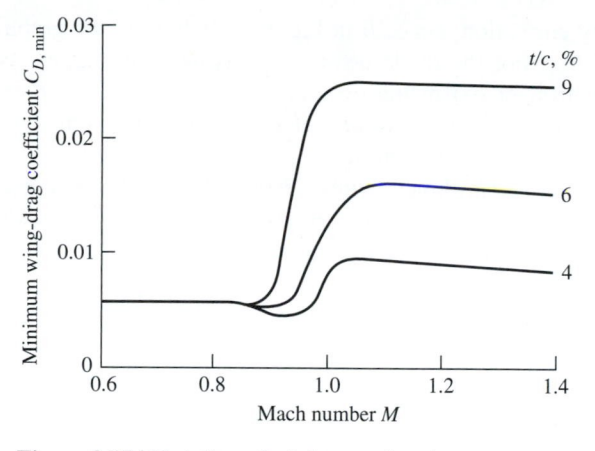

Figure 9.17 | Variation of minimum wing drag coefficient versus Mach number with airfoil thickness ratio as a parameter. The wing is swept, with a sweep angle of 47 degrees. (From Loftin, *Quest for Performance,* NASA SP 468, 1985.)

For purposes of discussion, consider the wing of an airplane. In most cases, if something is done during the design of the wing to increase M_{cr}, then usually the value of $M_{drag-divergence}$ also increases. This is a good thing, because the wing can fly closer to Mach one before the large drag rise is encountered. In airplane design, there have been two classic features employed to increase M_{cr}, hence, $M_{drag-divergence}$. The first simply is to make the wing thinner. As already discussed, a thinner airfoil will have a higher M_{cr} than a thicker airfoil, everything else being equal. This is reinforced by the wind tunnel data shown in Fig. 9.17, where the drag coefficient is plotted versus free-stream Mach number for three wings with three different thicknesses. Note the particularly large drag rise encountered by the wing with 9 percent thickness-to-chord ratio, and that it occurs at a value of $M_{drag-divergence}$ of about 0.88. By reducing the wing thickness to 6 and 4 percent, the magnitude of the drag rise is progressively reduced, and the value of $M_{drag-divergence}$ is progressively increased, moving closer to Mach one.

The other classic design feature used to increase M_{cr} is to sweep the wing. To see how wing sweep increases the critical Mach number of the wing, first consider a straight wing, a portion of which is sketched in Fig. 9.18a. We define a straight wing as one for which the midchord line is perpendicular to the free stream; this is certainly the case for the rectangular planform shown in Fig. 9.18a. Assume the straight wing has an airfoil section with a thickness-to-chord ratio of 0.15, as shown at the left of Fig. 9.18a. Streamline AB flowing over this wing sees the airfoil with $t_1/c_1 = 0.15$. Now consider the same wing swept back through the angle $\Lambda = 45°$, as shown in Fig. 9.18b. Streamline CD, which flows over this wing (ignoring any three-dimensional curvature effects), sees an effective airfoil shape with the same

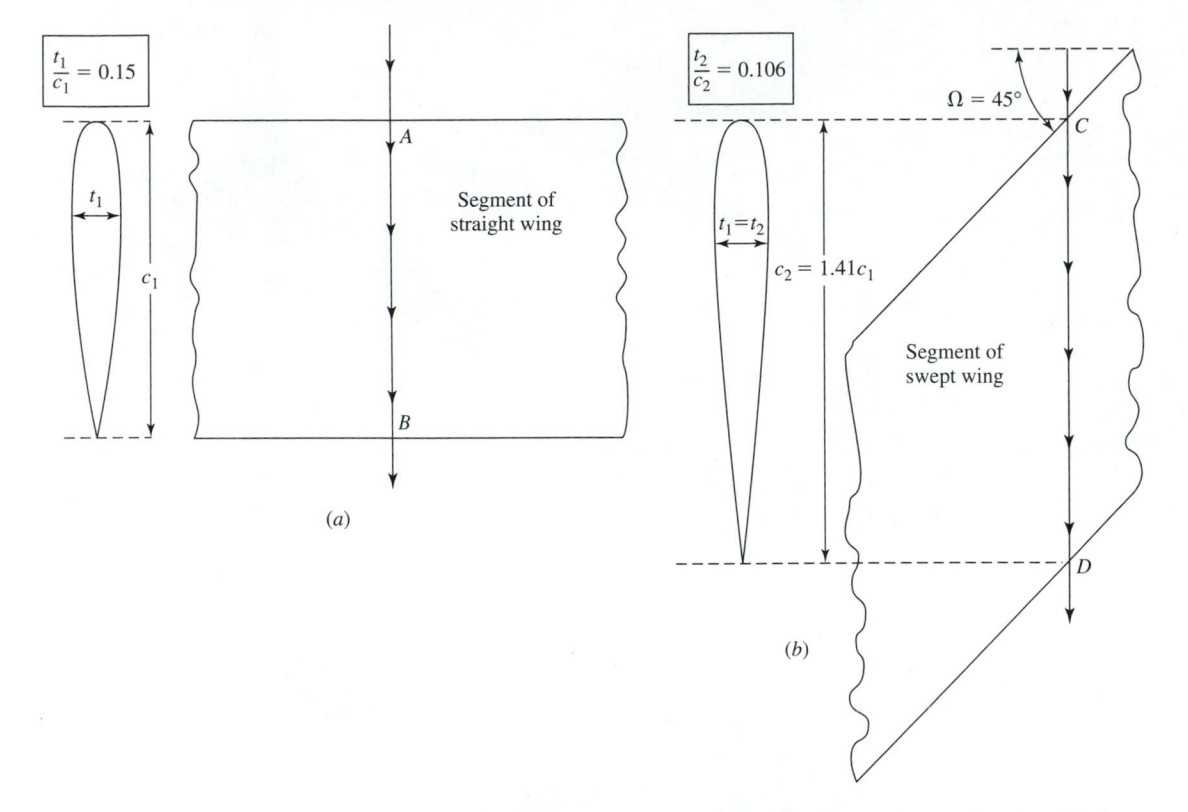

Figure 9.18 | By sweeping the wing, a streamline effectively sees a thinner airfoil, hence increasing the critical Mach number of the wing.

thickness as before ($t_2 = t_1$), but the effective chord length c_2 is longer by a factor of 1.41 (i.e., $c_2 = 1.41c_1$). This makes the effective thickness-to-chord ratio seen by streamline CD equal to $t_2/c_2 = 0.106$—thinner by almost one-third compared to the straight-wing case. Hence, by sweeping the wing, the flow behaves as if the airfoil section is thinner, with a consequent increase in the critical Mach number of the wing. Everything else being equal, a swept wing has a larger critical Mach number, hence a large drag-divergence Mach number than a straight wing. For this reason, most high-speed airplanes designed since the middle 1940s have swept wings. (The only reason why the Bell X-1, shown in Fig. 1.9, had straight wings is because its design commenced in 1944 before any knowledge or data about swept wings was available in the United States. Later, when such swept-wing data flooded into the United States from Germany in mid-1945, the Bell designers were conservative, and stuck with the straight wing.) A wonderful example of an early swept-wing fighter is the North American F-86 of Korean War vintage, shown in Fig. 9.19.

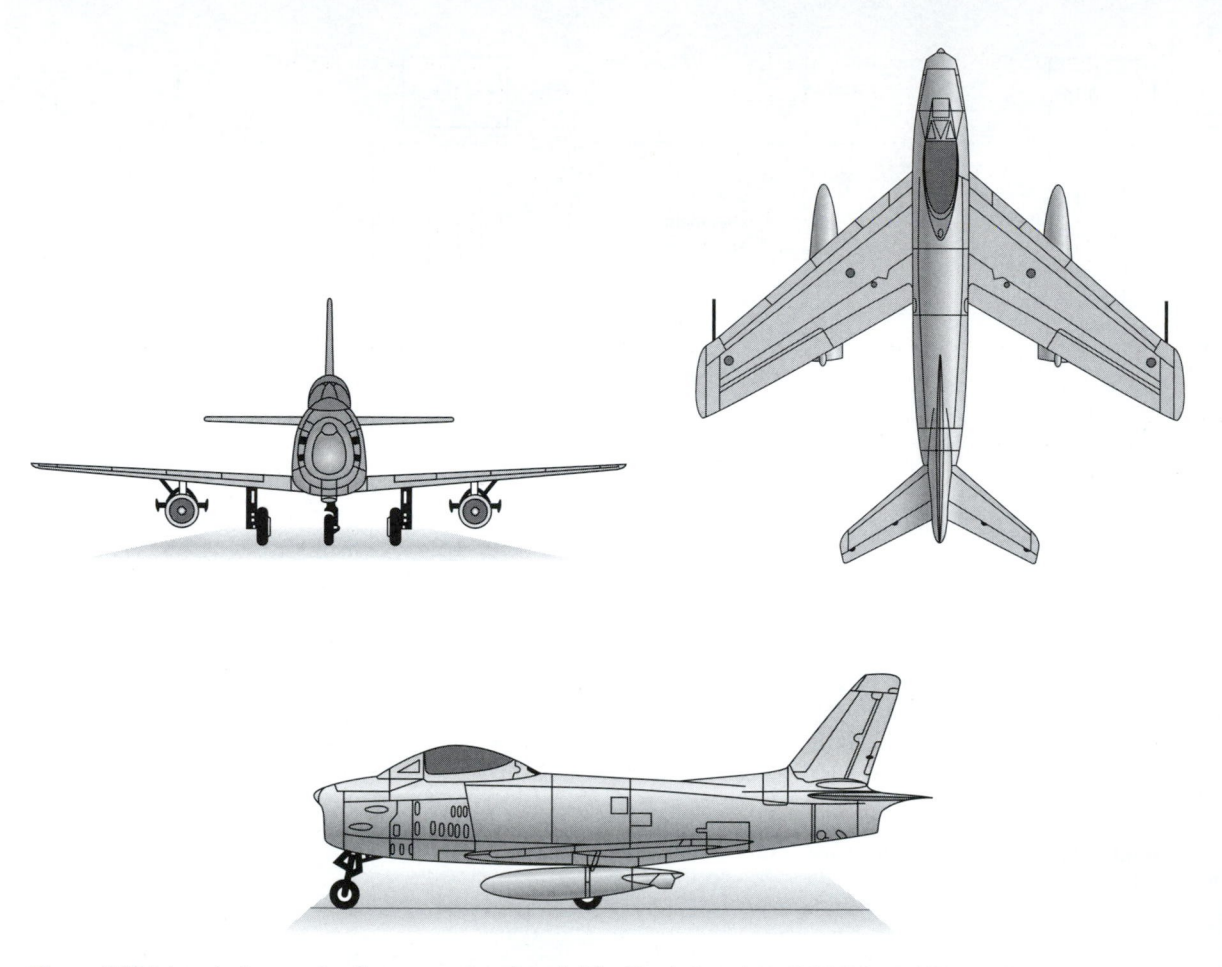

Figure 9.19 | A typical example of a swept-wing aircraft. The North American F-86 Sabre of Korean War fame.

9.8 | SUMMARY

This chapter has presented some of the technical aspects of subsonic and supersonic linearized flow for two-dimensional bodies and wall geometries. Closed-form analytical results have been obtained which illustrate important physical trends, and which dramatically contrast some fundamental differences between subsonic and supersonic flow. Although modern numerical techniques now exist for the accurate solution of flows with complex geometry (to be discussed in subsequent chapters), linearized solutions still play an important role in the whole spectrum of modern compressible flow.

Finally, it should be noted that linearized theory has also been applied to three-dimensional flows, yielding results for slender bodies of revolution at small angles of attack, and for finite wings. Although space will not be devoted in this book to such three-dimensional linearized flows, the reader is strongly encouraged to study this aspect in the classical literature. (See, for example, Refs. 5, 6, and 9.)

9.9 | HISTORICAL NOTE: THE 1935 VOLTA CONFERENCE—THRESHOLD TO MODERN COMPRESSIBLE FLOW; WITH ASSOCIATED EVENTS BEFORE AND AFTER

Some of the threads of the early history of compressible flow have already been established in previous chapters. We have seen in Sec. 3.10 how normal shock wave theory was well established by Rankine and Hugoniot in the latter half of the nineteenth century, and capped off by Rayleigh and Taylor in 1910. This work was extended to two dimensions by Prandtl and Meyer during the period from 1905 to 1908, when they developed and presented the fundamentals of both oblique shock and expansion wave theories for supersonic flow (see Sec. 4.16). Moreover, the basic properties of quasi-one-dimensional flow through supersonic nozzles were examined by de Laval in the 1880s and 1890s, and by Stodola and Prandtl in the first decade of the twentieth century. (See Secs. 4.16, 5.8, and 5.9.) However, at this time the only practical application of such work was in the design and analysis of steam turbines— supersonic wind tunnels, rocket engines, and high-speed aircraft were still far in the future.

The next major contribution to the advancement of compressible flow theory occurred in the 1920s. Although the flight speeds of all airplanes at that time were comfortably within the realm of incompressible flow (less than 100 m/s), the tip speeds of propellers regularly approached the speed of sound. This promoted an early interest in the effect of compressibility on propeller airfoils. As early as 1922, Prandtl is quoted as stating that the lift coefficient increased according to $(1 - M_\infty^2)^{-1/2}$; he mentioned this conclusion in his lectures at Göttingen, but without written proof. This result was mentioned again 6 years later by Jacob Ackeret, a colleague of Prandtl, in the famous German series *Handbuch der Physik,* again without proof. Subsequently, the concept was formally established by H. Glauert in 1928. Using only six pages in the *Proceedings of the Royal Society.* Glauert presented a derivation based on linearized small-perturbation theory (similar to that described in Sec. 9.4), which confirmed the $(1 - M_\infty^2)^{-1/2}$ variation. In this paper, entitled "The Effect of Compressibility on the Lift of an Airfoil," vol. 118, p. 113. Glauert derived the famous Prandtl–Glauert compressibility correction given here as Eqs. (9.36) and (9.37). This result was to stand alone, unaltered, for the next 10 years.

The next major advance in compressible flow theory involved the calculation of properties on a sharp right-circular cone in supersonic flow. (This will be the subject of Chap. 10.) In 1928, Adolf Busemann, a colleague of Prandtl's at Göttingen, arrived at a graphical solution for supersonic conical flows. However, in 1933 a more practical analytical formulation leading to the numerical solution of an ordinary differential equation for conical flow was given by G. I. Taylor and J. W. Maccoll in a paper entitled "The Air Pressure on a Cone Moving at High Speeds" which appeared in the *Proceedings of the Royal Society,* vol. 139A, 1933, pp. 278–311. We will develop and study this Taylor–Maccoll equation in Chap. 10 in a form that is virtually unchanged from the original formulation in 1933.

In addition, the 1920s also saw the development of linearized theory for two-dimensional supersonic flow by Jacob Ackeret. In 1925, Ackeret presented a paper entitled "Luftkrafte auf Flugel, die mit groserer als Schallgeschwingigkeit bewegt werden" ("Air Forces on Wings Moving at Supersonic Speeds") which appeared in *Zeitschrift für Flugtechnik und Motorluftschiffahrt,* vol. 16, 1925, p. 72. In this paper, Ackeret derived the $(M_\infty^2 - 1)^{-1/2}$ variation for a linearized pressure coefficient given above by Eq. (9.51) in Sec. 9.6. Ackeret's paper showed for the first time the now familiar decrease in pressure coefficient as the supersonic Mach number increases, as sketched in Fig. 9.11. Shortly thereafter, in 1929, Prandtl and Busemann developed for the first time in history exact nonlinear solutions for two-dimensional supersonic flow by means of the method of characteristics (a story to be told in Chap. 11). Busemann went on to apply this method of characteristics to the design of a supersonic nozzle, leading to the first practical supersonic wind tunnel in the mid-1930s. (See Sec. 11.17.)

In these paragraphs, a rather unexpected picture develops. Today we have a tendency to think of compressible flow as a very modern engineering science. This is because such material did not enter the majority of university engineering curricula until the 1950s, nor did industry require a substantial expertise in this field until about the same period. However, it is clear from the above sketch that the fundamentals of compressible flow were well established *before 1935.* This status is underscored by an article that appeared in 1934 in the monumental series *Aerodynamic Theory,* edited by W. F. Durand (see Ref. 22). Sponsored by the Guggenheim Fund for the Promotion of Aeronautics, *Aerodynamic Theory* is a six-volume compendium of the aerodynamic state of the art of that day (and still remains an important contemporary cornerstone for the study of aerodynamics). In Volume III of this series, G. I. Taylor and J. W. Maccoll authored a section entitled "The Mechanics of Compressible Fluids." This article takes only 41 pages out of a total of 2158 in the complete series, reflecting the relative practical unimportance of high-speed flow at that time. However, the material in those 41 pages could be used as a text for the standard compressible flow course of today. Taylor and Maccoll range from a discussion on acoustic theory and finite waves as we have presented in Chap. 7, to shock wave theory as given in Chaps. 3 and 4, to nozzle flows and the design of high-speed wind tunnels as we have discussed in Chap. 5, to potential theory and the Prandtl–Glauert relation as presented in this chapter, to conical flow as will be described in Chap. 10, and even to a brief introduction to the essence of characteristic theory (to be developed in Chap. 11). It is therefore remarkable that, as the world entered the year 1935 on a collision course with war and with airplanes still flying at Mach 0.3 or less, the foundation of theoretical compressible flow was securely laid. This foundation would finally see extensive use, beginning about 15 years later.

In light of the above, it is not surprising that 1935 was a fertile time for an international meeting of those few fluid mechanicians dealing with compressible flow. The time was right, and in Italy the circumstances were right. Since 1931 the Royal

Academy of Science in Rome had been conducting a series of important scientific conferences sponsored by the Alessandro Volta Foundation. (Alesandro Volta was an Italian physicist who invented the electric battery in 1800. The unit of electromotive force, the volt, is named in his honor.) The first conference dealt with nuclear physics, and then rotated between the sciences and the humanities on alternate years. The second Volta conference had the title "Europe," and in 1933 the third conference was the subject of immunology. This was followed by the subject "The Dramatic Theater" in 1934. During this period, the influence of Italian aeronautics was gaining momentum, led by General Arturo Crocco, an aeronautical engineer who had become interested in flight in 1903. He was also the father of Luigi Crocco, who distinguished himself as a leading aeronautical scientist in the midtwentieth century. [Luigi is responsible for Crocco's theorem embodied in Eq. (6.59).] General Crocco had become interested in ramjet engines in 1931, and therefore was well aware of the potential impact of compressible flow theory and experiment on future aviation. This led to the choice of the topic of the fifth Volta conference—"High Velocities in Aviation." Participation was by invitation only, and due to the prestige of the conference and the excitement of the subject matter, the participants paid special attention to the preparation of their papers. As a result, between September 30 and October 6, 1935, the major figures in the development of compressible flow gathered in Rome— Theodore von Karman and Eastman Jacobs from the United States, Prandtl and Busemann from Germany, Ackeret from Switzerland, G. I. Taylor from England, Crocco and Enrico Pistolesi from Italy, and many more. The fifth Volta conference was to become a major threshold, opening the established theory of compressible flow to practical applications in the decades to come.

The technical content of that Volta conference ranged from subsonic to supersonic flow, and from experimental to theoretical considerations. For example, Prandtl gave a general introduction and survey paper on compressible flow, showing many schlieren pictures (such as Figs. 4.41 and 4.42) for illustration. G. I. Taylor discussed supersonic conical flow theory, and von Karman presented research on minimum wave-drag shapes for axisymmetric bodies. The linearized Prandtl–Glauert relation was once again derived and presented by Enrico Pistolesi, along with several higher-order calculations for compressibility corrections. Eastman Jacobs presented new test results for compressibility effects on subsonic airfoils, obtained in several high-speed wind tunnels at the NACA Langley Aeronautical Laboratory in Virginia. Jakob Ackeret gave a paper on many different subsonic and supersonic wind tunnel designs. There were also presentations on propulsion techniques for high-speed flight, including rockets and ramjets. The meeting also included a field trip to the new Italian aerodynamic research center at Guidonia near Rome. Guidonia was equipped with several high-speed wind tunnels, subsonic and supersonic, all designed after the work of Ackeret and constructed under his consultation. This laboratory was to produce a large bulk of supersonic experimental data before and during World War II, and was to produce from its ranks a leading supersonic aerodynamicist, Antonio Ferri. (Much of the work performed at Guidonia is reflected in Ferri's book, Ref. 5.)

However, probably one of the most farsighted and important papers given at the fifth Volta conference was presented by Adolf Busemann (see Fig. 9.20). Entitled

Figure 9.20 | Adolf Busemann.

"Aerodynamischer Auftrieb bei Uberschallgeschwindigkeit" ("Aerodynamic Forces at Supersonic Speeds"), this paper introduced for the first time in history the concept of the swept wing as a mechanism for reducing the large drag increase encountered beyond the critical Mach number (see Sec. 9.7). Busemann reasoned that the flow over a wing is governed mainly by the component of velocity perpendicular to the leading edge. If the wing is swept this component will decrease, as illustrated in Fig. 9.21, which is taken directly from Busemann's original paper. Consequently, the free-stream Mach number at which the large rise in drag is encountered is increased. Therefore, airplanes with swept wings could fly faster before encountering the drag-divergence phenomena discussed in Sec. 9.7. This swept-wing concept of Busemann's is now reflected in the vast majority of high-speed aircraft in operation today.

It is interesting to note that the fifth Volta conference was given special significance by the Italian government. Its prestige was reflected in its location—it was held in an impressive Renaissance building that served as the city hall during the Holy Roman Empire. Moreover, the Italian dictator Benito Mussolini chose the conference to make his announcement that Italy had invaded Ethiopia. It is curious that such a political statement was saved for a technical meeting on high-speed flow.

The conference served to spread excitement about the future of high-speed flight, and provided the first major international exchange of information on compressible

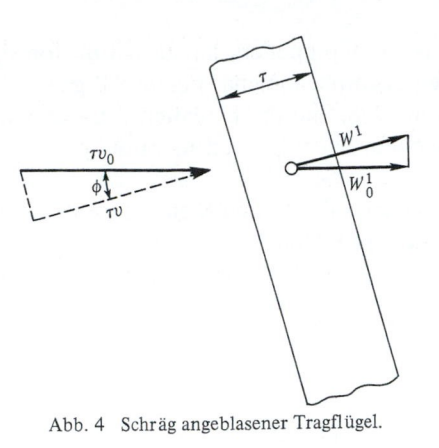

Abb. 4   Schräg angeblasener Tragflügel.

Figure 9.21 | The swept-wing concept as it appeared in Busemann's original paper in 1935.

flow. However, in many respects, it had a delayed impact. For example, Busemann's work on swept wings appeared to drop from sight. This was because the German Luftwaffe recognized its military significance, and classified the concept in 1936—one year after the conference. The Germans went on to produce a large bulk of swept-wing research during World War II, resulting in the design of the first operational jet airplane—the Me 262—which had a moderate degree of sweep. After the war, technical teams from the three allied nations. England, Russia, and the United States, swooped into the German research laboratories at Penemunde and Braunschweig, and gathered all the swept-wing data they could find. (The United States also gathered Adolf Busemann himself, who was moved to the NACA Langley Aeronautical Laboratory. Later, Busemann became a professor at the University of Colorado, and he now lives an active retired life in Boulder, Colorado.) Virtually all the modern high-speed airplanes of today can trace their lineage back to the original data obtained from Germany, and ultimately to Busemann's paper at the fifth Volta conference.

Strangely enough, the significance of Busemann's idea was lost on most attendees at the conference. Von Karman and Jacobs did not spread it upon their return to the United States. Indeed, 10 years later, when World War II was reaching its conclusion and jet airplanes were beginning to revolutionize aviation, the idea of swept wings was developed independently by R. T. Jones, an ingenious aerodynamicist at the NACA Langley Laboratory. When Jones made such a proposal to Jacobs and von Karman in 1945, neither man remembered Busemann's idea from the Volta conference. (See Ref. 134 for more historical details on the invention of the swept wing.)

On the positive side, however, the Volta conference did serve to spur highspeed research in the United States. Renewed efforts were made by the NACA to obtain data on compressibility effects on high-speed subsonic airfoils—this time prompted not only by high tip speeds of propellers, but also by the foresight that airplane wings would soon encounter such phenomena. Figure 9.8 gives some experimental data published by NACA in 1938. Shortly thereafter, von Karman and Tsien published a compressibility correction that improved upon the older Prandtl–Glauert relation (see Sec. 9.5).

Nevertheless, in general the United States reacted slowly to the stimulus provided by the Volta conference. Upon his return from Italy in late 1935, von Karman urged both the Army and the NACA to develop high-speed wind tunnels, including supersonic facilities. He encountered deaf ears. Finally, as the clouds of war enveloped the United States in 1941, such urging encountered more receptive attitudes. Von Karman established at Cal Tech the first major university curriculum in compressible flow in 1942; this course of study was highly populated by military officers. Finally, in 1944, the first operational supersonic wind tunnel in the United States was built at the Army Ballistics Research Laboratory in Aberdeen, Maryland. This tunnel was designed by von Karman and his colleagues at Cal Tech, and was operated by Cal Tech personnel at Aberdeen under contract from the Army. Twelve years after Busemann began to collect data in his supersonic tunnel in Germany, and 9 years after the fifth Volta conference and the construction of supersonic tunnels at Guidonia in Italy, the United States was finally seriously in the business of supersonic research.

9.10 | HISTORICAL NOTE: PRANDTL— A BIOGRAPHICAL SKETCH

The name of Ludwig Prandtl (see Fig. 9.22) pervades virtually all of twentieth century fluid mechanics, ranging from inviscid incompressible flow over airfoils and finite wings, to the ingenious idea of the boundary layer for viscous flows, and extending through the early development of high-speed subsonic and supersonic flows. We have already mentioned his impact on the advancement of compressible flow in Secs. 4.16 and 9.9. Who was this man who gathers so much respect, even bordering on reverence, from fluid mechanicians? Let us take a closer look.

Ludwig Prandtl was born on February 4, 1875, in Freising, Bavaria. His father was Alexander Prandtl, a professor of surveying and engineering at the agricultural college at Weihenstephan, near Freising. Although three children were born into the Prandtl family, two died at birth and Ludwig grew up as an only child. At an early age, Prandtl became interested in his father's books on physics, machinery, and instruments. Much of Prandtl's remarkable ability to intuitively go to the heart of a physical problem can be traced to his environment at home as a child, where his father, a great lover of nature, induced Ludwig to observe natural phenomena and to reflect upon them.

In 1894, Prandtl began his formal scientific studies at the Technische Hochschule in Munich, where his principal teacher was A. Foppl. Six years later, he graduated from the University of Munich with a Doctor's degree. However, by this time he was alone, his father having died in 1896 and his mother in 1898.

By 1900, Prandtl had not done any work nor shown any interest in fluid mechanics. Indeed, his doctor's thesis at Munich was in solid mechanics, dealing with

Figure 9.22 | Ludwig Prandtl (1875–1953).

unstable elastic equilibrium in which bending and distortion acted together. (It is not generally recognized by people in fluid dynamics that Prandtl continued his interest and research in solid mechanics through most of his life—this work is eclipsed, however, by his major contributions to the study of fluid flow.) However, soon after graduation from Munich, Prandtl had his first major encounter with fluid mechanics. Joining the Nuremburg works of the Maschinenfabrick Augsburg as an engineer. Prandtl worked in an office designing mechanical equipment for the new factory. He was made responsible for redesigning an apparatus for removing machine shavings by suction. Finding no reliable information in the scientific literature about the fluid mechanics of suction. Prandtl arranged his own experiments to answer a few fundamental questions about the flow. The result of this work was his new design for shavings cleaners. The apparatus was modified with pipes of improved shape and size, and carried out satisfactory operation at one-third its original power consumption. Prandtl's contributions in fluid mechanics had begun.

One year later, in 1901, he became Professor of Mechanics in the Mathematical Engineering Department at the Technische Hochschule in Hanover. (Please note that in Germany a "technical highschool" is equivalent to a technical university in the United States.) It was at Hanover that Prandtl enhanced and continued his new-found interest in fluid mechanics. It was here, and not at Göttingen, that Prandtl first developed his famous boundary layer theory. It was also here that he first became interested in the steam flow through Laval nozzles, in parallel with the pioneering work by Stodola (see Sec. 5.9).

In 1904, Prandtl delivered his famous paper on the concept of the boundary layer to the Third Congress of Mathematicians at Heidelberg. From this time on, the star of Prandtl was to rise meteorically. Later that year he moved to Göttingen to become Director of the Institute for Technical Physics, later to be renamed Applied Mechanics.

It should be noted that, at the turn of the century, no engineering curriculum existed in any pure university in Germany; such training was provided by the technische hochschules. However, at this time Felix Klein, a powerful mathematician, was director at the University of Göttingen. He recognized that, since the University provided no formal instruction in engineering, it consequently had little connection with industry and the rapidly increasing influence of technology on society. Attempting to rectify this situation, Klein established a series of professional chairs and institutes dedicated to the applied sciences. One of these was the Institute for Technical Physics, for which Prandtl was chosen as Director (at the age of 30) in 1904. This institute gave instruction in mechanics, thermodynamics, strength of materials, and hydraulics. Other institutes were in applied mathematics and applied electricity. Of course, in the meantime, Göttingen was maintaining and fostering its already excellent reputation in pure mathematics and physics (see Sec. 4.16). So it is no wonder that Prandtl flourished in this environment.

In the fall of 1909, Prandtl married Gertrude Foppl, a daughter of August Foppl, Prandtl's old professor from the Technische Hochschule in Munich. The marriage subsequently produced two daughters.

As described in Sec. 4.16, Prandtl made substantial contributions to the understanding of compressible flow during the period 1905 to 1910—this work on flow through Laval nozzles, and especially on oblique shock and expansion waves, was of particular note. During the period 1910 to 1920, his primary output shifted to low-speed airfoil and finite-wing theory, leading to the famous Prandtl lifting line and lifting surface theories for calculating lift and induced drag. About this time, after a long hiatus, researchers in England and the United States began to grasp the significance of Prandtl's boundary layer theory, and his work on wing theory quickly spread via various English language translations of his papers. By 1925, Prandtl had firmly established a worldwide reputation as the leader in aerodynamics. Students and colleagues flocked to Göttingen, and then fanned out to various international locations to establish centers of aerodynamic research. These included Jakob Ackeret in Zurich, Switzerland, Adolf Busemann in Germany, and Theodore von Karman at Cal Tech in the United States.

During the 1920s and 1930s, Prandtl's responsibilities at Göttingen expanded. In addition to the Institute for Applied Mechanics, he now was in charge of the newly established Kaiser Wilhelm Institute for Fluid Dynamics. (After World War II, the name was changed to the Max Planck Institute.) In these years, Prandtl continued his interest in high-speed flow, leading in part to the development of the Prandtl–Glauert compressibility correction (see Secs. 9.4 and 9.9). Moreover, a major aerodynamic laboratory—the Aerodynamische Versuchsanstalt—was established at Göttingen, containing a number of low- and high-speed wind tunnels and other expensive research equipment.

Shortly after the Nazis came to power in Germany in 1933, Göttingen experienced a major exodus of Jewish professors, causing the university to lose substantial expertise and prestige, especially in the area of pure mathematics and physics. However, Prandtl was not directly affected, and in fact the Air Ministry of the new German government began to provide major support to his aerodynamic research. Prandtl continued to work under these conditions until 1945, when the Americans passed through Göttingen during the last days of World War II. By all accounts, Prandtl was concerned about the fate of his Jewish colleagues, but he was a scientist without a major sense of political awareness. As a matter of dedication to his country, Prandtl subjugated personal misgivings to what he felt was obligation. Some insight into Prandtl's character and thinking during this period is given by von Karman in his autobiography entitled *The Wind and Beyond* (Little, Brown and Co., 1967). Von Karman's comments on Prandtl, his former teacher, are not particularly complimentary, and have been the source of some rebuttal from other colleagues of Prandtl. Nevertheless, von Karman's viewpoint is worth reading, and in fact the entire book is an excellent portrait of the growth of twentieth century fluid mechanics, with many interesting observations on the cast of characters by someone who himself played a large part in its development.

Prandtl's personal technical contributions during the last years of his life were not as potent as in his early days. However, his interests remained in fluid dynamics, although he published a few papers in his original field of solid mechanics, discussing nonelastic phenomena in more conventional terms. He also became interested in meteorological fluid dynamics, and was actively working in this area until the end of his life.

Prandtl died in 1953. He was clearly the father of modern aerodynamics—a monumental figure in fluid dynamics. Each day, around the world, his name will continue to be spoken for as long as we maintain and extend our technical society.

9.11 | HISTORICAL NOTE: GLAUERT— A BIOGRAPHICAL SKETCH

Equations (9.36) and (9.37) give the famous Prandtl–Glauert compressibility correction. Every student of fluid dynamics has some knowledge of Prandtl. But who was Glauert? Let us take a look.

Hermann Glauert was born in Sheffield, England, on October 4, 1892. He was well-educated, first at the King Edward VII School at Sheffield, and then later at Trinity College, Cambridge, where he received many honors for his high leadership in the classroom. For example, he was awarded the Ryson Medal for astronomy in 1913, an Isaac Newton Scholarship in 1914, and the Rayleigh Prize in 1915.

In 1916, as the second year of World War I waxed on, Glauert joined the staff of the Royal Aircraft Establishment in Farnborough. There, he quickly grasped the fundamentals of aerodynamics, and wrote numerous reports and memoranda dealing with airfoil and propeller theory, the performance, stability, and control of airplanes, and the theory of the autogyro. In 1926, he published a book entitled *The Elements of Aerofoil and Airscrew Theory;* this book was the single most important instrument for spreading Prandtl's airfoil and wing theory around the English-speaking world, and to this day is still used as a reference in courses dealing with incompressible flow.

Glauert did not collaborate with Prandtl on the development of the Prandtl–Glauert rule. As related in Sec. 9.9, Glauert worked independently and was the first person to derive the rule from established aerodynamic theory, publishing his results in 1928 in the *Proceedings of the Royal Society* (see Sec. 9.9).

By the early 1930s, Glauert was probably the leading theoretical aerodynamicist in England. He had also become the Principal Scientific Officer of the RAE, as well as Head of its Aerodynamics Department. However, on August 4, 1934, Glauert was strolling through a small park called Fleet Common at Farnborough. It was a pleasant day, and he stopped to watch some Royal Engineers who were blowing up tree stumps. Suddenly, from 8 yards away, a blast tore a stump to pieces, hurling fragments of wood in all directions. One hit Glauert squarely on the forehead; he died a few hours later. England, and the world, were suddenly and prematurely deprived of one of its best aerodynamicists.

9.12 | SUMMARY

In addition to the intermediate summary comments made in Sec. 9.8, we give a more specific summary of the basic results from linearized theory here.

For an irrotational, inviscid, compressible flow, the continuity, momentum, and energy equations reduce to one equation with one dependent variable, namely, the velocity potential Φ, defined as $\mathbf{V} = \nabla \Phi$. The full velocity potential equation is

$$\left(1 - \frac{\Phi_x^2}{a^2}\right)\Phi_{xx} + \left(1 - \frac{\Phi_y^2}{a^2}\right)\Phi_{yy} + \left(1 - \frac{\Phi_z^2}{a^2}\right)\Phi_{zz}$$

$$- \frac{2\Phi_x\Phi_y}{a^2}\Phi_{xy} - \frac{2\Phi_x\Phi_z}{a^2}\Phi_{xz} - \frac{2\Phi_y\Phi_z}{a^2}\Phi_{yz} = 0 \qquad (8.17)$$

This is an exact equation for irrotational flow; it holds for the flow over arbitrary bodies, thin or thick, at arbitrary angles of attack, small or large. However, defining a perturbation velocity potential ϕ as $\Phi(x, y, z) = V_\infty x + \phi(x, y, z)$ and assuming small perturbations, Eq. (8.17) reduces to a simpler form, applicable to subsonic and supersonic flow, but not applicable to transonic or hypersonic flow:

$$\left(1 - M_\infty^2\right)\frac{\partial^2 \phi}{\partial x^2} + \frac{\partial^2 \phi}{\partial y^2} + \frac{\partial^2 \phi}{\partial z^2} = 0 \qquad (9.6)$$

This is the linearized small-perturbation velocity potential equation. Since Eq. (9.6) is linear, it is much more amenable to analytic solution than the full velocity potential equation given by Eq. (8.17). However, to obtain this advantage with Eq. (9.6), we trade accuracy; Eq. (9.6) is an *approximate* relation that holds only for small perturbations (thin bodies at small angles of attack) and only for subsonic or supersonic flow.

For the linearized solution of both subsonic and supersonic compressible flows, Eq. (9.6) represents one important tool. Two additional necessary tools are the form of the pressure coefficient consistent with small perturbations,

$$C_p = -\frac{2u'}{V_\infty} \tag{9.17}$$

and the boundary condition

$$v' = V_\infty \theta \tag{9.19}$$

For subsonic compressible flow, these tools lead to the *Prandtl–Glauert* rule,

$$C_p = \frac{C_{p_o}}{\sqrt{1 - M_\infty^2}} \tag{9.36}$$

where C_{p_o} is the pressure coefficient at low speeds (incompressible flow). Also,

$$C_L = \frac{C_{L_o}}{\sqrt{1 - M_\infty^2}} \tag{9.37a}$$

and

$$C_M = \frac{C_{M_o}}{\sqrt{1 - M_\infty^2}} \tag{9.37b}$$

where C_L and C_M are the lift and moment coefficients.

For supersonic flow, the preceding tools lead to an expression for the pressure coefficient given by

$$C_p = \frac{2\theta}{\sqrt{M_\infty^2 - 1}} \tag{9.51}$$

As derived in the homework problems, Eq. (9.51) when applied to a flat plate at an angle of attack α yields

$$C_L = \frac{4\alpha}{\sqrt{M_\infty^2 - 1}}$$

$$C_D = \frac{4\alpha^2}{\sqrt{M_\infty^2 - 1}}$$

$$C_{M_{c/4}} = \frac{-\alpha}{\sqrt{M_\infty^2 - 1}}$$

where C_L, C_D, and $C_{M_{c/4}}$ are the lift, drag, and moment coefficients, respectively. Here, $C_{M_{c/4}}$ is taken about the quarter-chord point (a point 0.25 of the chord length from the leading edge).

PROBLEMS

9.1   Show that this nonlinear equation is valid for *transonic* flow with small perturbations:

$$\left(1 - M_\infty^2\right) \frac{\partial^2 \phi}{\partial x^2} + \frac{\partial^2 \phi}{\partial y^2} + \frac{\partial^2 \phi}{\partial z^2} = M_\infty^2 \left[\frac{\gamma + 1}{V_\infty} \frac{\partial \phi}{\partial x} \right] \frac{\partial^2 \phi}{\partial x^2}$$

9.2   The low-speed lift coefficient for an NACA 2412 airfoil at an angle of attack of 4° is 0.65. Using the Prandtl–Glauert rule, calculate the lift coefficient for $M_\infty = 0.7$.

9.3   In low-speed flow, the pressure coefficient at a point on an airfoil is −0.9. Calculate the value of C_p at the same point for $M_\infty = 0.6$ by means of

   a.   The Prandtl–Glauert rule

   b.   Laitone's correction

   c.   The Karman–Tsien rule

9.4   Consider a flat plate with chord length c at an angle of attack α to a supersonic free stream of Mach number M_∞. Let L and D be the lift and drag per unit span, and S be the planform area of the plate per unit span, $S = c(1)$. Using linearized theory, derive the following expressions for the lift and drag coefficients (where $C_L \equiv L/\frac{1}{2}\rho_\infty V_\infty^2 S$ and $C_D \equiv D/\frac{1}{2}\rho_\infty V_\infty^2 S$):

   a. $C_L = \dfrac{4\alpha}{\sqrt{M_\infty^2 - 1}}$

   b. $C_D = \dfrac{4\alpha^2}{\sqrt{M_\infty^2 - 1}}$

9.5   For the flat plate in Problem 9.4, the quarter-chord point is located, by definition, at a distance equal to $c/4$ from the leading edge. Using linearized theory, derive the following expression for the moment coefficient about the quarter-chord point for supersonic flow

$$C_{M_{c/4}} = \frac{-\alpha}{\sqrt{M_\infty^2 - 1}}$$

where $C_{M_{c/4}} \equiv M_{c/4}/\frac{1}{2}\rho_\infty V_\infty^2 Sc$, and as usual in aeronautical practice, a positive moment by convention is in the direction of increasing angle of attack.

9.6   Consider a flat plate at an angle of attack of 4°.

   a.   Calculate C_L and $C_{M_{c/4}}$ for $M_\infty = 0.03$ (essentially incompressible flow). (*Hint:* Consult a book, such as Reference 104, for the aerodynamic properties of a flat plate using incompressible flow thin airfoil theory.)

   b.   Apply the Prandtl–Glauert rule to the results of part (*a*), and calculate C_L and $C_{M_{c/4}}$ for $M_\infty = 0.6$.

9.7 Consider a diamond-shaped airfoil such as that sketched in Fig. 4.35. The half-angle is ε, thickness is t, and chord is c. For supersonic flow, use linearized theory to derive the following expression for C_D at $\alpha = 0$:

$$C_D = \frac{4}{\sqrt{M_\infty^2 - 1}} \left(\frac{t}{c}\right)^2$$

9.8 Supersonic linearized theory predicts that, for a thin airfoil of arbitrary shape and thickness at angle of attack α, $C_L = 4\alpha/\sqrt{M_\infty^2 - 1}$, independent of the shape and thickness. Prove this result.

9.9 Repeat Prob. 4.17, except using linearized theory. Plot the linearized results on top of the same graphs produced for Prob. 4.17 in order to assess the differences between linear theory (which is approximate) and shock-expansion theory (which is exact). From this comparison, over what angle-of-attack range would you feel comfortable in applying linear theory?

9.10 Linear supersonic theory predicts that the curve of wave drag versus Mach number has a minimum point at a certain value of $M_\infty > 1$.

a. Calculate this value of M_∞.

b. Does it make physical sense for the wave drag to have a minimum value at some supersonic value of M_∞ above 1? Explain. What does this say about the validity of linear theory for certain Mach number ranges?

9.11 At $\alpha = 0°$, the minimum pressure coefficient for an NACA 0009 airfoil in low-speed flow is -0.25. Calculate the critical Mach number for this airfoil using

a. The Prandtl–Glauert rule

b. The (more accurate) Karman–Tsien rule

CHAPTER 10

Conical Flow

Gas dynamics as a branch of physics and applied mathematics has grown with the growth of high speed flight.
Howard W. Emmons, 1958

This chapter is all about supersonic flow over cones. In Sec. 4.4 we contrasted the flow over wedges and cones. The flow over a wedge is two-dimensional and can be analyzed using oblique shock theory and algebraic relationships. Although side views of both a wedge and a cone look identical (see Fig. 4.11), the flow over a cone takes place in three-dimensional space, and is analyzed using the governing flow equations in partial differential form. After a discussion about the differences between wedge flows and cone flows, we stated in the last sentence of the main text in Sec. 4.4 that "because of these differences, the study in this book of supersonic flow over cones will be delayed until Chap. 10." Well, here we are at Chap. 10, and we are ready to study cone flows.

A cone is a basic geometric shape, and the results of this chapter find frequent application in the aerodynamics of supersonic missiles, inlet diffusers with conical centerbodies for supersonic airplanes, and basic research experiments on the physics of supersonic and hypersonic flows, among others. Also, the flow over a cone is frequently used as the generating flowfield for the design of hypersonic waverider vehicles, an interesting configuration that is discussed for example in Refs. 135 and 136. So in addition to being a fundamental flowfield in the study of compressible flow, the supersonic flow over cones is a useful foundation in many applications.

This chapter is short, and the intellectual path is straightforward. So we do not need a roadmap. However, return to the overall roadmap for this book, Fig. 1.7. We are now back to the center column, namely, box 11, where, with conical flow, we commence a series of interesting studies of more complex and important flows. We begin a long cruise straight down to the end of the center column.

10.1 | INTRODUCTION

In contrast to the linearized two-dimensional flows considered in Chap. 9, this chapter deals with the exact nonlinear solution for a special degenerate case of three-dimensional flow—the axisymmetric supersonic flow over a sharp cone at zero angle of attack to the free stream. Consider a body of revolution (a body generated by rotating a given planar curve about a fixed axis) at zero angle of attack as shown in Fig. 10.1. A cylindrical coordinate system (r, ϕ, z) is drawn, with the z axis as the axis of symmetry aligned in the direction of V_∞. By inspection of Fig. 10.1, the flowfield must be symmetric about the z axis, i.e., all properties are

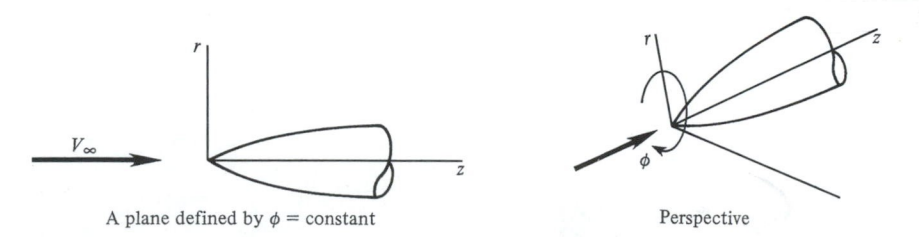

A plane defined by ϕ = constant   Perspective

Figure 10.1 | Cylindrical coordinate system for an axisymmetric body.

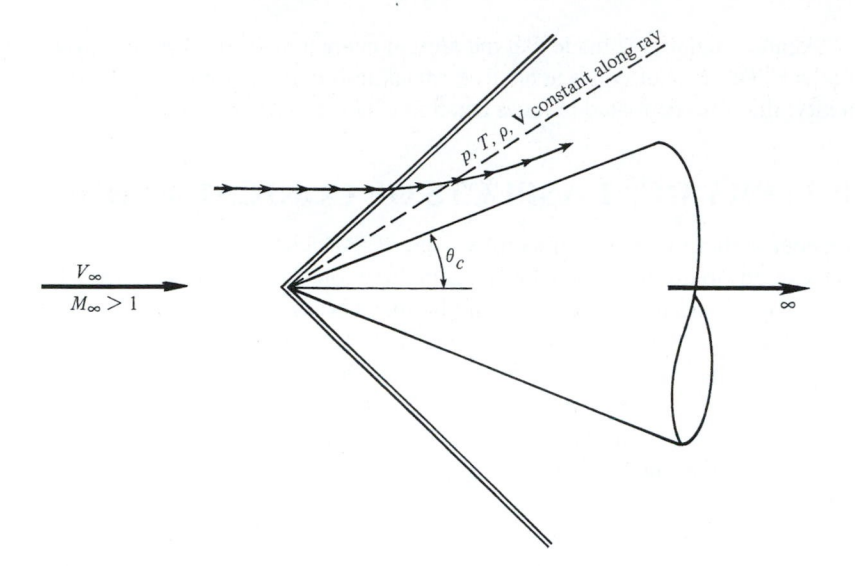

Figure 10.2 | Supersonic flow over a cone.

independent of ϕ:

$$\frac{\partial}{\partial \phi} \equiv 0$$

The flowfield depends only on r and z. Such a flow is defined as *axisymmetric flow*. It is a flow that takes place in three-dimensional space; however, because there are only two independent variables, r and z, axisymmetric flow is sometimes called "quasi-two-dimensional" flow.

In this chapter, we will further specialize to the case of a sharp right-circular cone in a supersonic flow, as sketched in Fig. 10.2. This case is important for three reasons:

1. The equations of motion can be solved exactly for this case.
2. The supersonic flow over a cone is of great practical importance in applied aerodynamics; the nose cones of many high-speed missiles and projectiles are approximately conical, as are the nose regions of the fuselages of most supersonic airplanes.
3. The first solution for the supersonic flow over a cone was obtained by A. Busemann in 1929, long before supersonic flow became fashionable (see Ref. 26). This solution was essentially graphical, and illustrated some of the important physical phenomena. A few years later, in 1933, G. I. Taylor and J. W. Maccoll (see Ref. 27) represented a numerical solution that is a hallmark in the evolution of compressible flow. Therefore, the study of conical flow is of historical significance.

Again, emphasis is made that the present chapter deals with cones at zero angle of attack. The case of cones at angle of attack introduces additional geometric complexity; this case is treated in more detail in Chap. 13.

10.2 | PHYSICAL ASPECTS OF CONICAL FLOW

Consider a sharp cone of semivertex angle θ_c, sketched in Fig. 10.2. Assume this cone extends to infinity in the downstream direction (a semi-infinite cone). The cone is in a supersonic flow, and hence an oblique shock wave is attached at the vertex. The shape of this shock wave is also conical. A streamline from the supersonic free stream discontinuously deflects as it traverses the shock, and then curves continuously downstream of the shock, becoming parallel to the cone surface asymptotically at infinity. Contrast this flow with that over a two-dimensional wedge (Chap. 4) where all streamlines behind the shock are immediately parallel to the wedge surface.

Because the cone extends to infinity, distance along the cone becomes meaningless: If the pressure were different at the 1- and 10-m stations along the surface of the cone, then what would it become at infinity? This presents a dilemma that can be reconciled only by assuming that the pressure is constant along the surface of the cone, as well as that all other flow properties are also constant. Since the cone surface is simply a ray from the vertex, consider other such rays between the cone surface and the shock wave, as illustrated by the dashed line in Fig. 10.2. It only makes sense to assume that the flow properties are constant along these rays as well. Indeed, the *definition of conical flow is where all flow properties are constant along rays from a given vertex*. The properties vary from one ray to the next. This aspect of conical flow has been experimentally proven. Theoretically, it results from the lack of a meaningful scale length for a semi-infinite cone.

10.3 | QUANTITATIVE FORMULATION (AFTER TAYLOR AND MACCOLL)

Consider the superimposed cartesian and spherical coordinate systems sketched in Fig. 10.3a. The z axis is the axis of symmetry for the right-circular cone, and V_∞ is oriented in the z direction. The flow is axisymmetric; properties are independent of ϕ. Therefore, the picture can be reoriented as shown in Fig. 10.3b, where r and θ are the two independent variables and V_∞ is now horizontal. At any point e in the flow-field, the radial and normal components of velocity are V_r and V_θ, respectively. Our objective is to solve for the flowfield between the body and the shock wave. Recall that for axisymmetric conical flow

$$\frac{\partial}{\partial \phi} \equiv 0 \quad \text{(axisymmetric flow)}$$

$$\frac{\partial}{\partial r} \equiv 0 \quad \text{(flow properties are constant along a ray from the vertex)}$$

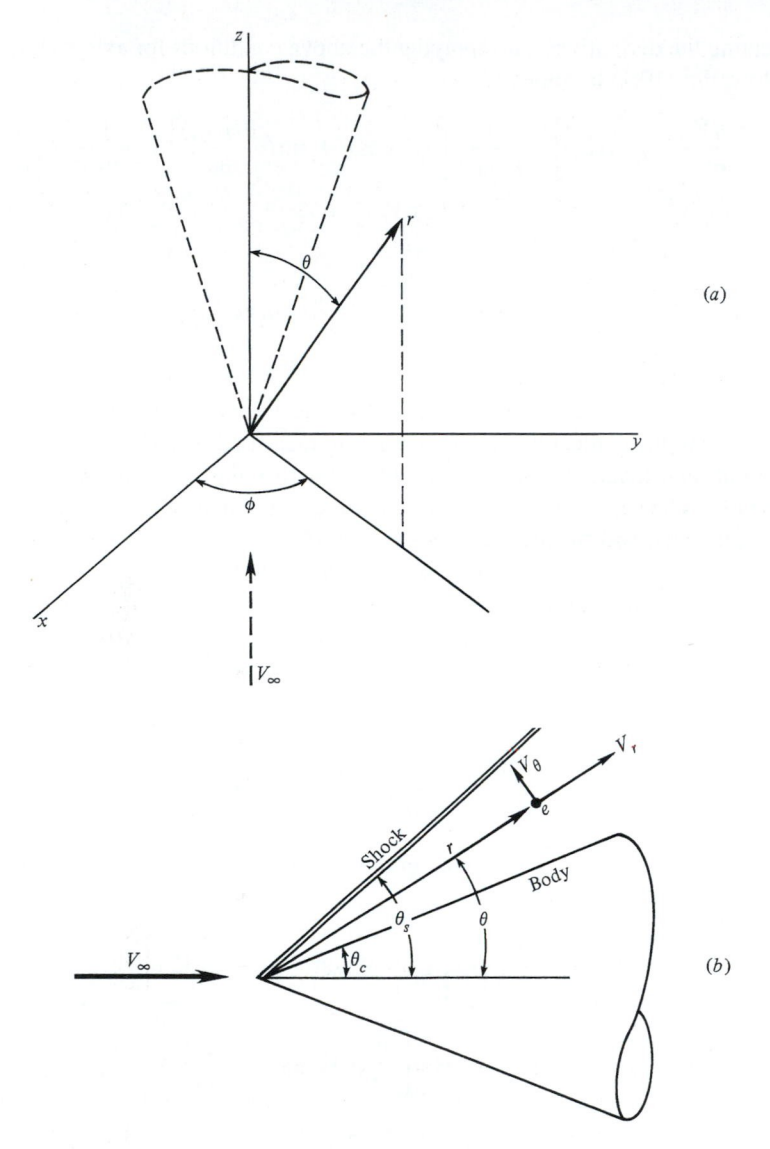

(a)

(b)

Figure 10.3 | Spherical coordinate system for a cone.

The continuity equation for steady flow is Eq. (6.5),

$$\nabla \cdot (\rho \mathbf{V}) = 0$$

In terms of spherical coordinates, Eq. (6.5) becomes

$$\nabla \cdot \rho(\mathbf{V}) = \frac{1}{r^2} \frac{\partial}{\partial r}(r^2 \rho V_r) + \frac{1}{r \sin \theta} \frac{\partial}{\partial \theta}(\rho V_\theta \sin \theta) + \frac{1}{r \sin \theta} \frac{\partial(\rho V_\phi)}{\partial \phi} = 0$$

$$(10.1)$$

Evaluating the derivatives, and applying the above conditions for axisymmetric conical flow, Eq. (10.1) becomes

$$\frac{1}{r^2}\left[r^2\frac{\partial(\rho V_r)}{\partial r}+\rho V_r(2r)\right]+\frac{1}{r\sin\theta}\left[\rho V_\theta\cos\theta+\sin\theta\frac{\partial(\rho V_\theta)}{\partial\theta}\right]+\frac{1}{r\sin\theta}\frac{\partial(\rho V_\phi)}{\partial\phi}=0$$

$$\frac{2\rho V_r}{r}+\frac{\rho V_\theta}{r}\cot\theta+\frac{1}{r}\left(\rho\frac{\partial V_\theta}{\partial\theta}+V_\theta\frac{\partial\rho}{\partial\theta}\right)=0$$

$$2\rho V_r+\rho V_\theta\cot\theta+\rho\frac{2V_\theta}{\partial\theta}+V_\theta\frac{\partial\rho}{\partial\theta}=0$$

$$(10.2)$$

Equation (10.2) is the continuity equation for axisymmetric conical flow.

Return to the conical flowfield sketched in Figs. 10.2 and 10.3. The shock wave is straight, and hence the increase in entropy across the shock is the same for all streamlines. Consequently, throughout the conical flowfield, $\nabla s = 0$. Moreover, the flow is adiabatic and steady, and hence Eq. (6.45) dictates that $\Delta h_o = 0$. Therefore, from Crocco's equation, Eq. (6.60), we find that $\nabla \times \mathbf{V} = 0$, i.e., the conical flowfield is *irrotational*. Since Croco's theorem is a combination of the momentum and energy equations (see Sec. 6.6), then $\nabla \times \mathbf{V} = 0$ can be used in place of either one. In spherical coordinates,

$$\nabla\times\mathbf{V}=\frac{1}{r^2\sin\theta}\begin{vmatrix}\mathbf{e}_r & r\mathbf{e}_\theta & (r\sin\theta)\mathbf{e}_\phi\\\frac{\partial}{\partial r} & \frac{\partial}{\partial\theta} & \frac{\partial}{\partial\phi}\\V_r & rV_\theta & (r\sin\theta)V_\phi\end{vmatrix}=0 \qquad (10.3)$$

where $\mathbf{e}_r$, $\mathbf{e}_\theta$, and $\mathbf{e}_\phi$ are unit vectors in the r, θ, and ϕ directions, respectively. Expanded, Eq. (10.3) becomes

$$\nabla\times\mathbf{V}=\frac{1}{r^2\sin\theta}\left\{\mathbf{e}_r\left[\frac{\partial}{\partial\theta}(rV_\phi\sin\theta)-\frac{\partial}{\partial\phi}(rV_\theta)\right]\right.$$

$$-r\mathbf{e}_\theta\left[\frac{\partial}{\partial r}(rV_\phi\sin\theta)-\frac{\partial}{\partial\phi}(V_r)\right]$$

$$\left.+(r\sin\theta)\mathbf{e}_\phi\left[\frac{\partial}{\partial r}(rV_\theta)-\frac{\partial V_r}{\partial\theta}\right]\right\}=0 \qquad (10.4)$$

Applying the axisymmetric conical flow conditions, Eq. (10.4) dramatically simplifies to

$$V_\theta=\frac{\partial V_r}{\partial\theta} \qquad (10.5)$$

Equation (10.5) is the irrotationality condition for axisymmetric conical flow.

Since the flow is irrotational, we can apply Euler's equation in any direction in the form of Eq. (8.7):

$$dp=-\rho V\,dV$$

where

$$V^2=V_r^2+V_\theta^2$$

Hence, Eq. (8.7) becomes

$$dp = -\rho(V_r \, dV_r + V_\theta \, dV_\theta) \tag{10.6}$$

Recall that, for isentropic flow,

$$\frac{dp}{d\rho} \equiv \left(\frac{\partial p}{\partial \rho}\right)_s = a^2$$

Thus, Eq. (10.6) becomes

$$\frac{d\rho}{\rho} = -\frac{1}{a^2}(V_r \, dV_r + V_\theta \, dV_\theta) \tag{10.7}$$

From Eq. (6.45), and defining a new reference velocity V_{max} as the maximum theoretical velocity obtainable from a fixed reservoir condition (when $V = V_{max}$, the flow has expanded theoretically to zero temperature, hence $h = 0$), we have

$$h_o = \text{const} = h + \frac{V^2}{2} = \frac{V_{max}^2}{2}$$

Note that V_{max} is a constant for the flow and is equal to $\sqrt{2h_o}$. For a calorically perfect gas, the above becomes

$$\frac{a^2}{\gamma - 1} + \frac{V^2}{2} = \frac{V_{max}^2}{2}$$

or

$$a^2 = \frac{\gamma - 1}{2}\left(V_{max}^2 - V^2\right) = \frac{\gamma - 1}{2}\left(V_{max}^2 - V_r^2 - V_\theta^2\right) \tag{10.8}$$

Substitute Eq. (10.8) into (10.7):

$$\frac{d\rho}{\rho} = -\frac{2}{\gamma - 1}\left(\frac{V_r \, dV_r + V_\theta \, dV_\theta}{V_{max}^2 - V_r^2 - V_\theta^2}\right) \tag{10.9}$$

Equation (10.9) is essentially Euler's equation in a form useful for studying conical flow.

Equations (10.2), (10.5), and (10.9) are three equations with three dependent variables: ρ, V_r, and V_θ. Due to the axisymmetric conical flow conditions, there is only one independent variable, namely θ. Hence, the partial derivatives in Eqs. (10.2) and (10.5) are more properly written as ordinary derivatives. From Eq. (10.2),

$$2V_r + V_\theta \cot\theta + \frac{dV_\theta}{d\theta} + \frac{V_\theta}{\rho}\frac{d\rho}{d\theta} = 0 \tag{10.10}$$

From Eq. (10.9),

$$\frac{d\rho}{d\theta} = -\frac{2\rho}{\gamma - 1}\left(\frac{V_r\dfrac{dV_r}{d\theta} + V_\theta\dfrac{dV_\theta}{d\theta}}{V_{max}^2 - V_r^2 - V_\theta^2}\right) \tag{10.11}$$

Substitute Eq. (10.11) into Eq. (10.10):

$$2V_r + V_\theta \cot\theta + \frac{dV_\theta}{d\theta} - \frac{2V_\theta}{\gamma - 1} \left(\frac{V_r \dfrac{dV_r}{d\theta} + V_\theta \dfrac{dV_\theta}{d\theta}}{V_{max}^2 - V_r^2 - V_\theta^2} \right) = 0$$

or $\quad \dfrac{\gamma - 1}{2} \left(V_{max}^2 - V_r^2 - V_\theta^2 \right) \left(2V_r + V_\theta \cot\theta + \dfrac{dV_\theta}{d\theta} \right) - V_\theta \left(V_r \dfrac{dV_r}{d\theta} + V_\theta \dfrac{dV_\theta}{d\theta} \right) = 0$

$$(10.12)$$

Recall from Eq. (10.5) that

$$V_\theta = \frac{dV_r}{d\theta}$$

Hence, $\qquad\qquad\qquad\qquad \dfrac{dV_\theta}{d\theta} = \dfrac{d^2 V_r}{d\theta^2}$

Substituting this result into Eq. (10.12), we have

$$
\boxed{
\begin{aligned}
&\frac{\gamma - 1}{2} \left[V_{max}^2 - V_r^2 - \left(\frac{dV_r}{d\theta} \right)^2 \right] \left[2V_r + \frac{dV_r}{d\theta} \cot\theta + \frac{d^2 V_r}{d\theta^2} \right] \\
&- \frac{dV_r}{d\theta} \left[V_r \frac{dV_r}{d\theta} + \frac{dV_r}{d\theta} \left(\frac{d^2 V_r}{d\theta^2} \right) \right] = 0
\end{aligned}
}
$$

$$(10.13)$$

Equation (10.13) is the *Taylor-Maccoll* equation for the solution of conical flows. Note that it is an ordinary differential equation, with only one dependent variable, V_r. Its solution gives $V_r = f(\theta)$; V_θ follows from Eq. (10.5), namely,

$$V_\theta = \frac{dV_r}{d\theta} \qquad\qquad (10.14)$$

There is no closed-form solution to Eq. (10.13); it must be solved *numerically*. To expedite the numerical solution, define the nondimensional velocity V' as

$$V' \equiv \frac{V}{V_{max}}$$

Then, Eq. (10.13) becomes

$$
\begin{aligned}
&\frac{\gamma - 1}{2} \left[1 - V_r'^2 - \left(\frac{dV_r'}{d\theta} \right)^2 \right] \left[2V_r' + \frac{dV_r'}{d\theta} \cot\theta + \frac{d^2 V_r'}{d\theta^2} \right] \\
&- \frac{dV_r'}{d\theta} \left[V_r' \frac{dV_r'}{d\theta} + \frac{dV_r'}{d\theta} \frac{d^2 V_r'}{d\theta^2} \right] = 0
\end{aligned}
$$

$$(10.15)$$

The nondimensional velocity V' is a function of Mach number only. To see this more clearly recall that

$$h + \frac{V^2}{2} = \frac{V_{max}^2}{2}$$

$$\frac{a^2}{\gamma - 1} + \frac{V^2}{2} = \frac{V_{max}^2}{2}$$

$$\frac{1}{\gamma - 1}\left(\frac{a}{V}\right)^2 + \frac{1}{2} = \frac{1}{2}\left(\frac{V_{max}}{V}\right)^2$$

$$\frac{2}{\gamma - 1}\left(\frac{1}{M}\right)^2 + 1 = \left(\frac{V_{max}}{V}\right)^2$$

$$\frac{V}{V_{max}} \equiv V' = \left[\frac{2}{(\gamma - 1)M^2} + 1\right]^{-1/2} \tag{10.16}$$

Clearly, from Eq. (10.16), $V' = f(M)$; given M, we can always fine V', or vice versa.

10.4 | NUMERICAL PROCEDURE

For the numerical solution of the supersonic flow over a right-circular cone, we will employ an *inverse* approach. By this, we mean that a given shock wave will be assumed, and the particular cone that supports the given shock will be calculated. This is in contrast to the *direct* approach, where the cone is given and the flowfield and shock wave are calculated. The numerical procedure is as follows:

1. Assume a shock wave angle θ_s and a free-stream Mach number M_∞, as sketched in Fig. 10.4. From this, the Mach number and flow deflection angle, M_2 and δ, respectively, immediately behind the shock can be found from the oblique shock relations (see the discussion of three-dimensional shocks in Sec. 4.13). Note that, contrary to our previous practice, the flow deflection angle is here denoted by δ so as not to confuse it with the polar coordinate θ.

2. From M_2 and δ, the radial and normal components of flow velocity, V_r' and V_θ', respectively, directly behind the shock can be found from the geometry of Fig. 10.4. Note that V' is obtained by inserting M_2 into Eq. (10.16).

3. Using the above value of V_r' directly behind the shock as a boundary value, solve Eq. (10.15) for V_r' numerically in steps of θ, marching away from the shock. Here, the flowfield is divided into incremental angles $\Delta\theta$, as sketched in Fig. 10.4. The ordinary differential equation (10.15) can be solved at each $\Delta\theta$ using any standard numerical solution technique, such as the Runge-Kutta method.

4. At each increment in θ, the value of V_θ' is calculated from Eq. (10.14). At some value of θ, namely $\theta = \theta_c$, we will find $V_\theta' = 0$. The normal component of velocity at an impermeable surface is zero. Hence, when $V_\theta' = 0$ at $\theta = \theta_c$, then θ_c must represent the surface of the particular cone which supports the shock

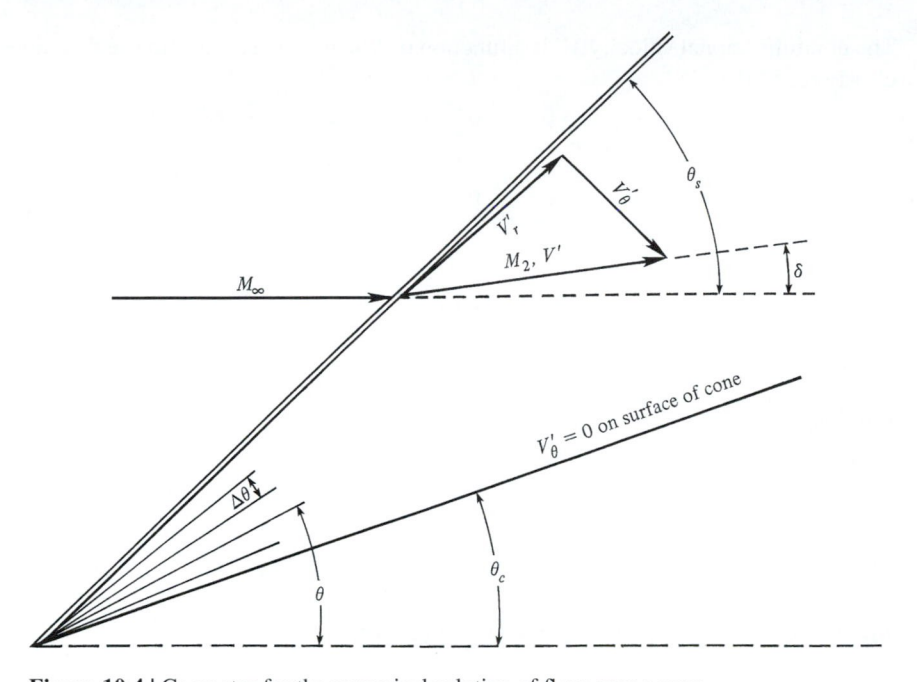

Figure 10.4 | Geometry for the numerical solution of flow over a cone.

wave of given wave angle θ_s at the given Mach number M_∞ as assumed in step 1 on the previous page. That is, the cone angle compatible with M_∞ and θ_s is θ_c. The value of V'_r at θ_c gives the Mach number along the cone surface via Eq. (10.16).

5.   In the process of steps 1 through 4 here, the complete velocity flowfield between the shock and the body has been obtained. Note that, at each point (or ray), $V' = \sqrt{(V'_r)^2 + (V'_\theta)^2}$ and M follows from Eq. (10.16). The pressure, density, and temperature along each ray can then be obtained from the isentropic relations, Eqs. (3.28), (3.30), and (3.31).

If a different value of M_∞ and/or θ_s is assumed in step 1, a different flowfield and cone angle θ_c will be obtained from steps 1 through 5. By a repeated series of these calculations, tables or graphs of supersonic cone properties can be generated. Such tables exist in the literature, the most common being those by Kopal (Ref. 28) and Sims (Ref. 29).

10.5 | PHYSICAL ASPECTS OF SUPERSONIC FLOW OVER CONES

Some typical numerical results obtained from the solution in Sec. 10.4 are illustrated in Fig. 10.5, which gives the shock wave angle θ_s as a function of cone angle θ_c, with M_∞ as a parameter. Figure 10.5 for cones is analogous to Fig. 4.8 for two-dimensional wedges; the two figures are qualitatively similar, but the numbers are different.

Examine Fig. 10.5 closely. Note that, for a given cone angle θ_c and given M_∞, there are two possible oblique shock waves—the strong- and weak-shock solutions.

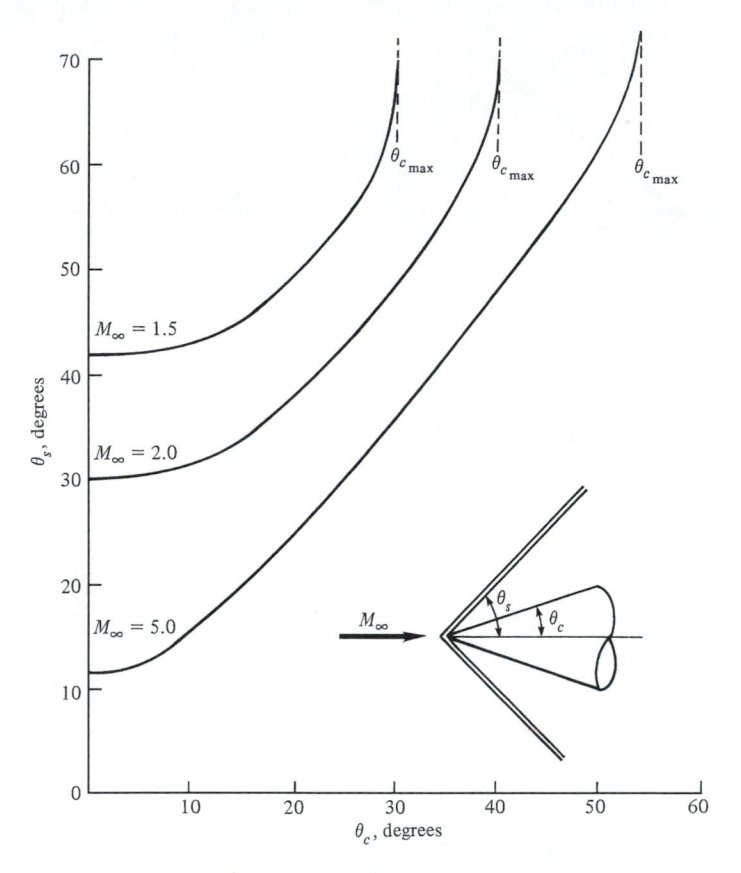

Figure 10.5 | θ_c-θ_s-M diagram for cones in supersonic flow. (The top portion of the curves curl back for the strong shock solution, which is not shown here.)

This is directly analogous to the two-dimensional case discussed in Chap. 4. The weak solution is almost always observed in practice on real finite cones; however, it is possible to force the strong-shock solution by independently increasing the back-pressure near the base of the cone.

Also note from Fig. 10.5 that, for a given M_∞, there is a maximum cone angle $\theta_{c_{max}}$, beyond which the shock becomes detached. This is illustrated in Fig. 10.6. When $\theta_c > \theta_{c_{max}}$, there exists no Taylor-Maccoll solution as given here; instead, the flowfield with a detached shock must be solved by techniques such as those discussed in Chap. 12.

In comparison to the two-dimensional flow over a wedge, the three-dimensional flow over a cone has an extra dimension in which to expand. This "three-dimensional relieving effect" was discussed in Sec. 4.4, which should now be reviewed by the reader. In particular, recall from Fig. 4.11 that the shock wave on a cone of given angle is weaker than the shock wave on a wedge of the same angle. It therefore follows that the cone experiences a lower surface pressure, temperature, density, and entropy than the wedge. It also follows that, for a given M_∞, the maximum allowable cone angle

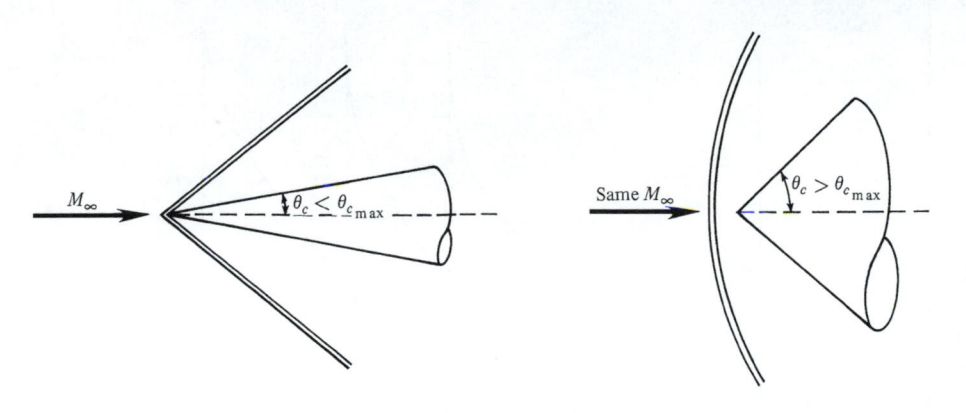

Figure 10.6 | Attached and detached shock waves on cones.

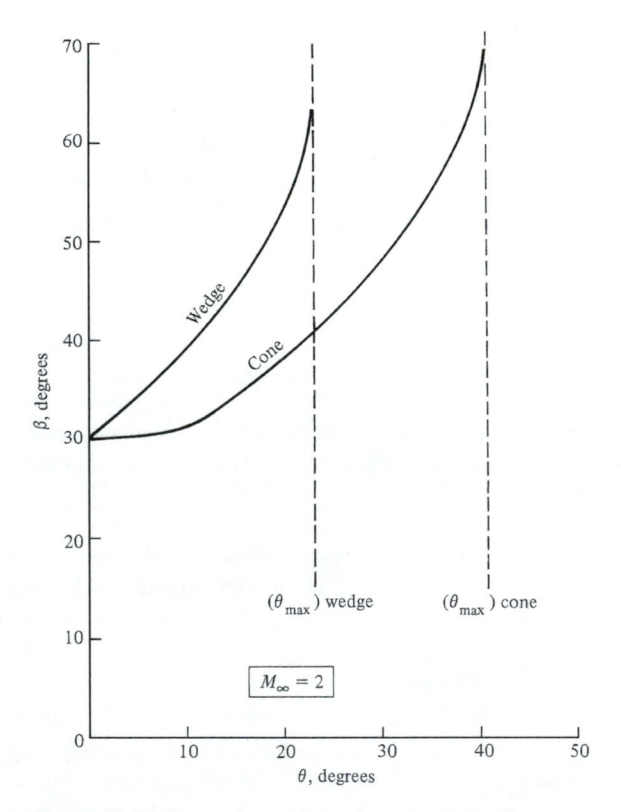

Figure 10.7 | Comparison of shock wave angles for wedges and cones at Mach 2.

for an attached shock solution is greater than the maximum wedge angle. This is clearly demonstrated in Fig. 10.7.

Finally, the numerical results show that any given streamline between the shock wave and cone surface is curved, as sketched in Fig. 10.8, and asymptotically becomes parallel to the cone surface at infinity. Also, for most cases, the complete

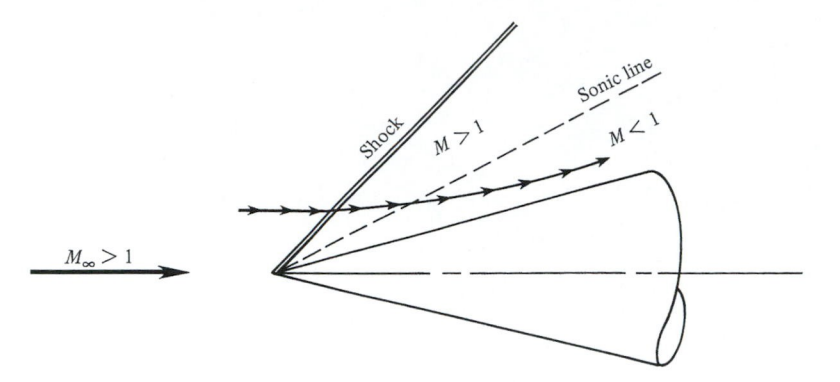

Figure 10.8 | Some conical flowfields are characterized by an isentropic compression to subsonic velocities near the cone surface.

flowfield between the shock and the cone is supersonic. However, if the cone angle is large enough, but still less than $\theta_{c_{max}}$, there are some cases where the flow becomes subsonic near the surface. This case is illustrated in Fig. 10.8, where one of the rays in the flowfield becomes a sonic line. In this case, we see one of the few instances in nature where a supersonic flowfield is actually *isentropically* compressed from supersonic to subsonic velocities. A transition from supersonic to subsonic flow is almost invariably accompanied by shock waves, as discussed in Chap. 5. However, flow over a cone is an exception to this observation.

PROBLEMS

(For these problems, use any of the existing tables and charts for conical flow.)

10.1 Consider a 15° half-angle cone at 0° angle of attack in a free stream at standard sea level conditions with $M_\infty = 2.0$. Obtain:

a. The shock wave angle

b. p, T, ρ, and M immediately behind the shock wave

c. p, T, ρ, and M on the cone surface

10.2 For the cone in Prob. 10.1, below what value of M_∞ will the shock wave be detached? Compare this with the analogous value for a wedge.

10.3 The drag coefficient for a cone can be defined as $C_D = D/q_\infty A_b$, where A_b is the area of the base of the cone. For a 15° half-angle cone, plot the variation of C_D with M_∞ over the range $1.5 \leq M_\infty \leq 7.0$. Assume the base pressure p_b is equal to free-stream pressure. (*Note:* You will not find C_D in the tables. Instead, derive a formula for C_D in terms of the surface pressure p_c, and use the tables to find p_c.)

CHAPTER 11

Numerical Techniques for Steady Supersonic Flow

It might be remarked that mathematics is undergoing a renaissance similar to that caused in physics by the discovery of the electron. This has been brought about by the advent of electronic computers of such fantastic speed and memory compared to their human counterparts that nonintegrable equations can be solved by numerical integration in a reasonably short space of time. This is having far-reaching effects in aerodynamics, where most problems are non-linear in nature, and exact analytical solutions are the exception rather than the rule.

William F. Hilton, 1951

PREVIEW BOX

Figure 11.1 shows the computed three-dimensional pressure coefficient contours over the surface of a generic modern fighter aircraft flying at Mach 0.85 at an angle of attack of 10° and an angle of yaw of 30°. (Pressure contours are the locus of constant pressure in the flow; along each contour shown in Fig. 11.1, the pressure is constant, but it is a different constant value from one contour to the next.) The results shown in Fig. 11.1 are dramatic. They were obtained by means of a numerical solution of the governing inviscid flow equations,

Eqs. (6.5), (6.11)–(6.13), and (6.17), carried out on a high-speed digital computer. The equations solved were the full nonlinear equations for inviscid flow without any mathematical simplification or physical approximations, solved for a highly three-dimensional flow field over a complex geometry, namely, a complete airplane. These results are an example of a relatively new discipline in fluid dynamics—*computational fluid dynamics*. In this chapter, we are introduced to the basic concept of computational fluid dynamics (CFD). But it is only an

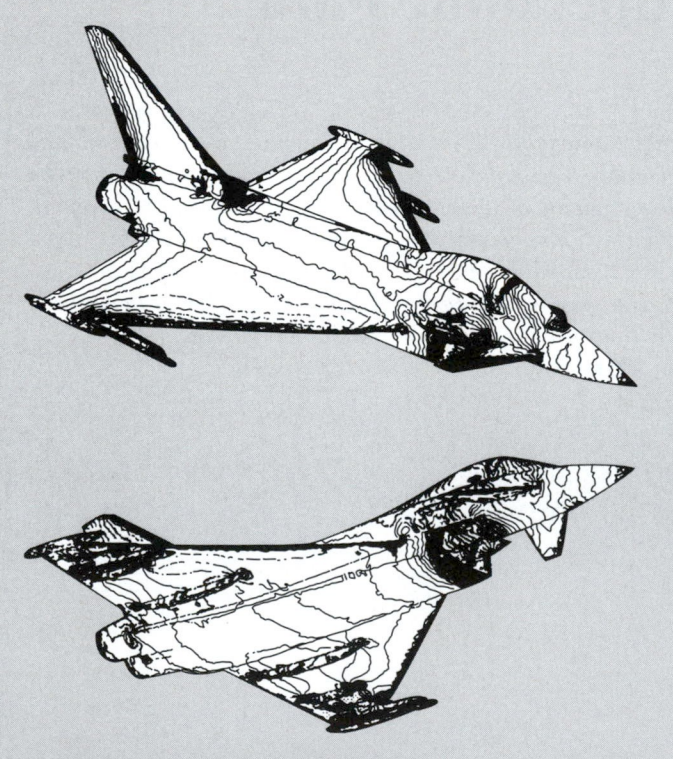

Figure 11.1 | Computed three-dimensional pressure coefficient contours over the surface of a generic fighter aircraft. Mach number = 0.85, angle of attack = 10 degrees, angle of yaw = 30 degrees. (From Selmin, Hettena and Formaggia, "An Unstructured Node Centered Scheme for the Simulation of 3-D Inviscid Flows," in C. Hirsch, J. Periaux and W. Kordulla (eds), *Computational Fluid Dynamics, 92,* vol. 2, Elsevier, Amsterdam, 1992, pp. 823–828.

introduction. Computational fluid dynamics is a sophisticated discipline, steeped in applied mathematics, and the subject of a number of recent books. It is a subject by itself, well beyond the scope of this book. Also, the main thrust of this book is to emphasize the *physical* aspects of compressible flow, and not to get deeply immersed in detailed mathematical or computational discourses. However, this is also a book on "modern" compressible flow, and in this modern world of fluid dynamics in general, it is virtually impossible not to be touched or affected by some aspects of CFD. Numerical solutions of the full governing continuity, momentum, and energy equations for complex geometries are now an integral part of fluid dynamics, albeit still not routine in any sense of the word. The intent of the present chapter is to give you some idea of the nature of CFD for compressible flow applications. The present chapter deals with steady, supersonic flow, but we will see examples of CFD applications in other compressible flow problems in subsequent chapters.

The roadmap for the present chapter is shown in Fig. 11.2. Here we deal with two, somewhat separate

numerical techniques for solving steady supersonic flows. Starting with the left-hand column, we discuss the method of characteristics, a rather classical and elegant numerical approach to the solution of such flows. The method of characteristics was applied to steady supersonic flow as early as the 1920s (see Sec. 11.17), but because of its intensive computational load, it did not really come into its own until the advent of the high-speed digital computer in the 1950s. As indicated in Fig. 11.2, the details of the application of the method of characteristics depend on the nature of the flow; they are somewhat different for two-dimensional irrotational flow, axisymmetric irrotational flow, rotational flow, and three-dimensional flow. We will examine each of these situations. Of great importance, and perhaps the most frequent use of the method of characteristics, is the design of the contour (the *shape*) of a supersonic nozzle for smooth, shock-free flow. We will discuss this nozzle design in some detail. Then we move to the right-hand column in Fig. 11.2 and present some of the basic ideas of the finite-difference technique. We will introduce the concept of downstream marching for

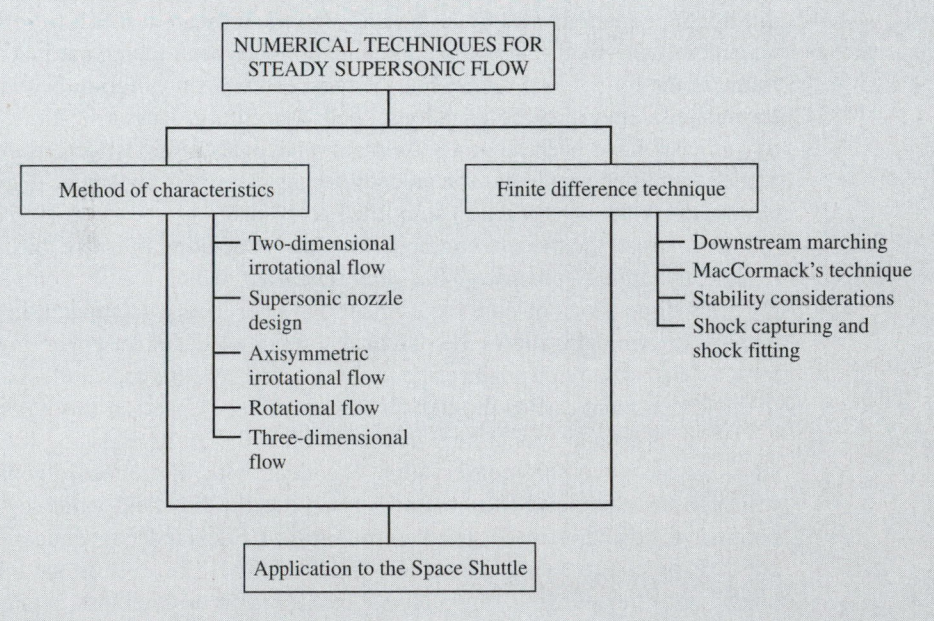

Figure 11.2 | Roadmap for Chapter 11.

(*continued on next page*)

(*continued from page 379*)

finite-difference solutions, and will describe one of the most straightforward (and student-friendly) finite-difference methods, namely MacCormack's technique. We will mention the idea of stability criteria—what is necessary to ensure stable numerical calculations. The two philosophies for dealing with shock waves in these numerical solutions—shock capturing and shock fitting—will be discussed. Finally, we end the chapter with an illustration of the application of both the method of characteristics and the finite-difference technique to the calculation of the flow over the Space Shuttle.

Note: For a very introductory presentation of computational fluid dynamics, see Anderson, *Computational Fluid Dynamics: The Basics with Applications,* McGraw-Hill, 1995. This book is written to be used before you study any of the other, more advanced books on CFD.

11.1 | AN INTRODUCTION TO COMPUTATIONAL FLUID DYNAMICS

As we have seen from the previous chapters, the cornerstone of theoretical fluid dynamics is a set of conservation equations that describe the physics of fluid motion; these equations speak words, such as: (1) mass is conserved; (2) $\mathbf{F} = m\mathbf{a}$ (Newton's second law); and (3) energy is conserved. These equations also describe the variations of fluid pressure, temperature, density, velocity, etc., throughout space and time. In their most general form, they are integral equations (see Chap. 2) or partial differential equations (see Chap. 6), and consequently are difficult to solve. Indeed, no general analytical solution to these equations has been found, nor is it likely to be found in the foreseeable future. For the two centuries since Bernoulli and Euler first formulated some of these equations in St. Petersburg, Russia, in the 1730s, fluid dynamicists have been laboring to obtain analytical solutions for certain restricted and/or simplified problems. The preceding chapters of this book have dealt primarily with such (relatively speaking) simplified problems.

In contrast, the modern engineer of today is operating in a new third dimension in fluid dynamics—*computational fluid dynamics,* which readily complements the previous dimensions of pure experiment and pure theory. Computational fluid dynamics, in principle, allows the practical solution of the exact governing equations for a myriad of applied engineering problems, and it is this aspect that is introduced in this chapter and carried through all the remaining chapters of this book.

What is computational fluid dynamics? It is the art of replacing the individual terms in the governing conservation equations with discretized algebraic forms, which in turn are solved to obtain *numbers* for the flowfield variables at discrete points in time and/or space. The end product of CFD is indeed a collection of numbers, in contrast to a closed-form analytical solution. However, in the long run, the objective of most engineering analyses, closed form or otherwise, is a quantitative description of the problem, i.e., *numbers*. If the governing conservation equations are given in integral form, the integral terms themselves are replaced with discrete algebraic expressions involving the flowfield variables at discrete grid points distributed throughout the flow. This is called the *finite-volume technique*. If the equations are given in partial differential equation form, the partial derivative terms are replaced with discrete algebraic difference quotients involving the flowfield variables at discrete grid points. This is called the *finite-difference technique*. In this book, our utilization of CFD will involve the finite-difference technique.

Perhaps the first major example of computational fluid dynamics applied to a practical engineering problem was the work of Kopal (Ref. 28), who in 1947 compiled massive tables of the supersonic flow over sharp cones by numerically solving the governing Taylor-Maccoll differential equation [see Chap. 10, and specifically Eqs. (10.13) and (10.15)]. The solutions were carried out on a primitive digital computer at the Massachusetts Institute of Technology. However, the first major generation of computational fluid-dynamic solutions appeared during the 1950s and early 1960s, spurred by the simultaneous advent of efficient, high-speed computers and the need to solve the high-velocity, high-temperature reentry body problem. High temperatures necessitated the inclusion of molecular vibrational energies and chemical reactions in flow problems, sometimes in equilibrium and at other times in nonequilibrium. As we shall see in Chaps. 16 and 17, such high-temperature physical phenomena generally cannot be solved analytically, even for the simplest flow geometry. Therefore, numerical solutions of the governing equations on a high-speed computer were an absolute necessity. Even though it was not fashionable at the time to describe such high-temperature gasdynamic calculations as "computational fluid dynamics," they nevertheless represented the first generation of the discipline.

The second generation of computational fluid-dynamic solutions, those that today are generally descriptive of the discipline, involve the application of the general equations of motion to applied fluid-dynamic problems that are in themselves so complicated (without the presence of chemical reactions, etc.) that a computer must be utilized. Examples of such inherently difficult problems are mixed subsonic-supersonic flows such as the supersonic blunt body problem (to be discussed in Chap. 12), and viscous flows which are not amenable to the boundary layer approximation, such as separated and recirculating flows. In the latter case, the full Navier–Stokes equations are required for an exact solution. Such viscous flows are outside the scope of this book; here we will deal with inviscid flows only.

Two major numerical techniques for the solution of completely supersonic, steady inviscid flows are introduced in this chapter—the method of characteristics and finite-difference methods. The method of characteristics is older and more developed, and is limited to inviscid flows, whereas finite-difference techniques (along with finite-volume techniques) are still evolving as computational fluid dynamics grows and matures, and have much more general application to inviscid and viscous flows. In this chapter, only some flavor and general guidance on finite-difference solutions can be given. Computational fluid dynamics is an extensive subject on its own, and its detailed study is beyond the scope of this book. Some early surveys of CFD can be found in Refs. 30 through 33. Some excellent modern textbooks on CFD at the graduate level are now available; see for examples Refs. 102 and 137 through 142. For a text written specifically for an elementary introduction to CFD, intended to be read before studying some of the more advanced texts, see Ref. 18. The reader is strongly encouraged to examine this literature in order to develop a more substantial understanding of CFD. In addition to the introduction given in the present chapter, all the remaining chapters of this book deal to a greater or lesser extent with computational techniques. However, in all cases our discussions will be self-contained; you are not expected to be familiar with the details of CFD. Indeed, the main thrust of this book is to emphasize the *physical* fundamentals of compressible flow, not to constitute a study of detailed mathematical or computational methods. But if this material wets your appetite to look further into CFD, you now know where to look.

Finally, the numerical techniques discussed in the remainder of this chapter have three aspects in common:

1. They involve the calculation of flowfield properties at discrete points in the flow. For example, consider an xy coordinate space that is divided into a rectangular grid, as sketched in Fig. 11.3. The solid circles denote *grid points* at which the flow properties are either known or to be calculated. The points are indexed by the letters i in the x direction and j in the y direction. For example, the point directly in the middle of the grid is denoted by (i, j), the point immediately to its right is $(i + 1, j)$, and so forth. It is not necessary to always deal with a rectangular grid as shown in Fig. 11.3, although such grids are preferable for finite-difference solutions. For the method of characteristics solutions, we will deal with a nonrectangular grid.

2. They are predicated on the ability to expand the flowfield properties in terms of a Taylor's series. For example, if $u_{i,j}$ denotes the x component of velocity known at point (i, j), then the velocity $u_{i+1,j}$ at point $(i + 1, j)$ can be obtained from

$$u_{i+1, j} = u_{i, j} + \left(\frac{\partial u}{\partial x}\right)_{i, j} \Delta x + \left(\frac{\partial^2 u}{\partial x^2}\right)_{i, j} \frac{(\Delta x)^2}{2} + \cdots \qquad (11.1)$$

Equation (11.1) will be useful in the subsequent sections.

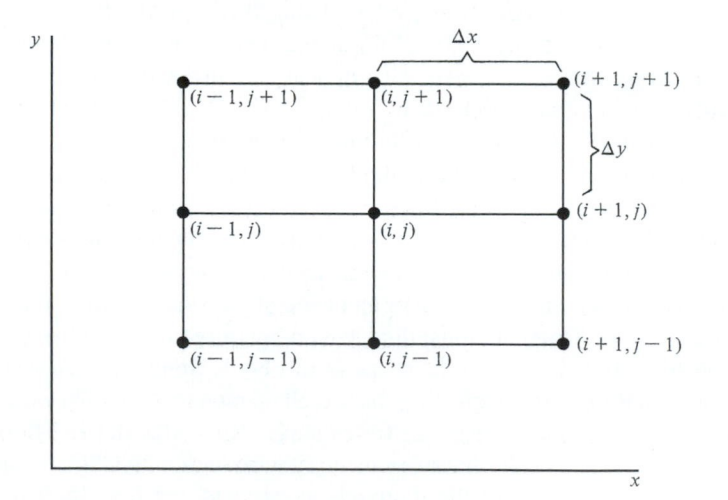

Figure 11.3 | Rectangular finite-difference grid.

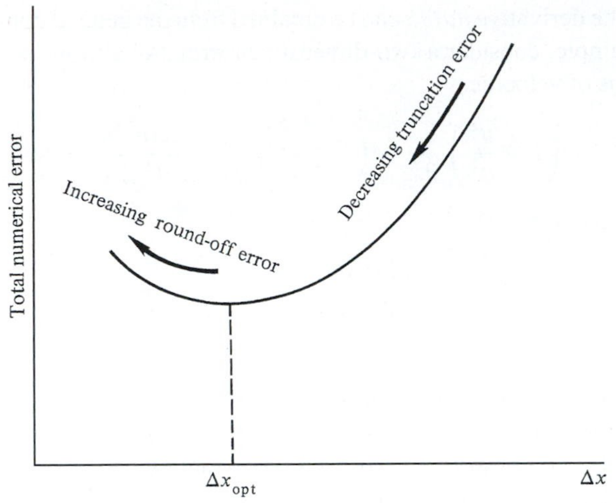

Figure 11.4 | Schematic of the effect of grid size on numerical error.

3. In the theoretical limit of an infinite number of grid points (i.e., Δx and $\Delta y \to 0$ in Fig. 11.3), the solutions are *exact*. Since all practical calculations obviously utilize a finite number of grid points, such numerical solutions are subject to *truncation error*, due to neglect of the higher-order terms in Eq. (11.1). Moreover, because all digital computers round off each number to a certain significant figure, the flowfield calculations are also subject to *round-off error*. By reducing the value of Δx in Eq. (11.1), the truncation error is reduced; however, the number of steps required to calculate a certain distance in x is correspondingly increased, therefore increasing the round-off error. This trend is illustrated in Fig. 11.4, which shows the total numerical error as a function of step size, Δx. Note that there is an optimum value $(\Delta x)_{\text{opt}}$ at which maximum accuracy is obtained; it does *not* correspond to $\Delta x \to 0$. Although all computations are subject to these numerical errors, this author feels that, as long as the full nonlinear equations of motion are being solved along with the exact boundary conditions, such solutions are properly designated as *exact solutions*. Therefore, an important advantage of computational fluid dynamics is its inherent ability to provide exact solutions to difficult, nonlinear problems.

11.2 | PHILOSOPHY OF THE METHOD OF CHARACTERISTICS

Let us begin to obtain a feeling for the method of characteristics by considering again Fig. 11.3 and Eq. (11.1). Neglect the second-order term in Eq. (11.1), and write

$$u_{i+1, j} = u_{i, j} + \left(\frac{\partial u}{\partial x} \right)_{i, j} \Delta x + \cdots \tag{11.2}$$

The value of the derivative $\partial u/\partial x$ can be obtained from the general conservation equations. For example, consider a two-dimensional irrotational flow, so that Eq. (8.17) yields, in terms of velocities,

$$\left(1 - \frac{u^2}{a^2}\right)\frac{\partial u}{\partial x} + \left(1 - \frac{v^2}{a^2}\right)\frac{\partial v}{\partial y} - \frac{2uv}{a^2}\frac{\partial u}{\partial y} = 0 \qquad (11.3)$$

Solve Eq. (11.3) for $\partial u/\partial x$:

$$\frac{\partial u}{\partial x} = \frac{\dfrac{2uv}{a^2}\dfrac{\partial u}{\partial y} - \left(1 - \dfrac{v^2}{a^2}\right)\dfrac{\partial v}{\partial y}}{(1 - u^2/a^2)} \qquad (11.4)$$

Now assume the velocity $\mathbf{V}$, and hence u and v, is known at each point along a vertical line, $x = x_o$, as sketched in Fig. 11.5. Specifically, the values of u and v are known at point (i, j), as well as above and below, at points $(i, j + 1)$ and $(i, j - 1)$. Hence, the y derivatives, $\partial u/\partial y$ and $\partial v/\partial y$, are known at point (i, j). (They can be calculated from finite-difference quotients, to be discussed later.) Consequently, the right-hand side of Eq. (11.4) yields a number for $(\partial u/\partial x)_{i, j}$, which can be substituted into Eq. (11.2) to calculate $u_{i+1, j}$. However, *there is one notable exception:* If the denominator of Eq. (11.4) is zero, then $\partial u/\partial x$ is at least indeterminate, and may even be discontinuous. The denominator is zero when $u = a$, i.e., when the component of flow velocity perpendicular to $x = x_o$ is sonic, as shown in Fig. 11.5. Moreover, from the geometry of Fig. 11.5, the angle μ is defined by $\sin \mu = u/V = a/V = 1/M$, i.e., μ is the *Mach angle*. The orientation of the x and y axes with respect to $\mathbf{V}$ in Fig. 11.5 is arbitrary; the germane aspect of this discussion is that a line that makes a Mach angle with respect to the streamline direction at a point is also a

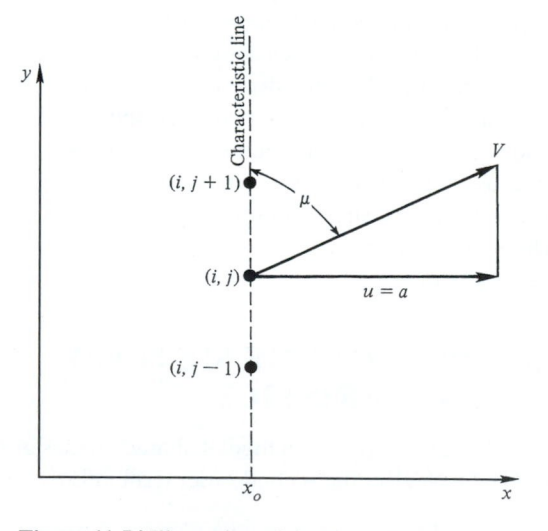

Figure 11.5 | Illustration of the characteristic direction.

line along which the derivative of u is indeterminate, and across which it may be discontinuous. We have just demonstrated that such lines exist, and that they are Mach lines. The choice of u was arbitrary in the above discussion. The derivatives of the other flow variables, p, ρ, T, v, etc., are also indeterminate along these lines. Such lines are defined as *characteristic lines*.

With this in mind, we can now outline the general philosophy of the method of characteristics. Consider a region of steady, supersonic flow in xy space. (For simplicity, we will initially deal with two-dimensional flow; extensions to three-dimensional flows will be discussed later.) This flowfield can be solved in three steps, as follows:

Step 1. Find some particular lines (directions) in the xy space where *flow variables* $(p, \rho, T, u, v$, etc.) are *continuous,* but along which the *derivatives* ($\partial p / \partial x, \partial u / \partial y$, etc.) are *indeterminate,* and in fact across which the derivatives may even sometimes be discontinuous. As already defined, such lines in the xy space are called *characteristic lines*.

Step 2. Combine the partial differential conservation equations in such a fashion that ordinary differential equations are obtained that hold *only along the characteristic lines.* Such ordinary differential equations are called the *compatibility equations.*

Step 3. Solve the compatibility equations step by step *along* the characteristic lines, starting from the given initial conditions at some point or region in the flow. In this manner, the complete flowfield can be mapped out along the characteristics. In general, the characteristic lines (sometimes referred to as the "characteristics net") depend on the flowfield, and the compatibility equations are a function of geometric location along the characteristic lines; hence, the characteristics and the compatibility equations must be constructed and solved simultaneously, step by step. An exception to this is two-dimensional irrotational flow, for which the compatibility equations become algebraic equations explicitly independent of geometric location. This will be made clear in subsequent sections.

As an analog to this discussion, the above philosophy is clearly exemplified in the unsteady, one-dimensional flow discussed in Chap. 7. Consider a centered expansion wave traveling to the left, as sketched in Fig. 11.6. In Chap. 7, the governing partial differential equations were reduced to ordinary differential equations (compatibility equations) which held only along certain lines in the xt plane that had slopes of $dx/dt = u \pm a$. The compatibility equations are Eqs. (7.65) and (7.66), and the lines were defined as characteristic lines in Sec. 7.6. These characteristics are sketched in Fig. 11.6a. However, in Chap. 7, we did not explicitly identify such characteristic lines with indeterminate or discontinuous derivatives. Nevertheless, this identification can be made by examining Eq. (7.89), which gives $u = u(x, t)$. Consider a given time $t = t_1$, which is illustrated by the dashed horizontal line in Fig. 11.6a. At time t_1, the head of the wave is located at x_b, and the tail at x_e. Equation (7.89) for the mass motion u is evaluated at time t_1, as sketched in Fig. 11.6b. Note that at x_b the velocity is continuous, but $\partial u / \partial x$ is discontinuous across the leading characteristic. Similarly, at x_e, u is continuous but $\partial u / \partial x$ is discontinuous across

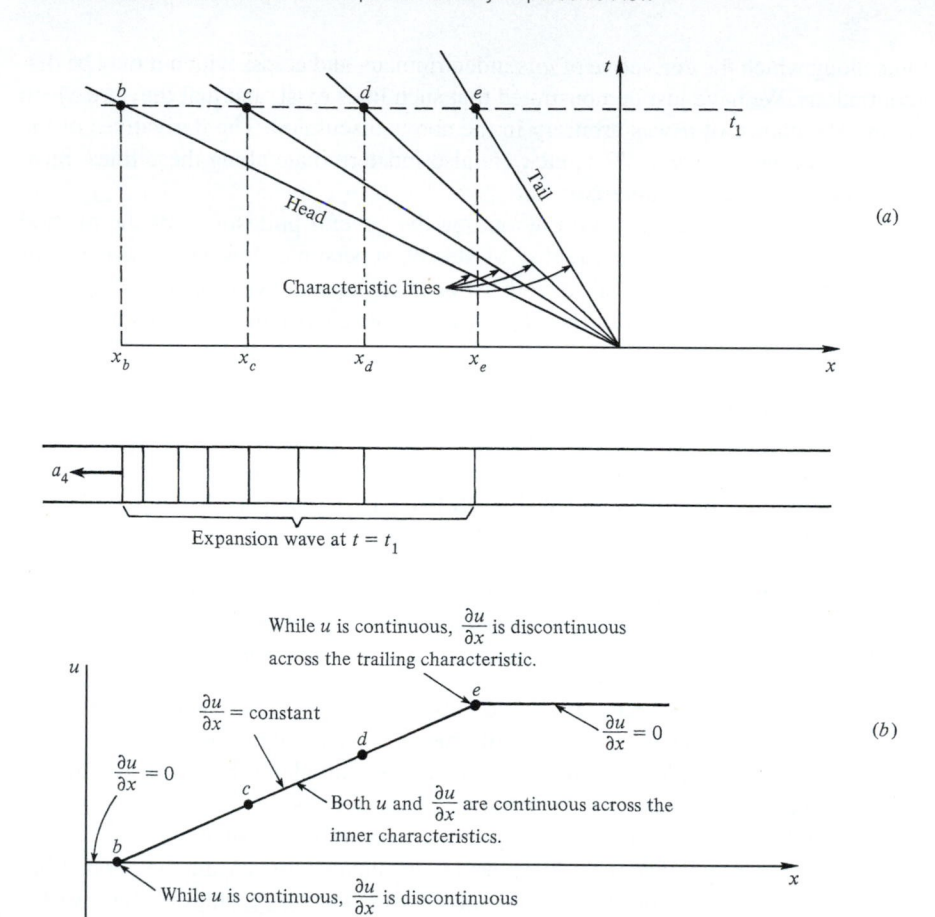

Figure 11.6 | Relationship of characteristics in unsteady one-dimensional flow.

the trailing characteristic. Hence, by examining Fig. 11.6a and b, we see that the characteristic lines identified in Chap. 7 are indeed consistent with the definition of characteristics given in the present chapter.

11.3 | DETERMINATION OF THE CHARACTERISTIC LINES: TWO-DIMENSIONAL IRROTATIONAL FLOW

At the beginning of Sec. 11.2, Mach lines in the flow were identified as characteristic lines in a somewhat heuristic fashion. Are there other characteristic lines in the flow? Is there a more deterministic approach to identifying characteristic lines? Those questions are addressed in this section.

To begin with, consider steady, adiabatic, two-dimensional, irrotational super-sonic flow. Other types of flow will be considered in subsequent sections. The governing nonlinear equations are Eqs. (8.17) and (8.18). For two-dimensional flow, Eq. (8.17) becomes

$$\left(1 - \frac{\Phi_x^2}{a^2}\right)\Phi_{xx} + \left(1 - \frac{\Phi_y^2}{a^2}\right)\Phi_{yy} - \frac{2\Phi_x\Phi_y}{a^2}\Phi_{xy} = 0 \qquad (11.5)$$

Note that Φ is the full-velocity potential, *not* the perturbation potential. In fact, in all of our work in this chapter, we are not using perturbations in any way. Hence,

$$\Phi_x = u \qquad \Phi_y = v \qquad \mathbf{V} = u\mathbf{i} + v\mathbf{j}$$

Recall that $\Phi_x = f(x, y)$; hence,

$$d\Phi_x = \frac{\partial \Phi_x}{\partial x} dx + \frac{\partial \Phi_x}{\partial y} dy = \Phi_{xx} dx + \Phi_{xy} dy \qquad (11.6)$$

$$d\Phi_y = \frac{\partial \Phi_y}{\partial x} dx + \frac{\partial \Phi_y}{\partial y} dy = \Phi_{xy} dx + \Phi_{yy} dy \qquad (11.7)$$

Recopying these equations,

From Eq. (11.5) $\qquad \left(1 - \dfrac{u^2}{a^2}\right)\Phi_{xx} - \dfrac{2uv}{a^2}\Phi_{xy} + \left(1 - \dfrac{v^2}{a^2}\right)\Phi_{yy} = 0$

From Eq. (11.6) $\qquad\qquad\qquad\qquad (dx)\Phi_{xx} + (dy)\Phi_{xy} = du$

From Eq. (11.7) $\qquad\qquad\qquad\qquad (dx)\Phi_{xy} + (dy)\Phi_{yy} = dv$

These equations can be treated as a system of simultaneous, linear, algebraic equations in the variables Φ_{xx}, Φ_{yy}, and Φ_{xy}. For example, using Cramer's rule, the solution for Φ_{xy} is

$$\Phi_{xy} = \frac{\begin{vmatrix} 1 - \dfrac{u^2}{a^2} & 0 & 1 - \dfrac{v^2}{a^2} \\ dx & du & 0 \\ 0 & dv & dy \end{vmatrix}}{\begin{vmatrix} 1 - \dfrac{u^2}{a^2} & -\dfrac{2uv}{a^2} & 1 - \dfrac{v^2}{a^2} \\ dx & dy & 0 \\ 0 & dx & dy \end{vmatrix}} = \frac{N}{D} \qquad (11.8)$$

Now consider point A and its surrounding neighborhood in an arbitrary flow-field, as sketched in Fig. 11.7. The derivative of the velocity potential, Φ_{xy}, has a specific value at point A. Equation (11.8) gives the solution for Φ_{xy} at point A for an arbitrary choice of dx and dy, i.e., for an arbitrary *direction* away from point A defined by the choice of dx and dy. For the chosen dx and dy, there are corresponding values of the change in velocity du and dv. No matter what values are chosen for dx and dy, the corresponding values of du and dv will always yield the same number for Φ_{xy} from Eq. (11.8), *with one exception*. If dx and dy are chosen such that $D = 0$

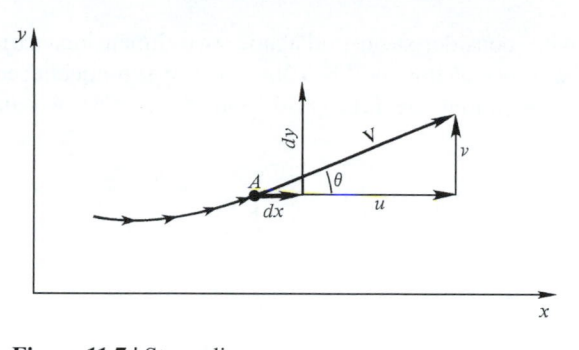

Figure 11.7 | Streamline geometry.

in Eq. (11.8), then Φ_{xy} is not defined in that particular direction dictated by dx and dy. However, we know that Φ_{xy} has a specific finite value at point A, even though it is not uniquely determined when the direction through point A is defined by this particular choice of dx and dy, which yields $D = 0$ in Eq. (11.8). Clearly, an infinite value of Φ_{xy} is physically inconsistent. For example, return to Fig. 11.6b. At points b and e, $\partial u/\partial x$ is not uniquely determined, but we have to say that its value should be somewhere between zero and the constant value given by the slope between points b and e. As a consequence, if the direction from A (dx and dy) is chosen so that $D = 0$ in Eq. (11.8), then to keep Φ_{xy} finite, $N = 0$ in Eq. (11.8) also:

$$\Phi_{xy} = \frac{N}{D} = \frac{0}{0}$$

That is, $\Phi_{xy} = \partial u/\partial y = \partial v/\partial x$ is *indeterminate*. We have previously defined the directions in the flowfield along which the derivatives of the flow properties are indeterminate and across which they may be discontinuous as characteristic directions. Therefore, the lines in xy space for which $D = 0$ (and hence $N = 0$) are characteristic lines.

This now provides a means to calculate the equations of the characteristic lines. In Eq. (11.8) set $D = 0$. This yields

$$\left(1 - \frac{u^2}{a^2}\right)(dy)^2 + \frac{2uv}{a^2}\,dx\,dy + \left(1 - \frac{v^2}{a^2}\right)(dx)^2 = 0$$

or $\qquad \left(1 - \frac{u^2}{a^2}\right)\left(\frac{dy}{dx}\right)^2_{\text{char}} + \frac{2uv}{a^2}\left(\frac{dy}{dx}\right)_{\text{char}} + \left(1 - \frac{v^2}{a^2}\right) = 0 \qquad$ (11.9)

In Eq. (11.9), $(dy/dx)_{\text{char}}$ is the slope of the characteristic lines. Using the quadratic formula, Eq. (11.9) yields

$$\left(\frac{dy}{dx}\right)_{\text{char}} = \frac{-2uv/a^2 \pm \sqrt{(2uv/a^2)^2 - 4[1 - (u^2/a^2)][1 - (v^2/a^2)]}}{2[1 - (u^2/a^2)]}$$

or $\qquad \boxed{\left(\frac{dy}{dx}\right)_{\text{char}} = \frac{-uv/a^2 \pm \sqrt{[(u^2 + v^2)/a^2] - 1}}{[1 - (u^2/a^2)]}} \qquad$ (11.10)

Equation (11.10) defines the characteristic curves in the physical xy space.

Examine Eq. (11.10) more closely. The term inside the square root is

$$\frac{u^2 + v^2}{a^2} - 1 = \frac{V^2}{a^2} - 1 = M^2 - 1$$

Hence, we can state

1. If $M > 1$, there are two real characteristics through each point of the flowfield. Moreover, for this situation, Eq. (11.5) is defined as a *hyperbolic* partial differential equation.
2. If $M = 1$, there is one real characteristic through each point of the flow. By definition, Eq. (11.5) is a *parabolic* partial differential equation.
3. If $M < 1$, the characteristics are imaginary, and Eq. (11.5) is an *elliptic* partial differential equation.

Therefore, we see that steady, inviscid supersonic flow is governed by hyperbolic equations, sonic flow by parabolic equations, and subsonic flow by elliptic equations. Moreover, because two real characteristics exist through each point in a flow where $M > 1$, the method of characteristics becomes a practical technique for solving supersonic flows. In contrast, because the characteristics are imaginary for $M < 1$, the method of characteristics is not used for subsonic solutions. (An exception is transonic flow, involving mixed subsonic-supersonic regions, where solutions have been obtained in the complex plane using imaginary characteristics.) Also, it is worthwhile mentioning that the unsteady one-dimensional flow in Chap. 7 is hyperbolic, and hence two real characteristics exist through each point in the xt plane, as we have already seen. Indeed, *unsteady* inviscid flow is hyperbolic for two and three spatial dimensions, and for any speed regime—subsonic, transonic, supersonic, or hypersonic. This feature of unsteady flow underlies the strength of the time-dependent numerical technique to be described in Chap. 12.

Concentrating on steady, two-dimensional supersonic flow, let us examine the real characteristic lines given by Eq. (11.10). Consider a streamline as sketched in Fig. 11.7. At point A, $u = V \cos \theta$ and $v = V \sin \theta$. Hence, Eq. (11.10) becomes

$$\left(\frac{dy}{dx}\right)_{char} = \frac{-\dfrac{V^2 \cos \theta \sin \theta}{a^2} \pm \sqrt{\dfrac{V^2}{a^2}(\cos^2 \theta + \sin^2 \theta) - 1}}{1 - \dfrac{V^2}{a^2}\cos^2 \theta} \tag{11.11}$$

Recall that the Mach angle μ is given by $\mu = \sin^{-1}(1/M)$, or $\sin \mu = 1/M$. Thus, $V^2/a^2 = M^2 = 1/\sin^2 \mu$, and Eq. (11.11) becomes

$$\left(\frac{dy}{dx}\right)_{char} = \frac{-\dfrac{\cos \theta \sin \theta}{\sin^2 \mu} \pm \sqrt{\dfrac{\cos^2 \theta + \sin^2 \theta}{\sin^2 \mu} - 1}}{1 - \dfrac{\cos^2 \theta}{\sin^2 \mu}} \tag{11.12}$$

From trigonometry,

$$\sqrt{\frac{\cos^2\theta + \sin^2\theta}{\sin^2\mu} - 1} = \sqrt{\frac{1}{\sin^2\mu} - 1} = \sqrt{\csc^2\mu - 1} = \sqrt{\cot^2\mu} = \frac{1}{\tan\mu}$$

Thus, Eq. (11.12) becomes

$$\left(\frac{dy}{dx}\right)_{\text{char}} = \frac{-\cos\theta\sin\theta/\sin^2\mu \pm 1/\tan\mu}{1 - (\cos^2\theta/\sin^2\mu)} \tag{11.13}$$

After more algebraic and trigonometric manipulation, Eq. (11.13) reduces to

$$\boxed{\left(\frac{dy}{dx}\right)_{\text{char}} = \tan(\theta \mp \mu)} \tag{11.14}$$

A graphical interpretation of Eq. (11.14) is given in Fig. 11.8, which is an elaboration of Fig. 11.7. At point A in Fig. 11.8, the streamline makes an angle θ with the x axis. Equation (11.14) stipulates that there are two characteristics passing through point A, one at the angle μ above the streamline, and the other at the angle μ below the streamline. Hence, *the characteristic lines are Mach lines.* This fact was deduced in Sec. 11.2; however, the derivation given here is more rigorous. Also, the characteristic given by the angle $\theta + \mu$ is called a C_+ characteristic; it is a left-running

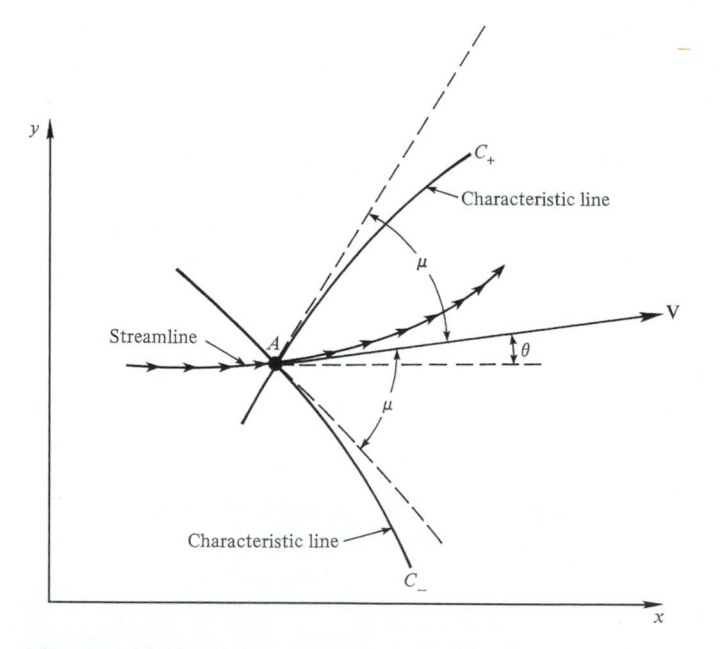

Figure 11.8 | Illustration of left- and right-running characteristic lines.

characteristic analogous to the C_+ characteristics used in Chap. 7. The characteristic in Fig. 11.8 given by the angle $\theta - \mu$ is called a C_- characteristic; it is a right-running characteristic analogous to the C_- characteristics used in Chap. 7. Note that the characteristics are curved in general, because the flow properties (hence θ and μ) change from point to point in the flow.

11.4 | DETERMINATION OF THE COMPATIBILITY EQUATIONS

In essence, Eq. (11.8) represents a combination of the continuity, momentum, and energy equations for two-dimensional, steady, adiabatic, irrotational flow. In Sec. 11.3, we derived the characteristic lines by setting $D = 0$ in Eq. (11.8). In this section, we will derive the compatibility equations by setting $N = 0$ in Eq. (11.8).

When $N = 0$, the numerator determinant yields

$$\left(1 - \frac{u^2}{a^2}\right) du\, dy + \left(1 - \frac{v^2}{a^2}\right) dx\, dv = 0$$

or

$$\frac{dv}{du} = \frac{-[1 - (u^2/a^2)]}{[1 - (v^2/a^2)]} \frac{dy}{dx} \tag{11.15}$$

Keep in mind that N is set to zero only when $D = 0$ in order to keep the flowfield derivatives finite, albeit of the indeterminate form $0/0$. When $D = 0$, we are restricted to considering directions only along the characteristic lines, as explained in Sec. 11.3. Hence, when $N = 0$, we are held to the same restriction. Therefore, *Eq. (11.15) holds only along the characteristic lines*. Therefore, in Eq. (11.15),

$$\frac{dy}{dx} \equiv \left(\frac{dy}{dx}\right)_{\text{char}}$$

Substituting Eq. (11.10) into (11.15), we have

$$\frac{dv}{du} = -\frac{\left(1 - \dfrac{u^2}{a^2}\right)}{\left(1 - \dfrac{v^2}{a^2}\right)} \left[\frac{-\dfrac{uv}{a^2} \pm \sqrt{\dfrac{u^2 + v^2}{a^2} - 1}}{\left(1 - \dfrac{u^2}{a^2}\right)}\right]$$

which simplifies to

$$\frac{dv}{du} = \frac{\dfrac{uv}{a^2} \mp \sqrt{\dfrac{u^2 + v^2}{a^2} - 1}}{1 - \dfrac{v^2}{a^2}} \tag{11.16}$$

Recall that $u = V \cos\theta$ and $v = V \sin\theta$. Then, Eq. (11.16) becomes

$$\frac{d(V \sin\theta)}{d(V \cos\theta)} = \frac{M^2 \cos\theta \sin\theta \mp \sqrt{M^2 - 1}}{1 - M^2 \sin^2\theta}$$

which, after some algebraic manipulations, reduces to

$$d\theta = \mp\sqrt{M^2 - 1}\frac{dV}{V} \qquad (11.17)$$

Equation (11.17) is the *compatibility equation,* i.e., the equation that describes the variation of flow properties *along* the characteristic lines. From a comparison with Eq. (11.14), we note that

$$d\theta = -\sqrt{M^2 - 1}\frac{dV}{V} \qquad \text{(applies along the } C_- \text{ characteristic)} \qquad (11.18)$$

$$d\theta = \sqrt{M^2 - 1}\frac{dV}{V} \qquad \text{(applies along the } C_+ \text{ characteristic)} \qquad (11.19)$$

Compare Eq. (11.17) with Eq. (4.35) for Prandtl–Meyer flow. They are identical. Hence, Eq. (11.17) can be integrated to give the Prandtl–Meyer function $\nu(M)$ as displayed in Eq. (4.44). Therefore, Eqs. (11.18) and (11.19) are replaced by the *algebraic compatibility equations:*

$$\theta + \nu(M) = \text{const} = K_- \qquad \text{(along the } C_- \text{ characteristic)} \qquad (11.20)$$

$$\theta - \nu(M) = \text{const} = K_+ \qquad \text{(along the } C_+ \text{ characteristic)} \qquad (11.21)$$

In Equations (11.20) and (11.21), K_- and K_+ are constants along their respective characteristics, and are analogous to the Riemann invariants J_- and J_+ for unsteady flow as defined in Chap. 7.

The compatibility equations (11.20) and (11.21) relate velocity magnitude and direction along the characteristic lines. For this reason, they are sometimes identified in the literature as "hodograph characteristics." Plots of the hodograph characteristics are useful for graphical solutions or hand calculations using the method of characteristics. The reader is encouraged to read the classic texts by Ferri (Ref. 5) and Shapiro (Ref. 16) for further discussions of the hodograph approach. We shall not take a graphical approach here. Rather, Eqs. (11.20) and (11.21) are in a sufficient form for direct numerical calculations; they are the most useful form for modern computer calculations.

It is important to note that the compatibility equations (11.20) and (11.21) have no terms involving the spatial coordinates x and y. Hence, they can be solved without requiring knowledge of the geometric location of the characteristic lines. This geometrical independence of the compatibility equations is peculiar only to the present case of two-dimensional irrotational flow. For all other cases, the compatibility equations are dependent upon the spatial location, as will be discussed later.

11.5 | UNIT PROCESSES

In Sec. 11.2, the philosophy of the method of characteristics was given as a three-step process. Step 1—the determination of the characteristic lines—was carried out in Sec. 11.3. Step 2—the determination of the compatibility equations which hold along the characteristics—was carried out in Sec. 11.4. Step 3—the solution of the compatibility equations point by point along the characteristics—is discussed in this section. The machinery for *applying* the method of characteristics is a series of specific computations called "unit processes," which vary depending on whether the points at which calculations are being made are internal to the flowfield, on a solid or free boundary, or on a shock wave.

11.5.1   Internal Flow

If we know the flowfield conditions at two points in the flow, then we can find the conditions at a third point, as sketched in Fig. 11.9. Here, the values of v_1 and θ_1 are known at point 1, and v_2 and θ_2 are known at point 2. Point 3 is located by the intersection of the C_- characteristic through point 1 and the C_+ characteristic through point 2. Along the C_- characteristic through point 1, Eq. (11.20) holds:

$$\theta_1 + v_1 = (K_-)_1 \quad \text{(known value along } C_-)$$

Also along the C_+ characteristic through point 2, Eq. (11.21) holds:

$$\theta_2 - v_2 = (K_+)_2 \quad \text{(known value along } C_+)$$

Hence, at point 3, from Eq. (11.20),

$$\theta_3 + v_3 = (K_-)_3 = (K_-)_1 \qquad (11.22)$$

and from Eq. (11.21),

$$\theta_3 - v_3 = (K_+)_3 = (K_+)_2 \qquad (11.23)$$

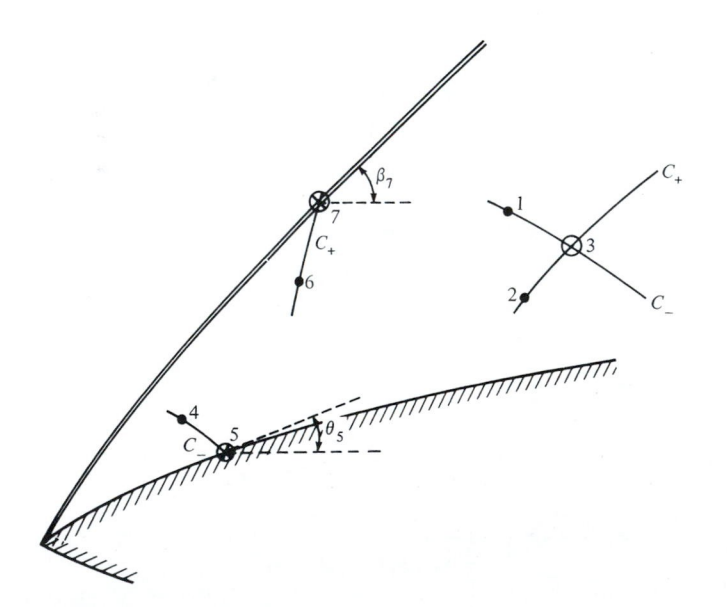

Figure 11.9 | Unit processes for the steady-flow, two-dimensional, irrotational method of characteristics.

Solving Eqs. (11.22) and (11.23), we obtain θ_3 and ν_3 in terms of the known values of K_+ and K_-:

$$\theta_3 = \tfrac{1}{2}[(K_-)_1 + (K_+)_2] \tag{11.24}$$

$$\nu_3 = \tfrac{1}{2}[(K_-)_1 - (K_+)_2] \tag{11.25}$$

Thus, the flow conditions at point 3 are now determined from the known values at points 1 and 2. Recall that ν_3 determines M_3 through Eq. (4.44), and that M_3 determines the pressure, temperature, and density through the isentropic flow relations, Eqs. (3.28), (3.30), and (3.31).

The *location* of point 3 in space is determined by the intersection of the C_- characteristic through point 1 and the C_+ characteristic through point 2, as shown in Fig. 11.9. However, the C_- and C_+ characteristics are generally curved lines, and all we know are their directions at points 1 and 2. How can we then locate point 3? An approximate but usually sufficiently accurate procedure is to assume the characteristics are straight-line segments between the grid points, with slopes that are average values. For example, consider Fig. 11.10. Here, the C_- characteristic through point 1 is drawn as a straight line with an average slope angle given by

$$\left[\tfrac{1}{2}(\theta_1 + \theta_3) - \tfrac{1}{2}(\mu_1 + \mu_3)\right]$$

The C_+ characteristic through point 2 is drawn as a straight line with an average slope angle given by $[\tfrac{1}{2}(\theta_2 + \theta_3) + \tfrac{1}{2}(\mu_2 + \mu_3)]$. Their intersection locates point 3.

11.5.2 Wall Point

If we know conditions at a point in the flow near a solid wall, we can find the flow variables *at* the wall as follows. Consider point 4 in Fig. 11.9, at which the flow is known. Hence, along the C_- characteristic through point 4, the value K_- is known:

$$(K_-)_4 = \theta_4 + \nu_4 \quad \text{(known)}$$

The C_- characteristic intersects the wall at point 5. Hence, at point 5,

$$(K_-)_4 = (K_-)_5 = \theta_5 + \nu_5 \tag{11.26}$$

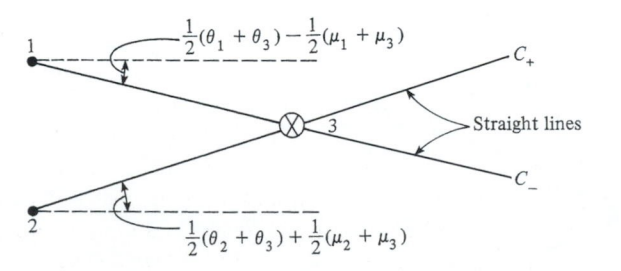

Figure 11.10 | Approximation of characteristics by straight lines.

However, the shape of the wall is known, and since the flow must be tangent at the wall, θ_5 is known. Thus, in Eq. (11.26), v_5 is the only unknown, and can be written as

$$v_5 = v_4 + \theta_4 - \theta_5$$

11.5.3   Shock Point

If we know conditions at a point in the flow near a shock wave, we can find the flow variables immediately behind the shock as well as the local shock angle as follows. Consider point 6 in Fig. 11.9, at which the flow is known. Hence, along the C_+ characteristic through point 6, the value K_+ is known:

$$(K_+)_6 = \theta_6 - v_6 \quad \text{(known)}$$

The C_+ characteristic intersects the shock at point 7. Hence, at point 7,

$$(K_+)_6 = (K_+)_7 = \theta_7 - v_7 \tag{11.27}$$

For a given free-stream Mach number M_∞, find the value of the local shock angle β_7 which yields the value of $\theta_7 - v_7$ immediately behind the shock that agrees with the number obtained in Eq. (11.27). This is a trial-and-error process using the oblique shock relations developed in Chap. 4. Then, given β_7 and M_∞, all other flow properties at point 7 are known from the oblique shock relations.

11.5.4   Initial Data Line

The unit processes discussed in this section must *start* somewhere. In order to implement the method of characteristics, we must have a line in the locally supersonic flow along which the flowfield properties are known. Then the method of characteristics can be carried out as described here, marching downstream from the initial data line. Such a downstream-marching method is mathematically a property of hyperbolic and parabolic partial differential equations. For the calculation of an internal flow, such as a nozzle flow, the initial data line is taken at or downstream of the limiting characteristic, which is slightly downstream of the sonic line. (The concept of limiting characteristics is described in Sec. 12.3.) The properties along this initial data line must be obtained from an independent calculation, such as the time-marching method discussed in Chap. 12. An alternative for starting a nozzle calculation is simply to assume that the sonic line in the nozzle throat is straight, and to assume a centered expansion emanating from the wall of the nozzle in the throat region (see Example 11.1 in Sec. 11.7). For the calculation of an external flow, such as the flow over a sharp-nosed airfoil shape, the initial data line can be established by assuming wedge flow at the sharp leading edge, and using wedge-flow properties along a line across the flow between the body and the shock wave just a small distance downstream of the leading edge. In any event, we repeat that the method of characteristics solution for a steady supersonic flow must start from a given initial data line, and then the calculation can be marched downstream from the line.

11.6 | REGIONS OF INFLUENCE AND DOMAINS OF DEPENDENCE

Our discussion on characteristic lines leads to the conclusion that in a steady supersonic flow disturbances are felt only in limited regions. This is in contrast to a subsonic flow where disturbances are felt everywhere throughout the flowfield. (This distinction was clearly made in the contrast between subsonic and supersonic linearized flow discussed in Chap. 9.) To better understand the propagation of disturbances in a steady supersonic flow, consider point A in a uniform supersonic stream, as sketched in Fig. 11.11a. Assume that two needlelike probes are introduced upstream of point A. The probes are so thin that their shock waves are essentially Mach waves. In the sketch shown, the tips of the probes at points B and C are located such that point A is outside the Mach waves. Hence, even though the probes are upstream of point A, their presence is not felt at point A. The disturbances introduced by the probes are confined within the Mach waves. On the other hand, if another probe is introduced at point D upstream of point A such that point A falls inside the Mach wave (see Fig. 11.11b), then obviously the presence of the probe is felt at point A.

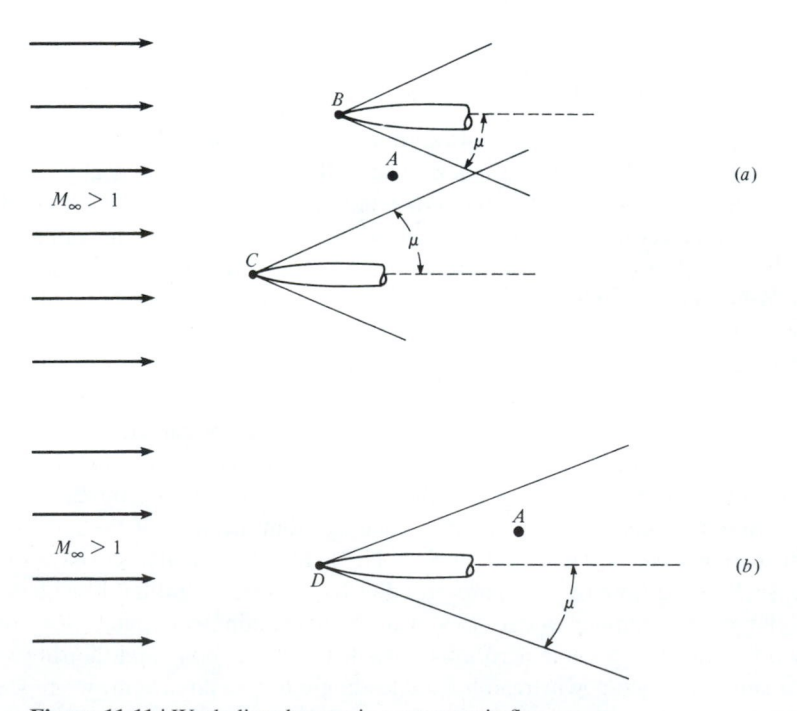

Figure 11.11 | Weak disturbances in a supersonic flow.

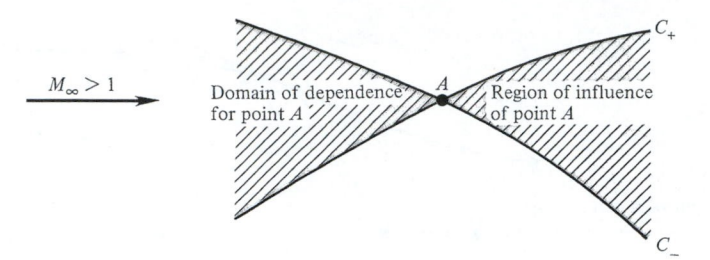

Figure 11.12 | Domain of dependence and region of influence.

The above simple picture leads to the definition of two zones associated with point A, as illustrated in Fig. 11.12. Consider the left- and right-running characteristics through point A. The area between the two upstream characteristics is defined as the *domain of dependence* for point A. Properties at point A "depend" on any disturbances or information in the flow within this upstream region. The area between the two downstream characteristics is defined as the *region of influence* of point A. This region is "influenced" by any action that is going on at point A. Clearly, disturbances that are generated at point A do *not* propagate upstream. This is a general and important behavior of steady supersonic flow—*disturbances do not propagate upstream*. (However, keep in mind from Chap. 7 that, in an *unsteady* supersonic flow, compression waves can propagate upstream.)

11.7 | SUPERSONIC NOZZLE DESIGN

In order to expand an internal steady flow through a duct from subsonic to supersonic speed, we established in Chap. 5 that the duct has to be convergent-divergent in shape, as sketched in Fig. 11.13a. Moreover, we developed relations for the local Mach number, and hence the pressure, density, and temperature, as functions of local area ratio A/A^*. However, these relations assumed quasi-one-dimensional flow, whereas, strictly speaking, the flow in Fig. 11.13a is two-dimensional. Moreover, the quasi-one-dimensional theory tells us nothing about the proper *contour* of the duct, i.e., what is the proper variation of area with respect to the flow direction $A = A(x)$. If the nozzle contour is not proper, shock waves may occur inside the duct.

The method of characteristics provides a technique for properly designing the contour of a supersonic nozzle for shockfree, isentropic flow, taking into account the multidimensional flow inside the duct. The purpose of this section is to illustrate such an application.

The subsonic flow in the convergent portion of the duct in Fig. 11.13a is accelerated to sonic speed in the throat region. In general, because of the multidimensionality of the converging subsonic flow, the sonic line is gently curved. However, for most applications, we can assume the sonic line to be straight, as illustrated by the straight dashed line from a to b in Fig. 11.13a. Downstream of the sonic line, the duct diverges. Let θ_w represent the angle of the duct wall with respect to the x direction.

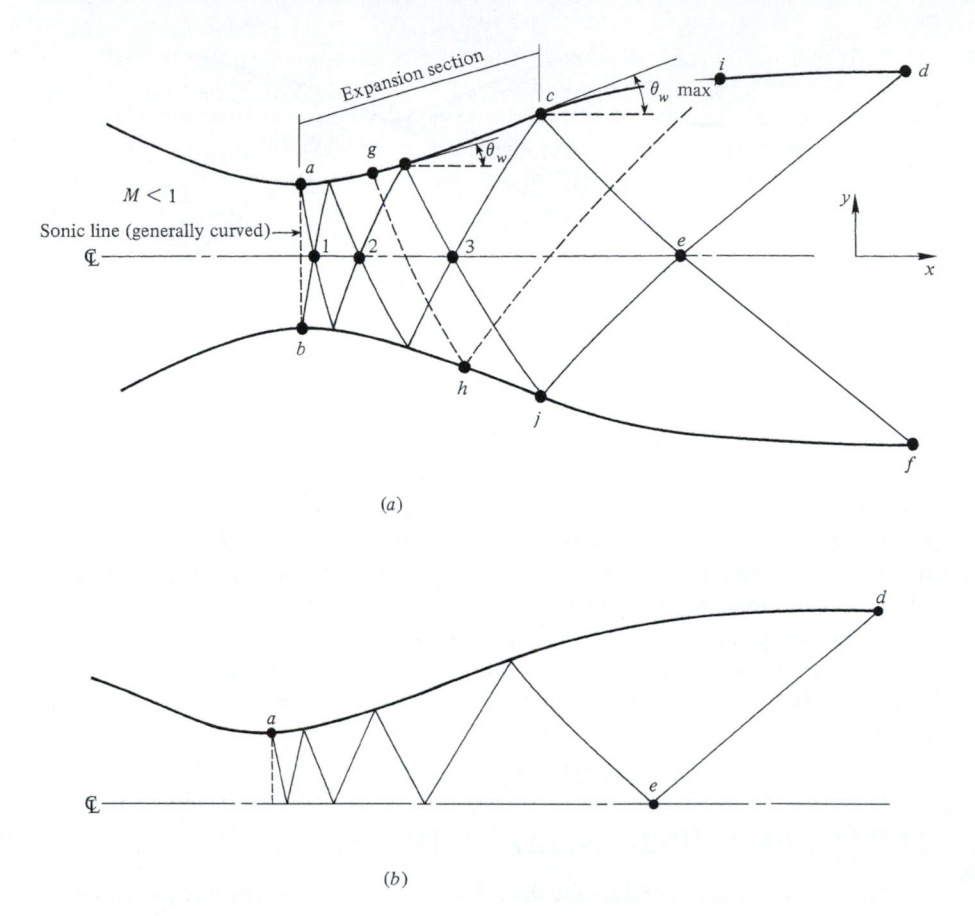

Figure 11.13 | Schematic of supersonic nozzle design by the method of characteristics.

The section of the nozzle where θ_w is increasing is called the *expansion* section; here, expansion waves are generated and propagate across the flow downstream, reflecting from the opposite wall. Point c is an inflection point of the contour, where $\theta_w = \theta_{w_{\max}}$. Downstream of point c, θ_w decreases until the wall becomes parallel to the x direction at points d and f. The section from c to d is a "straightening" section specifically designed to cancel all the expansion waves generated by the expansion section. For example, as shown by the dashed line in Fig. 11.13a, the expansion wave generated at g and reflected at h is canceled at i. Also shown in Fig. 11.13a are the characteristic lines going through points d and f at the nozzle exit. These characteristics represent infinitesimal expansion waves in the nozzle, i.e., Mach waves. Tracing these two characteristics upstream, we observe multiple reflections up to the throat region. The area $acejb$ is the expansion region of the nozzle, covered with both left- and right-running characteristics. Such a region with waves of both families is defined as a *nonsimple region* (analogous to the nonsimple waves described for unsteady one-dimensional flow in Sec. 7.7). In this region, the characteristics are curved lines. In contrast, the regions *cde* and *jef* are covered by waves of only one family because the other family is cancelled at the wall. Hence, these are *simple regions*, where the characteristic lines are straight. Downstream of *def,* the flow is uniform and parallel, at the desired Mach number. Finally, due to the symmetry of the nozzle flow, the waves (characteristics) generated from the top wall act as if they are "reflected" from the centerline. This geometric ploy due to symmetry allows us to consider in our calculations only the flow above the centerline, as sketched in Fig. 11.13*b*.

Supersonic nozzles with gently curved expansion sections as sketched in Fig. 11.13*a* and *b* are characteristic of wind tunnel nozzles where high-quality, uniform flow is desired in the test section (downstream of *def*). Hence, wind tunnel nozzles are long, with a relatively slow expansion. By comparison, rocket nozzles are short in order to minimize weight. Also, in cases where rapid expansions are desirable, such as the nonequilibrium flow in modern gasdynamic lasers (see Ref. 21), the nozzle length is as short as possible. In such *minimum-length nozzles,* the expansion section in Fig. 11.13*a* is shrunk to a point, and the expansion takes place through a centered Prandtl–Meyer wave emanating from a sharp-corner throat with an angle $\theta_{w_{\max}, M_L}$, as sketched in Fig. 11.14*a*. The length of the supersonic nozzle, denoted as L in Fig. 11.14*a* is the minimum value consistent with shockfree, isentropic flow. If the contour is made shorter than L, shocks will develop inside the nozzle.

Assume that the nozzles sketched in Figs. 11.13*a* and 11.14*a* are designed for the same exit Mach numbers. For the nozzle in Fig. 11.13*a* with an arbitrary expansion contour *ac*, multiple reflections of the characteristics (expansion waves) occur from the wall along *ac*. A fluid element moving along a streamline is constantly accelerated while passing through these multiple reflected waves. In contrast, for the minimum-length nozzle shown in Fig. 11.14*a*, the expansion contour is replaced by a sharp corner at point *a*. There are no multiple reflections and a fluid element encounters only two systems of waves—the right-running waves emanating from point *a* and the left-running waves emanating from point *d*. As a result, $\theta_{w_{\max}, M_L}$ in Fig. 11.14*a* must be larger than $\theta_{w_{\max}}$ in Fig. 11.13*a*, although the exit Mach numbers are the same.

Let ν_M be the Prandtl–Meyer function associated with the design exit Mach number. Hence, along the C_+ characteristic *cb* in Fig. 11.14*a*, $\nu = \nu_M = \nu_c = \nu_b$. Now consider the C_- characteristic through points *a* and *c*. At point *c*, from Eq. (11.20),

$$\theta_c + \nu_c = (K_-)_c \tag{11.28}$$

However, $\theta_c = 0$ and $\nu_c = \nu_M$. Hence, from Eq. (11.28),

$$(K_-)_c = \nu_M \tag{11.29}$$

At point *a*, along the same C_- characteristic *ac*, from Eq. (11.20),

$$\theta_{w_{\max}, M_L} + \nu_a = (K_-)_a \tag{11.30}$$

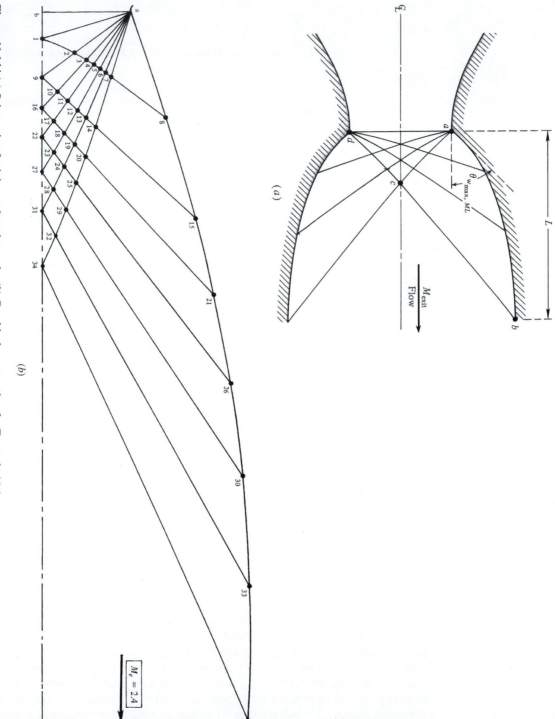

Figure 11.14 | (*a*) Schematic of minimum-length nozzle. (*b*) Graphical construction for Example 11.1.

400

Since the expansion at point a is a Prandtl–Meyer expansion from initially sonic conditions, we know from Sec. 4.14 that $\nu_a = \theta_{w_{max}, M_L}$. Hence, Eq. (11.30) becomes

$$\theta_{w_{max}, M_L} = \tfrac{1}{2}(K_-)_a \tag{11.31}$$

However, along the same C_- characteristic, $(K_-)_a = (K_-)_c$; hence, Eq. (11.31) becomes

$$\theta_{w_{max}, M_L} = \tfrac{1}{2}(K_-)_c \tag{11.32}$$

Combining Eqs. (11.29) and (11.32), we have

$$\boxed{\theta_{w_{max}, M_L} = \frac{\nu_M}{2}} \tag{11.33}$$

Equation (11.33) demonstrates that, *for a minimum-length nozzle the expansion angle of the wall downstream of the throat is equal to one-half the Prandtl–Meyer function for the design exit Mach number.* For other nozzles such as that sketched in Fig. 11.11a, the maximum expansion angle is *less* than $\nu_M/2$.

The shape of the finite-length expansion section in Fig. 11.13a can be somewhat arbitrary (within reason). It is frequently taken to be a circular arc with a diameter larger than the nozzle throat height. However, once the shape of the expansion section is chosen, then its length and $\theta_{w_{max}}$ are determined by the design exit Mach number. These properties can be easily found by noting that the characteristic line from the end of the expansion section intersects the centerline at point e, where the local Mach number is the same as the design exit Mach number. Hence, to find the expansion section length and $\theta_{w_{max}}$, simply keep track of the centerline Mach number (at points 1, 2, 3, etc.) as you construct your characteristics solution starting from the throat region. When the centerline Mach number equals the design exit Mach number, this is point e. Then the expansion section is terminated at point c, which fixes both its length and the value of $\theta_{w_{max}}$.

EXAMPLE 11.1

Compute and graph the contour of a two-dimensional minimum-length nozzle for the expansion of air to a design exit Mach number of 2.4.

■ Solution

The results of this problem are given in Fig. 11.14b. To begin with, the sonic line at the throat, ab, is assumed to be straight. The first characteristic $(a - 1)$ emanating from the sharp throat is chosen as inclined only slightly from the normal sonic line. ($\Delta\theta = 0.375°$; hence $\theta + \nu = 0.75°$ and $dy/dx = \theta - \mu = -73.725°$.) The remainder of the expansion fan is divided into six increments with $\Delta\theta = 3°$. The total corner angle $\theta_{w_{max}} = \nu/2 = 36.75°/2 = 18.375°$. The values of K_+, K_-, θ, and ν are tabulated in Table 11.1 for all grid points. The

Table 11.1

Point no.	$K_- = $ $\theta + v$	$K_+ = $ $\theta - v$	$\theta = $ $\frac{1}{2}(K_- + K_+)$	$v = $ $\frac{1}{2}(K_- - K_+)$	M	μ	Comments
1	0.75	0	$0.375^\dagger$	$0.375^\dagger$	1.04	74.1	
2	6.75	0	$3.375^\dagger$	$3.375^\dagger$	1.19	57.2	
3	12.75	0	$6.375^\dagger$	$6.375^\dagger$	1.31	49.8	
4	18.75	0	$9.375^\dagger$	$9.375^\dagger$	1.41	45.2	
5	24.75	0	$12.375^\dagger$	$12.375^\dagger$	1.52	41.1	
6	30.75	0	$15.375^\dagger$	$15.375^\dagger$	1.62	38.1	
7	36.75	0	$18.375^\dagger$	$18.375^\dagger$	1.72	35.6	
8	$36.75^\dagger$	$0^\dagger$			$1.72^\dagger$	$35.6^\dagger$	Same as point 7
9	$6.75^\dagger$	-6.75	$0^\dagger$	6.75	1.32	49.3	
10	$12.75^\dagger$	$-6.75^\dagger$	3	9.75	1.43	44.4	
11	$18.75^\dagger$	$-6.75^\dagger$	6	12.75	1.53	40.8	
12	$24.75^\dagger$	$-6.75^\dagger$	9	15.75	1.63	37.8	
13	$30.75^\dagger$	$-6.75^\dagger$	12	18.75	1.73	35.3	
14	$36.75^\dagger$	$-6.75^\dagger$	15	21.75	1.84	32.9	
15	$36.75^\dagger$	$-6.75^\dagger$	$15^\dagger$	$21.75^\dagger$	$1.84^\dagger$	$32.9^\dagger$	Same as point 14
16	$12.75^\dagger$	-12.75	$0^\dagger$	12.75	1.53	40.8	
17	$18.75^\dagger$	$-12.75^\dagger$	3	15.75	1.63	37.8	
18	$24.75^\dagger$	$-12.75^\dagger$	6	18.75	1.73	35.3	
19	30.75^*	$-12.75^\dagger$	9	21.75	1.84	32.9	
20	$36.75^\dagger$	$-12.75^\dagger$	12	24.75	1.94	31.0	
21	$36.75^\dagger$	$-12.75^\dagger$	$12^\dagger$	$24.75^\dagger$	$1.94^\dagger$	$31.0^\dagger$	Same as point 20
22	$18.75^\dagger$	-18.75	$0^\dagger$	18.75	1.73	35.3	
23	$24.75^\dagger$	$-18.75^\dagger$	3	21.75	1.84	32.9	
24	$30.75^\dagger$	$-18.75^\dagger$	6	24.75	1.94	31.0	
25	$36.75^\dagger$	$-18.75^\dagger$	9	27.75	2.05	29.2	
26	$36.75^\dagger$	$-18.75^\dagger$	$9^\dagger$	$27.75^\dagger$	$2.05^\dagger$	$29.2^\dagger$	Same as point 25
27	$24.75^\dagger$	-24.75	$0^\dagger$	24.75	1.94	31.0	
28	$30.75^\dagger$	$-24.75^\dagger$	3	27.75	2.05	29.2	
29	$36.75^\dagger$	$-24.75^\dagger$	6	30.75	2.16	27.6	
30	$36.75^\dagger$	$-24.75^\dagger$	$6^\dagger$	$30.75^\dagger$	$2.16^\dagger$	$27.6^\dagger$	Same as point 29
31	$30.75^\dagger$	-30.75	$0^\dagger$	30.75	2.16	27.6	
32	$36.75^\dagger$	-30.75	3	33.75	2.28	26.0	
33	$36.75^\dagger$	$-30.75^\dagger$	$3^\dagger$	$33.75^\dagger$	$2.28^\dagger$	$26.0^\dagger$	Same as point 32
34	$36.75^\dagger$	-36.75	$0^\dagger$	36.75	2.4	24.6	
35	$36.75^\dagger$	$-36.75^\dagger$	$0^\dagger$	$36.75^\dagger$	$2.4^\dagger$	$24.6^\dagger$	Same as point 34

†Known quantities at beginning of each step.

nozzle contour is drawn by starting at the throat corner (where $\theta_a = \theta_{w\max} = 18.375°$), drawing a straight line with an average slope, $\frac{1}{2}(\theta_a + \theta_8)$, and defining point 8 on the contour as the intersection of this straight line with the left-running characteristic 7–8. Point 15 is located by the intersection of a straight line through point 8 having a slope of $\frac{1}{2}(\theta_8 + \theta_{15})$ with the left-running characteristic 14–15. This process is repeated to generate the remainder of the contour, points 21, 26, etc.

For this example, the computed area ratio $A_e/A^* = 2.33$. This is within 3 percent of the value $A_e/A^* = 2.403$ from Table A.1. This small error is induced by the graphical construction of Fig. 11.14*b*, and by the fact that only seven increments are chosen for the corner expansion fan. For a more accurate calculation, finer increments should be used, resulting in a more closely spaced characteristic net throughout the nozzle.

Note that a small inconsistency is involved with the properties at point 1 in Fig. 11.14, as listed in the first line of Table 11.1. The entry in Table 11.1 for θ at point 1 is a nonzero (but small) number, namely 0.375°. This is inconsistent with the physical picture in Fig. 11.14, which shows point 1 on the nozzle centerline where $\theta = 0$. This inconsistency is due to the necessity of *starting* the calculations with the straight characteristic line, $a-1$, along which the value of θ is constant and equal to 0.375°. In reality, the characteristic $a-1$ is curved because of the nonuniform flow inside the region $a-b-1$ in Fig. 11.14, but we have no way of knowing what that nonuniform flow is for this problem. In Sec. 12.7, we will show that a finite-difference calculation in the throat region can provide such information. However, within the framework of the method of characteristics in the present section, we must live with this inconsistency. As long as the first characteristic line $a-1$ is taken as close as possible to the assumed straight sonic line, this inconsistency will be minimized.

11.8 | METHOD OF CHARACTERISTICS FOR AXISYMMETRIC IRROTATIONAL FLOW

For axisymmetric irrotational flow, the philosophy of the method of characteristics is the same as discussed earlier; however, some of the details are different, principally the compatibility equations. The purpose of this section is to illustrate those differences.

Consider a cylindrical coordinate system, as sketched in Fig. 11.15. The cylindrical coordinates are r, ϕ, and x, with corresponding velocity components v, w,

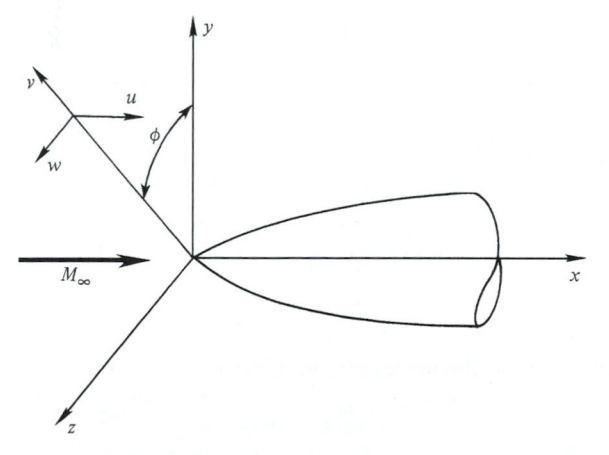

Figure 11.15 | Superposition of rectangular and cylindrical coordinate systems for axisymmetric flow.

and u, respectively. In these cylindrical coordinates, the continuity equation

$$\nabla \cdot (\rho \mathbf{V}) = 0$$

becomes

$$\frac{\partial(\rho u)}{\partial x} + \frac{\partial(\rho v)}{\partial r} + \frac{1}{r}\frac{\partial(\rho w)}{\partial \phi} + \frac{\rho v}{r} = 0 \tag{11.34}$$

Recalling from Sec. 10.1 that axisymmetric flow implies $\partial/\partial \phi = 0$, Eq. (11.34) becomes

$$\frac{\partial(\rho u)}{\partial x} + \frac{\partial(\rho v)}{\partial r} + \frac{\rho v}{r} = 0 \tag{11.35}$$

From Euler's equation for irrotational flow, Eq. (8.7),

$$dp = -\rho V\, dV = \frac{-\rho}{2} d(V^2) = -\frac{\rho}{2} d(u^2 + v^2 + w^2) \tag{11.36}$$

However, the speed of sound $a^2 = (\partial p/\partial \rho)_s = dp/d\rho$. Hence, along with $w = 0$ for axisymmetric flow, Eq. (11.36) becomes

$$d\rho = -\frac{\rho}{a^2}(u\, du + v\, dv) \tag{11.37}$$

from which follows

$$\frac{\partial \rho}{\partial x} = -\frac{\rho}{a^2}\left(u\frac{\partial u}{\partial x} + v\frac{\partial v}{\partial x}\right) \tag{11.38}$$

$$\frac{\partial \rho}{\partial r} = -\frac{\rho}{a^2}\left(u\frac{\partial u}{\partial r} + v\frac{\partial v}{\partial r}\right) \tag{11.39}$$

Substituting Eqs. (11.38) and (11.39) into Eq. (11.35), we obtain, after factoring,

$$\left(1 - \frac{u^2}{a^2}\right)\frac{\partial u}{\partial x} - \frac{uv}{a^2}\frac{\partial v}{\partial x} - \frac{uv}{a^2}\frac{\partial u}{\partial r} + \left(1 - \frac{v^2}{a^2}\right)\frac{\partial v}{\partial r} = -\frac{v}{r} \tag{11.40}$$

The condition of irrotationality is

$$\nabla \times \mathbf{V} = 0$$

which in cylindrical coordinates can be written as

$$\nabla \times \mathbf{V} = \frac{1}{r}\begin{vmatrix} \mathbf{e}_r & r\mathbf{e}_\phi & \mathbf{e}_x \\ \dfrac{\partial}{\partial r} & \dfrac{\partial}{\partial \phi} & \dfrac{\partial}{\partial x} \\ v & rw & u \end{vmatrix} = 0 \tag{11.41}$$

For axisymmetric flow, Eq. (11.41) yields

$$\frac{\partial u}{\partial r} = \frac{\partial v}{\partial x} \tag{11.42}$$

Substituting Eq. (11.42) into (11.40), we have

$$\left(1 - \frac{u^2}{a^2}\right)\frac{\partial u}{\partial x} - 2\frac{uv}{a^2}\frac{\partial v}{\partial x} + \left(1 - \frac{v^2}{a^2}\right)\frac{\partial v}{\partial r} = -\frac{v}{r} \tag{11.43}$$

Keeping in mind that $u = u(x, r)$ and $v = v(x, r)$, we can also write

$$du = \frac{\partial u}{\partial x}dx + \frac{\partial u}{\partial r}dr = \frac{\partial u}{\partial x}dx + \frac{\partial v}{\partial x}dr \tag{11.44}$$

and

$$dv = \frac{\partial v}{\partial x}dx + \frac{\partial v}{\partial r}dr \tag{11.45}$$

Equations (11.43), (11.44), and (11.45) are three equations which can be solved for the three derivatives $\partial u/\partial x$, $\partial v/\partial x$, and $\partial v/\partial r$.

The reader should by now suspect that we are on the same track as in our previous development of the characteristic equations. Equations (11.43) through (11.45) for axisymmetric flow are analogous to Eqs. (11.5) through (11.7) for two-dimensional flow. To determine the characteristic lines and compatibility equations, solve Eqs. (11.43) through (11.45) for $\partial v/\partial x$ as follows:

$$\frac{\partial v}{\partial x} = \frac{\begin{vmatrix} 1 - \dfrac{u^2}{a^2} & -\dfrac{v}{r} & 1 - \dfrac{v^2}{a^2} \\ dx & du & 0 \\ 0 & dv & dr \end{vmatrix}}{\begin{vmatrix} 1 - \dfrac{u^2}{a^2} & -2\dfrac{uv}{a^2} & 1 - \dfrac{v^2}{a^2} \\ dx & dr & 0 \\ 0 & dx & dr \end{vmatrix}} = \frac{N}{D} \tag{11.46}$$

The characteristic directions are found by setting $D = 0$. This yields

$$\left(\frac{dr}{dx}\right)_{\text{char}} = \frac{-uv/a^2 \pm \sqrt{[(u^2 + v^2)/a^2] - 1}}{1 - (u^2/a^2)} \tag{11.47}$$

Equation (11.47) is identical to Eq. (11.10). The discussion following Eq. (11.10), leading to Eq. (11.14), also holds here. Consequently,

$$\boxed{\left(\frac{dr}{dx}\right)_{\text{char}} = \tan(\theta \mp \mu)} \tag{11.48}$$

and we see that *for axisymmetric irrotational flow, the characteristic lines are Mach lines*. The C_+ and C_- characteristics are the same as those sketched in Fig. 11.6.

The compatibility equations that hold along these characteristic lines are found by setting $N = 0$ in Eq. (11.46). The result is

$$\frac{dv}{du} = \frac{-\left(1 - \dfrac{u^2}{a^2}\right) - \dfrac{v}{r}\dfrac{dx}{du}}{\left(1 - \dfrac{v^2}{a^2}\right)\dfrac{dx}{dr}}$$

or

$$\frac{dv}{du} = -\frac{\left(1 - \dfrac{u^2}{a^2}\right)}{\left(1 - \dfrac{v^2}{a^2}\right)}\frac{dr}{dx} - \frac{\dfrac{v}{r}\dfrac{dr}{du}}{\left(1 - \dfrac{v^2}{a^2}\right)} \tag{11.49}$$

In Eq. (11.49), the term dr/dx is the characteristic direction given by Eq. (11.47). Hence, substituting Eq. (11.47) into (11.49), we have

$$\frac{dv}{du} = \frac{\dfrac{uv}{a^2} \mp \sqrt{\dfrac{u^2 + v^2}{a^2} - 1}}{\left(1 - \dfrac{v^2}{a^2}\right)} - \frac{\dfrac{v}{r}\dfrac{dr}{du}}{\left(1 - \dfrac{v^2}{a^2}\right)} \tag{11.50}$$

Note that Eq. (11.50) for axisymmetric flow differs from Eq. (11.16) for two-dimensional flow by the additional term involving dr/r. Referring again to Fig. 11.6, we make the substitution $u = V\cos\theta$ and $v = V\sin\theta$ into Eq. (11.50), which after algebraic manipulation becomes

$$d\theta = \mp\sqrt{M^2 - 1}\,\frac{dV}{V} \pm \frac{1}{\sqrt{M^2 - 1} \mp \cot\theta}\frac{dr}{r} \tag{11.51}$$

The first term on the right-hand side of Eq. (11.51) is the differential of the Prandtl–Meyer function, dv (see Sec. 4.14). Hence, the final form of the compatibility equation is

$$\boxed{d(\theta + v) = \frac{1}{\sqrt{M^2 - 1} - \cot\theta}\frac{dr}{r} \quad \text{(along a } C_- \text{ characteristic)}} \tag{11.52}$$

$$\boxed{d(\theta - v) = -\frac{1}{\sqrt{M^2 - 1} + \cot\theta}\frac{dr}{r} \quad \text{(along a } C_+ \text{ characteristic)}} \tag{11.53}$$

Equations (11.52) and (11.53) are the compatibility equations for axisymmetric irrotational flow. Compare them with the analogous results for two-dimensional irrotational flow given by Eqs. (11.20) and (11.21). For axisymmetric flow, we note the following:

1. The compatibility equations are *differential* equations, not algebraic equations as before.
2. The quantity $\theta + \nu$ is no longer constant along a C_- characteristic. Instead, its value depends on the spatial location in the flowfield as dictated by the dr/r term in Eq. (11.52). The same qualification is made for $\theta - \nu$ along a C_+ characteristic.

For the actual numerical computation of an axisymmetric flowfield by the method of characteristics, the differentials in Eqs. (11.52) and (11.53) are replaced by finite differences (which are to be discussed later). The flow properties and their location are found by a step-by-step solution of Eqs. (11.52) and (11.53) coupled with the construction of the characteristics net using Eq. (11.48).

11.9 | METHOD OF CHARACTERISTICS FOR ROTATIONAL (NONISENTROPIC AND NONADIABATIC) FLOW

The assumption of irrotationality in the previous sections allows a great simplification. For example, Eq. (11.5) for two-dimensional irrotational flow contains only three velocity derivatives, namely $\Phi_{xx} = \partial u/\partial x$, $\Phi_{yy} = \partial v/\partial y$, and $\Phi_{xy} = \partial u/\partial y = \partial v/\partial x$. The irrotationality condition allows the use of the velocity potential and, in particular, eliminates one of the possible velocity derivatives as an unknown via $\partial u/\partial y = \partial v/\partial x$. Along with Eqs. (11.6) and (11.7), we have a system of equations with three unknown velocity derivatives, which can be solved by means of three-by-three determinants, Eq. (11.8). Similarly, for axisymmetric irrotational flow, the irrotationality condition, Eq. (11.42), allows the derivation of a governing equation, Eq. (11.43), which contains only three unknown velocity derivatives. This again leads to a system of three-by-three determinants, namely, Eq. (11.46).

In contrast, rotational flow is more complex, although the philosophy of the method of characteristics remains the same. Only a brief outline of the rotational method of characteristics will be given here; the reader is referred to Shapiro (Ref. 16) for additional details.

Crocco's theorem, Eq. (6.60), repeated here,

$$T\nabla s = \nabla h_o - \mathbf{V} \times (\nabla \times \mathbf{V})$$

tells us that rotational flow occurs when nonisentropic and/or nonadiabatic conditions are present. An example of the former is the flow behind a curved shock wave (see Fig. 4.29), where the entropy increase across the shock is different for different streamlines. An example of the latter is a shock layer within which the static temperature is high enough for the gas to lose a substantial amount of energy due to thermal radiation.

Without the simplification afforded by the irrotationality condition, it is not possible to obtain a system of three independent equations with three unknown derivatives for the flow variables. Instead, for a rotational flow, the conservation equations as well as auxiliary relations [such as Eqs. (11.44) and (11.45)] lead to a minimum of eight equations with eight unknown derivatives. The characteristic lines and corresponding compatibility equations are then found by evaluating eight-by-eight determinants. Obviously, we will not take the space to go through such an evaluation. The results for two-dimensional and axisymmetric rotational flows show that there are *three* sets of characteristics—the left- and right-running *Mach lines,* and the *streamlines* of the flow. The compatibility equations along the Mach lines are of the form

$$dV = f\left(d\theta, ds, dh_o, \frac{dr}{r}\right) \tag{11.54}$$

and along the streamlines, from Eqs. (6.43) and (6.49),

$$dh_o = \dot{q} \tag{11.55}$$

$$T\,ds = de + p\,d\frac{1}{\rho} \tag{11.56}$$

In Eq. (11.54), ds and dh_o denote changes in entropy and total enthalpy along the Mach lines; in Eqs. (11.55) and (11.56) the respective changes ds and dh_o are along the streamlines.

Equations (11.54) through (11.56), along with the characteristics net of Mach lines and streamlines, must be solved in a step-by-step coupled fashion. A typical unit process is illustrated in Fig. 11.16. Here, all properties are known at points 1 and 2. Point 3 is located by the intersection of the C_- characteristic through point 1 and the C_+ characteristic through point 2. The streamline direction θ_3 at point 3 is first estimated by assuming an average of θ_1 and θ_2. This streamline is traced upstream until it intersects at point 4 the known data plane through points 1 and 2. The values of s_4 and h_{o4} are interpolated from the known values at points 1 and 2. Then the values of s_3 and h_{o3} are obtained from the compatibility equations along the streamline, Eqs. (11.55) and (11.56). Once s_3 and h_{o3} are found as above, the compatibility equation along the Mach lines, Eq. (11.54), yields values of V_3 and θ_3. The whole unit process is then repeated in an iterative sense until the desired accuracy is obtained at point 3.

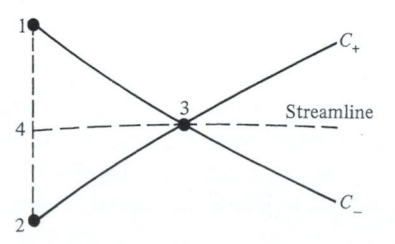

Figure 11.16 | Characteristic directions for a nonisentropic flow.

The reader is cautioned that the above discussion is purely illustrative; the details of a given problem obviously depend on the specific physical phenomena being treated (the thermodynamics of the gas, the form of energy loss, etc.). However, the major purpose of this section is to underscore that, for a general two-dimensional or axisymmetric flow, the streamlines are characteristics, and the derivation of the appropriate compatibility equations is more complex than for the irrotational case discussed in Secs. 11.3 through 11.8.

11.10 | THREE-DIMENSIONAL METHOD OF CHARACTERISTICS

The general conservation equations for three-dimensional inviscid flow were derived in Chap. 6. These equations can be used, for example, to solve the three-dimensional flow over a body at angle of attack, as sketched in Fig. 11.17. For supersonic three-dimensional flow, these equations are hyperbolic. Hence, the method of characteristics can be employed, albeit in a much more complex form than for the two-dimensional or axisymmetric cases treated earlier. Again, only the general results will be given here; the reader is urged to consult Refs. 34 through 38 for examples of detailed solutions.

Consider point b in a general supersonic three-dimensional flow, as sketched in Fig. 11.17. Through this point, the characteristic directions generate two sets of three-dimensional *surfaces*—a *Mach cone* with its vertex at point b and with a half-angle equal to the local Mach angle μ, and a *stream* surface through point b. The intersections of these surfaces establish a complex three-dimensional network of grid points. Moreover, as if this were not complicated enough, the compatibility equations along arbitrary rays of the Mach cone contain cross derivatives that have to be

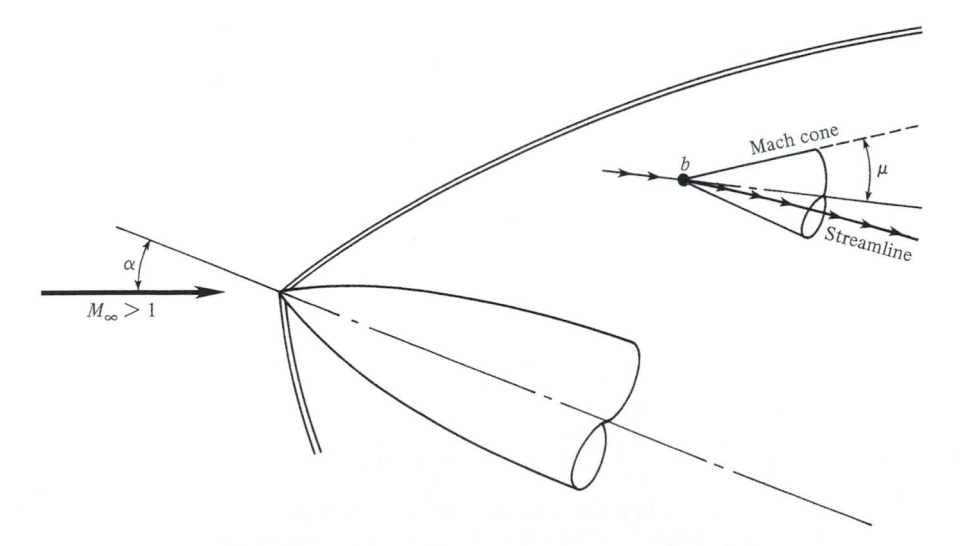

Figure 11.17 | Illustration of the Mach cone in three-dimensional flow.

evaluated in directions not along the characteristics. Nevertheless, such solutions can be obtained (see Refs. 34 through 38).

Rakich (Refs. 37 and 38) has utilized a modification of the above philosophy, which somewhat simplifies the calculations. In this approach, which is sometimes labeled "semicharacteristics" or the "reference plane method," the three-dimensional flowfield is divided into an arbitrary number of planes containing the centerline of the body. This is sketched in Fig. 11.18, which is a front view of the body and shock. One of these planes, say $\phi = \phi_2$, is projected on Fig. 11.19. In this particular reference plane, a series of grid points are established along arbitrarily spaced straight lines locally perpendicular to the body surface. Assume that the flowfield properties

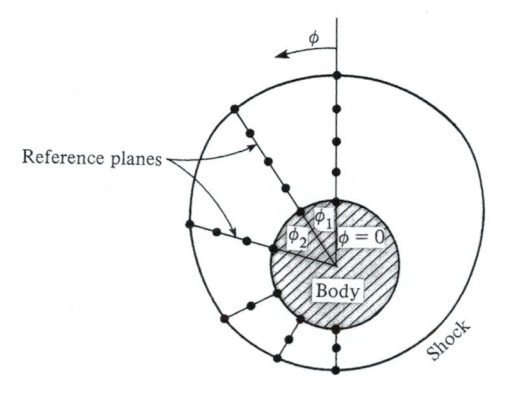

Figure 11.18 | Grid network in a cross-sectional plane for an axisymmetric body at angle of attack; three-dimensional method of characteristics.

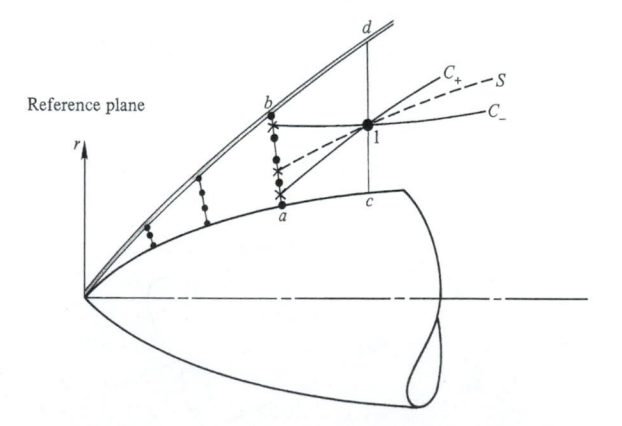

Figure 11.19 | Grid network in the meridional plane for an axisymmetric body at angle of attack; three-dimensional method of characteristics.

known at the grid points denoted by solid circles along the straight line *ab*. Furthermore, arbitrarily choose point 1 on the next downstream line, *cd*. Let $C_+, C_-,$ and S denote the projection in the reference plane of the Mach cone and streamline through point 1. Extend these characteristics upstream until they intersect the data line *ab* at the cross marks. Data at these intersections are obtained by interpolating between the known data at the solid circles. Then, the flowfield properties at point 1 are obtained by solving these compatibility equations along the characteristics:

$$\frac{\beta}{\rho V^2} \frac{dp}{dC_+} + \cos\psi \frac{d\theta}{dC_+} = (f_1 + \beta f_2) \sin\mu$$

$$\frac{\beta}{\rho V^2} \frac{dp}{dC_-} - \cos\psi \frac{d\theta}{dC_-} = (f_1 - \beta f_2) \sin\mu$$

$$\frac{d\psi}{dS} = f_3$$

where $\psi =$ the cross-flow angle defined by $\sin\psi = w/V$

$$\beta = \sqrt{M^2 - 1}$$

$$f_1 = -\frac{\cos\psi \sin\theta}{r}$$

$$f_2 = -\frac{\sin^2\psi \cos\theta}{r}$$

$$f_3 = -\frac{\sin\psi \sin\theta}{r}$$

It is beyond the scope of this book to describe the details of such an analysis. Again, the reader is referred to Refs. 37 and 38 for further elaboration. The major point made here is that the method of characteristics can be used for three-dimensional supersonic flows, and several modern techniques have been devised for its implementation.

11.11 | INTRODUCTION TO FINITE DIFFERENCES

The method of characteristics, discussed in the previous sections, is a numerical solution of the governing conservation equations wherein the grid points and computations are made along the characteristic lines. Following the characteristic lines is sometimes a numerical inconvenience, and at high Mach numbers the characteristics net can become particularly elongated and distorted, causing inordinate numerical error in the calculations. In contrast, the finite-difference approach discussed in this and subsequent sections is inherently more straightforward than the method of characteristics, and has the advantage that essentially arbitrary computational grids can be employed. Indeed, it is quite common to use simple rectangular grids for finite-difference methods, as shown in Fig. 11.3. It is for reasons such as these that finite-difference solutions of the governing conservation equations have become popular in modern compressible flow, supplanting characteristics solutions in many cases. Moreover, finite-difference methods have a much wider range of applicability; they are useful for subsonic and mixed subsonic-supersonic (transonic) flows where the method of characteristics is at best impractical. Finite-difference solutions for purely supersonic steady flows will be discussed in the remainder of this chapter. This will be followed in Chap. 12 with a presentation of the powerful time-marching finite-difference technique that has provided a major breakthrough in the analysis of mixed subsonic-supersonic flows.

The philosophy of finite-difference solutions is to replace the partial derivatives appearing in the conservation equations (see Chap. 6) with algebraic difference quotients, yielding algebraic equations for the flowfield variables at the specified grid points. The type of finite difference that is used to replace the partial derivatives can be selected from a number of different forms, depending on the desired accuracy of the solution, convergence behavior, stability, and convenience. However, the most common forms in current use are *forward, rearward,* and *central differences,* all of which stem from the Taylor's series given by Eq. (11.1). For example, assume that we write the conservation equations in cartesian coordinates, and we wish to replace the derivative $\partial u/\partial x$ in these equations with a finite difference at the grid point (i, j). In its present form, Eq. (11.1) is of "second-order accuracy" because terms involving $(\Delta x)^3$, $(\Delta x)^4$, etc., have been assumed small and can be neglected. If we are interested in only first-order accuracy, then Eq. (11.1) can be written as

$$u_{i+1,j} = u_{i,j} + \left(\frac{\partial u}{\partial x}\right)_{i,j} \Delta x + \cdots \tag{11.57}$$

From Eq. (11.57), we can form a *forward difference* for the derivative $\partial u/\partial x$,

$$\boxed{\left(\frac{\partial u}{\partial x}\right)_{i,j} = \frac{u_{i+1,j} - u_{i,j}}{\Delta x}} \tag{11.58}$$

which is of first-order accuracy. Similarly, if Eq. (11.1) is written for a minus value of Δx, we have

$$u_{i-1,j} = u_{i,j} + \left(\frac{\partial u}{\partial x}\right)_{i,j} (-\Delta x) + \left(\frac{\partial^2 u}{\partial x^2}\right)_{i,j} \frac{(-\Delta x)^2}{2} + \cdots \tag{11.59}$$

which, for first-order accuracy, can be written as

$$u_{i-1,j} = u_{i,j} - \left(\frac{\partial u}{\partial x}\right)_{i,j} \Delta x + \cdots \tag{11.60}$$

From Eq. (11.60), we can form a *rearward difference* for the derivative $\partial u/\partial x$,

$$\boxed{\left(\frac{\partial u}{\partial x}\right)_{i,j} = \frac{u_{i,j} - u_{i-1,j}}{\Delta x}} \tag{11.61}$$

which is of first-order accuracy. Finally, we can obtain a second-order-accurate finite difference for $\partial u/\partial x$ by subtracting Eq. (11.59) from Eq. (11.1), both of which contain $(\Delta x)^2$ and hence are of second-order accuracy. After subtraction, we have

$$u_{i+1,j} - u_{i-1,j} = 0 + 2\left(\frac{\partial u}{\partial x}\right)_{i,j} \Delta x + 0 + \cdots \qquad (11.62)$$

Solving Eq. (11.62) for $(\partial u/\partial x)_{i,j}$, we obtain the central difference

$$\boxed{\left(\frac{\partial u}{\partial x}\right)_{i,j} = \frac{u_{i+1,j} - u_{i-1,j}}{2\,\Delta x}} \qquad (11.63)$$

which is of second-order accuracy.

In summary, Eqs. (11.58) and (11.63) define forward, rearward, and central differences, respectively, for the derivative $\partial u/\partial x$. Analogous expressions exist for derivatives in the y direction. For example, returning to Fig. 11.3, we can write

$$\left(\frac{\partial u}{\partial y}\right)_{i,j} = \frac{u_{i,j+1} - u_{i,j}}{\Delta y} \qquad \text{(forward difference)}$$

$$\left(\frac{\partial u}{\partial y}\right)_{i,j} = \frac{u_{i,j} - u_{i,j-1}}{\Delta y} \qquad \text{(rearward difference)}$$

$$\left(\frac{\partial u}{\partial y}\right)_{i,j} = \frac{u_{i,j+1} - u_{i,j-1}}{2\,\Delta y} \qquad \text{(central difference)}$$

Finite-difference expressions for higher-order derivatives, such as $\partial^2 u/\partial x^2$, can also be constructed from Eq. (11.1). However, note from Chap. 6 that the conservation equations for inviscid compressible flow contain only first-order derivatives of the flowfield properties. Hence, in this book we need only be concerned with finite differences for first-order derivatives. This would not be true if we were dealing with viscous flows, where second-order derivatives are present in the momentum and energy equations.

Equations (11.58), (11.61), and (11.63) are finite-difference representations of the first partial derivative. When these difference quotients are used to replace the partial differentials in an equation, then a *difference equation* results. For example, consider the continuity equation given by Eq. (6.5), repeated here.

$$\frac{\partial \rho}{\partial t} + \nabla \cdot (\rho \mathbf{V}) = 0 \qquad (6.5)$$

For steady, two-dimensional flow, Eq. (6.5) becomes

$$\frac{\partial(\rho u)}{\partial x} + \frac{\partial(\rho v)}{\partial y} = 0 \qquad (11.64)$$

Defining $F = \rho u$ and $G = \rho v$, Eq. (11.64) is written as

$$\frac{\partial F}{\partial x} + \frac{\partial G}{\partial y} = 0 \qquad (11.65)$$

Replacing the x derivative in Eq. (11.65) with a forward difference [Eq. (11.58)], and the y derivative with a central difference [the y equivalent of Eq. (11.63)], we have

$$\frac{F_{i+1,j} - F_{i,j}}{\Delta x} = \frac{G_{i,j+1} - G_{i,j-1}}{2\,\Delta y} \tag{11.66}$$

or
$$F_{i+1,j} = F_{i,j} + \frac{\Delta x}{2\,\Delta y}(G_{i,j+1} - G_{i,j-1}) \tag{11.67}$$

Equation (11.66), or Eq. (11.67), is the *difference equation* that replaces the original partial differential equation, namely Eq. (11.65). Equation (11.66) is an approximation for Eq. (11.65); Eq. (11.66) contains a truncation error which is a combination of the truncation errors from the difference quotients in Eq. (11.58) and the y equivalent of Eq. (11.63).

A distinction between various finite-difference solutions is that of explicit versus implicit approaches. Let us make the distinction by way of an example. Assume we have a two-dimensional flowfield over which we place a rectangular grid, as sketched in Fig. 11.3. Assume the general direction of the flow is from left to right. Furthermore, assume that the flowfield properties are known at all the grid points along the vertical line through point (i, j). We wish to calculate the value of F at all the downstream grid points along the vertical line through point $(i + 1, j)$. Equation (11.67) allows us to calculate F at point $(i + 1, j)$ explicitly from the known values along the vertical line through point (i, j). By repeated application of Eq. (11.67) at all points on the upstream vertical line, $(i, j + 1)$, $(i, j - 1)$, etc., the values of F at all points along the downstream vertical line can be calculated one at a time. This type of approach, wherein the flowfield at a given downstream point is evaluated strictly in terms of the known upstream values, is defined as an explicit finite-difference solution. In contrast, let us construct an approach that assumes the y derivative in Eq. (11.65) is the average between the two vertical lines through points (i, j) and $(i + 1, j)$ in Fig. 11.3, i.e., let us form a difference equation for Eq. (11.65) as follows.

$$\frac{F_{i+1,j} - F_{i,j}}{\Delta x} = \frac{1}{2}\left[\left(\frac{G_{i,j+1} - G_{i,j-1}}{2\Delta y} + \frac{G_{i+1,j+1} - G_{i+1,j-1}}{2\Delta y}\right)\right]$$

or
$$F_{i+1,j} = F_{i,j} + \frac{\Delta x}{4\Delta y}(G_{i,j+1} + G_{i+1,j+1} - G_{i,j-1} - G_{i+1,j-1}) \tag{11.68}$$

In order to calculate $F_{i+1,j}$ from Eq. (11.68), knowing the flowfield at the upstream vertical line is not enough. The right-hand side of Eq. (11.68) also contains the unknown quantities $G_{i+1,j+1}$ and $G_{i+1,j-1}$ along the downstream vertical line. If Eq. (11.68) is applied at all points along the upstream vertical line, a system of simultaneous equations for $G_{i+1,j}$, $F_{i+1,j}$, $G_{i+1,j+1}$, etc., along the downstream vertical line is obtained. These unknowns must be solved simultaneously. Moreover, additional equations (momentum, etc.) are required because there are more unknowns than equations provided by Eq. (11.68). This type of approach, wherein the flowfield at a given downstream point is evaluated in terms of both known upstream values and unknown downstream values, is defined as an implicit finite-difference solution.

The advantage of explicit methods is that they are relatively simple to set up and program. The disadvantage is that the spatial increments Δx and Δy are limited due to stability constraints associated with explicit methods. For a given Δy, Δx is constrained to be less than a certain value dictated by numerical stability considerations. (Such stability analyses are discussed at length in Refs. 18 and 102.) In turn, if Δx is constrained to be too small, the computer time required to calculate the flow over a prescribed downstream distance can be large.

The advantage of implicit methods is that stability can be maintained over much larger values of Δx, hence using considerably fewer steps to make calculations over a prescribed downstream distance. A disadvantage of implicit methods is that they are more complicated to set up and program in comparison to explicit methods. Moreover, massive matrix manipulations are usually required at each spatial step to solve the simultaneous algebraic equations, hence the computer time per step is larger for the implicit approach. However, on the whole, implicit methods frequently result in smaller total computer times for a given flowfield calculation. Whether this continues to be the case is a matter of current research; for example, explicit methods are readily vectorizable for use on a vector-type supercomputer, and frequently can take much better advantage of the computer architecture than implicit methods.

Today, both implicit and explicit methods are in wide use. However, for the sake of simplicity, we will deal only with explicit methods in the remainder of this chapter. For details on both methods, see Refs. 18, 102, and 137–142.

A favorite form of the governing flow equations in use by many computational fluid dynamicists today is the conservation form; both conservation and nonconservation forms were derived in Secs. 6.2 and 6.4, respectively. Writing the conservation form of the governing equations for steady, three-dimensional flow, we have from Eqs. (6.5), (6.11) through (6.13), and (6.17),

Continuity:
$$\frac{\partial(\rho u)}{\partial x} + \frac{\partial(\rho v)}{\partial y} + \frac{\partial(\rho w)}{\partial z} = 0$$

x momentum:
$$\frac{\partial}{\partial x}(\rho u^2 + p) + \frac{\partial(\rho uv)}{\partial y} + \frac{\partial(\rho uw)}{\partial z} = \rho f_x$$

y momentum:
$$\frac{\partial(\rho vu)}{\partial x} + \frac{\partial}{\partial y}(\rho v^2 + p) + \frac{\partial(\rho uw)}{\partial z} = \rho f_y$$

z momentum:
$$\frac{\partial(\rho wu)}{\partial x} + \frac{\partial(\rho wv)}{\partial y} + \frac{\partial}{\partial z}(\rho w^2 + p) = \rho f_z$$

Energy:
$$\frac{\partial}{\partial x}\left[\rho\left(e + \frac{V^2}{2}\right)u + pu\right] + \frac{\partial}{\partial y}\left[\rho\left(e + \frac{V^2}{2}\right)v + pv\right]$$
$$+ \frac{\partial}{\partial z}\left[\rho\left(e + \frac{V^2}{2}\right)w + pw\right]$$
$$= \rho\dot{q} + \rho(uf_x + vf_y + wf_z)$$

These equations can be expressed in a single, generic form as

$$\frac{\partial F}{\partial x} + \frac{\partial G}{\partial y} + \frac{\partial H}{\partial z} + J = 0 \tag{11.69}$$

where F, G, H, and J are column vectors given by

$$F = \left\{ \begin{array}{c} \rho u \\ \rho u^2 + p \\ \rho v u \\ \rho w u \\ \rho(e + V^2/2)u + pu \end{array} \right\} \qquad G = \left\{ \begin{array}{c} \rho v \\ \rho u v \\ \rho v^2 + p \\ \rho w u \\ \rho(e + V^2/2)v + pv \end{array} \right\}$$

$$H = \left\{ \begin{array}{c} \rho w \\ \rho u w \\ \rho v w \\ \rho w^2 + p \\ \rho(e + V^2/2)w + pw \end{array} \right\} \qquad J = \left\{ \begin{array}{c} 0 \\ \rho f_x \\ \rho f_y \\ \rho f_z \\ \rho \dot{q} + \rho(u f_x + v f_y + w f_z) \end{array} \right\}$$

Here, J is called the *source term*. The governing equations in the form of Eq. (11.69) are called the *strong* conservation form, in contrast to Eqs. (6.5), (6.11) through (6.13), and (6.17) which are classified as the *weak* conservation form. In various applications of computational fluid mechanics, the form used for the governing equations can make a difference in the numerical solution; this distinction is particularly important for problems that involve shock waves, and has to do with the choice of the shock-capturing or shock-fitting approaches—to be discussed in Sec. 11.15.

It seems clear from this discussion that the finite-difference philosophy is inherently straightforward; just replace the partial derivatives in the governing equations with algebraic difference quotients, and grind away to obtain solutions of these algebraic equations at each grid point. However, this impression is misleading. For any given application, there is no guarantee that such calculations will be accurate, or even stable. Moreover, the *boundary conditions* for a given problem dictate the solution, and therefore the proper treatment of boundary conditions within the framework of a particular finite-difference technique is vitally important. For these reasons, general finite-difference solutions are by no means routine. Indeed, much of computational fluid dynamics today is still more of an art than a science; each different problem usually requires special thought and originality in its solution. The reader is strongly urged to study Refs. 39 through 45 in order to gain more appreciation for this state of affairs. These references, written early in the development of CFD, only scratch the surface of the finite-difference literature, but they represent a reasonable introduction to some of the problems. These, along with Refs. 18, 102, and 137–142, and the flavor given in this and subsequent sections, should provide the reader with an understanding of the power and usefulness of finite-difference solutions to compressible flow problems. It is beyond the scope of this book to provide the minute details of any given finite-difference solution; however, the purpose of this and subsequent chapters is to provide a roadmap from which the reader can make excursions into the literature as desired.

11.12 | MACCORMACK'S TECHNIQUE

Although a myriad of finite-difference schemes have been utilized for numerous problems, one specific algorithm gained wide use and acceptance in the 1970s and 1980s. This is a technique developed by Robert MacCormack at the NASA Ames Research Center, first published in 1969 in the context of a time-marching solution to the unsteady equations of motion (see Ref. 39). A discussion of such time-marching techniques will be deferred until Chap. 12. However, MacCormack's technique has also been applied to steady supersonic flows (see Refs. 40 through 44). MacCormack's technique has been supplanted by more modern algorithms in recent years. However, it is straightforward, very "student friendly," and works well for a number of applications. Therefore, it is highlighted in this section.

Let us consider the solution of a steady, two-dimensional, supersonic, inviscid flowfield in (x, y) space. The flow is assumed to be known along an initial data line, and the finite-difference calculation will march downstream from this initial data line, in the same fashion as described for the method of characteristics in Sec. 11.5. Once again, we note that this downstream-marching approach is consistent with the properties of hyperbolic or parabolic equations. For supersonic flow, Eq. (11.69) is hyperbolic. Let us rewrite Eq. (11.69) for two-dimensional flow with no source terms as

$$\frac{\partial F}{\partial x} = -\frac{\partial G}{\partial y} \tag{11.70}$$

Consider again the grid illustrated in Fig. 11.3. MacCormack's solution of Eq. (11.70) on the grid of Fig. 11.3 takes the form of a predictor-corrector technique, using forward differences on the predictor step and rearward differences on the corrector step. By using this two-step process, although the differences are of first-order accuracy in each step, the overall result is of second-order accuracy. Specifically, referring to Fig. 11.3, the flowfield is known at all points along the vertical lines through $(i - 1)$ and (i). Hence, $F_{i+1,j}$ can be calculated from a Taylor's series expansion in terms of x:

$$F_{i+1,j} = F_{i,j} + \left(\frac{\partial F}{\partial x}\right)_{\text{ave}} \Delta x \tag{11.71}$$

In Eq. (11.71), $F_{i,j}$ is known, and $(\partial F/\partial x)_{\text{ave}}$ is an average of the x derivative of F between points (i, j) and $(i + 1, j)$. A numerical value of this average derivative is obtained in two steps as we see next.

Predictor Step. First, predict the value of $F_{i+1,j}$ by using a Taylor's series where $\partial F/\partial x$ is evaluated at point (i, j). Denote this predicted value as $\bar{F}_{i+1,j}$:

$$\bar{F}_{i+1,j} = F_{i,j} + \left(\frac{\partial F}{\partial x}\right)_{i,j} \Delta x \tag{11.72}$$

In Eq. (11.72), $(\partial F/\partial x)_{i,j}$ is obtained from Eq. (11.70) using a forward difference for the y derivative:

$$\left(\frac{\partial F}{\partial x}\right)_{i,j} = -\left(\frac{\partial G}{\partial y}\right)_{i,j} = -\frac{G_{i,j+1} - G_{i,j}}{\Delta y} \tag{11.73}$$

In Eq. (11.73), $G_{i,j+1}$ and $G_{i,j}$ are known; hence, the calculated value of $(\partial F/\partial x)_{i,j}$ from Eq. (11.73) is substituted into Eq. (11.72) to yield the predicted value, $\bar{F}_{i+1,j}$. This process is repeated to obtain $\bar{F}_{i+1,j}$ at *all* values of j, i.e., at all grid points along the vertical line through $i+1, j$ in Fig. 11.3.

Corrector Step. The value of $\bar{F}_{i+1,j}$ obtained from the predictor step really represents individual numbers for the *flux variables* $(\bar{\rho}\bar{u})_{i+1,j}, (\bar{\rho}\bar{u}^2 + \bar{p})_{i+1,j},$ $(\bar{\rho}\bar{v}\bar{u})_{i+1,j}$, and $[\bar{\rho}(\bar{e} + \bar{V}^2/2)\bar{u} + \bar{p}\bar{u}]_{i+1,j}$, as displayed in Eq. (11.69). In turn, these numbers can be solved for the *primitive variables,* $\bar{\rho}_{i+1,j}, \bar{u}_{i+1,j}, \bar{v}_{i+1,j}$, and $\bar{e}_{i+1,j}$. These predicted primitive variables are then used to calculate numbers for $\bar{G}_{i+1,j}$. These predicted values of G are then used to calculate a predicted value of the derivative $(\overline{\partial F/\partial x})_{i+1,j}$ by using a rearward difference in Eq. (11.70):

$$\left(\overline{\frac{\partial F}{\partial x}}\right)_{i+1,j} = -\frac{\bar{G}_{i+1,j} - \bar{G}_{i+1,j-1}}{\Delta y} \tag{11.74}$$

In turn, the results from Eqs. (11.73) and (11.74) allow the calculation of the average derivative

$$\left(\frac{\partial F}{\partial x}\right)_{ave} = \frac{1}{2}\left[\left(\frac{\partial F}{\partial x}\right)_{i,j} + \left(\overline{\frac{\partial F}{\partial x}}\right)_{i+1,j}\right] \tag{11.75}$$

Finally, the average derivative calculated by Eq. (11.75) allows the calculation of the corrected value $F_{i+1,j}$ from Eq. (11.71). By simply marching downstream in steps of x, our algorithm allows the calculation of the complete flowfield downstream of a given initial data line. This is made possible because the equations for steady inviscid supersonic flow are hyperbolic. The above technique cannot be employed in subsonic regions; indeed, if an embedded subsonic region is encountered while marching downstream, the calculations will generally become unstable. However, such mixed subsonic and supersonic flows can be treated by the time-marching technique described in Chap. 12.

Finally, note that MacCormack's scheme is an explicit finite-difference technique. As mentioned earlier, it is of second-order accuracy. A generalization of MacCormack's scheme to third-order accuracy is described in Ref. 42.

11.13 | BOUNDARY CONDITIONS

Consider the flow in the vicinity of a solid wall, as sketched in Fig. 11.20. The algorithm described in Sec. 11.12 applies to grid points internally in the flowfield, such as point 1. Here it is possible to form both the required forward and rearward differences in the y direction. However, on the wall at point 2, it is not possible to form a rearward difference, since there are no points inside the wall. Various methods have been developed to calculate the flow at a wall boundary point, all with mixed degrees of success. Some methods work better than others, depending on the character of the specific flow problem and the slope of the boundary. An authoritative review of such

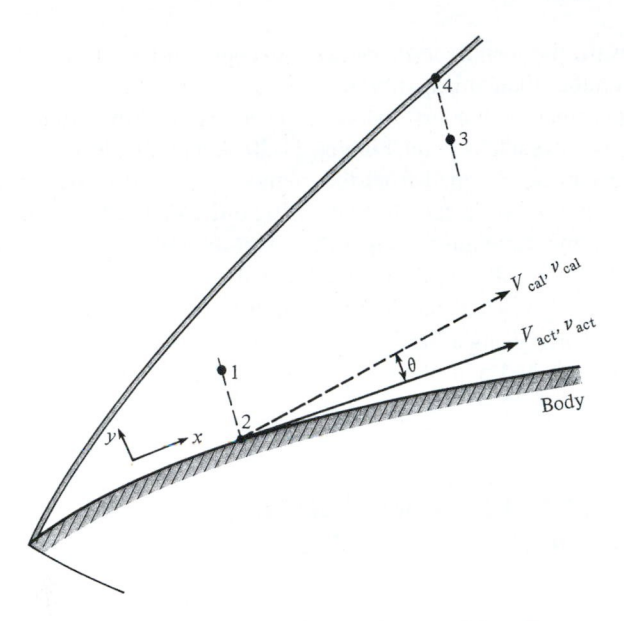

Figure 11.20 | Shock and wall boundary conditions for supersonic steady-flow finite-difference solutions.

boundary conditions is given in Ref. 46. We emphasize that the proper treatment of boundary conditions can make or break a flowfield calculation.

A generally accepted method for accurately dealing with a solid-wall boundary condition for inviscid steady supersonic flow is that due to Abbett (see Ref. 46). Abbett's method is in wide use; moreover, it is simple and accurate. Refer again to Fig. 11.20. First calculate values of the flowfield variables at point 2 using the internal flow algorithm described in Sec. 11.12, but incorporating forward derivatives in both the predictor and corrector steps. This will yield a calculated velocity V_{cal} at point 2, as well as calculated values of pressure, temperature, etc. In general the direction of V_{cal} will not be tangent to the wall due to inaccuracies in the calculational procedure. Figure 11.20 shows V_{cal} above the wall by the angle θ. However, the necessary boundary conditions at the wall for an inviscid flow dictate that the flow velocity be tangent to the wall. Therefore, Abbett suggests that the calculated velocity direction at point 2 be rotated by means of a Prandtl–Meyer expansion through the known angle θ. This yields the actual velocity at point 2, V_{act}, which is tangent to the wall. The Mach number (hence ultimately the velocity magnitude) at point 2 is obtained from the actual Prandtl–Meyer function, ν_{act}, where

$$\nu_{act} = \nu_{cal} + \theta$$

Analogously, the actual pressure and temperature at point 2 are obtained from the originally calculated values, modified by an isentropic expansion from ν_{cal} to ν_{act}. Figure 11.20 shows the case when V_{cal} is pointed away from the wall; when V_{cal} is toward the wall, the technique is the same except that the Prandtl–Meyer turn is a compression rather than an expansion.

Another common boundary condition in supersonic flow is that immediately behind a shock wave, such as point 4 in Fig. 11.20. Again, the flow properties at the interior point 3 can be obtained from the method discussed in Sec. 11.12. The flow properties at point 4 can be calculated by using one-sided differences (all forward or all rearward) in the same interior algorithm. The strength (hence angle) of the shock wave at point 4 then follows from the oblique shock relations described in Chap. 4. In Ref. 46, Abbett gives several alternative approaches to the shock boundary condition, including some using a local characteristics technique projected from the internal points and matched with the oblique shock relations. Such an approach will be detailed in Chap. 12.

11.14 | STABILITY CRITERION: THE CFL CRITERION

The rectangular grid shown in Fig. 11.3 does not always involve purely arbitrary spacing for Δx and Δy. Indeed, the ratio $\Delta x / \Delta y$ must be less than a certain value in order for the explicit finite-difference procedure described in Sec. 11.12 to be computationally stable. On the other hand, for implicit methods $\Delta x / \Delta y$ can be much larger—some implicit methods are unconditionally stable for any value of $\Delta x / \Delta y$ no matter how large. In these cases, however, the *accuracy* of the solution can become poor at large $\Delta x / \Delta y$ simply because the truncation errors, which depend on Δx and Δy, become large.

In this book, we are dealing primarily with explicit methods for simplicity. Moreover, MacCormack's method described in Sec. 11.12 is an explicit method; this method has been widely adopted, and because of its simplicity, MacCormack's method, in this author's experience, is very "student friendly." Therefore, in the present section, let us examine more closely the stability criterion associated with such an explicit method.

It is difficult to obtain from mathematical analysis a precise condition for $\Delta x / \Delta y$ that holds exactly for a governing system of *nonlinear* equations, such as the flow equations that we use in gasdynamics. However, we can use as guidance the stability criterion for a model equation that is linear, and that has many of the same mathematical properties as the nonlinear system. For the steady, supersonic, inviscid flows discussed in this chapter, the governing nonlinear equations are *hyperbolic,* as discussed in Sec. 11.3. A linear, hyperbolic equation can be used as a model for this system in terms of stability considerations. One example of a standard stability analysis of hyperbolic linear equations is the Von Neumann stability method, discussed at length in Refs. 18, 102, 128, and 137–142. The result of this analysis is the following stability criterion:

$$\Delta x \le \frac{\Delta y}{|\tan(\theta \pm \mu)|_{\max}} \tag{11.76}$$

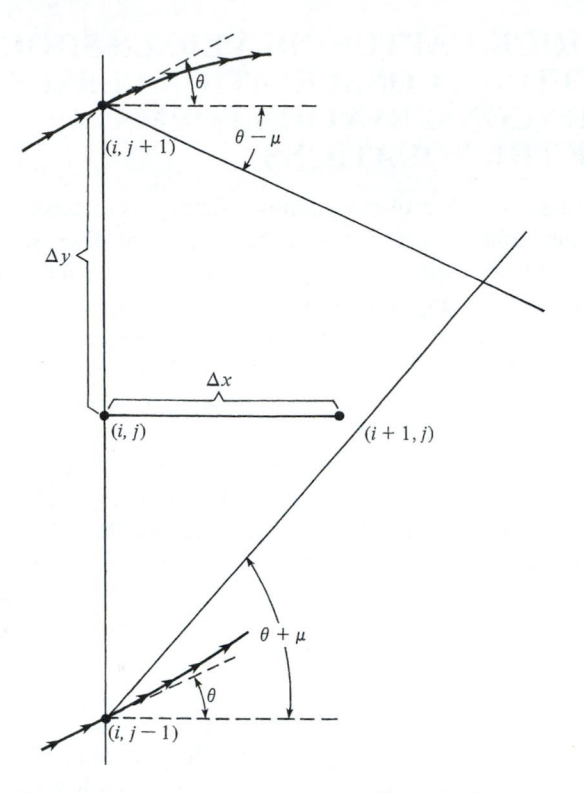

Figure 11.21 | Illustration of the stability criterion for steady two-dimensional supersonic flow.

Equation (11.76) is called the Courant–Friedrichs–Lewy criterion, the so-called CFL criterion. The interpretation of this criterion is shown in Fig. 11.21. Here, a vertical column of grid points $(i, j - 1), (i, j), (i, j + 1)$, etc., is considered, with Δy the spacing between adjacent points. Characteristic lines with angles $\theta + \mu$ and $\theta - \mu$ are drawn through points $(i, j - 1)$ and $(i, j + 1)$, respectively. The value of Δx allowed by Eq. (11.76) falls within the domain defined by these characteristic lines. If Δx is larger than stipulated by Eq. (11.76), then grid point $(i + 1, j)$ falls outside the domain of these characteristics, and the numerical computation will be unstable.

Note that, from Eq. (11.76), there can be a different value of Δx associated with each vertically arrayed grid point, i.e., a different Δx reaching downstream from each of points $(i, j - 1), (i, j), (i, j + 1)$, etc. However, the value actually used for Δx should be the same for each of these points so that we have a uniformly spaced grid in the x direction for the next column of grid points, i.e., the spacing between points $(i, j - 1)$ and $(i + 1, j - 1)$ should be the same as between (i, j) and $(i + 1, j)$, and so forth. Hence, in Eq. (11.76), the particular constant value of Δx to be used for all the vertically arrayed grid points is that associated with the maximum value of $|\tan(\theta \pm \mu)|$ in Eq. (11.76); this is the reason for the subscript max in Eq. (11.76).

11.15 | SHOCK CAPTURING VERSUS SHOCK FITTING; CONSERVATION VERSUS NONCONSERVATION FORMS OF THE EQUATIONS

Consider the supersonic flow over a sharp-nosed body, as sketched in Fig. 11.22. The downstream-marching, explicit finite-difference method discussed in the previous sections can readily be used to calculate the supersonic flowfield between the body and the shock wave, starting from a line of initial data near the nose. These initial data are usually obtained by assuming the nose of the body to be a sharp wedge, and using the results of Chap. 4 for starting conditions. If the body is three-dimensional, the nose can usually be assumed to be a cone, and the results of Chap. 10 can be used for the initial data. In Fig. 11.22, the body represents one set of boundary conditions, and the shock wave constitutes a second set. The methods discussed in Sec. 11.13 can be used for these boundaries. Because the shock wave in Fig. 11.22 is assumed to be a discontinuity, it is used as one of the boundaries of the flowfield and is determined by matching the oblique shock relations with the interior flowfield. This approach is defined as *shock fitting,* in contrast with an alternative approach, sketched in Fig. 11.23. Here, the finite-difference grid is extended far ahead of and above the body, and free-stream conditions are assumed along the outer boundaries. Again applying the algorithm in Sec. 11.12, the flowfield over the finite-difference grid can be calculated. The shock wave will automatically appear within the grid as a region of large gradients smeared over several grid points (the grid is in reality much finer than sketched in Fig. 11.23). Consequently, shock waves do not have to be explicitly assumed; they will appear at those locations in the flowfield where they belong. Such an approach is called *shock capturing.* An obvious advantage of shock-capturing techniques is that no *a priori* knowledge about the number or location of shock waves is needed. A disadvantage is that the shock is numerically smeared rather than

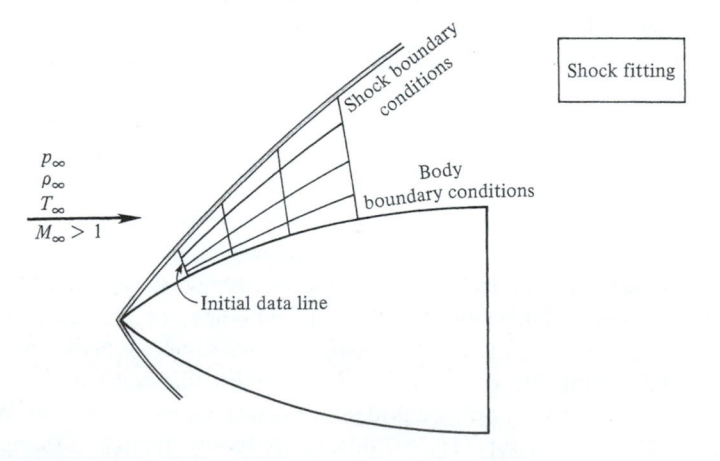

Figure 11.22 | Mesh for the shock-fitting finite-difference approach.

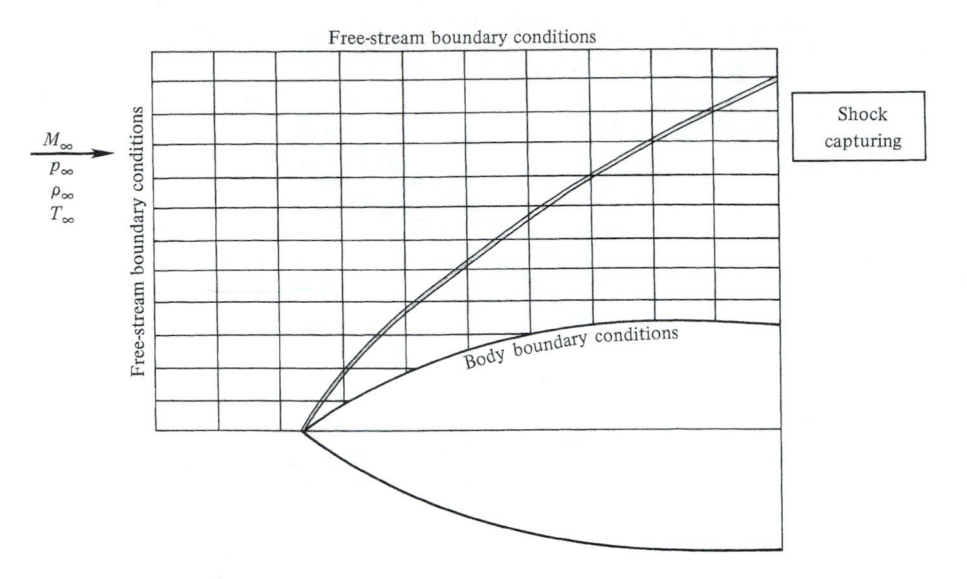

Figure 11.23 | Mesh for the shock-capturing finite-difference approach.

discontinuous; also, the grid points in the free stream are essentially wasted insofar as useful flowfield information is concerned.

Connected with the above considerations is the form of the governing equations. In Chap. 6, both conservation and nonconservation forms of the partial differential equations were obtained. It is generally acknowledged that the equations must be used in conservation form for the shock-capturing approach; this is to ensure conservation of the flux of mass, momentum, and energy across the shock waves within the grid. However, for the shock-fitting approach, either the conservation or nonconservation form of the equations can be used—MacCormack's technique discussed in Sec. 11.12 applies to both systems. The nonconservation form has a numerical advantage: The primitive variables ρ, u, v, p, T, etc., are calculated directly from the equations. In contrast, when the conservation form is used, the fluxes $\rho u, \rho v, \rho u^2$, etc., are calculated directly from the equations, and the primitive variables must be backed out; this causes extra computation and computer time. However, beyond these considerations, there is no reason to favor one form over the other; the choice is up to the user.

11.16 | COMPARISON OF CHARACTERISTICS AND FINITE-DIFFERENCE SOLUTIONS WITH APPLICATION TO THE SPACE SHUTTLE

It is suitable to conclude the technical portion of this chapter with a direct comparison of the method of characteristics with the finite-difference approach. The calculation of the flowfield around a three-dimensional body closely approximating NASA's

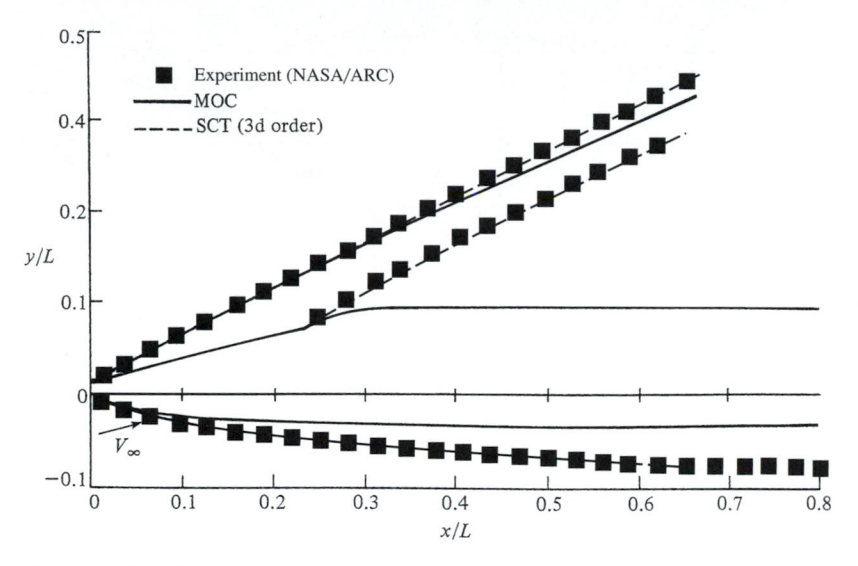

Figure 11.24 | Shock waves on a space shuttle configuration; comparison between method of characteristics and finite-different calculations (after Rakich and Kutler). $M_\infty = 7.4, \alpha = 15.3°$.

Space Shuttle is used as an example. The results given here are obtained from the work of Rakich and Kutler, which is described in detail in Ref. 45.

The body is illustrated in Fig. 11.24. The calculations are made for an angle of attack of 15.3°. In the immediate vicinity of the blunt nose, the flow is a mixed subsonic-supersonic region which is calculated by a blunt body method such as will be described in Chap. 12. Downstream of this region the flow is completely supersonic. Here, two sets of calculations are made: (1) a three-dimensional semicharacteristics calculation (MOC) as described in Sec. 11.10, and (2) a third-order-accurate shock-capturing finite-difference version of MacCormack's technique (SCT) based on the philosophy presented in Secs. 11.12 through 11.15. In Fig. 11.24, the shock waves emanating from the nose and canopy regions are shown for both sets of calculations; in addition, experimental data obtained at the NASA Ames Research Center are also shown. Even though the shape of the wind tunnel model in the canopy region varied slightly from the shape fed into the computer calculations, in general the agreement is quite good. A front view of the body and the corresponding shock waves is given in Fig. 11.25. Again, reasonable agreement is obtained. The slight discrepancy that occurs further downstream is due to numerical problems with the method of characteristics on the leaward (upper) side of the body—slight inaccuracies caused by the interpolation for data on the C_+ characteristic. The surface pressure distributions along the top ($\phi = 180°$) and bottom ($\phi = 0°$) of the vehicle are shown in Fig. 11.26. Again, good agreement is obtained between the two sets of calculations and experiment.

With regard to computer time for the two sets of calculations, Rakich and Kutler report that, on a single point basis, the time required for the elaborate three-dimensional

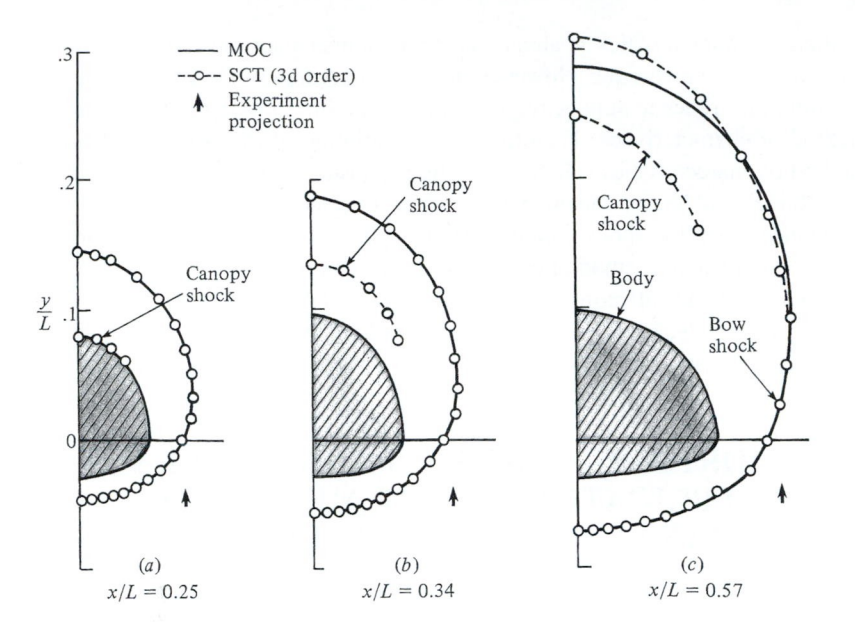

Figure 11.25 | Circumferential shock shape on a space shuttle configuration (after Rakich and Kutler), $M_\infty = 7.4$, $\alpha = 15.3°$.

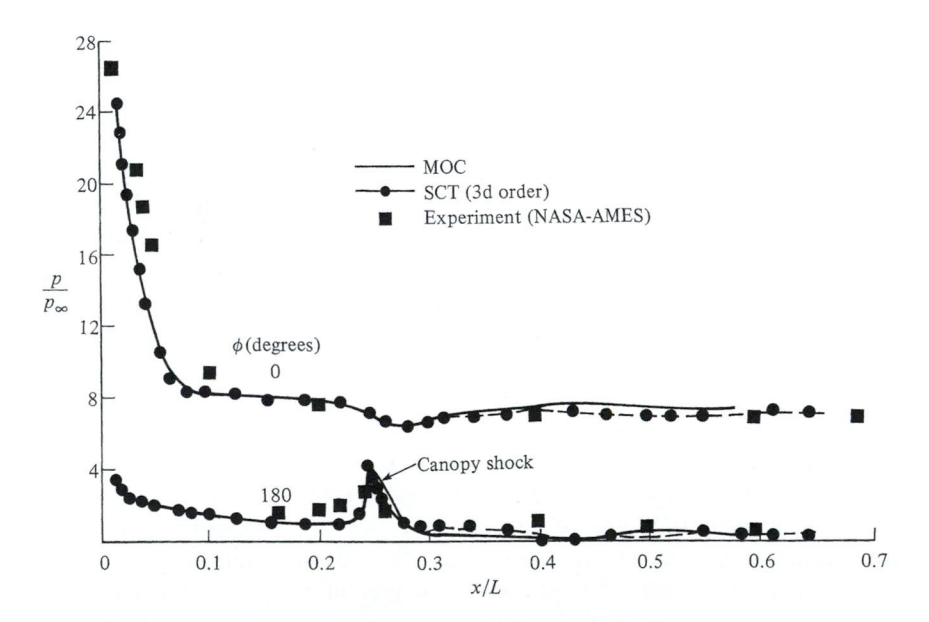

Figure 11.26 | Longitudinal surface pressure distribution on a space shuttle configuration (after Rakich and Kutler). $M_\infty = 7.4$, $\alpha = 15.3°$.

method of characteristics is about four times longer than the more straightforward finite-difference technique. However, in order to accurately capture the shock waves, the finite-difference technique required almost six times more grid points than did the method of characteristics. Therefore, for the solution of the complete flowfield, the method of characteristics solution was slightly faster. However, in their final evaluation, Rakich and Kutler conclude that, "when considering its versatility and computational efficiency, the shock-capturing (finite-difference) technique seems to have the edge on the present method of characteristics program." This is not a general conclusion to be applied to all cases; however, it is clear that the method of characteristics and finite-difference techniques are on reasonably equal footing for the numerical solution of steady, inviscid, supersonic flows.

11.17 | HISTORICAL NOTE: THE FIRST PRACTICAL APPLICATION OF THE METHOD OF CHARACTERISTICS TO SUPERSONIC FLOW

Ludwig Prandtl and Adolf Busemann—two names that occur with regularity throughout the history of compressible flow (see Secs. 4.16 and 9.9)—are responsible for the first successful implementation of the method of characteristics to supersonic flow problems. The *theory* of characteristics was developed by mathematicians to solve general systems of partial differential equations of the first order. Primarily responsible for this mathematical development were the French mathematician Jacques Salomon Hadamard in 1903 and the Italian mathematician Tullio Levi-Civita in 1932. However, in 1929, Prandtl and Busemann coauthored a classical paper in which the *method* of characteristics was applied for the first time to the calculation of two-dimensional supersonic flow. Entitled "Naherungsverfahren zur Zeichnerischen Ermittlung von Ebenen Stromungen mit Uberschallgeschwindigkeit" ("Procedure for the Graphical Determination of Plane Supersonic Flows") and published in *Stodola Festschrigt,* p. 499 (1929), this work provided graphs of the characteristics in the hodograph plane for two-dimensional flow with $\gamma = 1.4$. Furthermore, they showed that the physical characteristics (Mach lines) are perpendicular to the hodograph characteristics and can be obtained from the latter with the aid of a right triangle. This graphical construction was then used by Prandtl and Busemann to construct a contoured nozzle, as illustrated in Fig. 11.27. The approach given by Prandtl and Busemann was a major contribution to the development of compressible flow, and the graphical technique laid out in their paper is still taught today in standard university classes on compressible flow. (In our discussion of the method of characteristics in this chapter, however, we have chosen a numerical rather than a graphical approach for the convenience of computer implementation.)

The experience gained from this work was utilized a few years later by Busemann to design a contoured supersonic nozzle for the first practical supersonic wind tunnel in history, shown in Fig. 11.28. Designed during the early 1930s, this tunnel represented the epitome of the compressible flow research that revolved around Prandtl and his colleagues at Göttingen during the first half of the twentieth century.

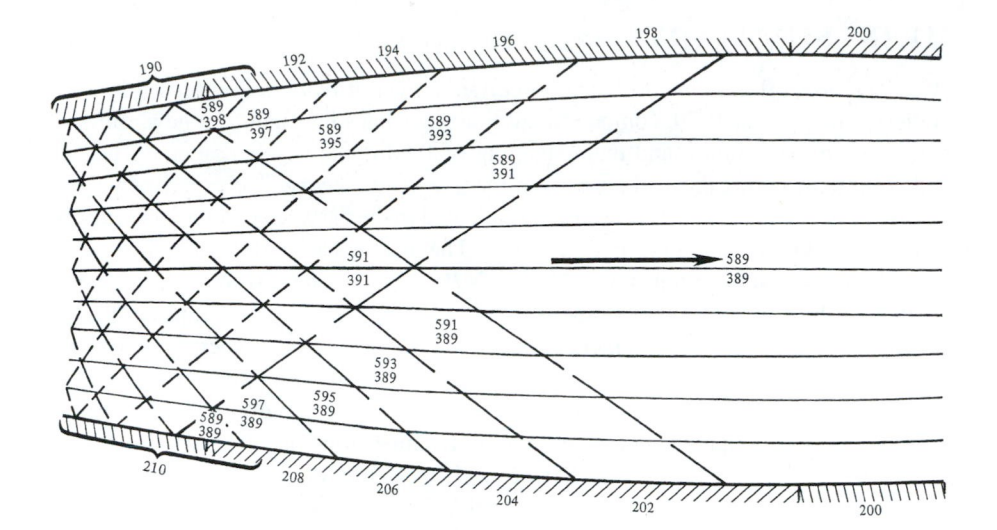

Figure 11.27 | Nozzle contour designed by means of the method of characteristics, after Prandtl and Busemann, 1929.

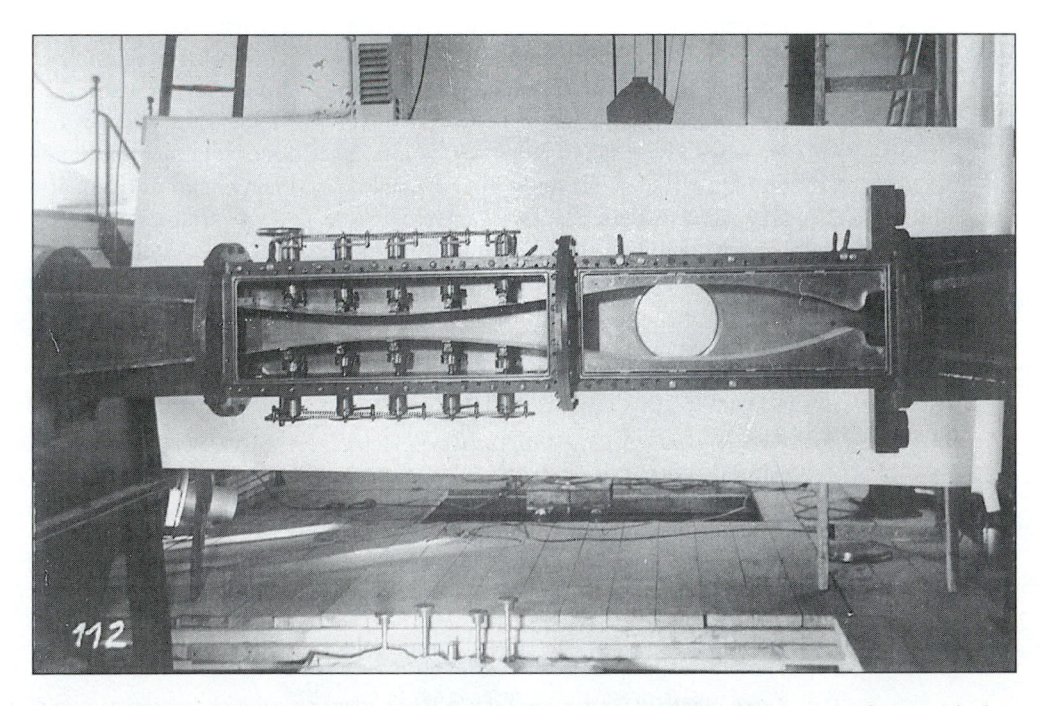

Figure 11.28 | Busemann's supersonic wind tunnel from the early 1930s. This was the first practical supersonic wind tunnel in history. The nozzle was designed by the method of characteristics as developed by Prandtl and Busemann in 1929.

11.18 | SUMMARY

Computational fluid dynamics is an important aspect of *modern* compressible flow; indeed, since about 1970, computational fluid dynamics has opened a new, third dimension in the solution and understanding of fluid dynamic phenomena. The two other dimensions are those of pure experiment and pure theory. The experimental tradition in physical science was solidly established in the early seventeenth century by the work of Galileo and his contemporaries. The methods and use of pure theory had their fundamental beginnings with Newton's *Principia* in 1687, with major advancements in fluid dynamics by Bernoulli and Euler in the early and mid-eighteenth century. Virtually all advancements in physical science and engineering since then were products of the two dimensions of pure theory and pure experiment working together. Today, computational fluid dynamics constitutes a new, third dimension, which directly complements the two previous dimensions of pure experiment and pure theory. The purpose of this chapter has been to introduce the basic philosophy and a small amount of the methodology of this new third dimension.

The method of characteristics, which had its origins somewhat earlier and independent from that of modern computational fluid dynamics, takes this tact:

1. Find those directions in space along which the flowfield derivatives are indeterminate and across which they may be discontinuous. These are called the characteristic curves (or surfaces, in three dimensions).
2. Find the equations, obtained from a proper treatment of the continuity, momentum, and energy equations, which hold along the characteristic lines (or surfaces). These are called the compatibility equations. These equations have the advantage of being in one less space dimension than the actual flow problem. That is, for three-dimensional flows, the compatibility equations are partial differential equations in two independent variables; for a two-dimensional flow, the compatibility equations are ordinary differential equations (in one independent variable). Furthermore, if the flow is two dimensional *and* irrotational, the compatibility equations reduce one step further, namely, to algebraic equations.

To be more precise, we have discussed these four cases:

1. *Two-dimensional, irrotational flow.* Here, there are two characteristic lines through any given point, the right- and left-running Mach lines (the C_- and C_+ characteristics, respectively). The compatibility equations are the algebraic relations:

$$\theta + \nu = K_- \quad \text{(along the } C_- \text{ characteristic)}$$

$$\theta - \nu = K_+ \quad \text{(along the } C_+ \text{ characteristic)}$$

2. *Asixymmetric, irrotational flow.* Here, there are two characteristic lines, again the right- and left-running Mach lines. The compatibility equations are ordinary differential equations given by Eqs. (11.52) and (11.53).

3. *Two-dimensional rotational flow.* Here, there are three characteristic lines through any given point, namely, the right- and left-running Mach waves and the streamline. The compatibility equations are ordinary differential equations represented by Eqs. (11.54)–(11.56).

4. *Three-dimensional flow.* Here, the characteristics are three-dimensional surfaces. At any given point, they are the Mach cones emanating from that point and a stream surface through the point. The compatibility equations are partial differential equations. However, using the method of "semicharacteristics" introduced by Rakich, the problem can be solved by means of the solution of ordinary differential equations (see Sec. 11.10).

In finite-difference methods, the partial derivatives in the governing continuity, momentum, and energy equations are replaced by algebraic difference quotients written in terms of the flowfield variables at distinct grid points in the flow. The problem then reduces to the solution of vast numbers of algebraic equations where the unknowns are the flowfield variables at the grid points. All finite-difference methods have as their source a Taylor series expansion. One particular method that has been widely used is MacCormack's method, described in Sec. 11.12. There are many different variations of finite-difference solutions in use; some are explicit and others are implicit; some use shock capturing and others use shock fitting. These concepts are discussed in Secs. 11.11 and 11.15.

The field of computational fluid dynamics is rapidly evolving at this time of writing. New advances are being made that improve on both the accuracy of solution and the speed of computation. Finite-volume and finite-element methods are becoming widespread, in some cases supplanting the older finite-difference methods. Improvements in smoothing the numerical results are being made with such schemes as the *total variation diminishing* (TVD) approach. Shock waves are being made sharper and better defined by means of *upwind differencing*. We have not discussed these matters here; they are the purview of more advanced books and papers. The reader is encouraged to consult the current literature for more details.

Finally, we note that the problems treated in this chapter are steady flows where the Mach number is supersonic at every point in the flow. For this type of flow, both the method of characteristics and the finite difference methods are *downstream marching*. That is, for the solution of a given problem whether it be an internal flow through a duct or an external flow over a supersonic body, the solution begins at an initial data line along which the flow properties are known and the unknown steady flowfield variables are calculated by moving in progressive increments in the downstream direction.

PROBLEMS

11.1 Using the method of characteristics, compute and graph the contour of a two-dimensional minimum-length nozzle for the expansion of air to a design exit Mach number of 2.

11.2 Repeat Prob. 11.1, except consider a nozzle with a finite expansion section which is a circular arc with a diameter equal to three throat heights. Compare this nozzle contour and total length with the minimum-length nozzle of Prob. 11.1.

11.3 Consider the external supersonic flow over the pointed body sketched in Fig. 11.22. Outline in detail how you would set up a method-of-characteristics solution for this flow.

CHAPTER 12

The Time-Marching Technique: With Application to Supersonic Blunt Bodies and Nozzles

Bodies in going through a fluid communicate their motion to the ambient fluid by little and little, and by that communication lose their own motion and by losing it are retarded.

Roger Coats, 1713, in the preface to the Second Edition of Newton's Principia

PREVIEW BOX

The Lockheed F-104, shown in the three-view in Fig. 12.1, was the first fighter designed for sustained flight at Mach 2. This airplane embodies excellent supersonic aerodynamics—slender body, pointed nose, thin wings with a sharp leading edge, low-aspect-ratio wings—all designed to minimize the strength of the shock waves on the airplane and hence to reduce wave drag at supersonic speeds. Extrapolating this philosophy to the design of much faster, hypersonic aircraft designed to fly at, say, Mach 20, you might think that such aircraft would be extreme examples of very slender bodies, with very thin wings, supersharp leading edges, etc. However, examine Fig. 12.2, which shows the Space Shuttle, one of today's most common hypersonic vehicles. Notice the blunt nose, thick body, and thick wings with blunt leading edges. Clearly, the design philosophy used for the Space Shuttle is almost the antithesis of that for the F-104. The difference is caused by aerodynamic heating, which becomes severe at hypersonic speeds. The design of hypersonic vehicles is dominated by the need to reduce aerodynamic heating to

the body; drag usually becomes a less significant consideration. Aerodynamic heating is dramatically less for blunt bodies compared to that for slender bodies, and that is why all hypersonic vehicles designed to date have blunt noses, blunt leading edges, etc. A qualitative discussion as to *why* blunt bodies minimize aerodynamic heating, and the history of the origin of this revolutionary design concept, is given in Chap. 1 of Ref. 104. Quantitative theoretical proof that aerodynamic heating varies inversely as the square root of the nose radius is given in Ref. 119. In short, blunt-nosed bodies have become important configurations for very high speed vehicles.

The qualitative aspects of the flow over a supersonic blunt body are discussed in Sec. 4.12. When the blunt body concept was first introduced for hypersonic vehicles in the early 1950s, there existed no theoretical solutions to such a flow field. At that time, the "supersonic blunt body problem" became a subject of intense research, and for the next 15 years platoons of researchers and many millions of dollars were devoted

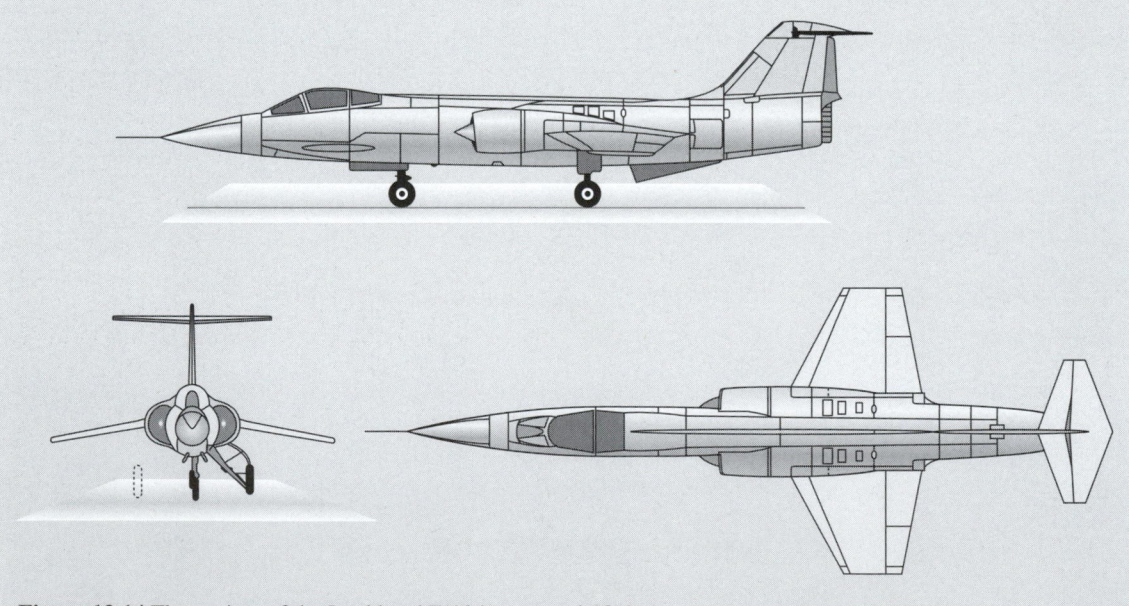

Figure 12.1 | Three-view of the Lockheed F-104 supersonic fighter.

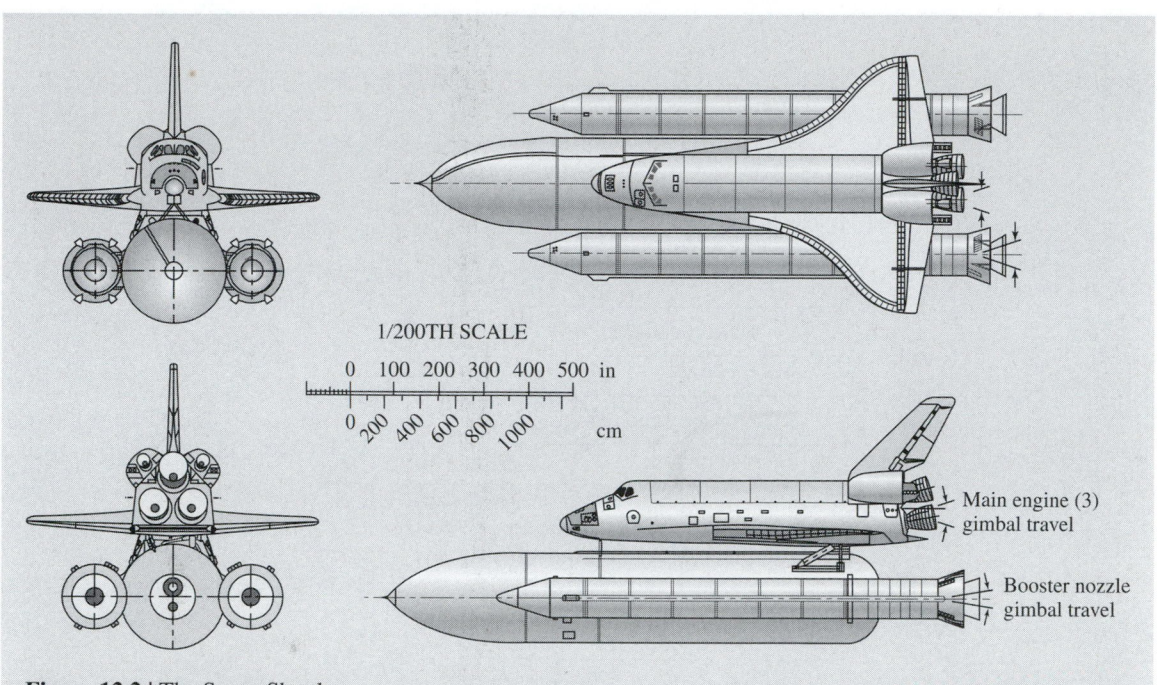

Figure 12.2 | The Space Shuttle.

to the theoretical solution of the flowfield over a blunt body moving at supersonic or hypersonic speeds—without any reasonable success. The problem was related to the mixed subsonic-supersonic nature of the flow behind the curved, detached shock wave over the body, as shown in Fig. 4.29. Whatever technique that would work in the subsonic region would fall apart in the supersonic region, and whatever method was good for the supersonic region (such as the methods of characteristics) did not apply to the subsonic region.

Then, in the mid-1960s, a breakthrough was achieved. The supersonic blunt body problem was solved by means of a time-marching numerical solution. The time-marching aspect makes all the difference. It allows for the straightforward calculation of both the subsonic and supersonic regions by a single uniform technique. This breakthrough was so dramatic that the calculation of the flow over a blunt body moving at supersonic or hypersonic speeds is routine today. The time-marching solution of the blunt body flow is now the industry standard.

The present chapter introduces the concept of time-marching solutions, and then discusses in detail the time-marching solution of the blunt body problem. This material is particularly important because modern computational fluid dynamics uses time-marching to solve many types of problems, not just the blunt body problem. Indeed, time-marching is one of the dominant features of modern CFD.

The roadmap for this chapter is given in Fig. 12.3. We first introduce the philosophy of time-marching solutions by way of application to a familiar problem, namely, the quasi-one-dimensional nozzle flow discussed in Chap. 5. This is followed by a discussion of the stability criterion for time-marching solutions. Then, in preparation for the blunt body problem, we define the limiting characteristic curves in the blunt body flow, and take a side excursion to consider Newtonian theory for the prediction of pressure coefficient on the surface of a body in a flow. Finally, we deal with the main aspect of this chapter—the application of the time-marching method to the supersonic blunt body problem.

(*continued on next page*)

(continued from page 433)

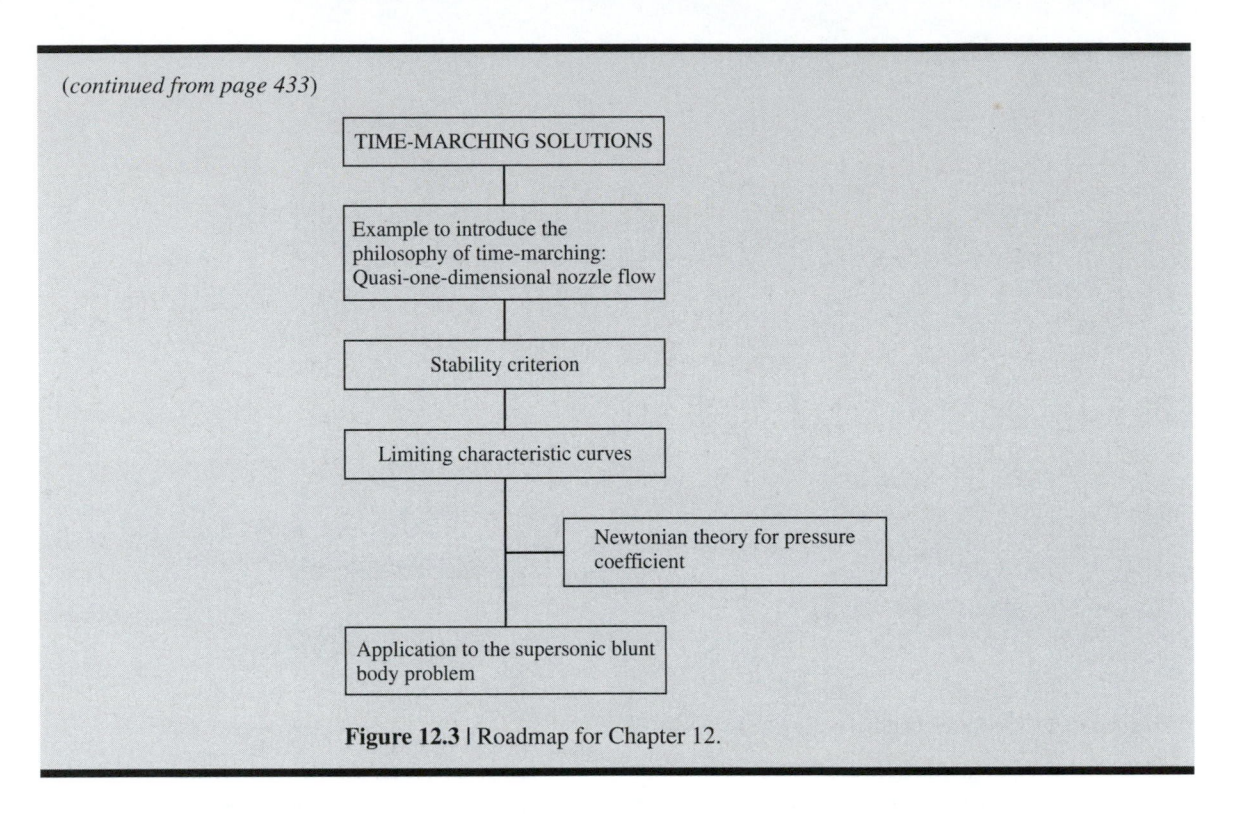

Figure 12.3 | Roadmap for Chapter 12.

12.1 | INTRODUCTION TO THE PHILOSOPHY OF TIME-MARCHING SOLUTIONS FOR STEADY FLOWS

We have seen from Chap. 11 that steady supersonic flowfields are governed by hyperbolic differential equations, whereas steady subsonic flowfields are described by elliptic differential equations. There are many applications where flowfields contain both subsonic and supersonic regions, such as the flow over a blunt body moving at supersonic velocity as sketched in Fig. 12.4a and the expansion to supersonic speeds through a convergent-divergent nozzle, as sketched in Fig. 12.4b. Both of these examples are mixed subsonic-supersonic flows, where the sonic line divides the two regions. The fact that the nature of the governing equations changes from elliptic to hyperbolic across the sonic line causes severe mathematical and numerical difficulties—so much so that steady-flow solutions of the subsonic and supersonic regions are usually treated separately and differently, and then somehow patched in the transonic region near the sonic line. So far, no practical steady-flow technique exists that can uniformly treat both the subsonic and supersonic regions of a general flowfield of arbitrary extent. Compounding this problem was the discovery in the early 1950s that high-speed missiles should have blunt noses to reduce aerodynamic heating. Almost overnight the supersonic blunt body problem, with its mixed subsonic-supersonic flowfield, became a central focus in theoretical and experimental aerodynamics. During the period

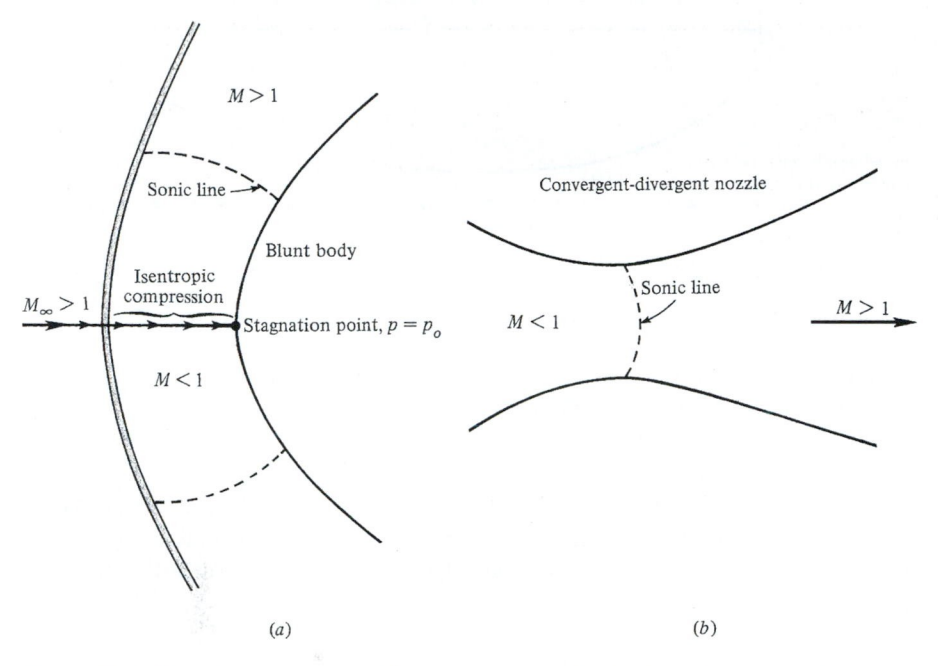

Figure 12.4 | Illustrations of mixed subsonic-supersonic flowfields, with curved sonic lines.

between 1955 and 1965, numerous blunt body solutions were advanced, all with several greater or lesser disadvantages or defects. (During this period it was common to have complete sessions during meetings of the Institute of Aeronautical Sciences—now the AIAA—just to discuss the blunt body problem.) Then, in the mid-1960s, a breakthrough occurred. The time-marching technique for the solution of steady flows was developed, and in 1966, Moretti and Abbett published the first truly practical solution for the supersonic blunt body problem (see Ref. 47). Since then, time-marching (sometimes called "time-dependent") solutions have become an important segment of computational fluid dynamics. The purpose of this chapter is to introduce the philosophy, approach, and some results of this very powerful technique.

The philosophy and approach of the time-marching technique is best described in the context of a simple example. Consider the quasi-one-dimensional flow of a calorically perfect gas through a given convergent-divergent nozzle, as studied in Chap. 5. The reservoir conditions p_o, ρ_o, T_o are given and held constant with time. Split the nozzle into a number of grid points in the flow direction, as sketched in Fig. 12.5. Arbitrarily *assume* values for all the flow variables, p, ρ, u, etc., at all the grid points except the first, which is associated with the fixed reservoir conditions. These are by no means the correct solutions (unless you are a magician at making the correct guess). Consider these guessed values as *initial* conditions throughout the flow. Then advance the flowfield variables at each grid point *in steps of time* by means of the Taylor's series

$$g(t + \Delta t) = g(t) + \left(\frac{\partial g}{\partial t}\right)_{\text{ave}} \Delta t + \cdots \tag{12.1}$$

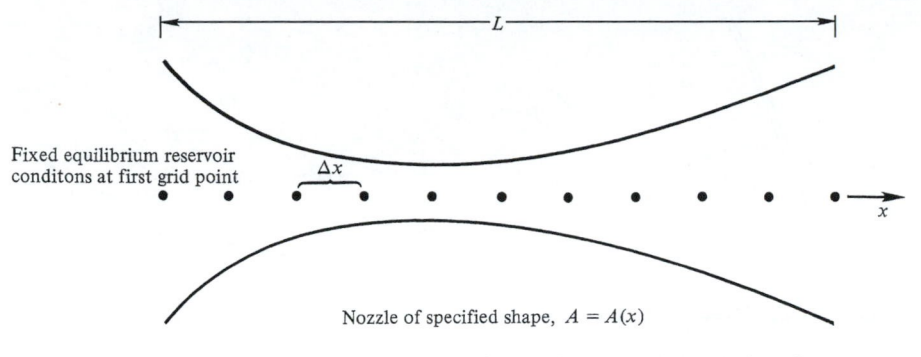

Figure 12.5 | Coordinate system and grid points for the time-marching solution of quasi-one-dimensional flow through a nozzle.

where g denotes p, ρ, T, or u, and Δt is a small increment in time chosen to satisfy certain stability criteria to be discussed in Sec. 12.2. For example, if at a given grid point we know u at time t, then we can calculate u at time $t + \Delta t$ at the same grid point from Eq. (12.1) *if* we can find a value for the time derivative $(\partial u/\partial t)_{\text{ave}}$. Let us pause for a moment and consider in the next paragraph where this time derivative comes from.

Obviously, Eq. (12.1) is just mathematics; the physics of the problem enters in the calculation of the time derivatives, which are obtained from the *unsteady* conservation equations. For the quasi-one-dimensional problem considered here, the governing *unsteady* equations can be obtained by applying the fundamental integral equations of continuity, momentum, and energy, Eqs. (2.2), (2.11), and (2.20), respectively, to an infinitesimally small control volume of variable area, as sketched in Fig. 5.7. For example, Eq. (2.2),

$$\frac{\partial}{\partial t} \iiint\limits_{\mathcal{V}} \rho \, d\mathcal{V} = - \oiint\limits_{S} \rho \mathbf{V} \cdot d\mathbf{S}$$

when applied to the control volume in Fig. 5.7 yields (noting that $d\mathcal{V} = A\, dx$)

$$\frac{\partial}{\partial t}(\rho A \, dx) = -[-\rho u A + (\rho + d\rho)(u + du)(A + dA)] \tag{12.2}$$

Ignoring products of differentials in Eq. (12.2), the result is

$$\frac{\partial(\rho A)}{\partial t} \, dx = -d(\rho u A) \tag{12.3}$$

Equation (12.3) can be more formally written as

$$\frac{\partial(\rho A)}{\partial t} = -\frac{\partial(\rho u A)}{\partial x} \tag{12.4}$$

Furthermore, since $A = A(x)$ does not depend on time, Eq. (12.4) becomes

$$\frac{\partial \rho}{\partial t} = -\frac{1}{A} \frac{\partial (\rho u A)}{\partial x} \tag{12.5}$$

Equation (12.4) or (12.5) represents the continuity equation for unsteady quasi-one-dimensional flow in partial differential equation form. By similar applications of Eqs. (2.11) and (2.20) to Fig. 5.7, and with some manipulation, we find (you should demonstrate this to yourself)

$$\frac{\partial u}{\partial t} = -\frac{1}{\rho} \left(\frac{\partial p}{\partial x} + \rho u \frac{\partial u}{\partial x} \right) \tag{12.6}$$

$$\frac{\partial e}{\partial t} = -\frac{1}{\rho} \left[p \frac{\partial u}{\partial x} + \rho u \frac{\partial e}{\partial x} + pu \frac{\partial \ln A}{\partial x} \right] \tag{12.7}$$

Equations (12.6) and (12.7) are the momentum and energy equations, respectively, for unsteady quasi-one-dimensional flow. Along with the perfect gas relations

$$p = \rho R T \tag{12.8}$$

$$e = c_v T \tag{12.9}$$

these equations are sufficient for calculating the flow we are considering.

Now return to the line of thought embodied in Eq. (12.1). The time derivative in Eq. (12.1) can now be obtained from Eqs. (12.5) through (12.7). Note that, in these equations, the time derivatives on the left-hand sides are given in terms of the spatial derivatives on the right-hand sides. These spatial derivatives are known—they can be expressed as finite differences from the known flowfield values at time t. Hence, Eqs. (12.5) through (12.7), along with (12.8) and (12.9), allow the calculation of $(\partial g/\partial t)$, evaluated at time t. If we desired first-order accuracy, then this value of $(\partial g/\partial t)_t$ in Eq. (12.1) would be enough to calculate $g(t + \Delta t)$. However, for second-order accuracy, $(\partial g/\partial t)_{\text{ave}}$ in Eq. (12.1) must be an average between t and $t + \Delta t$. This average derivative can be calculated by means of MacCormack's technique, first introduced in Sec. 11.12. For the present time-marching technique, MacCormack's predictor-corrector scheme is as follows.

Predictor Step. Calculate $(\partial g/\partial t)_t$ from Eqs. (12.5) through (12.9), using forward spatial differences on the right-hand sides from the known flowfield at time t. Use this value to obtain a *predicted* value of g at time $t + \Delta t$ from

$$\bar{g}(t + \Delta t) = g(t) + \left(\frac{\partial g}{\partial t} \right)_t \Delta t$$

Corrector Step. Using rearward spatial differences, insert the above values of $\bar{g}$ into Eqs. (12.5) through (12.9) to calculate a predicted value of $\overline{\partial g/\partial t}$. Then form the average derivative as

$$\left(\frac{\partial g}{\partial t} \right)_{\text{ave}} = \frac{1}{2} \left[\left(\frac{\partial g}{\partial t} \right)_t + \left(\overline{\frac{\partial g}{\partial t}} \right) \right] \tag{12.10}$$

Finally, insert the average derivative from Eq. (12.10) into Eq. (12.1) to obtain the corrected value, $g(t + \Delta t)$. This is the desired second-order-accurate value of g at time $t + \Delta t$.

Now we come to the *crux* of the time-marching technique. Using Eq. (12.1), with $(\partial g/\partial t)_{\text{ave}}$ calculated as outlined above, values of g at each grid point in Fig. 12.5 can be calculated in steps of time, starting from the guessed, arbitrary initial conditions. The values of the flowfield variables (represented by g) will change for each step in time. However, after a number of time steps, these changes will become smaller and smaller, finally asymptotically approaching a steady value. *It is this steady flowfield we are interested in as our solution—the time-marching technique is simply a means to achieve this end.* For example, Fig. 12.6 gives the temperature distribution for a nozzle with an area variation given by $A/A^* = 1 + 2.2(x - 1.5)^2$. Here, the nozzle throat is at $x = 1.5$; $x < 1.5$ is the subsonic section and $x > 1.5$ is the supersonic section. The dashed line in Fig. 12.6 represents the guessed initial temperature distribution at time $t = 0$. It is arbitrarily taken as a linear variation. The solid curves in Fig. 12.6 give the transient distributions after 8, 16, 32, 120, and 744 time steps, using the time-marching procedure described above. By the 744th time step, the distribution has become sufficiently invariant with time for this to be taken as the final steady state. This final steady state agrees with the classical results obtained from Chap. 5. This behavior is further illustrated in Fig. 12.7, which shows the variation of mass flow $\rho u A$ through the

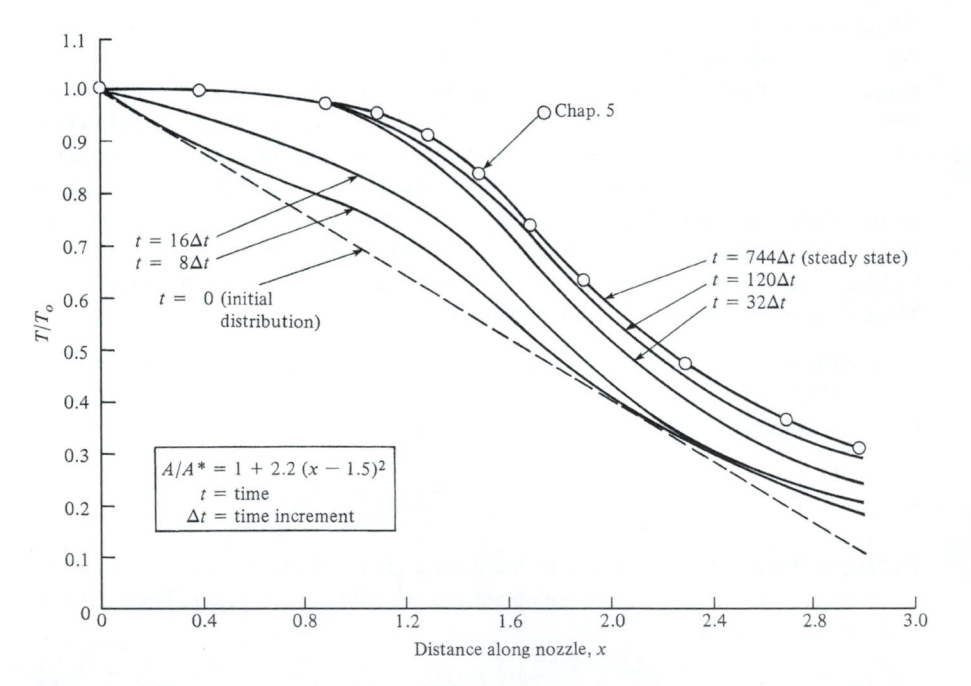

Figure 12.6 | Transient and final steady state temperature distributions for a calorically perfect gas obtained from the time-marching technique.

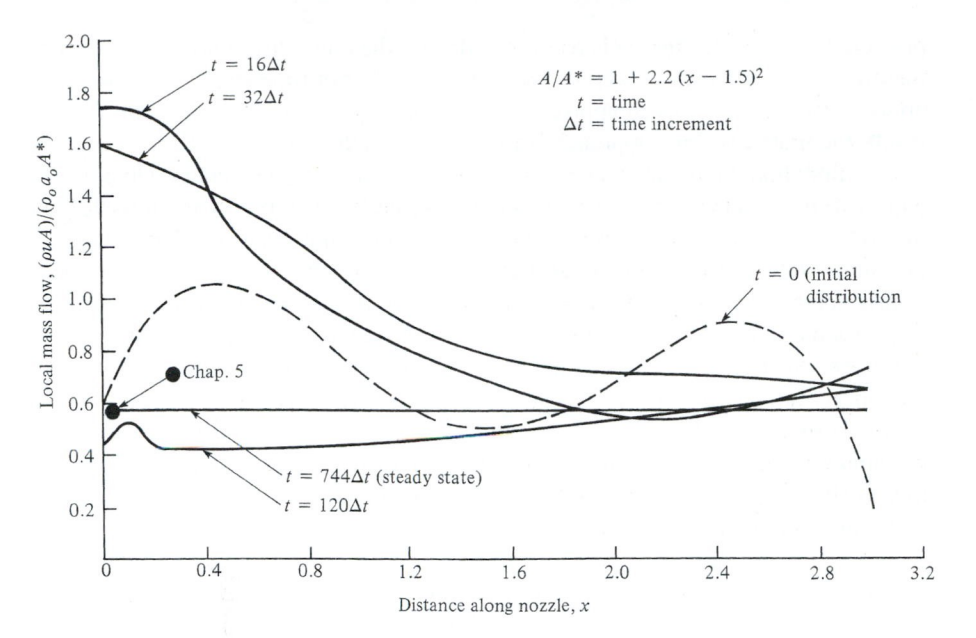

Figure 12.7 | Transient and final steady state mass-flow distributions for a calorically perfect gas obtained from the time-marching technique.

nozzle at different times. Again, the dashed line is the initial distribution due to the assumed flowfield values at time zero. The solid curves show the intermediate distributions after 16, 32, 120, and 744, time steps. Note that, at $t = 744\Delta t$, the mass flow distribution has become a straight, horizontal line—$\rho u A$ has become a constant throughout the nozzle, as it should be for steady flow as discussed in Chap. 5. Furthermore, it is the correct value as shown by comparison with the classical results from Chap. 5.

Please note that the above time-marching solution for the quasi-one-dimensional flow of a calorically perfect gas was chosen simply to illustrate the time-dependent technique. The closed-form algebraic solutions for nozzle flows given in Chap. 5 are considerably simpler than this finite-difference solution. However, for quasi-one-dimensional nozzle flows of nonequilibrium gases, such as may occur in high-temperature chemically reacting flows, the time-dependent technique present here has definite advantages; the classical results of Chap. 5 are no longer valid for such high-temperature flows. This situation will be addressed in Chap. 17. For further background and details on the application of the time-marching technique to quasi-one-dimensional nozzle flows, see Refs. 48 through 50.

Let us recapitulate. The essence of the time-marching technique to solve steady flows is as follows. For a given flow problem with prescribed steady boundary conditions, set down some arbitrary initial values of the flowfield at each grid point. Then advance these flow properties in steps of time using Eq. (12.1), where the time derivatives are obtained from the unsteady equations of motion. MacCormack's predictor-corrector technique is recommended for the finite-difference calculation of these time derivatives, as given above. After a number of time steps, the flow properties at each grid point will approach a steady state. This steady state is the desired result; the time-marching approach is just a means to this end.

At first thought, the introduction of time as an extra independent variable would appear to be an unnecessary complication for a steady-flow problem. However, on the contrary, for some problems it becomes a striking simplification. Consider again the blunt body and two-dimensional nozzle flows sketched in Fig. 12.4. As mentioned before, there is virtually no satisfactory, uniformly valid, steady state technique for the solution of these mixed flows—the mixed nature of the elliptic subsonic region and the hyperbolic supersonic region essentially rules out such a solution. On the other hand, the unsteady equations of motion [such as Eqs. (12.5) through (12.7)] are *hyperbolic with respect to time,* regardless of whether the flow is locally subsonic or supersonic. Hence, the complete flowfields shown in Fig. 12.4 lend themselves to a well-posed initial value problem with respect to time. Therefore, the time-marching technique becomes a very powerful tool for the solution of such mixed flows, being uniformly valid throughout the flowfield. (Note that the unsteady wave motion discussed in Chap. 7 is another example of a system which is hyperbolic in time, a fact which we took advantage of with our characteristics solutions for one-dimensional finite wave motion. However, in Chap. 7 we were concerned with the time variations themselves, whereas in the present chapter we are concerned with the final steady state as an asymptotic convergence of the transient flow.)

12.2 | STABILITY CRITERION

The time-marching technique with the use of MacCormack's approach as outlined here is explicit. For such an explicit solution, the value of Δt in Eq. (12.1) cannot be any arbitrary value; indeed, it must be less than or equal to some maximum value. This maximum value is usually estimated from a stability analysis performed on a set of approximate, linear equations after Courant, Friedrichs, and Lewy (Ref. 51)—the so-called CFL criterion. Without going into the mathematics, the physical significance of the CFL criterion is that Δt must be less than or at most equal to the time required for a sound wave to propagate between two adjacent grid points. Consider a two-dimensional rectangular grid such as shown in Fig. 11.3, where at any grid point the flowfield velocity is V, with x and y components u and v, respectively. The velocity of propagation of a sound wave in the x direction is $u + a$, and the time of propagation is

$$\Delta t_x = \frac{\Delta x}{u + a} \tag{12.11}$$

Similarly, in the y direction,

$$\Delta t_y = \frac{\Delta y}{v + a} \tag{12.12}$$

The CFL criterion can then be expressed as

$$\Delta t \leq \min(\Delta t_x, \Delta t_y) \tag{12.13}$$

where the number chosen on the right-hand side is the smaller of the values obtained from Eqs. (12.11) and (12.12). Experience has shown that the choice of the equals sign in Eq. (12.13) usually yields a Δt too large for stability of the nonlinear system associated with the flow problems of interest. Hence, in practice, Δt is chosen such that

$$\boxed{\Delta t = K[\min(\Delta t_x, \Delta t_y)]} \tag{12.14}$$

where K is less than unity, typically on the order of 0.5 to 0.8. A particular value of K suited to a particular application is usually determined by trial and error.

Note from Eqs. (12.11) through (12.14) that Δt is proportional to the grid spacing Δx or Δy. For coarse grids, Δt can be large, and the ensuing computer time correspondingly short. However, if the number of grid points is essentially quadrupled by halving both Δx and Δy in Fig. 11.3, the number of calculations at each time step will increase by a factor of 4. Moreover, the value of Δt will be halved, and twice as many time steps will be necessary to compute to a given value of time t. Hence, because of the coupling between Δt and the grid size as given in Eqs. (12.11) through (12.14), reducing the grid spacing by a factor of 2 results in a factor of 8 increase in computer execution time for a given time-marching solution. Therefore, this stability criterion can be very stringent.

Finally, contemporary work on *implicit* time-marching finite-difference techniques indicates that the stability criterion presented here can be relaxed considerably. It is beyond the scope of this book to review such current research; instead, the reader is encouraged to consult and follow the recent literature.

12.3 | THE BLUNT BODY PROBLEM—
QUALITATIVE ASPECTS AND
LIMITING CHARACTERISTICS

Some of the physical aspects concerning the flowfield over a supersonic blunt body were introduced in Sec. 4.12, which should be reviewed by the reader before progressing further.

It is again emphasized that the steady flowfield over the blunt body in Fig. 12.4 is a mixed subsonic-supersonic flow described by elliptic equations in the subsonic region and hyperbolic equations in the supersonic region. The sonic line divides these two regions. Moreover, we have emphasized in Chap. 11 that disturbances cannot propagate upstream in a steady supersonic flow. Hence, by examining Fig. 12.4*a*, we might assume that the subsonic region and the shape of the sonic line are governed by only that portion of the body shape between the two sonic lines. However, this is not completely valid. Consider the low supersonic flow (say $M_\infty < 2$) over a sphere as sketched in Fig. 12.8*a*. Point *a* is the intersection of the sonic line and the body. On

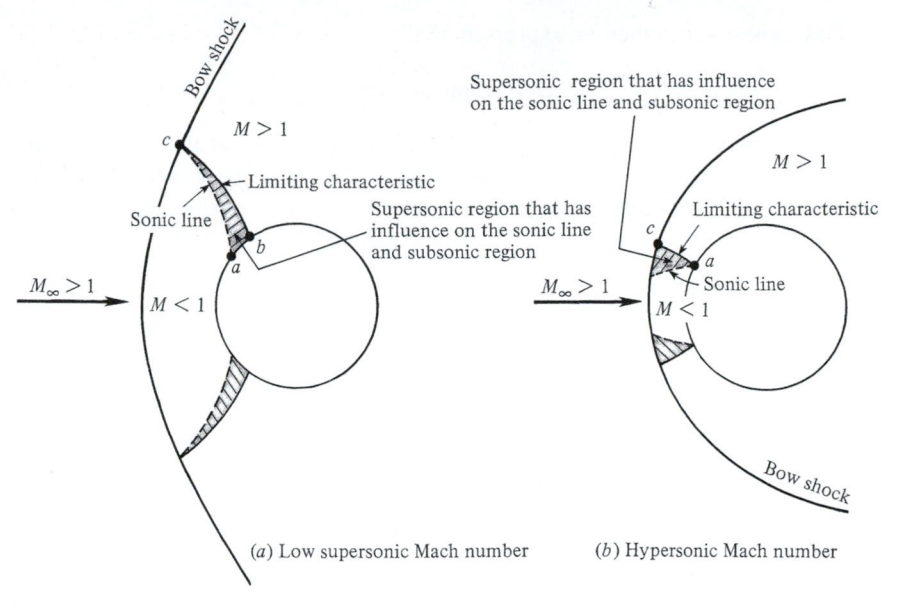

Figure 12.8 | Illustration of limiting characteristics.

the body downstream of point a the flow is supersonic. Consider the left-running characteristic lines that emanate from the body downstream of point a. The particular characteristic that emanates from the body at point b, and just exactly intersects the shock wave at the point where the sonic line also intersects the shock (point c), is defined as the *limiting characteristic*. Any characteristic line emanating from the body between points a and b will intersect the sonic line; any characteristic emanating downstream of point b will not. Hence, any disturbance originating within the shaded supersonic region between the sonic line and limiting characteristic in Fig. 12.8a will propagate along a left-running characteristic, will intersect the sonic line, and hence will be felt throughout the subsonic region. In particular, the shape of the body between points a and b will influence the shape of the sonic line and the subsonic flow even though the local flow between a and b is supersonic. Therefore, if the method of characteristics (see Chap. 11) is to be employed for calculating the supersonic region over a blunt body, the initial data line can be chosen no further upstream than the limiting characteristic; to use the sonic line as initial data improperly ignores the influence of the shaded regions. At higher Mach numbers, such as those sketched in Fig. 12.8b, the shock wave moves closer to the body and the sonic point behind the shock moves considerably downward, whereas the sonic point on the body moves downward only slightly. Hence, the shape of the sonic line is quite different at higher Mach numbers. Here, the right-running characteristic through point c on the shock intersects the sonic line at point a on the body. This characteristic is the *limiting characteristic* for such a case. Any disturbance in the supersonic shaded region in Fig. 12.8b will propagate along a right-running characteristic, will intersect the sonic line, and will influence the subsonic regions.

Again, it is emphasized that any theoretical or numerical solution of the blunt body flowfield must be valid not only in the subsonic region, but must carry downstream to at least the limiting characteristic. Then, the methods described in Chap. 11 can be used downstream of the limiting characteristic.

For an elaborate and detailed description of the blunt body flowfield and various early approaches to its solution, see the authoritative book by Hayes and Probstein (Ref. 52). In this chapter, we will emphasize the time-marching solution of blunt body flows.

12.4 | NEWTONIAN THEORY

As noted on our roadmap in Fig. 12.3, this section is a slight diversion from our main emphasis in this chapter on time-marching solutions. Here, we will obtain a simple expression for the pressure distribution over the surface of a blunt body, which will be useful in subsequent discussions.

In Propositions 34 and 35 of his *Principia,* Isaac Newton considered that the force of impact between a uniform stream of particles and a surface is obtained from the loss of momentum of the particles normal to the surface. For example, consider a stream of particles with velocity V_∞ incident on a flat surface inclined at the angle θ with respect to the velocity, as shown in Fig. 12.9a. Upon impact with the surface, Newton assumed that the normal momentum of the particles is transferred to the surface, whereas the tangential momentum is preserved. Hence, after collision with the surface, the particles move along the surface, as sketched in Fig. 12.9a. The change in normal velocity is simply $V_\infty \sin\theta$. Now consider Fig. 12.9b. The mass flux of particles incident on a surface of area A is $\rho V_\infty A \sin\theta$. Hence, the time rate of change of momentum of this mass flux, from Newton's reasoning, is

$$\text{Mass flux} \times \text{velocity change}$$

or
$$(\rho V_\infty A \sin\theta)(V_\infty \sin\theta) = \rho V_\infty^2 A \sin^2\theta$$

And in turn, from Newton's second law, this time rate of change of momentum is equal to the force F on the surface:

$$F = \rho V_\infty^2 A \sin^2\theta \tag{12.15}$$

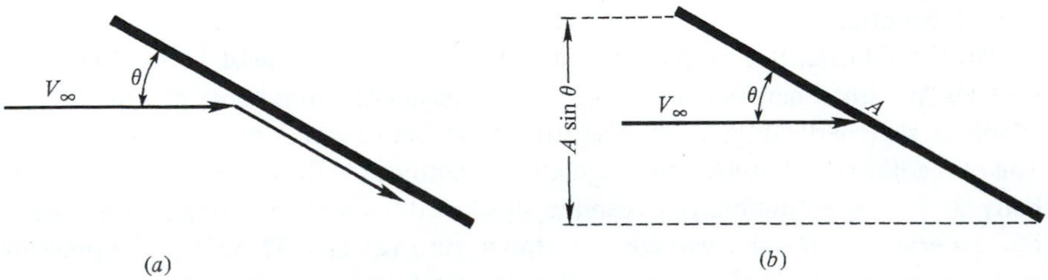

(a) (b)

Figure 12.9 | Schematic for newtonian impact theory.

In turn, the pressure is force per unit area, which from Eq. (12.15) is

$$\frac{F}{A} = \rho V_\infty^2 \sin^2 \theta \tag{12.16}$$

Newton assumed the stream of particles in Fig. 12.9*b* to be linear, i.e., he assumed that the individual particles do not interact with each other, and have no random motion. Since modern science recognizes that static pressure is due to the random motion of the particles, and since Eq. (12.16) considers only the linear, directed motion of the particles, the value of F/A in Eq. (12.16) must be interpreted as the pressure *difference* above static pressure, namely, $F/A = p - p_\infty$. Therefore, from Eq. (12.16), and recalling from Chap. 9 the definition of the pressure coefficient, $C_p = (p - p_\infty)/\frac{1}{2}\rho V_\infty^2$, we have

$$p - p_\infty = \rho V_\infty^2 \sin^2 \theta$$

$$\frac{p - p_\infty}{\frac{1}{2}\rho V_\infty^2} = 2 \sin^2 \theta$$

$$\boxed{C_p = 2 \sin^2 \theta} \tag{12.17}$$

Equation (12.17) is the newtonian "sine-squared" law for the pressure distribution on a surface inclined at an angle θ with respect to the free stream. Of course, the physical picture used to derive Eq. (12.17) in no way describes a realistic flow—subsonic or supersonic. Newton did not have the advantage of our knowledge in the twentieth century; he did not know about shock waves, nor did he have the proper image of fluid mechanics. However, at high supersonic and hypersonic Mach numbers, the shock wave moves closer to the body, and the flowfield begins to resemble some of the characteristics sketched in Fig. 12.9*a*, namely, a uniform flow ahead of the shock wave, and a flow reasonably parallel to the body in the shock layer between the body and the shock. Therefore, particularly at hypersonic Mach numbers, the newtonian theory provides reasonable results for the pressure distribution over an inclined surface, with increasing accuracy as M_∞ and θ increase. Also, it is interesting to note that the exact shock wave relations (see Chap. 4) approached the newtonian result as γ approaches unity. Therefore, Eq. (12.17) is more accurate for dissociating and ionizing flow (where the "effective" γ is low) than for monatomic gases such as helium (where $\gamma = 1.67$). A discussion of such chemically reacting flows and the consequent effect on γ is given in Chaps. 16 and 17.

In 1955, Lester Lees, a professor at the California Institute of Technology, proposed a "modified newtonian" pressure law. Consider a blunt body at zero angle of attack, as sketched in Fig. 12.4*a*. The streamline that passes through the normal portion of the bow shock is also the stagnation streamline. At the stagnation point of the body, $V = 0$ by definition. Between the shock and the body, the stagnation streamline experiences an *isentropic* compression to zero velocity. Therefore, the pressure at the stagnation point is simply equal to the total pressure behind a normal shock wave at M_∞—a quantity easily calculated from the results of Chap. 3. Moreover, this is the maximum pressure on the body; away from the stagnation point, the pressure decreases as indicated by Eq. (12.17). Therefore, the surface pressure coefficient attains its maximum value at the stagnation point, namely, $C_{p_{max}} = (p_o - p_\infty)/\frac{1}{2}\rho_\infty V_\infty^2$. Lees suggested that Eq. (12.17) be modified by replacing the coefficient 2 with $C_{p_{max}}$. Hence,

$$C_p = C_{p_{max}} \sin^2 \theta \qquad (12.18)$$

This modified newtonian pressure law is now in wide use for estimating pressure distributions over blunt surfaces at high Mach numbers. It is more accurate than Eq. (12.17).

Further elaboration on the use of newtonian theory for hypersonic flows is given in Sec. 15.4.

12.5 | TIME-MARCHING SOLUTION OF THE BLUNT BODY PROBLEM

Let us now consider the detailed time-marching solution of the blunt body flowfield. Assume that we are given the free-stream Mach number M_∞, and the body shape, as sketched in Fig. 12.10. This approach, where the body shape is specified, and the shock wave shape and flowfield are to be calculated, is called the *direct problem.* This is in contrast with the *inverse problem,* where the shock shape is specified and the body shape that supports the given shock is to be calculated. Numerous steady-flow solutions in the past have taken the inverse approach. However, the direct problem is usually the one encountered in practice, and the time-marching approach described here is the only technique available at present that allows the exact solution of the direct problem. (Note that the inverse approach can be iterated until a desired body shape is converged upon; in this sense the direct problem can be solved by an iterative repetition of the inverse approach.)

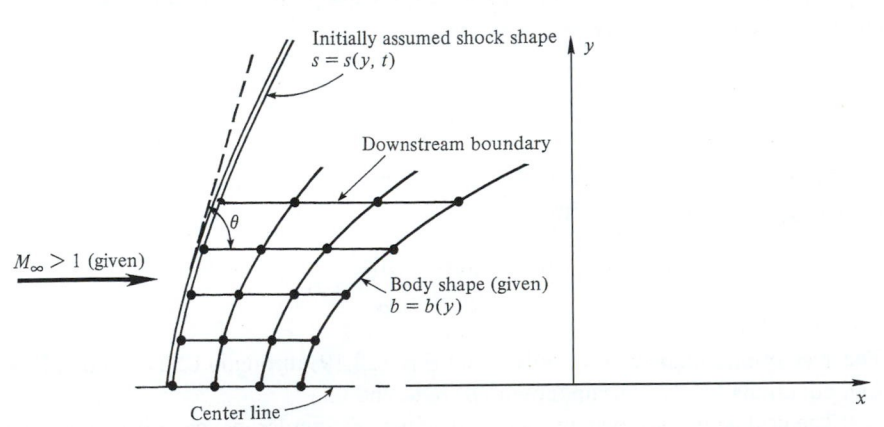

Figure 12.10 | Finite-difference grid in physical space for the blunt body problem.

In the time-marching approach, the initial shock wave shape is assumed at time $t = 0$. The abscissa of the shock and body are denoted by s and b, respectively. The flowfield between the assumed shock wave and the specified body is divided into a number of grid points, as shown in Fig. 12.10. Here, a two-dimensional coordinate system is illustrated; the case for axisymmetric flow is similar. At each grid point, values of all the flowfield variables are arbitrarily set. Then, starting from these guessed values and using time-marching machinery analogous to that developed in Sec. 12.1, new values of the flowfield variables, shock detachment distance, and shock shape are calculated in steps of time. After a number of time steps, the flowfield converges to the proper steady state value; this steady state is the desired result, and the time-marching approach is just a means to that end. Some details of this solution are discussed next.

The governing equations for two-dimensional or axisymmetric isentropic flow are, from Chap. 6:

Continuity:
$$\frac{\partial \rho}{\partial t} + \frac{\partial (\rho u)}{\partial x} + \frac{\partial (\rho v)}{\partial y} + K\frac{\rho v}{y} = 0 \tag{12.19}$$

x momentum:
$$\rho\frac{\partial u}{\partial t} + \rho u\frac{\partial u}{\partial x} + \rho v\frac{\partial u}{\partial y} = -\frac{\partial p}{\partial x} \tag{12.20}$$

y momentum:
$$\rho\frac{\partial v}{\partial t} + \rho u\frac{\partial v}{\partial x} + \rho v\frac{\partial v}{\partial y} = -\frac{\partial p}{\partial y} \tag{12.21}$$

Energy:
$$\frac{Ds}{Dt} = 0 \tag{12.22}$$

where $K = 0$ for two-dimensional flow, and $K = 1$ for axisymmetric flow. Here, the energy equation is stated in the form of the isentropic assumption. Moreover, for isentropic conditions, $p/\rho^\gamma = $ const for a calorically perfect gas. Hence, Eq. (12.22) can be written alternatively as

$$\frac{D(p/\rho^\gamma)}{Dt} = 0 \tag{12.23}$$

Expanding the derivative in Eq. (12.23), and defining $\psi = \ln p - \gamma \ln \rho$, we obtain the energy equation in the form

$$\frac{\partial \psi}{\partial t} + u\frac{\partial \psi}{\partial x} + v\frac{\partial \psi}{\partial y} = 0 \tag{12.24}$$

The system of equations to be solved are Eqs. (12.19) through (12.21), and (12.24), four equations for the four unknowns $p, \rho, u,$ and v.

The grid network shown in Fig. 12.10 is not rectangular; hence, it is inconvenient for the formation of finite differences. To transform the shock layer into a rectangular

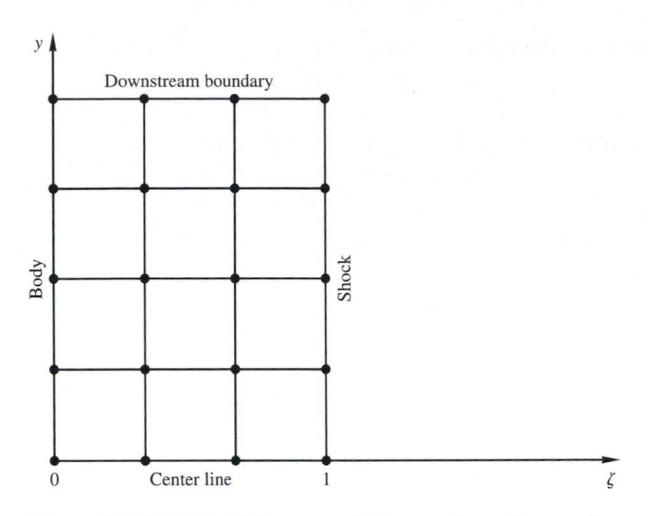

Figure 12.11 | Finite-difference grid in transformed space for the blunt body problem.

grid, define a new independent variable ζ such that

$$\zeta \equiv \frac{x - b}{\delta}$$

where $\delta = s - b$, i.e., δ is the local shock detachment distance. In this fashion, ζ always varies between 0 (at the body) and 1 (at the shock), and the grid now appears as sketched in Fig. 12.11. In addition, let $W = ds/dt$ be the x component of the shock wave velocity (note that the shock wave will be in motion until the final steady state is reached), and let θ represent the angle between the tangent to the shock and the x axis. In addition, these variables are defined:

$$C \equiv (\zeta - 1)\frac{db}{dy} - \zeta \cot\theta$$

$$P = \ln p$$

$$R = \ln \rho$$

$$B = \frac{u - W\zeta + vC}{\delta}$$

where, in terms of the earlier material,

$$\psi = \ln p - \gamma \ln \rho = P - \gamma R$$

Finally, nondimensionalize all the variables as follows: Divide p and ρ by their free-stream values; divide the velocities by $(p_\infty/\rho_\infty)^{1/2}$; divide the lengths by a characteristic length L; and obtain a nondimensional time by dividing the dimensional time by $L/(p_\infty/\rho_\infty)^{1/2}$. The resulting equations, where now the symbols represent the nondimensional values, are

Continuity:

$$\frac{\partial R}{\partial t} = -\left[B\frac{\partial R}{\partial \zeta} + \frac{1}{\delta}\frac{\partial u}{\partial \zeta} + \frac{C}{\delta}\frac{\partial v}{\partial \zeta} + \frac{\partial v}{\partial y} + v\frac{\partial R}{\partial y} + K\frac{v}{y} \right] \tag{12.25}$$

x momentum:

$$\frac{\partial u}{\partial t} = -\left[B\frac{\partial u}{\partial \zeta} + v\frac{\partial u}{\partial y} + \frac{P}{\rho\delta}\frac{\partial P}{\partial \zeta} \right] \tag{12.26}$$

y momentum:

$$\frac{\partial v}{\partial t} = -\left[B\frac{\partial v}{\partial \zeta} + v\frac{\partial v}{\partial y} + \frac{pC}{\rho\delta}\frac{\partial P}{\partial \zeta} + \frac{p}{\rho}\frac{\partial P}{\partial y} \right] \tag{12.27}$$

Energy: $\quad\dfrac{\partial \psi}{\partial t} = -\left[B\dfrac{\partial \psi}{\partial \zeta} + v\dfrac{\partial \psi}{\partial y} \right]$ $\tag{12.28}$

The form of these equations is useful because the nondimensional variables are usu-ally of the order of magnitude of unity—a convenient ploy used by some people in helping to examine and interpret the results from the computer.

To advance the flowfield in steps of time, the grid points in Fig. 12.11 are treated as four distinct sets: interior points (all grids points except on the shock, body, and downstream boundary); the shock points ($\zeta = 1$); the body points ($\zeta = 0$); and the downstream boundary points. The flowfield at the interior points is advanced in time by means of MacCormack's predictor-corrector method discussed in Sec. 12.1. Note that Eqs. (12.25) through (12.28) are written with the known spatial derivatives on the right-hand side; these derivatives are replaced with forward differences on the predictor step and rearward differences on the corrector step. This allows the calcu-lation of the time derivatives that appear on the left-hand side of Eqs. (12.25) through (12.28). In turn, these time derivatives ultimately lead to the advancement of the flowfield in steps of time via Eq. (12.1).

For the shock points, the values of the flow variables behind the shock at time $t + \Delta t$ can be obtained from the Rankine–Hugoniot relations for a moving shock wave (see Chap. 7). However, this implies that a value of the shock wave velocity, $W(t + \Delta t)$, must first be assumed, since it is not known at the beginning of each time step in the computations. Therefore, an iterative process must be established wherein the values of the flowfield variables at the shock grid points must be obtained from some independent calculation, and then compared with those obtained from the shock relations for the assumed W. In the analysis of Moretti and Abbett (Ref. 47), this in-dependent calculation is made via a characteristic technique utilizing information from the interior points. Specifically, the characteristic equations are obtained from the two-dimensional unsteady governing equations written for a (ξ, η, t) coordinate frame, where ξ and η are cartesian coordinates locally normal and tangential, respec-tively, to the shock wave. This is illustrated in Fig. 12.12. The assumption is made in

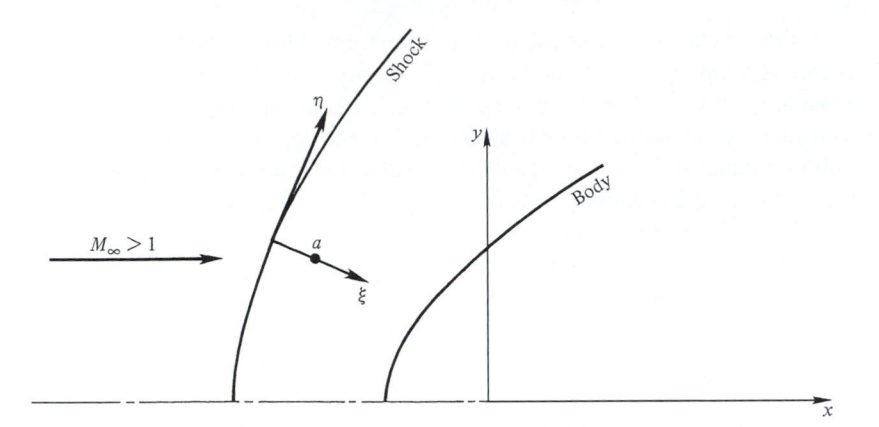

Figure 12.12 | Shock-oriented coordinates for the characteristic treatment of the shock boundary conditions.

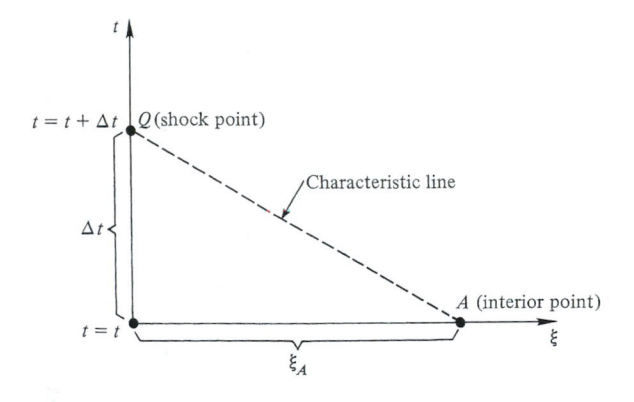

Figure 12.13 | Construction of the one-dimensional unsteady characteristic at the shock wave.

obtaining the compatibility equations that the governing equations can be written as quasi-one-dimensional in the ξ direction, modified by "forcing terms" containing derivatives in the tangential direction. That is, the characteristic directions are drawn in the (ξ, t) plane only, as shown in Fig. 12.13. This characteristic line, along with the compatibility equations, allows the flowfield to be calculated at $\xi = 0$ (the shock point) at time $t + \Delta t$, i.e., point Q in Fig. 12.13, from the known flowfield at point A at time t. Since the location of point A in Figs. 12.12 and 12.13 generally will not correspond to one of the interior grid points in Fig. 12.11, the information at A must be obtained by spatial interpolation. Finally, the information at point Q obtained from this characteristics approach is compared with the information calculated from the shock relations for the assumed W, and if agreement is not obtained, new values of W are assumed until the iteration converges. For more details, including the form of the characteristic equations, see Ref. 47.

A similar characteristics calculation is performed for the body points. Here, the analysis is simpler because the body is stationary, and also because the entropy is known at the body. (Note that $Ds/Dt = 0$, i.e., the entropy of a given fluid element is constant, even when the flow is unsteady. The entropy at the stagnation point must be chosen equal to its proper steady state value for a normal shock obtained from Chap. 3, because this entropy wets the entire body surface and is constant throughout the time-marching calculation.)

Finally, the flowfield values at the downstream boundary shown in Figs. 12.10 and 12.11 are obtained by simple linear extrapolation from the upstream grid points. They can also be obtained alternatively from MacCormack's technique using only one-sided differences in both the predictor and corrector steps. In either event, as long as the downstream boundary points are in a locally supersonic flow, the rest of the flowfield is not strongly influenced by slight inaccuracies at the downstream boundary. However, it is important to make certain that the downstream boundary points are in a supersonic region; experience has shown that the calculations will become unstable if the above extrapolation or one-sided differencing is performed in a subsonic region.

For more details concerning this solution, the reader is urged to consult Refs. 47 and 53.

12.6 | RESULTS FOR THE BLUNT BODY FLOWFIELD

Let us now examine some typical results obtained with the time-marching blunt body solution described in the previous sections; these results are presented in more detail in Ref. 53. The purpose of this section is twofold: (1) to further illustrate the nature and behavior of time-marching solutions of steady state flowfields, and (2) to describe some fluid dynamic aspects of the supersonic blunt body flowfield.

First, consider the two-dimensional flow over a parabolic cylinder as shown in Fig. 12.14. The free-stream Mach number is $M_\infty = 4$. The assumed shock shape is labeled $0\,\Delta t$. All flow properties between this assumed shock and the prescribed body are also given arbitrary values. Starting from these assumed initial conditions, the flowfield is calculated in steps of time. Note that after 100 time steps, the shock wave has moved considerably forward of its initially assumed position, and has changed shape. However, also note that its movement has slowed, and that after 300 time steps, its location has become essentially stationary. Moreover, the flowfield properties between the shock and the body do not materially change after 300 time steps— the steady state has been obtained. This time-varying behavior is further illustrated in Fig. 12.15, which gives the time variation of the stagnation point pressure. The assumed initial value (at $t = 0$) is the proper steady-state value known in advance (we know the steady-state shock velocity should be zero, and the stagnation point pressure should be that behind a stationary normal shock wave with $M_\infty = 4$). Note from Figs. 12.14 and 12.15 that (1) the most extreme transients occur at early times where the "driving potential" toward the steady state is the strongest, and (2) the steady state is, for all practical purposes, achieved at large values of time. The reader is

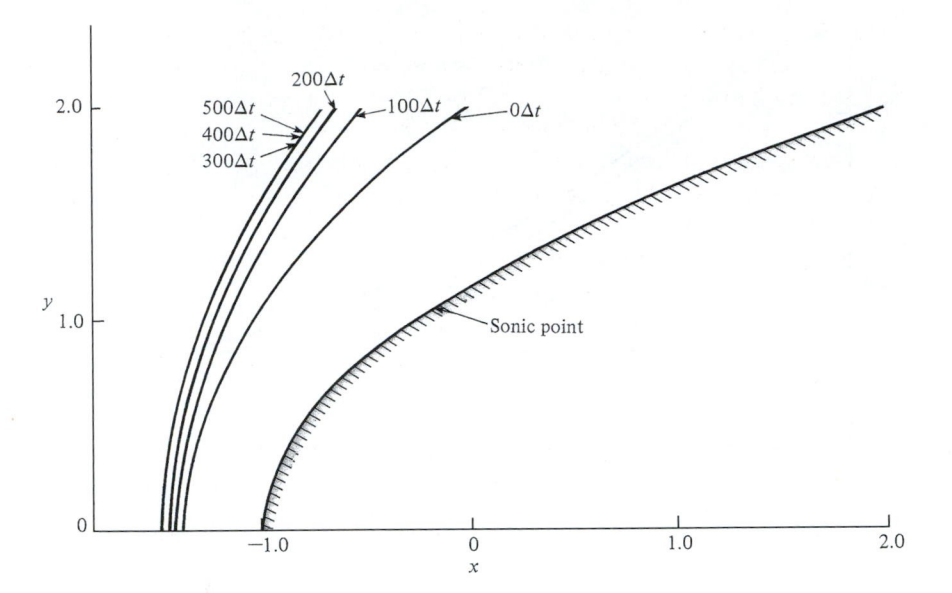

Figure 12.14 | Time-marching shock wave motion, parabolic cylinder, $M_\infty = 4$.

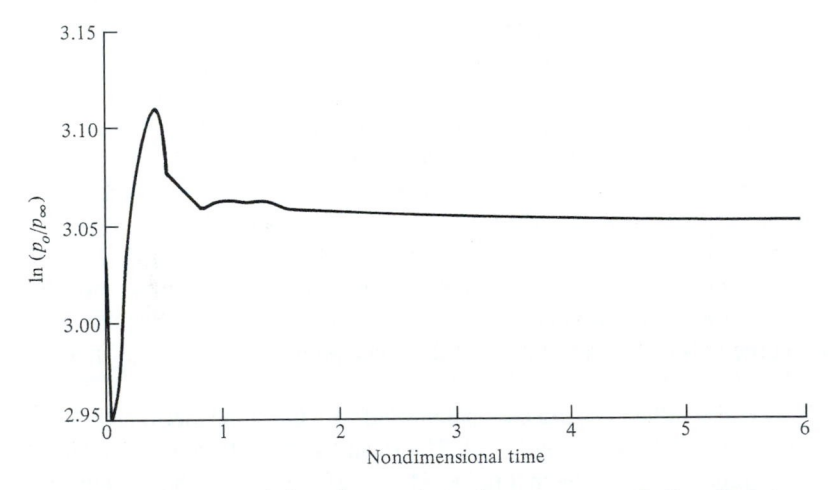

Figure 12.15 | Time-variation of stagnation point pressure, parabolic cylinder, $M_\infty = 4$.

again reminded that this steady state is the desired result of the calculations; the transient behavior shown in Figs. 12.14 and 12.15 is simply a means to that end.

For the remainder of this section, we will concentrate on the final steady-state flowfields. For example, Fig. 12.16 gives the steady-state surface pressure distribution, normalized with respect to stagnation point pressure, around the parabolic cylinder. The pressure distributions are calculated for two Mach numbers, $M_\infty = 4$ and 8. The solid lines are exact results from the time-marching solution; these are compared with

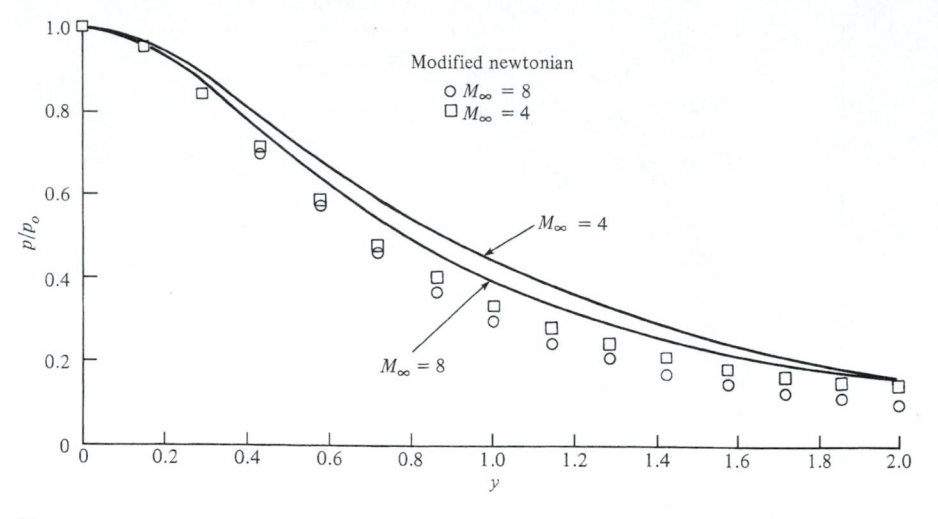

Figure 12.16 | Surface pressure distributions, parabolic cylinder.

results from the modified newtonian formula given by Eq. (12.18). Note that the modified newtonian distribution underestimates the actual pressure distribution, and that better agreement is obtained at $M_\infty = 8$ rather than $M_\infty = 4$. This is consistent with our discussion in Sec. 12.4, where it was argued that the assumptions underlying newtonian theory are more closely approached at hypersonic Mach numbers than at low Mach numbers. The relative lack of agreement with modified newtonian results, as given in Fig. 12.16, is typical of two-dimensional blunt bodies; in practice, pressure distributions over axisymmetric bodies agree more closely with modified newtonian results than do two-dimensional pressure distributions, as will be shown later.

The steady-state shock shapes and sonic lines are shown in Fig. 12.17 for the parabolic cylinder at $M_\infty = 4$ and 8. Note that, as the Mach number increases, the shock detachment distance decreases and the sonic line shifts downward. The sonic point on the shock moves farther than the sonic point on the body, as mentioned in Sec. 12.3.

Now consider an axisymmetric paraboloid with the same meridian cross section as the parabolic cylinder in Fig. 12.14. The steady-state surface pressure distribution for the paraboloid is given in Fig. 12.18 for $M_\infty = 4$. The solid line gives the exact results from the time-marching solution. The open squares are results from Eq. (12.18); note that, in contrast to the earlier two-dimensional comparison, the agreement with newtonian theory is excellent for the axisymmetric case, even for a long distance along the body. In Fig. 12.18, a comparison is also made with an inverse steady-state method developed by Lomax and Inouye (see Ref. 54). Again, reasonable agreement is obtained. However, like all steady-state techniques for the blunt body problem, Lomax and Inouye's results are valid only up to the sonic region and limiting characteristic; steady-state techniques usually become unstable downstream of this region. In contrast, the time-marching technique is uniformly valid in both the subsonic and supersonic regions, and can given results for any desired distance downstream, as clearly shown in Fig. 12.18.

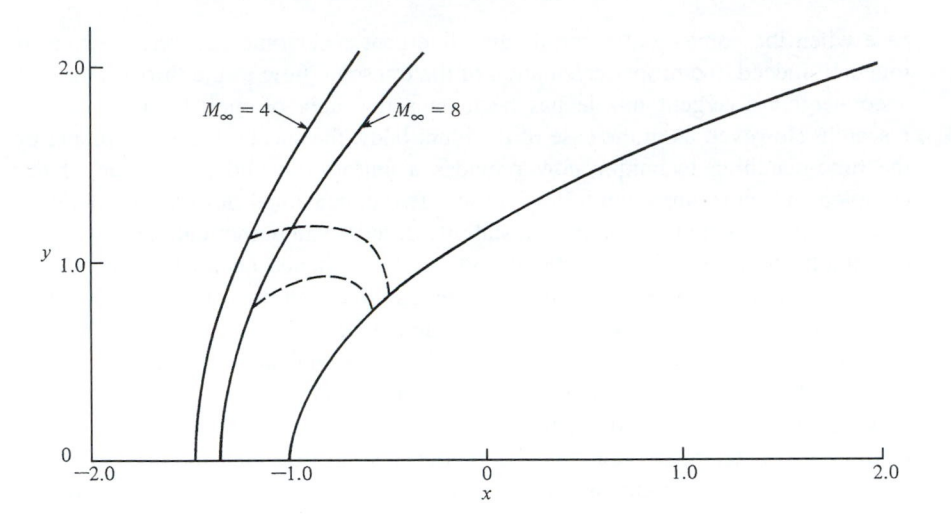

Figure 12.17 | Shock shapes and sonic lines, parabolic cylinder.

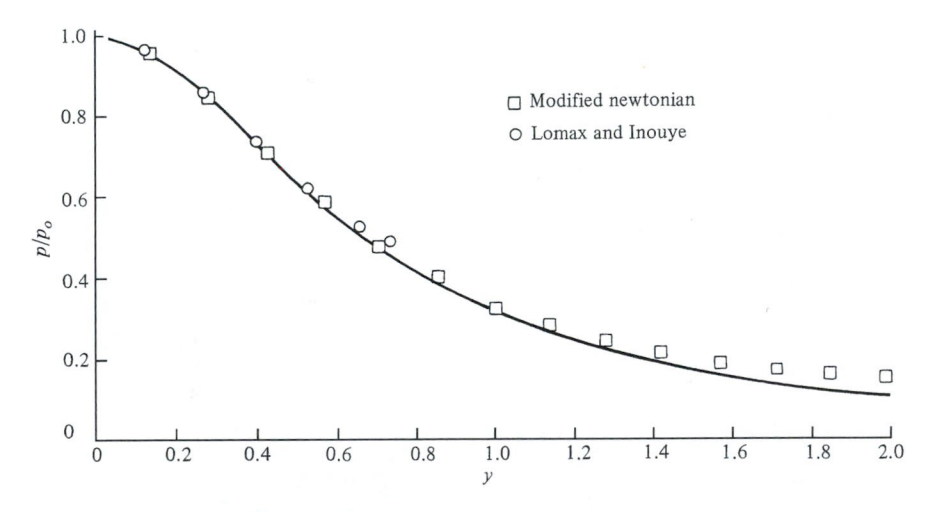

Figure 12.18 | Surface pressure distribution, paraboloid, $M_\infty = 4$.

12.7 | TIME-MARCHING SOLUTION OF TWO-DIMENSIONAL NOZZLE FLOWS

Return to Fig. 12.4, which illustrates the similarities between the flow over a supersonic blunt body and the flow through a two-dimensional (or axisymmetric) convergent-divergent supersonic nozzle. Both cases are mixed subsonic-supersonic flows, with curved sonic lines. Indeed, the flow around the blunt body in Fig. 12.4*a* can be visualized as a series of streamtubes with the general features of Fig. 12.4*b*. Therefore, the difficulties in developing a uniformly valid steady-state technique for the solution of blunt body flows also occur with the two-dimensional nozzle case when the convergent subsonic and divergent supersonic sections are treated together. Indeed, the proper calculation of the transonic flow in the throat region of a convergent-divergent nozzle has been an active area of modern aerodynamic research. However, as in the case of the blunt body, the successful development of the time-marching technique now provides a uniformly valid calculation of the complete subsonic-supersonic flowfield in a two-dimensional nozzle. The philosophy, equations, boundary conditions, stability criteria, and numerical machinery are essentially the same as for the blunt body problem; hence no further elaboration will be made here. For an example of a time-marching solution of two-dimensional nozzle flows, the reader is encouraged to examine Ref. 55.

Note that the method of characteristics discussed in Chap. 11 is a standard approach to the calculation of the *supersonic* region of a convergent-divergent nozzle; however, it cannot be used for the subsonic or transonic regions. Moreover, the method of characteristics requires prior knowledge of the sonic line, or, more precisely, the limiting characteristics for the nozzle throat region. For our applications in Chap. 11, the sonic line was assumed to be a straight line—a common assumption for many practical characteristics solutions. However, in general, the sonic line in the throat region of a convergent-divergent nozzle is curved, and its curvature becomes more pronounced as the convergence of the subsonic section is made more rapid. Therefore, for short, rapid-expansion nozzles, it is preferable to start a characteristics solution from the limiting characteristics associated with the more accurate curved sonic line rather than assuming a straight sonic line. The curved sonic line can be computed from a time-marching technique as illustrated in Ref. 55.

Some steady-state results for Mach number contours in the throat region of a convergent-divergent nozzle are given in Fig. 12.19. The solid lines are results from

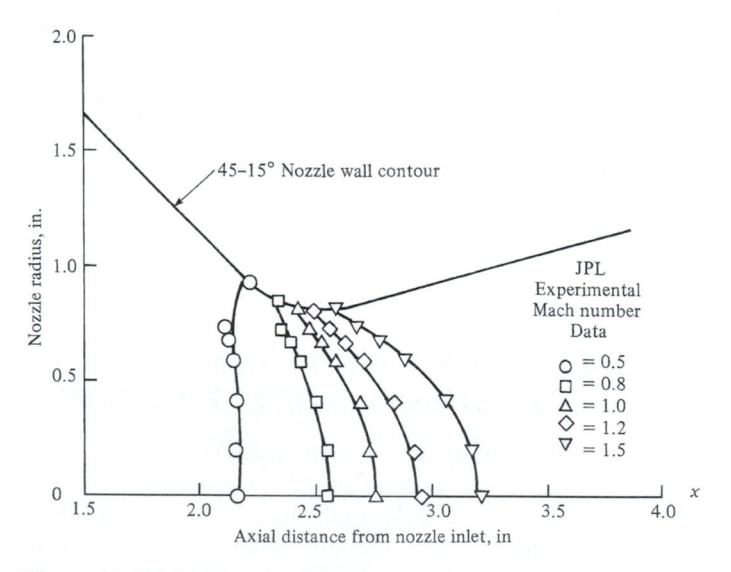

Figure 12.19 | Constant Mach number lines in a 45° to 15° conical nozzle; results from the time-marching calculations of Serra (Ref. 55).

the time-marching technique described in Ref. 55. The open symbols are experimental measurements from the Jet Propulsion Laboratory. Agreement between theory and experiment is quite satisfactory. Note especially that the sonic line ($M = 1$ contour) is highly curved due to the rapid convergence of the 45° subsonic section.

12.8 | OTHER ASPECTS OF THE TIME-MARCHING TECHNIQUE; ARTIFICIAL VISCOSITY

A virtue of the time-marching technique is its relative simplicity, in spite of the complexity of the steady-state flow that is being solved. Moreover, the time-marching technique is straightforward to program on a digital computer, thus minimizing the labor invested to set up the solution. However, the reader is cautioned that the technique is not yet (and may never be) routine. Like all computational fluid dynamic applications, solutions are frequently more of an art than a science. For example, throughout this chapter we have stated that time-marching calculations begin with "arbitrary" initial conditions for the flowfield. For a physical problem that has a unique solution, this is conceptually true. However, in practice, the initial conditions usually cannot be completely arbitrary, rather, they must be prescribed within a certain latitude. A case in point is the blunt body solution described in Secs. 12.5 and 12.6. Here, the initial shock wave must not be assumed too close or too far away from the body. If the shock detachment distance is initially too large or too small, the shock wave tends to accelerate too rapidly, thus producing strong gradients of the flowfield variables behind the wave. Consequently, the finite-difference scheme using a fixed grid becomes inaccurate, ultimately causing some aspect of the calculations to collapse. Other applications are frequently plagued by analogous situations. Therefore, it is wise to choose initial conditions intelligently, using any existing *a priori* knowledge about the flow to guide your choice. Also keep in mind that the closer the initial conditions are to the final steady state, the faster the program will converge to this steady state, hence conserving computer time.

Another problem of time-marching solutions is that small inaccuracies introduced at the boundaries can propagate as short-wavelength disturbances throughout the flowfield, sometimes focusing on a certain region of the flow and causing the calculation to become unstable. This is why the proper treatment of boundary conditions is so important. If the flow is physically viscous, these unwanted disturbances tend to dissipate, and frequently do not cause problems. On the other hand, for inviscid flows, there are applications and techniques where the calculations must be artificially damped by the addition of a mathematical quantity called *artificial viscosity*. The concept of artificial viscosity can be introduced as follows.

First, consider a quantity G which is a function of both x and t. A finite-difference expression for the *second* partial derivative with respect to x can be obtained from the Taylor's series expansion:

$$G_{i+1} = G_i + \left(\frac{\partial G}{\partial x}\right)_i \Delta x + \left(\frac{\partial^2 G}{\partial x^2}\right)_i \frac{(\Delta x)^2}{2} + \cdots \qquad (12.29)$$

where $i + 1$ and i are two neighboring grid points in the x direction. In Eq. (12.29), replace the term $(\partial G/\partial x)_i$ with a central difference,

$$\left(\frac{\partial G}{\partial x}\right)_i = \frac{G_{i+1} - G_{i-1}}{2\,\Delta x}$$

thus yielding for Eq. (12.29)

$$G_{i+1} = G_i + \tfrac{1}{2}(G_{i+1} - G_{i-1}) + \left(\frac{\partial^2 G}{\partial x^2}\right)_i \frac{(\Delta x)^2}{2} + \cdots \qquad (12.30)$$

Solving Eq. (12.30) for the second partial derivative, we obtain

$$\boxed{\left(\frac{\partial^2 G}{\partial x^2}\right)_i = \frac{G_{i+1} - 2G_i + G_{i-1}}{(\Delta x)^2}} \qquad (12.31)$$

Equation (12.31) is a central second difference of second-order accuracy.

Now consider another quantity F, which is also a function of x and t and which is related to G through the simple partial differential equation

$$\frac{\partial G}{\partial t} = \frac{\partial F}{\partial x} \qquad (12.32)$$

Let us finite-difference this equation by using a central difference for F,

$$\left(\frac{\partial F}{\partial x}\right)_i^k = \frac{F_{i+1}^k - F_{i-1}^k}{2\Delta x} \qquad (12.33)$$

In Eq. (12.33), the superscript k has been added to denote evaluation at the kth time step. Also, let us represent $\partial G/\partial t$ in Eq. (12.32) by a finite-difference expression introduced by Lax (see Ref. 56). Lax's technique has been used in several computational fluid dynamic applications, particularly during the mid-1960s. According to Lax, the time derivative is based on an average value of G between points $(i + 1)$ and $(i - 1)$, i.e.,

$$\left(\frac{\partial G}{\partial t}\right)_i^k = \frac{G_i^{k+1} - \tfrac{1}{2}\left[G_{i+1}^k + G_{i-1}^k\right]}{\Delta t} \qquad (12.34)$$

Substitute Eqs. (12.33) and (12.34) into Eq. (12.32):

$$G_i^{k+1} = \tfrac{1}{2}\left[G_{i+1}^k + G_{i-1}^k\right] + \left(F_{i+1}^k - F_{i-1}^k\right)\frac{\Delta t}{2\Delta x} \qquad (12.35)$$

Subtract G_i^k from both sides of Eq. (12.35), and divide by Δt:

$$\frac{G_i^{k+1} - G_i^k}{\Delta t} = \frac{G_{i+1}^k - 2G_i^k + G_{i-1}^k}{2\,\Delta t} + \frac{F_{i+1}^k - F_{i-1}^k}{2\,\Delta x} \qquad (12.36)$$

Multiply the numerator and denominator of the first term on the right-hand side of Eq. (12.36) by $(\Delta x)^2$:

$$\frac{G_i^{k+1} - G_i^k}{\Delta t} = \left[\frac{G_{i+1}^k - 2G_i^k + G_{i-1}^k}{(\Delta x)^2}\right]\frac{(\Delta x)^2}{2\,\Delta t} + \frac{F_{i+1}^k - F_{i-1}^k}{2\,\Delta x} \qquad (12.37)$$

Look closely at Eq. (12.37). Recalling the central second difference from Eq. (12.31), taking the limit of Eq. (12.37) as Δx and Δt go to zero, and utilizing the mathematical definition of a derivative, Eq. (12.37) becomes

$$\frac{\partial G}{\partial t} = \frac{\partial F}{\partial x} + \frac{(\Delta x)^2}{2\,\Delta t}\frac{\partial^2 G}{\partial x^2} \tag{12.38}$$

Note that Eq. (12.38) is *different* than Eq. (12.32), with which we first started. We applied Lax's finite-difference procedure to Eq. (12.32), obtained a difference equation (12.35), and then found that, by applying the definition of the derivative to the difference equation, we recovered a partial differential equation (12.38) that is different than the one we started with. In particular, Eq. (12.38) now contains a term involving a second derivative $\partial^2 G/\partial x^2$ multiplied by a coefficient $\nu = (\Delta x)^2/2\,\Delta t$. This is analogous to the viscous terms in the Navier–Stokes equations for flow with friction, where second-order derivatives are multiplied by the physical viscosity. However, in Eq. (12.38), the second-order derivative is simply a mathematical consequence of the differencing procedure, and its coefficient $\nu \equiv (\Delta x)^2/2\,\Delta t$ is called the *artificial viscosity*.

In Lax's technique, the artificial viscosity is implicit in the finite-difference algorithm. However, in other numerical techniques, terms such as $\nu(\partial^2 G/\partial x^2)$ are explicitly added to the inviscid equations of motion *before* the finite-differencing procedure is implemented. This idea for damping the calculations by explicitly adding dissipative terms to the equations of motion is due to Von Neumann and Richtmyer (see Ref. 57), who were the first to employ a time-dependent technique on a practical problem. They were concerned with the calculation of properties across a shock wave; the main motivation of artificial viscosity was to provide some mathematical dissipation analogous to the real viscous effects inside a shock wave. In this fashion, the inviscid equations could be used to calculate the jump conditions across a shock wave. The shock structure was spread over several grid points, analogous to the shock-capturing approach described in Sec. 11.15. However, the shock thickness produced by the artificial viscosity bears no relation to the actual shock thickness produced by the physical viscosity, although Von Neumann and Richtmyer did obtain the correct jump conditions for properties across the shock wave.

Virtually all computational fluid dynamic techniques contain artificial viscosity to some degree, either implicitly or explicitly. MacCormack's predictor-corrector technique highlighted here and in Chap. 11 has some slight implicit artificial viscosity. As long as the amount is small, the accuracy of the numerical results is not compromised. However, if a large amount of damping is necessary for ensuring numerical stability, the artificial viscosity will materially increase the entropy of the flowfield and will cause inaccuracies. Moreover, heavy numerical damping may obscure other inconsistencies in the technique, producing results that may be stable but not valid. It is wise to avoid explicitly using artificial viscosity as much as possible.

Finally, note that the time-marching technique described in this chapter is a valid solution of the *unsteady* equations of motion. The transient approach to the steady state flow is physically meaningful—it follows nature, if nature were starting from the assumed initial conditions. Therefore, even though the main thrust of this chapter has been the solution of steady state flows by means of the time-marching technique, the technique itself can be readily applied to study transient flows in their own right. One such example is the time-varying flowfield inside a reciprocating internal combustion engine (see Ref. 58).

12.9 | HISTORICAL NOTE: NEWTON'S SINE-SQUARED LAW—SOME FURTHER COMMENTS

Sections 6.7 and 12.4 relate Isaac Newton's interest in fluid mechanics. This interest was focused on the calculation of the force on a body moving through a fluid, culminating in the famous sine-squared law we derived in Sec. 12.4. In Newton's day, there was a high interest in such force calculations, spurred by the development of naval architecture and its attendant practical need to calculate the flow resistance of ship hulls. However, it is interesting to note that Newton's fluid mechanics work was also driven by a more philosophical reason. Many scholars of that day still held the belief of Aristotle that the planets and stars moved through space which was occupied by a continuous medium, i.e., they assumed that space was not a vacuum. However, contemporary astronomical data of that day, including his own, convinced Newton that such was not the case. If space were occupied by a continuous medium, the heavenly bodies would encounter a resistance that would affect their motion. Observations of celestial motion did not show any such effects. Therefore, Newton was motivated to establish the laws of resistance of a body in a fluid medium in order to show that, indeed, such a resistance existed, and that it invalidated the Aristotelian philosophy. His conclusions were that the force of resistance was finite, that it depended on the fluid density, velocity, and shape of the body, and that it varied as $\sin^2 \theta$, where θ is the angle of incidence between the surface and the velocity direction.

It is also interesting to note that, like the complete scientist he was, Newton carried out experiments to check his theory. Using pendulums, and falling bodies in both air and water, Newton was able to establish that "all agree with the theory."

However, it was later recognized by others that all did *not* agree with the theory. For example, a series of experiments were carried out by d'Alembert in 1777 under the support of the French government in order to measure the resistance of ships in canals. The results showed that "the rule that for oblique planes resistance varies with the sine square of the angle of incidence holds good only for angles between $50°$ and $90°$ and must be abandoned for lesser angles." Also, in 1781, Euler pointed out the physical inconsistency of Newton's model consisting of a linear, rectilinear stream impacting without warning on a surface. In contrast to this model, Euler noted that the fluid moving toward a body "before reaching the latter, bends its direction and its velocity so that when it reaches the body it flows past it along the surface, and exercises no other force on the body except the pressure corresponding to the single points of contact." Euler went on to present a formula for resistance that attempted to take into account the shear stress distribution along the surface as well as the pressure distribution. This expression for large incidence angles became proportional to $\sin^2 \theta$, whereas at small incidence angles it was proportional to $\sin \theta$. Euler noted that such a variation was in reasonable agreement with the experiments by d'Alembert and others.

None of this early work produced expressions for aerodynamic forces with the accuracy and fundamental integrity that we are accustomed to today. In particular, Newton's sine-squared law produced such inconsistencies and inaccuracies that some aspects of fluid mechanics were actually set back by its use. For example, the lift on a surface at very small incidence angles—a few degrees—was grossly under-predicted by Newton's law. In 1799, Sir George Cayley in England first proposed the fundamental concept of the modern airplane, with a fixed wing at small incidence angle to provide lift. However, some responsible scientists of the nineteenth century used the sine-squared law to show that the wing area would have to be so large to support the airplane's weight as to be totally impractical. For this reason, some historians feel that Newton actually hindered the advancement toward powered flight in the nineteenth century.

However, Newton's sine-squared law came into its own in the last half of the twentieth century. Shortly after World War II and the development of the atomic bomb, the major world powers scrambled to develop an unmanned vehicle that could deliver the bomb over large distances. This led to the advent of the intercontinental ballistic missile in the 1950s. These missiles were to be launched over thousands of miles, with the trajectory of the warhead carrying it far beyond the outer limits of the atmosphere, and then entering the atmosphere at Mach numbers above 20. At such hypersonic speeds, the bow shock wave on these entry vehicles closely approaches the surface of the body, leaving only a very thin shock layer between the body and the shock. Consequently, as sketched in Fig. 12.20, the physical picture of hypersonic

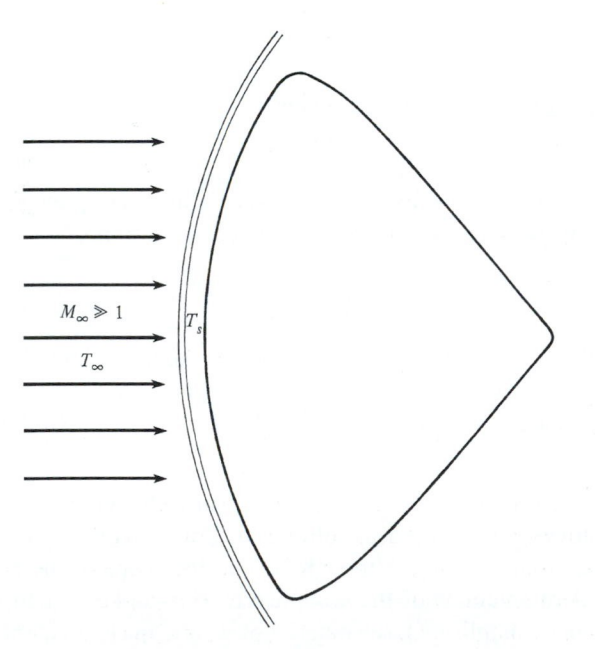

Figure 12.20 | Schematic of the thin shock layer on a hypersonic body. This picture approximates fairly reasonably the model considered by Isaac Newton.

flow over blunt bodies actually closely approximates the model used by Newton—a uniform stream impacting the surface, and then flowing along the surface. Indeed, such newtonian impact theory yields good results for pressure distributions and forces at these speeds, as discussed in previous sections of this chapter. Therefore, 250 years after its inception, Newton's sine-squared law finally found an application for which it was reasonably suited.

12.10 | SUMMARY

Flowfields encompassing mixed regions of subsonic and supersonic flow are best solved by the time-marching philosophy as described in this chapter. A particularly important example is the solution of the supersonic blunt body problem. In this case, an initial guess is made for the flowfield, which is treated as the initial condition at time zero. Then, the unsteady flow equations are solved numerically in steps of time. One method for this solution (but by no means the only method) is the predictor-corrector explicit method of MacCormack. The flowfield properties change from one time step to another; however, after a large enough number of time steps, the flowfield changes become negligibly small, i.e., a steady state is approached. This steady state is the desired flowfield and the time marching is just the means to that end.

Certain physical characteristics of the steady flow over a blunt body moving at supersonic speeds are:

1. As the free-stream Mach number increases, the bow shock wave becomes more curved and the shock detachment distance becomes smaller.
2. As the free-stream Mach number increases, the sonic line becomes more curved and moves closer to the centerline. The sonic point on the shock moves down faster than the sonic point on the body, i.e., the sonic line rotates toward the body as M_∞ is increased.
3. Modified newtonian theory provides a simple means of predicting the surface pressure distribution over the blunt nose, obtained from

$$C_p = C_{p_{\max}} \sin^2 \theta \qquad (12.18)$$

Modified newtonian results are reasonably accurate for blunt bodies at hypersonic free-stream Mach numbers; the accuracy seems better for axisymmetric and three-dimensional bodies than for two-dimensional shapes.

The time-marching philosophy is a powerful technique in computational fluid dynamics, allowing the numerical solution of flowfields that previously were not solvable by any other means. Although beyond the scope of the present book, we note that modern solutions of the complete Navier–Stokes equations for viscous flows, including complicated separated flows, are made tractable by the time-marching approach. The reader is encouraged to consult the current literature for such matters.

PROBLEMS

12.1   Consider a convergent-divergent nozzle of length L with an area-ratio variation given by $A/A^* = 1 + 10|x/L|$, where $-0.5 \leq x/L \leq 0.5$. Assume quasi-one-dimensional flow and a calorically perfect gas with $\gamma = 1.4$.

    a.   Write a computer program to calculate the variation of p/p_o, T/T_o, ρ/ρ_o, u/a_o, and M as a function of x/L by means of the time-dependent finite-difference technique. Plot some results at intermediate times, as well as the final steady state results. Use Fig. 12.6 as a model for your plots.

    b.   On the same plots, compare your steady state numerical results with the answers obtained from Table A.1.

12.2   Consider the two-dimensional, subsonic-supersonic flow in a convergent-divergent nozzle.

    a.   If the sonic line is straight, sketch the limiting characteristics.

    b.   If the sonic line is curved, sketch the limiting characteristics.

12.3   Consider a 15° half-angle right-circular cone. Using newtonian theory, calculate the drag coefficient for $1.5 \leq M_\infty \leq 7$, assuming the base pressure is equal to p_∞. Plot these results on the same graph as you prepared for Prob. 10.3. From the comparison, what can you conclude about the use of newtonian theory for small- and moderate-angle cones?

12.4   Consider a blunt axisymmetric body at an angle of attack α in a supersonic stream. Assume a calorically perfect gas. Outline in detail how you would carry out a time-dependent, finite-difference solution of this flowfield. Point out the differences between this problem and the solution for $\alpha = 0$ discussed in Sec. 12.5.

12.5   Consider a hemisphere with a flat base in a hypersonic flow at 0° angle of attack (the hemispherical portion faces into the flow). Assuming that the base pressure is equal to free-stream static pressure, use modified newtonian theory to derive an expression for the drag coefficient $C_D = D/q_\infty \pi R^2$ as a function of $C_{p_{max}}$.

12.6   This problem, as well as Probs. 12.7 and 12.8, are related to the discussion on computational fluid dynamics contained in Appendix B. In that discussion, the time-dependent (time-marching) solution of isentropic subsonic-supersonic quasi-one-dimensional flow is given, albeit under rather controlled conditions, such as the use of qualitatively proper initial conditions. Using the computer program you wrote for Prob. 12.1, and the same nozzle shape, explore the effect of different initial conditions on the behavior of the time-marching process. Specifically for one exploration, feed in constant property initial conditions, i.e., assume density, velocity, and temperature are constant through the nozzle at time zero, equal to their reservoir values. Compare the time-marching behavior with that from Prob. 12.1. Do not be surprised if you cannot get a solution (i.e., if the attempted solution "blows up" on the computer). What can you say about the importance of the selection of initial conditions?

12.7 Using the computer program and nozzle shape from Prob. 12.1, calculate the purely subsonic isentropic flow through the nozzle for the case when the ratio of exit static pressure to reservoir pressure is held fixed at 0.996. (Do not be surprised if you have difficulty. For help, consult the discussion on this type of CFD solution in the author's book *Computational Fluid Dynamics: The Basics with Applications,* McGraw-Hill, 1995. Read why the use of the governing equations completely in conservation form might be helpful.)

12.8 Using your computer program from Prob. 12.1, solve the flow described in Prob. 5.11 involving a normal shock wave inside the nozzle. (Again, do not be surprised if you have difficulty, because the conservation form of the equations with artificial viscosity is usually employed for this type of flow. See *Computational Fluid Dynamics: The Basics with Applications,* for a detailed discussion of the CFD solution of this type of flow.)

CHAPTER 13

Three-Dimensional Flow

There is no royal road to geometry.

Proclus (410–485 A.D.) an Athenian philospher, commenting on the works of Euclid

PREVIEW BOX

For decades after the advent of high-speed flight, the accurate calculation of three-dimensional flows was the Holy Grail of compressible flow—constantly sought after but not achievable. Until recently, our understanding of the physical aspects of complex three-dimensional flows, what there was of it, was based on experimental data. Note that in the previous chapters of this book we dealt with one- or two-dimensional flows. Even the supersonic flow over a right-circular cone at zero angle of attack (as studied in Chap. 11) is described by two independent variables, although the flow takes place in three-dimensional space. Such flows are called "quasi-two-dimensional." A general three-dimensional flow, for example, in cartesian space a flow where the properties are a function of x, y, and z, has resisted analytical solution.

However, with the advent of computational fluid dynamics, calculations of such three-dimensional flows can be, and are being made. In the present chapter, we emphasize the physical aspects of three-dimensional compressible inviscid flows. In many three-dimensional flows, nature has some surprises for us—new physical phenomena that we would not ordinarily expect based on our previous experience with two-dimensional flows. In the course of our discussions, we will highlight several computational techniques used to calculate three-dimensional flows. Even with the power of modern CFD, such calculations are today still not routine. They are labor intensive, usually requiring long run times on the computer. However, complex three-dimensional flows are being successfully calculated, and they are now an inherent part of computational fluid dynamics.

The material in this chapter is exciting. It is part of the *modern* compressible flow that we enjoy today. So, read and enjoy.

The intellectual organization of this chapter is straightforward; we first study cones at angle of attack, and then blunt bodies at angle of attack. Therefore, there is no need for a roadmap to help keep us on course.

13.1 | INTRODUCTION

For a moment, return to Fig. 1.9. Here you see the Bell XS-1, the first aircraft to fly faster than the speed of sound in level flight. What you see is a geometrically three-dimensional object at an angle of attack; hence, the flowfield over the XS-1 is *three-dimensional*. Indeed, the flowfields associated with all practical flight vehicles are three-dimensional. In contrast, the vast majority of flow problems treated in this book are either one- or two-dimensional. Why? The answer is straightforward—*for simplicity*. We have used these simpler problems to great advantage in the study of the fundamentals of compressible flow, which can be readily demonstrated by a myriad of different one- and two-dimensional applications. Moreover, these simpler flows have practical applications on their own. The one-dimensional and quasi-one-dimensional flows discussed in Chaps. 3, 5, and 7 have direct application to flows in ducts and streamtubes, and such one-dimensional analyses are used extensively in fluids engineering, propulsion, and aerodynamics. The two-dimensional flows discussed in Chaps. 4 and 9 are applied locally to those parts of a body where the flow is essentially two-dimensional, such as straight wings, control surfaces (such as ailerons on a wing), and for any object that has a long span in one direction perpendicular to the flow. Also, in Sec. 4.13, we demonstrated that the flow properties behind any point on a three-dimensional shock wave surface are determined by the conventional two-dimensional oblique shock relations applied locally at that point. Even the flow over a sharp, right-circular cone at zero angle of attack is "one-dimensional" in the sense that the conical flowfield depends only on one independent variable, namely the polar angle θ as described in Chap. 10. In short, the one- and two-dimensional flows treated thus far have served us well in our study of compressible flow.

On the other hand, the vast majority of practical problems in compressible flow, especially those involving external flows over aerodynamic bodies, are three-dimensional. We may be concerned with the calculation of the flow over a complete airplane configuration, such as the Bell XS-1 shown in Fig. 1.9. Or, we may be interested in the flowfield around a simple missile-like body, but with the body at angle of attack—another example of a three-dimensional flowfield. We have already briefly touched on the analysis of three-dimensional flows, such as in Sec. 11.10 on the three-dimensional method of characteristics, and in Sec. 11.16 where an example of the three-dimensional flow over a space-shuttle configuration was calculated by both the method of characteristics and the finite-difference method. However, for the most part, we have not dealt squarely with the calculation of three-dimensional flows. Because of the importance of such flows, it is now appropriate for us to devote a chapter to such matters.

In general, the addition of a "third dimension" in aerodynamic analyses causes at least an order-of-magnitude increase in the amount of work and thought necessary to obtain a solution. In fact, in terms of pure analysis, there are very few analytical, three-dimensional flowfield solutions in existence. Indeed, before the late 1960s, the calculation of three-dimensional flows was a major state-of-the-art research area—very few solutions existed. Since the early 1970s, the solution of three-dimensional flows over very complex shapes has become more attainable through the methods of computational fluid dynamics (CFD). However, even today, modern numerical calculations of three-dimensional flows require a great deal more time to program and execute than their two-dimensional counterparts. And of course, the large amount of numerical data produced in the course of a three-dimensional CFD solution is sometimes overwhelming, and can be made tractable only by the intelligent use of sophisticated computer graphics.

In light of this, the present chapter will be long on philosophy and methodology, but short on details. The subject of three-dimensional flows deserves a book all its own. To paraphrase Proclus' quotation at the beginning of the chapter, there is no royal road to three-dimensional flow methods. Our purpose in the present chapter is to introduce some of the physical aspects that distinguish three-dimensional flows from their one- and two-dimensional counterparts, and to discuss some of the methods, both old and new, for the calculation of such flows. Finally, we hope to provide the reader with some intuitive understanding of three-dimensional compressible flows in general. Because of the dominant role played by three-dimensional flow problems in modern aerodynamics, along with the advanced numerical methods presently used for calculating such flows, the material in this chapter is essential to the study of *modern* compressible flow.

13.2 | CONES AT ANGLE OF ATTACK: QUALITATIVE ASPECTS

Chapter 10 was devoted to the study of supersonic flow over a right-circular cone at zero angle of attack. Referring to Fig. 10.3, we saw that the flow was conical (independent of distance along a conical ray r from the vertex of the cone) and axisymmetric (independent of the azimuthal angle ϕ). Hence, the flow properties are functions only of the polar angle θ. In this sense, the problem, which involves a three-dimensional geometric body (the cone), is, from the point of view of the governing flow equations, a special type of "one-dimensional" flow in that the dependent variables are functions of only one independent variable. Mathematically, this means that the flow is described by an ordinary differential equation, namely, the Taylor–Maccoll equation, Eq. (10.13). In our later discussions, it will be useful to describe the conical flowfield as projected on a spherical surface generated by rays from the cone vertex of constant length r. For the case of the cone at zero angle of attack, this is sketched in Fig. 13.1. The flow in any azimuthal plane ($\phi = $ const) is shown in Fig. 13.1a. Consider streamline ab between the shock and the body, and the two conical rays that go through points a and b, respectively, on the streamline.

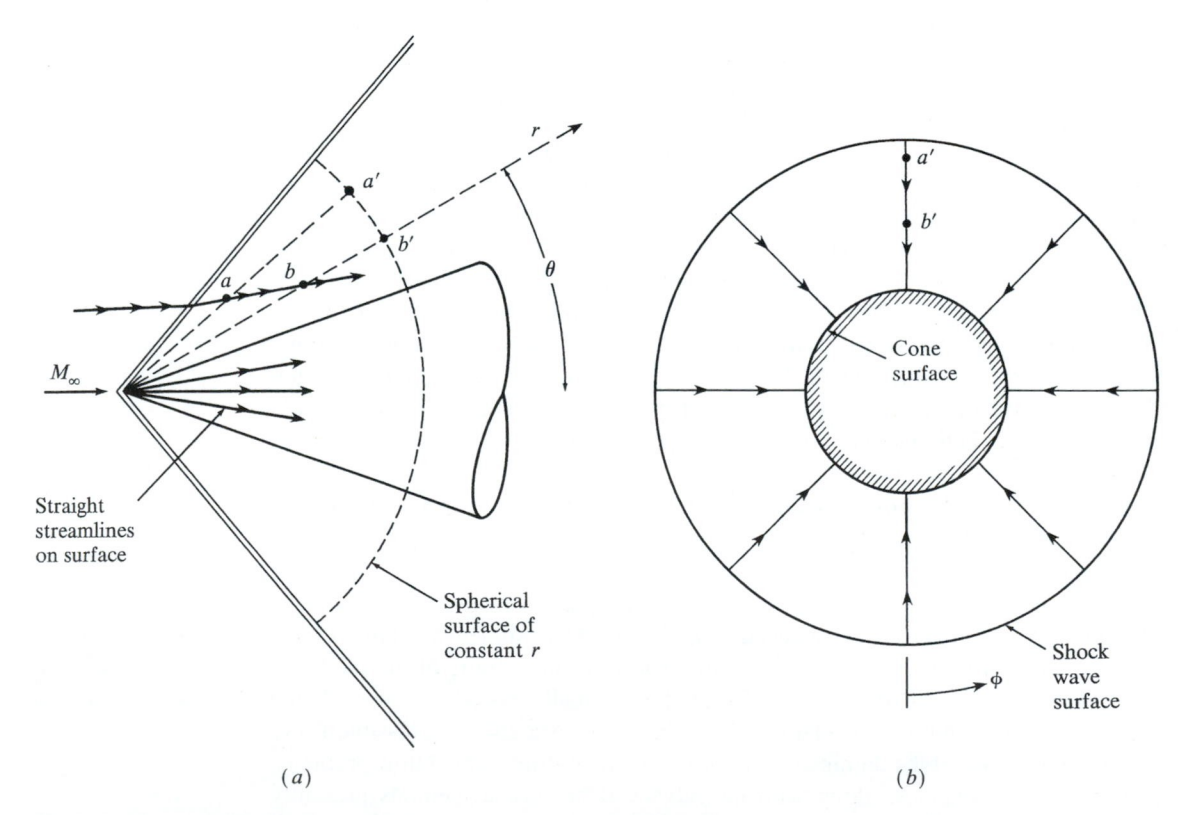

(a) (b)

Figure 13.1 | (a) Right-circular cone at zero angle of attack; (b) Projection of the body, shock wave, and streamlines on a spherical surface.

Points a and b are projected along their respective rays, and appear as points a' and b', respectively, on the spherical surface generated by a constant length along all the conical rays. A front view of this spherical surface is shown in Fig. 13.1b. Here, both the cone surface and shock surface project as concentric circles. Moreover, the streamline ab projects as a straight line, as shown by the line $a'b'$ in Fig. 13.1b. Indeed, for each meridian plane defined by $\phi = $ const, the streamlines project as straight lines on the spherical surface, and therefore the conical flow in Fig. 13.1a is seen on the spherical surface in the manner shown in Fig. 13.1b. Also, the streamlines on the surface of the cone are straight lines emanating from the cone vertex, as shown in Fig. 13.1a.

Now, consider a right-circular cone at angle of attack α, as sketched in Fig. 13.2. The same spherical coordinate system is used here as was shown earlier in Fig. 10.3a, with the z axis along the centerline of the cone. The free-stream velocity vector $\mathbf{V}_\infty$ lies in the yz plane at an angle α to the z axis. Relative to the xyz cartesian axes, we draw spherical coordinates, $r, \theta,$ and ϕ, where θ is measured from the z axis and ϕ is the azimuthal angle in the xy plane. The flow velocity components in the spherical coordinates are shown as V_r, V_θ, and V_ϕ, corresponding to the directions of increasing $r, \theta,$ and ϕ, respectively. The flowfield as it would appear in the yz plane is sketched in Fig. 13.3. Here, θ_s is the shock angle measured from the cone centerline. Just as in the zero angle-of-attack case, this flowfield is conical, i.e., flow properties are constant along rays from the cone vertex—the presence of an angle of attack does *not* destroy the conical nature of the flow. However, this is the only similarity with the zero angle-of-attack case. In all other respects, the flowfield in Fig. 13.3 is markedly different from the zero-α case. For example:

1. The flowfield shown in Fig. 13.3 is a function of *two* independent variables, θ and ϕ, in contrast to the zero-α case where θ is the only independent variable.

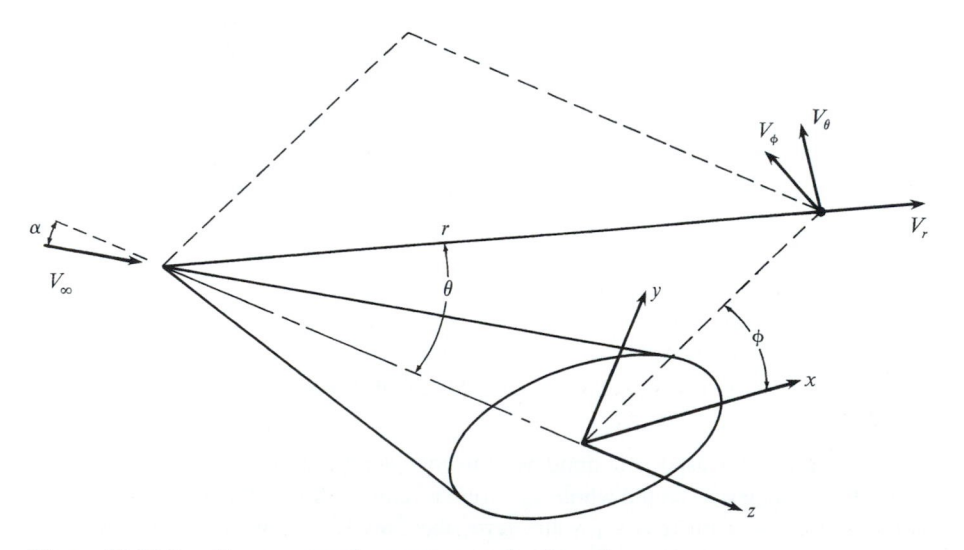

Figure 13.2 | Coordinate system for a cone at angle of attack.

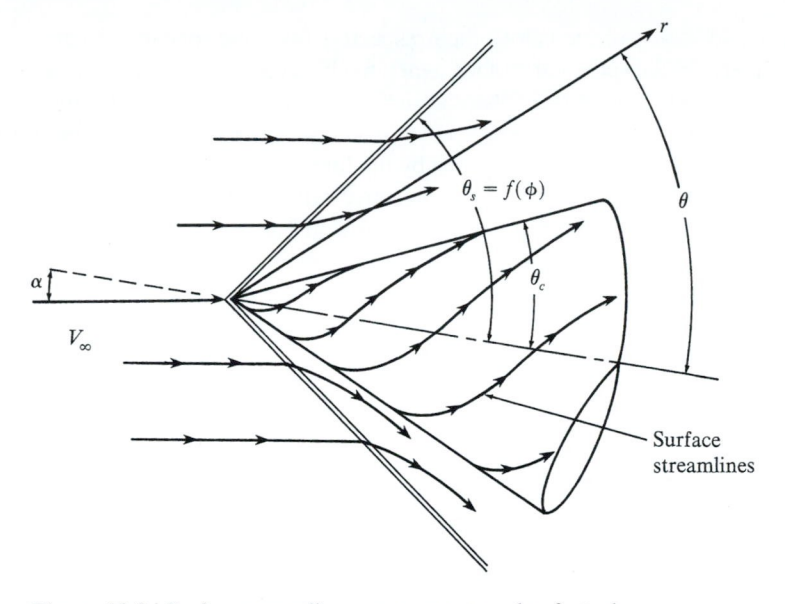

Figure 13.3 | Surface streamlines on a cone at angle of attack.

2. The shock wave angle θ_s is different for each meridional plane, i.e., θ_s is a function of ϕ.

3. The streamlines along the cone surface are now *curved* streamlines which curl around the body from the bottom of the cone (called the *windward* surface) to the top of the cone (called the *leeward* surface). However, each of the curved streamlines along the surface emanates from the vertex of the cone. Only two surface streamlines are straight—those along the very top and bottom rays.

4. The streamlines in the flow between the shock wave and the body are no longer planar; they are curved in the three-dimensional space between the shock and the body.

5. Because the flow is adiabatic and inviscid, the entropy is constant along a given streamline between the shock and the body. However, streamlines that pass through different points on the shock wave experience different increases in entropy across the shock, because the shock wave angle θ_s is different. Hence, the flow between the shock and body has finite gradients in entropy perpendicular to the streamlines. An important consequence of these entropy gradients is that the flow is *rotational,* as seen from Crocco's theorem, given by Eq. (6.60). In this sense, the supersonic flow over a cone at angle of attack is analogous to the flow over a supersonic blunt body discussed in Secs. 12.3 through 12.6.

With the above aspects in mind, we can consider the zero angle-of-attack case as almost a "singularity" in the whole spectrum of conical flows. It has singular behavior because, as α decreases toward zero, the flow does not uniformly approach the zero angle-of-attack case in all respects. For example, as the limit of $\alpha \to 0$ is reached, the flow changes discontinuously from rotational to irrotational. Also, the

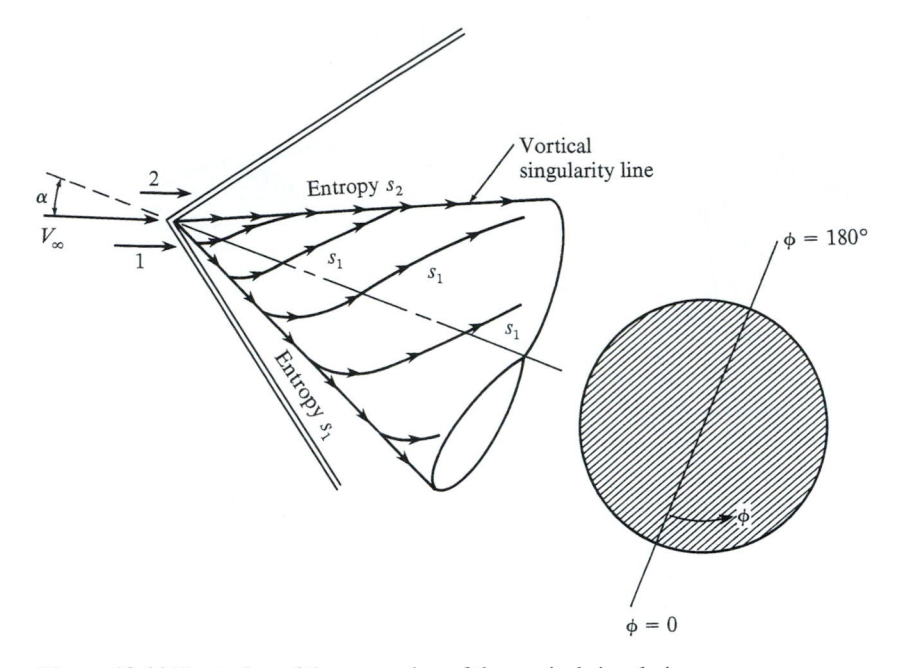

Figure 13.4 | Illustration of the generation of the vortical singularity on a cone at angle of attack.

number of independent variables drops from two to one. This means the system of equations necessary for the flow analysis changes when $\alpha \to 0$. Also, the qualitative flow picture changes. For example, the curved streamlines along the surface shown in Fig. 13.3 become straight at zero angle of attack.

There is another important aspect of the angle of attack case which does not exist at $\alpha = 0$, namely, the existence of a *vortical singularity* on the leeward surface of the cone at angle of attack. The nature of this vortical singularity can be seen in Fig. 13.4, which shows a cone at angle of attack α, along with a cross section of the cone body that identifies the most windward streamline with $\phi = 0$ and the most leeward streamline with $\phi = 180°$. At the cone vertex, the streamline at $\phi = 0$, identified as streamline 1, crosses the shock wave, and acquires entropy s_1. In turn, the flow through this point wets the entire body surface, and hence all the curved streamlines shown along the body also have entropy s_1. In contrast, the streamline at the vertex at $\phi = 180°$, identified as streamline 2, crosses a weaker portion of the shock wave, and acquires a smaller entropy s_2. In the sketch shown in Fig. 13.4, where α is less than θ_c, streamline 2 flows downstream along the top of the cone, where $\phi = 180°$. However, all of the streamlines along the surface that are curving upward from the windward side of the cone are also converging along the ray $\phi = 180°$. Therefore, the ray along the cone surface at $\phi = 180°$ has a multivalued entropy—s_2 and s_1, as well as other values as we will soon see. This line is a *vortical singularity,* and was first defined by Ferri in 1950 (see Refs. 81 and 82).

It is useful to examine the angle-of-attack flows projected on a spherical surface, such as shown for the zero angle-of-attack case in Fig. 13.1b. When $\alpha < \theta_c$, the flow

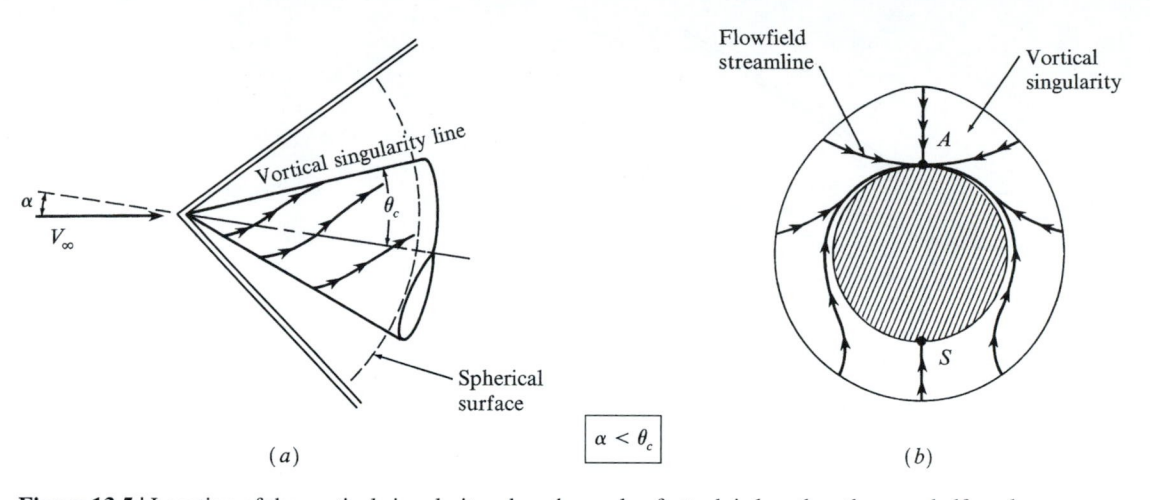

(a)

$\boxed{\alpha < \theta_c}$

(b)

Figure 13.5 | Location of the vortical singularity when the angle of attack is less than the cone half-angle.

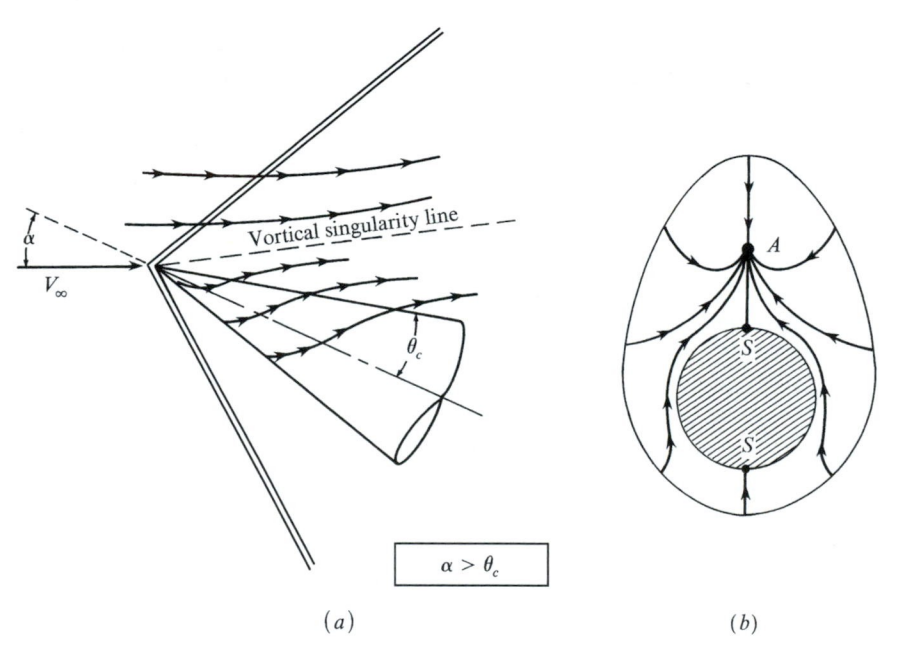

(a)

$\boxed{\alpha > \theta_c}$

(b)

Figure 13.6 | Location of the vortical singularity when the angle of attack is greater than the cone half-angle.

is such as illustrated in Fig. 13.5. The vortical singularity lies along the top of the cone as shown in Fig. 13.5a, and projects into the spherical surface as point A in Fig. 13.5b. The curved streamlines in the flowfield project onto the spherical surface also as curves, and they all converge at the vortical singularity A. Hence, the vortical singularity is truly multivalued, with values of entropy ranging from the lowest to the highest within the flowfield. When $\alpha > \theta_c$, the flow is different, as sketched in Fig. 13.6. Here, the vortical singularity lifts off the surface, and is located at point A

away from the surface as shown in Fig. 13.6b. It is also observed in both Figs. 13.5b and 13.6b that the streamlines with different values of entropy are closely squeezed together near the cone surface. Hence, an *entropy layer* exists adjacent to the surface of a cone at angle of attack, which is characterized by large gradients in entropy normal to the streamlines.

In describing three-dimensional flows, the type of pictures shown in Figs. 13.5b and 13.6b are called *cross flows*. To be more specific, the xy plane shown in Fig. 13.2 is called the cross-flow plane, and the velocity given by the vector addition of $\mathbf{V}_\theta$ and $\mathbf{V}_\phi$ is called the cross-flow velocity at a given point in the cross-flow plane. Any point where $V_\theta^2 + V_\phi^2 = 0$ is called a stagnation point in the cross-flow plane; such stagnation points are labeled by S in Figs. 13.5b and 13.6b. The vortical singularity A is also a cross-flow stagnation point. Note that S and A are not *true* stagnation points, because the radial velocity V_r is finite at these points. Indeed, there are no points in the inviscid conical flowfield where $V = 0$, i.e., there are no true stagnation points in this flowfield. For more details on cross-flow stagnation points and vortical singularities, see the work by Melnik (Ref. 83).

As the angle of attack increases, the cross-flow velocity also increases. When it becomes supersonic, i.e., when $V_\theta^2 + V_\phi^2 > a^2$, then embedded shock waves can occur in the leeward portion of the flow, as sketched in Fig. 13.7. These shocks are usually relatively weak and appear in most cases when the angle of attack is larger than the cone half-angle, i.e., when $\alpha > \theta_c$. Modern computational fluid dynamic solutions of the inviscid flow over cones at angle of attack have shown weak embedded shocks in the results, as will be discussed in the next section.

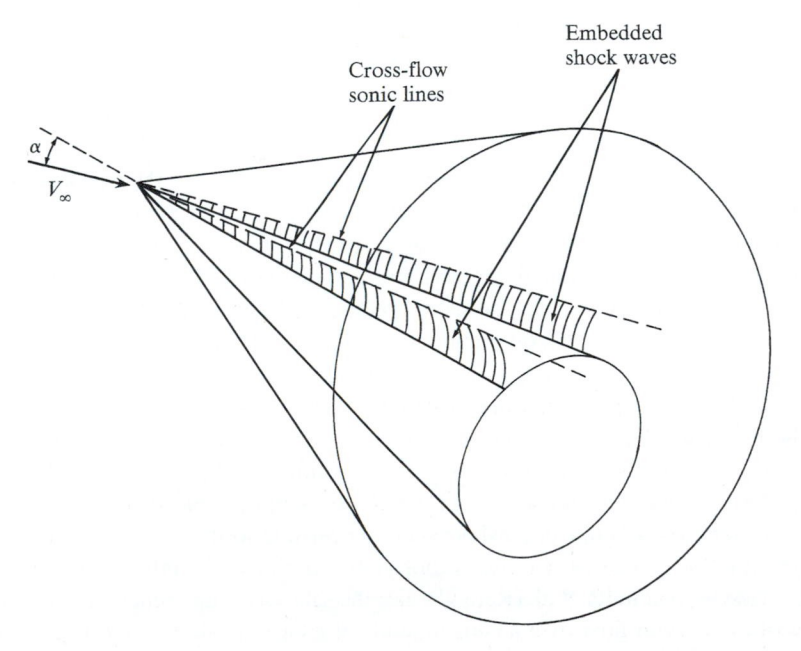

Figure 13.7 | Schematic of embedded shocks on the leeward surface of a cone at angle of attack.

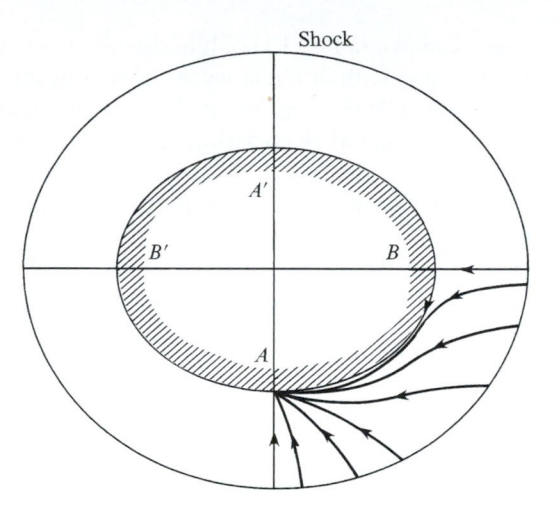

Figure 13.8 | Flowfield around an elliptic cone
at $\alpha = 0$, as projected onto a spherical surface
defined by $r = $ const.

It should be noted that flows over cones that do not have circular cross sections
are also conical flows (constant properties along r). For many high-speed applications, cones with elliptical cross sections are attractive. The flow over such elliptic
cones at angle of attack exhibits many of the same features as for the right-circular
cone, with the flow variables depending on both θ and ϕ. However, unlike the right-
circular cone, the flow over an eliptic cone at zero angle of attack still depends on
both θ and ϕ, and has cross-flow stagnation points and vortical singularities even at
$\alpha = 0$. The flow over an elliptic cone at zero angle of attack is shown in Fig. 13.8,
where points A and A' are vortical singularities and B and B' are cross-flow stagnation points.

Finally, we emphasize that all of the qualitative features of the flows over cones
at angle of attack discussed herein are for *inviscid* flows. Experimental measurements of real flows over cones at angle of attack show that the windward region is accurately described by inviscid analysis, but that the leeward region is characterized
by flow separation. The surface pressure gradient in the circumferential direction (the
direction of increasing ϕ) is favorable on the windward side. However, for angles of
attack greater than θ_c, the pressure gradient on the leeward side becomes unfavor-
able; the circumferential pressure distribution attains a local minimum somewhere
on the leeward side, and the boundary layer separates from the cone surface along a
constant ray just downstream of this pressure minimum. Associated with this separated flow are primary and secondary separation vortices, and if the cross-flow velocity is supersonic, embedded shocks will occur due to the abrupt change in flow
direction in the separated regions. A thorough experimental study of such flows has
been made by Feldhuhn et al. (Ref. 84). For the sake of completeness, the major features of the *viscous* flow over a cone at angle of attack are shown in Fig. 13.9, taken
from Feldhuhn et al. In Fig. 13.9, where the flow is again projected on a spherical

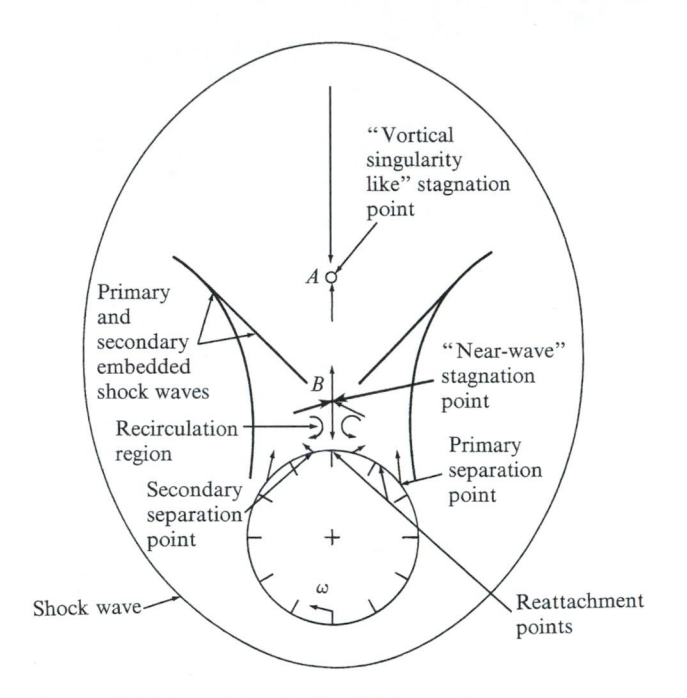

Figure 13.9 | A model of the flowfield around a cone at large angle of attack based on the experimental data of Feldhuhn et al. (Ref. 84).

surface, we see the flow separation points, separation vortices, and the embedded shocks. In spite of the viscous effects, some of the flowfield on the leeward side exhibits familiar inviscid properties, such as the vortical singularity at point A. Since this book deals with inviscid flows, we will not pursue these viscous properties any further. The interested reader is referred to Ref. 84 for more details. However, we note in passing an aspect of separated flows that is a current state-of-the-art research topic in aerodynamics. Modern computational fluid dynamic calculations of inviscid rotational flowfields are yielding results that simulate flow separation without any direct accounting of the local viscous effects. As a result, there is a growing number of researchers who feel that separated flows are dominated by inviscid phenomena, and that the actual viscosity plays only a secondary role. Because of the present controversial nature of this theory, we will not elaborate here; instead, we will wait for a resolution at some future date. In the spirit of the present section, we note the recent work by Marconi (Ref. 85) on the calculation of separated flows over cones and cone-cylinders at angle of attack, where the calculation involved the solution of the Euler equations, i.e., dealing with an inviscid flow only. Results were obtained which included a separated flow such as sketched in Fig. 13.10, which shows a vortex sheet leaving the body surface along the separation line. A typical surface streamline pattern from Marconi's calculations is shown in Fig. 13.11 for a cone-cylinder at 36° angle of attack in a Mach 2.3 free stream. The distinction between the attached flow on the windward side and the separated region on the leeward side is striking.

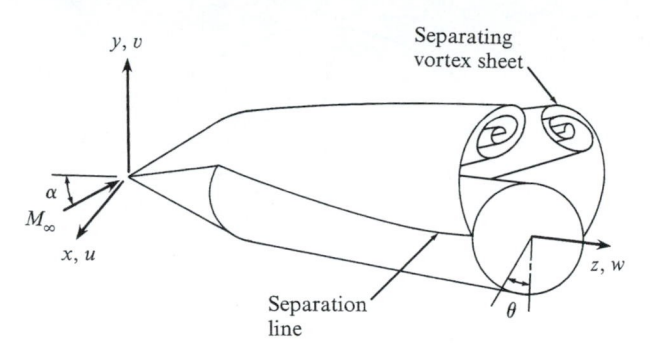

Figure 13.10 | A model of the separated flowfield over an axisymmetric body at angle of attack based on the assumption of inviscid flow. (After Marconi, Ref. 85.)

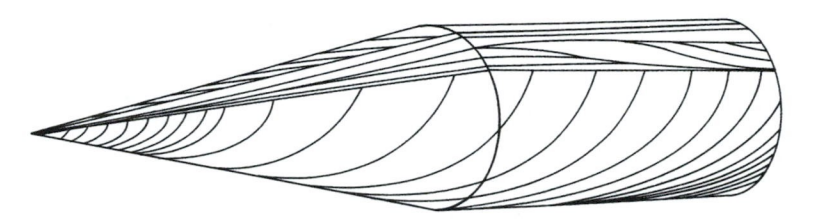

Figure 13.11 | Surface streamlines on a cone-cylinder at angle of attack, from the inviscid flow solutions of Marconi (Ref. 85). $M_\infty = 2.3$ and $\alpha = 36°$. Flow separation on the leeward side is modeled as part of the inviscid solution.

13.3 | CONES AT ANGLE OF ATTACK: QUANTITATIVE ASPECTS

Early work on the calculation of flows over cones at angle of attack expressed the flow variables in terms of series expansions in α around the zero angle-of-attack case. (See, for example, the work of Kopal in Ref. 86, and that of Sims in Ref. 87, which complement the zero angle-of-attack tables in Refs. 28 and 29 by those same authors, respectively.) These analyses are approximate, and are limited to small angle of attack. They will not be discussed here because they have essentially been superseded by the techniques of modern computational fluid dynamics that allow the exact inviscid solution to be obtained.

Before discussing the exact numerical solution, let us examine an aspect of the governing equations for conical flow that mathematically allows the existence of a vortical singularity, as described qualitatively in Sec. 13.2. Because the flow is isentropic along a given streamline, Eq. (6.51) holds:

$$\frac{Ds}{Dt} = 0 \qquad (6.51)$$

For a steady flow, this is

$$\mathbf{V} \cdot \nabla s = 0 \tag{13.1}$$

In terms of the spherical coordinates shown in Fig. 13.2, we have

$$\nabla s = \frac{\partial s}{\partial r} \mathbf{e}_r + \frac{1}{r} \frac{\partial s}{\partial \theta} \mathbf{e}_\theta + \frac{1}{r \sin \theta} \frac{\partial s}{\partial \phi} \mathbf{e}_\phi \tag{13.2}$$

and

$$\mathbf{V} = V_r \mathbf{e}_r + V_\theta \mathbf{e}_\theta + V_\phi \mathbf{e}_\phi \tag{13.3}$$

where $\mathbf{e}_r$, $\mathbf{e}_\theta$, and $\mathbf{e}_\phi$ are the unit vectors in the r, θ, and ϕ directions, respectively. Combining Eqs. (13.1) through (13.3), we have

$$\mathbf{V} \cdot \nabla s = V_r \frac{\partial s}{\partial r} + \frac{V_\theta}{r} \frac{\partial s}{\partial \theta} + \frac{V_\phi}{r \sin \theta} \frac{\partial s}{\partial \phi} = 0 \tag{13.4}$$

For conical flow, $\partial s / \partial r = 0$, and Eq. (13.4) becomes

$$\frac{V_\theta}{r} \frac{\partial s}{\partial \theta} + \frac{V_\phi}{r \sin \theta} \frac{\partial s}{\partial \phi} = 0 \tag{13.5}$$

or

$$\frac{\partial s}{\partial \phi} = -\frac{V_\theta}{V_\phi} \sin \theta \frac{\partial s}{\partial \theta} \tag{13.6}$$

Keep in mind that Eq. (13.6) holds along a streamline in the flow; more appropriately, it holds along the projection of a streamline in the spherical surface defined by $r = $ const, as discussed in Sec. 13.2. The *shape* of this streamline in the spherical surface can be found by noting that the entropy is a function of θ and ϕ in the spherical surface. Thus,

$$ds = \frac{\partial s}{\partial \theta} d\theta + \frac{\partial s}{\partial \phi} d\phi \tag{13.7}$$

Along a streamline, $ds = 0$, and Eq. (13.7) gives

$$\frac{d\phi}{d\theta} = -\frac{\partial s / \partial \theta}{\partial s / \partial \phi} \tag{13.8}$$

Substituting Eq. (13.6) into (13.8), we have

$$\frac{d\phi}{d\theta} = \frac{V_\phi}{V_\theta \sin \theta} \tag{13.9}$$

Equation (13.9) gives the shape of the streamlines as projected on the spherical surface in terms of the velocity field. This equation, in conjunction with Eq. (13.6), conceptually allows the solution of the entropy distribution over the spherical surface defined by $r = $ const. There is one exception, however, namely, at any point where both V_θ and V_ϕ are zero, i.e., at a cross-flow stagnation point. At such a point, Eqs. (13.6) and (13.9) are indeterminant forms, which allow the possibility of a multivalued entropy at that point—namely, a vortical singularity. Hence, the governing flow equations predict that such vortical singularities may exist. In Sec. 13.2 we have already shown on the basis of physical reasoning that not all cross-flow stagnation points are vortical singularities, but that all vortical singularities are cross-flow stagnation points. Equations (13.6) and (13.9) simply show that the existence of vortical singularities are compatible with the mathematics.

Modern solutions to the flowfield over cones at angle of attack usually involve a finite-difference solution to the governing partial differential equations of three-dimensional, inviscid, adiabatic compressible flow. These equations have been derived and discussed at length in Chap. 6. For example, repeating Eqs. (6.5), (6.29), and (6.44), we have

Continuity:
$$\frac{\partial \rho}{\partial t} + \nabla \cdot (\rho \mathbf{V}) = 0 \tag{6.5}$$

Momentum:
$$\rho \frac{D\mathbf{V}}{Dt} = -\nabla p + \rho \mathbf{f} \tag{6.29}$$

Energy:
$$\rho \frac{Dh_o}{Dt} = \frac{\partial p}{\partial t} + \rho (\mathbf{f} \cdot \mathbf{V}) \tag{6.44}$$

Written in terms of spherical coordinates, and specialized to a steady flow with no body forces, these equations become:

Continuity:
$$\frac{\partial \rho}{\partial r} = -\frac{1}{V_r} \left[\rho \frac{\partial V_r}{\partial r} + \frac{2\rho V_r}{r} + \frac{1}{r \sin\theta} \frac{\partial}{\partial \phi}(\rho V_\theta \sin\theta) + \frac{1}{r \sin\theta} \frac{\partial(\rho V_\phi)}{\partial \phi} \right] \tag{13.10}$$

Momentum in r direction:
$$\frac{\partial V_r}{\partial r} = -\frac{1}{V_r} \left[\frac{V_\theta}{r} \frac{\partial V_r}{\partial \theta} + \frac{V_\phi}{r \sin\theta} \frac{\partial V_r}{\partial \phi} - \frac{V_\theta^2 + V_\phi^2}{r} + \frac{1}{\rho} \frac{\partial p}{\partial r} \right] \tag{13.11}$$

Momentum in θ direction:
$$\frac{\partial V_\theta}{\partial r} = -\frac{1}{V_r} \left[\frac{V_\theta}{r} \frac{\partial V_\theta}{\partial \theta} + \frac{V_\phi}{r \sin\theta} \frac{\partial V_\theta}{\partial \phi} + \frac{V_r V_\theta}{r} - \frac{V_\phi^2 \cot\theta}{r} + \frac{1}{r\rho} \frac{\partial p}{\partial \theta} \right] \tag{13.12}$$

Momentum in φ direction:
$$\frac{\partial V_\theta}{\partial r} = -\frac{1}{V_r} \left[\frac{V_\theta}{r} \frac{\partial V_\phi}{\partial \theta} + \frac{V_\phi}{r \sin\theta} \frac{\partial V_\phi}{\partial \theta} + \frac{V_r V_\phi}{r} + \frac{V_\theta V_\phi \cot\theta}{r} + \frac{1}{\rho r \sin\theta} \frac{\partial p}{\partial \theta} \right] \tag{13.13}$$

Energy: $h_o = h_\infty + \dfrac{V_\infty^2}{2} = h + \dfrac{1}{2}\left(V_r^2 + V_\theta^2 + V_\phi^2\right)$ (13.14)

In addition to these flow equations, we also have the perfect gas equation of state:

$$p = \rho RT \tag{13.15}$$

and the state relation for a calorically perfect gas:

$$h = c_p T \tag{13.16}$$

Equations (13.10) through (13.16) are seven equations for the seven unknowns, ρ, V_r, V_θ, V_ϕ, p, T, and h. They are the equations, written in spherical coordinates, for a steady, adiabatic, inviscid, compressible flow, and therefore are applicable for the solution of the flow over a cone at angle of attack in a supersonic stream.

Note that Eqs. (13.10) through (13.13) have been written such that the r derivatives are on the left-hand side, and the θ and ϕ derivatives are on the right-hand side. This hints strongly of a finite-difference solution that marches in the r direction, directly analogous to the time-marching solutions discussed in Chap. 12. Indeed, a novel approach to the solution of the cone problem using the principle of marching in the r direction was first set forth by Moretti in Ref. 88. The general philosophy of Moretti's approach is illustrated in Fig. 13.12. We are interested in calculating the flowfield over a cone with half-angle θ_c at an angle of attack α in a free stream at M_∞. Start with an *assumed* flowfield on the spherical surface given by $r = r_o$, where

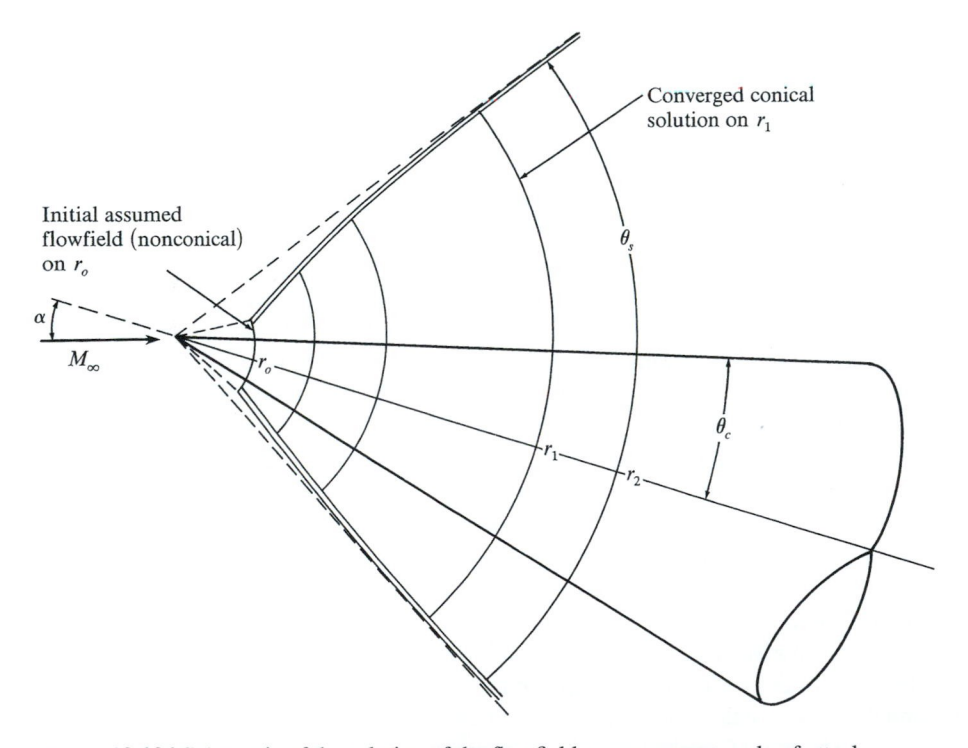

Figure 13.12 | Schematic of the solution of the flowfield over a cone at angle of attack, starting with an initially assumed nonconical flowfield, and marching downstream until convergence is obtained.

the spherical surface is bounded between the body and the shock. This will be a non-conical flow, and of course is not the correct flow solution for the cone. The assumed flow on $r = r_o$ can be somewhat arbitrary, but must have at each point the local total enthalpy equal to the free-stream value, and the integrated mass flow through $r = r_o$ must equal the free-stream mass flow intercepted by the spherical surface. Since the flowfield properties are now specified on $r = r_o$, the θ and ϕ derivatives that appear on the right-hand side of Eqs. (13.10) through (13.13) can be expressed in terms of known finite differences. This immediately allows the calculation of $\partial \rho / \partial r$, $\partial V_r / \partial r$, $\partial V_\theta / \partial r$, and $\partial V_\phi / \partial r$ from Eqs. (13.10) through (13.13). In turn, the r derivatives are used to calculate the flow over the next downstream spherical surface located at $r_o + \Delta r$. For this purpose, MacCormack's technique, as discussed in Sec. 11.12, can be used. For example, if the flow is known over the spherical surface located at r, then the density at $r + \Delta r$ can be obtained from

$$\rho(r + \Delta r) = \rho(r) + \left(\frac{\partial \rho}{\partial r} \right)_{ave} \Delta r \qquad (13.17)$$

In Eq. (13.17), the average value of $\partial \rho / \partial r$ is obtained from the predictor-corrector approach directly analogous to that described in Sec. 11.12. That is, a predicted value $\partial \rho / \partial r$ is obtained from Eq. (13.10) using forward differences in θ and ϕ. Then a corrected value $(\overline{\partial \rho / \partial r})$ is obtained from Eq. (13.10) using rearward differences in θ and ϕ with predicted values $\bar{\rho}$, $\bar{V}_r$, $\bar{V}_\theta$, and $\bar{V}_\phi$. Then the average r derivative is formed as

$$\left(\frac{\partial \rho}{\partial r} \right)_{ave} = \frac{1}{2} \left[\left(\frac{\partial \rho}{\partial r} \right) + \left(\overline{\frac{\partial \rho}{\partial r}} \right) \right] \qquad (13.18)$$

Finally, the value of ρ at $r + \Delta r$ is obtained from Eq. (13.17). Of course, to allow the proper formulation of the finite differences, the flowfield shown in Fig. 13.12 should be transformed such that it is a rectangular shape in the transformed plane. Along with this, the governing equations (13.10) through (13.13) should also be transformed to the computational space. Since our purpose here is to present the general philosophy of the method, we will not clutter our discussion with details.

As the finite-difference solution marches downstream to subsequent spherical surfaces, the flowfield changes from one value of r to another. In the process, the shock wave shape and location change as we march downstream. For details on the calculation of the shock shape, as well as the numerical formulation of the boundary conditions behind the shock and along the body, see Ref. 88. However, as we progress far enough downstream, the flowfield properties begin to approach a converged value, i.e., $\partial \rho / \partial r$ becomes smaller and smaller, until we reach some spherical surface, denoted by $r = r_1$ in Fig. 13.12, where there is virtually no change in the flowfield in the r direction. That is, at $r = r_1$, $(\partial / \partial r) \approx 0$ for all the flow variables, and therefore the flow variables over the next downstream surface r_2 are virtually unchanged from r_1. Clearly, when this convergence is achieved, then, by definition, the flowfield has become conical, and the flowfield solution over the spherical surface $r = r_1$ is indeed the solution of the flow over the given cone at the given angle of attack. The nonconical flow computed between r_o and r_1 was just a means to obtain the final conical flow solution. In this vein, the present technique is directly analogous to the time-marching method for the solution of the flow over a blunt body discussed in Chap. 12, where the calculated transient flowfield is just a means to an end, namely, obtaining the final steady flow over the body at large times. Here, we have replaced the time marching of Chap. 12 with spatial marching in the r direction, leading to a converged conical flow at large values of r.

Some typical results obtained by Moretti are shown in Figs. 13.13 and 13.14, for a free-stream Mach number of 7.95, and a 10° half-angle cone at 8° angle of attack. These figures illustrate the mechanics of the downstream-marching philosophy. Figure 13.13 illustrates the rate of change of the average calculated shock coordinate with distance r. If $\bar{\theta}_s$ represents the average polar coordinate of the shock at a given r [note that $\theta_s = f(\phi)$ for a given r], then $\partial\bar{\theta}_s/\partial r$ is an indicator of convergence. When $\partial\bar{\theta}_s/\partial r$ becomes small, the correct conical flow is approached. This variation is given in Fig. 13.13, which is a plot of $\partial\bar{\theta}_s/\partial r$ as a function of radial location referenced to the initial value $r = r_o$. This shows that a downstream-marching distance of more than $100r_o$ was necessary before the converged conical flow was obtained.

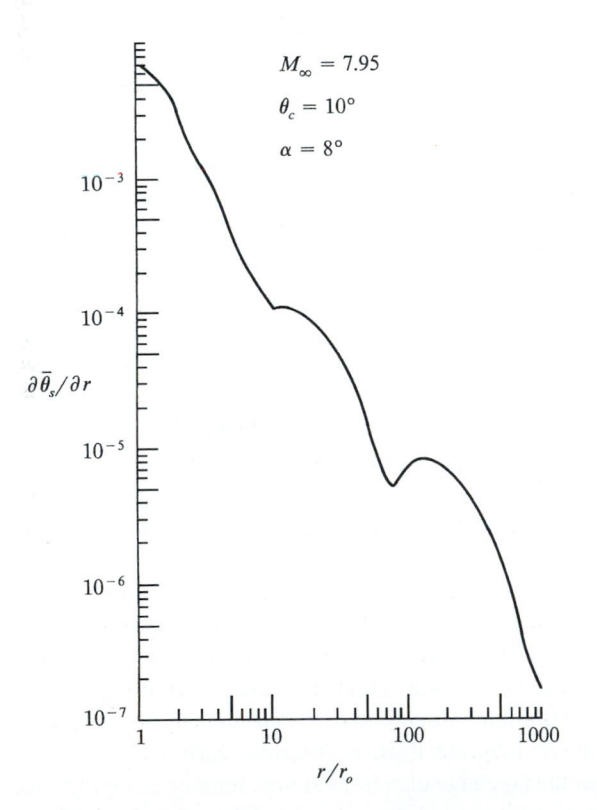

Figure 13.13 | Rate of convergence, as indicated by the spatial rate of change of the mean shock angle with downstream distance r. (From Moretti, Ref. 88.)

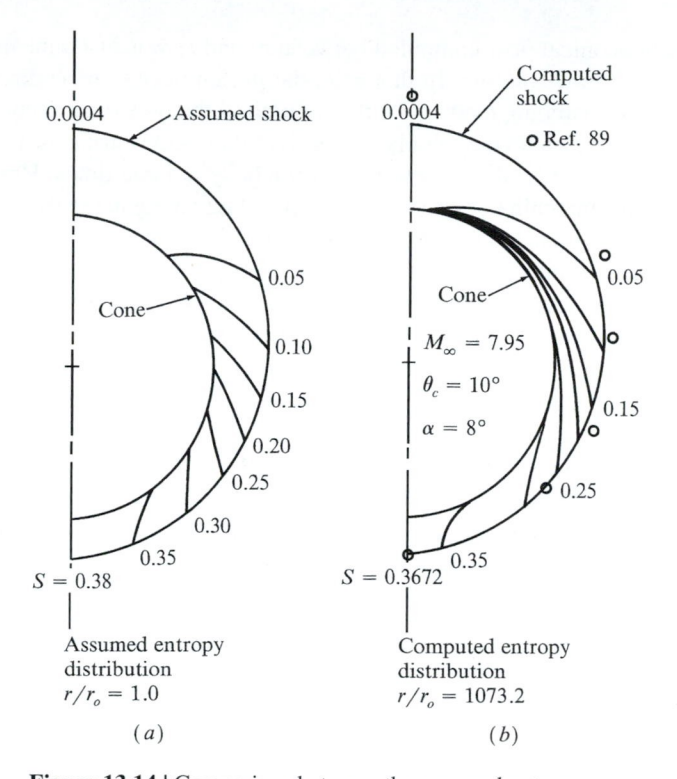

Figure 13.14 | Comparison between the assumed entropy distribution at $r = r_o$ (a) and the converged entropy distribution at $r = 1073.2r_o$ (b). (From Moretti, Ref. 88.)

In Fig. 13.14, the initially assumed entropy distribution at $r = r_o$ (Fig. 13.14a) is compared with the final converged result at $r = 1073.2r_o$ (Fig. 13.14b). Starting with the assumed nonconical flow in Fig. 13.14a, the converged conical flow shown in Fig. 13.14b is obtained. The answer to the problem is Fig. 13.14b. Concentrating on Fig. 13.14b, we see a projection of the computed lines of constant nondimensional entropy on the spherical surface. Note that these lines all converge at the top of the cylinder, showing that the numerical solution is predicting the expected vortical singularity. Also, since the entropy is constant along a given streamline, then the curves shown in Fig. 13.14b are also traces of the streamline shapes. The computed shock wave shape is also shown, and is compared with experimental data for the shock shape obtained from Tracy (Ref. 89) denoted by the open circles. The expected good agreement between calculation and experiment is seen on the windward side, but the measured shock is slightly higher than the calculated shock on the leeward side—an effect due to the real viscous flow, as described earlier.

In terms of the language of computational fluid dynamics (as described in Chaps. 11 and 12), Moretti's cone solution is a *shock-fitting* method. Moreover, the governing equations given by Eqs. (13.10) through (13.13) are in the *nonconservation form,* which is appropriate in conjunction with a shock-fitting method. However, this leads to a restriction on the range of problems that can be solved by the specific method described above. The nonconservation form of the equations is not appropriate for capturing shock waves; hence, any embedded shocks that might be present in the leeward flowfield will not be properly calculated. This limited the cone solutions carried out by Moretti in Ref. 88 to cases where $\alpha < \theta_c$. However, this in no way compromises the overall philosophy presented by Moretti, namely, that the proper conical flow solution can be obtained by marching downstream from an assumed initial nonconical flow.

Using Moretti's downstream-marching philosophy, Kutler and Lomax (Refs. 40 and 90) removed the restrictions mentioned above by computing the flow with a *shock-capturing* method using the conservation form of the governing equations. In this approach, the finite-difference grid reaches beyond the conical shock wave, which in turn is captured internally within the grid in the same vein as discussed in Sec. 11.15 and pictured in Fig. 11.23. Embedded shock waves on the leeward side, if present, are also captured within the grid. Kutler and Lomax, with an eye toward applications to nonconical bodies, did not use a spherical coordinate system; rather, they employed a body-oriented system, with x measured along the surface, y perpendicular to the surface, and ϕ as the meridional angle. This coordinate system and grid is shown in Fig. 13.15, obtained from Ref. 40. The governing steady-flow equations in conservation form are written in the body-oriented coordinates, and the solution is marched downstream in the x direction using MacCormack's technique

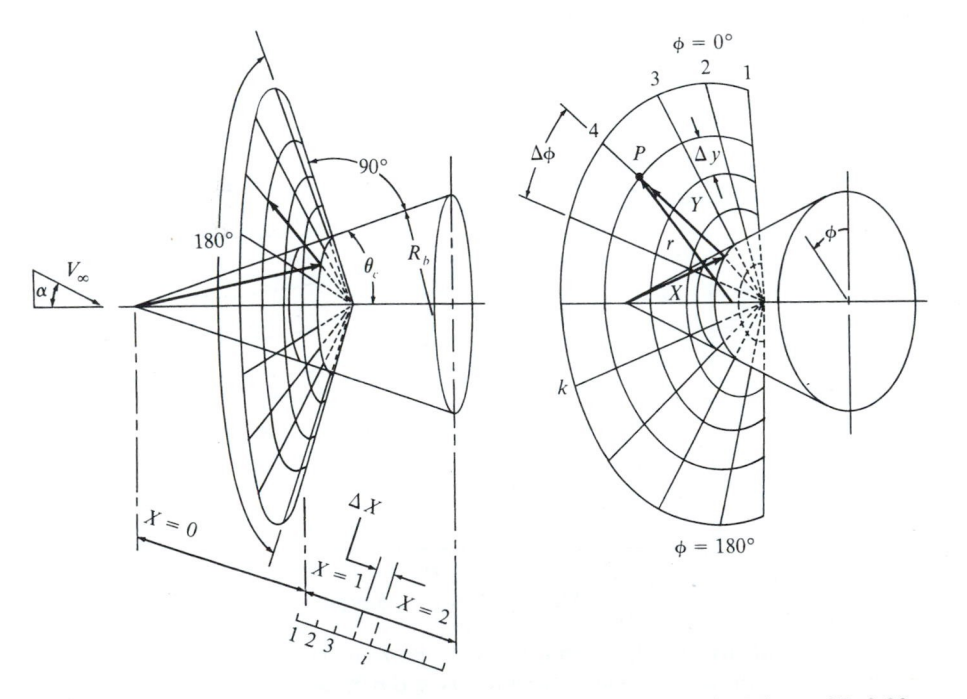

Figure 13.15 | Coordinate system and finite-difference mesh for the calculations of Ref. 90.

until a converged, conical flow is obtained. For more details on the computational method, see Ref. 40.

The circumferential pressure distributions around a 10° half-angle cone at Mach 5 are shown in Fig. 13.16 for angles of attack ranging from 0° to 15°, as calculated by Kutler and Lomax. Note that for $\alpha < 10°$, the pressure distributions monotonically decrease from the windward to the leeward side. However, for $\alpha > 10°$,

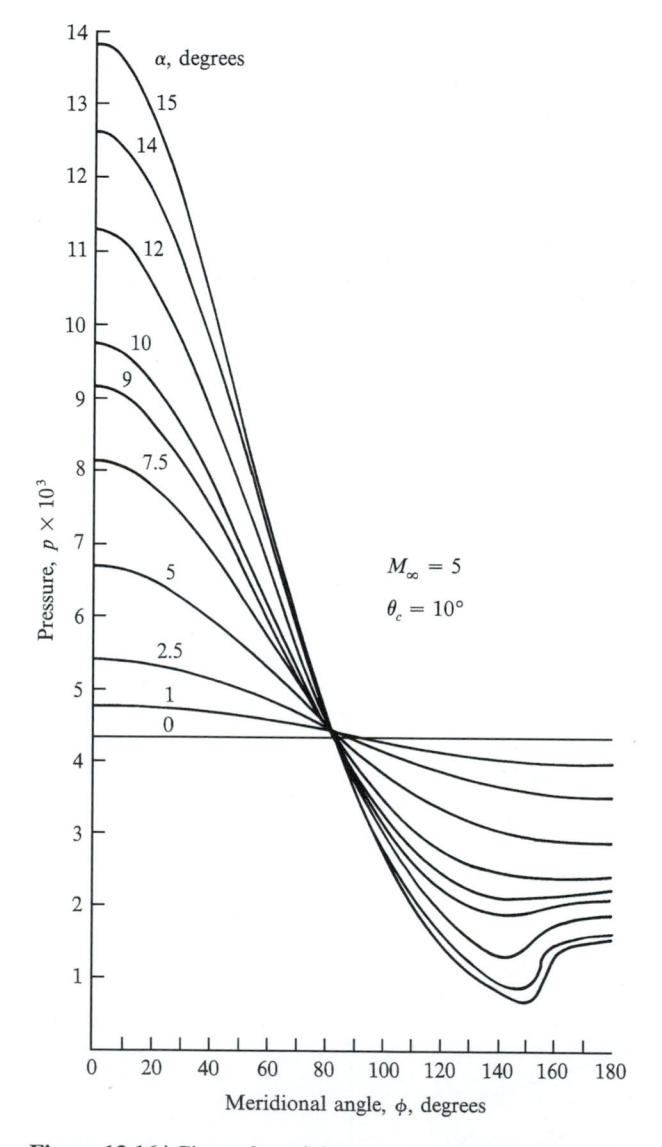

Figure 13.16 | Circumferential pressure distributions around a 10° cone at various angles of attack. (From Kutler and Lomax, Ref. 90.)

the pressure first decreases, reaches a local minimum value partway around the leeward side, and then increases to the top of the cone. Kutler and Lomax found weak embedded shock waves on the leeward side corresponding to the region of adverse pressure gradients. In a separate analysis of flows over cones at angle of attack, Fletcher (Ref. 91) also observed embedded shocks. Fletcher's approach utilized the same downstream-marching philosophy as described earlier, along with shock fitting of the primary shock wave. The flowfield calculations were carried out using a hybrid numerical and analytical method; see Ref. 91 for details. His results for a Mach 7.95 flow over a 10° cone at $\alpha = 16°$ are shown in Fig. 13.17. Here, we see, projected on a spherical surface, the calculated shock wave shape, the vortical singularity (denoted by VS), the calculated embedded shocks (labeled NN), and the sonic lines in the windward and leeward regions. Also shown are some experimental data from Tracy (Ref. 89). Note that the outer primary shock shape agrees well with experiment, but that the experimentally measured embedded shocks EE lie outside of the numerically computed shocks.

With this, we end our discussion of the flow over a cone at angle of attack in a supersonic flow. This problem, which prior to 1965 was very different to solve for large angles of attack, has been made almost routine by the modern methods of computational fluid dynamics. Our purpose has been twofold: (1) to achieve some

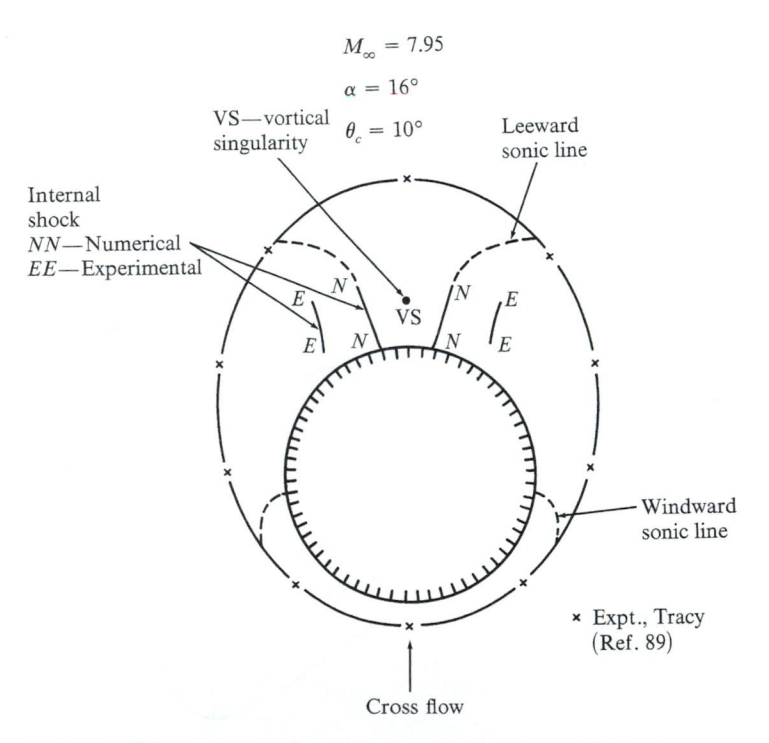

Figure 13.17 | Comparison between experiment and numerical calculations for flow over a cone at angle of attack. (From Fletcher, Ref. 91.)

overall understanding of the various computational techniques for the solution of this problem, and (2) to study the physical aspects of such flows as an example of a classic three-dimensional flowfield.

13.4 | BLUNT-NOSED BODIES AT ANGLE OF ATTACK

Recall that the flow over a cone at zero angle of attack—a three-dimensional geometric object—is "one-dimensional" in the sense that the conical flowfield depends only on one independent variable, namely the polar angle θ as described in Chap. 10. Similarly, the flow over a cone at angle of attack is "two-dimensional" in the sense that the flowfield, which is still conical, depends only on two independent variables, namely, θ and ϕ, as discussed in Secs. 13.2 and 13.3.

In this section, and for the remainder of this chapter, we discuss flowfields that are truly three-dimensional in the sense that they depend on three spatial independent variables. An important example of such a flow is the supersonic blunt body at angle of attack. The supersonic blunt body at zero angle of attack was studied in Sec. 12.5, where a time-marching method was used to obtain the steady flow in the limit of large time. The first practical zero angle-of-attack blunt body solution—indeed, made practical by the time-marching philosophy—was carried out by Moretti and Abbett in Ref. 47. This work was quickly extended to the angle-of-attack case by Moretti in Ref. 92. Since we followed Moretti's approach in Sec. 12.5, let us do the same here for the angle-of-attack case.

Consider a blunt body at angle of attack as shown in Fig. 13.18. A cylindrical coordinate system, r, ϕ, z, is drawn with the z axis along the centerline of the body.

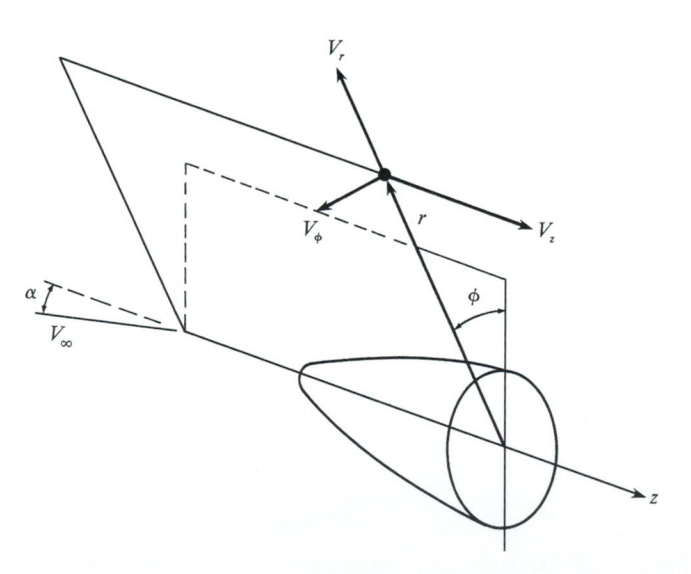

Figure 13.18 | Cylindrical coordinate system in physical space for the angle-of-attack blunt body problem.

The governing three-dimensional flowfield equations, analogous to the two-dimensional equations given by Eqs. (12.19) through (12.22), are, in cylindrical coordinates,

Continuity:
$$\frac{\partial \rho}{\partial t} + \frac{1}{r}\frac{\partial}{\partial r}(\rho r V_r) + \frac{1}{r}\frac{\partial(\rho V_\phi)}{\partial \phi} + \frac{\partial(\rho V_z)}{\partial z} = 0 \qquad (13.19)$$

Momentum in r direction:

$$\rho\frac{\partial V_r}{\partial t} + \rho V_r\frac{\partial V_r}{\partial r} + \frac{\rho V_\phi}{r}\frac{\partial V_r}{\partial \phi} - \frac{\rho V_\phi^2}{r} + \rho V_z\frac{\partial V_r}{\partial z} = -\frac{\partial p}{\partial r} \qquad (13.20)$$

Momentum in ϕ direction:

$$\rho\frac{\partial V_\phi}{\partial t} + \rho V_r\frac{\partial V_\phi}{\partial r} + \frac{\rho V_\phi}{r}\frac{\partial V_\phi}{\partial \theta} - \frac{\rho V_r V_\phi}{r} + \rho V_z\frac{\partial V_\phi}{\partial z} = -\frac{1}{r}\frac{\partial p}{\partial \phi} \qquad (13.21)$$

Momentum in z direction:

$$\rho\frac{\partial V_z}{\partial t} + \rho V_r\frac{\partial V_z}{\partial r} + \frac{\rho V_\phi}{r}\frac{\partial V_z}{\partial \phi} + \rho V_z\frac{\partial V_z}{\partial z} = -\frac{\partial p}{\partial z} \qquad (13.22)$$

Energy:
$$\frac{\partial s}{\partial t} + V_r\frac{\partial s}{\partial r} + \frac{V_\phi}{r}\frac{\partial s}{\partial \phi} + V_z\frac{\partial s}{\partial z} = 0 \qquad (13.23)$$

Recall that Eq. (13.23) is really the entropy equation, and it states that the entropy of a given fluid element is constant during its motion in the shock layer between the shock wave and the body—a ramification of the flow being inviscid and adiabatic. Following Moretti and Bleich, Eqs. (13.19) through (13.23) are nondimensionalized and transformed as follows. For simplicity, assume the body is axisymmetric (this is *not* a necessary aspect of the method). Hence, the body shape is given by

$$z = b(r)$$

The shock wave shape is given by

$$z = S(r, \phi, t)$$

Let $\delta = S - b$. Then a new set of independent variables is defined as

$$\zeta = \frac{z - b}{\delta} \qquad Y = r \qquad X = \phi \quad t = t \qquad (13.24)$$

With the relations given in Eq. (13.24), the three-dimensional flowfield between the shock and body transforms to the right parallelepiped in the ζ-X-Y space as shown in Fig. 13.19. In turn, this is used as the computational space in which finite-difference quotients are formed. The dependent variables were transformed in Ref. 92 as

$$P = \ln p \qquad R = \ln \rho \qquad \psi = P - \gamma R \qquad (13.25)$$

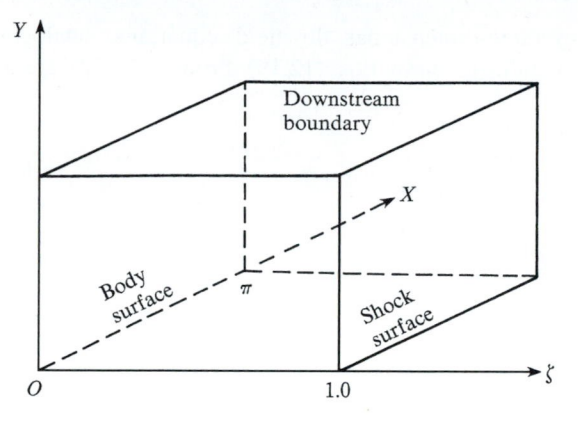

Figure 13.19 | Transformed coordinate system in computational space for the angle-of-attack blunt body problem.

With the relations defined in Eqs. (13.24) and (13.25), the governing flow equations given by Eqs. (13.19) through (13.23) become

Continuity:

$$\frac{\partial R}{\partial t} = -\left[V_r \frac{\partial R}{\partial Y} + A \frac{\partial R}{\partial X} + B \frac{\partial R}{\partial \zeta} + \frac{\partial V_r}{\partial Y} + E \frac{\partial V_r}{\partial \zeta} + \left(\frac{\partial V_\infty}{\partial X} + V_r \right) \middle/ Y \right.$$

$$\left. - F \frac{\partial V_\phi}{\partial \zeta} + \frac{1}{\delta} \frac{\partial V_z}{\partial \zeta} \right] \tag{13.26}$$

Momentum in r direction:

$$\frac{\partial V_r}{\partial t} = -\left[V_r \frac{\partial V_r}{\partial Y} + A \frac{\partial V_r}{\partial X} + B \frac{\partial V_r}{\partial \zeta} - A V_r + G \left(\frac{\partial P}{\partial Y} + E \frac{\partial P}{\partial \zeta} \right) \right] \tag{13.27}$$

Momentum in φ direction:

$$\frac{\partial V_\phi}{\partial t} = -\left[V_r \frac{\partial V_\phi}{\partial Y} + A \frac{\partial V_\phi}{\partial X} + B \frac{\partial V_\phi}{\partial \zeta} + A V_r + G \left(\frac{1}{Y} \frac{\partial P}{\partial x} - F \frac{\partial P}{\partial \zeta} \right) \right] \tag{13.28}$$

Momentum in z direction:

$$\frac{\partial V_z}{\partial t} = -\left[V_r \frac{\partial V_z}{\partial Y} + A \frac{\partial V_z}{\partial X} + B \frac{\partial V_z}{\partial \zeta} + \frac{G}{\delta} \frac{\partial P}{\partial \zeta} \right] \tag{13.29}$$

Energy (entropy):

$$\frac{\partial \psi}{\partial t} = -\left[V_r \frac{\partial \psi}{\partial Y} + A \frac{\partial \psi}{\partial X} + B \frac{\partial \psi}{\partial \zeta} \right] \tag{13.30}$$

where

$$A = \frac{V_\phi}{Y} \qquad C = \frac{db}{dr}(\zeta - 1) - \zeta\frac{\partial S}{\partial r}$$

$$B = \frac{1}{\delta}\left[V_z + CV_r - \zeta\left(\frac{\partial S}{\partial t} + A\frac{\partial S}{\partial \phi}\right)\right]$$

$$E = \frac{C}{\delta} \qquad F = \frac{\zeta}{Y\delta}\frac{\partial S}{\partial \phi} \qquad G = \frac{p}{\rho}$$

Note that Eqs. (13.26) through (13.30) are written with the time derivatives on the left-hand side and the spatial derivatives on the right-hand side. Assuming that the flowfield is known at time t, these spatial derivatives can be replaced with finite-difference expressions evaluated in the ζ-X-Y computational space shown in Fig. 13.19. This allows the calculation of the time derivatives of R, V_r, V_ϕ, V_z and ψ from Eqs. (13.26) through (13.30), from which new values of the flowfield variables are obtained at time $(t + \Delta t)$. The actual time-marching method can be carried out using MacCormack's technique as given in Sec. 12.5 for the two-dimensional blunt body problem, i.e., by using a predictor-corrector approach where the ζ, X, and Y derivatives are replaced by forward differences on the predictor step, and by rearward differences on the corrector step. The boundary conditions along the shock and body can be treated numerically by using a locally one-dimensional method of characteristics analysis matched to the calculation of the interior flowfield, exactly as described in Sec. 12.5 for the two-dimensional blunt body problem. See Ref. 92 for more details.

Typical results obtained by Moretti and Bleich are shown in Figs. 13.20 through 13.23. In Fig. 13.20, the time-dependent motion of the bow shock wave is shown for the flow over a blunt body consisting of an ellipsoidal nose with a major-to-minor axis ratio of 1.5, blending into a 14° half-angle cone downstream; the body is at a 30° angle of attack, and $M_\infty = 8$. The assumed initial shock shape at $t = 0$ is shown; for simplicity, it is initially chosen as an axisymmetric shape. During the course of the time-marching solution, the shock wave changes shape and location, and of course all the flow variables between the shock and the body are changing with time. Results for the transient shock wave are shown after 100, 200, 300, and 400 time steps. The 400th step is essentially the converged steady state result—the desired answer—yielding a nonaxisymmetric shock. Figure 13.21 gives the calculated steady-state Mach number distribution around the surface of the body for the symmetry plane $\phi = 0$, plotted as a function of r. (The r-body coordinates on the windward and leeward side of the body are illustrated in Fig. 13.20.) At the left side of Fig. 13.21, the Mach number plot is started at a value of r at a downstream location on the windward side. As we move from left to right along the horizontal axis in Fig. 13.21, we are moving along the windward body surface toward the nose. The value $r = 0$ corresponds to the nose tip. Then, we continue to move over the top of

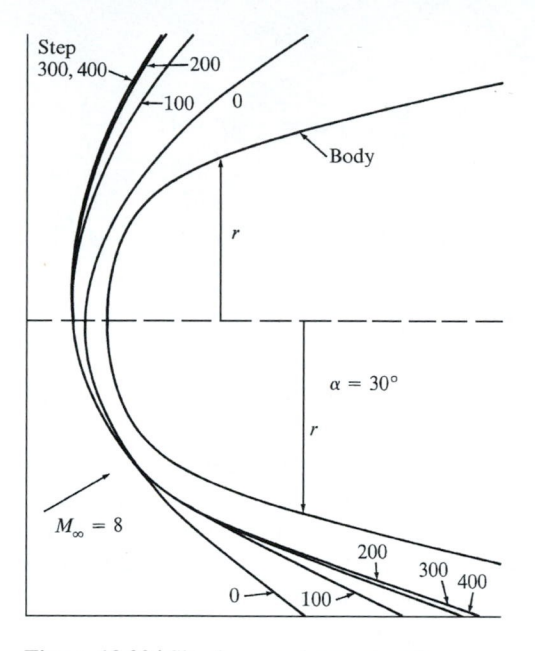

Figure 13.20 | Shock wave shapes at various times during time marching toward the steady state. (From Moretti and Bleich, Ref. 92.)

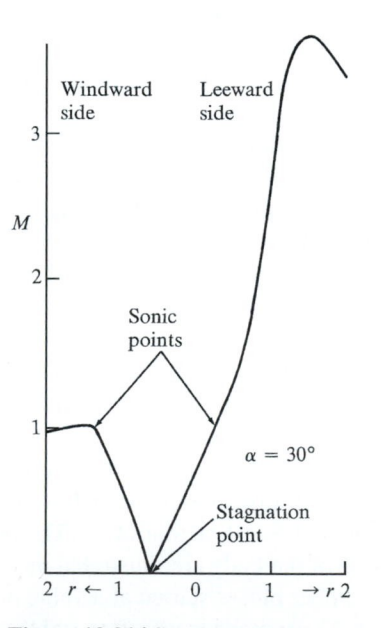

Figure 13.21 | Steady-state Mach number distribution along the surface of the body shown in Fig. 13.20 (Ref. 92).

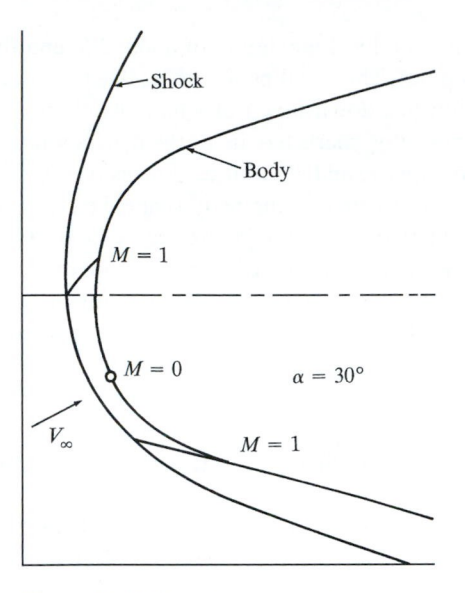

Figure 13.22 | Steady-state shock wave, sonic lines, and stagnation point in the symmetry plane for the flow problem in Fig. 13.20 (Ref. 92).

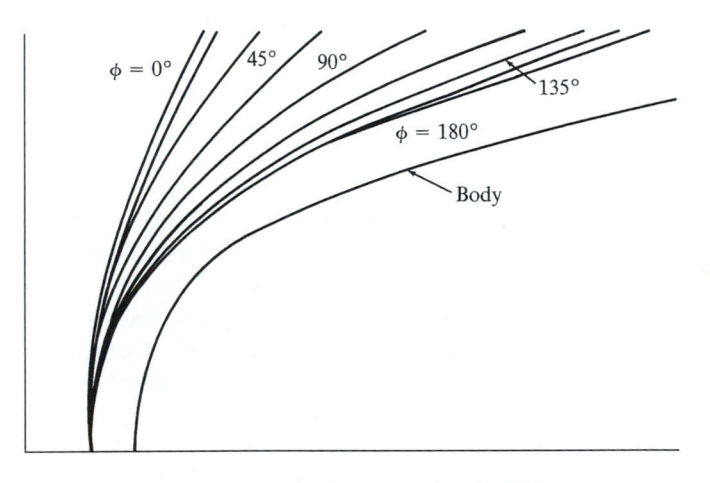

Figure 13.23 | Steady-state shock wave shapes in different meridional planes, $M_\infty = 8.0$ and $\alpha = 30°$ (Ref. 92).

the body away from the nose over the leeward side. Note that the Mach number M at the left of Fig. 13.21 is essentially sonic, determining the sonic point on the lower section of the body. As we move closer to the nose, M decreases to zero, thus locating the stagnation point, which occurs on the leeward side. Then, moving away from the stagnation point, M increases toward the nose tip, continues to increase over the leeward side to a local maximum of about 2.6, and then slightly decreases downstream of this point. This local peak in M is due to a local "overexpansion" of the flow in the region just downstream of where the ellipsoid nose mates with the cone. This overexpansion is characteristic of the hypersonic inviscid flow (note that $M_\infty = 8$) over axisymmetric and other three-dimensional bodies that have a discontinuous change in the *derivative* of the body shape, i.e., a discontinuity in d^2b/dr^2, such as the case shown here, even though the slopes themselves (db/dr for the ellipsoid and db/dr for the cone) are matched at the juncture of the two geometric shapes. In Fig. 13.22, the steady-state shock wave shape is shown along with the upper and lower sonic lines, and the stagnation point location. These are all typical of a blunt body at angle of attack. Finally, the steady-state shock shape in different meridional planes defined by different values of ϕ is given in Fig. 13.23, starting with $\phi = 0$ at the top of the body, and ending with $\phi = 180°$ at the bottom of the body. The fact that the shock is highly three-dimensional (highly nonaxisymmetric) is clearly evident here.

The work of Moretti and Bleich in Ref. 92 has been greatly extended in recent years. An example of a more recent application is described by Weilmuenser (Ref. 93), who calculated the inviscid flow over a space-shuttle-like vehicle at high angle of attack. The body shape and finite-difference grid is shown in Fig. 13.24. A spherical coordinate system is used in the nose region, patched to a cylindrical coordinate system downstream of the nose. The governing unsteady flow equations in spherical coordinates are given by Eqs. (13.10) through (13.13) for continuity and momentum; the unsteady energy equation (entropy equation) in spherical coordinates is given by

$$\frac{\partial s}{\partial t} = -\left[V_r \frac{\partial s}{\partial r} + \frac{V_\theta}{r} \frac{\partial s}{\partial \theta} + \frac{V_\phi}{r \sin\theta} \frac{\partial s}{\partial \phi} \right] \tag{13.31}$$

The unsteady flow equations in cylindrical coordinates are given by Eqs. (13.19) through (13.23). These are the governing equations for the time-marching solution of the inviscid flowfield over the body shown in Fig. 13.24. The approach used by Weilmuenser follows the shock-fitting philosophy pioneered by Moretti and the explicit time-marching predictor-corrector technique of MacCormack. Both of these concepts have already been discussed elsewhere in this book, and hence no further elaboration is given here. Typical results from Ref. 93 are shown in Fig. 13.25. Here the steady-state three-dimensional shock wave shape over the shuttle-like body is given for the case of $M_\infty = 16.25$ and $\alpha = 39.8°$. Of course, the entire steady flowfield between the shock and the body is also calculated. Figure 13.25 illustrates an advanced capability for the calculation of three-dimensional flowfields. Such calculations do not come cheap, however. For the solution shown in Fig. 13.25, nearly 100,000 grid points are used, and a supercomputer is necessary for the calculations.

The shuttle vehicle at high angle of attack, such as shown in Fig. 13.25, has a large region of subsonic flow over the lower compression surface. This is why a

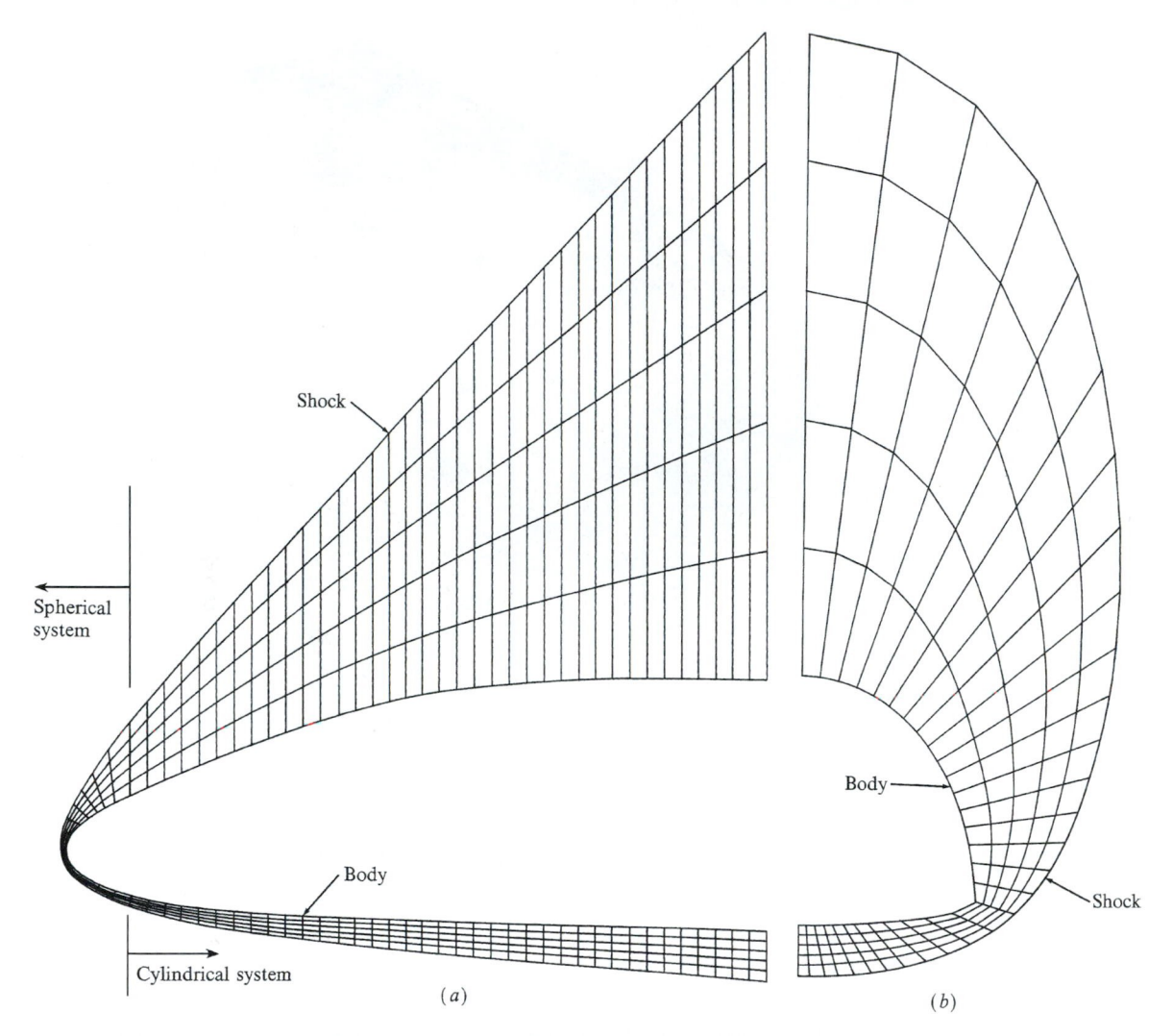

Figure 13.24 | (*a*) Physical grid in the symmetry plane for the calculation of the flow over a shuttle-like vehicle (Ref. 93). (*b*) Physical grid in the cross-flow plane.

time-marching method is used to calculate the entire flowfield. However, there are numerous applications involving blunt-nosed bodies at small enough angles of attack where a large region of locally supersonic flow exists downstream of the blunt nose. One such example has already been discussed in Sec. 11.16, where the inviscid flowfield over the space shuttle is calculated by Rakich and Kutler (Ref. 45), comparing results obtained from a downstream-marching finite-difference solution and a three-dimensional method of characteristics solution. Both of these solutions

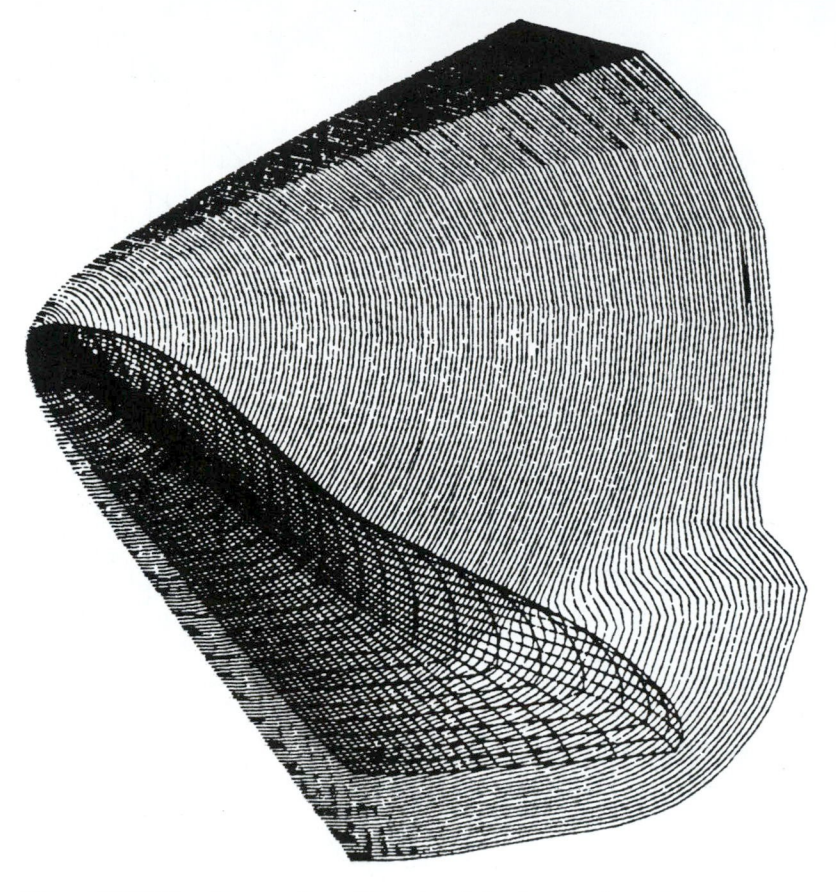

Figure 13.25 | Steady state, three-dimensional shock wave shape over a shuttle-like vehicle. $M_\infty = 16.25$ and $\alpha = 39.8°$. (From Weilmuenser, Ref. 93.)

are started from an initial data plane generated from a time-marching blunt body solution in the nose region. (It is instructional to reread Sec. 11.16 before progressing further.) A modern example of a three-dimensional flowfield calculation using the downstream-marching method is given by the work of Newberry et al. in Ref. 94, where the inviscid flow over the hypersonic entry research vehicle configuration shown in Fig. 13.26 is calculated. Here, a highly efficient downstream-marching method by Chakravarthy et al. (Refs. 95 through 97) is used, again starting from an initial data surface obtained from a time-marching blunt body calculation. Typical results for the Mach number distribution throughout the flowfield are shown by the computer graphics representations in Fig. 13.27. Here, the Mach number contours (lines of constant Mach number) are shown in six different cross-sectional planes corresponding to six streamwise locations along the body. The free-stream Mach number is 16, and the angle of attack is 8°.

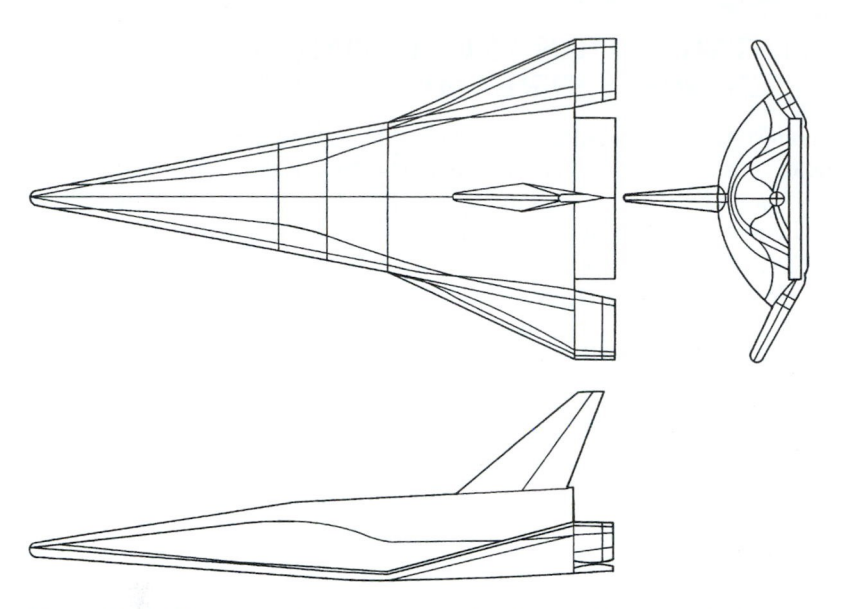

Figure 13.26 | The generic hypersonic research vehicle used for the calculations of Newberry et al. (From Ref. 94).

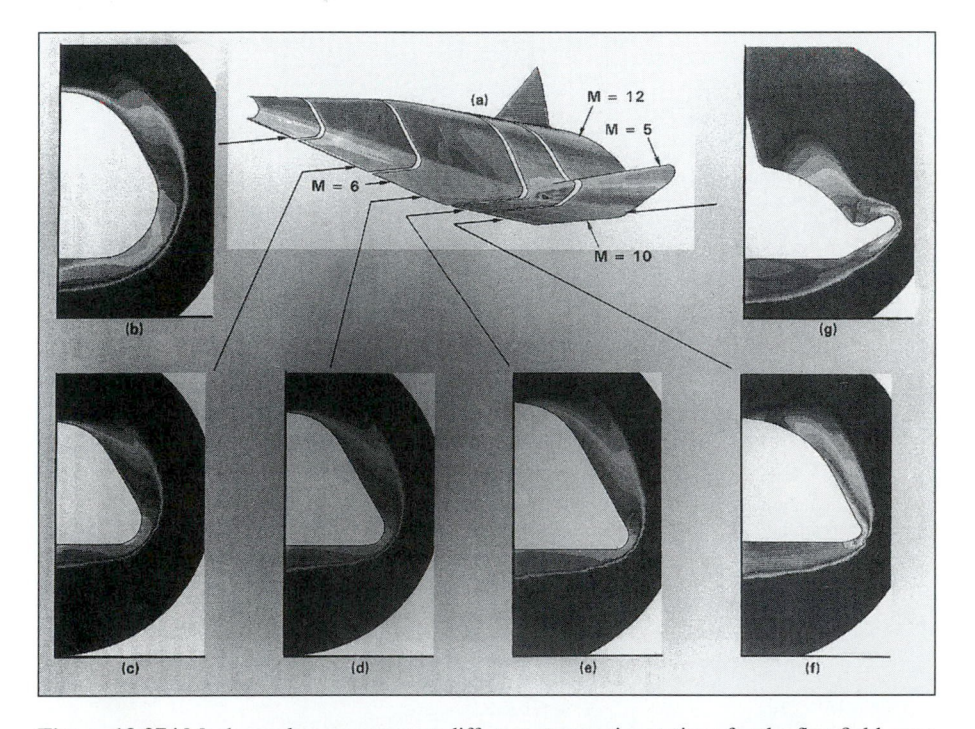

Figure 13.27 | Mach numbers contours at different streamwise stations for the flowfield over the generic hypersonic research vehicle shown in Fig. 13.26. The location of each station is identified by the arrows in the diagram (from Newberry et al., Ref. 94).

13.5 | STAGNATION AND MAXIMUM ENTROPY STREAMLINES

An interesting physical aspect of the three-dimensional flow over a blunt body at an angle of attack to a supersonic free stream is that the streamline going through the stagnation point is *not* the maximum entropy streamline. For a symmetric body at zero angle of attack, the stagnation streamline and the stagnation point are along the centerline, as sketched in Fig. 13.28a. This streamline crosses the bow shock wave at precisely the point where the wave angle is $90°$, that is, it crosses a normal shock, and hence the entropy of the stagnation streamline between the shock and the body is the maximum value. In contrast, consider the *asymmetric* cases shown in Figs. 13.28b and c; an asymmetric flow can be produced by a nonsymmetric body, an angle of attack, or both. In these cases, the shape and location of the stagnation streamline,

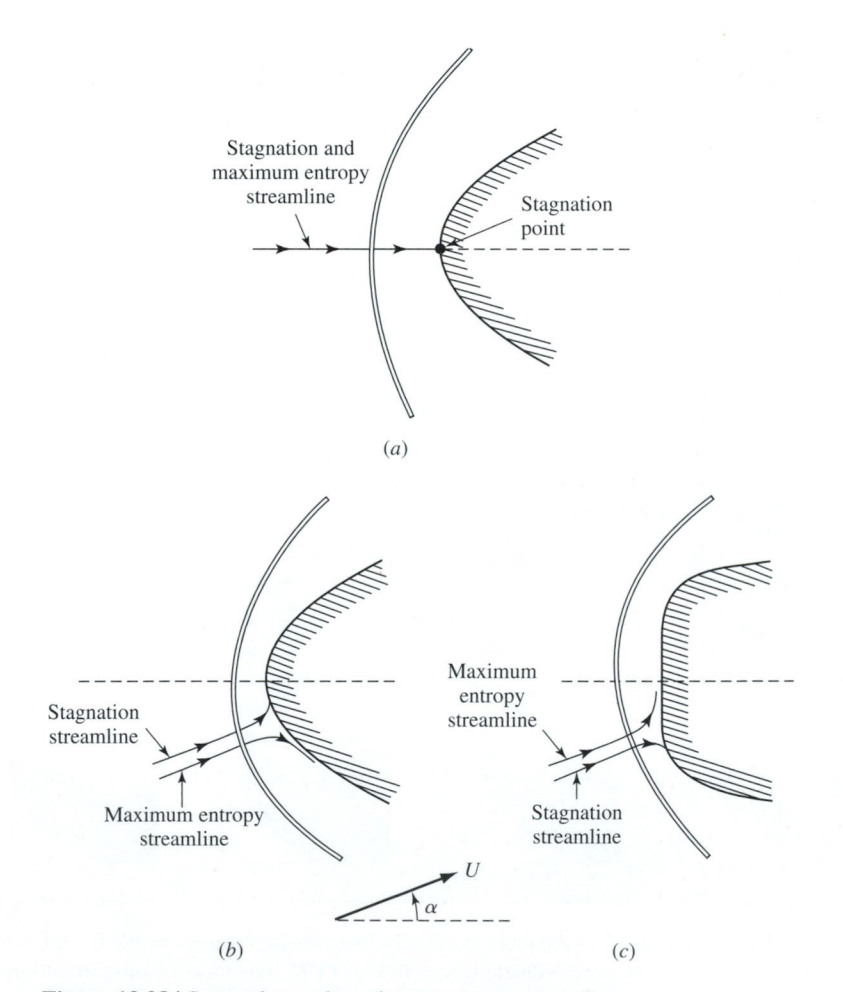

Figure 13.28 | Stagnation and maximum entropy streamlines.

and hence of the stagnation point, are not known in advance; they must be obtained as part of the numerical solution. Moreover, the stagnation streamline does not pass through the normal portion of the bow shock wave, and hence it is not the maximum entropy streamline. The relative locations of the stagnation streamline and the maximum entropy streamline for two nose shapes is shown in Figs. 13.28*b* and *c*. Note that the stagnation streamline is always attracted to that portion of the body with maximum curvature, whereas the maximum entropy streamline will turn in the direction of decreasing body curvature. More details on this matter can be found in Ref. 52.

13.6 | COMMENTS AND SUMMARY

The calculations shown in Figs. 13.24 through 13.27, in their time, represented the state of the art for inviscid three-dimensional flowfields over supersonic and hypersonic bodies. They were among the first of their kind, and therefore are classic in the field of CFD. This is why we discuss them here. Today such calculations are made with more modern numerical techniques utilizing much more sophisticated grids and algorithms. Because this chapter has emphasized the physical aspects of three-dimensional flow, and these aspects are nicely illustrated by the classical CFD calculations, we have chosen not to highlight more recent calculations from the current generation of CFD.

The purpose of this chapter has been to give the reader a basic familiarity with some of the features of three-dimensional flows over supersonic bodies. Emphasis has been placed on the physical aspects of such flows, along with a general understanding of several computational methods for calculating these flows. In particular, we have studied these cases.

1. Flows over elliptic cones and cones at angle of attack. These are three-dimensional geometries that, by virtue of the conical nature of the flow, generate flowfields that are "two-dimensional," i.e., that depend on only two independent variables, such as θ and ϕ, in a spherical coordinate system centered at the vertex of the cone. These flows exhibit vortical singularities, i.e., points where the entropy is multivalued. Also, embedded shocks may appear in the leeward region when the cross-flow velocity becomes supersonic, which usually occurs approximately when the angle of attack is greater than the cone half-angle. The calculational method for obtaining the "two-dimensional" conical flows uses a downstream-marching philosophy, starting with an initial nonconical flow and approaching the correct conical flow in the limit of large distances downstream.

2. Flows over blunt bodies at angle of attack. These are truly three-dimensional flows, involving three independent spatial variables, such as r, θ, and z, in a cylindrical coordinate system. Moreover, the numerical solution of such flows involves a time-marching philosophy; hence, t becomes a fourth independent variable, which is made necessary by virtue of the calculational method itself.

The desired steady three-dimensional flowfield solution is approached in the limit of large times.

3. Flows over slender blunt-nosed bodies at angle of attack, such as the vehicle shown in Fig. 13.26. Here, the flow in the blunt-nosed region is calculated by means of a time-marching method. When the steady state is achieved in this region, a plane of data located in the supersonic region just downstream of the limiting characteristic surface is chosen as the initial data plane, from which a three-dimensional steady downstream-marching procedure is used to calculate the remainder of the supersonic flowfield. This downstream marching can be carried out using the three-dimensional method of characteristics, or which is more usually the case today, a finite-difference or finite-volume solution of the steady-flow equations. However, if and when a pocket of locally subsonic flow is encountered during this downstream marching, we must revert back to a time-marching solution for this locally subsonic region. (See, for example, Ref. 95.)

In summary, the types of flowfields encountered in the vast majority of practical aerodynamic applications are three-dimensional. Unfortunately, the analysis of such three-dimensional flows has been extremely difficult in the past; indeed, exact solutions of such flows were only dreams in the minds of aerodynamicists during most of this century. It has been a state-of-the-art research problem since the beginning of rational fluid dynamics with Leonhard Euler in the eighteenth century. However, since the late 1960s, the advent of computational fluid dynamics has changed this situation; as we have seen in this chapter, numerical techniques now exist for the computation of general three-dimensional flowfields, and many such computations have successfully been completed. The solution of three-dimensional flows is still a state-of-the-art problem today, but only from the point of view as to improvements in the numerical accuracy, the efficiency of solution (the quest to reduce the computer time necessary to obtain solutions), and the proper methods for presenting, studying and interpreting the large amount of numerical data, generated by such solutions (a problem in computer graphics).

CHAPTER 14

Transonic Flow

We call the speed range just below and just above the sonic speed—Mach number nearly equal to 1—the transonic range. Dryden (Hugh Dryden, well-known fluid dynamicist and past administrator of the National Advisory Committee for Aeronautics, now NASA) and I invented the word "transonic." We had found that a word was needed to denote the critical speed range of which we were talking. We could not agree whether it should be written with one s or two. Dryden was logical and wanted two s's. I thought it wasn't necessary always to be logical in aeronautics, so I wrote it with one s. I introduced the term in this form in a report to the Air Force. I am not sure whether the general who read it knew what it meant, but his answer contained the word, so it seemed to be officially accepted . . . I well remember this period (about 1941) when designers were rather frantic because of the unexpected difficulties of transonic flight. They thought the troubles indicated a failure in aerodynamic theory.

Theodore von Karman, in a lecture given at Cornell University, 1953

PREVIEW BOX

The transonic flight regime is one that supersonic and hypersonic vehicles spend as little time as possible flying in during acceleration or deceleration through Mach 1. This is because of the drag-divergence phenomena, the rapid shift of center of pressure, and the unsteady and somewhat unpredictable effect of shock waves on control surfaces, all of which are undesirable aspects of transonic flight. Current jet transports nudge this regime by cruising near or slightly above the critical Mach number, but never beyond the drag-divergence Mach number. Typical cruise Mach numbers of jet transports

range from 0.75 to about 0.83. However, the quest to push this envelope, to increase the drag-divergence Mach number closer to 1, has been active for decades. Doing this requires a fundamental understanding of transonic flow.

At the time of this writing, a graphic example of pushing the envelope is the Boeing Aircraft Company's new concept for a transonic jet transport, to cruise at Mach 0.95. An artist's sketch of a possible configuration is shown in Fig. 14.1. Compare this configuration with that for the Boeing 777 shown in Fig. 1.4. You see in

Figure 14.1 | A transonic airplane concept from Boeing.

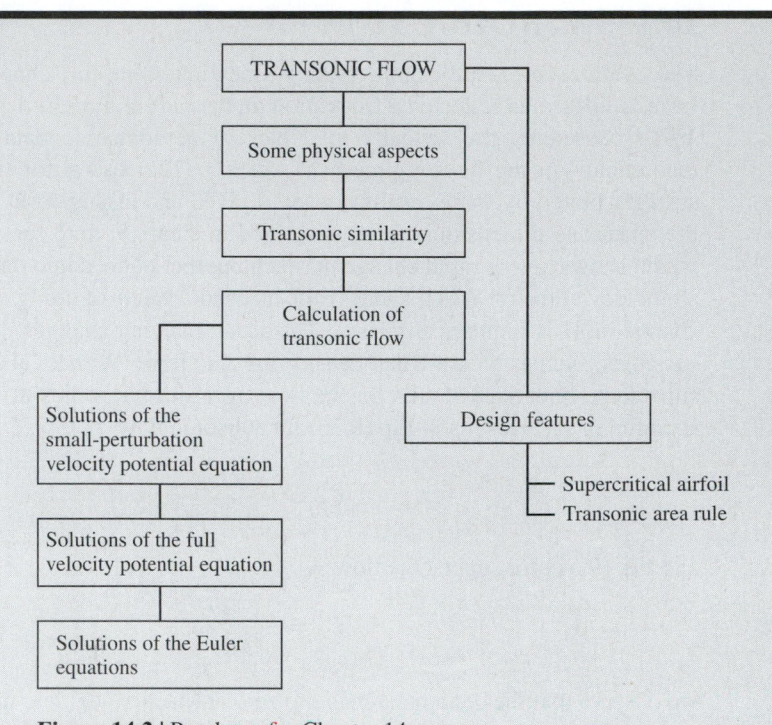

Figure 14.2 | Roadmap for Chapter 14.

Fig. 1.4 the standard configuration used by designers of most current jet transports since Boeing introduced the pioneering 707 in the late 1950s—characterized by a relatively high aspect ratio swept wing with engines mounted in pods located underneath the wing, or in some cases on the rear portion of the fuselage. The configuration for Boeing's "sonic cruiser" in Fig. 14.1 is a radical departure from this standard configuration. Whether this or some other configuration is finally developed by Boeing is not germane here. What is important is that some serious effort is being made to design an airplane to cruise in the transonic flight regime. More than ever this requires a fundamental understanding of the physical properties of the gasdynamics in the transonic regime, and the ability to accurately calculate such flows. This is the subject of the present chapter.

Transonic flow has always been important. It is now more so than ever. The material in this chapter will give you a fundamental understanding of some of the problems to be faced in the design of a transonic transport such that sketched in Fig. 14.1. This is important material, and the future applications are exciting.

The roadmap for this chapter is given in Fig. 14.2. We begin with a discussion of the physical aspects of transonic flow. We follow with the theoretical aspects of transonic similarity, identifying the transonic similarity principle and the transonic similarity parameters. This is classic transonic theory. The remainder of the chapter is devoted mostly to the numerical calculation of inviscid transonic flow. Such calculations have historically evolved in three steps, involving chronologically the numerical solution of (1) the small-perturbation velocity potential equation, (2) the full velocity potential equation, and finally (3) the Euler equations. As you might expect, the complexity of the solutions increase with each of these three steps. We end the chapter with a discussion of two important design features for transonic aircraft, the supercritical airfoil and the transonic area rule. This discussion is integrated within an extensive historical note on transonic flight.

A quick glance at the overall roadmap for the book in Fig. 1.7 shows that we are now at box 15, almost at the end of the center column.

14.1 | INTRODUCTION

The "failure in aerodynamic theory" mentioned in the chapter-opening quote from von Karman reflected a frustration on the part of aircraft designers in the early 1940s caused by the virtually total lack of aerodynamic data—experimental or theoretical—in the flow regime near Mach 1. The reason for this lack of data is strongly hinted by some results already derived and discussed in this book. Witness the quasi-one-dimensional flows described in Chap. 5; note for example Fig. 5.13, which shows a very rapid change in Mach number at the sonic throat, i.e., for a very slight deviation of A/A^* away from its sonic value of unity, the corresponding change in M is dramatically large. The accompanying changes in all the other flow variables, such as pressure and density, are also large. Witness also the subsonic and supersonic linearized results for the flow over slender bodies discussed in Chap. 9. Examining such results in Eq. (9.36) for subsonic flow

$$C_p = \frac{C_{p_o}}{\sqrt{1 - M_\infty^2}} \qquad (9.36)$$

and Eq. (9.51) for supersonic flow

$$C_p = \frac{2\theta}{\sqrt{M_\infty^2 - 1}} \qquad (9.51)$$

we observe that the denominators go to zero at Mach 1, yielding infinitely large pressure coefficients—an obvious physical impossibility. These examples from our previous discussion wave a red flag about flow near or at Mach 1. Certainly such transonic flow is *extremely sensitive* to slight changes, hence presenting experimental difficulties in obtaining good transonic data in wind tunnels. Also, in Chap. 9 we saw that subsonic and supersonic flows involving small perturbations can be described by *linear* theory, providing such useful results as Eqs. (9.36) and (9.51) listed here. We also saw that flow in the transonic regime is described by *nonlinear* theory—a much more difficult situation, and hence presenting theoretical difficulties in obtaining good transonic information. In short, transonic flow historically has been an exceptionally challenging problem in aerodynamics, yielding its secrets only slowly and grudgingly over the years. Today, the use of slotted-throat wind tunnels (test sections with holes or longitudinal slots in the walls to relieve the sensitivity of transonic flows to slight changes, and to attenuate waves from the test model which propagate outward at nearly right angles to the flow and impinge on the tunnel walls) has created a revolution in the accurate experimental measurement of transonic flows. Also, the power of computational fluid dynamics has created a similar revolution in the ability to calculate and predict the nature of transonic flows. However, in spite of these "revolutions," transonic flow today still stands as a challenging state-of-the-art problem in modern compressible flow, and this is one of the two reasons why we are devoting a chapter to it here. The other reason is because of the importance of transonic flow for engineering applications. For example, almost all the existing commercial jet transports today cruise at free-stream Mach numbers around 0.8—penetrating the lower side of the transonic regime. Also, almost all air combat among modern supersonic fighter planes takes place at or near Mach 1—no matter what the top speed of the aircraft. Of course, all supersonic and hypersonic aircraft—including the space shuttle—must pass through the transonic regime on their way up and down. Hence, in the world of modern compressible flow, it is important to have some feeling for the nature of transonic flow, and some understanding of the analysis of such flows.

The purpose of this chapter is to provide such a "feeling," and nothing more. Transonic flow is a subject that dictates a book almost by itself, such as given by Ref. 98. In this chapter we will only examine the major aspects of the subject; in this fashion, as in Chap. 13, the present chapter will be intentionally long on philosophy and methodology, but short on details.

14.2 | SOME PHYSICAL ASPECTS OF TRANSONIC FLOWS

A general physical picture of transonic flows is discussed in Sec. 1.3, and sketched in Figs. 1.10b and c; this material should be reviewed at this stage before progressing further. Also, the concept of the *critical* Mach number M_{cr} is discussed in Sec. 9.7. The critical Mach number is that free-stream Mach number at which sonic flow is first obtained on a body; in this sense, the transonic regime begins when the critical Mach number is reached. The material in Sec. 9.7 should also be reviewed before progressing further.

As discussed in Sec. 1.3, transonic flow is characterized by mixed regions of locally subsonic and supersonic flow that occur over a body moving at Mach numbers near unity. Also, the general three-dimensional flow in the throat region of supersonic nozzles is transonic. The physical characteristics of transonic flow are nicely illustrated by the series of schlieren photographs shown in Fig. 14.3, obtained from Ref. 99. Here, we see the flow over three different airfoils for different values of the free-stream Mach number. Moving from bottom to top, you can see the influence of increasing free-stream Mach number from $M_\infty = 0.79$ to $M_\infty = 1.0$. Going from left to right, you can observe the effect of increasing airfoil thickness, ranging from the NACA 64A006 airfoil of 6 percent thickness to the NACA 64A012 airfoil of 12 percent thickness. The schlieren photographs show the various shock wave patterns as well as regions of the flow separation. Superimposed on each photograph are the measured pressure coefficient distributions over the top (solid curve) and bottom (dashed curve) surfaces of the airfoil. The scale for the magnitude of the pressure coefficient is shown at the left of each row; as is usual in aeronautical practice, negative values of C_p are given above the horizontal axis, and positive values below. Also, the short horizontal dashed line at the left of each row gives the value of the critical pressure coefficient $C_{p_{cr}}$ corresponding to the specific value of M_∞ listed at the right of each row. (See Sec. 9.7 for the definition and significance of $C_{p_{cr}}$.) Starting with the lower left-hand photograph for the NACA 64A006 airfoil at $M_\infty = 0.79$, we observe a pocket of supersonic flow extending from just downstream of the leading edge to about 35 percent of the chord length, where it is terminated by a nearly normal shock

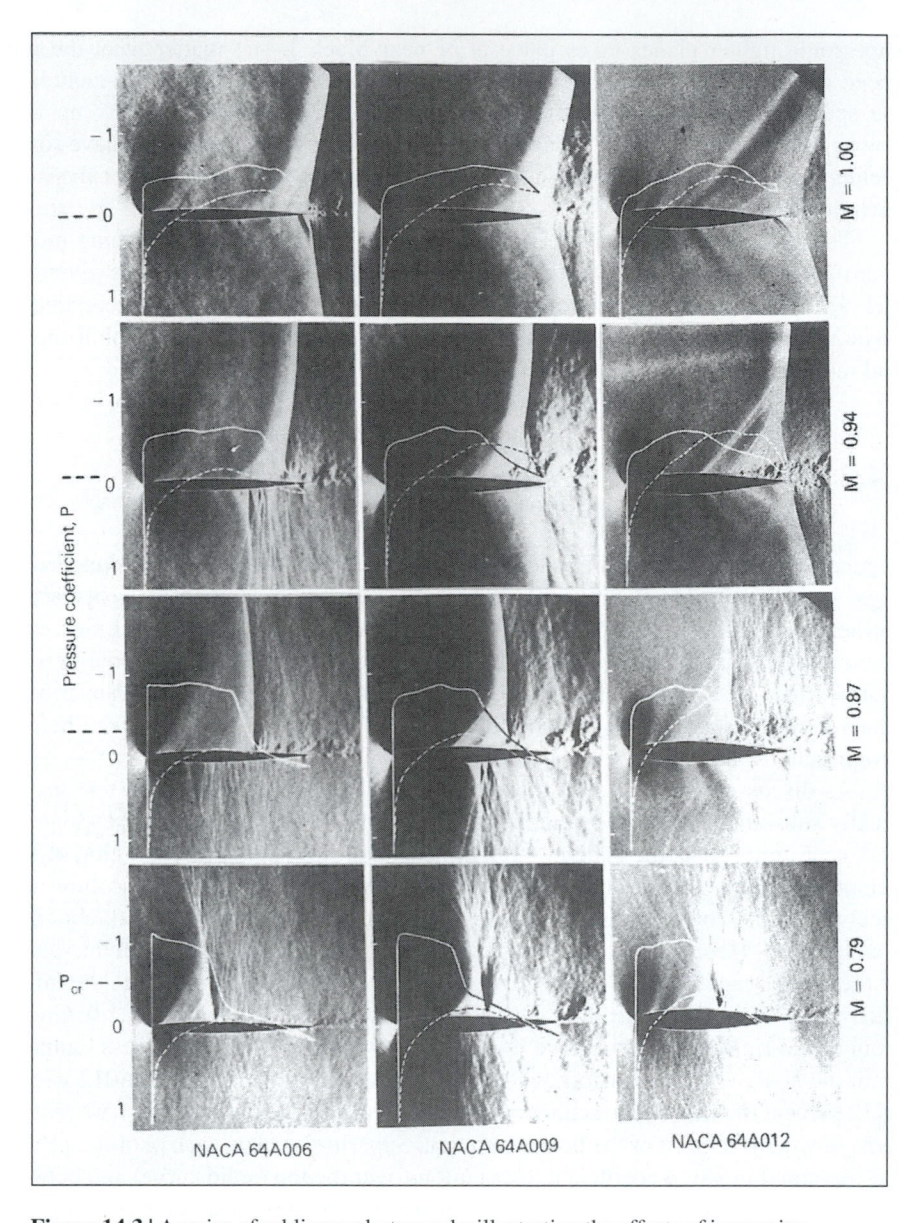

Figure 14.3 | A series of schlieren photographs illustrating the effects of increasing free-stream Mach number (from bottom to top in the figure) and increasing airfoil thickness (from left to right in the figure) on the transonic flow over airfoils. (From Ref. 99.)

wave. This supersonic pocket is identified by the nearly white region in the photograph; the supersonic flow has weak expansion waves propagating from the airfoil surface, and terminating at the sonic line above the airfoil or at the shock wave itself. (The optical nature of the schlieren method applied here causes regions of decreasing density such as expansion waves to appear light and regions of increasing density such as shock waves to appear dark.) This picture illustrates the type of flow characteristics sketched earlier in Fig. 1.10b. Note that the magnitude of the measured pressure coefficient along the top surface substantially exceeds $C_{p_{cr}}$ for a distance of about 35 percent of the chord length downstream of the leading edge, further confirming the existence of locally supersonic flow in that region. Note that the measured C_p almost discontinuously drops to a value below $C_{p_{cr}}$ behind the shock, heralding the region of locally subsonic flow downstream of the shock. Note also that C_p along the bottom surface does not exceed $C_{p_{cr}}$; hence, the flow over the bottom surface is completely subsonic. Now move to the next photograph directly above. Here, for the same NACA 64A006 airfoil, the free-stream Mach number has been increased to $M_\infty = 0.87$. For this case we observe a greatly enlarged region of supersonic flow over the top surface, and the shock wave has moved downstream, closer to the trailing edge of the airfoil. The shock is now stronger, and this causes the viscous boundary layer to separate from the surface in the region where the shock impinges on the surface. The separated boundary layer can be seen as a region of intense vorticity trailing downstream of the shock impingement point. The flow is still subsonic along the lower surface. Moving to the next photograph directly above (for $M_\infty = 0.94$), we see virtually the entire upper surface immersed in a locally supersonic flow, and the shock wave has almost reached the trailing edge. There is now a small pocket of supersonic flow under the bottom surface as well, as indicated by the weak waves shown in the schlieren photograph; this is also indicated by the values of C_p on the lower surface that slightly exceed $C_{p_{cr}}$ over a small portion of the bottom surface. When M_∞ is increased to 1.0, as shown in the top photograph, the flow is supersonic over the entire top surface, and is supersonic over a substantial portion of the bottom surface. The shock waves have moved to the trailing edge itself, and the mechanism for forming the leading-edge bow shock wave is beginning to appear. In this sense, this photograph shows the beginning of the type of flowfield sketched in Fig. 1.10c. Now, as we move from left to right in Fig. 14.3, we see the effect of increasing the airfoil thickness. Note that the increased thickness causes a larger perturbation of the flow; the flow will expand to a greater degree over a thicker airfoil, and hence the transonic effects are stronger for thicker airfoils. The local Mach numbers inside the supersonic regions become larger, which in turn causes the terminating shock waves to be stronger. Note that the regions of separated flow induced by the impingement of these shock waves on the viscous boundary layer also become more extensive. Scanning along the top photographs in Fig. 14.3, namely those of $M_\infty = 1.0$, we note that both the upper and lower shocks are now at the trailing edge, and for this case the region of separated flow is greatly diminished.

The separated flow associated with the shock wave/boundary layer interaction shown in Fig. 14.3 is caused by the following mechanism. The pressure increases almost discontinuously across the shock wave. This represents an extremely large *adverse* pressure gradient. (An adverse pressure gradient is one where the pressure *increases* in the flow direction.) It is well-known that boundary layers readily separate from the surface in regions of adverse pressure gradients. When the shock wave impinges on the surface, the boundary layer encounters an extremely large adverse pressure gradient, and it will almost always separate. This shock wave/boundary layer interaction is one of the most important aspects of transonic flow. Along with the total pressure losses (entropy increases) caused by the shock waves themselves, the shock-induced separated flows create a large rise in drag on the airfoil—the *drag-divergence* phenomenon that is always associated with flight in the transonic regime. (See Ref. 1 for a basic description of the drag-divergence behavior of airfoils.) This drag-divergence phenomenon is illustrated in Fig. 14.4, taken from Ref. 100. This is a plot of drag coefficient versus M_∞ for an NACA 2315 airfoil; the different curves correspond to different angles of attack. Note the extremely rapid rise in drag coefficient as the Mach number approaches 1. This is perhaps the most significant consequence of the transonic regime.

In the present chapter, we deal with inviscid flows only; hence, the shock wave/boundary layer interaction will not be discussed further. As we will see, modern

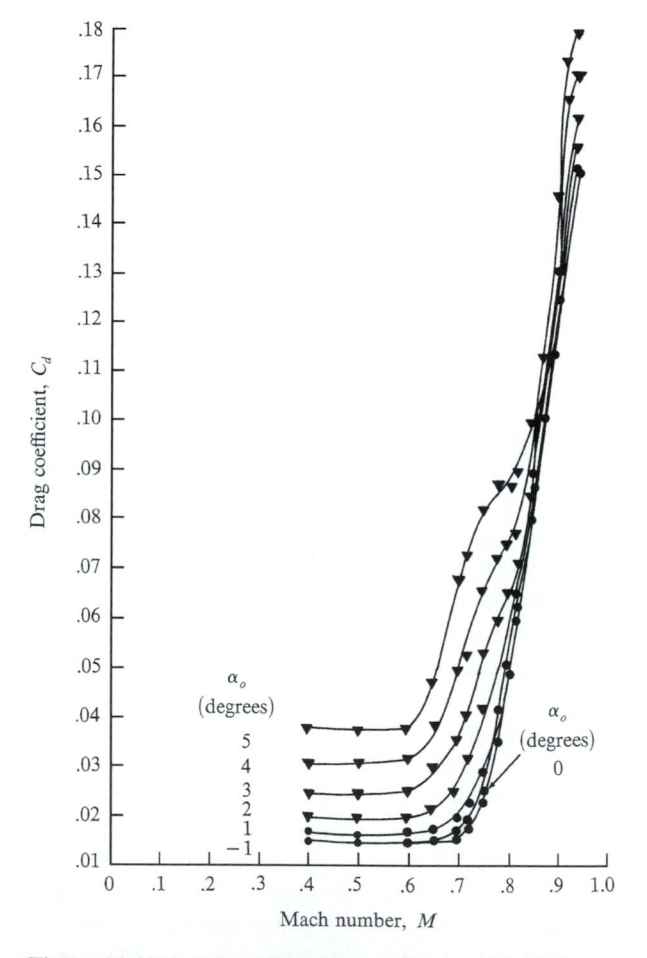

Figure 14.4 | Variation of the drag coefficient with Mach number for an NACA 2315 airfoil, illustrating the drag-divergence phenomenon as Mach 1 is approached. Experimental results are given for angles of attack ranging from −1° to 5°. (From Loftin, Ref. 100.)

computational solutions of inviscid transonic flows can predict many aspects of transonic flows, including the strength and location of the shock waves. From these solutions, the drag-rise phenomenon shown in Fig. 14.4 can be modeled to some extent. However, for the most accurate analysis, a viscous flow solution is necessary. Such solutions for viscous transonic flows are now focusing on numerical solutions of the complete Navier–Stokes equations—a state-of-the-art problem that is far beyond the scope of this book.

14.3 | SOME THEORETICAL ASPECTS OF TRANSONIC FLOWS; TRANSONIC SIMILARITY

Inviscid transonic flows are governed by the partial differential equations derived in Chap. 6, namely the *Euler equations,* repeated here:

Continuity:
$$\frac{\partial \rho}{\partial t} + \nabla \cdot (\rho \mathbf{V}) = 0 \tag{14.1}$$

Momentum:
$$\rho \frac{D\mathbf{V}}{Dt} = -\nabla p \tag{14.2}$$

Energy:
$$\rho \frac{Dh_o}{Dt} = \frac{\partial p}{\partial t} \tag{14.3}$$

In these equations, we are assuming an inviscid, adiabatic flow with no body forces. For numerical solutions of inviscid transonic flows, the Euler equations are conceptually the most accurate equations. Entropy gradients are present in transonic flows due to the presence of the shock waves seen in Fig. 14.3; in turn, these flows are rotational as demonstrated by Crocco's theorem (see Sec. 6.6). The Euler equations given by Eqs. (14.1) through (14.3) are applicable whether or not the flow is rotational.

We have stated that the transonic flows shown in Fig. 14.3 are rotational—but to what degree? Are the shock waves that appear in such flows weak enough to allow us to neglect the rotationality of the flow in some cases? Let us address this question further. Return to Eq. (3.60) for the entropy change across a normal shock wave, repeated below:

$$s_2 - s_1 = c_p \ln\left\{\left[1 + \frac{2\gamma}{\gamma+1}\left(M_1^2 - 1\right)\right]\left[\frac{2 + (\gamma-1)M_1^2}{(\gamma+1)M_1^2}\right]\right\}$$

$$- R \ln\left[1 + \frac{2\gamma}{\gamma+1}\left(M_1^2 - 1\right)\right] \tag{3.60}$$

Recall that $c_p = \gamma R/(\gamma - 1)$. Then Eq. (3.60) becomes

$$\frac{s_2 - s_1}{R} = \frac{1}{\gamma-1}\ln\left[1 + \frac{2\gamma}{\gamma+1}\left(M_1^2 - 1\right)\right] + \frac{\gamma}{\gamma-1}\ln\left[\frac{2 + (\gamma-1)M_1^2}{(\gamma+1)M_1^2}\right]$$

$$\tag{14.4}$$

For convenience, let $m = M_1^2 - 1$. Then the first term in square brackets in Eq. (14.4) becomes

$$1 + \frac{2\gamma}{\gamma + 1}\left(M_1^2 - 1\right) = 1 + \frac{2\gamma}{\gamma + 1}m \tag{14.5}$$

and the second term in square brackets becomes

$$
\begin{aligned}
\frac{2 + (\gamma - 1)M_1^2}{(\gamma + 1)M_1^2} &= \frac{1}{M_1^2}\left[\frac{2 + (\gamma - 1)M_1^2}{\gamma + 1}\right] \\
&= \frac{1}{M_1^2}\left[\frac{2 + (\gamma - 1)M_1^2 + (\gamma + 1) - (\gamma + 1)}{\gamma + 1}\right] \\
&= \frac{1}{M_1^2}\left[\frac{(\gamma - 1)M_1^2 - (\gamma - 1)}{\gamma + 1} + 1\right] \\
&= \frac{1}{M_1^2}\left[\frac{(\gamma - 1)\left(M_1^2 - 1\right)}{\gamma + 1} + 1\right] \\
&= \frac{1}{m + 1}\left[\frac{\gamma - 1}{\gamma + 1}m + 1\right]
\end{aligned}
\tag{14.6}
$$

Substituting Eqs. (14.5) and (14.6) into Eq. (14.4), we have

$$
\begin{aligned}
\frac{s_2 - s_1}{R} ={}& \frac{1}{\gamma - 1}\ln\left(1 + \frac{2\gamma}{\gamma + 1}m\right) - \frac{\gamma}{\gamma - 1}\ln(m + 1) \\
&+ \frac{\gamma}{\gamma - 1}\ln\left(\frac{\gamma - 1}{\gamma + 1}m + 1\right)
\end{aligned}
\tag{14.7}
$$

For transonic flows, $M_1 \approx 1$, hence $m \ll 1$. Thus, each logarithmic term in Eq. (14.7) is of the form $(1 + \varepsilon)$, where $\varepsilon \ll 1$. Recall the series expansion:

$$\ln(1 + \varepsilon) = \varepsilon - \varepsilon^2/2 + \varepsilon^3/3 + \cdots$$

With this, Eq. (14.7) is given by

$$
\begin{aligned}
\frac{s_2 - s_1}{R} ={}& \frac{1}{\gamma - 1}\left[\frac{2\gamma}{\gamma + 1}m - \left(\frac{2\gamma}{\gamma + 1}\right)^2\frac{m^2}{2} + \left(\frac{2\gamma}{\gamma + 1}\right)^3\frac{m^3}{3} + \cdots\right] \\
&- \frac{\gamma}{\gamma - 1}\left[m - \frac{m^2}{2} + \frac{m^3}{3} + \cdots\right] \\
&+ \frac{\gamma}{\gamma - 1}\left[\frac{\gamma - 1}{\gamma + 1}m - \left(\frac{\gamma - 1}{\gamma + 1}\right)^2\frac{m^2}{2} + \left(\frac{\gamma - 1}{\gamma + 1}\right)^3\frac{m^3}{3} + \cdots\right]
\end{aligned}
\tag{14.8}
$$

Note that the terms involving m and m^2 in Eq. (14.8) cancel, yielding

$$\frac{s_2 - s_1}{R} = \frac{2\gamma}{3(\gamma + 1)^2} m^3 + \cdots$$

or

$$\frac{s_2 - s_1}{R} \approx \frac{2\gamma}{3(\gamma + 1)^2} \left(M_1^2 - 1\right)^3$$

(14.9)

The result in Eq. (14.9) states that the entropy increase across a weak shock is of *third order* in terms of $(M_1^2 - 1)$; when $(M_1^2 - 1) \ll 1$ as for transonic flows, then the entropy increase across the shock is *very* small. [Note from Eq. (3.57) that the *strength* of a shock as indicated by the ratio $(p_2 - p_1)/p_1$ is proportional to $M_1^2 - 1$; hence, Eq. (14.9) states that the entropy increase across the shock is of third order in the shock strength.] Therefore, for the transonic flows shown in Fig. 14.1, we can *assume* that the flow is essentially *isentropic,* the actual increase in entropy being of third order in shock strength and hence negligible for the case of transonic flow. In turn, we can *assume* that the flow is essentially *irrotational.* This answers the question asked at the beginning of the paragraph. (Keep in mind that this is an approximation only; for the high end of the transonic range, say for $M_1 = 1.2$, the entropy changes may be too large to ignore. We have already explored this matter in Example 3.9, which you should review before proceeding further.)

If we make the assumption that the transonic flow is irrotational on the basis of very small entropy changes as discussed above, then a velocity potential Φ can be defined such that $\mathbf{V} = \nabla\Phi$, and the governing Euler equations, Eqs. (14.1) through (14.3), cascade to a single equation in terms of Φ, as described in Chap. 8. This equation was derived as Eq. (8.17), repeated here:

$$\left(1 - \frac{\Phi_x^2}{a^2}\right)\Phi_{xx} + \left(1 - \frac{\Phi_y^2}{a^2}\right)\Phi_{yy} + \left(1 - \frac{\Phi_z^2}{a^2}\right)\Phi_{zz}$$

$$- \frac{2\Phi_x\Phi_y}{a^2}\Phi_{xy} - \frac{2\Phi_x\Phi_z}{a^2}\Phi_{xz} - \frac{2\Phi_y\Phi_z}{a^2}\Phi_{yz} = 0 \qquad (8.17)$$

The advantages of Eq. (8.17) for a flowfield analysis, as long as the flow is irrotational, were described in Chap. 8; it is strongly recommended that you review the material in Chap. 8 before proceeding further, especially concerning the derivation of Eq. (8.17) and the theoretical advantages obtained by using Eq. (8.17). For an irrotational, isentropic flow, Eq. (8.17) is an exact relation. Its use for the analysis of the transonic flows shown in Fig. 14.3 is only approximate, but as argued here, the approximation appears to be reasonable.

Equation (8.17) holds for any body shape, thick or thin, at any angle of attack. If we are concerned with the transonic flow over a slender body at small angle of attack, then we can make the assumption of small perturbations, as described in Chap. 9. This leads to the definition of a *perturbation* velocity potential ϕ, defined as $\Phi = V_\infty x + \phi$, and Eq. (8.17) is written in terms of ϕ, yielding the perturbation-velocity potential equation given by Eq. (9.1). This equation is still exact for an irrotational, isentropic flow. It can be reduced to a simpler form if the assumption of *small perturbations* is made, as explained in Sec. 9.2. The result of this reduction leads to Eq. (9.6), which

$$\tau = \frac{b}{c}$$

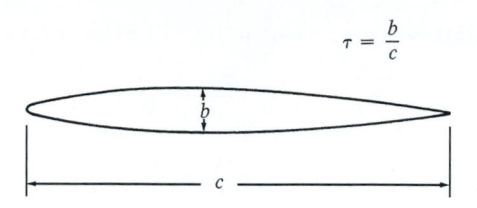

Figure 14.5 | Definition of slenderness ratio τ.

holds for subsonic and supersonic flow. However, as noted in Sec. 9.2, for transonic flow an extra term appears in the reduced, small-perturbation equation, yielding

$$\left(1 - M_\infty^2\right)\phi_{xx} + \phi_{yy} + \phi_{zz} = M_\infty^2\left[(\gamma + 1)\frac{\phi_x}{V_\infty}\right]\phi_{xx} \qquad (14.10)$$

Equation (14.10) is the *transonic* small-perturbation equation. Make certain to review Sec. 9.2 to understand how this equation is obtained.

Equation (14.10) is a dimensional equation; a particularly interesting result can be obtained by nondimensionalizing this equation, as follows. Let τ be the slenderness ratio of the body; $\tau = b/c$, where b and c are the maximum thickness and length of the body, respectively, as sketched in Fig. 14.5. Note that, for flow with small perturbations, τ must be small. Also observe from Fig. 14.3 that the disturbances in a transonic flow reach far above and below the airfoil, i.e., the *lateral* extent of the disturbances is large compared to the streamwise extent. Hence, in Eq. (14.10) the physical domain where nonzero values of the perturbation potential ϕ are concentrated extends to large values of y and z, but are limited to the streamwise region of $x \approx c$. This motivates a transformation of (x, y, z) into $(\bar{x}, \bar{y}, \bar{z})$, where $\bar{x}$, $\bar{y}$, and $\bar{z}$ are all of the same order of magnitude. This can be achieved by defining the following nondimensional independent variables:

$$\bar{x} = \frac{x}{c} \qquad \bar{y} = \frac{y\tau^{1/3}}{c} \qquad \bar{z} = \frac{z\tau^{1/3}}{c}$$

At the same time, consider a nondimensional perturbation velocity potential defined by

$$\bar{\phi} = \frac{\phi}{cV_\infty\tau^{2/3}}$$

To nondimensionalize Eq. (14.10) according to the definitions just presented, we first write it as

$$\left(1 - M_\infty^2\right)\frac{\partial^2(\phi/cV_\infty\tau^{2/3})}{\partial(x/c)^2}\left(\frac{cV_\infty\tau^{2/3}}{c^2}\right) + \frac{\partial^2(\phi/cV_\infty\tau^{2/3})}{\partial(y\tau^{1/3}/c)^2}\left(\frac{cV_\infty\tau^{2/3}}{c^2/\tau^{2/3}}\right)$$

$$+ \frac{\partial^2(\phi/cV_\infty\tau^{2/3})}{\partial(z\tau^{1/3}/c)^2}\left(\frac{cV_\infty\tau^{2/3}}{c^2/\tau^{2/3}}\right)$$

$$= M_\infty^2\left[\frac{(\gamma + 1)}{V_\infty}\frac{\partial(\phi/cV_\infty\tau^{2/3})}{\partial(x/c)}\left(\frac{cV_\infty\tau^{2/3}}{c}\right)\right]\frac{\partial^2(\phi/cV_\infty\tau^{2/3})}{\partial(x/c)^2}\left(\frac{cV_\infty\tau^{2/3}}{c^2}\right)$$

or $\qquad \left(1 - M_\infty^2\right)\frac{\partial^2\bar{\phi}}{\partial\bar{x}^2} + \tau^{2/3}\frac{\partial^2\bar{\phi}}{\partial\bar{y}^2} + \tau^{2/3}\frac{\partial^2\bar{\phi}}{\partial\bar{z}^2} = \tau^{2/3}M_\infty^2(\gamma + 1)\left(\frac{\partial\bar{\phi}}{\partial\bar{x}}\right)\frac{\partial^2\bar{\phi}}{\partial\bar{x}^2}$

Combining terms, we obtain

$$\left[\frac{(1 - M_\infty^2)}{\tau^{2/3}} - M_\infty^2(\gamma + 1)\bar{\phi}_{\bar{x}} \right] \bar{\phi}_{\bar{x}\bar{x}} + \bar{\phi}_{\bar{y}\bar{y}} + \bar{\phi}_{\bar{z}\bar{z}} = 0 \qquad (14.11)$$

Let us define the *transonic similarity parameter K* as

$$K = \frac{1 - M_\infty^2}{\tau^{2/3}} \qquad (14.12)$$

Then Eq. (14.11) is written as

$$\left[K - M_\infty^2(\gamma + 1)\bar{\phi}_{\bar{x}} \right] \bar{\phi}_{\bar{x}\bar{x}} + \bar{\phi}_{\bar{y}\bar{y}} + \bar{\phi}_{\bar{z}\bar{z}} = 0 \qquad (14.13)$$

Finally, assuming that the Mach numbers are near unity for transonic flow, replace M_∞ in Eq. (14.13) by unity, obtaining

$$\left[K - (\gamma + 1)\bar{\phi}_{\bar{x}} \right] \bar{\phi}_{\bar{x}\bar{x}} + \bar{\phi}_{\bar{y}\bar{y}} + \bar{\phi}_{\bar{z}\bar{z}} = 0 \qquad (14.14)$$

Equation (14.14) is the *transonic similarity equation;* it is essentially another form of the transonic small-perturbation equation given by Eq. (14.10). However, Eq. (14.14) contains a special message. Consider two flows at different values of M_∞ (but both transonic) over two bodies with different values of τ, but with M_∞ and τ for both flows such that the transonic similarity parameter K is the same for both flows. Then Eq. (14.14) states that the solution for both flows in terms of the nondimensional quantities $\bar{\phi}(\bar{x}, \bar{y}, \bar{z})$ will be the same. This is the essence of the *transonic similarity principle.* In turn, the pressure coefficients for the two flows are related such that $C_p/\tau^{2/3}$ is the same between the two flows, i.e.,

$$\frac{C_p}{\tau^{2/3}} = -2\bar{\phi}_{\bar{x}} = f(K, \bar{x}, \bar{y}, \bar{z}) \qquad (14.15)$$

The proof of Eq. (14.15) is left as a homework problem. Keep in mind that transonic similarity is an approximate theory, good only for flows over slender bodies at small angles of attack, and where the transonic shock waves are weak enough to assume an isentropic, irrotational flow.

In summary, there are three echelons of transonic inviscid flow theory.

1. Solutions of the Euler equations, given by Eqs. (14.1) through (14.3). These are the exact solutions, since the Euler equations contain no special assumptions in regard to the inviscid flow.

2. Solutions of the potential equation, given by Eq. (8.17). These solutions are approximate, because they assume the shock wave present in the transonic flowfield is weak enough to justify treating the flow as isentropic and irrotational. This is frequently a good assumption, because the entropy change across a shock wave at transonic speeds is only of third order in the shock strength.

3. Solutions of the small-perturbation potential equation, in the form of Eq. (14.10) or Eq. (14.14). These solutions are a further approximation, good only for the flows over slender bodies at small angles of attack. It is within this framework that the transonic similarity principle holds, as derived here.

It is important to note that all three levels of equations for the analysis of transonic flow—the Euler equations, the full potential equation, and the perturbation potential equation—are *nonlinear equations*. Any type of transonic theory is *nonlinear* theory. This important aspect of transonic flow was first noted in Sec. 9.2, and is plainly evident in the equations discussed in the present section. The nonlinearity of transonic flows has made such flows very difficult to solve in the past; this is essentially responsible for the "failure in aerodynamic theory" expressed in von Karman's quote at the beginning of this chapter. However, the advent of computational fluid dynamics has changed this situation in recent years. Successful numerical solutions to all three echelons of equations itemized above have been obtained for a variety of applications. These numerical solutions are the subject of the rest of this chapter.

14.4 | SOLUTIONS OF THE SMALL-PERTURBATION VELOCITY POTENTIAL EQUATION: THE MURMAN AND COLE METHOD

In the present section, we will address the solution of Eq. (14.10), or equivalently, Eq. (14.14). This class of transonic flowfield solutions is best exemplified by the work of Murman and Cole (Ref. 101), which has become a classic in the field. We will outline their approach in this section.

To illustrate the method, we will consider the airfoil in physical space shown at the left of Fig. 14.6. For simplicity, the angle of attack is zero and the airfoil is

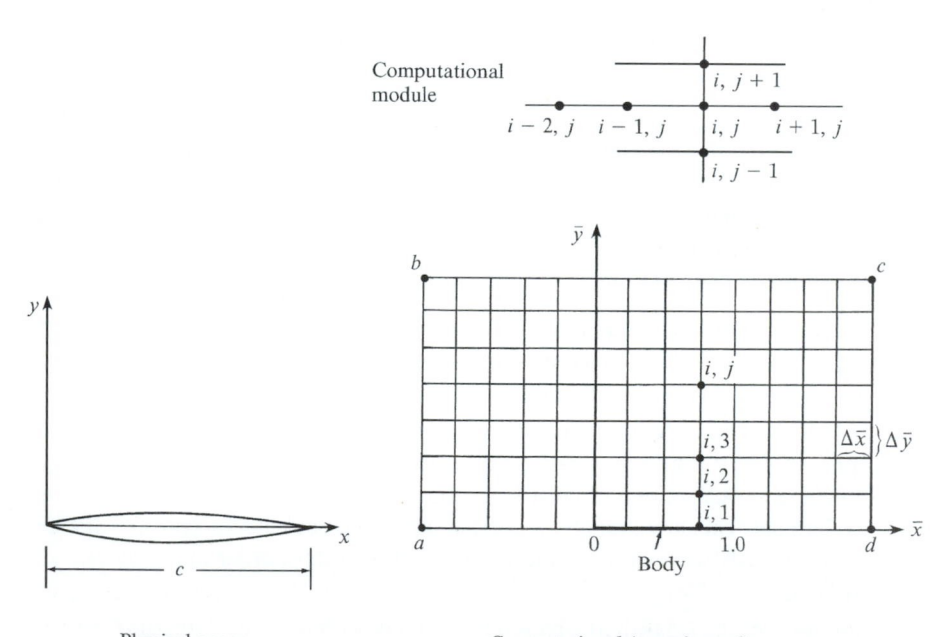

Figure 14.6 | Physical and computational spaces.

symmetric, hence a zero-lift case is considered. (However, this is not necessary; small-perturbation solutions can be obtained for thin nonsymmetric airfoils at small angle of attack.) We wish to obtain the two-dimensional, inviscid, transonic flowfield over this airfoil as governed by Eq. (14.10) written in (x, y) space. The numerical solution itself is carried out in the transformed $(\bar{x}, \bar{y})$ space shown at the right of Fig. 14.6, using the transformed equivalent of Eq. (14.10), namely Eq. (14.14). In particular, Eq. (14.14) is replaced by a finite-difference equation evaluated over the rectangular grid in $(\bar{x}, \bar{y})$ space. A computational module [a segment of the grid, showing the grid points used for the finite-difference representations at the grid point (i, j)] is drawn above the grid. The airfoil is represented by the line from 0 to 1.0 along the $\bar{y} = 0$; axis; the surface tangency boundary condition along the body is evaluated at $\bar{y} = 0$, consistent with the small-perturbation assumption. This boundary condition is given by Eq. (9.19), where the shape of the body is expressed as $y = f(x)$. That is,

$$\frac{df}{dx} = \frac{v'}{V_\infty} \tag{14.16}$$

However, from the transformation defined in Sec. 14.3, we have

$$v' = \frac{\partial \phi}{\partial y} = \tau V_\infty \frac{\partial \bar{\phi}}{\partial \bar{y}} \tag{14.17}$$

Combining Eqs. (14.16) and (14.17), the surface boundary condition becomes

$$\bar{\phi}_{\bar{y}}(\bar{x}, 0) = \frac{1}{\tau} \frac{df}{dx} \tag{14.18}$$

where df/dx is a known function of x, hence $\bar{x}$. Equation (14.18) represents the boundary condition for $0 \le \bar{x} \le 1$ along the $\bar{y} = 0$ axis, as shown by the heavy line in the grid drawn in Fig. 14.6. For all other values of $\bar{x}$ along the line $\bar{y} = 0$, the flow symmetry condition, $\bar{\phi}_{\bar{y}} = 0$, is used. An appropriate, second-order one-sided difference for $\bar{\phi}_{\bar{y}}$ at the surface is (see Ref. 18)

$$\bar{\phi}_{\bar{y}} = \frac{\partial \bar{\phi}}{\partial \bar{y}} = \frac{1}{2\,\Delta \bar{y}}(-3\bar{\phi}_{i,1} + 4\bar{\phi}_{i,2} - \bar{\phi}_{i,3}) \tag{14.19}$$

where grid point $(i, 1)$ is along the $\bar{y} = 0$ axis, and points $(i, 2)$ and $(i, 3)$ are directly above it, as shown in Fig. 14.6. Hence, a finite-difference expression for the surface boundary condition is, from Eqs. (14.18) and (14.19),

$$\frac{1}{\tau}\left(\frac{df}{dx}\right)_i = \frac{1}{2\,\Delta \bar{y}}(-3\bar{\phi}_{i,1} + 4\bar{\phi}_{i,2} - \bar{\phi}_{i,3}) \tag{14.20}$$

For the boundary conditions along ab, bc, and cd which form the left, upper, and right boundaries of the grid in Fig. 14.6, it is tempting to apply free-stream conditions. However, keep in mind that, in a subsonic flow (albeit near Mach 1), disturbances reach out to infinity in all directions away from the body. Therefore, we should apply the free-stream conditions only if the outer boundaries of the grid were an infinite distance away, which is certainly not the practical case shown in Fig. 14.6. Instead, a more appropriate "far-field" boundary condition—not the free-stream conditions—should be applied along ab, bc, and cd. This "far-field" boundary condition is expressed in terms of the far field associated with a doublet singularity. It takes the form of

$$\bar{\phi}(\bar{x}, \bar{y}) = \frac{1}{2\pi K^{1/2}} \frac{\mathscr{D}\bar{x}}{\bar{x}^2 + K\bar{y}^2} \tag{14.21}$$

where $\mathscr{D}$ is the effective doublet strength, obtained as part of the solution, and $\bar{x}$ and $\bar{y}$ are the coordinates along ab, bc, and cd. The arguments surrounding the development of Eq. (14.21) as well as the calculation of $\mathscr{D}$ are too lengthy to relate here; the reader is encouraged to study Ref. 101 for the details. Equation (14.21) is given here only for the sake of illustration in our discussion of the boundary conditions.

For the remainder of the flowfield over the grid in Fig. 14.6, Eq. (14.14) is used. The proper finite-difference form of the $\bar{x}$ derivatives in Eq. (14.14) depends on whether the flow is locally subsonic or supersonic, and it is this aspect where Murman and Cole in Ref. 101 make a fundamental contribution to the state of the art of transonic flowfield calculations. If the flow is locally subsonic, then information at point (i, j) can come from both upstream and downstream, and an appropriate finite-difference representation is the standard second-order central difference formula:

$$(\bar{\phi}_{\bar{x}})_{i,j} = \left(\frac{\partial\bar{\phi}}{\partial\bar{x}}\right)_{i,j} = \frac{\bar{\phi}_{i+1,j} - \bar{\phi}_{i-1,j}}{2\,\Delta\bar{x}} \tag{14.22}$$

and

$$(\bar{\phi}_{\bar{x}\bar{x}})_{i,j} = \left(\frac{\partial^2\bar{\phi}}{\partial\bar{x}^2}\right)_{i,j} = \frac{\bar{\phi}_{i+1,j} - 2\bar{\phi}_{i,j} + \bar{\phi}_{i-1,j}}{(\Delta\bar{x})^2} \tag{14.23}$$

However, if the flowfield is locally supersonic, then information at point (i, j) can only come from upstream. This motivates the use of *upwind differences*, namely,

$$(\bar{\phi}_{\bar{x}})_{i,j} = \left(\frac{\partial\bar{\phi}}{\partial\bar{x}}\right)_{i,j} = \frac{\bar{\phi}_{i,j} - \bar{\phi}_{i-2,j}}{2\,\Delta\bar{x}} \tag{14.24}$$

and

$$(\bar{\phi}_{\bar{x}\bar{x}})_{i,j} = \left(\frac{\partial^2\bar{\phi}}{\partial\bar{x}^2}\right)_{i,j} = \frac{\bar{\phi}_{i,j} - 2\bar{\phi}_{i-1,j} + \bar{\phi}_{i-2,j}}{(\Delta\bar{x})^2} \tag{14.25}$$

In both the locally subsonic and supersonic cases, the $\bar{y}$ derivative is replaced by central differences, as follows:

$$(\bar{\phi}_{\bar{y}\bar{y}})_{i,j} = \left(\frac{\partial^2\bar{\phi}}{\partial\bar{y}^2}\right)_{i,j} = \frac{\bar{\phi}_{i,j+1} - 2\bar{\phi}_{i,j} + \bar{\phi}_{i,j-1}}{(\Delta\bar{y})^2} \tag{14.26}$$

The grid points which are used in Eqs. (14.22) through (14.26) are shown in the computational module in Fig. 14.6. Using the above finite-difference quotients, let us obtain the difference equation which results from Eq. (14.14). First, consider locally subsonic flow, where the derivatives are expressed by Eqs. (14.22), (14.23), and (14.26). By direct substitution into Eq. (14.14), we have

$$\left[K - (\gamma + 1)\left(\frac{\bar{\phi}_{i+1,j} - \bar{\phi}_{i-1,j}}{2\,\Delta\bar{x}} \right) \right] \frac{\bar{\phi}_{i+1,j} - 2\bar{\phi}_{i,j} + \bar{\phi}_{i-1,j}}{(\Delta\bar{x})^2}$$

$$+ \frac{\bar{\phi}_{i,j+1} - 2\bar{\phi}_{i,j} + \bar{\phi}_{i,j-1}}{(\Delta\bar{y})^2} = 0 \qquad (14.27)$$

In the case of locally supersonic flow, where the derivatives are expressed by Eqs. (14.24) through (14.26), the difference form of Eq. (14.14) is

$$\left[K - (\gamma + 1)\left(\frac{\bar{\phi}_{i,j} - \bar{\phi}_{i-2,j}}{2\,\Delta\bar{x}} \right) \right] \frac{\bar{\phi}_{i,j} - 2\bar{\phi}_{i-1,j} + \bar{\phi}_{i-2,j}}{(\Delta\bar{x})^2}$$

$$+ \frac{\bar{\phi}_{i,j+1} - 2\bar{\phi}_{i,j} + \bar{\phi}_{i,j+1}}{(\Delta\bar{y})^2} = 0 \qquad (14.28)$$

Equations (14.27) and (14.28) can be solved by the rather standard relaxation technique, also called the iterative technique, which is described at length in most numerical analysis texts; in particular, see Ref. 102 for details. The relaxation technique is carried out as follows. Examining the computational grid shown in Fig. 14.6, first *assume* values for $\bar{\phi}$ at all grid points. Now concentrate on the grid point (i, j). Test to see if the flow is locally subsonic or supersonic at (i, j); if it is subsonic, use Eq. (14.27), and if it is supersonic, use Eq. (14.28). In either Eq. (14.27) or (14.28), as the case may be, treat $\bar{\phi}_{i,j}$ as the *unknown* variable, and use the assumed (specified) values for the $\bar{\phi}$'s at other grid points. In this manner, Eq. (14.27) is expressed as

$$A\bar{\phi}_{i,j} = B \qquad (14.29)$$

where A and B are known numbers, and Eq. (14.28) is expressed as

$$C(\bar{\phi}_{i,j})^2 + D\bar{\phi}_{i,j} = E \qquad (14.30)$$

where C, D, and E are known numbers. Solve either Eq. (14.29) or (14.30), as the case may be, at each internal grid point throughout the computational grid in Fig. 14.6. Now, use the new set of $\bar{\phi}$'s just obtained above to calculate new values for A, B, C, D, and E, and again solve Eq. (14.29) or (14.30) at each grid point. Continue this process until the values of $\bar{\phi}_{i,j}$ relax to the same values from one computational step to another, i.e., until the solution converges.

The simple relaxation procedure discussed above can be somewhat lengthy in terms of the computer time required for obtaining convergence. The convergence can be accelerated by using successive *line* relaxation (see Ref. 102). In this modification of the simple relaxation method, the values of $\bar{\phi}$ along a vertical line of grid points in Fig. 14.6 are singled out to be treated as the unknowns in Eq. (14.27) or (14.28). That is, in these equations, $\bar{\phi}_{i,j+1}$, $\bar{\phi}_{i,j}$, and $\bar{\phi}_{i,j-1}$ are treated as unknowns; all the other $\bar{\phi}$'s that appear in these equations are given the known value obtained from the previous relaxation step (or the previous line relaxation). When Eq. (14.27) or (14.28) is applied at each grid point in the vertical line $(i, 1)$, $(i, 2)$, ..., (i, j), ..., a system of simultaneous algebraic equations is obtained; these equations must be solved together. In each of these equations there are three unknowns. For example, at grid point $(i, 3)$, the unknowns are $\bar{\phi}_{i,2}$, $\bar{\phi}_{i,3}$, and $\bar{\phi}_{i,4}$. At grid point (i, j), the unknowns are $\bar{\phi}_{i,j-1}$, $\bar{\phi}_{i,j}$, and $\bar{\phi}_{i,j+1}$. And so forth. When expressed in terms of matrix representation, these equations for the single vertical line of grid points result in a tridiagonal matrix, which can be easily treated by standard techniques. After all the unknowns are solved along the vertical row of grid points as described above, we move to the right in Fig. 14.6, and now treat the next vertical row of points $(i + 1, 1)$, $(i + 1, 2)$, $(i + 1, 3)$, ..., $(i + 1, j)$, ..., in the same manner. In this fashion, all the $\bar{\phi}$'s for one relaxation step are calculated by solving the unknowns along each vertical line, sweeping from left to right in the grid shown in Fig. 14.6. When this sweep is finished, return to the vertical line of grid points at the extreme left, and start the next relaxation step.

This description is intended to provide only a "feeling" for the numerical technique used to solve Eq. (14.14). For more details on the numerical approach, consult Ref. 102, and for details on the complete solution of the transonic small-disturbance solutions, see Murman and Cole (Ref. 101).

Typical results obtained by Murman and Cole are shown in Fig. 14.7. Here, the surface pressure coefficient distributions are given for a symmetric circular arc airfoil at zero angle of attack for two different values of the transonic similarity parameter K. The solid line represents the calculations from Murman and Cole, and the open circles are experimental data obtained from Knechtel (Ref. 103). In Fig. 14.7a, K_s is a modified transonic similarity parameter, defined as

$$K_s = \left(1 - M_\infty^2\right) / \left(M_\infty^2 \tau\right)^{2/3}$$

The value of $K_s = 3$ pertains to a free-stream Mach number below M_{cr}; hence, the flow is completely subsonic. Note the smooth, symmetric pressure distribution for this case. In Fig. 14.7b, the value $K_s = 1.3$ pertains to a free-stream Mach number above M_{cr}; hence, the flow is mixed subsonic-supersonic. That portion of the flow where $|C_p| > |C_{p_{cr}}|$ is locally supersonic. Note the unsymmetrical pressure distribution as well as the rapid increase in pressure at about $\bar{x} \approx 0.8$. This rapid pressure change is indicative of a shock wave at that location; the drop from supersonic to subsonic flow at about $\bar{x} \approx 0.8$ is another indication of the presence of the shock wave. Note that the pressure jump across the shock wave is relatively sharp in the calculations, but that it is somewhat diffused in the experimental data. This is most likely due to the effect of shock wave/boundary layer interaction in the experimental results, creating a locally separated flow at the surface. Such viscous effects are, of course, not included in the inviscid calculations.

The value of small-perturbation solutions of transonic flows is demonstrated by the results in Fig. 14.7. For the subcritical case (Fig. 14.7a), excellent agreement

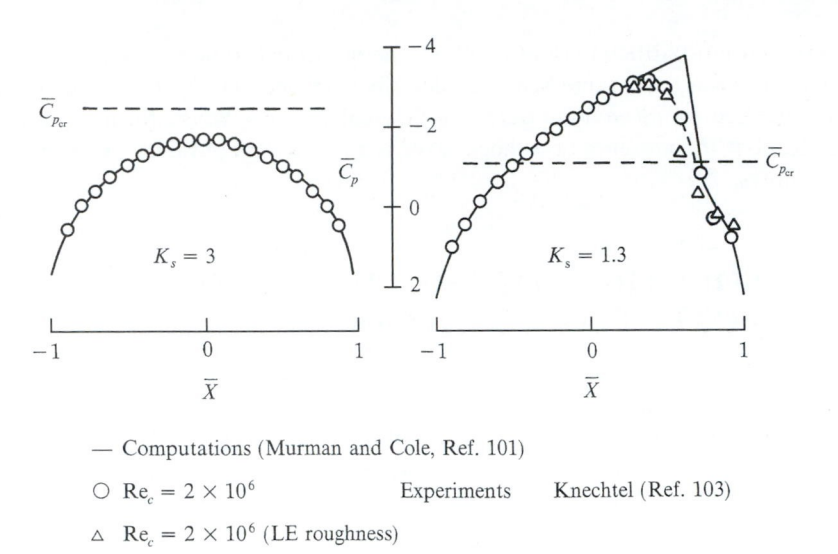

— Computations (Murman and Cole, Ref. 101)

○ $Re_c = 2 \times 10^6$ Experiments   Knechtel (Ref. 103)

△ $Re_c = 2 \times 10^6$ (LE roughness)

(*a*)   (*b*)

$$\bar{x} = \frac{x}{c}$$

$$\tau = \frac{b}{c}$$

$$K_s = \frac{(1 - M_\infty^2)}{(M_\infty^2 \tau)^{2/3}}$$

Figure 14.7 | Pressure coefficient distributions for a circular arc airfoil; comparison between experiment and calculation. (From Murman and Cole, Ref. 101.) (*a*) Free-stream Mach number below M_{cr} (subcritical case). (*b*) Free-stream Mach number above M_{cr} (supercritical case).

between computation and experiment is obtained. For the supercritical case (Fig. 14.7*b*), excellent agreement is also obtained, except in the vicinity of the shock wave. Hence, small-perturbation solutions of transonic flows—the simplest of the hierarchy of techniques described in Sec. 14.3—can give useful results.

Returning again to Fig. 14.7*b*, we repeat that the numerical calculations give results that are indicative of a shock wave in the flow (as we would expect, on the basis of our physical considerations discussed in Sec. 14.2). However, this leads to the following question: Since the transonic small-perturbation equation assumes an isentropic flow, how can a shock wave be predicted by such an equation? The answer rests in the artificial viscosity which is present in the numerical solution. As discussed in Sec. 12.8, the truncation error in a numerical solution can give rise to an inherent artificial viscosity in the numerics, and this "numerical dissipation" acts mathematically to create a shock wave in the same sense as friction and thermal conduction act to create the internal structure of a real shock front. Hence, even though a governing equation is being used that assumes isentropic flow [Eq. (14.14)], the presence of artificial viscosity allows the numerical solution to *capture* a shock wave in exactly the same sense as described in Sec. 11.15. This is a fortunate circumstance for *all* inviscid transonic flow calculations, where for many practical applications the presence of a shock wave is an important physical characteristic of such flows.

14.5 | SOLUTIONS OF THE FULL VELOCITY POTENTIAL EQUATION

The small-perturbation solutions described in Sec. 14.4 have certain limitations. As always, they are limited to thin bodies at small angle of attack. This is done to ensure that the perturbation velocities in the flow are indeed small. However, even for these cases there are regions where the perturbations are not small. For example, no matter how thin the airfoil, the flow velocity at the stagnation point near the leading edge will go to zero—hardly a "small" perturbation. The same can be said about the sharp, acute-angle trailing edge, where in subsonic flow the Kutta condition stipulates $V = 0$. (See Ref. 104 for a discussion of the Kutta condition in aerodynamics.) In spite of this, the small-perturbation solutions give good results in both the leading- and trailing-edge regions, as already seen in Fig. 14.7. This agreement is most likely fortuitous; as theorized by Caughey (Ref. 105), and supported by the work of Keyfitz et al. (Ref. 106), in the leading- and trailing-edge regions the error associated with the small-perturbation assumption is compensated by the truncation error in the numerical solution due to the finite grid size. Finally, examining Fig. 14.7, the changes in flow properties across the shock wave are not small, and there might be some inaccuracy in the shock location and shock properties when the small-perturbation equation is used.

The concerns raised in the previous paragraph are obviated by solving the full potential equation for transonic flows, namely Eq. (8.17). As stated in Chap. 8, and repeated in Sec. 14.3, Eq. (8.17) deals with the full velocity potential Φ, and hence allows for large changes in the flowfield variables. In particular, Eq. (8.17) can be applied to any size body at any angle of attack. However, the use of Eq. (8.17) still assumes the flow to be irrotational and isentropic. Solutions of Eq. (8.17) represent the next step in our discussion of the hierarchy of transonic flow analysis, the first step being the small-perturbation solutions discussed in Sec. 14.4.

The numerical solution of Eq. (8.17) can be carried out by means of the relaxation technique discussed in Sec. 14.4. However, exemplifying the adage that "you cannot get something for nothing," the increased accuracy associated with the use of the full potential equation is accompanied by increased complexity of the numerical solution. This increased complexity is associated with the body surface boundary condition. In the small-perturbation solution, the body boundary condition, namely Eq. (14.18), was applied along the $\bar{x}$ axis, i.e., at $\bar{y} = 0$. In contrast, for the full potential solution, the body boundary condition should be applied on the body surface itself, i.e.,

$$\frac{\partial \Phi}{\partial n} = 0 \quad \text{on } y = f(x) \tag{14.31}$$

where $f(x)$ is the shape of the body in the (x, y) plane, and n is the direction locally normal to the surface. If a rectangular finite-difference grid is used in the physical plane, it becomes difficult to numerically apply the boundary condition at the surface of the body, Eq. (14.31). First, very few (if any) of the regularly spaced rectangular grid points would fall on the body surface, and therefore a complex system of interpolation has to be used to place oddly spaced grid points on the body surface. Such a rectangular grid, along with its complexity for the surface boundary condition, was used and described by Magnus and Yoshihara in Ref. 107. This grid involves a fine grid embedded in a coarse grid, which finally switches to a polar coordinate grid in the far field, as shown in Fig. 14.8a. A detail of the grid at the body surface is shown in Fig. 14.8b, along with the points required to apply the boundary

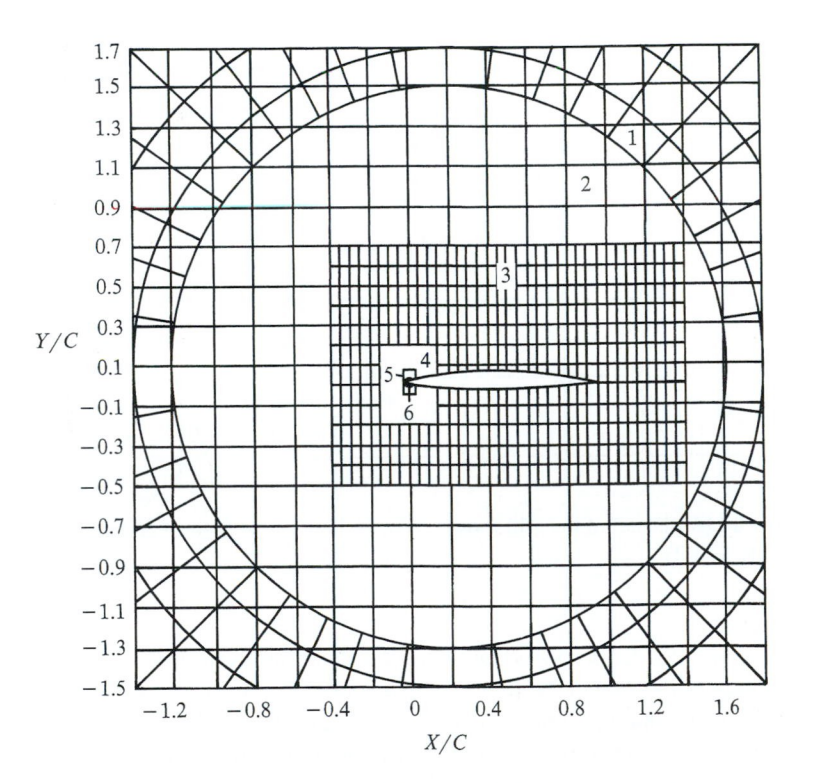

Figure 14.8a | The patching of six different grids for the numerical calculation of the transonic flow over an airfoil; an approach circa 1970 before the advent of curvilinear grid generation. (From Magnus and Yoshihara, Ref. 107.)

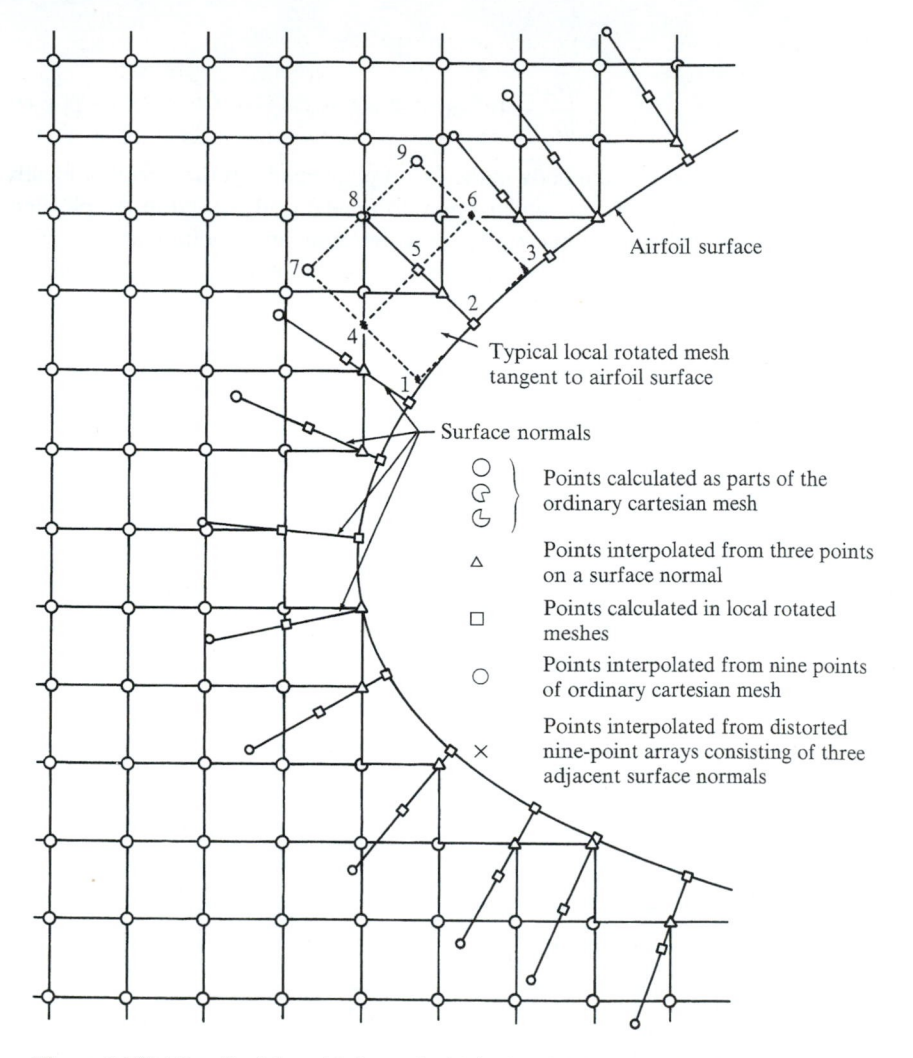

Figure 14.8*b* | Detail of the grid shown in (*a*) in the vicinity of the leading edge (Ref. 107).

condition, namely the derivative of Φ normal to the surface. One glance at Fig. 14.8 quickly impresses upon us the complexity associated with a rectangular grid. In spite of this, Magnus and Yoshihara successfully used such a grid for the solution of the Euler equations for a transonic flow; these solutions will be discussed in the next section.

The grid problem was made much more tractable in 1974 when Thompson et al. (Ref. 108) developed an ingenious method for constructing a boundary-fitted coordinate system around a body of arbitrary shape. In this method, the body surface becomes a coordinate line in physical space, and other coordinate lines away from the body are generated by means of the solution of two elliptic partial differential equations. To be more specific, a transformation is constructed to map the curvilinear, boundary-fitted grid in physical space to a rectangular grid in the computational space. That is, the physical (x, y) space is transformed into (ξ, η) space via a set of elliptic partial differential equations such as

$$A\frac{\partial^2 x}{\partial \xi^2} - 2B\frac{\partial^2 x}{\partial \xi\, \partial \eta} + C\frac{\partial^2 x}{\partial \eta^2} = f(\xi, \eta) \tag{14.32}$$

and

$$A\frac{\partial^2 y}{\partial \xi^2} - 2B\frac{\partial^2 y}{\partial \xi\, \partial \eta} + C\frac{\partial^2 y}{\partial \eta^2} = g(\xi, \eta) \tag{14.33}$$

Figure 14.9 illustrates this transformation. The physical (x, y) space is shown in Fig. 14.9a, along with the boundary-fitted coordinate system for an airfoil. Note in the physical plane that the airfoil surface is a coordinate line, namely $\eta = \text{const} = c_1$. All the grid points along $\eta = c_1$ fall on the airfoil surface. In Fig. 14.9a, c_1 is set to zero; hence $\eta = 0$ is the coordinate of the airfoil surface. The next coordinate curve away from the airfoil surface is $\eta = \text{const} = c_2$. The furthest curve away from the body is $\eta = \text{const} = c_n$. Fanning out from the body are a second series of coordinate lines, $\xi = \text{const}$. The (ξ, η) grid in the transformed space, Fig. 14.9b, is a rectangular grid. The relationship of this rectangular grid to the analogous curvilinear grid in the physical space, Fig. 14.9a, is set by the transformation in Eqs. (14.32) and (14.33). That is, Eqs. (14.32) and (14.33) are solved to give the (x, y) coordinates in physical space which correspond to the (ξ, η) coordinates in the transformed space. Note in Eqs. (14.32) and (14.33) that x and y are the dependent variables, and that a solution of Eqs. (14.32) and (14.33) gives

$$x = x(\xi, \eta)$$

and

$$y = y(\xi, \eta)$$

From this solution, any grid point in the rectangular grid in (ξ, η) space in Fig. 14.9 can be located in the curvilinear grid in (x, y) space in Fig. 14.9a. Equations (14.32) and (14.33) are elliptic partial differential equations which can be solved numerically by a relaxation method. These equations, and their solution, are associated with the generation of the curvilinear, boundary-fitted coordinate system in Fig. 14.9a; they have absolutely *nothing* to do with the physics of the flowfield itself. Equations (14.32) and (14.33) are simply the definition of a grid transformation, and nothing else. Because the transformation is defined by a set of elliptic partial differential equations, it is called an *elliptic grid transformation*. See Ref. 108 for a detailed discussion of elliptic grid generation. Also, an extensive but elementary treatment is contained in Ref. 18. An actual example of a boundary-fitted curvilinear grid for an airfoil is shown in Fig. 14.10; this is an elliptically generated grid from Ref. 110, obtained using the technique of Ref. 108. In this grid, the points near the body surface are so close together that the graphics show them essentially as a continuous black

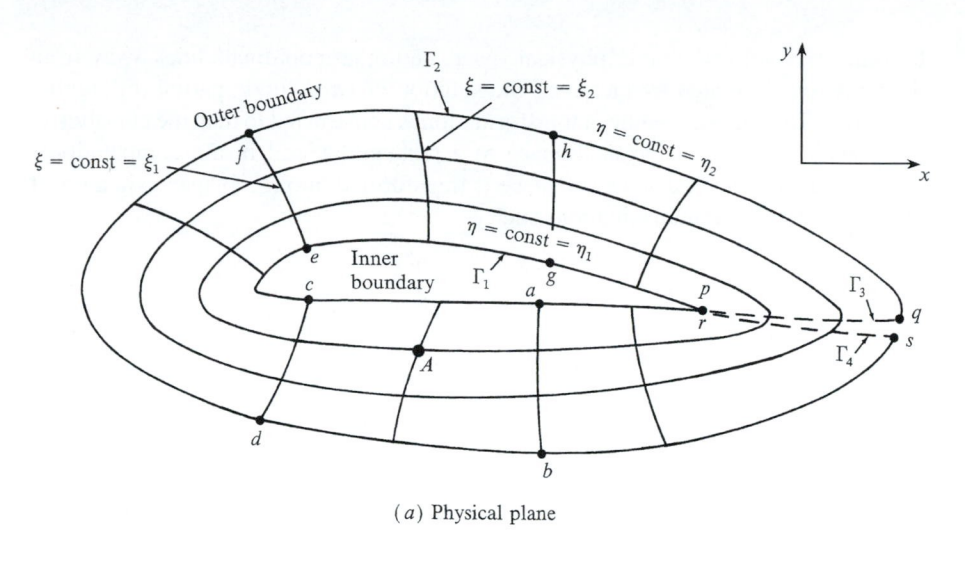

(a) Physical plane

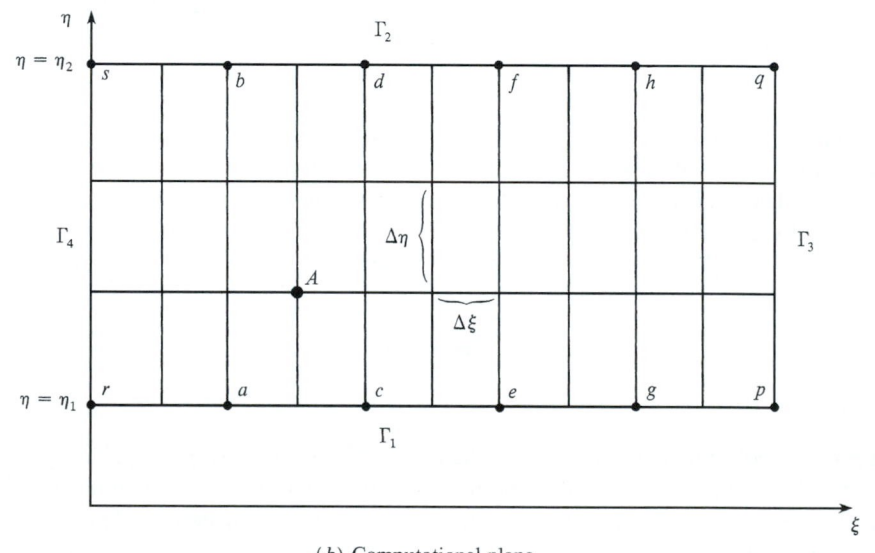

(b) Computational plane

Figure 14.9 | (a) Schematic of a boundary-fitted curvilinear grid in the physical (x, y) space. (b) Schematic of a rectangular grid in the computational (ξ, η) space, obtained from the grid in (a) by means of a suitable transformation. (From Ref. 18.)

area. Fig. 14.10b shows that portion of the grid near the airfoil; in reality, the full grid reaches much further away from the body such as shown in Fig. 14.10a. This grid was constructed for a viscous flow solution; hence, it requires a number of finely spaced points near the body. For the inviscid flow discussed here, the actual grid may not have to be so finely spaced.

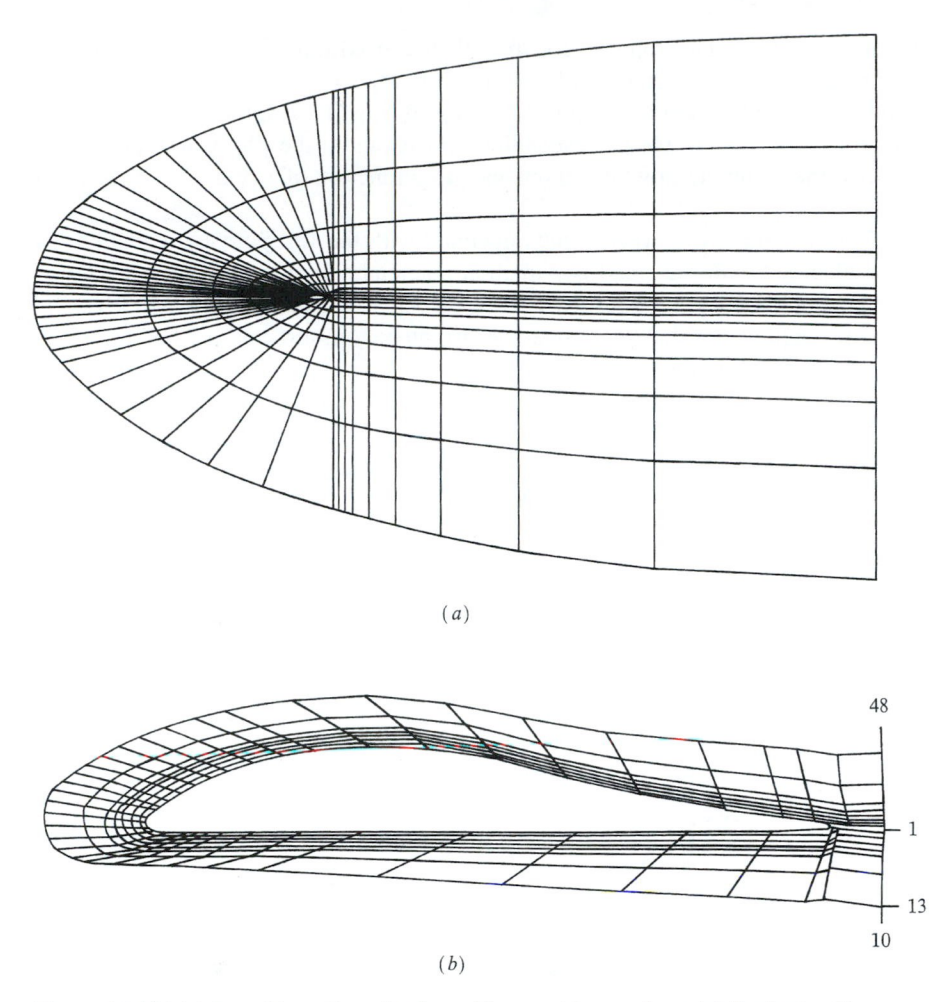

Figure 14.10 | (*a*) Actual boundary-fitted curvilinear grid around an airfoil, obtained by an elliptical grid generation technique patterned after Thompson et al. (Ref. 108), and carried out by Kothari and Anderson in Ref. 110. The airfoil is the small speck in the center of the grid. (*b*) Detail of the boundary-fitted grid in the vicinity of the airfoil. (From Ref. 110.)

For a given problem, the curvilinear grid is constructed first, independent of the flowfield solution itself. After this grid is formed, then the flowfield is solved using the full potential equation, namely Eq. (8.17). This equation is solved in the rectangular grid in (ξ, η) space shown in Fig. 14.9*b*. To this end, Eq. (8.17) must be transformed into (ξ, η) space. The details of this transformation are straight-forward, but lengthy; see Refs. 102 and 109 for a complete description of the general transformation. Finally, the transformed version of Eq. (8.17) in terms of $\partial^2 \Phi / \partial \eta^2$, $\partial \Phi / \partial \eta$, $\partial^2 \Phi / \partial \xi$, $\partial \Phi / \partial \xi$, etc., is solved. [The derivation of the transformed version of Eq. (8.17) is left as a homework problem.] These derivatives are replaced by the finite-difference expressions shown in Eqs. (14.22) through (14.25), except now in terms of ξ and η. The solution for Φ is then carried out by a relaxation method using the transformed version of Eq. (8.17). After the Φ and the corresponding flow variables are calculated in the transformed grid (Fig. 14.9b), these same variables are carried directly to the corresponding grid points in the physical plane; in this manner, the complete flowfield is obtained as a function of x and y in the physical plane.

The differences between results obtained with the full potential equation and those obtained with the small-perturbation potential equation are graphically illustrated in Fig. 14.11, which shows data calculated by Keyfitz et al. (Ref. 106). Here, the pressure coefficient distributions over the top and bottom surfaces of a Joukowski airfoil are shown; only the leading-edge region is shown, where $0 \leq x/c \leq 0.1$. (Note that, in contrast to the usual aerodynamic convention in Fig. 14.11 *positive* values of

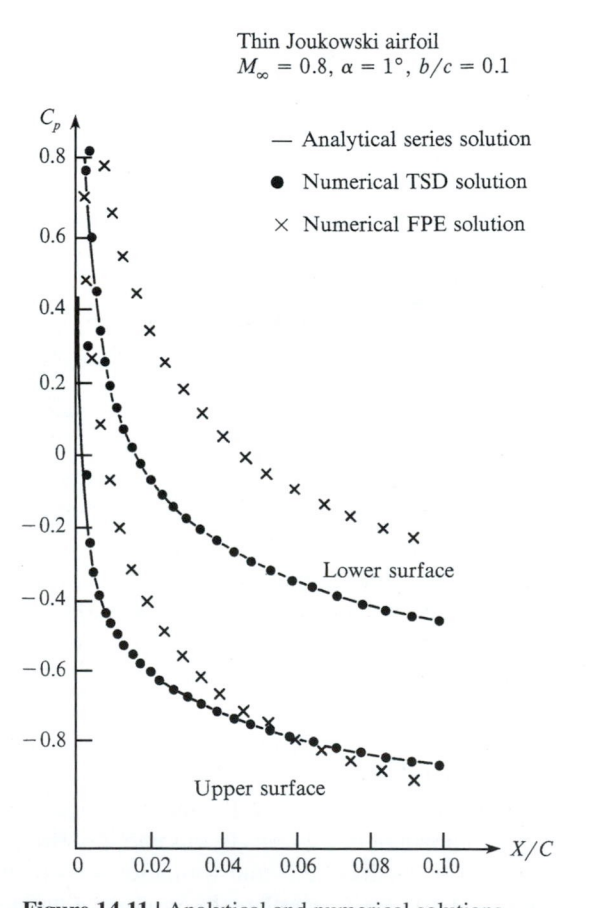

Figure 14.11 | Analytical and numerical solutions for the pressure coefficient distributions near the leading edge of a thin Joukowski airfoil. (By Keyfitz et al., Ref. 106.)

C_p are plotted in the upper quadrant.) As described earlier, it is this leading-edge region of the airfoil where the assumption of small perturbations is least accurate. In Fig. 14.11, TSD stands for transonic small disturbance (solution of the small-perturbation potential equation as discussed in Sec. 14.4), and FPE stands for full potential equation (solution of the full velocity potential equation as discussed in the present section). Also, the solid line in Fig. 14.11 represents an analytical solution to the small-perturbation potential equation in the leading-edge region, as reported in Ref. 106; this analytical solution agrees well with the TSD numerical solution. However, the primary message conveyed by Fig. 14.11 is that the more accurate FPE solution is quite different from the TSD solution in the leading-edge region; note that, for the most part, the TSD solution underpredicts the pressure, and shows a more rapid rise in pressure as the leading edge is approached. It should be noted that Keyfitz et al. examined the effect of mesh size on the results, and found that the TSD results in the nose region were very sensitive to the fineness of the grid in that region. The results shown in Fig. 14.11 were obtained with a mesh fine enough such that the results are relatively grid-independent.

Numerical solutions to both the small-perturbation and full potential equations in transonic flows have been extensively developed since the early 1970s, including the calculation of three-dimensional flows. Such a three-dimensional calculation is illustrated in Fig. 14.12. Here, the inviscid, transonic flow over a three-dimensional finite wing is illustrated. The free-stream Mach number is 0.9, and the wing is at an angle of attack such that the lift coefficient is 0.5. The airfoil section of the wing is a modern supercritical airfoil shape. Cordwise pressure coefficient distributions at three different spanwise stations are shown in Fig. 14.12. Two sets of calculations are displayed: (1) The dashed lines are numerical solutions of the small-perturbation potential equation using the computer code developed by Bailey and Ballhaus (Ref. 111), and (2) The solid curves are numerical solutions of the full potential equation using what has now become a relatively standard computer code called FLO-22 developed by Jameson and Caughey as reported in Ref. 112. The circles are experimental data points obtained by Hinson and Burdges (Ref. 113). Indeed, the comparisons shown in Fig. 14.12 were first made in Ref. 113, and then commented upon by Caughey in Ref. 105. Examining Fig. 14.12, we make these observations.

1. There is a substantial difference between the small-perturbation and full potential results, including a difference in the shock location.

2. On the whole, the full potential results agree better with the experimental data than the small-perturbation results.

3. The full potential results more accurately predict the shock wave location (the shock wave is evidenced by the rapid change in C_p, which occurs toward the back of the airfoil section). However, the effect of the artificial viscosity seems to spread the calculated shock jump over a wider region than shown by experiment. It is interesting that, although the small-perturbation results do not accurately predict the shock location, they do provide a qualitatively sharper shock jump than the full potential results.

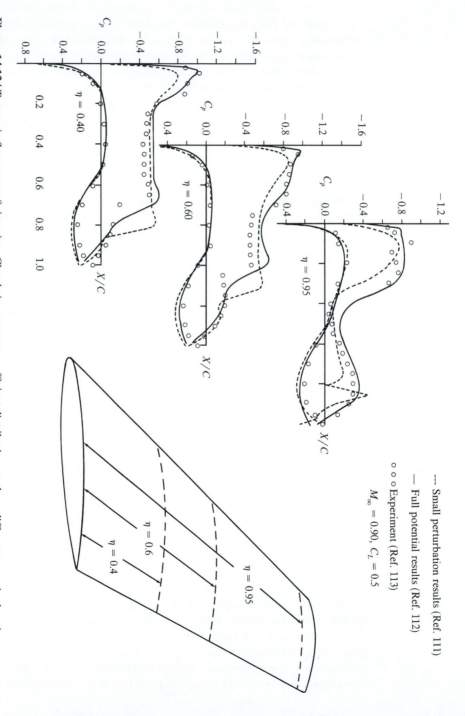

Figure 14.12 | Transonic flow over a finite wing. Chordwise pressure coefficient distributions at three different spanwise locations. Comparison between small-perturbation and full potential results with experiment. (From Caughey, Ref. 105.)

--- Small perturbation results (Ref. 111)

— Full potential results (Ref. 112)

o o o Experiment (Ref. 113)

$M_\infty = 0.90$, $C_L = 0.5$

In summary, the full potential solutions are more accurate than the small-perturbation results—no surprise, because the full potential equation itself [Eq. (8.17)] is more accurate than the small-perturbation potential equation [Eq. (14.14)]. On the other hand, the full potential solutions require more work and effort, principally due to the treatment of the boundary condition. In modern transonic flow calculations, the proper application of the surface boundary condition is carried out in concert with the generation of a curvilinear, boundary-fitted coordinate system, thus requiring the solution of the velocity potential equation in the transformed (ξ, η) space, which is rectangular. The advantage obtained with the full potential solutions is frequently worth this extra effort.

14.6 | SOLUTIONS OF THE EULER EQUATIONS

The use of the small-perturbation velocity potential equation (Sec. 14.4) and the full velocity potential equation (Sec. 14.5) both assume irrotational flow. The results obtained seem to justify this assumption; however, note that all the results given in Secs. 14.4 and 14.5 apply to the low end of the transonic regime, i.e., for subsonic free-stream Mach numbers, for which the shock wave at the end of the pocket of supersonic flow is relatively weak. For transonic applications that involve stronger shock waves, especially those situations where the free-stream Mach number is above unity, the assumption of irrotational flow becomes much less accurate. Consequently, attention to transonic flow analyses in recent times has shifted to the solution of the Euler equations, given by Eqs. (14.1) through (14.3). These equations hold for both rotational and irrotational flows; as discussed in Chap. 6, the only assumptions contained in Eqs. (14.1) through (14.3) are inviscid, adiabatic flow with no body forces. This also implies isentropic flow along a streamline. However, as discussed in Sec. 11.15, numerical solutions of the Euler equations also allow the capturing of shock waves in the flow, with the proper jump conditions across the shock wave including a discontinuous increase in entropy across the shock. This is the role of artificial viscosity in the numerical solution since some degree of numerical dissipation is necessary to generate the shock. Of course, the flow along a streamline is isentropic in front of the shock with one constant value of entropy, and it is isentropic behind the shock with another, but higher, constant value of entropy. The entropy change at the shock wave can be different from one streamline to another; thus, numerical solutions of the Euler equations allow for entropy gradients normal to the streamlines. Indeed, this is precisely the same physical mechanism actually occurring in transonic flows with shocks; hence, within the assumption of an inviscid flow, a solution of the Euler equations represents essentially an "exact" approach to the analysis of transonic flow. Hence, Euler solutions are the third and final echelon of the solution of transonic flows as discussed in Sec. 14.3. Such Euler solutions are the subject of this section.

Transonic flows are mixed regions of locally subsonic and supersonic flows; hence, the mathematical nature of such flows in the steady state is a mixed elliptic-hyperbolic problem. This is exactly the same problem associated with the steady flow over a supersonic blunt body as described in Chap. 12. As discussed in Chap. 12, this mixed-flow problem is circumvented by carrying out a time-marching solution, approaching the proper steady state in the limit of large times. In the same vein, solutions of the Euler equations for transonic flow problems are also time-marching solutions, beginning at some initially assumed starting point, and advancing the flowfield in steps of time using a numerical solution of the Euler equations for unsteady flow [i.e., using Eqs. (14.1) through (14.3) with the time derivatives included] until a steady-state result is obtained in the limit of large time. The time-marching philosophy and approach is discussed at length in Chap. 12, hence no further elaboration is given here.

The first time-marching solution of the Euler equations for transonic flow was carried out by Magnus and Yoshihara (Ref. 107). Using the rectangular grid shown previously in Fig. 14.8, they set up an algorithm vaguely similar, but different in detail, to the MacCormack method discussed in Chap. 12. See Ref. 107 for such details. The application treated in Ref. 107 was the flow over an NACA 64A410 airfoil at a 4° angle of attack in a Mach 0.72 free stream. The calculated pressure coefficient distributions over the top and bottom surfaces of the airfoil are compared with experimental measurements by Stivers (Ref. 114) in Fig. 14.13. Good agreement

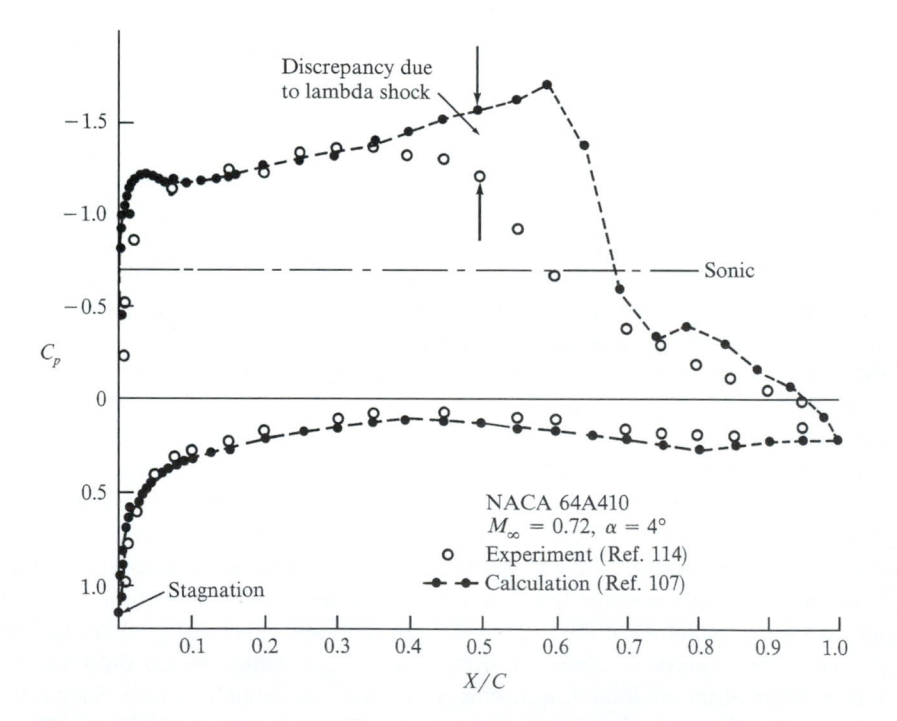

Figure 14.13 | An early finite-difference solution of the complete Euler equations for transonic flow, circa 1970 by Magnus and Yoshihara (Ref. 107). Pressure coefficient distribution for an NACA 64A410 airfoil.

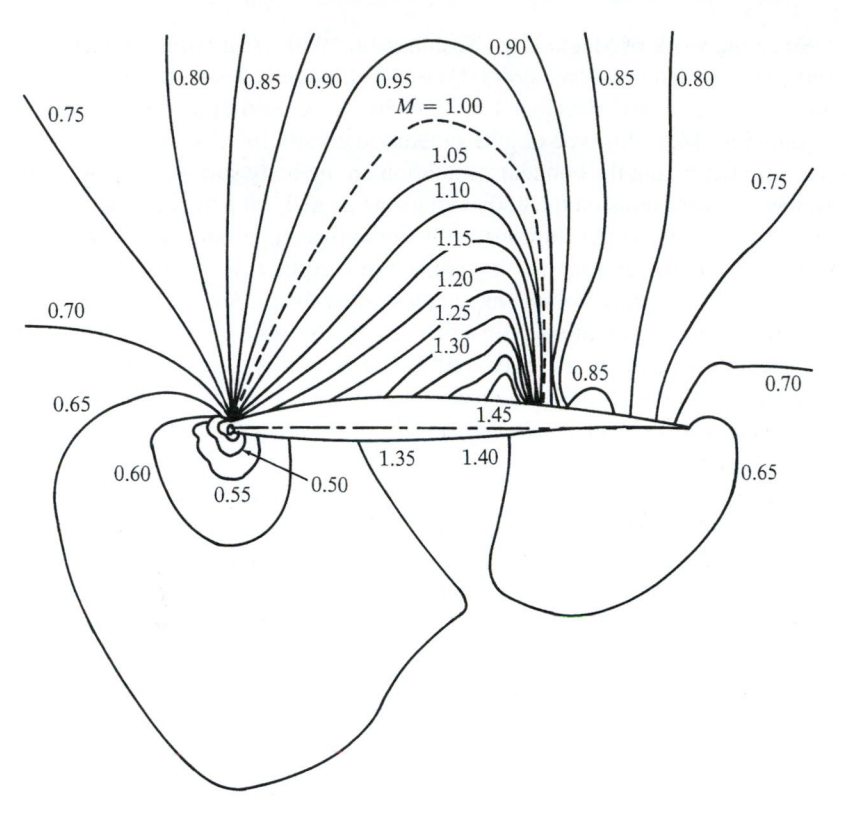

Figure 14.14 | Calculated Mach number contours for an NACA 64A410 airfoil $M_\infty = 0.72, \alpha = 4°$ (Ref. 107).

between the time-marching solution and the experimental data is obtained over the bottom surface of the airfoil and for a substantial portion of the upper surface. However, the region in the vicinity of the shock wave is not predicted well; Magnus and Yoshihara explain this difference as due to the shock wave/boundary layer interaction which is obviously not included in the Euler solution. Mach number contours are shown in Fig. 14.14. The sonic line is highlighted by the dashed curve. Note the large region of supersonic flow over the top surface, reaching far above the airfoil. Also note the value of the maximum Mach number in this region, about $M = 1.45$, even though the free-stream Mach number is only 0.72. This relatively large maximum Mach number is due to the angle of attack, causing the flow to expand rapidly over the top surface.

By today's standards, the technique developed in Ref. 107 is somewhat outdated, both in regard to the grid employed as well as the details of the algorithm. However, this work was pioneering because it was the first solution of the complete Euler equations for a transonic flow, and it introduced the time-marching approach for such flows.

Since the work of Magnus and Yoshihara in 1970, great strides have been made in Euler solutions to transonic flows. First, the elliptically generated, boundary-fitted coordinate system was developed in 1974 by Thompson et al. (Ref. 108), as discussed in Sec. 14.5; this type of grid generation greatly increased the ease and accuracy of implementing the boundary condition on the body surface simply by placing a number of grid points on the body surface as an integral and consistent part of the entire grid. Second, major improvements in the time-marching approach have been made which greatly shorten the computer time required to obtain the final steady state. In particular, finite-volume techniques rather than finite-difference approaches have certain advantages, along with a fine grid/coarse grid coupling technique called "multigrid." Such aspects are far beyond the scope of this chapter. A major developer of improved Euler solutions to transonic flow has been Tony Jameson of Princeton University; for further details of such modern solutions, see the extensive surveys by Jameson in Refs. 115 and 116.

To complete this section, we will present a few results which are examples of modern Euler solutions to transonic flow. To begin with, let us consider the flow over a circular cylinder; this is a classic configuration in aerodynamic theory. The solution for the inviscid incompressible flow over a circular cylinder can be obtained from exact potential theory for incompressible flow, and is constructed by superimposing the flows associated with a doublet and a uniform free stream; see Ref. 104 for details on this solution. Such an incompressible flow solution theoretically corresponds to $M_\infty = 0$, and leads to the exact formula for the pressure coefficient:

$$C_p = 1 - 4\sin^2\theta$$

where θ is the polar angle measured along the surface from the front stagnation point. This incompressible flow result, labeled as $M_\infty = 0$, is given in Fig. 14.15. For the compressible flow over a cylinder, because the circular shape is a "blunt" body, the flow very rapidly expands over the top and bottom surfaces. For this reason, the critical Mach number for a circular cylinder is quite low. Indeed, it is interesting to note the critical Mach number for both a circular cylinder and a sphere, obtained from Ref. 16, as

Circular cylinder: $M_{cr} = 0.404$

Sphere: $M_{cr} = 0.57$

The higher critical Mach number for the sphere is yet another example of the three-dimensional relieving effect discussed in previous chapters. Note that the transonic flow occurs over cylinders and spheres even though the free-stream Mach number is quite low. Euler solutions for the transonic flow over a circular cylinder were obtained by Jameson in Ref. 115. These results are labeled as $M_\infty = 0.35$ and $M_\infty = 0.45$ in Fig. 14.15—free-stream Mach numbers just below and just above M_{cr}, respectively. For $M_\infty = 0.35$, the flow is completely subsonic, and a smooth, symmetrical C_p distribution is obtained. Note that the peak (negative) C_p at the top of the cylinder is about -3.4, larger in magnitude than the incompressible result

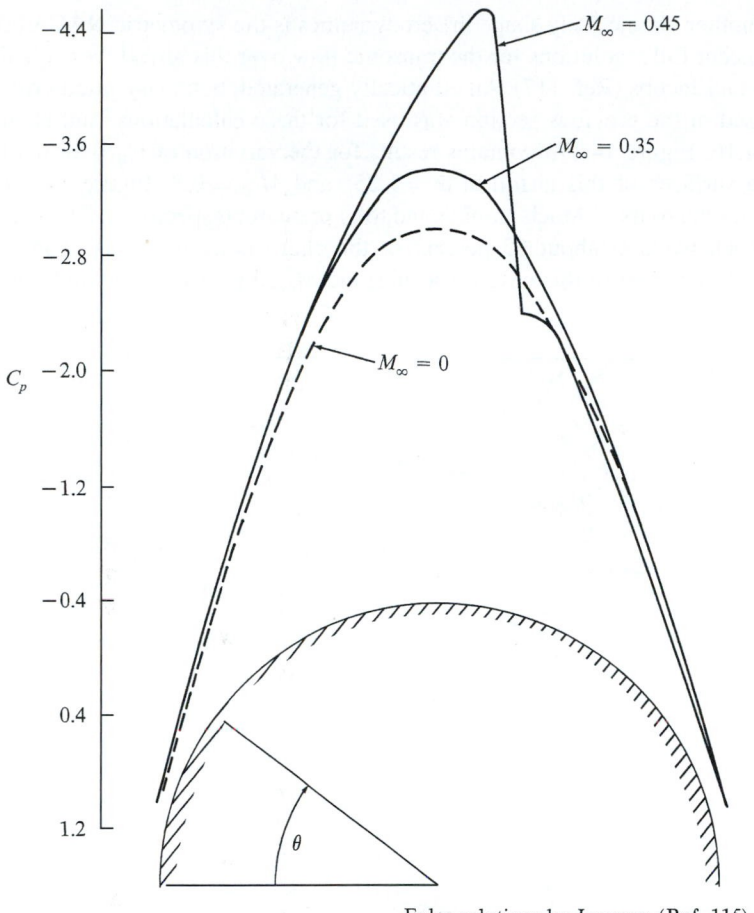

— Euler solutions by Jameson (Ref. 115)

--- Incompressible flow; $C_p = 1 - 4\sin^2\theta$

Figure 14.15 | Transonic flow over a circular cylinder; finite-volume solutions of the Euler equations by Jameson (Ref. 115). $M_\infty = 0.35$ is a subcritical case, and $M_\infty = 0.45$ is a supercritical case. Comparison with classical incompressible results ($M_\infty = 0$).

of -3.0. This is consistent with the effect of compressibility on C_p as discussed in Secs. 9.4 and 9.5 [Applying the simple Prandtl–Glauert correction from Eq. (9.36), we obtain $C_p = -3.2$, it is no surprise that Eq. (9.36) underpredicts C_p because the Prandtl–Glauert theory is based on small-perturbation theory, and hence is applicable to slender bodies only.] For $M_\infty = 0.45$, the flow over the cylinder is partly supersonic. Note the dramatic qualitative and quantitative changes in C_p; the pressure distribution is no longer symmetrical, and a shock wave occurs slightly downstream of the $\theta = 90°$ location.

Another classic body shape in aerodynamics is the symmetric NACA 0012 airfoil. Recent Euler solutions for the transonic flow over this airfoil were obtained by Reddy and Jacobs (Ref. 117). An elliptically generated, boundary-fitted grid such as discussed in the previous section was used for these calculations, and is shown in Fig. 14.16. Figure 14.17a contains results for the variation of C_p over the top and bottom surfaces of this airfoil at $\alpha = 1.25°$ and $M_\infty = 0.8$. Figure 14.17b and c illustrates contours of Mach number and total pressure, respectively. The nearly normal shock wave at about 65 percent of the chord is clearly evident in all these figures. In contrast to this case for a subsonic M_∞, Fig. 14.18a, b, and c gives the

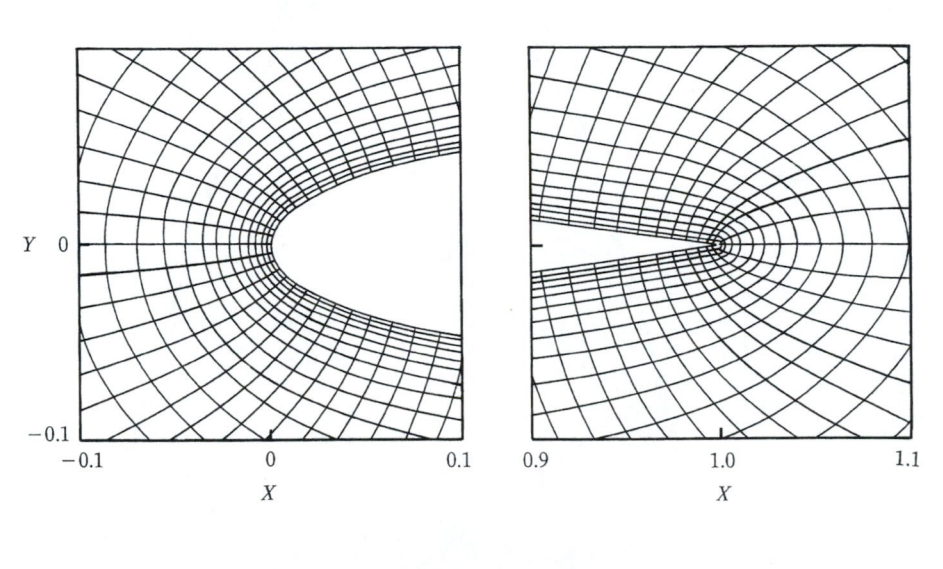

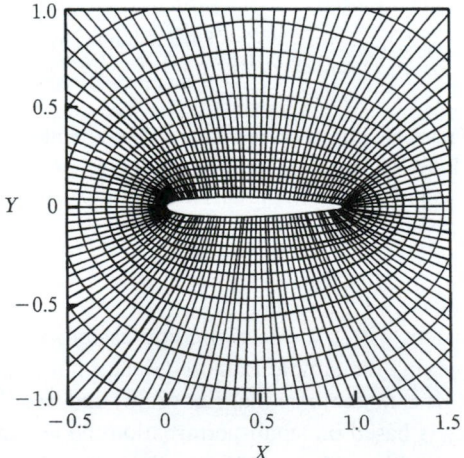

Figure 14.16 | Boundary-fitted curvilinear grid for the Euler solutions by Reddy and Jacobs (Ref. 117).

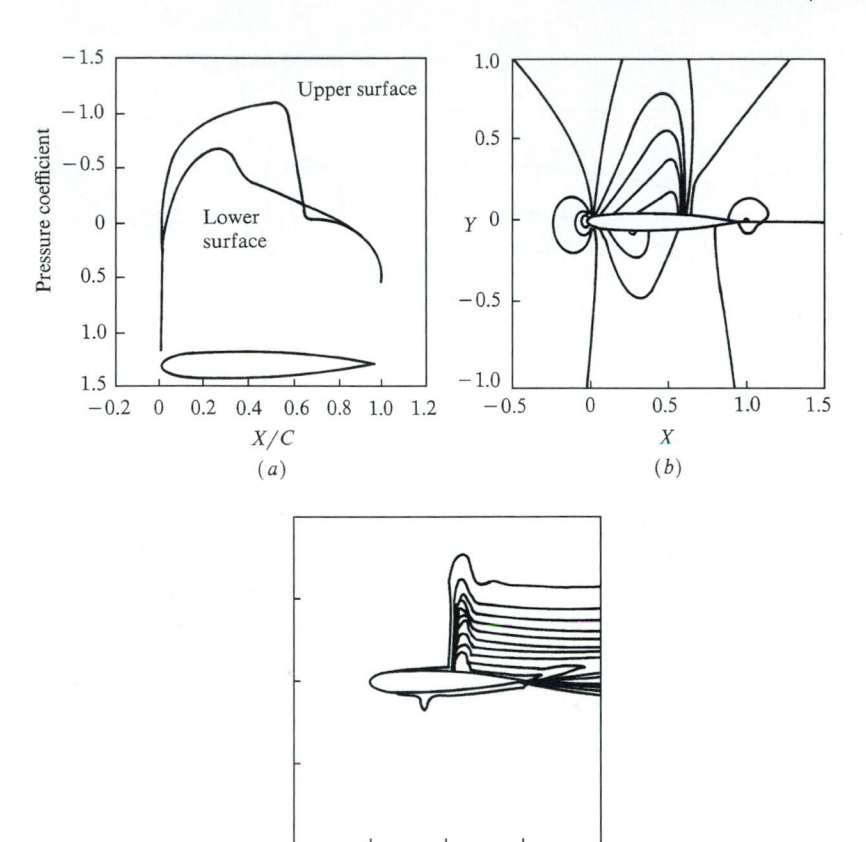

Figure 14.17 | Transonic flow over an NACA 0012 airfoil with a subsonic free-stream Mach number of 0.8 and an angle of attack of 1.25°, from the calculations of Reddy and Jacobs (Ref. 117). (*a*) Pressure coefficient distributions. (*b*) Mach number contours. (*c*) Stagnation pressure contours.

same information for the case of a supersonic M_∞; in particular, for Fig. 14.18, $M_\infty = 1.2$ and $\alpha = 7.0°$. Comparing Figs. 14.17 and 14.18, note the dramatic differences between subsonic and supersonic values of M_∞. For the supersonic case, Fig. 14.18*a* shows a constantly decreasing pressure along both the top and bottom surfaces from the leading to the trailing edge. The Mach number contours in Fig. 14.18*b* fan out in an almost "Mach wave" pattern away from the body, in comparison to the closed loops seen in Fig. 14.17*b*. The total pressure contours in Fig. 14.18*c* clearly show an oblique shock wave at the trailing edge; the nearly normal bow shock upstream of the nose occurs far ahead of the nose, and is off to the left side of the graph.

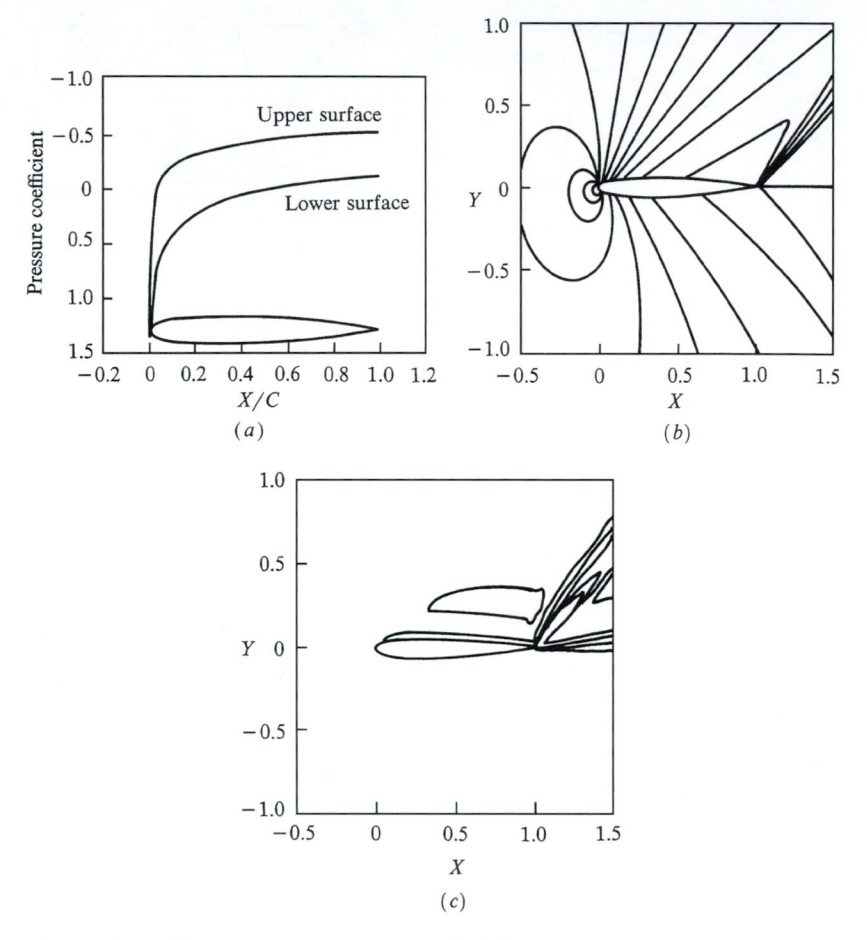

Figure 14.18 | Transonic flow over an NACA 0012 airfoil with a supersonic free-stream Mach number of 1.2 and an angle of attack of $7.0°$, from the calculations of Reddy and Jacobs (Ref. 117). (*a*) Pressure coefficient distributions. (*b*) Mach number contours. (*c*) Stagnation pressure contours.

14.7 | HISTORICAL NOTE: TRANSONIC FLIGHT— ITS EVOLUTION, CHALLENGES, FAILURES, AND SUCCESSES

Return to Fig. 1.9 for a moment and examine the picture of the Bell XS-1 in flight, circa late 1947. This is a photograph of aeronautical engineering poetry in motion— an aircraft that stretched the contemporary aerodynamic state of the art to the limit and whose design represented a voyage into previously uncharted regions of transonic flow. When Chuck Yeager nudged the XS-1 to a Mach number of 1.06 on October 14, 1947, the Bell XS-1 became the first manned aircraft to fly faster than sound in level flight. As noted in Sec. 1.1, this flight was one of the high-water marks in the engineering application of compressible flow. The success of the XS-1 was the culmination of a number of aerodynamic projects over the preceding 30 years— projects undertaken to lay bare the secrets of flows at or very near Mach 1, i.e., transonic flows. Let us reach back over these years (and in some cases, much earlier) and examine the pioneering work that was ultimately highlighted by the XS-1 in Fig. 1.9.

The major obstacle to transonic and supersonic flight is the large drag rise that occurs when the free-stream Mach number exceeds the drag-divergence Mach number (recall the trend shown so dramatically in Fig. 14.4). The variation of drag with flow velocity has always been of great interest as far back as the fifteenth century, when Leonardo da Vinci guessed incorrectly that flow resistance was proportional to the first power of velocity. This same tenant was held by Galileo a century later. However, two experimentalists, Edme Mariotte and Christiaan Huygens, both members of the Paris Academy of Sciences, within the space of 20 years of each other determined the *velocity-squared* law, which today we take almost for granted. Specifically, in 1673, Mariotte gave a paper at the Academy, where he described a series of tests involving water impinging on one end of a beam supported at the middle. By adjusting weights on the other end of the beam, Mariotte found that the force was proportional to V^2. In 1690, Huygens published a paper that made the same claim, but based on an entirely different set of experiments involving falling bodies through air and other media. Of course, today we know that the drag coefficient C_D is relatively constant with velocity (Mach number) for a body moving at subsonic speeds and hence the drag D varies as V^2 through the familiar relation $D = \frac{1}{2}\rho_\infty V_\infty^2 S C_D$. However, in the late seventeenth century, the independent results of Mariotte, and then of Huygens, represented a tremendous advancement in aerodynamics. On the other hand, neither of these gentlemen had the remotest idea of what happens when the speed of sound is approached. Indeed, we might be inclined to think that knowledge of the transonic drag rise is a twentieth century event—but not so! The transonic drag rise was first noted in the early eighteenth century by the well-known English mathematician and ballistician, Benjamin Robins. Robins invented the ballistic pendulum, and by firing high-speed projectiles into the pendulum, he noted that the drag of a projectile was a function of V^2 for most cases. However, at high speeds the drag exhibited a stronger velocity variation, more nearly proportional to V^3. Moreover, in his paper entitled "Resistance of the Air and Experiments Relating to Air Resistance" in the *Philosophical Transactions, London,* dated 1746, he states

> that the velocity at which the moving body shifts resistance is nearly the same with which sound is propagated through the air

Clearly, Benjamin Robins was the first person to appreciate the existence of the transonic drag rise near Mach 1, and this was 30 years before the Declaration of Independence by the colonies in America. Gun-fired projectiles were routinely reaching the speed of sound and faster, by that time. Hence, as early as 1746, investigators in the field of ballistics knew that an unusually large increase in drag occurred near the speed of sound; they simply did not understand why. The first quantitative graph showing the actual variation of drag coefficient versus velocity for a projectile, with velocities ranging from 300 to 1000 m/s at sea level, appeared in Germany in 1910.

In the journal *Artillerische Monatshefte,* Hauptman Bensberg and C. Cranz published a graph that clearly showed a constant C_D below 300 m/s, a large increase in C_D in the region between 300 and 400 m/s, and then a gradual decrease in C_D as the velocity increases above 400 m/s. Since the speed of sound at standard sea level is 341 m/s, we know the large peak in C_D observed by Bensberg and Cranz in the velocity range 300 to 400 m/s is the now familiar transonic drag rise. This graph by Bensberg and Cranz is the first of its kind in history, the first to *quantify* the drag rise near Mach 1, and the first to show that C_D actually *decreases* with increasing speed above Mach 1. In short, long before aerodynamicists were probing the transonic region, ballisticians knew what was happening. This provides some poetic justice to the fact that the fuselage of the Bell XS-1 (see again Fig. 1.9) is exactly the shape of a 50-caliber machine gun bullet.

Transonic aerodynamics in the twentieth century evolved through three distinct phases: (1) the knowledge that something different was happening at or near Mach 1, (2) a physical understanding of *why* these differences occurred, and (3) the ability to measure and compute these differences. Let us examine these phases in more detail.

14.7.1   Something Different

We have already seen that, for 150 years before the twentieth century, ballisticians knew that the drag on a projectile rapidly increased when its velocity approaches the speed of sound. However, in the world of airplane aerodynamics, this was of little concern in the early days of flight. When the Wright brothers successfully flew for the first time in 1903, their flight speed was only 35 mph; the compressibility problems associated with flight near Mach 1 never entered their minds. However, by the end of World War I, in 1918, compressibility problems forced themselves onto the aerodynamic community in a somewhat unexpected manner. By then, the forward speed of high performance fighters, such as the Spad and Nieuport, had increased sufficiently (to about 120 mph) that, in combination with the relative velocity due to the rotation of the propeller, the *propeller tip speeds* were approaching, and even slightly exceeding, Mach 1. By 1919, British researchers had already observed the loss in thrust and large increase in blade drag for a propeller with tip speeds up to 1180 ft/s—slightly above the speed of sound. To examine this effect further, F. W. Caldwell and E. N. Fales, both engineers at the U.S. Army Engineering Division at McCook Field near Dayton, Ohio (the forerunner of the massive Air Force research and development facilities at Wright-Patterson Air Force Base today), conducted a series of high-speed airfoil tests. They designed and built the first high-speed wind tunnel in the United States—a facility with a 14-in.-diameter test section capable of velocities up to 675 ft/s. In 1918, they conducted the first wind-tunnel test involving the high-speed flow over a stationary airfoil. Their results showed large decreases in lift coefficient and major increases in drag coefficient for the thicker airfoils at angle of attack. These were the first measured "compressibility effects" on an airfoil in history. Caldwell and Fales noted that such changes occurred at a certain air velocity, which they denoted as the *critical speed*—a term that was to evolve into the critical Mach number at a later date. Because of the importance of these adverse effects on the overall propeller performance, additional investigations were carried out at the National Bureau of Standards (NBS) in the early and mid-1920s by Lyman J. Briggs and Hugh Dryden. After designing and building a high-speed wind tunnel with a 12-in. diameter test section, capable of producing Mach 0.95 at the nozzle exit, these researchers observed the same phenomena as Caldwell and Fales. In fact, in their report on these experiments, entitled "Aerodynamic Characteristics of Airfoils at High Speeds" (NACA Report No. 207; published in 1925), Briggs and Dryden observed:

> We may suppose that the speed of sound represents an upper limit beyond which an additional loss of energy takes place. If at any point along the wing the velocity of sound is reached the drag will increase. From our knowledge of the flow around airfoils at ordinary speeds we know that the velocity near the surface is much higher than the general stream velocity . . . the increase being greater for the larger angles and thicker sections. This corresponds very well with the earlier flow breakdown for the thicker wings and all of the wings at high angles of attack.

Hence, by 1925 there was plenty of evidence that an airfoil section encounters some marked deleterious phenomena near Mach 1. Moreover, from the preceding quote by Briggs and Dryden, it was well recognized that thicker airfoils encountered such phenomena at lower free-stream Mach numbers. Even as early as 1922, Sylvanus A. Reed of the NACA published results showing that a propeller with a thin airfoil section at the tip did not encounter the same loss in performance as an equivalent propeller with a thick section at the top. Clearly, by 1925, the superiority of thin airfoil sections at near sonic speeds was appreciated; the only aspect that was lacking was the total understanding as to *why*. Indeed, as reflected in Briggs and Dryden's report, there was no physical understanding of the true mechanism prevailing in the high speed flow over an airfoil. To state as they did that "an additional loss of energy takes place" when the local flow velocity becomes sonic is simply begging the point.

14.7.2   A Physical Understanding

The work of Briggs and Dryden, although carried out by the National Bureau of Standards, was actually sponsored by a grant from the National Advisory Committee for Aeronautics (NACA). In the 1920s, the NACA mounted a program to explain the "why" of transonic flow over airfoils. An initial part of this program was the continued contractual support of Briggs and Dryden, who proceeded to build a new, small high-speed wind tunnel with a 2-in. diameter jet. Located at Edgewood Arsenal in Maryland, just north of Baltimore, this tunnel had a mildly converging-diverging nozzle, which produced Mach 1.08 at the exit. Using the same airfoils as in their earlier work, Briggs and Dryden examined the detailed pressure distributions over the airfoil surface. These tests were the first experiments in a supersonic flow carried out in the United States. Moreover, in NACA Report No. 255, published in 1927, Briggs and Dryden give the first inklings of the physical understanding of transonic airfoil flows. For example, they:

1. Deduced that the flow separated from the upper surface. However, they did not realize (as we do today) that the flow separation is induced by the presence of a shock wave interacting with the boundary layer on the upper surface.

2. Noted that the drag coefficient for the airfoil followed the same type of drag-divergence phenomena encountered by projectiles between about Mach 0.95 and 1.08.

3. Observed for the first time in history that the flow at Mach 1.08 involved a bow shock wave standing in front of the leading edge.

As the speeds of airplanes continued to increase through the 1920s, the loss of propeller performance when the tip speeds exceeded the speed of sound became a more serious problem. Spurred by this situation, the NACA initiated an in-house program to explore the "why" of transonic flow—a program that was to continue uninterrupted for 25 years, and which was to become one of the NACA's crowning accomplishments. A series of high-speed wind tunnels was constructed at the NACA Langley Memorial Laboratory, beginning with a rudimentary facility with a 12-in. diameter nozzle exit. With Eastman Jacobs as the tunnel director and John Stack (newly arrived after just graduating from MIT) as the chief researcher, a series of tests were run on various standard airfoil shapes. Frustrated by their continual lack of understanding about the flowfield, they turned to optical techniques, i.e., they assembled a crude schlieren system. Their first tests using the schlieren system dealt with flow over a cylinder. Recall from our earlier discussion that the critical Mach number for a cylinder is about 0.4. Hence, their results were spectacular. Shock waves were seen, along with the resulting flow separation. Visitors flocked to the wind tunnel to observe the results, including Theodore Theodorsen, one of the ranking NACA theoretical aerodynamicists of that period. An indicator of the psychology at that time is given by Theodorsen's comment that since the freestream flow was subsonic, what appeared as shock waves in the schlieren pictures must be an "optical illusion." However, Eastman Jacobs and John Stack knew differently. They proceeded with a major series of airfoil testing, using standard NACA sections. Their schlieren pictures, along with detailed pressure measurements, revealed the secrets of flow over the airfoils at Mach numbers above the critical Mach number. Quickly, a second high-speed tunnel was built at Langley, this one with a 24-in. diameter nozzle exit. The transonic airfoil work continued at a rapid pace. In 1935, Jacobs traveled to Italy, where he presented results of the NACA high-speed airfoil research at the fifth Volta Conference (see Sec. 9.9). This is the first time in history that photographs of the transonic flow field over standard-shaped airfoils were presented in a large public forum. One of these original photographs is shown and discussed in Ref. 134, which should be consulted for more details. These photographs were much like those shown in Fig. 14.3 (which are more recent in origin, dating from 1949).

During the course of such work in the 1930s, the incentive for high-speed aerodynamic research shifted from propeller applications to concern about the airframe of the airplane itself. By the mid-1930s, the possibility of the 550 mi/h airplane was more than a dream—reciprocating engines were becoming powerful enough to consider such a speed regime for propeller-driven aircraft. In turn, the entire airplane itself (wings, cowling, tail, etc.) would encounter compressibility effects. This led to the construction of a large 8-ft high-speed tunnel at Langley, capable of test section velocities above 500 mi/h. This tunnel, along with the two earlier tunnels, established the NACA's dominance in high-speed subsonic research in the late 1930s. In the process, by 1940, the high-speed flow over airfoils was relatively well understood, certainly on a firm qualitative basis, and for free-stream Mach numbers on the subsonic side of transonics, say for M_∞ less than about 0.95, on a firm quantitative basis as obtained experimentally in the Langley wind tunnels. Although experimental transonic airfoil research continues today, not only with NASA (the successor of the NACA) but also at many locations throughout the world, the basic physical understanding of such flows was essentially in hand by the early 1940s due to the pioneering work of Eastman Jacobs, John Stack, and their colleagues at the NACA Langley Memorial Laboratory. For more historical details, see Ref. 134.

14.7.3 Measuring and Computing

The measurement of transonic flows below $M_\infty = 0.95$ and above $M_\infty = 1.1$ was carried out with reasonable accuracy in the early NACA high-speed wind tunnels. However, the data obtained between Mach 0.95 and 1.1 were of questionable accuracy; for these Mach numbers very near unity, the flow was quite sensitive and if a model of any reasonable cross-sectional area were placed in the tunnel, the flow became choked. This choking phenomenon was one of the most difficult aspects of high-speed tunnel research. Small models had to be used; for example, Fig. 14.19

Figure 14.19 | Wind tunnel model of the Bell XS-1 in the Langley 8-ft tunnel, circa 1947. (From Ref. 99.)

shows a small model of the Bell XS-1 mounted in the Langley 8-ft high-speed tunnel in 1947—one year before Yeager's history-making flight. The wing span was slightly over 1 ft whereas the test-section diameter was much larger, namely, 8 ft. In spite of this small model size, valid data could not be obtained at free-stream Mach numbers above 0.92 due to choking of the tunnel at higher Mach numbers.

The Mach number gap between 0.95 and 1.1, in which valid data could not be obtained in the existing high-speed wind tunnels in the late 1940s, contributed much to the aerodynamic uncertainties that dominated the Bell XS-1 program, up to its first supersonic flight on October 14, 1947. Moreover, the advancement of basic aerodynamics in the transonic range was greatly hindered by this situation. Throughout the late 1930s and 1940s, NACA engineers attempted to rectify this choking problem in their high-speed tunnels. Various test section designs were tried—closed test sections, totally open test sections, a bump on the test section wall to tailor the flow constrictions, as well as various methods of supporting models in the test section to minimize blockage. None of these ideas solved the problem. Thus, the stage was set for a technical breakthrough, which came in the late 1940s—the slotted-throat transonic tunnel, as described below.

In 1946, Ray H. Wright, a theoretician at NACA Langley, carried out an analysis that indicated that if the test section contained a series of long, thin rectangular slots parallel to the flow direction that resulted in about 12 percent of the test section periphery being open, then the blockage problem might be greatly alleviated. This idea met with some skepticism, but it was almost immediately accepted by John Stack, who by that time was a highly placed administrator at Langley. A decision was made to slot the test section of the small 12-in. high-speed tunnel, which resulted in greatly improved performance in early 1947. However, this was simply an experiment, and much skepticism still prevailed. On the surface the NACA made no plans to implement this development. On the other hand, Stack confided privately to his colleagues that he favored slotting the large 16-ft high-speed tunnel. Without fanfare, this work began in the spring of 1948, buried in a larger project to increase the horsepower of the tunnel. Almost simultaneously, Stack made the decision to slot the 8-ft tunnel as well. The work on the 8-ft tunnel proceeded faster than on its larger counterpart, and on October 6, 1950, it became operational for research. By December of that same year, the modified 16-ft tunnel also became operational. Subsequent operation of these facilities proved that the slotted-throat concept allowed the smooth transition of the tunnel flow through Mach 1 simply by the increase of the tunnel power—the problem of blockage was basically solved. In this respect, these tunnels became the first truly *transonic* wind tunnels, and for this accomplishment, John Stack and his colleagues at NACA Langley were awarded the prestigious Collier Trophy in 1951. The measurement of transonic flows in the laboratory was now well in hand.

The same could not be said at that time for the computation of transonic flows. As emphasized earlier in this chapter, transonic flow is nonlinear flow, and the analysis of such flows was, therefore, exceptionally difficult in the period before the development of the high-speed digital computer. In 1951, as Stack and the Langley engineers were being awarded the Collier Trophy, there was virtually no useful aerodynamic method for the calculation of transonic flows. Transonic similarity was known and understood (see Sec. 14.3) at that time, but similarity concepts are useful only for relating one solution or set of measurements to another situation; it is not a solution of the flow per se. Also known at that time was the approximate means of estimating the critical Mach number of an airfoil using the Prandtl–Glauert rule, or any other compressibility correction, as was described in Sec. 9.7. Indeed, the method described in Sec. 9.7 was first developed by Eastman Jacobs and John Stack in the late 1930s. Clearly, in 1950 the practical analysis of transonic flow fields themselves was lagging greatly behind the experimental progress. This situation prevailed until the advent of modern computational fluid dynamics and, in particular, the pioneering method advanced by Murman and Cole (see Sec. 14.4). In this sense, the work described in Sec. 14.4 and the subsequent sections speaks for itself as an historical chronology of modern transonic flow analysis. Today, with a few exceptions, we can finally make a statement analogous to that given above about the experimental status in 1950, namely, that by the 1980s, the calculation of transonic flow is now well in hand.

14.7.4   The Transonic Area Rule and the Supercritical Airfoil

We would be remiss in this discussion of the historical aspects of transonic flight if we did not mention two major configuration breakthroughs that have made transonic flight practical—the area rule and the supercritical airfoil. Both of these advancements were a product of the transonic wind tunnels at Langley and both were driven by the same person—Richard Whitcomb. Let us examine these two matters more closely.

First, on a technical basis, the area rule and the supercritical airfoil both have the same objective, namely, to reduce drag in the transonic regime. However, this drag reduction is accomplished in different ways. Consider the qualitative sketch of drag coefficient versus Mach number given in Fig. 14.20 for a transonic body. The variation for a standard body shape without area rule and without a supercritical airfoil is given by the solid curve. Now, let us consider the area rule by itself. First, the area rule is a simple statement that the cross-sectional area of the body should have a smooth variation with longitudinal distance along the body; there should be no rapid or discontinuous changes in the cross-sectional area distribution. For example, a conventional wing-body combination will have a sudden cross-sectional area increase in the region where the wing cross section is added to the body cross section. The area rule says that to compensate, the body cross section should be decreased in the vicinity of the wing, producing a wasp-like or coke-bottle shape for the body. The aerodynamic advantage of the area rule is shown in Fig. 14.20, where the drag variation of the area-ruled body is given by the dashed curve. Simply stated, the area rule reduces the peak transonic drag by a considerable amount. The supercritical airfoil, on the other hand, acts in a different fashion. A supercritical airfoil is shaped somewhat flat on the top surface in order to reduce the local Mach number inside the supersonic region below what it would be for a conventional airfoil under the same flight conditions. As a result, the shock wave strength is lower, the boundary

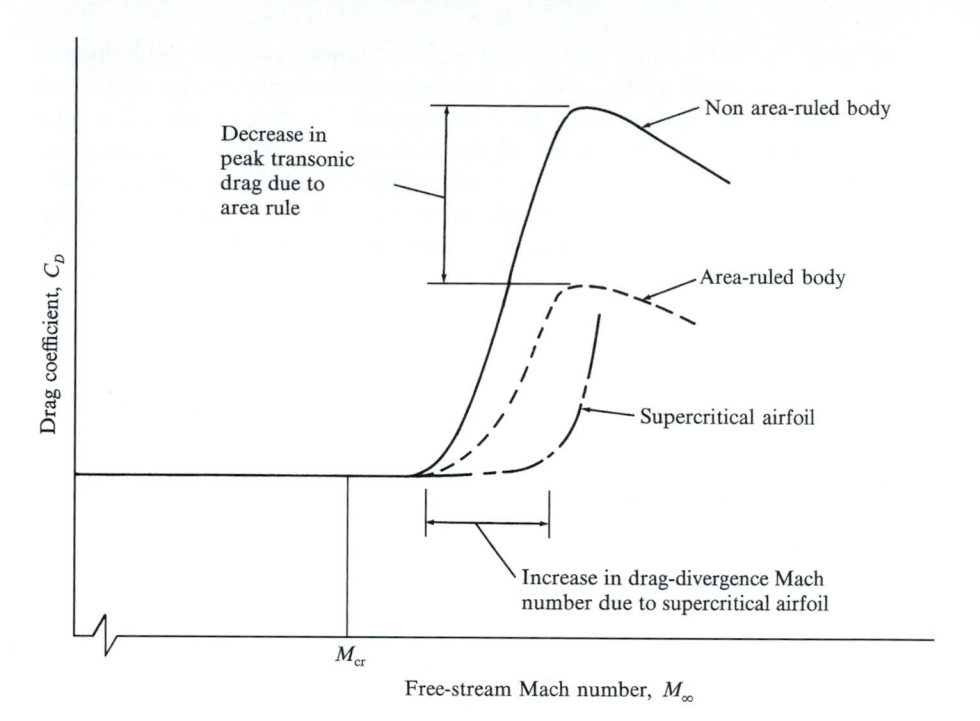

Figure 14.20 | Illustration of the separate effects of the area rule and the supercritical airfoil.

layer separation is less severe, and hence the free-stream Mach number can be higher before the drag-divergence phenomenon sets in. The drag variation for a supercritical airfoil is sketched in Fig. 14.20, shown by the broken curve. Here, the role of a supercritical airfoil is clearly shown; although the supercritical airfoil and an equivalent standard airfoil may have the same critical Mach number, the drag-divergence Mach number for the supercritical airfoil is much larger. That is, the supercritical airfoil can tolerate a much larger increase in the free-stream Mach number above the critical value before drag divergence is encountered. In this fashion, such airfoils are designed to operate far above the critical Mach number—hence the label "supercritical" airfoils.

The area rule was introduced in a most spectacular fashion in the early 1950s. Although there had been some analysis that obliquely hinted about the area rule, and although workers in the field of ballistics had known for years that projectiles with sudden changes in cross-sectional area exhibited high drag at high speeds, the importance of the area rule was not fully appreciated until a series of wind tunnel tests on various transonic bodies were conducted in the slotted-throat 8-ft wind tunnel at Langley by Richard Whitcomb. These data, and an appreciation of the area rule, came just in time to save a new airplane program at Convair. In 1951, Convair was designing one of the new "century series" fighters intended to fly at supersonic speeds. Designated the YF-102, this aircraft had a delta-wing and was powered by

(*a*)   (*b*)

Figure 14.21 | (*a*) The Convair YF-102, no area ruling. (*b*) The Convair YF-102A, with area ruling. Note the wasp-like shape of the fuselage in comparison with the YF-102 shown in (*a*).

the Pratt and Whitney J-57 turbojet—the most powerful engine in the United States at that time. A photograph of the YF-102 is given in Fig. 14.21*a*. Aeronautical engineers at Convair expected the YF-102 to easily fly supersonically. On October 24, 1953, flight tests of the YF-102 began at Muroc Air Force Base (now Edwards), while a production line was forming at the San Diego plant of Convair. However, as the flight tests progressed, it became painfully clear that the YF-102 could not fly faster than sound—the transonic drag rise was simply too large, even for the powerful J-57 engine to overcome. After consultation with the NACA aerodynamicists and inspection of the area rule results that had been obtained in the Langley 8-ft tunnel, the Convair engineers designed a modified airplane—the YF-102A—with an area-ruled fuselage. A photograph of the YF-102A, with its coke bottle-shaped fuselage is given in Fig. 14.21*b*. Wind tunnel data for the YF-102A looked promising. Figure 14.22 was obtained from that data; it shows the variation of drag coefficient with free-stream Mach number for both the YF-102 and YF-102A. In the upper left of Fig. 14.22, the cross-sectional area distribution of the YF-102 is shown, including how it is built up from the different body components. Note the irregular and bumpy nature of the total cross-sectional area distribution. At the bottom right, given by the dashed line, is the cross-sectional area distribution for the YF-102A—a much smoother variation than that for the YF-102. The data shown in Fig. 14.22 are obtained from Reference 100. The comparison between the drag coefficients for the conventional YF-102 (solid curve) and the area-ruled YF-102A (dashed curve) dramatically illustrates the tremendous transonic drag reduction to be obtained with

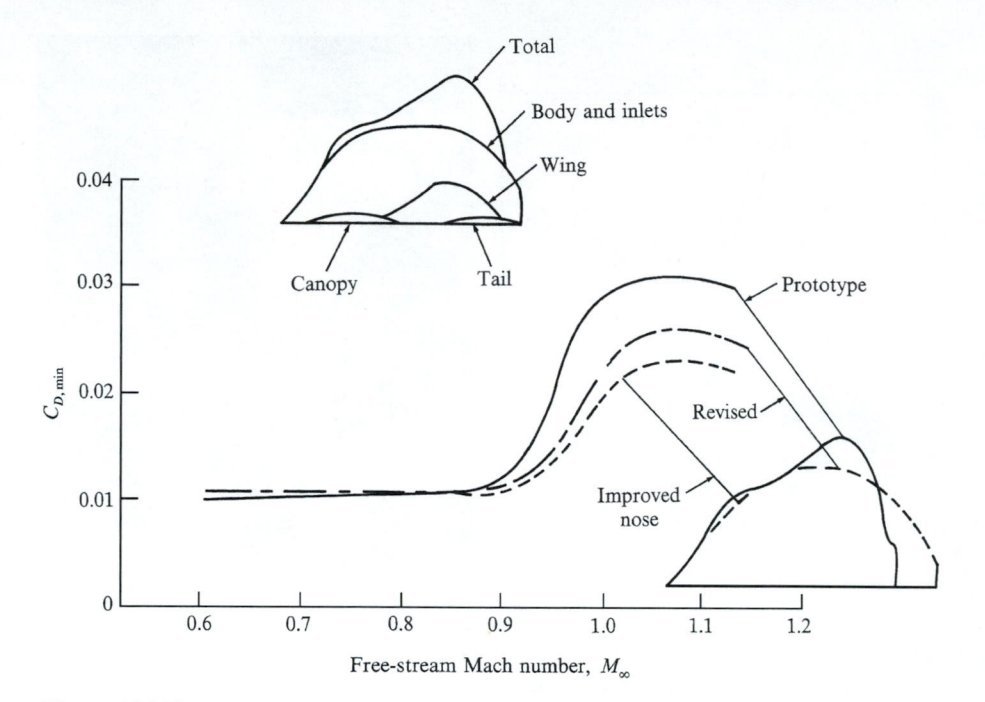

Figure 14.22 | The effect of the area rule modifications made on the original non-area-ruled Convair YF-102 (labeled prototype) and the resulting area-ruled YF-102A (labeled revised and improved nose). (From Ref. 100.)

the use of the area rule. (Recall from Fig. 14.20 that the function of the area rule is to decrease the peak transonic drag; Fig. 14.22 quantifies this function.) Encouraged by these wind tunnel results, the Convair engineers began a flight test program for the YF-102A. On December 20, 1954, the prototype YF-102A left the ground at Lindbergh field, San Diego—it broke the speed of sound while still climbing. The use of the area rule had increased the top speed of the airplane by 25 percent. The production line rolled, and 870 F-102As were built for the Air Force. The area rule had been ushered in with dramatic style.

The supercritical airfoil, also pioneered by Richard Whitcomb, based on data obtained in the 8-ft wind tunnel, was a development of the 1960s. Recall from Fig. 14.20 that the function of the supercritical airfoil is to increase the increment between the critical Mach number and the drag-divergence Mach number. The data in the Langley tunnel indicated a possible 10 percent increase in cruise Mach number due to a supercritical wing. NASA introduced the technical community to the supercritical airfoil data in a special conference in 1972. Since that time, the supercritical airfoil concept has been employed on virtually all new commercial aircraft and some military airplanes. Physical data for a supercritical airfoil and for the standard NACA 64-A215 airfoil are compared in Figs. 14.23 and 14.24, along with a comparison

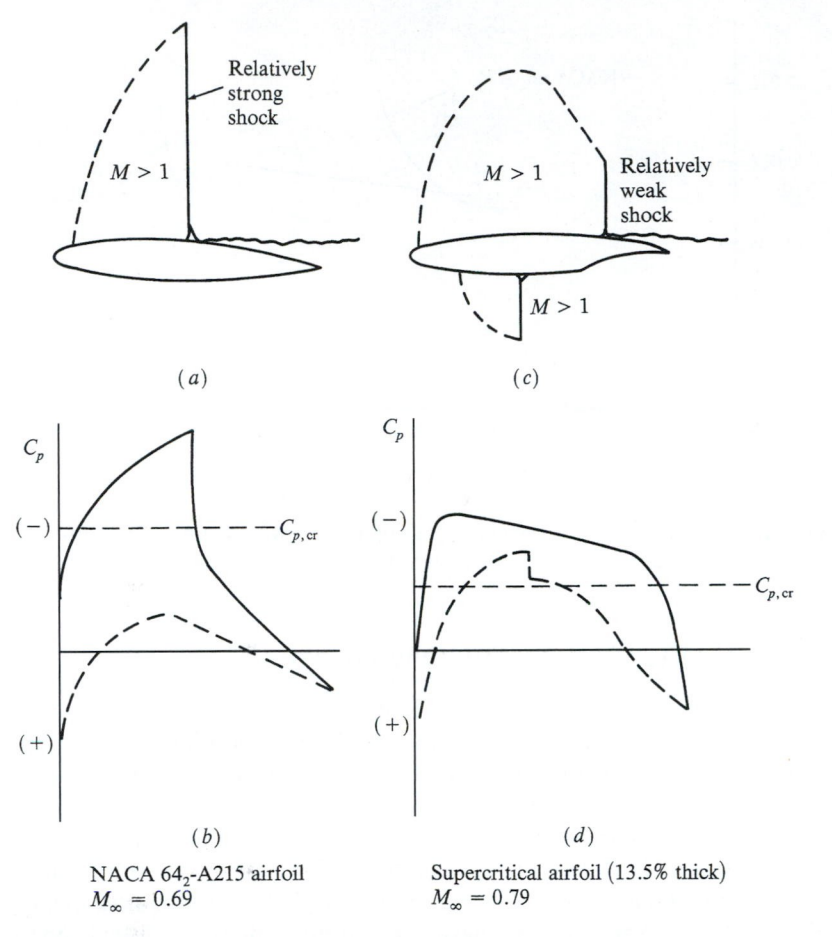

Figure 14.23 | Standard NACA 64-series airfoil compared with a supercritical airfoil at cruise lift conditions. (From R. T. Whitcomb and L. R. Clark, "An Airfoil Shape for Efficient Flight At Supercritical Mach Numbers," NASA TMX-1109, July 1965.)

of their shapes. The performances advantage of the supercritical airfoil is clearly evident.

With this, we end this rather lengthy historical note on transonic flight. Our purpose has been to provide just the flavor of what constitutes one of the most exciting chapters from the annals of aerodynamics and aeronautical engineering. We have seen how the secrets of transonic flow were slow to be revealed, how a concerted, intelligent attack on this problem eventually led to useful wind tunnel data as well as modern methods of computation for transonic flows, and finally how this transonic data ultimately resulted in two of the major aerodynamic breakthroughs in the latter half of the twentieth century—the area rule and the supercritical airfoil.

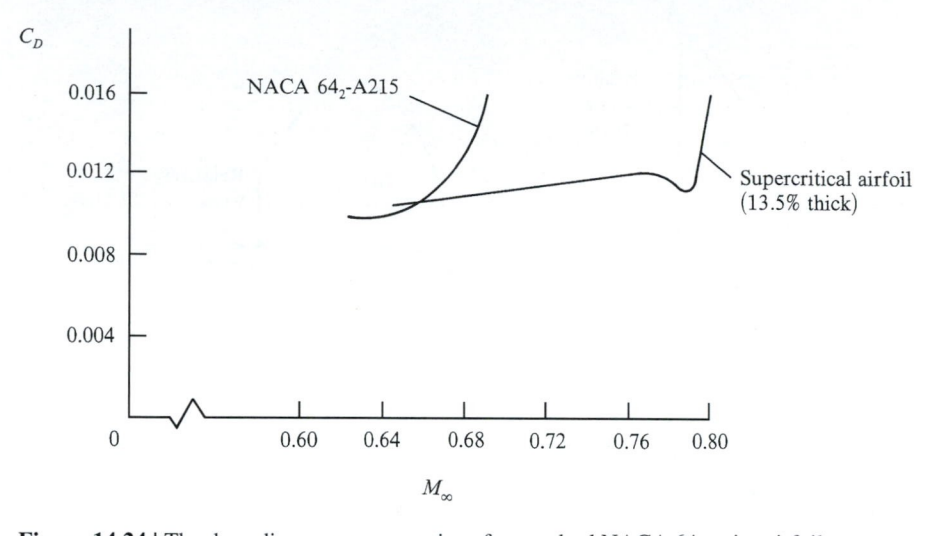

Figure 14.24 | The drag-divergence properties of a standard NACA 64-series airfoil and a supercritical airfoil. (From NASA TMX-1109, as in Fig. 14.23.)

14.8 | SUMMARY AND COMMENTS

In this chapter, we have covered some of the essential physical and theoretical aspects of transonic flow. If this chapter had been written 30 years ago, it would have been completely different. First, it would have been much shorter, and it would have emphasized only a few specialized theories. One such theory is called the hodograph method, and uses the transonic small-perturbation equation in the hodograph plane, for which some shock-free exact solutions can be obtained. Such solutions are discussed, for example, in Shapiro (see Ref. 16). In the more modern treatment of transonic flows given here, we have intentionally not covered such hodograph techniques. Instead, we have concentrated on the main echelons of transonic flow numerical solutions, namely,

1. Small-perturbation solutions
2. Full potential solutions
3. Euler solutions

These solutions are listed in order of increasing accuracy, and as life would have it, also of increasing difficulty and effort. The small-perturbation solutions assume irrotational flow, and slender bodies at small angles of attack. The full potential solutions also assume irrotational flow, but pertain to any body of arbitrary thickness and angle of attack. In both cases, the assumption of irrotational flow is motivated by the change in entropy across a weak shock, which is of third order in shock strength and hence is small. The Euler solutions make no such assumptions, and hence represent "exact" solutions of inviscid transonic flow.

Modern, state-of-the-art research in transonic flow is now concentrating on numerical solutions of the complete Navier-Stokes equations in order to properly include the viscous effects, particularly those effects associated with the shock wave/boundary layer interaction region. Since the present book deals with inviscid flow only, such matters are beyond our scope. However, these viscous effects can play a strong role in transonic flows, and the interested reader is encouraged to read the modern literature on such transonic viscous flows. The *AIAA Journal,* the *Journal of Aircraft, Computers and Fluids,* and the *Journal of Computational Physics* are good sources of such literature.

CHAPTER 15

Hypersonic Flow

Almost everyone has their own definition of the term hypersonic. If we were to conduct something like a public opinion poll among those present, and asked everyone to name a Mach number above which the flow of a gas should properly be described as hypersonic there would be a majority of answers round about five or six, but it would be quite possible for someone to advocate, and defend, numbers as small as three, or as high as 12.

P. L. Roe, comment made in a lecture at the von Karman Institute, Belgium, January 1970

PREVIEW BOX

The Space Shuttle is shown in Fig. 12.2. Take another look at this figure—it shows one of the most important hypersonic flight vehicles to date. After the end of each mission in space, the shuttle enters the earth's atmosphere at Mach 26, and flies most of its flight path at hypersonic speeds until it returns to the earth's surface. The design of the Space Shuttle benefited from a massive hypersonic research program in the 1950s and 1960s, driven by the necessities of the intercontinental ballistic missile and the manned space program. Now take a look at Fig. 15.1, which shows the X-43 unmanned hypersonic research vehicle developed by NASA, and expected to make its first flight in late 2002 or early 2003. The X-43 is representative of a new class of hypersonic vehicles designed to cruise within the atmosphere using airbreathing propulsion as the prime mover. The engine itself is pioneering—a supersonic combustion ramjet engine (scramjet). Such engines are at the very frontier of propulsion development. One of the research goals of the X-43 is to demonstrate for the first time in actual flight the performance of a scramjet.

The blunt-nosed space shuttle in Fig. 12.2, and the rather slender X-43 hypersonic cruise vehicle in Fig. 15.1, along with its novel hypersonic propulsion, scream *hypersonic aerodynamics.* Hypersonic flow is the subject of this chapter. The material presented here is a short introduction to a flow regime that will take on increased importance in the twenty-first century. I predict that many of the readers of this book will have the opportunity to help push forward the frontier of hypersonic flight, in regard to both the aerodynamic design of new vehicles and the pioneering application of compressible flow to the development of exotic and novel engines to power these vehicles. So pay special attention to this chapter—it describes some basic aspects of hypersonic flow that for some of you will be the wave of the future.

The roadmap for this chapter is given in Fig. 15.2. We start out with a basic physical description of hypersonic flow, addressing the question: hypersonic flow—what is it? Then we treat four specific theoretical aspects of hypersonic flow, starting at the left in Fig. 15.2 with

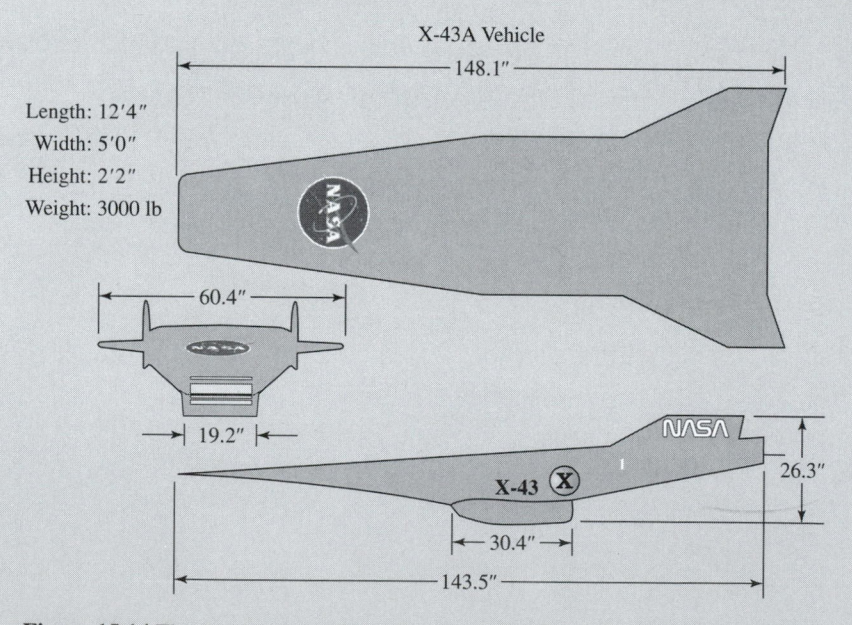

Figure 15.1 | The X-43A Hyper-X hypersonic research vehicle (NASA).

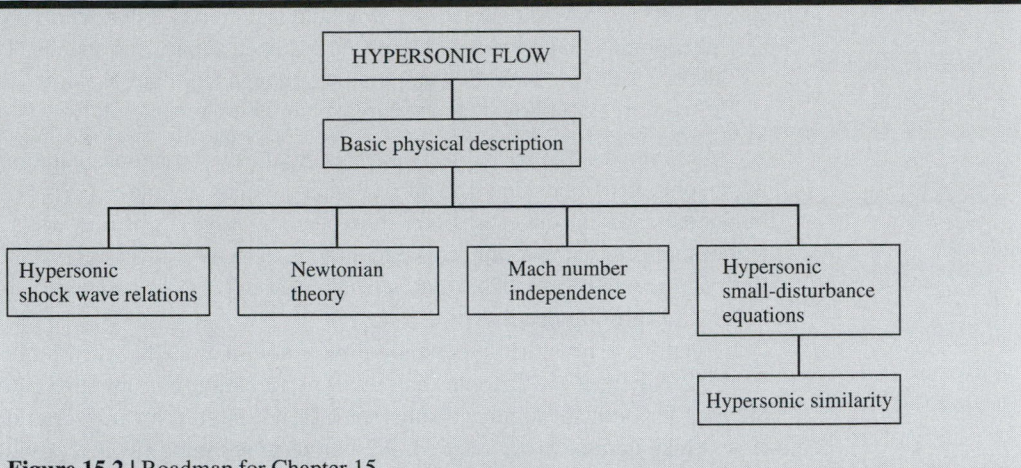

Figure 15.2 | Roadmap for Chapter 15.

a discussion of the simplification of the shock wave relations afforded by the assumption of high Mach numbers. We then discuss Newtonian theory, a special approach to quickly estimate pressure distributions on hypersonic shapes. This is followed by a demonstration that pressure coefficients, lift and drag coefficients, and shock wave shapes in hypersonic flow do not change very much with increasing Mach number—a phenomenon called Mach number independence. Finally we develop the hypersonic small-disturbance equations, which in turn lead to the principle of hypersonic similarity. Although not explicitly shown in Fig. 15.2, we briefly address the matter of CFD solutions to hypersonic flows at the end of the chapter. All aspects treated in this chapter assume a calorically perfect gas (constant specific heats). High-temperature effects that are so important to hypersonic flow, and that dramatically change the thermodynamics, are discussed as an integral part of Chaps. 16 and 17, dealing with high-temperature gas dynamics.

Finally, refer to the roadmap for the book given in Fig. 1.7. With the present chapter we reach the end of the center column of the roadmap.

15.1 | INTRODUCTION

When the space shuttle enters the earth's atmosphere from near-earth orbit, it is flying at Mach 25. When the Apollo spacecraft returned from the moon, it entered the atmosphere at Mach 36. These very high Mach numbers are associated with the extreme, high-Mach-number portion of the flight spectrum which is labeled as *hypersonic* flight. The hypersonic flow regime was briefly described in Sec. 1.3; this short discussion should be reviewed before progressing further.

There are two reasons for singling out hypersonic flow for a separate chapter in this book, as follows.

1. Hypersonic flight is of extreme interest today because of new vehicle concepts designed to fly at very high Mach numbers. Hypersonic aerodynamics is an important part of the entire flight spectrum, and therefore it is an integral part of any study of modern compressible flow.

2. At very high Mach numbers, a flowfield is dominated by certain physical phenomena that are not so important at lower, supersonic speeds. These special aspects of hypersonic flow are distinct enough from our previous discussions of compressible flow that a separate chapter on hypersonic flow is necessary.

As in the case of the subjects covered by the two previous chapters, the topic of hypersonic flow considered in this chapter justifies an entire book by itself. Such books exist; an introductory book in hypersonic flow is given by Ref. 119, and the reader interested in this subject is encouraged to study Ref. 119 closely. Our scope in this chapter will be much like that of Chaps. 13 and 14—long on philosophy and concepts, and short on details.

Finally, we note that hypersonic flow is *nonlinear*. This was first brought out in Sec. 9.2, where it was shown that small-perturbation considerations lead to linear theories for both subsonic and supersonic flows, but not for transonic or hypersonic flow. Make certain to review Sec. 9.2 before progressing further, paying special attention to the effect of hypersonic Mach numbers.

15.2 | HYPERSONIC FLOW—WHAT IS IT?

There is a conventional rule of thumb that defines hypersonic aerodynamics as those flows where the Mach number is greater than 5. However, this is no more than just a rule of thumb; when a flow is accelerated from $M = 4.99$ to $M = 5.01$, there is no "clash of thunder" and the flow does not "instantly turn from green to red." Rather, hypersonic flow is best defined as that regime where certain physical flow phenomena become progressively more important as the Mach number is increased to higher values. In some cases, one or more of these phenomena may become important above Mach 3, whereas in other cases they may not be compelling until Mach 7 or higher. The purpose of this section is to describe briefly these physical phenomena; in some sense this entire section will constitute a "definition" of hypersonic flow.

15.2.1   Thin Shock Layers

Recall from oblique shock theory (see Chap. 4) that, for a given flow deflection angle, the density increase across the shock wave becomes progressively larger as the Mach number is increased. At higher density, the mass flow behind the shock can more easily "squeeze through" smaller areas. For flow over a hypersonic body, this means that the distance between the body and the shock wave can be small. The flowfield between the shock wave and the body is defined as the shock layer, and for hypersonic speeds this shock layer can be quite thin. For example, consider the Mach 36 flow of a calorically perfect gas with a ratio of specific heats, $\gamma = c_p/c_v = 1.4$, over a wedge of 15° half-angle. From standard oblique shock theory the shock wave angle will be only 18° as shown in Fig. 15.3. If high-temperature, chemically reacting effects are included, the shock wave angle will be even smaller. Clearly, this shock layer is thin. It is a basic characteristic of hypersonic flows that

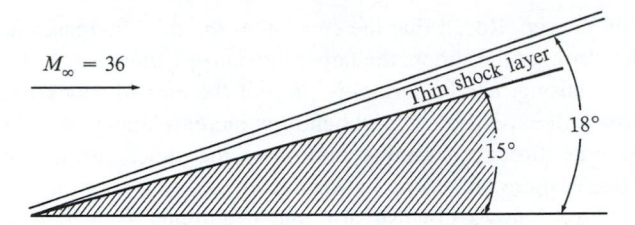

Figure 15.3 | Illustration of a thin shock layer at hypersonic Mach numbers.

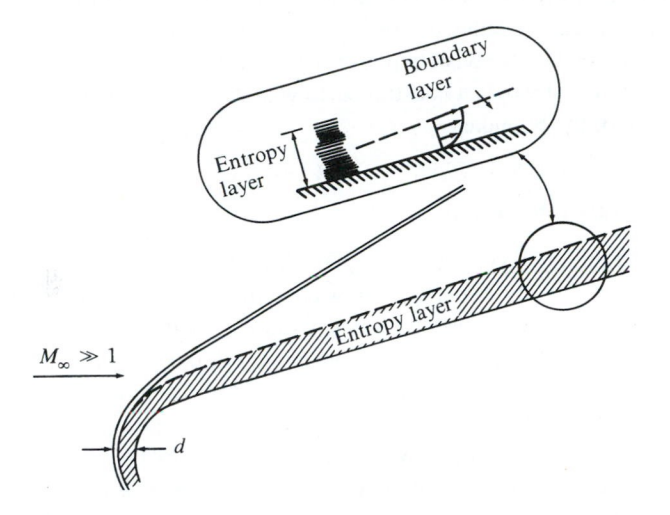

Figure 15.4 | Illustration of the entropy layer of a blunt-nosed slender body at hypersonic speeds.

shock waves lie close to the body, and that the shock layer is thin. In turn, this can create some physical complications, such as the merging of the shock wave itself with a thick, viscous boundary layer growing from the body surface—a problem which becomes important at low Reynolds numbers. However, at high Reynolds numbers, where the shock layer is essentially inviscid, its thinness can be used to theoretical advantage, leading to a general analytical approach called "thin shock layer theory" (see Ref. 119). In the extreme, a thin shock layer approaches the fluid dynamic model postulated by Isaac Newton in 1687; such "newtonian theory" is simple and straightforward, and is frequently used in hypersonic aerodynamics for approximate calculations (to be discussed in Sec. 15.4).

15.2.2 Entropy Layer

Consider the wedge shown in Fig. 15.3, except now with a blunt nose, as sketched in Fig. 15.4. At hypersonic Mach numbers, the shock layer over the blunt nose is also very thin, with a small shock detachment distance d. In the nose region, the shock wave is highly curved. Recall that the entropy of the flow increases across a shock wave, and the stronger the shock, the larger the entropy increase. A streamline passing through the strong, nearly normal portion of the curved shock near the centerline of the flow will experience a larger entropy increase than a neighboring streamline which passes through a weaker portion of the shock further away from the centerline. Hence, there are strong entropy gradients generated in the nose region; this "entropy layer" flows downstream, and essentially wets the body for large distances from the nose, as shown in Fig. 15.4. The boundary layer along the surface grows inside this entropy layer, and is affected by it. Since the entropy layer is also a region of strong vorticity, as related through Crocco's theorem (see Sec. 6.6), this interaction is sometimes called a "vorticity interaction." The entropy layer causes analytical problems when we wish to perform a standard boundary layer calculation on the surface, because there is a question as to what the proper conditions should be at the outer edge of the boundary layer.

15.2.3 Viscous Interaction

Consider a boundary layer on a flat plate in a hypersonic flow, as sketched in Fig. 15.5. A high-velocity, hypersonic flow contains a large amount of kinetic energy; when this flow is slowed by viscous effects within the boundary layer, the lost kinetic energy is transformed (in part) into internal energy of the gas—this is called viscous dissipation. In turn, the temperature increases within the boundary layer; a typical temperature profile within the boundary layer is also sketched in Fig. 15.5. The characteristics of hypersonic boundary layers are dominated by such temperature increases. For example, the viscosity coefficient increases with temperature, and this by itself will make the boundary layer thicker. In addition, because the pressure p is constant in the normal direction through a boundary layer, the increase in temperature T results in a decrease in density ρ through the equation of state $\rho = p/RT$. In order to pass the required mass flow through the boundary layer at reduced density, the boundary layer thickness must be larger. Both of these phenomena combine to

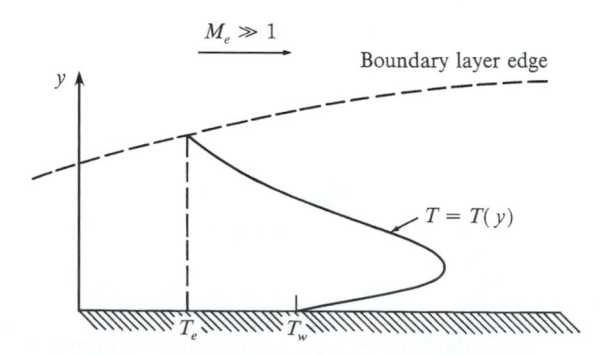

Figure 15.5 | Schematic of a temperature profile in a hypersonic boundary layer.

make hypersonic boundary layers grow more rapidly than at slower speeds. Indeed, the flat plate compressible laminar boundary layer thickness δ grows essentially as

$$\delta \propto \frac{M_\infty^2}{\sqrt{\mathrm{Re}_x}}$$

where M_∞ is the free-stream Mach number, and Re_x is the local Reynolds number. (See Ref. 119 for a derivation of this relation.) Clearly, since δ varies as the square of M_∞, it can become inordinately large at hypersonic speeds.

The thick boundary layer in hypersonic flow can exert a major displacement effect on the inviscid flow outside the boundary layer, causing a given body shape to appear much thicker than it really is. Due to the extreme thickness of the boundary layer flow, the outer inviscid flow is greatly changed; the changes in the inviscid flow in turn feed back to affect the growth of the boundary layer. This major interaction between the boundary layer and the outer inviscid flow is called viscous interaction. Viscous interactions can have important effects on the surface pressure distribution, hence lift, drag, and stability on hypersonic vehicles. Moreover, skin friction and heat transfer are increased by viscous interaction. For example, Fig. 15.6 illustrates the viscous interaction on a sharp, right-circular cone at zero angle of attack. Here, the pressure distribution on the cone surface p is given as a function of distance from the tip. These are experimental results obtained from Ref. 120. If there were no viscous interaction, as discussed in Chap. 10, the inviscid surface pressure would be constant, equal to p_c (indicated by the horizontal dashed line in Fig. 15.6). However, due to the viscous interaction, the pressure near the nose is considerably greater; the surface pressure distribution decays further downstream, ultimately approaching the inviscid value far downstream.

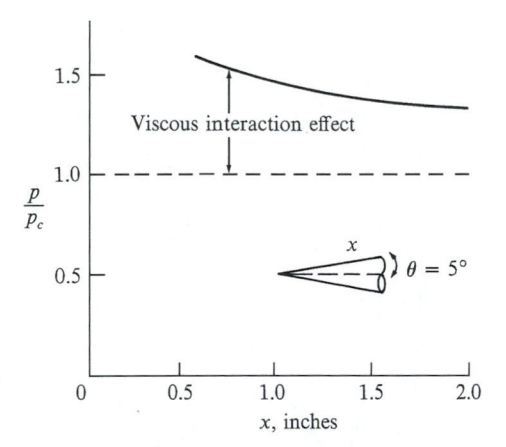

Figure 15.6 | Viscous interaction effect. Induced pressure on a sharp cone at $M_\infty = 11$ and $\mathrm{Re} = 1.88 \times 10^5$ per foot. (From Ref. 120.)

15.2.4 High-Temperature Flows

As discussed previously, the kinetic energy of a high-speed, hypersonic flow is dissipated by the influence of friction within a boundary layer. The extreme viscous dissipation that occurs within hypersonic boundary layers can create very high temperatures—high enough to excite vibrational energy internally within molecules, and to cause dissociation and even ionization within the gas. If the surface of a hypersonic vehicle is protected by an ablative heat shield, the products of ablation are also present in the boundary layer, giving rise to complex hydrocarbon chemical reactions. On both accounts, we see that the surface of a hypersonic vehicle can be wetted by a chemically reacting boundary layer.

The boundary layer is not the only region of high-temperature flow over a hypersonic vehicle. Consider the nose region of a blunt body, as sketched in Fig. 15.7. The bow shock wave is normal, or nearly normal, in the nose region, and the gas temperature behind this strong shock wave can be enormous at hypersonic speeds. The magnitudes of these temperatures, as well as the physical consequences of such temperatures, are discussed at length in Sec. 16.1.

High-temperature chemically reacting flows can have an influence on lift, drag, and moments on a hypersonic vehicle. For example, such effects have been found to be important for estimating the amount of body-flap deflection necessary to trim the space shuttle during high-speed reentry. However, by far the most dominant aspect of high temperatures in hypersonics is the resultant high heat-transfer rates to the surface. Aerodynamic heating dominates the design of all hypersonic machinery, whether it be a flight vehicle, a ramjet engine to power such a vehicle, or a wind tunnel to test the vehicle. This aerodynamic heating takes the form of heat transfer from the hot boundary layer to the cooler surface—called convective heating, and denoted

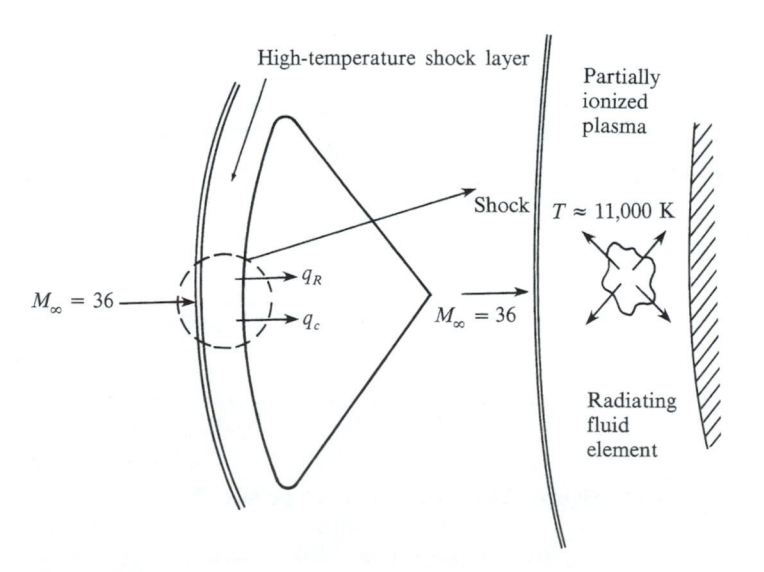

Figure 15.7 | Illustration of a high-temperature shock layer on a blunt body moving at hypersonic speeds.

by q_c in Fig. 15.7. Moreover, if the shock layer temperature is high enough, the thermal radiation emitted by the gas itself can become important, giving rise to a radiative flux to the surface—called radiative heating, and denoted by q_r in Fig. 15.7. (In the winter, when you warm yourself beside a roaring fire in the fireplace, the warmth you feel is not hot air blowing out of the fireplace, but rather radiation from the flame itself. Imagine how "warm" you would feel standing next to the gas behind a strong shock wave at Mach 36, where the temperature is 11,000 K—about twice the surface temperature of the sun.) For example, for Apollo reentry, radiative heat transfer was more than 30 percent of the total heating. For a space probe entering the atmosphere of Jupiter, the radiative heating will be more than 95 percent of the total heating.

Another consequence of high-temperature flow over hypersonic vehicles is the "communications blackout" experienced at certain altitudes and velocities during atmospheric entry, where it is impossible to transmit radio waves either to or from the vehicle. This is caused by ionization in the chemically reacting flow, producing free electrons that absorb radio-frequency radiation. Therefore, the accurate prediction of electron density within the flowfield is important.

Clearly, high-temperature effects can be a dominant aspect of hypersonic aerodynamics. Because of this importance to hypersonic applications, as well as to many other problems dealing with compressible flow, the chemistry and physics of high-temperature gases, and their application to gasdynamic flows, are discussed in Chaps. 16 and 17.

In summary, hypersonic flow is best defined as that regime where all or some of the above physical phenomena become important as the Mach number is increased to high values. Note that viscous effects, such as viscous interactions and aerodynamic heating, are particularly important aspects of hypersonic flow; since we focus on inviscid flows in this book, such matters will not be addressed here. The high-temperature aspects of hypersonic flow are also very important. Chapters 16 and 17 cover the gasdynamics of high-temperature flows—a vital part of modern compressible flow in general, and of hypersonic flow in particular. Therefore, in the present chapter we will deal with inviscid hypersonic flow of a calorically perfect gas. The question we address here is simply: What happens to our conventional compressible flow already discussed in this book when the Mach number becomes very large? For a discussion of the full range of hypersonic flow problems—inviscid, viscous, and high temperature—see the book by Anderson (Ref. 119).

15.3 | HYPERSONIC SHOCK WAVE RELATIONS

The basic oblique shock relations are derived and discussed in Chap. 4. These are *exact* shock relations, and hold for all Mach numbers greater than unity, supersonic or hypersonic (assuming a calorically perfect gas). However, some interesting approximate and simplified forms of these shock relations are obtained in the limit of high Mach number. These limiting forms are called the hypersonic shock relations; they are obtained below.

Consider the flow through a straight oblique shock wave, as sketched in Fig. 15.8. Upstream and downstream conditions are denoted by subscripts 1 and 2, respectively. For a calorically perfect gas, the classical results for changes across the

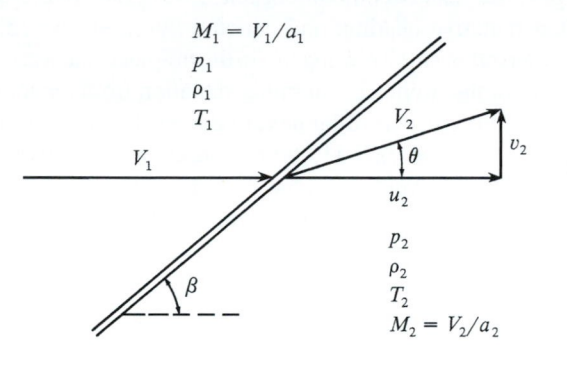

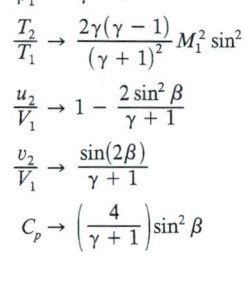

$$\frac{p_2}{p_1} \rightarrow \frac{2\gamma}{\gamma+1} M_1^2 \sin^2 \beta$$

$$\frac{\rho_2}{\rho_1} \rightarrow \frac{\gamma+1}{\gamma-1}$$

$$\frac{T_2}{T_1} \rightarrow \frac{2\gamma(\gamma-1)}{(\gamma+1)^2} M_1^2 \sin^2 \beta$$

$$\frac{u_2}{V_1} \rightarrow 1 - \frac{2\sin^2\beta}{\gamma+1}$$

$$\frac{v_2}{V_1} \rightarrow \frac{\sin(2\beta)}{\gamma+1}$$

$$C_p \rightarrow \left(\frac{4}{\gamma+1}\right)\sin^2\beta$$

In the hypersonic limit and for small θ:

$$\beta \rightarrow \frac{\gamma+1}{2}\theta$$

Figure 15.8 | Oblique shock wave geometry.

shock are given in Chap. 4. To begin with, the exact oblique shock relation for pressure ratio across the wave is given by Eq. (4.9), repeated here:

Exact:
$$\frac{p_2}{p_1} = 1 + \frac{2\gamma}{\gamma+1}\left(M_1^2 \sin^2 \beta - 1\right) \tag{4.9}$$

where β is the wave angle shown in Fig. 15.8. In the limit as M_1 goes to infinity, the term $M_1^2 \sin^2 \beta \gg 1$, and hence Eq. (4.9) becomes

as $M_1 \rightarrow \infty$:
$$\boxed{\frac{p_2}{p_1} = \frac{2\gamma}{\gamma+1}M_1^2 \sin^2 \beta} \tag{15.1}$$

In a similar vein, the density and temperature ratios are given by Eqs. (4.8) and (4.11), respectively:

Exact:
$$\frac{\rho_2}{\rho_1} = \frac{(\gamma+1)M_1^2 \sin^2 \beta}{(\gamma-1)M_1^2 \sin^2 \beta + 2} \tag{4.8}$$

as $M_1 \rightarrow \infty$:
$$\boxed{\frac{\rho_2}{\rho_1} = \frac{\gamma+1}{\gamma-1}} \tag{15.2}$$

$$\frac{T_2}{T_1} = \frac{(p_2/p_1)}{(\rho_2/\rho_1)} \qquad \text{(from the equation of state: } p = \rho RT\text{)}$$

as $M_1 \rightarrow \infty$:
$$\boxed{\frac{T_2}{T_1} = \frac{2\gamma(\gamma-1)}{(\gamma+1)^2}M_1^2 \sin^2 \beta} \tag{15.3}$$

Returning to Fig. 15.8, note that u_2 and v_2 are the components of the flow velocity behind the shock wave parallel and perpendicular to the upstream flow (not parallel and perpendicular to the shock wave itself, as is frequently done, and as was done in Chap. 4). With this in mind, it can be shown that

Exact:
$$\frac{u_2}{V_1} = 1 - \frac{2\left(M_1^2 \sin^2 \beta - 1\right)}{(\gamma + 1)M_1^2} \tag{15.4}$$

as $M_1 \to \infty$:
$$\boxed{\frac{u_2}{V_1} = 1 - \frac{2 \sin^2 \beta}{\gamma + 1}} \tag{15.5}$$

Exact:
$$\frac{v_2}{V_1} = \frac{2\left(M_1^2 \sin^2 \beta - 1\right) \cot \beta}{(\gamma + 1)M_1^2} \tag{15.6}$$

For large M_1, Eq. (15.6) can be approximated by

$$\frac{v_2}{V_1} = \frac{2\left(M_1^2 \sin^2 \beta\right) \cot \beta}{(\gamma + 1)M_1^2} = \frac{2 \sin \beta \cos \beta}{\gamma + 1} \tag{15.7}$$

Since $2 \sin \beta \cos \beta = \sin 2\beta$, then, from Eq. (15.7),

as $M_1 \to \infty$:
$$\boxed{\frac{v_2}{V_1} = \frac{\sin 2\beta}{\gamma + 1}} \tag{15.8}$$

In this equation, the choice of velocity components parallel and perpendicular to the upstream flow direction rather than to the shock wave is intentional. Equations (15.5) and (15.8) are useful in studying various aspects of the velocity field over a slender hypersonic body, as will be discussed later.

Note from Eqs. (15.1) and (15.3) that both p_2/p_1 and T_2/T_1 become infinitely large as $M_1 \to \infty$. In contrast, from Eqs. (15.2), (15.5), and (15.8), ρ_2/ρ_1, u_2/V_1, and v_2/V_1 approach limiting finite values as $M_1 \to \infty$.

In aerodynamics, pressure distributions are usually quoted in terms of the nondimensional pressure coefficient C_p, rather than the pressure itself. The pressure coefficient is defined as

$$C_p = \frac{p_2 - p_1}{q_1} \tag{15.9}$$

where p_1 and q_1 are the upstream (free-stream) static pressure and dynamic pressure, respectively. Recall from Sec. 9.3 that Eq. (15.9) can also be written as Eq. (9.10), repeated below:

$$C_p = \frac{2}{\gamma M_1^2} \left(\frac{p_2}{p_1} - 1\right) \tag{9.10}$$

Combining Eqs. (9.10) and (15.1), we obtain an exact relation for C_p behind an oblique shock wave as follows:

Exact:
$$C_p = \frac{4}{\gamma + 1} \left(\sin^2 \beta - \frac{1}{M_1^2}\right) \tag{15.10}$$

In the hypersonic limit,

as $M_1 \to \infty$:

$$\boxed{C_p = \left(\frac{4}{\gamma + 1}\right) \sin^2 \beta} \tag{15.11}$$

The relationship between Mach number M_1, shock angle β, and deflection angle θ is expressed by the so-called θ-β-M relation given by Eq. (4.17), repeated below:

Exact:

$$\tan \theta = 2 \cot \beta \left[\frac{M_1^2 \sin^2 \beta - 1}{M_1^2(\gamma + \cos 2\beta) + 2}\right] \tag{15.12}$$

This relation is plotted in Fig. 4.8, which is a standard plot of wave angle versus deflection angle, with Mach number as a parameter. Returning to Fig. 4.8, we note that, in the hypersonic limit, where θ is small, β is also small. Hence, in this limit, we can insert the usual small-angle approximations into Eq. (15.12):

$$\sin \beta \approx \beta$$

$$\cos 2\beta \approx 1$$

$$\tan \theta \approx \sin \theta \approx \theta$$

resulting in

$$\theta = \frac{2}{\beta}\left[\frac{M_1^2 \beta^2 - 1}{M_1^2(\gamma + 1) + 2}\right] \tag{15.13}$$

Applying the high Mach number limit to Eq. (15.13), we have

$$\theta = \frac{2}{\beta}\left[\frac{M_1^2 \beta^2}{M_1^2(\gamma + 1)}\right] \tag{15.14}$$

In Eq. (15.14) M_1 cancels, and we finally obtain in both the small-angle and hypersonic limits:

as $M_1 \to \infty$ and θ hence β is small:

$$\boxed{\frac{\beta}{\theta} = \frac{\gamma + 1}{2}} \tag{15.15}$$

Note that for $\gamma = 1.4$,

$$\boxed{\beta = 1.2\theta} \tag{15.16}$$

It is interesting to observe that, in the hypersonic limit for a slender wedge, the wave angle is only 20 percent larger than the wedge angle—a graphic demonstration of a thin shock layer in hypersonic flow. (Check Fig. 15.3, drawn from exact oblique shock results, and note that the $18°$ shock angle is 20 percent larger than the $15°$ wedge angle at Mach 36—truly an example of the hypersonic limit.)

For your convenience, the limiting hypersonic shock relations obtained in this section are summarized in Fig. 15.8. These limiting relations, which are clearly simpler than the corresponding exact oblique shock relations, will be important for the development of some of our hypersonic aerodynamic techniques in subsequent sections.

15.4 | A LOCAL SURFACE INCLINATION METHOD: NEWTONIAN THEORY

Linearized supersonic theory leads to a simple relation for the surface pressure coefficient, namely Eq. (9.51), repeated here:

$$C_p = \frac{2\theta}{\sqrt{M_\infty^2 - 1}} \tag{9.51}$$

Note from Eq. (9.51) that C_p depends only on θ, the local surface inclination angle defined by the angle between a line tangent to the surface and the free-stream direction. In this sense, Eq. (9.51) is an example of a "local surface inclination method" for linearized supersonic flow. *Question:* Do any local surface inclination methods exist for hypersonic flow? The answer is yes, and this constitutes the subject of the present section.

The oldest and most widely used of the hypersonic local surface inclination methods is *newtonian theory*. This theory has already been developed and discussed in Sec. 12.4, leading to the famous newtonian "sine-squared" law in Eq. (12.17):

$$C_p = 2\sin^2\theta \tag{12.17}$$

Additional insight into the physical meaning of Eq. (12.17) can be obtained from an examination of the hypersonic oblique shock relations, as described below.

Temporarily discard any thoughts of newtonian theory, and simply recall the exact oblique shock relation for C_p as given by Eq. (15.10), repeated here (with free-stream conditions now denoted by a subscript ∞ rather than a subscript 1, as used in Chap. 2):

$$C_p = \frac{4}{\gamma + 1}\left[\sin^2\beta - \frac{1}{M_\infty^2}\right] \tag{15.10}$$

Equation (15.11) gave the limiting value of C_p as $M_\infty \to \infty$, repeated here:

as $M_\infty \to \infty$:
$$C_p \to \frac{4}{\gamma + 1}\sin^2\beta \tag{15.11}$$

Now take the additional limit of $\gamma \to 1.0$. From Eq. (15.11), in both limits as $M_\infty \to \infty$ and $\gamma \to 1.0$, we have

$$C_p \to 2\sin^2\beta \tag{15.17}$$

Equation (15.17) is a result from exact oblique shock theory; it has nothing to do with newtonian theory (as yet). Keep in mind that β in Eq. (15.17) is the wave angle, not the deflection angle.

Let us go further. Consider the exact oblique shock relation for ρ/ρ_∞, given by Eq. (4.8), repeated here (again with a subscript ∞ replacing the subscript 1):

$$\frac{\rho_2}{\rho_\infty} = \frac{(\gamma + 1)M_\infty^2\sin^2\beta}{(\gamma - 1)M_\infty^2\sin^2\beta + 2} \tag{4.8}$$

Equation (15.2) was obtained as the limit where $M_\infty \to \infty$, namely,

as $M_\infty \to \infty$:
$$\frac{\rho_2}{\rho_\infty} \to \frac{\gamma + 1}{\gamma - 1}$$
(15.2)

In the additional limit as $\gamma \to 1$, we find

as $\gamma \to 1$ and $M_\infty \to \infty$:
$$\boxed{\frac{\rho_2}{\rho_\infty} \to \infty}$$
(15.18)

i.e., the density behind the shock is infinitely large. In turn, mass flow considerations then dictate that the shock wave is coincident with the body surface. This is further substantiated by Eq. (15.15), which is good for $M_\infty \to \infty$ and small deflection angles

$$\frac{\beta}{\theta} \to \frac{\gamma + 1}{2}$$
(15.15)

In the additional limit as $\gamma \to 1$, we have:

as $\gamma \to 1$ and $M_\infty \to \infty$ and θ and β small:

$$\boxed{\beta = \theta}$$

i.e., the shock wave lies on the body. In light of this result, Eq. (15.17) is written as

$$\boxed{C_p = 2 \sin^2 \theta}$$
(15.19)

Examine Eq. (15.19). It is a result from exact oblique shock theory, taken in the combined limit of $M_\infty \to \infty$ and $\gamma \to 1$. However, it is also precisely the newtonian results given by Eq. (12.17). Therefore, we make the following conclusion. The closer the actual hypersonic flow problem is to the limits $M_\infty \to \infty$ and $\gamma \to 1$, the closer it should be physically described by newtonian flow. In this regard, we gain a better appreciation of the true significance of newtonian theory. We can also state that the application of newtonian theory to practical hypersonic flow problems, where γ is always greater than unity (for air flows where the local static temperature is less than 800 K, $\gamma = 1.4$) is theoretically not proper, and the agreement that is frequently obtained with experimental data has to be viewed as somewhat fortuitous. Nevertheless, the simplicity of newtonian theory along with its (sometimes) reasonable results (no matter how fortuitous) has made it a widely used and popular engineering method for the estimation of surface pressure distributions, hence lift and wave drag coefficients, for hypersonic bodies.

In the newtonian model of fluid flow, the particles in the free stream impact only on the frontal area of the body; they cannot curl around the body and impact on the back surface. Hence, for that portion of a body which is in the "shadow" of the incident flow, such as the shaded region sketched in Fig. 15.9, no impact pressure is felt. Hence, over this shadow region it is consistent to assume that $p = p_\infty$, and therefore $C_p = 0$, as indicated in Fig. 15.9.

It is instructive to examine newtonian theory applied to a flat plate, as sketched in Fig. 15.10. Here, a two-dimensional flat plate with chord length c is at an angle of

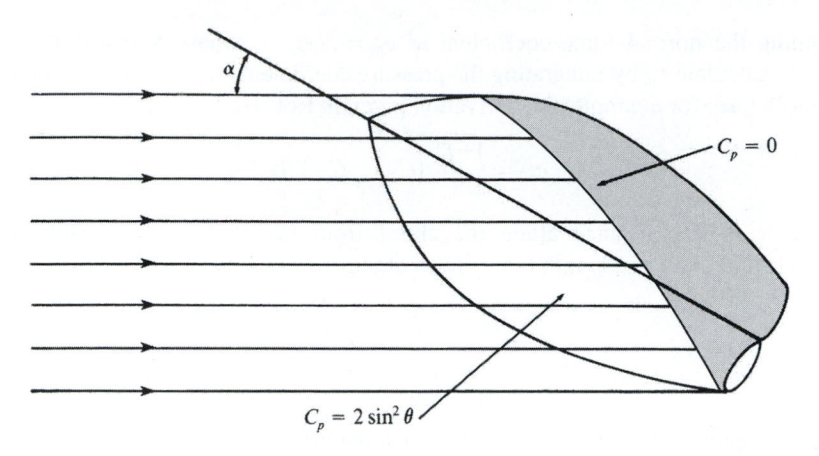

Figure 15.9 | Shadow region on the leeward side of a body, from
newtonian theory.

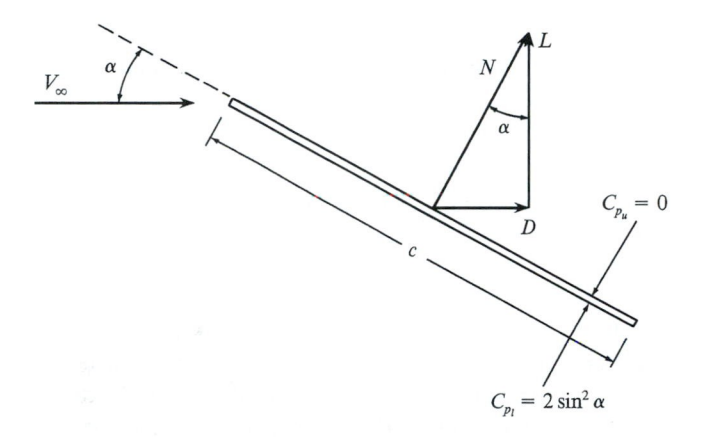

Figure 15.10 | Flat plate at angle of attack. Illustration
of aerodynamic forces.

attack α to the free stream. Since we are not including friction, and because surface
pressure always acts normal to the surface, the resultant aerodynamic force is perpendicular to the plate, i.e., in this case the normal force N is the resultant aerodynamic force. (For an infinitely thin flat plate, this is a general result which is not limited to newtonian theory, or even to hypersonic flow.) In turn, N is resolved into lift
and drag, denoted by L and D, respectively, as shown in Fig. 15.10. According to
newtonian theory, the pressure coefficient on the lower surface is

$$C_{p_l} = 2 \sin^2 \alpha \tag{15.20}$$

and that on the upper surface, which is in the shadow region, is

$$C_{p_u} = 0 \tag{15.21}$$

Defining the normal force coefficient as $c_n = N/q_\infty S$, where $S = (c)(l)$, we can readily calculate c_n by integrating the pressure coefficients over the lower and upper surfaces (see, for example, the derivation given in Ref. 104):

$$c_n = \frac{1}{c} \int_o^c \left(C_{p_l} - C_{p_u} \right) dx \tag{15.22}$$

where x is the distance along the chord from the leading edge. Substituting Eqs. (15.20) and (15.21) into (15.22), we obtain

$$c_n = \frac{1}{c}(2\sin^2\alpha)c$$

or

$$c_n = 2\sin^2\alpha \tag{15.23}$$

From the geometry of Fig. 15.10, we see that the lift and drag coefficients, defined as $c_l = L/q_\infty S$ and $c_d = D/q_\infty S$, respectively, where $S = (c)(l)$, are given by

$$c_l = c_n \cos\alpha \tag{15.24}$$

and

$$c_d = c_n \sin\alpha \tag{15.25}$$

Substituting Eq. (15.23) into Eqs. (15.24) and (15.25), we obtain

$$c_l = 2\sin^2\alpha\cos\alpha \tag{15.26}$$

and

$$c_d = 2\sin^3\alpha \tag{15.27}$$

Finally, from the geometry of Fig. 15.10, the lift-to-drag ratio is given by

$$\frac{L}{D} = \cot\alpha \tag{15.28}$$

[Note that Eq. (15.28) is a general result for inviscid supersonic or hypersonic flow over a flat plate. For such flows, the resultant aerodynamic force is the normal force N. From the geometry shown in Fig. 15.10, the resultant aerodynamic force makes the angle α with respect to lift, and clearly, from the right triangle between L, D, and N, we have $L/D = \cot\alpha$. Hence, Eq. (15.28) is not limited to newtonian theory.]

The results obtained here for the application of newtonian theory to an infinitely thin flat plate are plotted in Fig. 15.11. Here L/D, c_l, and c_d are plotted versus angle of attack α. From this figure, note these aspects:

1.   The value of L/D increases monotonically as α is decreased. Indeed, $L/D \to \infty$ as $\alpha \to 0$. However, this is misleading; when skin friction is added to this picture, D becomes finite at $\alpha = 0$, and then $L/D \to 0$ as $\alpha \to 0$.

2.   The lift curve peaks at about $\alpha \approx 55°$. (To be exact, it can be shown from newtonian theory that maximum c_l occurs at $\alpha = 54.7°$; the proof of this is left as a homework problem.) It is interesting to note that $\alpha \approx 55°$ for maximum lift is fairly realistic; the maximum lift coefficient for many practical hypersonic vehicles occurs at angles of attack in this neighborhood.

3.   Examine the lift curve at low angle of attack, say in the range of α from 0 to 15°. Note that the variation of c_l with α is very nonlinear. This is in direct

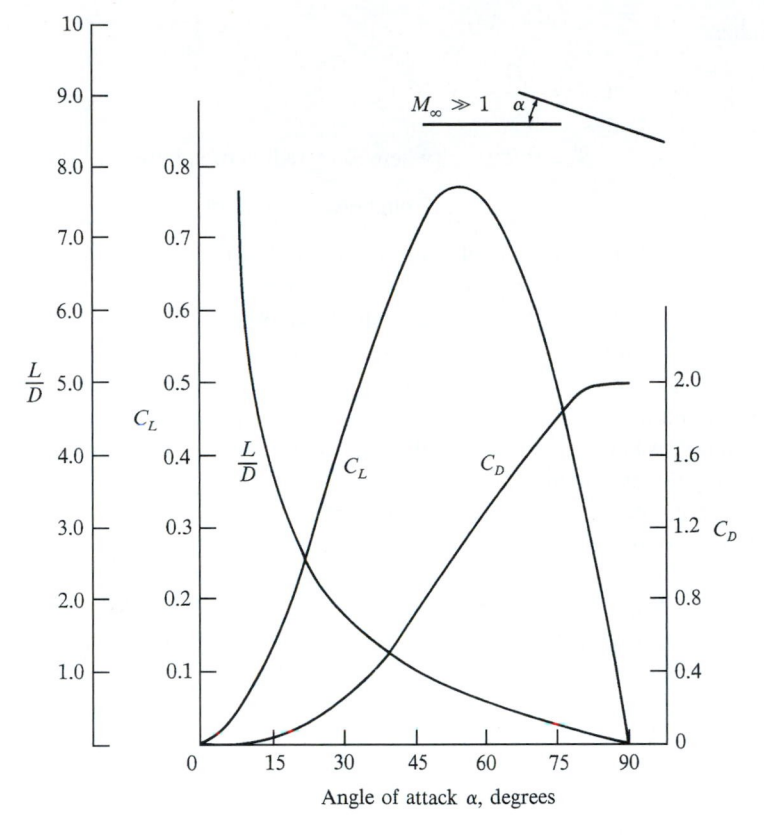

Figure 15.11 | Newtonian results for a flat plate.

contrast to the familiar results for subsonic and supersonic flow, where for thin bodies at small α, the lift curve is a linear function of α. (Recall, for example, that the theoretical lift slope from incompressible thin airfoil theory is 2π per radian.) Hence, the nonlinear lift curve shown in Fig. 15.11 is a graphic demonstration of the nonlinear nature of hypersonic flow.

Consider two other basic aerodynamic bodies; the circular cylinder of infinite span, and the sphere. Newtonian theory can be applied to estimate the hypersonic drag coefficients for these shapes; the results are

1. Circular cylinder of infinite span:

$$c_d = \frac{D}{q_\infty S}$$

$$S = 2R \qquad \text{(where } R = \text{radius of cylinder)}$$

$$c_d = \frac{4}{3} \qquad \text{(from newtonian theory)}$$

2.   Sphere

$$C_D = \frac{D}{q_\infty S}$$

$$S = \pi R^2 \qquad \text{(where } R = \text{ radius of sphere)}$$

$$C_D = 1 \qquad \text{(from newtonian theory)}$$

The derivations of these drag coefficient values are left for homework problems.

It is interesting to note that these results from newtonian theory do not explicitly depend on Mach number. Of course, they implicitly assume that M_∞ is high enough for hypersonic flow to prevail; outside of that, the precise value of M_∞ does not enter the calculations. This is compatible with the Mach number independence principle, to be discussed in the next section. In short, this principle states that certain aerodynamic quantities become relatively independent of Mach number if M_∞ is made sufficiently large. Newtonian results are the epitome of this principle.

As a final note on our discussion of newtonian theory, consider Fig. 15.12. Here, the pressure coefficients for a 15° half-angle wedge and a 15° half-angle cone are plotted versus free-stream Mach number for $\gamma = 1.4$. The exact wedge results are obtained from Eq. (15.10), and the exact cone results are obtained from the solution of the classical Taylor–Maccoll equation (see Chap. 10). Both sets of results are

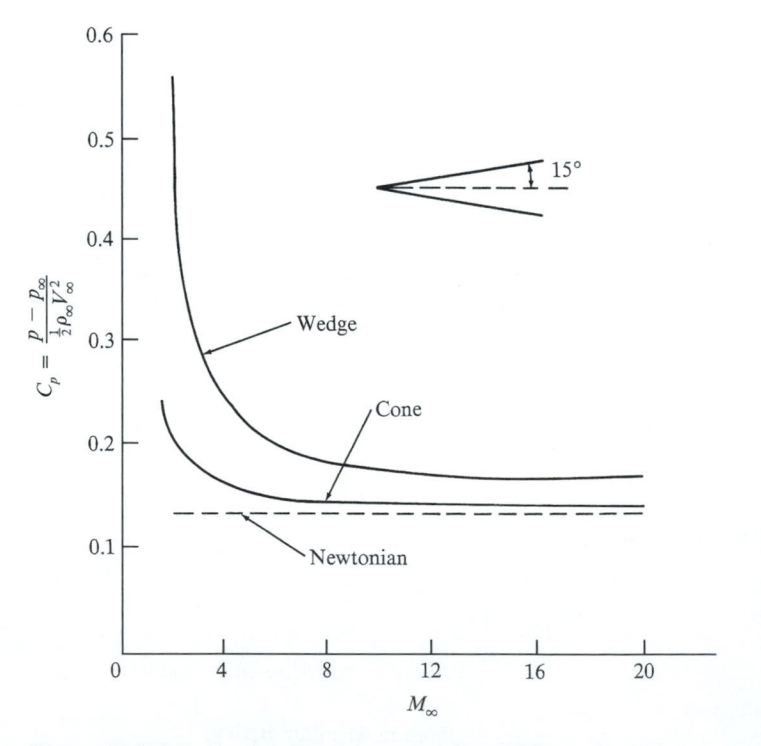

Figure 15.12 | Comparison between newtonian and exact results for the pressure coefficient on a sharp wedge and a sharp cone.

compared with newtonian theory, $C_p = 2\sin^2\theta$, shown as the dashed line in Fig. 15.12. This comparison demonstrates two general aspects of newtonian results:

1. The accuracy of newtonian results improves as M_∞ increases. This is to be expected from our previous discussion. Note from Fig. 15.12 that below $M_\infty = 5$, the newtonian results are not even close, but the comparison becomes much closer as M_∞ increases above 5.

2. Newtonian theory is usually more accurate for three-dimensional bodies (e.g., the cone) than for two-dimensional bodies (e.g., the wedge). This is clearly evident in Fig. 15.12 where the newtonian result is much closer to the cone results than to the wedge results.

This ends our discussion of the application of newtonian theory to hypersonic bodies. For more details, including the treatment of centrifugal force corrections to newtonian theory, see Ref. 119.

In addition to newtonian theory, there are three other local surface inclination methods that are frequently used for the estimation of pressure distributions over hypersonic bodies. These are the tangent wedge, tangent cone, and shock-expansion methods. There is not space in the present chapter to describe these methods; they are covered in detail in Ref. 119.

15.5 | MACH NUMBER INDEPENDENCE

Return again to Fig. 15.12, where values of C_p for both a 15° half-angle wedge and cone are plotted versus Mach number. Note that at low supersonic Mach numbers, C_p decreased rapidly as M_∞ was increased. However, at hypersonic speeds, the rate of decrease diminishes considerably, and C_p appears to reach a plateau as M_∞ becomes large, i.e., C_p becomes relatively independent of M_∞ at high Mach numbers. This is the essence of the Mach number independence principle; at high Mach numbers, certain aerodynamic quantities such as pressure coefficient, lift and wave-drag coefficients, and flowfield structure (such as shock wave shapes and Mach wave patterns) become essentially independent of Mach number. Indeed, newtonian theory (discussed in Sec. 15.4), gives results that are totally independent of Mach number, as clearly demonstrated by Eq. (15.19). The hypersonic Mach number independence principle is more than just an observed phenomenon; it has a mathematical foundation, which is the subject of this section. We will examine the roots of this Mach number independence more closely.

The governing partial differential equations for inviscid compressible flow are derived in Chap. 6; as before, we will refer to these equations as the Euler equations. Ignoring body forces, they can be expressed as Eqs. (6.5), (6.26) through (6.28), and (6.51), repeated here and renumbered for convenience:

Continuity:
$$\frac{\partial \rho}{\partial t} + \frac{\partial(\rho u)}{\partial x} + \frac{\partial(\rho v)}{\partial y} + \frac{\partial(\rho w)}{\partial z} = 0 \qquad (15.29)$$

x momentum:
$$\rho\frac{\partial u}{\partial t} + \rho u\frac{\partial u}{\partial x} + \rho v\frac{\partial u}{\partial y} + \rho w\frac{\partial u}{\partial z} = -\frac{\partial p}{\partial x} \qquad (15.30)$$

y momentum:
$$\rho\frac{\partial v}{\partial t} + \rho u\frac{\partial v}{\partial x} + \rho v\frac{\partial v}{\partial y} + \rho w\frac{\partial v}{\partial z} = -\frac{\partial p}{\partial y} \tag{15.31}$$

z momentum:
$$\rho\frac{\partial w}{\partial t} + \rho u\frac{\partial w}{\partial x} + \rho v\frac{\partial w}{\partial y} + \rho w\frac{\partial w}{\partial z} = -\frac{\partial p}{\partial z} \tag{15.32}$$

Energy:
$$\frac{\partial s}{\partial t} + u\frac{\partial s}{\partial x} + v\frac{\partial s}{\partial y} + w\frac{\partial s}{\partial z} = 0 \tag{15.33}$$

In reality, Eq. (15.33) is the "entropy equation"; for an inviscid, adiabatic flow, Eq. (15.33) can serve as the energy equation—indeed, it is fundamentally an energy equation as described in Sec. 6.5. Equation (15.33) simply states that the entropy of a fluid element is constant. For an isentropic process in a calorically perfect gas, $p/\rho^\gamma = $ const. Hence, if the entropy of a moving fluid element is constant as stated by Eq.(15.33), then the quantity p/ρ^γ is also constant for the moving fluid element, and for a calorically perfect gas Eq. (15.33) can be replaced by

$$\frac{\partial}{\partial t}\left(\frac{p}{\rho^\gamma}\right) + u\frac{\partial}{\partial x}\left(\frac{p}{\rho^\gamma}\right) + v\frac{\partial}{\partial y}\left(\frac{p}{\rho^\gamma}\right) + w\frac{\partial}{\partial z}\left(\frac{p}{\rho^\gamma}\right) = 0 \tag{15.34}$$

Let us nondimensionalize Eqs. (15.29) through (15.32) and (15.34) as follows. Define the nondimensional variables (the barred quantities) as

$$\bar{x} = \frac{x}{l} \qquad \bar{y} = \frac{y}{l} \qquad \bar{z} = \frac{z}{l}$$

$$\bar{u} = \frac{u}{V_\infty} \qquad \bar{v} = \frac{v}{V_\infty} \qquad \bar{w} = \frac{w}{V_\infty}$$

$$\bar{p} = \frac{p}{\rho_\infty V_\infty^2} \qquad \bar{\rho} = \frac{\rho}{\rho_\infty}$$

where l denotes a characteristic length of the flow, and ρ_∞ and V_∞ are the free-stream density and velocity, respectively. Assuming steady flow ($\partial/\partial t = 0$), we obtain from Eqs. (15.29) through (15.32) and (15.34)

$$\partial\frac{(\bar{\rho}\bar{u})}{\partial\bar{x}} + \partial\frac{(\bar{\rho}\bar{v})}{\partial\bar{y}} + \partial\frac{(\bar{\rho}\bar{w})}{\partial\bar{z}} = 0 \tag{15.35}$$

$$\bar{\rho}\bar{u}\frac{\partial\bar{u}}{\partial\bar{x}} + \bar{\rho}\bar{v}\frac{\partial\bar{u}}{\partial\bar{y}} + \bar{\rho}\bar{w}\frac{\partial\bar{u}}{\partial\bar{z}} = -\frac{\partial\bar{p}}{\partial\bar{x}} \tag{15.36}$$

$$\bar{\rho}\bar{u}\frac{\partial\bar{v}}{\partial\bar{x}} + \bar{\rho}\bar{v}\frac{\partial\bar{v}}{\partial\bar{y}} + \bar{\rho}\bar{w}\frac{\partial\bar{v}}{\partial\bar{z}} = -\frac{\partial\bar{p}}{\partial\bar{y}} \tag{15.37}$$

$$\bar{\rho}\bar{u}\frac{\partial\bar{w}}{\partial\bar{x}} + \bar{\rho}\bar{v}\frac{\partial\bar{w}}{\partial\bar{y}} + \bar{\rho}\bar{w}\frac{\partial\bar{w}}{\partial\bar{z}} = -\frac{\partial\bar{p}}{\partial\bar{z}} \tag{15.38}$$

$$\bar{u}\frac{\partial}{\partial\bar{x}}\left(\frac{\bar{p}}{\bar{\rho}^\gamma}\right) + \bar{v}\frac{\partial}{\partial\bar{y}}\left(\frac{\bar{p}}{\bar{\rho}^\gamma}\right) + \bar{w}\frac{\partial}{\partial\bar{z}}\left(\frac{\bar{p}}{\bar{\rho}^\gamma}\right) = 0 \tag{15.39}$$

Any particular solution of these equations is governed by the boundary conditions, which are discussed next.

The boundary condition for steady inviscid flow at a surface is simply the statement that the flow must be tangent to the surface. Let **n** be a unit normal vector at some point on the surface, and let **V** be the velocity vector at the same point. Then, for the flow to be tangent to the body,

$$\mathbf{V} \cdot \mathbf{n} = 0 \qquad (15.40)$$

Let n_x, n_y, and n_z be the components of **n** in the x, y, and z directions, respectively. Then, Eq. (15.40) can be written as

$$u n_x + v n_y + w n_z = 0 \qquad (15.41)$$

Recalling the definition of direction cosines from analytic geometry, note, in Eq. (15.41) that n_x, n_y, and n_z are also the direction cosines of n with respect to the x, y, and z axes, respectively. With this interpretation, n_x, n_y, and n_z may be considered dimensionless quantities, and the nondimensional boundary condition at the surface is readily obtained from Eq. (15.41) as

$$\bar{u} n_x + \bar{v} n_x + \bar{w} n_z = 0 \qquad (15.42)$$

Assume that we are considering the external flow over a hypersonic body, where the flowfield of interest is bounded on one side by the body surface, and on the other side by the bow shock wave. Equation (15.42) gives the boundary condition on the body surface. The boundary conditions right behind the shock wave are given by the oblique shock properties expressed by Eqs. (4.9), (4.8), (15.4), and (15.6), repeated here for convenience (replacing the subscript 1 with the subscript ∞ for free-stream properties):

$$\frac{p_2}{p_\infty} = 1 + \frac{2\gamma}{\gamma + 1} \left(M_\infty^2 \sin^2 \beta - 1 \right) \qquad (4.9)$$

$$\frac{\rho_2}{\rho_\infty} = \frac{(\gamma + 1) M_\infty^2 \sin^2 \beta}{(\gamma - 1) M_\infty^2 \sin^2 \beta + 2} \qquad (4.8)$$

$$\frac{u_2}{V_\infty} = 1 - \frac{2 \left(M_\infty^2 \sin^2 \beta - 1 \right)}{(\gamma + 1) M_\infty^2} \qquad (15.4)$$

$$\frac{v_2}{V_\infty} = \frac{2 \left(M_\infty^2 \sin \beta - 1 \right) \cot \beta}{(\gamma + 1) M_\infty^2} \qquad (15.6)$$

In terms of the nondimensional variables, and noting that for a calorically perfect gas

$$p_2/p_\infty = \bar{p}_2 \left(\rho_\infty V_\infty^2 \right) / p_\infty = \bar{p}_2 V_\infty^2 / R T_\infty = \bar{p}_2 \gamma V_\infty^2 / a_\infty^2 = \bar{p}_2 \gamma M_\infty^2$$

Eqs. (4.9), (4.8), (15.4), and (15.6) become

$$\bar{p}_2 = \frac{1}{\gamma M_\infty^2} + \frac{2}{\gamma + 1}\left(\sin^2 \beta - \frac{1}{M_\infty^2}\right) \tag{15.43}$$

$$\bar{\rho}_2 = \frac{(\gamma + 1)M_\infty^2 \sin^2 \beta}{(\gamma - 1)M_\infty^2 \sin^2 \beta + 2} \tag{15.44}$$

$$\bar{u}_2 = 1 - \frac{2\left(M_\infty^2 \sin^2 \beta - 1\right)}{(\gamma + 1)M_\infty^2} \tag{15.45}$$

$$\bar{v}_2 = \frac{2\left(M_\infty^2 \sin \beta - 1\right)\cot \beta}{(\gamma + 1)M_\infty^2} \tag{15.46}$$

In the limit of high M_∞, as $M_\infty \to \infty$, Eqs. (15.43) through (15.46) go to

$$\bar{p}_2 \to \frac{2\sin^2 \beta}{\gamma + 1} \tag{15.47}$$

$$\bar{\rho}_2 \to \frac{\gamma + 1}{\gamma - 1} \tag{15.48}$$

$$\bar{u}_2 \to 1 - \frac{2\sin^2 \beta}{\gamma + 1} \tag{15.49}$$

$$\bar{v}_2 \to \frac{\sin^2 \beta}{\gamma + 1} \tag{15.50}$$

Now consider a hypersonic flow over a given body. This flow is governed by Eqs. (15.35) through (15.39), with boundary conditions given by Eqs. (15.42) through (15.46).

Question: Where does M_∞ explicitly appear in these equations?

Answer: Only in the shock boundary conditions, Eqs. (15.43) through (15.46).

Now consider the hypersonic flow over a given body in the limit of large M_∞. The flow is again governed by Eqs. (15.35) through (15.39), but with boundary conditions given by Eqs. (15.42) and (15.47) through (15.50).

Question: Where does M_∞ explicitly appear in these equations?

Answer: No place!

Conclusion: At high M_∞, the solution is *independent* of Mach number.

Clearly, from this last consideration, we can see that the Mach number independence principle follows directly from the governing equations of motion with the appropriate boundary conditions written in the limit of high Mach number. Therefore, when the free-stream Mach number is sufficiently high, the nondimensional dependent variables in Eqs. (15.35) through (15.39) become essentially independent of Mach number; this trend applies also to any quantities derived from these nondimensional variables. For example, C_p can be easily obtained as a function of $\bar{p}$ only; in turn, the lift and wave-drag coefficients for the body, C_L and C_{D_w}, respectively, can be

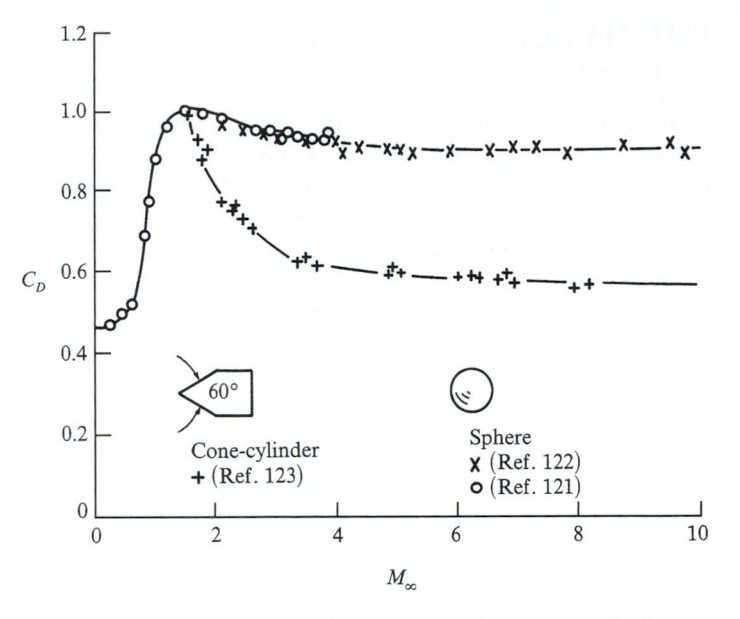

Figure 15.13 | Drag coefficient for a sphere and a cone-cylinder from ballistic range measurements; an illustration of Mach number independence. (From Ref. 124.)

expressed in terms of C_p integrated over the body surface (see, for example, Ref. 104). Therefore, C_p, C_L, and C_{D_w} also become independent of Mach number at high M_∞. This is demonstrated by the data shown in Fig. 15.13 obtained from Refs. 121 through 123 as gathered in Ref. 124. In Fig. 15.13, the measured drag coefficients for spheres and for a large-angle cone-cylinder are plotted versus Mach number, cutting across the subsonic, supersonic, and hypersonic regimes. Note the large drag rise in the subsonic regime associated with the drag-divergence phenomenon near Mach 1, and the decrease in C_D in the supersonic regime beyond Mach 1. Both of these variations are expected and well understood. (See, for example, Secs. 14.2 and 9.6, respectively.) For our purposes in the present section, note in particular the variation of C_D in the hypersonic regime; for both the sphere and cone-cylinder, C_D approaches a plateau, and becomes relatively independent of Mach number as M_∞ becomes large. Note also that the sphere data appear to achieve "Mach number independence" at lower Mach numbers than the cone-cylinder. This is to be expected, as follows. In Eqs. (15.43) through (15.46), the Mach number frequently appears in the combined form $M_\infty^2 \sin^2 \beta$; for a given Mach number, this quantity is larger for blunt bodies (β large) than for slender bodies (β small). Hence blunt body flows will tend to approach Mach number independence at lower M_∞ than will slender bodies.

Finally, keep in mind from the above analysis that it is the nondimensional variables that become Mach number independent. Some of the dimensional variables, such as p, are not Mach number independent; indeed, $p \to \infty$ as $M_\infty \to \infty$.

15.6 | THE HYPERSONIC SMALL-DISTURBANCE EQUATIONS

In Chap. 9, the concept of *perturbation* velocities was introduced. For irrotational flow, the Euler equations cascade to a single equation in terms of the perturbation velocities, u', v', and w', namely Eq. (9.4). In turn, then Eq. (9.4) reduces to the linear Eq. (9.5) which holds for subsonic and supersonic flows. On the other hand, we saw in Sec. 9.2 that Eq. (9.5) does not hold for transonic flow; this was reinforced in Chap. 14 where transonic flow was described as a basically *nonlinear* flow regime, even for small perturbations. The same is true for hypersonic flow, as noted in Sec. 9.2. At hypersonic Mach numbers, Eq. (9.5) does not hold. This raises the question: What equations do hold for hypersonic flow when the assumption of small perturbations is made? The answer to this question is the subject of this section. In particular, making the assumption that u', v', and w' are small, we will derive the *hypersonic small-disturbance equations*. In the following section, we will put these equations to work in order to obtain the principle of hypersonic similarity.

From the definition of the perturbation velocities as given in Chap. 9, we have

$$u = V_\infty + u'$$
$$v = v'$$
$$w = w'$$

In terms of these perturbation velocities, Eqs. (15.29) through (15.32) and (15.34) are written as

$$\frac{\partial[\rho(V_\infty + u')]}{\partial x} + \frac{\partial(\rho v')}{\partial y} + \frac{\partial(\rho w')}{\partial z} = 0 \tag{15.51}$$

$$\rho(V_\infty + u')\frac{\partial(V_\infty + u')}{\partial x} + \rho v'\frac{\partial(V_\infty + u')}{\partial y} + \rho w'\frac{\partial(V_\infty + u')}{\partial z} = -\frac{\partial p}{\partial x} \tag{15.52}$$

$$\rho(V_\infty + u')\frac{\partial v'}{\partial x} + \rho v'\frac{\partial v'}{\partial y} + \rho w'\frac{\partial v'}{\partial z} = -\frac{\partial p}{\partial y} \tag{15.53}$$

$$\rho(V_\infty + u')\frac{\partial w'}{\partial x} + \rho v'\frac{\partial w'}{\partial y} + \rho w'\frac{\partial w'}{\partial z} = -\frac{\partial p}{\partial z} \tag{15.54}$$

$$(V_\infty + u')\frac{\partial}{\partial x}\left(\frac{p}{\rho^\gamma}\right) + v'\frac{\partial}{\partial y}\left(\frac{p}{\rho^\gamma}\right) + w'\frac{\partial}{\partial z}\left(\frac{p}{\rho^\gamma}\right) = 0 \tag{15.55}$$

We wish to nondimensionalize Eqs. (15.51) through (15.55). Moreover, we wish to have nondimensional variables with an order of magnitude of unity, for reasons to be made clear later. To obtain a hint about reasonable nondimensionalizing quantities, consider the oblique shock relations in the limit as $M_\infty \to \infty$, obtained in Sec. 15.3. Also note that for a slender body at hypersonic speeds, both the shock wave angle β and the deflection angle θ are small; hence,

$$\sin\beta \approx \sin\theta \approx \theta \approx \frac{dy}{dx} \approx \tau$$

where $y = f(x)$ is the body shape, and τ is the slenderness ratio defined in Sec. 14.3. Thus, from Eq. (15.1), repeated below for convenience:

$$\frac{p_2}{p_\infty} \to \frac{2\gamma}{\gamma+1} M_\infty^2 \sin^2 \beta \tag{15.1}$$

we have the order-of-magnitude relationship:

$$\frac{p_2}{p_\infty} \to O\left[M_\infty^2 \tau^2\right] \tag{15.56}$$

This in turn implies that the pressure throughout the shock layer over the body will be on the order of $M_\infty^2 \tau^2 p_\infty$, and hence a reasonable definition for a nondimensional pressure which would be on the order of magnitude of unity is $\bar{p} = p/\gamma M_\infty^2 \tau^2 p_\infty$. (The reason for the γ will become clear later.) In regard to density, consider Eq. (15.2), repeated here:

$$\frac{\rho_2}{\rho_\infty} \to \frac{\gamma+1}{\gamma-1} \tag{15.2}$$

For $\gamma = 1.4$, $\rho_2/\rho_\infty \to 6$, which for our purposes is on the order of magnitude near unity. Hence, a reasonable nondimensional density is simply $\bar{\rho} = \rho/\rho_\infty$. In regard to velocities, first consider Eq. (15.5), repeated here:

$$\frac{u_2}{V_\infty} \to 1 - \frac{2\sin^2 \beta}{\gamma+1} \tag{15.5}$$

Define the *change* in the x component of velocity across the oblique shock as $\Delta u = V_\infty - u_2$. From Eq. (15.5), we have

$$\frac{\Delta u}{V_\infty} = \frac{V_\infty - u_2}{V_\infty} \to \frac{2\sin^2 \beta}{\gamma+1} \to O(\tau^2) \tag{15.57}$$

This implies that the nondimensional perturbation velocity $\bar{u}'$ (which is also a change in velocity in the x direction) should be defined as $\bar{u}' = u'/V_\infty \tau^2$ in order to be of an order of magnitude of unity. Finally, consider Eq. (15.8), repeated here:

$$\frac{v_2}{V_\infty} \to \frac{\sin^2 \beta}{\gamma+1} \tag{15.8}$$

From Eq. (15.8), we have

$$\frac{\Delta v}{V_\infty} = \frac{v_2}{V_\infty} \to \frac{\sin 2\beta}{\gamma+1} \to O(\tau) \tag{15.58}$$

This implies that the nondimensional perturbation velocity $\bar{v}'$ should be $\bar{v}' = v'/V_\infty \tau$, which is on the order of magnitude of 1.

[We pause to observe an interesting physical fact evidenced by Eqs. (15.57) and (15.58). Since we are dealing with slender bodies, τ is a small number, much less than unity. Hence, by comparing Eqs. (15.57) and (15.58), we see that Δu, which varies as τ^2, is much smaller than Δv, which varies as τ. Therefore, we conclude in the case of hypersonic flow over a slender body that the change in v dominates the flow, i.e., the changes in u and v are both small compared to V_∞, but that the change in v is large compared to the change in u.]

Based on these arguments, we define the following nondimensional quantities, all of which are on the order of magnitude of unity. Note that we add a third dimension in the z direction, and that y and z in the thin shock layer are much smaller than x:

$$\bar{x} = \frac{x}{l} \qquad\qquad \bar{y} = \frac{y}{l\tau} \qquad\qquad \bar{z} = \frac{z}{l\tau}$$

$$\bar{u}' = \frac{u'}{V_\infty \tau^2} \qquad\qquad \bar{v}' = \frac{v'}{V_\infty \tau} \qquad\qquad \bar{w}' = \frac{w'}{V_\infty \tau}$$

$$\bar{p} = \frac{p}{\gamma M_\infty^2 \tau^2 p_\infty} \qquad\qquad \bar{\rho} = \frac{\rho}{\rho_\infty}$$

(*Note:* The barred quantities here are different than the barred quantities used in Sec. 15.5, but since the present section is self-contained, there should be no confusion.) In terms of the nondimensional quantities defined here, Eqs. (15.51) through (15.55) can be written as shown next. From Eq. (15.51)

$$\frac{\partial}{\partial \bar{x}}\left[\bar{\rho}\left(\frac{1}{\tau^2}+\bar{u}'\right)\right][\rho_\infty V_\infty \tau^2] + \frac{\partial(\bar{\rho}\bar{v}')}{\partial \bar{y}}\left[\frac{\rho_\infty V_\infty \tau}{\tau}\right]$$

$$+\frac{\partial(\bar{\rho}\bar{w}')}{\partial \bar{z}}\left[\frac{\rho_\infty V_\infty^2 \tau}{\tau}\right] = 0 \tag{15.59}$$

From Eq. (15.52)

$$\bar{\rho}\left(\frac{1}{\tau^2}+\bar{u}'\right)\frac{\partial}{\partial \bar{x}}\left(\frac{1}{\tau^2}+u'\right)[\rho_\infty V_\infty^2 \tau^4] + \bar{\rho}\bar{v}'\frac{\partial}{\partial \bar{y}}\left(\frac{1}{\tau^2}+\bar{u}'\right)[\rho_\infty V_\infty^2 \tau^3]$$

$$+\bar{\rho}\bar{w}'\frac{\partial}{\partial \bar{z}}\left(\frac{1}{\tau^2}+\bar{u}'\right)\left[\frac{\rho_\infty V_\infty^2 \tau^3}{\tau}\right] = -\frac{\partial \bar{p}}{\partial \bar{x}}[\gamma M_\infty^2 \tau^2 p_\infty]$$

or, noting that

$$\rho_\infty V_\infty^2 = \frac{\gamma p_\infty}{\gamma p_\infty}\rho_\infty V_\infty^2 = \gamma p_\infty \frac{V_\infty^2}{a_\infty^2} = \gamma p_\infty M_\infty^2$$

we have

$$\bar{\rho}(1+\bar{u}'\tau^2)\frac{\partial \bar{u}'}{\partial \bar{x}'} + \bar{\rho}\bar{v}'\frac{\partial \bar{u}'}{\partial \bar{y}} + \bar{\rho}\bar{w}'\frac{\partial \bar{u}'}{\partial \bar{z}} = -\frac{\partial \bar{p}}{\partial \bar{x}} \tag{15.60}$$

From Eq. (15.53)

$$\bar{\rho}\left(\frac{1}{\tau^2}+\bar{u}'\right)\frac{\partial \bar{v}'}{\partial \bar{x}}[\rho_\infty V_\infty^2 \tau^3] + \bar{\rho}\bar{v}'\frac{\partial \bar{v}'}{\partial \bar{y}}\left[\frac{\rho_\infty V_\infty^2 \tau^2}{\tau}\right]$$

$$+\bar{\rho}\bar{w}'\frac{\partial \bar{v}'}{\partial \bar{z}'}\left[\frac{\rho_\infty V_\infty^2 \tau^2}{\tau}\right] = -\frac{\partial \bar{p}}{\partial \bar{y}}\left[\frac{\gamma M_\infty^2 \tau^2 p_\infty}{\tau}\right]$$

or

$$\bar{\rho}(1+\bar{u}'\tau^2)\frac{\partial \bar{v}'}{\partial \bar{x}} + \bar{\rho}\bar{v}'\frac{\partial \bar{v}'}{\partial \bar{y}} + \bar{\rho}\bar{w}'\frac{\partial \bar{v}'}{\partial \bar{z}} = -\frac{\partial \bar{p}}{\partial \bar{y}} \tag{15.61}$$

From Eq. (15.54), we similarly have

$$\bar{\rho}(1 + \bar{u}'\tau^2)\frac{\partial \bar{w}'}{\partial \bar{x}} + \bar{\rho}\bar{v}'\frac{\partial \bar{w}'}{\partial \bar{y}} + \bar{\rho}\bar{w}'\frac{\partial \bar{w}'}{\partial \bar{z}} = -\frac{\partial \bar{p}}{\partial \bar{z}} \tag{15.62}$$

From Eq. (15.55)

$$\left(\frac{1}{\tau^2} + \bar{u}'\right)\frac{\partial}{\partial \bar{x}}\frac{\bar{p}}{\bar{\rho}^\gamma}\left[V_\infty \tau^4 \gamma p_\infty M_\infty^2 \rho_\infty^\gamma\right] + \bar{v}'\frac{\partial}{\partial \bar{y}}\frac{\bar{p}}{\bar{\rho}^\gamma}\left[\frac{\gamma V_\infty \tau^3 p_\infty M_\infty^2 \rho_\infty^\gamma}{\tau}\right]$$

$$+ \bar{w}'\frac{\partial}{\partial \bar{z}}\frac{\bar{p}}{\bar{\rho}^\gamma}\left[\frac{\gamma V_\infty \tau^3 p_\infty M_\infty^2 \rho_\infty^\gamma}{\tau}\right] = 0$$

or

$$(1 + \tau^2 \bar{u}')\frac{\partial}{\partial \bar{x}}\left(\frac{\bar{p}}{\bar{\rho}^\gamma}\right) + \bar{v}'\frac{\partial}{\partial \bar{y}}\left(\frac{\bar{p}}{\bar{\rho}^\gamma}\right) + \bar{w}'\frac{\partial}{\partial \bar{z}}\left(\frac{\bar{p}}{\bar{\rho}^\gamma}\right) = 0 \tag{15.63}$$

Examine Eqs. (15.59) through (15.63) closely. Because of our choice of nondimensionalized variables, each term in these equations is of order of magnitude unity except for those multiplied by τ^2, which is very small. Therefore, the terms involving τ^2 can be ignored in comparison to the remaining terms, and Eqs. (15.59) through (15.63) can be written as

$$\frac{\partial \bar{\rho}}{\partial \bar{x}} + \frac{\partial(\bar{\rho}\bar{v}')}{\partial \bar{y}} + \frac{\partial(\bar{\rho}\bar{w}')}{\partial \bar{z}} = 0 \tag{15.64}$$

$$\bar{\rho}\frac{\partial \bar{u}'}{\partial \bar{x}} + \bar{\rho}\bar{v}'\frac{\partial \bar{u}'}{\partial \bar{y}} + \bar{\rho}\bar{w}'\frac{\partial \bar{u}'}{\partial \bar{z}} = -\frac{\partial \bar{p}}{\partial \bar{x}} \tag{15.65}$$

$$\bar{\rho}\frac{\partial \bar{v}'}{\partial \bar{x}} + \bar{\rho}\bar{v}'\frac{\partial \bar{v}'}{\partial \bar{y}} + \bar{\rho}\bar{w}'\frac{\partial \bar{v}'}{\partial \bar{z}} = -\frac{\partial \bar{p}}{\partial \bar{y}} \tag{15.66}$$

$$\bar{\rho}\frac{\partial \bar{w}'}{\partial \bar{x}} + \bar{\rho}\bar{v}'\frac{\partial \bar{w}'}{\partial \bar{y}} + \bar{\rho}\bar{w}'\frac{\partial \bar{w}'}{\partial \bar{z}} = -\frac{\partial \bar{p}}{\partial \bar{z}} \tag{15.67}$$

$$\frac{\partial}{\partial \bar{x}}\left(\frac{\bar{p}}{\bar{\rho}^\gamma}\right) + \bar{v}'\frac{\partial}{\partial \bar{y}}\left(\frac{\bar{p}}{\bar{\rho}^\gamma}\right) + \bar{w}'\frac{\partial}{\partial \bar{z}}\left(\frac{\bar{p}}{\bar{\rho}^\gamma}\right) = 0 \tag{15.68}$$

Equations (15.64) through (15.68) are the *hypersonic small-disturbance equations*. They closely approximate the hypersonic flow over slender bodies. They are limited to flow over slender bodies because we have neglected terms of order τ^2. They are also limited to hypersonic flow because some of the nondimensionalized terms are of order of magnitude unity only for high Mach numbers; we made certain of this in the argument that preceded the definition of the nondimensional quantities. Hence, the fact that each term in Eqs. (15.64) through (15.68) is of the order of magnitude unity [which is essential for dropping the τ^2 terms in Eqs. (15.59) through (15.63)] holds only for hypersonic flow.

Equations (15.64) through (15.68) exhibit an interesting property. Look for $\bar{u}'$ in these equations; you can find it only in Eq. (15.65). Therefore, in the hypersonic small-disturbance equations, $\bar{u}'$ is decoupled from the system. In principle, Eqs. (15.64) and (15.66) through (15.68) constitute four equations for the four unknowns, $\bar{\rho}$, $\bar{p}$, $\bar{v}'$, and $\bar{w}'$. After this system is solved, then $\bar{u}'$ follows directly from Eq. (15.65). This decoupling of $\bar{u}'$ from the rest of the system is another ramification of the fact already mentioned earlier, namely that the change in velocity in the flow direction over a hypersonic slender body is much smaller than the change in velocity perpendicular to the flow direction.

The hypersonic small-disturbance equations are used to obtain practical information about hypersonic flows over slender bodies. An example is given in the next section, dealing with hypersonic similarity.

(Note the importance of obtaining the limiting hypersonic shock relations in Sec. 15.3. We have already used these relations several times for important developments. For example, they were used to demonstrate Mach number independence in Sec. 15.5, and they were instrumental in helping to define the proper nondimensional variables in the hypersonic small-disturbance equations obtained in this section. So the work done in Sec. 15.3 was more than just an academic exercise; the specialized forms of the oblique shock relations in the hypersonic limit are indeed quite useful.)

The hypersonic small-disturbance equations, Eqs. (15.64) through (15.68), are the analog to Eq. (9.5) for subsonic and supersonic flow. However, unlike Eq. (9.5) which is linear, Eqs. (15.65) through (15.68) are *nonlinear*. Therefore, we have clearly demonstrated that, for hypersonic flow, the assumption of small perturbations does *not* lead to a linear theory; this is indeed just another ramification of the inherent nonlinearity of hypersonic flow.

15.7 | HYPERSONIC SIMILARITY

The concept of flow similarity is well entrenched in fluid mechanics. In general, two or more different flows are defined to be dynamically similar when: (1) the streamline shapes of the flows are geometrically similar, and (2) the variation of the flowfield properties is the same for the different flows when plotted in a nondimensional geometric space. Such dynamic similarity is ensured when: (1) the body shapes are geometrically similar, and (2) certain nondimensional parameters involving freestream properties and lengths, called similarity parameters, are the same between the different flows. See Ref. 104 for a more detailed discussion of flow similarity.

In the present section, we discuss a special aspect of flow similarity which applies to hypersonic flow over slender bodies. In the process, we will identify what is meant by hypersonic similarity, and will define a useful quantity called the hypersonic similarity parameter.

Consider a slender body at hypersonic speeds. The governing equations are Eqs. (15.64) through (15.68). To these equations must be added the boundary conditions at the body surface and behind the shock wave. At the body surface, the flow tangency condition is given by Eq. (15.41), repeated below:

$$un_x + vn_y + wn_z = 0 \qquad (15.41)$$

In terms of the perturbation velocities defined in Sec. 15.6, Eq. (15.41) becomes

$$(V_\infty + u')n_x + v'n_y + w'n_z = 0 \qquad (15.69)$$

In terms of the nondimensional perturbation velocities defined in Sec. 15.6, Eq. (15.69) becomes

$$\left(\frac{1}{\tau^2} + \bar{u}'\right)(V_\infty \tau^2)n_x + \bar{v}'(V_\infty \tau)n_y + \bar{w}'(V_\infty \tau)n_z = 0$$

or

$$(1 + \tau^2 \bar{u}')n_x + \bar{v}'\tau n_y + \bar{w}'\tau n_z = 0 \qquad (15.70)$$

In Eq. (15.70), the direction cosines n_x, n_y, and n_z are in the (x, y, z) space; these values are somewhat changed in the transformed space $(\bar{x}, \bar{y}, \bar{z})$ defined in Sec. 15.6. Letting $\bar{n}_x, \bar{n}_y$, and $\bar{n}_z$ denote the direction cosines in the transformed space, we have (within the slender body assumption)

$$n_x = \tau \bar{n}_x \qquad n_y = \bar{n}_y \qquad n_z = \bar{n}_z \qquad (15.71)$$

(See Ref. 119 for a more detailed discussion of the transformed direction cosines.) With the relations given in Eqs. (15.71), the boundary condition given by Eqs. (15.70) becomes

$$(1 + \tau^2 \bar{u}')\tau \bar{n}_x + \bar{v}'\tau \bar{n}_y + \bar{w}'\tau \bar{n}_z = 0$$

or

$$(1 + \tau^2 \bar{u}')\bar{n}_x + \bar{v}'\bar{n}_y + \bar{w}'n_z = 0 \qquad (15.72)$$

Consistent with the derivation of the hypersonic small-disturbance equations in Sec. 15.6, we neglect the term of order τ^2 in Eq. (15.72), yielding the final result for the surface boundary condition:

$$\boxed{\bar{n}_x + \bar{v}'\bar{n}_y + \bar{w}'\bar{n}_z = 0} \qquad (15.73)$$

The shock boundary conditions, consistent with the transformed coordinate system, can be obtained as follows. Consider Eq. (4.8), repeated here:

$$\frac{\rho_2}{\rho_\infty} = \bar{\rho}_2 = \frac{(\gamma + 1)M_\infty^2 \sin^2 \beta}{(\gamma - 1)M_\infty^2 \sin^2 \beta + 2} \qquad (4.8)$$

or

$$\bar{\rho}_2 = \left(\frac{\gamma + 1}{\gamma - 1}\right)\left[\frac{M_\infty^2 \sin^2 \beta}{M_\infty^2 \sin^2 \beta + 2/(\gamma - 1)}\right] \qquad (15.74)$$

For hypersonic flow over a slender body, β is small. Hence,

$$\sin \beta \approx \beta \approx \left(\frac{dy}{dx}\right)_s = \left(\frac{d\bar{y}}{d\bar{x}}\right)_s \tau$$

where $(d\bar{y}/d\bar{x})$ is the slope of the shock wave in the transformed space. Thus, Eq. (15.74) becomes

$$\boxed{\bar{\rho}_2 = \left(\frac{\gamma + 1}{\gamma - 1}\right)\left\{\frac{(d\bar{y}/d\bar{x})_s}{(d\bar{y}/d\bar{x})_s^2 + 2/(\gamma - 1)M_\infty^2 \tau^2}\right\}} \qquad (15.75)$$

Repeating Eq. (4.9),

$$\frac{p_2}{p_\infty} = 1 + \frac{2\gamma}{\gamma + 1} \left(M_\infty^2 \sin^2 \beta - 1 \right) \tag{4.9}$$

and recalling that $\bar{p} = p/\gamma M_\infty^2 \tau^2 \rho_\infty$, Eq. (4.9) becomes

$$\frac{p_2}{\gamma M_\infty^2 \tau^2 p_\infty} = \frac{1}{\gamma M_\infty^2 \tau^2} + \frac{2\gamma}{\gamma + 1} \left(M_\infty^2 \sin^2 \beta - 1 \right) \frac{1}{\gamma M_\infty^2 \tau^2}$$

$$\bar{p}_2 = \frac{1}{\gamma M_\infty^2 \tau^2} + \frac{2\gamma}{\gamma + 1} \left[M_\infty^2 \tau^2 \left(\frac{d\bar{y}}{d\bar{x}} \right)_s^2 - 1 \right] \frac{1}{\gamma M_\infty^2 \tau^2}$$

$$\bar{p}_2 = \frac{1}{\gamma M_\infty^2 \tau^2} + \frac{2(d\bar{y}/d\bar{x})_s^2}{\gamma + 1} - \frac{2}{(\gamma + 1) M_\infty^2 \tau^2}$$

$$\bar{p}_2 = \frac{2(d\bar{y}/d\bar{x})_s^2}{\gamma + 1} + \frac{(\gamma - 1) - 2\gamma}{\gamma(\gamma + 1) M_\infty^2 \tau^2}$$

$$\boxed{\bar{p}_2 = \frac{2}{\gamma + 1} \left[\left(\frac{d\bar{y}}{d\bar{x}} \right)_s^2 + \frac{1 - \gamma}{2\gamma M_\infty^2 \tau^2} \right]} \tag{15.76}$$

Repeating Eq. (15.4),

$$\frac{u_2}{V_\infty} = 1 - \frac{2 \left(M_\infty^2 \sin^2 \beta - 1 \right)}{(\gamma + 1) M_\infty^2} \tag{15.4}$$

and recalling that $u_2 = V_\infty + u_2' = \bar{u}_2'/V_\infty \tau^2$, Eq. (15.4) becomes

$$1 + \frac{u_2'}{V_\infty} = 1 - \frac{2 \left[M_\infty^2 \tau^2 (d\bar{y}/d\bar{x})_s^2 - 1 \right]}{(\gamma + 1) M_\infty^2}$$

$$\frac{u_2'}{V_\infty \tau^2} = \frac{2 \left[M_\infty^2 \tau^2 (d\bar{y}/d\bar{x})_s^2 - 1 \right]}{(\gamma + 1) M_\infty^2 \tau^2}$$

$$\boxed{\bar{u}_2' = -\frac{2}{\gamma + 1} \left[\left(\frac{d\bar{y}}{d\bar{x}} \right)_s^2 - \frac{1}{M_\infty^2 \tau^2} \right]} \tag{15.77}$$

Repeating Eq. (15.6),

$$\frac{v_2}{V_\infty} = \frac{2 \left(M_\infty^2 \sin^2 \beta - 1 \right) \cot \beta}{(\gamma + 1) M_\infty^2} \tag{15.6}$$

and recalling that $v_2 = v_2'$ and $\bar{v}_2' = \bar{v}_2'/V_\infty \tau$, Eq. (15.6) becomes

$$\frac{v_2'}{V_\infty \tau} = \frac{2}{\gamma + 1} \left[\beta^2 - \frac{1}{M_\infty^2} \right] \frac{1}{\beta \tau}$$

$$\bar{v}_2' = \frac{2}{\gamma + 1} \left[\left(\frac{d\bar{y}}{d\bar{x}} \right)_s^2 \tau^2 - \frac{1}{M_\infty^2} \right] \frac{1}{(d\bar{y}/d\bar{x})_s \tau^2}$$

$$\boxed{\bar{v}_2' = \frac{2}{\gamma + 1} \left[\left(\frac{d\bar{y}}{d\bar{x}} \right)_s^2 - \frac{1}{M_\infty^2 \tau^2} \right] \frac{1}{(d\bar{y}/d\bar{x})_s}} \qquad (15.78)$$

Equations (15.75) through (15.78) represent boundary conditions immediately be-hind the shock wave in terms of the transformed variables. Note that these equations were obtained from the exact oblique shock relations, making only the one assumption of small wave angle; nothing was said about very high Mach numbers; hence, Eqs. (15.75) through (15.78) should apply to moderate as well as large hypersonic Mach numbers.

Examine carefully the complete system of equations for hypersonic flow over a slender body—the governing flow equations [Eq. (15.64) through (15.68)], the surface boundary condition [Eq. (15.73)], and the shock boundary conditions [Eqs. (15.75) through (15.78)]. For this complete system, the free-stream Mach number M_∞ and the body slenderness ratio τ appear only as the product $M_\infty \tau$, and this appears only in the shock boundary conditions. The product $M_\infty \tau$ is identified as the hypersonic similarity parameter, which we will denote by K.

Hypersonic similarity parameter: $K \equiv M_\infty \tau$

Important: The meaning of the hypersonic similarity parameter becomes clear from an examination of the complete system of equations. Since $M_\infty \tau$ and γ are the only parameters that appear in these nondimensional equations, then solutions for two different flows over two different but affinely related bodies (bodies which have essentially the same mathematical shape, but which differ by a scale factor on one direction, such as different values of thickness) will be the same (in terms of the nondimensional variables, $\bar{u}'$, $\bar{v}'$, etc.) if γ and $M_\infty \tau$ are the same between the two flows. This is the principle of *hypersonic similarity*.

For affinely related bodies at a small angle of attack α, the principle of hypersonic similarity holds as long as, in addition to γ and $M_\infty \tau$, α/τ is also the same. For this case, the only modification to the above derivation occurs in the surface boundary condition, which is slightly changed; for small α, Eq. (15.73) is replaced by

$$\left(\bar{n}_x + \frac{\alpha}{\tau} \right) + \bar{v}' \bar{n}_y + \bar{w}' \bar{n}_z = 0 \qquad (15.79)$$

The derivation of Eq. (15.79) as well as an analysis of the complete system of equations for the case of small α, is left to the reader as a homework problem. In summary, including the effect of angle of attack, the solution of the governing equations along with the boundary conditions takes the functional form

$$\bar{p} = \bar{p}\left(\bar{x}, \bar{y}, \bar{z}, \gamma, M_\infty \tau, \frac{\alpha}{\tau}\right)$$

$$\bar{\rho} = \bar{\rho}\left(\bar{x}, \bar{y}, \bar{z}, \gamma, M_\infty \tau, \frac{\alpha}{\tau}\right)$$

etc.

Therefore, hypersonic similarity means that, if γ, $M_\infty \tau$, and α/τ are the same for two or more different flows over affinely related bodies, then the variation of the nondimensional dependent variables over the nondimensional space $\bar{p} = \bar{p}(\bar{x}, \bar{y}, \bar{z})$, etc., is clearly the same between the different flows.

Consider the pressure coefficient, defined as

$$C_p = \frac{p - p_\infty}{\frac{1}{2}\rho_\infty V_\infty^2} = \frac{p - p_\infty}{(\gamma/2)p_\infty M_\infty^2}$$

This can be written in terms of $\bar{p}$ as

$$C_p = \frac{2(p - p_\infty)\tau^2}{\gamma p_\infty M_\infty^2 \tau^2} = 2\tau^2\left(\bar{p} - \frac{1}{\gamma M_\infty^2 \tau^2}\right) \tag{15.80}$$

Since $\bar{p} = \bar{p}(\bar{x}, \bar{y}, \bar{z}, \gamma, M_\infty \tau, \alpha/\tau)$, then Eq. (15.80) becomes the following functional relation:

$$\boxed{\frac{C_p}{\tau^2} = f_1\left(\bar{x}, \bar{y}, \bar{z}, \gamma, M_\infty \tau, \frac{\alpha}{\tau}\right)} \tag{15.81}$$

From Eq. (15.81), we see another aspect of hypersonic similarity, namely, that flows over related bodies with the same values of γ, $M_\infty \tau$, and α/τ will have the same value of C_p/τ^2.

Since the lift and wave drag coefficients are obtained by integrating C_p over the body surface (see Ref. 104), then it is relatively straightforward to show that (see Ref. 119):

1. For a two-dimensional shape, referenced to planform area per unit span

$$\boxed{\begin{aligned} \frac{c_l}{\tau^2} &= f_2\left(\gamma, M_\infty \tau, \frac{\alpha}{\tau}\right) \\ \frac{c_d}{\tau^3} &= f_3\left(\gamma, M_\infty \tau, \frac{\alpha}{\tau}\right) \end{aligned}}$$ Referenced to planform area

2. For a three-dimensional shape, referenced to base area,

$$\boxed{\begin{aligned} \frac{C_L}{\tau} &= F_1\left(\gamma, M_\infty \tau, \frac{\alpha}{\tau}\right) \\ \frac{C_D}{\tau^2} &= F_2\left(\gamma, M_\infty \tau, \frac{\alpha}{\tau}\right) \end{aligned}}$$ Referenced to base area

Examine the results summarized in the two boxes above, namely the results for c_l and c_d for a two-dimensional low, and C_L and C_D for a three-dimensional flow. From these results, the principle of hypersonic similarity states that affinely related bodies with the same values of γ, $M_\infty \tau$, and α/τ will have: (1) the same values of c_l/τ^2 and c_d/τ^3 for two-dimensional flows, when referenced to planform area; and (2) the same values of C_L/τ and C_D/τ^2 for three-dimensional flows when referenced to base area.

The validity of the hypersonic similarity principle is verified by the results shown in Fig. 15.14, obtained from the work of Neice and Ehret (Ref. 125). Consider first Fig. 15.14a, which shows the variation of C_p/τ^2 as a function of distance downstream of the nose of a slender ogive-cylinder (as a function of $x = x/l$, expressed in percent of nose length). Two sets of data are presented, each for a different M_∞ and τ, but such that the product $K \equiv M_\infty \tau$ is the same value, namely 0.5. The data are exact calculations made by the method of characteristics. Hypersonic similarity states that the two sets of data should be identical, which is clearly the case shown in Fig. 15.14a.

A similar comparison is made in Fig. 15.14b, except for a higher value of the hypersonic similarity parameter, namely $K = 2.0$. The conclusion is the same; the data for two different values of M_∞ and τ, but with the same K, are identical. An interesting sideline is also shown in Fig. 15.14b. Two different methods of characteristics calculations are made—one assuming irrotational flow (the solid line), and the other treating rotational flow (the dashed line). There are substantial differences in implementing the method of characteristics for these two cases, as explained in Chap. 11. In reality, the flow over the ogive-cylinder is rotational because of the slightly curved shock wave over the nose. The effect of rotationality is to increase the value of C_p, as shown in Fig. 15.14b. However, Neice and Ehret state that no significant differences between the rotational-irrotational calculations resulted for the low value of $K = 0.5$ in Fig. 15.14a, which is why only one curve is shown. One can conclude from this comparison the almost intuitive fact that the effects of rotationality become more important as M_∞, τ, or both are progressively increased. However, the main reason for bringing up the matter of rotationality is to ask the question: Would we expect hypersonic similarity to hold for rotational flows? The question is rhetorical, because the answer is obvious. Examining the governing flow equations upon which hypersonic similarity is based, namely Eqs. (15.64) through (15.68), we note that they contain no assumption of irrotational flow—they apply to both cases. Hence, the principle of hypersonic similarity holds for both irrotational and rotational flows. This is clearly demonstrated in Fig. 15.14b, where the data calculated for irrotational flow for two different values of M_∞ and τ (but the same K) fall on the same curve, and the data calculated for rotational flow for the two different values of M_∞ and τ (but the same K) also fall on the same curve (but a different curve than the irrotational results).

Question: Over what range of values of $K \equiv M_\infty \tau$ does hypersonic similarity hold? The answer cannot be made precise. However, many results show that for very slender bodies (such as a 5° half-angle cone), hypersonic similarity holds for values of K ranging from less than 0.5 to infinitely large. On the other hand, for less slender

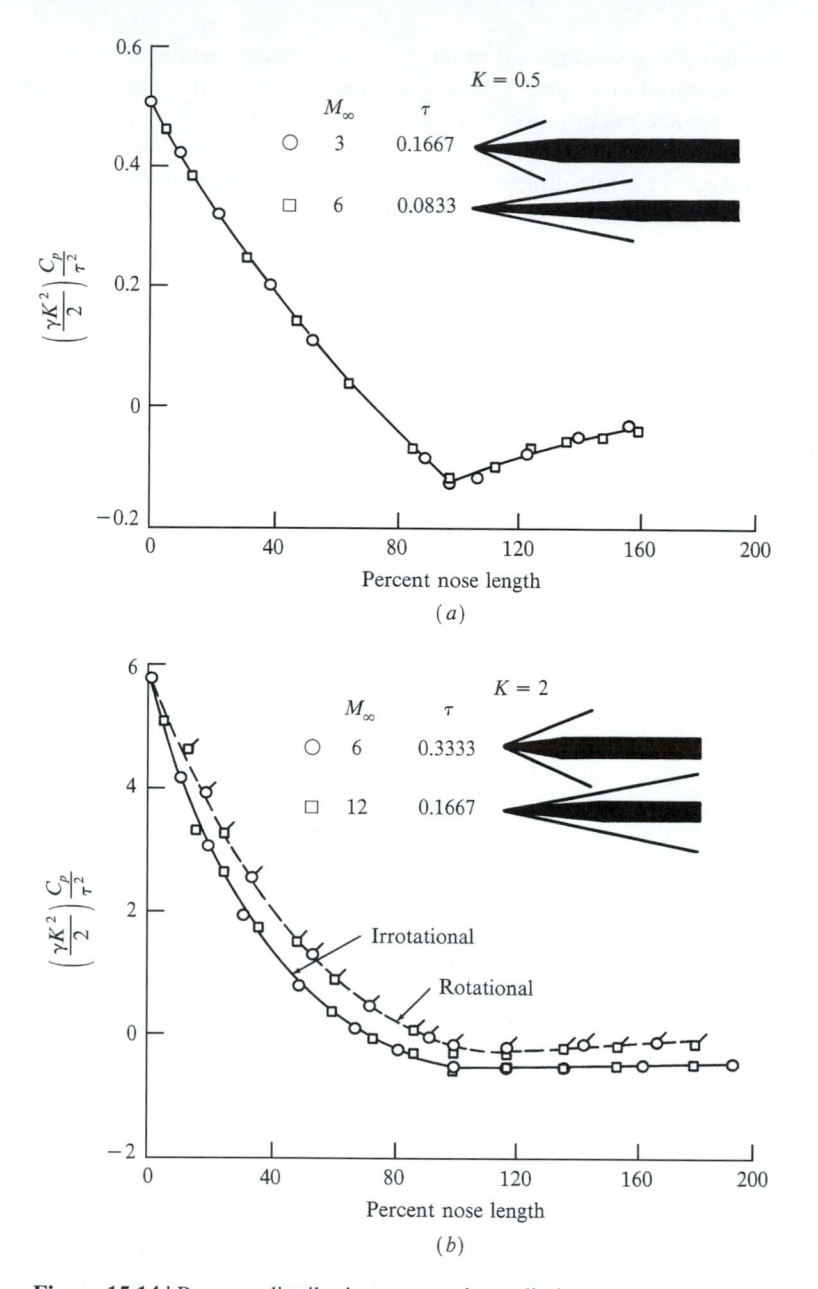

Figure 15.14 | Pressure distributions over ogive-cylinders; illustration of hypersonic similarity. (a) $K = 0.5$; (b) $K = 2.0$. (From Ref. 125.)

bodies (say, a $20°$ half-angle cone), the data do not correlate well until $K > 1.5$. However, always keep in mind that hypersonic similarity is based on the hypersonic small-disturbance equations, and we would expect the results to become more tenuous as the thickness of the body is increased.

An important historical note is in order here. The concept of hypersonic similarity was first developed by Tsien in 1946, and published in Ref. 126. In this paper, Tsien treated a two-dimensional potential (hence irrotational) flow. This work was further extended by Hayes (Ref. 127) who showed that Tsien's results applied to rotational flows as well. (As noted earlier, the development of hypersonic similarity in the present chapter started right from the beginning with the governing equations for rotational flow. There is no need to limit ourselves to the special case treated by Tsien.) However, of equal (or more) historical significance, Tsien's 1946 paper seems to be the source which coined the word hypersonic. After an extensive search of the literature, the present author could find no reference to the word "hypersonic" before 1946. Then, in his 1946 paper—indeed in the title of the paper—Tsien makes liberal use of the word "hypersonic," without specifically stating that he is coining a new word. In this sense, the word "hypersonic" seems to have entered our vocabulary with little or no fanfare.

15.8 | COMPUTATIONAL FLUID DYNAMICS APPLIED TO HYPERSONIC FLOW; SOME COMMENTS

The modern hypersonic aerodynamics of today is paced by computational fluid dynamics (CFD). Indeed, the impact of CFD on hypersonics has been the greatest of all the flight regimes discussed earlier in this book. This is due mainly to the lack of hypersonic ground test facilities for experimental studies, especially at the extreme ends of the spectrum where $M > 20$ and the stagnation temperatures are high enough to cause substantial chemical dissociation of the gas. In lieu of such high-performance facilities, the design of hypersonic vehicles must rely heavily on the results of computational fluid dynamics.

In this sense, the present section is essentially a summary section, because examples of computational fluid dynamics applied to flows with hypersonic Mach numbers can be found throughout this book. For example, in Sec. 11.16, there are examples of both the method of characteristics and an explicit finite-difference technique (essentially MacCormack's explicit technique) applied to a space-shuttle configuration at Mach 7.4. Blunt body solutions at Mach 8 are discussed in Sec. 12.6. Chapter 13 contains many results for three-dimensional flowfields at hypersonic speeds. These results, taken together, serve as our examples of the application of computational fluid dynamics to hypersonic flows. Only one additional example will be discussed here.

In Ref. 129, hypersonic flow over blunt-nosed cones at angle of attack is calculated. A blunt body solution (see Sec. 13.4) is used to obtain the initial data surface from which the three-dimensional method of characteristics (see Sec. 11.10)

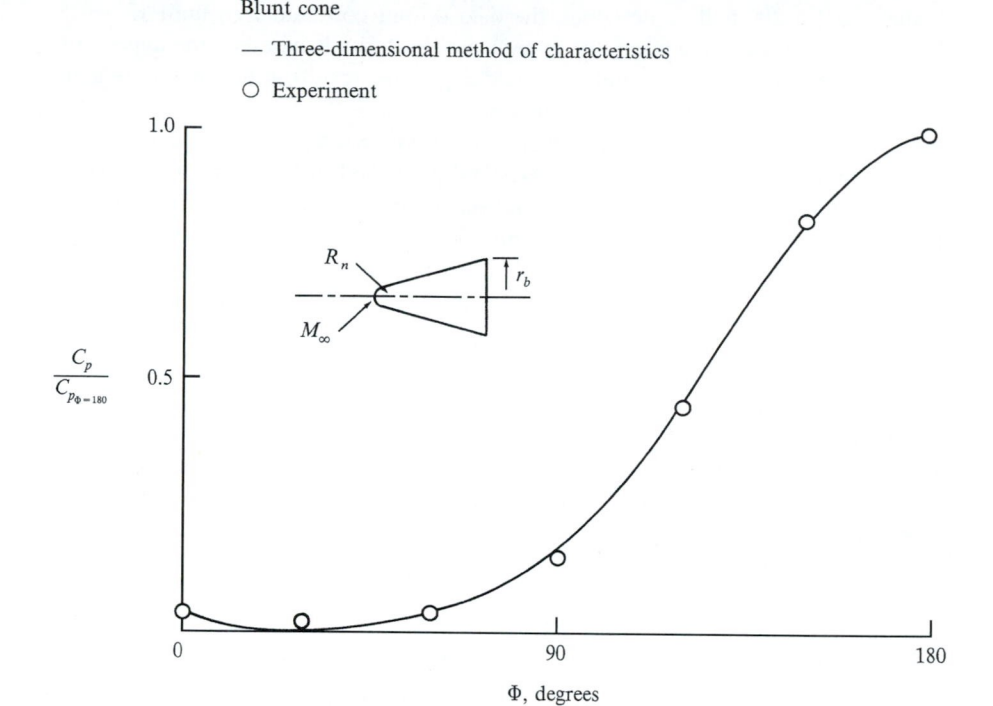

Figure 15.15 | Circumferential surface-pressure distribution at $x/R_n = 8$; comparison between theory and experiment in helium. $\theta_c = 15°$, $\alpha = 20°$, $M_\infty = 14.9$, $\gamma = 1.667$, Re $= 0.86 \times 10^6$. (From Ref. 129.)

is used to calculate the rest of the flowfield. Typical results are shown in Figs. 15.15 and 15.16. In Fig. 15.15, the circumferential pressure distribution around the conical surface at an axial location equal to eight nose radii downstream of the nose is shown. The most leeward location is $\Phi = 0$, and the most windward location is $\Phi = 180°$. The free-stream Mach number is 14.9. The circles are experimental data obtained in a hypersonic wind tunnel using helium as the test gas (for helium, $\gamma = 1.667$). The solid curve represents the calculation, also for $\gamma = 1.667$. Excellent agreement is obtained—a beautiful testimonial to the power of computational techniques applied to a rather complex hypersonic flow. In Fig. 15.16, the axial distributions of pressure coefficient are given for three different values of Φ (three different azimuthal locations). Again, $\Phi = 180°$ is the extreme windward location. Here, the circles represent experimental data obtained in air at Mach 10 (hence $\gamma = 1.4$). The solid curves are the computed results. Again, excellent agreement is obtained. Note that the results for $\Phi = 180°$ show a local overexpansion downstream of the nose, with a local recompression further downstream. This type of pressure variation is typical of the flow over blunt-nosed cones at hypersonic speeds. It appears that, in flowing over the blunt-nosed shape, the flow expands too far; after it reaches the conical part of the body, this overexpansion is then compensated by a local recompression.

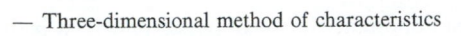

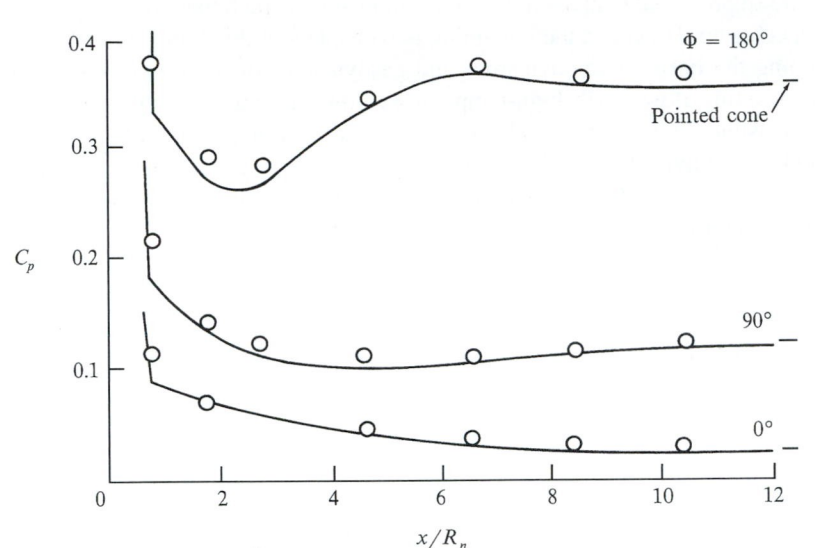

Figure 15.16 | Pressure distributions over a blunt-nosed cone; comparison between theory and experiment in air. $\alpha = 10°$, Re $= 0.6 \times 10^6$, $M_\infty = 10$, $\gamma = 1.4$. (From Ref. 129.)

Also note in Fig. 15.16 that the pressures far downstream approach the sharp-nosed cone results, given by the single dash at the end of each curve. However, the sharp cone results cannot be exactly achieved because of the presence of the entropy layer emanating from the blunt nose. This is in addition to the entropy layer that is always present on a cone—even a sharp-nosed cone—at angle of attack, as was discussed in Chap. 13.

Modern CFD applications to hypersonic flow abound in the literature. Such journals as the *AIAA Journal* and the *Journal of Propulsion and Power* are good sources of such literature.

15.9 | SUMMARY AND FINAL COMMENTS

In this chapter we have discussed some of the basic aspects of "classical" hypersonic aerodynamics, i.e., the hypersonic shock wave relations, newtonian flow, tangent-wedge and tangent-cone methods, Mach number independence, and hypersonic small-perturbation theory leading to the demonstration of hypersonic similarity. We have seen that hypersonic theory is *nonlinear,* even for small perturbations. The "modern" hypersonic aerodynamics is characterized by applications of computational fluid dynamics, as discussed in various sections throughout this book.

One of the most important aspects of hypersonic flow is the large temperature increases associated with such flows, and the resulting chemically reacting flowfields that are so prevalent in many hypersonic applications. Such matters are not discussed in this chapter. However, the remaining two chapters of this book are devoted to developing the basic thermodynamics and gasdynamics of high-temperature, chemically reacting flow. Such high-temperature flows are an essential part of *modern* compressible flow. The material in Chaps. 16 and 17 is applicable to many problems in addition to hypersonics. However, it is particularly applicable to hypersonic flow, and in that sense the following two chapters, although self-contained, can be visualized as a natural extension of the present chapter.

CHAPTER 16

Properties of
High-Temperature Gases

Science is eternal. It was started thousands of years ago and its progress is continuous. Principles that are deeply rooted are not likely to pass suddenly from the scene.

Theodore von Karman, 1963

PREVIEW BOX

Explosions, combustion, the searing temperatures associated with the very high speed flow of a gas—these are some examples of compressible flow where high-temperature effects must be taken into account. For these and other situations the assumption of a calorically perfect gas with constant specific heats, which has permeated all of the previous chapters, is simply not good enough. We have to take into account chemical reactions occurring in the flow, and other physical phenomena that cause the specific heats to be variable. This changes the complexion of our analyses and calculations completely. We have to deal with a combination of physical chemistry with gas dynamics—a combination that is exciting, important, and particularly interesting to study. In the present chapter we introduce some basic aspects of physical chemistry, which will then be combined with gas dynamics in the next chapter to study some fundamental high-temperature flows.

Return to the overall roadmap for the book given in Fig. 1.7. We now move to the extreme right, box 17, to study high-temperature flow. This will round out our overall study of modern compressible flow as represented by the entire map in Fig. 1.7.

The roadmap for the present chapter is given in Fig. 16.1. Here we will deal with two different states of gases, *equilibrium* as represented on the left of Fig. 16.1,

and *nonequilibrium* as represented on the right. These are two distinctly different situations; some high-temperature flows can be analyzed by assuming local equilibrium, but many others are dominated by nonequilibrium processes. So both sides of the roadmap in Fig. 16.1 are equally important. We will start on the left, and introduce some fundamental aspects of statistical thermodynamics (box 1), which in turn leads to the equilibrium thermodynamic properties of a single-species gas (box 2). Then we show how the equilibrium chemical composition of a chemically reacting gas can be calculated (box 3). This tells us how much of each chemical species is present in the gas. Boxes 2 and 3 are then combined to obtain the equilibrium thermodynamic properties of a chemically reacting gas. Finally, we move to the right-hand side of the roadmap, and deal with nonequilibrium processes. We emphasize two such processes: molecular vibrational nonequilibrium (box 5), and chemical nonequilibrium (box 6).

In short, view this chapter as a crash course in physical chemistry. We discuss only those aspects that are necessary for applications to compressible flows. We assume no prior knowledge of this material on your part; the concepts and equations are developed from first principles. View this chapter as an interesting and awarding adventure.

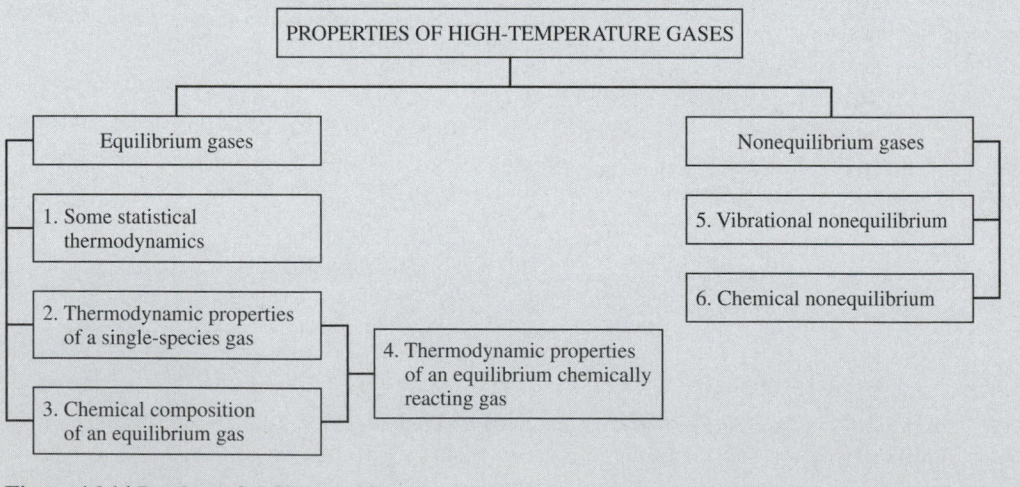

Figure 16.1 | Roadmap for Chapter 16.

16.1 | INTRODUCTION

Consider the atmospheric entry of the Apollo command vehicle upon return from the moon. At an altitude of approximately 53 km, the velocity of the vehicle is 11 km/s. As sketched in Fig. 12.20, a strong bow shock wave is wrapped around the blunt nose, and the shock layer between the shock and the body is relatively thin. This is a blunt body flowfield, as discussed in Chap. 12. Moreover, at a standard altitude of 53 km, the air temperature is 283 K and the resulting speed of sound is 338 m/s; hence the Mach number of the Apollo vehicle is 32.5—an extremely large hypersonic value. Using the theory developed in Chap. 3 for a calorically perfect gas, let us estimate the temperature in this shock layer. From Table A.2, for $M = 32.5$, $T_s/T_\infty = 206$, where T_s is the static temperature behind the normal portion of the bow shock wave. Hence, $T_s = (206)(283) = 58,300$ K. This is an extremely high temperature; it is also completely *incorrect*. Long before this temperature is reached, the air molecules will dissociate and ionize. Indeed, the shock layer becomes a partially ionized plasma, where the specific heat of the gas is a strong function of both pressure and temperature. The assumption of a calorically perfect gas made above is completely inaccurate; when this chemically reacting gas is properly calculated, the shock layer temperature is on the order of 11,600 K—still a high value, but a factor of 5 less than the temperature predicted on the basis of a calorically perfect gas. Figure 16.2 compares the variation of shock layer temperature as a function of flight velocity for both the cases of a calorically perfect gas and an equilibrium chemically reacting gas. Also noted are typical reentry velocities for various space vehicles such as an intermediate range ballistic missile (IRBM), intercontinental ballistic missile (ICBM), earth orbital vehicles (e.g., Mercury and Gemini), lunar return vehicles (e.g., Apollo) and Mars return vehicles. Clearly, for all such cases, the assumption of a calorically perfect gas is not appropriate; the effects of chemical reactions must be taken into account.

The remainder of this book will deal with the compressible flow of high-temperature gases. The importance of such a flow was illustrated above. In modern engineering applications, there are many other such examples: the flow through rocket engines, arc-driven hypersonic wind tunnels, high-performance shock tubes, high-energy gasdynamic and chemical lasers, and internal combustion engines, to name just a few. Therefore, as stated in Sec. 1.6, a study of modern compressible flow must include some discussion of high-temperature effects. To this end, the present chapter deals with the thermodynamics of high-temperature chemically reacting gases, providing the necessary foundation for the analyses of high-temperature flows to be given in Chap. 17. The objective of these remaining chapters is to give the reader some appreciation for the limitations of the calorically perfect gas results developed in previous chapters, as well as an ability to make modern compressible flow computations which properly include high-temperature effects.

There are two major physical characteristics which cause a high-temperature gas to deviate from calorically perfect gas behavior:

1. As the temperature of a diatomic or polyatomic gas is increased above standard conditions, the vibrational motion of the molecules will become important,

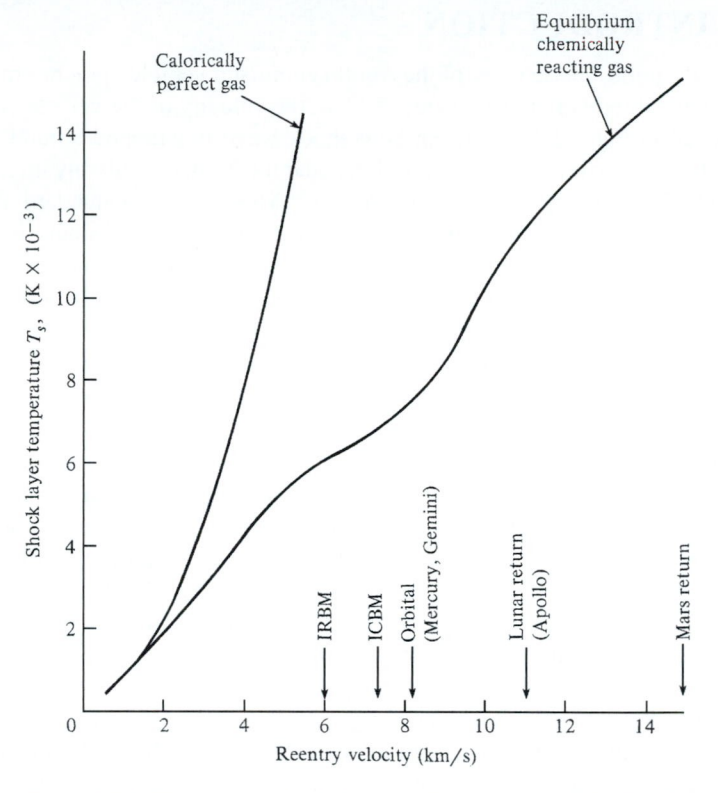

Figure 16.2 | Temperature behind a normal shock versus velocity for
air at a standard altitude of 52 km. Comparison between calorically
perfect and equilibrium gas results.

absorbing some of the energy which otherwise would go into the translational
and rotational molecular motion. As we shall soon see, the excitation of
vibrational energy causes the specific heat to become a function of temperature,
i.e., the gas gradually shifts from calorically perfect to thermally perfect (see
Sec. 1.4).

2. As the gas temperature is further increased, the molecules will begin to
    dissociate (the atoms constituting the various molecules will break away from
    the molecular structure) and even ionize (electrons will break away from the
    atoms). Under these conditions, the gas becomes chemically reacting, and the
    specific heat becomes a function of both temperature and pressure. If we
    consider air at 1 atm, the approximate temperatures at which various reactions
    will become important are illustrated in Fig. 16.3. If the gas is at lower
    pressure, these temperatures shift downward; later in this chapter we will
    learn why.

These physical effects—vibrational excitation and chemical reactions—will be
highlighted in the remainder of this book. The purpose of the present chapter is to

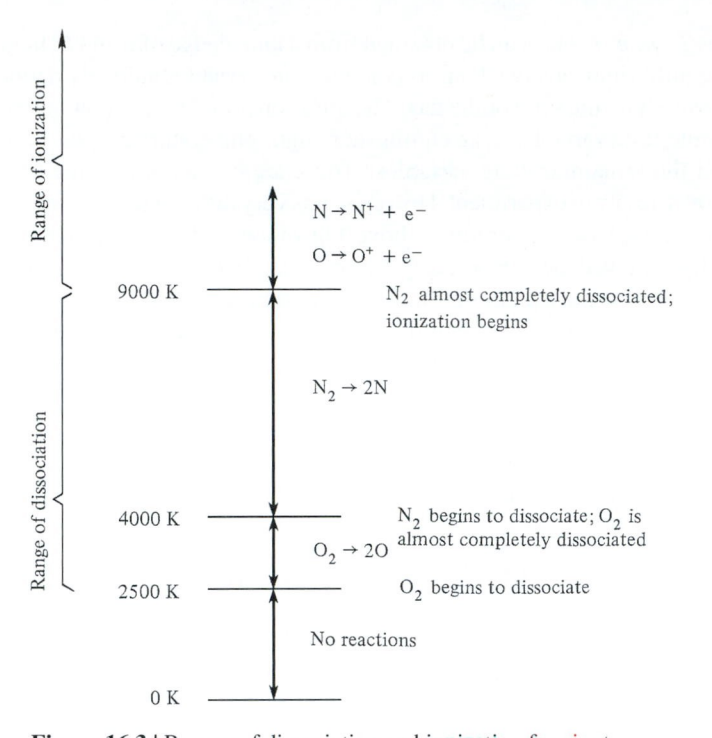

Figure 16.3 | Ranges of dissociation and ionization for air at approximately 1-atm pressure.

establish the thermodynamic behavior of such gases, much as Sec. 1.4 laid the basis for our previous flowfield analyses dealing principally with a calorically perfect gas. However, unlike Sec. 1.4, which was a review, the assumption is made here that the reader has not had a previous background in high-temperature thermodynamics. Therefore, some effort will be made to provide a background sufficiently thorough that the reader will feel comfortable with the results, to be used in Chap. 17.

To elaborate, an essential ingredient of any high-temperature flowfield analysis is the knowledge of the thermodynamic properties of the gas. For example, consider again the flowfield shown in Fig. 12.20. Assume that the gas is in local thermodynamic and chemical equilibrium (concepts to be defined later). The unknown flowfield variables, and how they can be obtained, are itemized as follows:

ρ—density $\left.\vphantom{\begin{matrix}1\\2\\3\end{matrix}}\right\}$ Obtained from a simultaneous solution
$\mathbf{V}$—velocity of the continuity, momentum, and
h—enthalpy energy equations

$T = T(\rho, h)$ $\left.\vphantom{\begin{matrix}1\\2\end{matrix}}\right\}$ Obtained from the equilibrium thermodynamic
$p = p(\rho, h)$ properties of high-temperature air

Here, we conceptually see that two thermodynamic variables ρ and h are obtained from the flowfield conservation equations, and that the remaining thermodynamic variables T, p, e, s, etc., can be obtained from a knowledge of ρ and h. In general, for a gas in equilibrium, any two thermodynamic state variables uniquely define the complete thermodynamic state of the gas. The question posed here is that, given two thermodynamic state variables in an equilibrium high-temperature gas, *how* do we obtain values of the remaining state variables? There are two answers. One is to *measure* these properties from experiment. However, it is very difficult to carry out accurate experiments on gases at temperatures above a few thousand degrees; such temperatures are usually achieved in the laboratory for only short periods of time in devices such as shock tubes, or by pulsed laser radiation absorption. The other answer is to *calculate* these properties. Fortunately, the powerful discipline of statistical mechanics developed over the last century, along with the advent of quantum mechanics in the early twentieth century, gives us a relatively quick and extremely accurate method of calculating equilibrium thermodynamic properties of high-temperature gases. These concepts form the basis of *statistical thermodynamics,* the elements of which will be developed and used in the following sections.

16.2 | MICROSCOPIC DESCRIPTION OF GASES

A molecule is a collection of atoms bound together by a rather complex intramolecular force. A simple concept of a diatomic molecule (two atoms) is the "dumbbell" model sketched in Fig. 16.4a. This molecule has several modes (forms) of energy:

1. It is moving through space, and hence it has translational energy $\varepsilon'_{\text{trans}}$, as sketched in Fig. 16.4b. The source of this energy is the translational kinetic energy of the center of mass of the molecule. Since the molecular translational velocity can be resolved into three components (such as V_x, V_y, and V_z in the xyz cartesian space shown in Fig. 16.4b), the molecule is said to have three "geometric degrees of freedom" in translation. Since motion along each coordinate direction contributes to the total kinetic energy, the molecule is also said to have three "thermal degrees of freedom."

2. It is rotating about the three orthogonal axes in space, and hence it has *rotational* energy $\varepsilon'_{\text{rot}}$, as sketched in Fig. 16.4c. The source of this energy is the rotational kinetic energy associated with the molecule's rotational velocity and its moment of inertia. However, for the diatomic molecule shown in Fig. 16.4c, the moment of inertia about the internuclear axis (the z axis) is very small, and therefore the rotational kinetic energy about the z axis is negligible in comparison to rotation about the x and y axis. Therefore, the diatomic molecule is said to have only two "geometric" as well as two "thermal" degrees of freedom. The same is true for a linear polyatomic molecule such as CO_2 shown in Fig. 16.4d. However, for nonlinear molecules, such as H_2O also shown in Fig. 16.4d, the number of geometric (and thermal) degrees of freedom in rotation are three.

3. The atoms of the molecule are vibrating with respect to an equilibrium location within the molecule. For a diatomic molecule, this vibration is modeled by a spring connecting the two atoms, as illustrated in Fig. 16.4e. Hence the

(a) Diatomic molecule

(b) Translational energy ϵ'_{trans}

Source

Translational kinetic energy of the center of mass (thermal degrees of freedom—3)

(c) Rotational energy ϵ'_{rot}

Rotational kinetic energy; (thermal degrees of freedom—2 for diatomic; 2 for linear polyatomic; and 3 for nonlinear polyatomic)

Rotational energy about the internuclear axis for a diatomic molecule is negligibly small.

(d)

CO_2; linear polyatomic molecule

H_2O; nonlinear polyatomic molecule

(e) Vibrational energy ϵ'_{vib}

1. kinetic energy
2. potential energy

(thermal degrees of freedom—2)

(f) Electronic energy ϵ'_{el}

Electrons

Nucleus

1. Kinetic energy of electrons in orbit
2. Potential energy of electrons in orbit

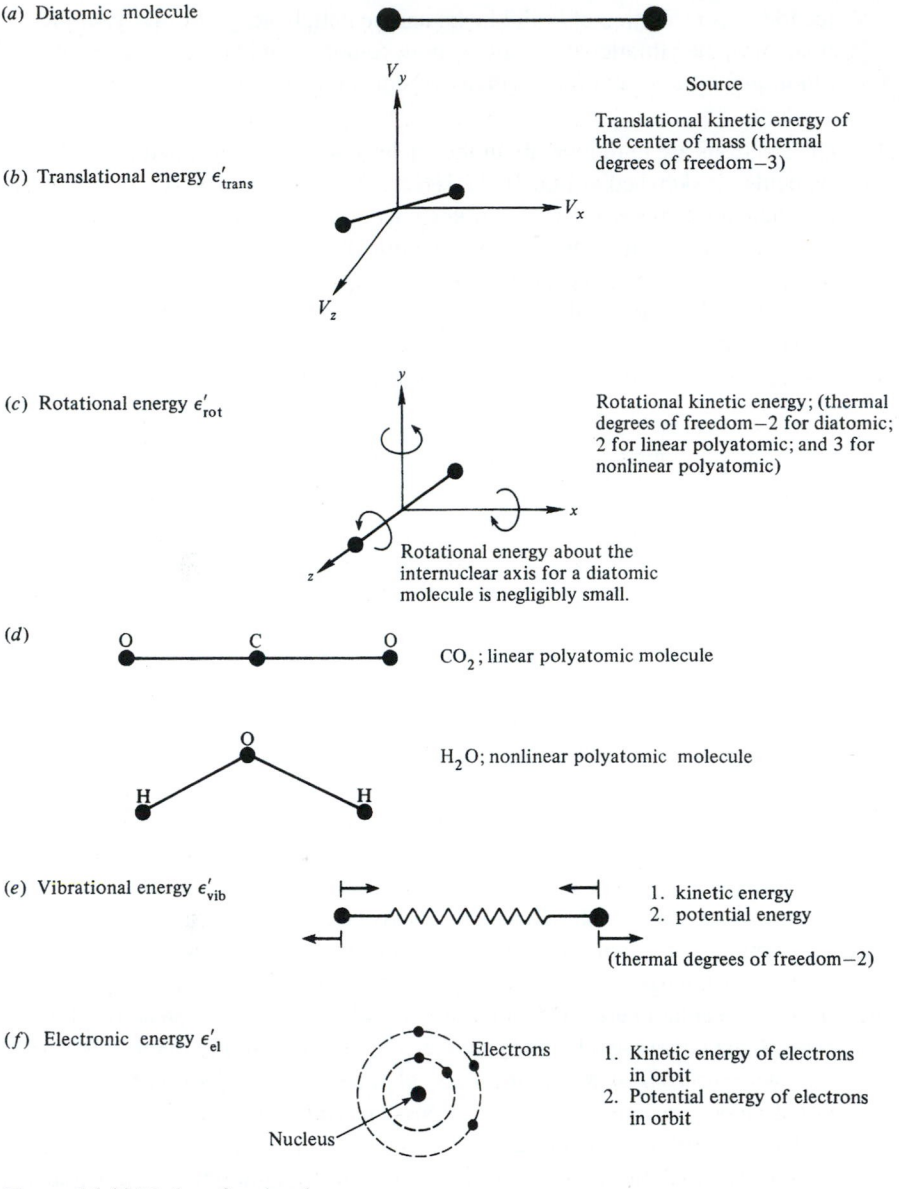

Figure 16.4 | Modes of molecular energy.

molecule has *vibrational* energy ε'_{vib}. There are two sources of this vibrational energy: the kinetic energy of the linear motion of the atoms as they vibrate back and forth, and the potential energy associated with the intramolecular force (symbolized by the spring). Hence, although the diatomic molecule has only one geometric degree of freedom (it vibrates only along one direction, namely, that of the internuclear axis), it has *two* thermal degrees of freedom due to the contribution of both kinetic and potential energy. For polyatomic molecules, the vibrational motion is more complex, and numerous fundamental vibrational modes can occur, with a consequent large number of degrees of freedom.

4. The electrons are in motion about the nucleus of each atom constituting the molecule, as sketched in Fig. 16.4f. Hence, the molecule has electronic energy ε'_{el}. There are two sources of electronic energy associated with each electron: kinetic energy due to its translational motion throughout its orbit about the nucleus, and potential energy due to its location in the electromagnetic force field established principally by the nucleus. Since the overall electron motion is rather complex, the concepts of geometric and thermal degrees of freedom are usually not useful for describing electronic energy.

Therefore, we see that the total energy of a molecule, ε', is the sum of its translational, rotational, vibrational, and electronic energies:

$$\varepsilon' = \varepsilon'_{trans} + \varepsilon'_{rot} + \varepsilon'_{vib} + \varepsilon'_{el} \qquad \text{(for molecules)}$$

For a single atom, only the translational and electronic energies exist:

$$\varepsilon' = \varepsilon'_{trans} + \varepsilon'_{el} \qquad \text{(for atoms)}$$

The results of quantum mechanics have shown that each of these energies is *quantized,* i.e., they can exist only at certain discrete values, as schematically shown in Fig. 16.5. This is a dramatic result. Intuition, based on our personal observations of nature, would tell us that at least the translational and rotational energies could be any value chosen from a continuous range of values (i.e., the complete real number system). However, our daily experience deals with the macroscopic, not the microscopic world, and we should not always trust our intuition when extrapolated to the microscopic scale of molecules. A major benefit of quantum mechanics is that it correctly describes microscopic properties, some of which are contrary to intuition. In the case of molecular energy, *all* modes are quantized, even the translational mode. These quantized energy levels are symbolized by the ladder-type diagram shown in Fig. 16.5, with the vertical height of each level as a measure of its energy. Taking the vibrational mode for example, the lowest possible vibrational energy is symbolized by $\varepsilon'_{0_{vib}}$. The next allowed quantized value is $\varepsilon'_{1_{vib}}$, then $\varepsilon'_{2_{vib}}, \ldots, \varepsilon'_{i_{vib}}, \ldots$. The energy of the ith vibrational energy level is $\varepsilon'_{i_{vib}}$, and so forth. Note that, as illustrated in Fig. 16.5, the spacing between the translational energy levels is very small, and if we were to look at this translational energy level diagram from across the room, it would look almost continuous. The spacings between rotational energy levels are much larger than between the translational energies; moreover, the spacing between two adjacent rotational levels increases as the energy increases (as we go up the ladder in Fig. 16.5). The spacings between vibrational levels are much larger than between rotational levels; also, contrary to rotation, adjacent vibrational energy levels become more closely spaced as the energy increases. Finally, the spacings between electronic

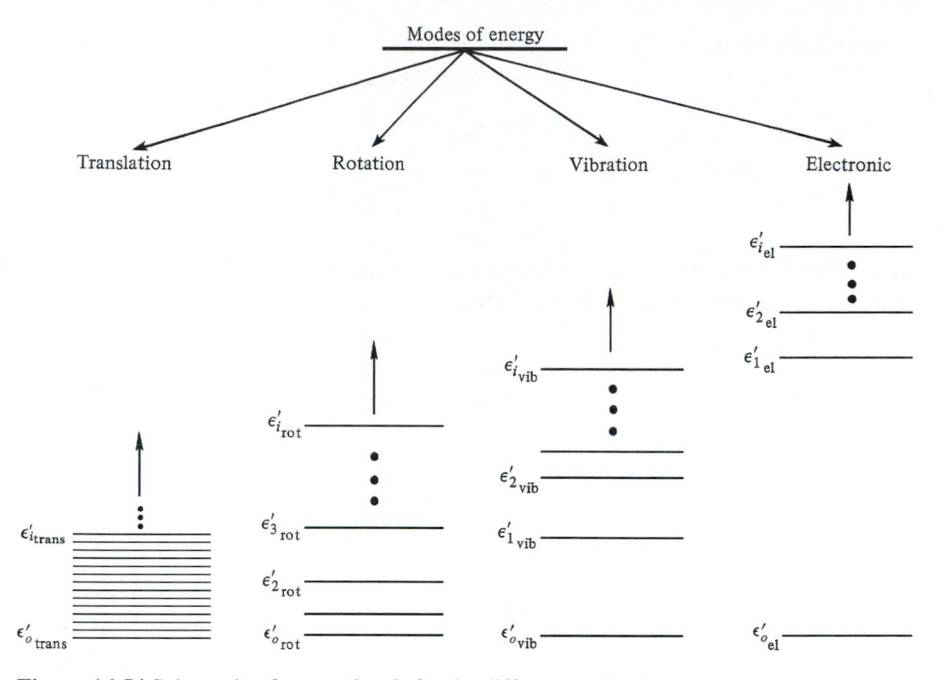

Figure 16.5 | Schematic of energy levels for the different molecular energy modes.

levels are considerably larger than between vibrational levels, and the difference between adjacent electronic levels decreases at higher electronic energies. The quantitative calculation of all these energies will be given in Sec. 16.7.

Again examining Fig. 16.5, note that the lowest allowable energies are denoted by $\varepsilon'_{0_{\text{trans}}}$, $\varepsilon'_{0_{\text{rot}}}$, $\varepsilon'_{0_{\text{vib}}}$, and $\varepsilon'_{0_{\text{el}}}$. These levels are defined as the *ground state* for the molecule. They correspond to the energy that the molecule would have if the gas were theoretically at a temperature of absolute zero; hence the values are also called the *zero-point energies* for the translational, rotational, vibrational, and electronic modes, respectively. It will be shown in Sec. 16.7 that the rotational zero-point energy is precisely zero, whereas the zero-point energies for translation, vibration, and electronic motion are not. This says that, if the gas were theoretically at absolute zero, the molecules would still have some finite translational motion (albeit very small) as well as some finite vibrational motion. Moreover, it only makes common sense that some electronic motion should theoretically exist at absolute zero, or otherwise the electrons would fall into the nucleus and the atom would collapse. Therefore, the total zero-point energy for a molecule is denoted by ε'_o, where

$$\varepsilon'_o = \varepsilon'_{0_{\text{trans}}} + \varepsilon'_{0_{\text{vib}}} + \varepsilon'_{0_{\text{el}}}$$

recalling that $\varepsilon'_{0_{\text{rot}}} = 0$.

It is common to consider the energy of a molecule as measured *above* its zero-point energy. That is, we can define the translational, rotational, vibrational, and electronic energies all *measured above the zero-point energy* as $\varepsilon_{j_{\text{trans}}}$, $\varepsilon_{k_{\text{rot}}}$, $\varepsilon_{l_{\text{vib}}}$, and $\varepsilon_{m_{\text{el}}}$, respectively, where

$$\varepsilon_{j_{\text{trans}}} = \varepsilon'_{j_{\text{trans}}} - \varepsilon'_{o_{\text{trans}}}$$

$$\varepsilon_{k_{\text{rot}}} = \varepsilon'_{k_{\text{rot}}}$$

$$\varepsilon_{l_{\text{vib}}} = \varepsilon'_{l_{\text{vib}}} - \varepsilon'_{o_{\text{vib}}}$$

$$\varepsilon_{m_{\text{el}}} = \varepsilon'_{m_{\text{el}}} - \varepsilon'_{o_{\text{el}}}$$

(Note that the *unprimed* values denote energy measured *above* the zero-point value.) In light of this, we can write the *total* energy of a molecule as ε'_i, where

$$\varepsilon'_i = \underbrace{\varepsilon_{j_{\text{trans}}} + \varepsilon_{k_{\text{rot}}} + \varepsilon_{l_{\text{vib}}} + \varepsilon_{m_{\text{el}}}}_{\substack{\text{All are measured above the zero-point} \\ \text{energy, thus all are equal to zero at} \\ T = 0\,\text{K}.}} + \underbrace{\varepsilon'_o}_{\substack{\text{This represents zero-point energy,} \\ \text{a fixed quantity for a given molecular} \\ \text{species that is equal to the energy} \\ \text{of the molecule at absolute zero.}}}$$

For an atom, the total energy can be written as

$$\varepsilon'_i = \varepsilon_{j_{\text{trans}}} + \varepsilon_{m_{\text{el}}} + \varepsilon'_o$$

If we examine a single molecule at some given instant in time, we would see that it simultaneously has a zero-point energy ε'_o (a fixed value for a given molecular species), a quantized electronic energy measured above the zero-point, $\varepsilon_{m_{\text{el}}}$, a quantized vibrational energy measured above the zero point, $\varepsilon_{l_{\text{vib}}}$, and so forth for rotation and translation. The total energy of the molecule at this given instant is ε'_i. Since ε'_i is the sum of individually quantized energy levels, then ε'_i itself is quantized. Hence, the allowable *total* energies can be given on a single energy level diagram, where $\varepsilon'_o, \varepsilon'_1, \varepsilon'_2, \ldots, \varepsilon'_i, \ldots$ are the quantized values of the total energy of the molecule.

In the above paragraphs, we have gone to some length to define and explain the significance of molecular energy *levels*. In addition to the concept of an energy level, we now introduce the idea of an energy *state*. For example, quantum mechanics identifies molecules not only with regard to their energies, but also with regard to angular momentum. Angular momentum is a vector quantity, and therefore has an associated direction. For example, consider the rotating molecule shown in Fig. 16.6. Three different orientations of the angular momentum vector are shown; in each orientation, assume the energy of the molecule is the same. Quantum mechanics shows

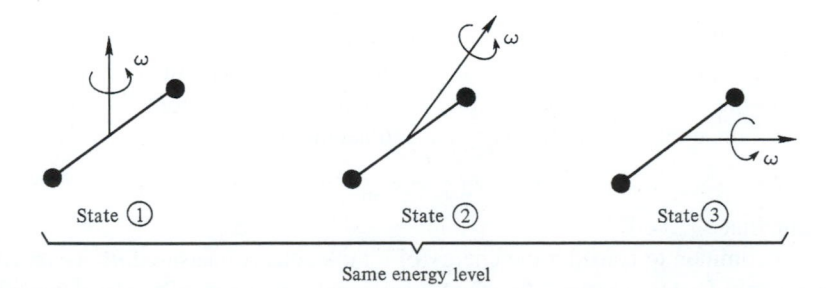

Figure 16.6 I Illustration of different energy states for the same energy level.

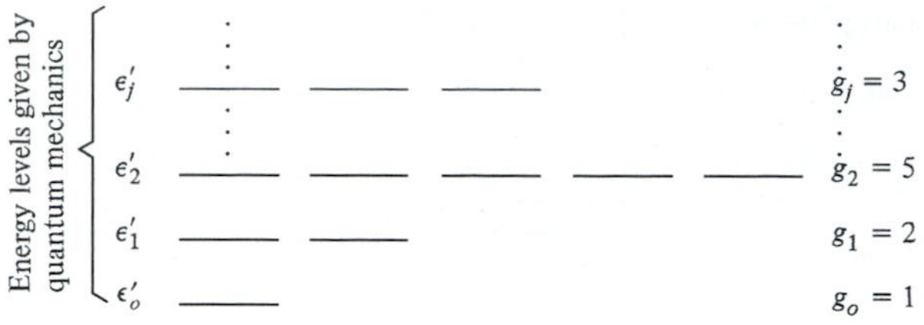

Figure 16.7 | Illustration of statistical weights.

that molecular orientation is also quantized, i.e., it can point only in certain directions. In all three cases shown in Fig. 16.6, the rotational energy is the same, but the rotational momentum has different *directions*. Quantum mechanics sees these cases as different and distinguishable *states*. Different states associated with the same energy level can also be defined for electron angular momentum, electron and nuclear spin, and the rather arbitrary lumping together of a number of closely spaced translational levels into one approximate "level" with many "states."

In summary, we see that, for any given energy level ε_i', there can be a number of different states that all have the same energy. This number of states is called the degeneracy or statistical weight of the given level ε_i', and is denoted by g_i. This concept is exemplified in Fig. 16.7, which shows energy levels in the vertical direction, with the corresponding states as individual horizontal lines arrayed to the right at the proper energy value. For example, the second energy level is shown with five states, all with an energy value equal to ε_2'; hence, $g_2 = 5$. The values of g_i for a given molecule are obtained from quantum theory and/or spectroscopic measurements.

Now consider a system consisting of a fixed number of molecules, N. Let N_j be the number of molecules in a given energy level ε_j'. This value N_j is defined as the population of the energy level. Obviously,

$$N = \sum_j N_j \qquad (16.1)$$

where the summation is taken over all energy levels. The different values of N_j associated with the different energy levels ε_j' form a set of numbers which is defined as the *population distribution*. If we look at our system of molecules at one instant in time, we will see a given set of N_j's, i.e., a certain population distribution over the energy levels. Another term for this set of numbers, synonomous with population distribution, is *macrostate*. Due to molecular collisions, some molecules will change from one energy level to another. Hence, when we look at our system at some later instant in time, there may be a different set of N_j's, and hence a different population distribution, or macrostate. Finally, let us denote the total energy of the system as E, where

$$E = \sum_j \varepsilon_j' N_j \qquad (16.2)$$

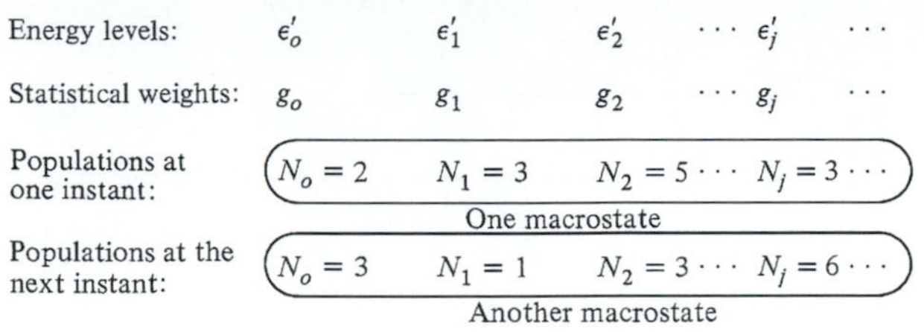

Energy levels:	ϵ'_o	ϵ'_1	ϵ'_2	$\cdots$	ϵ'_j	$\cdots$
Statistical weights:	g_o	g_1	g_2	$\cdots$	g_j	$\cdots$

Populations at one instant: $\left(N_o = 2 \quad N_1 = 3 \quad N_2 = 5 \cdots N_j = 3 \cdots \right)$

One macrostate

Populations at the next instant: $\left(N_o = 3 \quad N_1 = 1 \quad N_2 = 3 \cdots N_j = 6 \cdots \right)$

Another macrostate

Figure 16.8 | Illustration of macrostates.

The schematic in Fig. 16.8 reinforces the above definitions. For a system of N molecules and energy E, we have a series of quantized energy levels ε'_o, $\varepsilon'_1, \ldots, \varepsilon'_j, \ldots$, with corresponding statistical weights $g_o, g_1, \ldots, g_j, \ldots$. At some given instant, the molecules are distributed over the energy levels in a distinct way, $N_o, N_1, \ldots, N_j, \ldots$, constituting a distinct macrostate. In the next instant, due to molecular collisions, the populations of some levels may change, creating a different set of N_j's, and hence a different macrostate.

Over a period of time, one particular macrostate, i.e., one specific set of N_j's, will occur much more frequently than any other. This particular macrostate is called the *most probable macrostate* (or *most probable distribution*). It is the macrostate which occurs when the system is in *thermodynamic equilibrium*. In fact, this is the *definition* of thermodynamic equilibrium within the framework of statistical mechanics. The central problem of statistical thermodynamics, and the one to which we will now address ourselves, is as follows:

> Given a system with a fixed number of identical particles,
> $N = \sum_j N_j$, and a fixed energy $E = \sum_j \varepsilon'_j N_j$, find the most
> probable macrostate.

In order to solve the above problem, we need one additional definition, namely, that of a *microstate*. Consider the schematic shown in Fig. 16.9, which illustrates a given macrostate (for purposes of illustration, we choose $N_o = 2, N_1 = 5$, $N_2 = 3$, etc.). Here, we display each statistical weight for each energy level as a vertical array of boxes. For example, under ε'_1 we have $g_1 = 6$, and hence six boxes, one for each different energy *state* with the same energy ε'_1. In the energy level ε'_1, we have five molecules ($N_1 = 5$). At some instant in time, these five molecules individually occupy the top three and lower two boxes under g_1, with the fourth box left vacant (i.e., no molecules at that instant have the energy state represented by the fourth box). The way that the molecules are distributed over the available boxes defines a microstate of the system, say microstate I as shown in Fig. 16.9. At some later instant, the $N_1 = 5$ molecules may be distributed differently over the $g_1 = 6$ states, say leaving the second box vacant. This represents another, different microstate, labeled microstate II in Fig. 16.9. Shifts over the other vertical arrays of boxes between microstates I and II are also shown in Fig. 16.9. However, in both cases, N_o

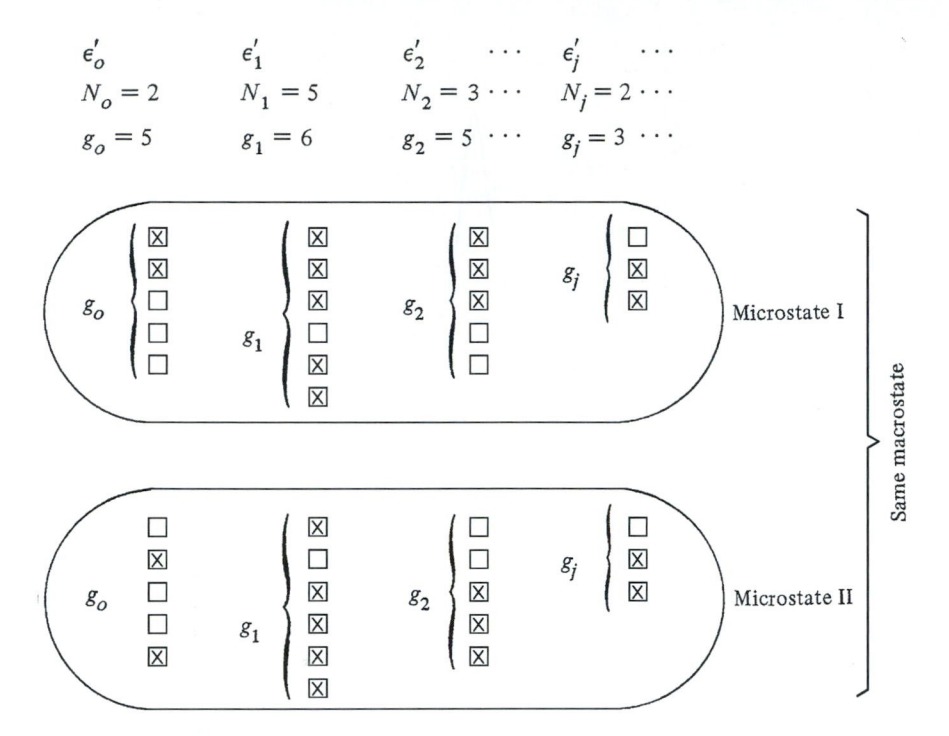

Figure 16.9 | Illustration of microstates.

still equals 2, N_1 still equals 5, etc.—i.e., the macrostate is still the same. Thus, any one macrostate can have a number of different microstates, depending on which of the degenerate states (the boxes in Fig. 16.9) are occupied by the molecules. In any given system of molecules, the microstates are constantly changing due to molecular collisions. Indeed, it is a central assumption of statistical thermodynamics that each microstate of a system occurs with equal probability. Therefore, it is easy to reason that *the most probable macrostate is that macrostate which has the maximum number of microstates.* If each microstate appears in the system with equal probability, and there is one particular macrostate that has considerably more microstates than any other, then that is the macrostate we will see in the system most of the time. This is indeed the situation in most real thermodynamic systems. Figure 16.10 is a schematic which plots the number of microstates in different macrostates. Note there is one particular macrostate, namely, macrostate D, that stands out as having by far the largest number of microstates. This is the *most probable macrostate;* this is the macrostate that is usually seen, and constitutes the situation of thermodynamic equilibrium in the system. Therefore, if we can count the number of microstates in any given macrostate, we can easily identify the most probable macrostate. This counting of microstates is the subject of Sec. 16.3. In turn, after the most probable macrostate is identified, the equilibrium thermodynamic properties of the system can be computed. Such thermodynamic computations will be discussed in subsequent sections.

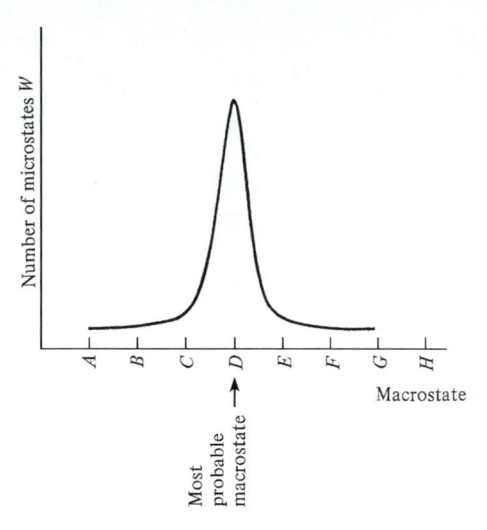

Figure 16.10 | Illustration of most probable macrostate as that macrostate that has the maximum number of microstates.

16.3 | COUNTING THE NUMBER OF MICROSTATES FOR A GIVEN MACROSTATE

Molecules and atoms are constituted from elementary particles—electrons, protons, and neutrons. Quantum mechanics makes a distinction between two different classes of molecules and atoms, depending on their number of elementary particles, as follows:

1. Molecules and atoms with an *even* number of elementary particles obey a certain statistical distribution called *Bose–Einstein statistics*. Let us call such molecules or atoms *Bosons*.

2. Molecules and atoms with an *odd* number of elementary particles obey a different statistical distribution called *Fermi–Dirac* statistics. Let us call such molecules or atoms *Fermions*.

There is an important distinction between the above two classes, as follows:

1. For *Bosons,* the number of molecules that can be in any one degenerate state (in any one of the boxes in Fig. 16.9) is *unlimited* (except, of course, that it must be less than or equal to N_j).

2. For *Fermions,* only one molecule may be in any given degenerate state at any instant.

This distinction has a major impact on the counting of microstates in a gas.

First, let us consider Bose–Einstein statistics. For the time being, consider one energy level by itself, say ε_j'. This energy level has g_j degenerate states and N_j

molecules. Consider the g_j states as the g_j containers diagrammed below.

$$
\overset{1}{\overbrace{\text{xxx}}} \mid \overset{2}{\overbrace{\text{xx}}} \mid \overset{3}{\overbrace{}} \mid \overset{4}{\overbrace{\text{x}}} \mid \cdots \mid \overset{g_j}{\overbrace{\text{xx}}}
$$

Distribute the N_j molecules among the containers, such as three molecules in the first container, two molecules in the second, etc., where the molecules are denoted by x in the above diagram. The vertical bars are partitions which separate one container from another. The distribution of molecules over these containers represents a distinct microstate. If a molecule is moved from container 1 to container 2, a different microstate is formed. To count the total number of different microstates possible, first note that the number of permutations between the symbols x and | is

$$
[N_j + (g_j - 1)]!
$$

This is the number of distinct ways that the N_j molecules and the $g_j - 1$ partitions can be arranged. However, the partitions are indistinguishable; we have counted them too many times. The $g_j - 1$ partitions can be permuted $(g_j - 1)!$ different ways. The molecules are also indistinguishable. They can be permuted $N_j!$ different ways without changing the picture drawn above. Therefore, there are $(g_j - 1)!N_j!$ different permutations which yield the identical picture drawn above, that is, the same microstate. Thus, the number of different ways N_j indistinguishable molecules can be distributed over g_j states is

$$
\frac{(N_j + g_j - 1)!}{(g_j - 1)!N_j!}
$$

This expression applies to one energy level ε_j' with population N_j, and gives the number of different microstates just due to the different arrangements within ε_j'. Consider now the whole set of N_j's distributed over the complete set of energy levels. (Keep in mind that the given set of N_j's defines a particular macrostate.) Letting W denote the total number of microstates for a given macrostate, the last expression, multiplied over all the energy levels, yields

$$
W = \prod_j \frac{(N_j + g_j - 1)!}{(g_j - 1)!N_j!} \tag{16.3}
$$

Note that W is a function of all the N_j values, $W = W(N_1, N_2, \ldots, N_j, \ldots)$. The quantity W is called the *thermodynamic probability,* and is a measure of the "disorder" of the system (as will be discussed later). In summary, Eq. (16.3) is the way to count the number of microstates in a given macrostate as long as the molecules are Bosons.

Next, let us consider Fermi–Dirac statistics. Recall that, for Fermions, only one molecule may be in any given degenerate state at any instant, i.e., there can be no more than one molecule per container. This implicitly requires that $g_j \geq N_j$. Consider the g_j containers. Take one of the molecules and put it in one of the containers. There will be g_j choices, or ways of doing this. Take the next particle, and put it in one of the *remaining* containers. However, there are now only $g_j - 1$ choices, because one of the containers is already occupied. Finally, placing the remaining molecules over the remaining containers, we find that the number of ways N_j particles can be distributed over g_j containers, with only one particle (or less) per container, is

$$g_j(g_j - 1)(g_j - 2) \cdots [g_j - (N_j - 1)] \equiv \frac{g_j!}{(g_j - N_j)!}$$

However, the N_j molecules are indistinguishable; they can be permutted $N_j!$ different ways without changing the above picture. Therefore, the number of different microstates just due to the different arrangements with ε_j' is

$$\frac{g_j!}{(g_j - N_j)!N_j!}$$

Considering all energy levels, the total number of microstates for a given macrostate for Fermions is

$$W = \prod_j \frac{g_j!}{(g_j - N_j)!N_j!} \tag{16.4}$$

In summary, if we are given a specific population distribution over the energy levels of a gas, i.e., a specific set of N_j's, i.e., a specific macrostate, Eqs. (16.3) or (16.4) allow us to calculate the number of microstates for that given macrostate for Bosons or Fermions, respectively. It is again emphasized that W is a function of the N_j's, and hence is a different number for different macrostates. Moreover, as sketched in Fig.16.10, there will in general be a certain macrostate, i.e., a certain distribution of N_j's, for which W will be considerably larger than for any other macrostate. This, by definition, will be the *most probable macrostate*. The precise solution for these N_j's associated with the most probable macrostate is the subject of Sec. 16.4.

16.4 | THE MOST PROBABLE MACROSTATE

The most probable macrostate is defined as that macrostate which contains the maximum number of microstates, i.e., which has W_{max}. Let us solve for the most probable macrostate, i.e., let us find the specific set of N_j's, which allows the maximum W.

First consider the case for Bosons. From Eq. (16.3) we can write

$$\ln W = \sum_j [\ln(N_j + g_j - 1)! - \ln(g_j - 1)! - \ln N_j!] \tag{16.5}$$

Recall that we are dealing with the combined translational, rotational, vibrational and electronic energies of a molecule, and that the closely spaced translational levels can be grouped into a number of degenerate states with essentially the same energy. Therefore, in Eq. (16.5), we can assume that $N_j \gg 1$ and $g_j \gg 1$, and hence that $N_j + g_j - 1 \approx N_j + g_j$ and $g_j - 1 \approx g_j$. Moreover, we can employ Sterling's formula

$$\ln a! = a \ln a - a \tag{16.6}$$

for the factorial terms in Eq. (16.5). Consequently, Eq. (16.5) becomes

$$\ln W = \sum_j [(N_j + g_j) \ln(N_j + g_j) - (N_j + g_j) - g_j \ln g_j + g_j - N_j \ln N_j + N_j]$$

Combining terms, this becomes

$$\ln W = \sum_j \left[N_j \ln\left(1 + \frac{g_j}{N_j}\right) + g_j \ln\left(\frac{N_j}{g_j} + 1\right) \right] \qquad (16.7)$$

Recall that $\ln W = f(N_j\text{'s}) = f(N_o, N_1, N_2, \ldots, N_j, \ldots)$. Also, to find the maximum value of W,

$$d(\ln W) = 0 \qquad (16.8)$$

From the chain rule of differentiation,

$$d(\ln W) = \frac{\partial(\ln W)}{\partial N_o} dN_o + \frac{\partial(\ln W)}{\partial N_1} dN_1 + \cdots + \frac{\partial(\ln W)}{\partial N_j} dN_j + \cdots \qquad (16.9)$$

Combining Eqs. (16.8) and (16.9),

$$d(\ln W) = \sum_j \frac{\partial(\ln W)}{\partial N_j} dN_j = 0 \qquad (16.10)$$

From Eq. (16.7),

$$\frac{\partial(\ln W)}{\partial N_j} = \ln\left(1 + \frac{g_j}{N_j}\right) \qquad (16.11)$$

Substituting Eq. (16.11) into Eq. (16.10),

$$d(\ln W) = \sum_j \left[\ln\left(1 + \frac{g_j}{N_j}\right) \right] dN_j = 0 \qquad (16.12)$$

In Eq. (16.12), the variation of N_j is not totally independent; dN_j is subject to two physical constraints, namely,

1. $N = \sum_j N_j = \text{const}$, and hence,

$$\sum_j dN_j = 0 \qquad (16.13)$$

2. $E = \sum_j \varepsilon'_j N_j = \text{const}$, and hence,

$$\sum_j \varepsilon'_j dN_j = 0 \qquad (16.14)$$

Letting α and β be two Lagrange multipliers (two constants to be determined later), Eqs. (16.13) and (16.14) can be written as

$$-\sum_j \alpha \, dN_j = 0 \qquad (16.15)$$

$$-\sum_j \beta \varepsilon'_j \, dN_j = 0 \qquad (16.16)$$

Adding Eqs. (16.12), (16.15), and (16.16), we have

$$\sum_j \left[\ln\left(1 + \frac{g_j}{N_j}\right) - \alpha - \beta\varepsilon_j' \right] dN_j = 0 \qquad (16.17)$$

From the standard method of Lagrange multipliers, α and β are defined such that each term in brackets in Eq. (16.17) is zero, i.e.,

$$\ln\left(1 + \frac{g_j}{N_j}\right) - \alpha - \beta\varepsilon_j' = 0$$

or

$$1 + \frac{g_j}{N_j} = e^\alpha e^{\beta\varepsilon_j'}$$

or

$$\boxed{N_j^* = \frac{g_j}{e^\alpha e^{\beta\varepsilon_j'} - 1}} \qquad (16.18)$$

The asterisk has been added to emphasize that N_j^* corresponds to the maximum value of W via Eq. (16.8), i.e., N_j^* corresponds to the most probable distribution of particles over the energy levels ε_j'. Equation (16.18) gives the *most probable macrostate* for Bosons. That is, the set of values obtained from Eq. (16.18) for all energy levels

$$N_o^*, \, N_1^*, \, N_2^*, \, \ldots, \, N_j^*, \, \ldots$$

is the most probable macrostate.

An analogous derivation for Fermions, starting from Eq. (16.4), yields for the most probable distribution

$$\boxed{N_j^* = \frac{g_j}{e^\alpha e^{\beta\varepsilon_j'} + 1}} \qquad (16.19)$$

which differs from the result for Bosons [Eq. (16.18)] only by the sign in the denominator. The details of that derivation are left to the reader.

16.5 | THE LIMITING CASE: BOLTZMANN DISTRIBUTION

At very low temperature, say less than 5 K, the molecules of the system are jammed together at or near the ground energy levels, and therefore the degenerate states of these low-lying levels are highly populated. As a result, the differences between Bose–Einstein statistics [Eq. (16.18)] and Fermi-Dirac statistics [Eq. (16.19)] are important. In contrast, at higher temperatures, the molecules are distributed over many energy levels, and therefore the states are generally sparsely populated, i.e.,

$N_j \ll g_j$. For this case, the denominators of Eqs. (16.18) and (16.19) must be very large,

$$e^{\alpha} e^{\beta \varepsilon'_j} - 1 \gg 1$$

and
$$e^{\alpha} e^{\beta \varepsilon'_j} + 1 \gg 1$$

Hence, in the high-temperature limit, the unity term in these denominators can be neglected, and both Eqs. (16.18) and (16.19) reduce to

$$\boxed{N_j^* = g_j e^{-\alpha} e^{-\beta \varepsilon'_j}} \tag{16.20}$$

This limiting case is called the *Boltzmann limit,* and Eq. (16.20) is termed the *Boltzmann distribution.* Since all gasdynamic problems generally deal with temperatures far above 5 K, the Boltzmann distribution is appropriate for all our future considerations. That is, in our future discussions, we will deal with Eq. (16.20) rather than Eqs. (16.18) or (16.19).

We still have two items of unfinished business with regard to the Boltzmann distribution, namely, α and β in Eq. (16.20). The link between classical and statistical thermodynamics is β. It can readily be shown (for example, see p. 434 of Ref. 119) that

$$\beta = \frac{1}{kT}$$

where k is the Boltzmann constant [see Eq. (1.15)] and T is the temperature of the system. Hence, Eq. (16.20) can be written as

$$N_j^* = g_j e^{-\alpha} e^{-\varepsilon'_j / kT} \tag{16.21}$$

To obtain an expression for α, recall that $N = \sum_j N_j^*$. Hence, from Eq. (16.21),

$$N = \sum_j g_j e^{-\alpha} e^{-\varepsilon'_j / kT} = e^{-\alpha} \sum_j g_j e^{-\varepsilon'_j / kT}$$

Hence
$$e^{-\alpha} = \frac{N}{\sum_j g_j e^{-\varepsilon'_j / kT}} \tag{16.22}$$

Substituting Eq. (16.22) into (16.21), we obtain

$$\boxed{N_j^* = N \frac{g_j e^{-\varepsilon'_j / kT}}{\sum_j g_j e^{-\varepsilon'_j / kT}}} \tag{16.23}$$

The Boltzmann distribution, given by Eq. (16.23), is important. It is the most probable distribution of the molecules over all the energy levels ε'_j of the system. Also, recall from Sec. 16.2 that ε'_j is the total energy, including the zero-point energy. However, Eq. (16.23) can also be written in terms of ε_j, the energy measured above the zero point, as follows. Since $\varepsilon_j' = \varepsilon_j + \varepsilon_o$, then

$$\frac{e^{-\varepsilon_j'/kT}}{\sum\limits_j g_j e^{-\varepsilon_j'/kT}} = \frac{e^{-(\varepsilon_j + \varepsilon_o)/kT}}{\sum\limits_j g_j e^{-(\varepsilon_j + \varepsilon_o)/kT}} = \frac{e^{-\varepsilon_o/kT} e^{-\varepsilon_j/kT}}{e^{-\varepsilon_o/kT} \sum\limits_j g_j e^{-\varepsilon_j/kT}}$$

$$= \frac{e^{-\varepsilon_j/kT}}{\sum\limits_j g_j e^{-\varepsilon_j/kT}}$$

Hence, Eq. (16.23) becomes

$$N_j^* = N \frac{g_j e^{-\varepsilon_j/kT}}{\sum\limits_j g_j e^{-\varepsilon_j/kT}} \tag{16.24}$$

where the energies are measured above the zero point. Finally, the *partition function Q* (or sometimes called the "state sum") is defined as

$$Q \equiv \sum_j g_j e^{-\varepsilon_j/kT}$$

and the Boltzmann distribution, from Eq. (16.24), can be written as

$$N_j^* = N \frac{g_j e^{-\varepsilon_j/kT}}{Q} \tag{16.25}$$

The partition function is a very useful quantity in statistical thermodynamics, as we will soon appreciate. Moreover, it is a function of the volume as well as the temperature of the system, as will be demonstrated later:

$$Q = f(T, V)$$

In summary, the Boltzmann distribution, given, for example, by Eq. (16.25), is extremely important. Equation (16.25) should be interpreted as follows. *For molecules or atoms of a given species, quantum mechanics says that a set of well-defined energy levels ε_j exists, over which the molecules or atoms can be distributed at any given instant, and that each energy level has a certain number of degenerate states, g_j. For a system of N molecules or atoms at a given T and V, Eq. (16.25) tells us how many such molecules or atoms, N_j^*, are in each energy level ε_j when the system is in thermodynamic equilibrium.*

16.6 | EVALUATION OF THERMODYNAMIC PROPERTIES IN TERMS OF THE PARTITION FUNCTION

The preceding formalism will now be cast in a form to yield practical thermodynamic properties for a high-temperature gas. In this section, properties such as internal energy will be expressed in terms of the partition function. In turn, in Sec. 16.7, the partition function will be developed in terms of T and V. Finally, in Sec. 16.8, the results will be combined to give practical expressions for the thermodynamic properties.

First consider the internal energy E, which is one of the most fundamental and important thermodynamic variables. From the microscopic viewpoint, for a system in equilibrium,

$$E = \sum_j \varepsilon_j N_j^* \tag{16.26}$$

Note that in Eq. (16.26) E is measured above the zero-point energy. Combining Eq. (16.26) with the Boltzmann distribution given by Eq. (16.25), we have

$$E = \sum_j \varepsilon_j N \frac{g_j e^{-\varepsilon_j/kT}}{Q} = \frac{N}{Q} \sum_j g_j \varepsilon_j e^{-\varepsilon_j/kT} \tag{16.27}$$

Recall from the previous section that

$$Q \equiv \sum_j g_j e^{-\varepsilon_j/kT} = f(V, T)$$

Hence

$$\left(\frac{\partial Q}{\partial T}\right)_V = \frac{1}{kT^2} \sum_j g_j \varepsilon_j e^{-\varepsilon_j/kT}$$

or

$$\sum_j g_j \varepsilon_j e^{-\varepsilon_j/kT} = kT^2 \left(\frac{\partial Q}{\partial T}\right)_V \tag{16.28}$$

Substituting Eq. (16.28) into (16.27),

$$E = \frac{N}{Q} kT^2 \left(\frac{\partial Q}{\partial T}\right)_V$$

or

$$\boxed{E = NkT^2 \left(\frac{\partial \ln Q}{\partial T}\right)_V} \tag{16.29}$$

This is the internal energy for a system of N molecules or atoms.

If we have 1 mol of atoms or molecules, then $N = N_A$, Avogadro's number. Also, $N_A k = \mathscr{R}$, the universal gas constant (see Sec. 1.4). Consequently, for the internal energy *per mole,* Eq. (16.29) becomes

$$E = \mathscr{R}T^2 \left(\frac{\partial \ln Q}{\partial T}\right)_V \tag{16.30}$$

In gasdynamics, a unit mass is a more fundamental quantity than a unit mole. Let M be the mass of the system of N molecules, and m be the mass of an individual molecule. Then $M = Nm$. From Eq. (16.29), the internal energy *per unit mass, e,* is

$$e = \frac{E}{M} = \frac{NkT^2}{Nm} \left(\frac{\partial \ln Q}{\partial T}\right)_V \tag{16.31}$$

However, $k/m = R$, the specific gas constant (again, see Sec. 1.4), and therefore Eq. (16.31) becomes

$$e = RT^2 \left(\frac{\partial \ln Q}{\partial T} \right)_V \tag{16.32}$$

The specific enthalpy is defined as

$$h = e + pv = e + RT$$

Hence, from Eq. (16.32),

$$h = RT + RT^2 \left(\frac{\partial \ln Q}{\partial T} \right)_V \tag{16.33}$$

Note that Eqs. (16.32) and (16.33) are "hybrid" equations, i.e., they contain a mixture of thermodynamic variables such as e, h, and T, and a statistical variable Q. Similar expressions for other thermodynamic properties can be obtained, as itemized next. For a system of N molecules or atoms, the entropy S is

$$S = Nk \left(\ln \frac{Q}{N} + 1 \right) + NkT \left(\frac{\partial \ln Q}{\partial T} \right)_V \tag{16.34}$$

and the pressure p is

$$p = NkT \left(\frac{\partial \ln Q}{\partial V} \right)_T \tag{16.35}$$

In all of these equations, Q is the key factor. If Q can be evaluated as a function of V and T, the thermodynamic state variables can then be calculated. This is the subject of Sec. 16.7.

16.7 | EVALUATION OF THE PARTITION FUNCTION IN TERMS OF T AND V

Since the partition function is defined as

$$Q \equiv \sum_j g_j e^{-\varepsilon_j/kT}$$

we need expressions for the energy levels ε_j in order to further evaluate Q. The quantized levels for translational, rotational, vibrational, and electronic energies are given by quantum mechanics. We state these results without proof here; see the classic books by Herzberg (Refs. 60 and 61) for details.

Recall that the total energy of a molecule is

$$\varepsilon' = \varepsilon'_{\text{trans}} + \varepsilon'_{\text{rot}} + \varepsilon'_{\text{vib}} + \varepsilon'_{\text{el}}$$

In this equation, from quantum mechanics,

$$\varepsilon'_{\text{trans}} = \frac{h^2}{8m}\left(\frac{n_1^2}{a_1^2} + \frac{n_2^2}{a_2^2} + \frac{n_3^2}{a_3^2}\right)$$

where n_1, n_2, n_3 are quantum numbers that can take the integral values 1, 2, 3, etc., and $a_1, a_2,$ and a_3 are linear dimensions which describe the size of the system. The values of $a_1, a_2,$ and a_3 can be thought of as the lengths of three sides of a rectangular box. (Also note here that h denotes Planck's constant, not enthalpy as before. In order to preserve standard nomenclature in both gasdynamics and quantum mechanics, we will live with this duplication. It will be clear which quantity is being used in our future expressions.) Also,

$$\varepsilon'_{\text{rot}} = \frac{h^2}{8\pi^2 I} J(J+1)$$

where J is the rotational quantum number, $J = 0, 1, 2$, etc., and I is the moment of inertia of the molecule. For vibration,

$$\varepsilon'_{\text{vib}} = h\nu\left(n + \tfrac{1}{2}\right)$$

where n is the vibrational quantum number, $n = 0, 1, 2$, etc., and ν is the fundamental vibrational frequency of the molecule. For the electronic energy, no simple expression can be written, and hence it will continue to be expressed simply as ε'_{el}.

In these expressions, I and ν for a given molecule are usually obtained from spectroscopic measurements; values for numerous different molecules are tabulated in Ref. 61, among other sources. Also note that $\varepsilon'_{\text{trans}}$ depends on the *size* of the system through $a_1, a_2,$ and a_3, whereas $\varepsilon'_{\text{rot}}, \varepsilon'_{\text{vib}},$ and ε'_{el} do not. Because of this spatial dependence of $\varepsilon'_{\text{trans}}$, Q depends on V as well as T. Finally, note that the lowest quantum number defines the zero-point energy for each mode, and from the above expressions, the zero-point energy for rotation is precisely zero, whereas it is a finite value for the other modes. For example,

$$\varepsilon'_{\text{trans}_o} = \frac{h^2}{8m}\left(\frac{1}{a_1^2} + \frac{1}{a_2^2} + \frac{1}{a_3^2}\right)$$

$$\varepsilon'_{\text{rot}_o} = 0$$

$$\varepsilon'_{\text{vib}_o} = \tfrac{1}{2}h\nu$$

In these equations, $\varepsilon'_{\text{trans}_o}$ is very small, but it is finite. In contrast, $\varepsilon'_{\text{vib}_o}$ is a larger finite value and $\varepsilon'_{\text{el}_o}$, although we do not have an expression for it, is larger yet.

Let us now consider the energy measured above the zero point:

$$\varepsilon_{\text{trans}} = \varepsilon'_{\text{trans}} - \varepsilon_{\text{trans}_o} \approx \frac{h^2}{8m}\left(\frac{n_1^2}{a_1^2} + \frac{n_2^2}{a_2^2} + \frac{n_3^2}{a_3^2}\right)$$

(Here, we are neglecting the small but finite value of $\varepsilon_{\text{trans}_o}$.)

$$\varepsilon_{\text{rot}} = \varepsilon'_{\text{rot}} - \varepsilon_{\text{rot}_o} = \frac{h^2}{8\pi^2 I} J(J+1)$$

$$\varepsilon_{\text{vib}} = \varepsilon'_{\text{vib}} - \varepsilon_{\text{vib}_o} = nh\nu$$

$$\varepsilon_{\text{el}} = \varepsilon'_{\text{el}} - \varepsilon_{\text{el}_o}$$

Therefore, the total energy is

$$\varepsilon' = \varepsilon_{\text{trans}} + \varepsilon_{\text{rot}} + \varepsilon_{\text{vib}} + \varepsilon_{\text{el}} + \varepsilon_o$$

Now, let us consider the *total* energy measured above the zero point, ε, where

$$\varepsilon = \underbrace{\varepsilon' - \varepsilon_o}_{\substack{\text{\textit{Sensible} energy,}\\ \text{i.e., energy}\\ \text{measured \textit{above}}\\ \text{zero-point energy.}}} = \underbrace{\varepsilon_{\text{trans}} + \varepsilon_{\text{rot}} + \varepsilon_{\text{vib}} + \varepsilon_{\text{el}}}_{\substack{\text{All measured \textit{above} the zero-point}\\ \text{energy. Thus, all are equal to zero at}\\ T = 0 \text{ K.}}}$$

Recall from Eqs. (16.24) and (16.25) that Q is defined in terms of the sensible energy, i.e., the energy measured above the zero point:

$$Q \equiv \sum_j g_j e^{-\varepsilon_j / kT}$$

where

$$\varepsilon_j = \varepsilon_{i_{\text{trans}}} + \varepsilon_{J_{\text{rot}}} + \varepsilon_{n_{\text{vib}}} + \varepsilon_{l_{\text{el}}}$$

Hence,

$$Q = \sum_i \sum_J \sum_n \sum_l g_i g_J g_n g_l \exp\left[-\frac{1}{kT}\left(\varepsilon_{i_{\text{trans}}} + \varepsilon_{J_{\text{rot}}} + \varepsilon_{n_{\text{vib}}} + \varepsilon_{l_{\text{el}}}\right)\right]$$

or

$$Q = \left[\sum_i g_i \exp\left(-\frac{\varepsilon_{i_{\text{trans}}}}{kT}\right)\right]\left[\sum_J g_J \exp\left(-\frac{\varepsilon_{J_{\text{rot}}}}{kT}\right)\right]$$

$$\times \left[\sum_n g_n \exp\left(-\frac{\varepsilon_{n_{\text{vib}}}}{kT}\right)\right]\left[\sum_l g_l \exp\left(-\frac{\varepsilon_{l_{\text{el}}}}{kT}\right)\right] \tag{16.36}$$

Note that the sums in each of the parentheses in Eq. (16.36) are partition functions for *each mode* of energy. Thus, Eq. (16.36) can be written as

$$Q = Q_{\text{trans}} Q_{\text{rot}} Q_{\text{vib}} Q_{\text{el}} \tag{16.37}$$

The evaluation of Q now becomes a matter of evaluating individually Q_{trans}, Q_{rot}, Q_{vib}, and Q_{el}.

First, consider Q_{trans}:

$$Q_{\text{trans}} = \sum_i g_{i_{\text{trans}}} \exp\left(-\frac{\varepsilon_{i_{\text{trans}}}}{kT}\right)$$

In the above, the summation is over all energy *levels,* each with g_i states. Therefore, the sum can just as well be taken over all energy states, and written as

$$Q_{\text{trans}} = \sum_j \exp\left(-\frac{\varepsilon_{j\text{trans}}}{kT}\right)$$

$$= \sum_{n_1=1}^{\infty}\sum_{n_2=1}^{\infty}\sum_{n_3=1}^{\infty} \exp\left[-\frac{h^2}{8mkT}\left(\frac{n_1^2}{a_1^2} + \frac{n_2^2}{a_2^2} + \frac{n_3^2}{a_3^2}\right)\right]$$

$$= \left[\sum_{n_1=1}^{\infty}\exp\left(-\frac{h^2}{8mkT}\frac{n_1^2}{a_1^2}\right)\right]\left[\sum_{n_2=1}^{\infty}\exp\left(-\frac{h^2}{8mkT}\frac{n_2^2}{a_2^2}\right)\right]$$

$$\times \left[\sum_{n_3=1}^{\infty}\exp\left(-\frac{h^2}{8mkT}\frac{n_3^2}{a_3^2}\right)\right]$$

If each of the terms in each summation above were plotted versus n, an almost continuous curve would be obtained because of the close spacings between the translational energies. As a result, each summation can be replaced by an *integral,* resulting in

$$Q_{\text{trans}} = a_1 \frac{\sqrt{2\pi mkT}}{h} a_2 \frac{\sqrt{2\pi mkT}}{h} a_3 \frac{\sqrt{2\pi mkT}}{h}$$

or

$$\boxed{Q_{\text{trans}} = \left(\frac{2\pi mkT}{h^2}\right)^{3/2} V} \qquad (16.38)$$

where $V = a_1 a_2 a_3 = $ volume of the system.

To evaluate the rotational partition function, we use the quantum mechanical result that $g_J = 2J + 1$. Therefore,

$$Q_{\text{rot}} = \sum_J g_J \exp\left(-\frac{\varepsilon_J}{kT}\right) = \sum_{J=0}^{\infty}(2J + 1)\exp\left[-\frac{h^2}{8\pi^2 IkT}J(J + 1)\right]$$

Again, if the summation is replaced by an integral,

$$\boxed{Q_{\text{rot}} = \frac{8\pi^2 IkT}{h^2}} \qquad (16.39)$$

To evaluate the vibrational partition function, results from quantum mechanics give $g_n = 1$ for all energy levels of a diatomic molecule. Hence,

$$Q_{\text{vib}} = \sum_n g_n e^{-\varepsilon_n/kT} = \sum_{n=0}^{\infty}e^{-nh\nu/kT}$$

This is a simple geometric series, with a closed-form expression for the sum:

$$\boxed{Q_{\text{vib}} = \frac{1}{1 - e^{-h\nu/kT}}} \qquad (16.40)$$

To evaluate the electronic partition function, no closed-form expression analogous to the above results is possible. Rather, the definition is used, namely,

$$Q_{el} \equiv \sum_{l=0}^{\infty} g_l e^{-\varepsilon_l/kT} = g_o + g_1 e^{-\varepsilon_1/kT} + g_2 e^{-\varepsilon_2/kT} + \cdots \qquad (16.41)$$

where spectroscopic data for the electronic energy levels ε_1, ε_2, etc., are inserted directly in the above terms. Usually, ε_l for the higher electronic energy levels is so large that terms beyond the first three shown in Eq. (16.41) can be neglected for $T \leq 15{,}000$ K.

Many results have been packed into this section, and the reader without previous exposure to quantum mechanics may feel somewhat uncomfortable. However, the purpose of this section has been to establish results for the partition function in terms of T and V; Eqs. (16.38) through (16.41) are those results. The discussion surrounding these equations removes, we hope, some of the mystery about their origin.

16.8 | PRACTICAL EVALUATION OF THERMODYNAMIC PROPERTIES FOR A SINGLE SPECIES

We now arrive at the focus of all the preceding discussion in this chapter, namely, the evaluation of the high-temperature thermodynamic properties of a single-species gas. We will emphasize the specific internal energy e; other properties are obtained in an analogous manner.

First, consider the translational energy. From Eq. (16.38),

$$\ln Q_{trans} = \frac{3}{2} \ln T + \frac{3}{2} \ln \frac{2\pi mk}{h^2} + \ln V$$

Therefore,

$$\left(\frac{\ln Q_{trans}}{\partial T} \right)_V = \frac{3}{2} \frac{1}{T} \qquad (16.42)$$

Substituting Eq. (16.42) into (16.32), we have

$$e_{trans} = RT^2 \frac{3}{2} \frac{1}{T}$$

$$\boxed{e_{trans} = \tfrac{3}{2} RT} \qquad (16.43)$$

Considering the rotational energy, we have from Eq. (16.39)

$$\ln Q_{rot} = \ln T + \ln \frac{8\pi^2 I k}{h^2}$$

Thus,

$$\frac{\partial \ln Q_{rot}}{\partial T} = \frac{1}{T} \qquad (16.44)$$

Substituting Eq. (16.44) into (16.32), we obtain

$$\boxed{e_{\text{rot}} = RT} \tag{16.45}$$

Considering the vibrational energy, we have from Eq. (16.40)

$$\ln Q_{\text{vib}} = -\ln(1 - e^{-h\nu/kT})$$

Thus,

$$\frac{\partial \ln Q_{\text{vib}}}{\partial T} = \frac{h\nu/kT^2}{e^{h\nu/kT} - 1} \tag{16.46}$$

Substituting Eq. (16.46) into (16.32), we obtain

$$\boxed{e_{\text{vib}} = \frac{h\nu/kT}{e^{h\nu/kT} - 1} RT} \tag{16.47}$$

Let us examine these results in light of a classical theorem from kinetic theory, the "theorem of equipartition of energy." Established before the turn of the century, this theorem states that each thermal degree of freedom of the molecule contributes $\frac{1}{2}kT$ to the energy of each molecule, or $\frac{1}{2}RT$ to the energy per unit mass of gas. For example, in Sec. 16.2, we demonstrated that the translational motion of a molecule or atom contributes three thermal degrees of freedom; hence, due to equipartition of energy, the translational energy per unit mass should be $3(\frac{1}{2}RT) = \frac{3}{2}RT$. This is precisely the result obtained in Eq. (16.43) from the modern principles of statistical thermodynamics. Similarly, for a diatomic molecule, the rotational motion contributes two thermal degrees of freedom; therefore, classically, $e_{\text{rot}} = 2(\frac{1}{2}RT) = RT$, which is in precise agreement with Eq. (16.45).

At this stage, you might be wondering why we have gone to all the trouble of the preceding sections if the principal of equipartition of energy will give us the results so simply. Indeed, extending this idea to the vibrational motion of a diatomic molecule, we recognize that the two vibrational thermal degrees of freedom should result in $e_{\text{vib}} = 2(\frac{1}{2}RT) = RT$. However, this is not confirmed by Eq. (16.47). Indeed, the factor $(h\nu/kT)/(e^{h\nu/kT} - 1)$ is less than unity except when $T \to \infty$, when it approaches unity; thus, in general, $e_{\text{vib}} < RT$, in conflict with classical theory. This conflict was recognized by scientists at the turn of the century, but it required the development of quantum mechanics in the 1920s to resolve the problem. Classical results are based on our macroscopic observations of the physical world, and they do not necessarily describe phenomena in the microscopic world of molecules. This is a major distinction between classical and quantum mechanics. As a result, the equipartition of energy principal is misleading. Instead, Eq. (16.47), obtained from quantum considerations, is the proper expression for vibrational energy.

In summary, we have for atoms:

$$\underbrace{e}_{\substack{\text{Internal energy per} \\ \text{unit mass measured} \\ \text{above zero-point} \\ \text{energy (sensible energy)}}} = \underbrace{\frac{3}{2}RT}_{\substack{\text{Translational} \\ \text{energy}}} + \underbrace{e_{\text{el}}}_{\substack{\text{Electronic energy,} \\ \text{obtained directly} \\ \text{from spectroscopic} \\ \text{measurements}}} \tag{16.48}$$

and for molecules:

$$
\underbrace{e}_{\substack{\text{Sensible}\\\text{energy}}} = \underbrace{\tfrac{3}{2}RT}_{\substack{\text{Translational}\\\text{energy}}} + \underbrace{RT}_{\substack{\text{Rotational}\\\text{energy}}} + \underbrace{\frac{h\nu/kT}{e^{h\nu/kT} - 1}RT}_{\substack{\text{Vibrational}\\\text{energy}}} + \underbrace{e_{\text{el}}}_{\substack{\text{Electronic}\\\text{energy}}}
$$

(16.49)

In addition, recalling the specific heat at constant volume, $c_v \equiv (\partial e/\partial T)_v$, Eq. (16.48) yields for atoms

$$
c_v = \frac{3}{2}R + \frac{\partial e_{\text{el}}}{\partial T}
$$

(16.50)

and Eq. (16.49) yields for molecules

$$
c_v = \frac{3}{2}R + R + \frac{(h\nu/kT)^2 e^{h\nu/kT}}{(e^{h\nu/kT} - 1)^2}R + \frac{\partial e_{\text{el}}}{\partial T}
$$

(16.51)

In light of the above results, we are led to the following important conclusions:

1. From Eqs. (16.48) through (16.51), we note that both e and c_v are functions of T only. This is the case for a *thermally perfect,* nonreacting gas, as defined in Sec. 1.4, i.e.,

$$
e = f_1(T) \quad \text{and} \quad c_v = f_2(T).
$$

This result, obtained from statistical thermodynamics, is a consequence of our assumption that the molecules are independent (no intermolecular forces) during the counting of microstates, and that each microstate occurs with equal probability. If we included intermolecular forces, such would not be the case.

2. For a gas with only translational and rotational energy, we have

$$
c_v = \tfrac{3}{2}R \quad \text{(for atoms)}
$$
$$
c_v = \tfrac{5}{2}R \quad \text{(for diatomic molecules)}
$$

That is, c_v is constant. This is the case of a *calorically perfect gas,* as also defined in Sec. 1.4. For air at or around room temperature, $c_v = \tfrac{5}{2}R$, $c_p = c_v + R = \tfrac{7}{2}R$, and hence $\gamma = c_p/c_v = \tfrac{7}{5} = 1.4 = $ const. So we see that air under normal conditions has translational and rotational energy, but no significant vibrational energy, and that the results of statistical thermodynamics predict $\gamma = 1.4 = $ const—which we have assumed in all the preceding chapters. However, when the air temperature reaches 600 K or higher, vibrational energy is no longer negligible. Under these conditions, we say that "vibration is excited"; consequently $c_v = f(T)$ from Eq. (16.51), and γ is no longer constant. For air at such temperatures, the "constant γ" results from the previous chapters are no longer strictly valid. Instead, we have to redevelop our gas dynamics using results for a thermally perfect gas such as Eq. (16.51). This will be the subject of Chap. 17.

3. In the theoretical limit of $T \rightarrow \infty$, Eq. (16.51) predicts $c_v \rightarrow \frac{7}{2}R$, and again we would expect c_v to be a constant. However, long before this would occur, the gas would dissociate and ionize due to the high temperature, and c_v would vary due to chemical reactions. This case will be addressed in subsequent sections.

4. Note that Eqs. (16.48) and (16.49) give the internal energy measured above the zero point. Indeed, statistical thermodynamics can only calculate the *sensible* energy or enthalpy; an absolute calculation of the total energy is not possible because we cannot in general calculate values for the zero-point energy. The zero-point energy remains a useful theoretical concept especially for chemically reacting gases, but not one for which we can obtain an absolute numerical value. This will also be elaborated upon in subsequent sections.

5. The theoretical variation of c_v for air as a function of temperature is sketched in Fig. 16.11. This sketch is qualitative only, and is intended to show that, at very low temperatures (below 1 K), only translation is fully excited, and hence $c_v = \frac{3}{2}R$. (We are assuming here that the gas does not liquefy at low temperatures.) Between 1 K and 3 K, rotation comes into play, and above 3 K rotation and translation are fully excited, where $c_v = \frac{5}{2}R$. Then, above 600 K, vibration comes into play, and c_v is a variable until approximately 2000 K. Above that temperature, chemical reactions begin to occur, and c_v experiences large variations, as will be discussed later. The shaded region in Fig. 16.11 illustrates the regime where all our previous gasdynamic results assuming a calorically perfect gas are valid. The purpose of this chapter, as well as Chap. 17, is to explore the high-temperature regime where γ is no longer constant, and where vibrational and chemical reaction effects become important.

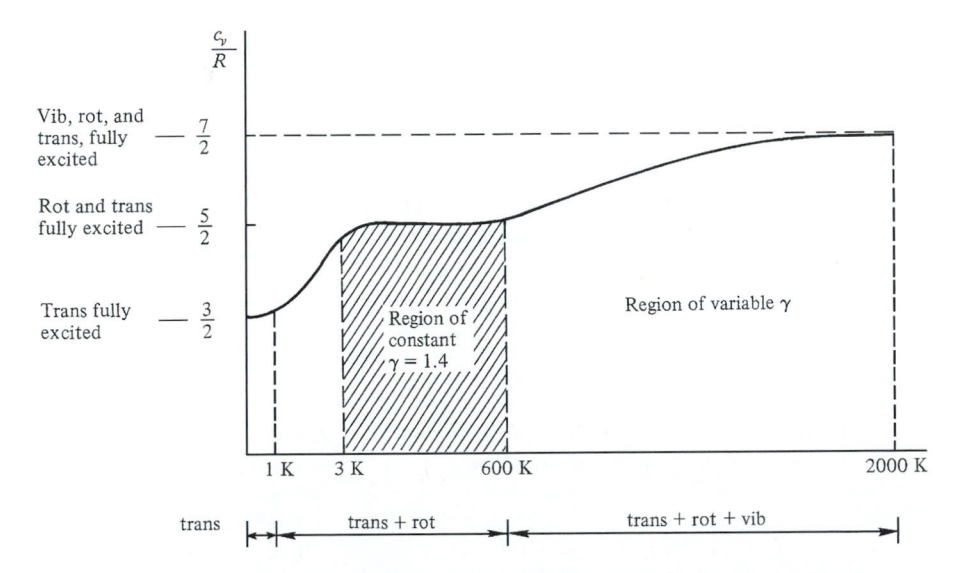

Figure 16.11 | Schematic of the temperature variation of the specific heat for a diatomic gas.

16.9 | THE EQUILIBRIUM CONSTANT

The theory and results obtained in the previous sections apply to a single chemical species. However, most high-temperature gases of interest are mixtures of several species. Let us now consider the statistical thermodynamics of a *mixture* of gases; the results obtained in this section represent an important ingredient for our subsequent discussions on equilibrium chemically reacting gases.

First, consider a gas mixture composed of three arbitrary chemical species A, B, and AB. The chemical equation governing a reaction between these species is

$$AB \rightleftharpoons A + B$$

Assume that the mixture is confined in a given volume at a given constant *pressure* and *temperature*. (We will soon appreciate that p and T are important variables in dealing with chemically reacting mixtures.) We assume that the system has existed long enough for the composition to become fixed, i.e., the above reaction is taking place an equal number of times to both the right and left (the forward and reverse reactions are balanced). This is the case of *chemical equilibrium*. Therefore, let N^{AB}, N^A, and N^B be the number of AB, A, and B particles, respectively, in the mixture at chemical equilibrium. Moreover, the A, B, and AB particles each have their own set of energy levels, populations, and degeneracies:

$$\varepsilon'_o, \varepsilon'^A_1, \varepsilon'^A_2, \ldots, \varepsilon'^A_j, \ldots \qquad \varepsilon'^B_o, \varepsilon'^B_1, \varepsilon'^B_2, \ldots, \varepsilon'^B_j, \ldots$$

$$N^A_o, N^A_1, N^A_2, \ldots, N^A_j, \ldots \qquad N^B_o, N^B_1, N^B_2, \ldots, N^B_j, \ldots$$

$$g^A_o, g^A_1, g^A_2, \ldots, g^A_j, \ldots \qquad g^B_o, g^B_1, g^B_2, \ldots, g^B_j, \ldots$$

$$\varepsilon'^{AB}_o, \varepsilon'^{AB}_1, \varepsilon'^{AB}_2, \ldots, \varepsilon'^{AB}_j, \ldots$$

$$N^{AB}_o, N^{AB}_1, N^{AB}_2, \ldots, N^{AB}_j, \ldots$$

$$g^{AB}_o, g^{AB}_1, g^{AB}_2, \ldots, g^{AB}_j, \ldots$$

A schematic of the energy levels is given in Fig. 16.12. Recall that, in most cases, we do not know the absolute values of the zero-point energies, but in general we know that $\varepsilon'^A_o \neq \varepsilon'^B_o \neq \varepsilon'^{AB}_o$. Therefore, the three energy-level ladders shown in Fig. 16.12 are at different heights. However, it is possible to find the *change* in zero-point energy for the reaction

$$\underbrace{AB}_{\text{Reactant}} \rightarrow \underbrace{A + B}_{\text{Products}}$$

$$\begin{bmatrix} \text{Change in zero-} \\ \text{point energy} \end{bmatrix} \equiv \begin{bmatrix} \text{Zero-point energy} \\ \text{of products} \end{bmatrix} - \begin{bmatrix} \text{Zero-point energy} \\ \text{of reactants} \end{bmatrix}$$

$$\Delta\varepsilon_o \qquad = \qquad \left(\varepsilon'^A_o + \varepsilon'^B_o\right) \qquad - \qquad \varepsilon'^{AB}_o$$

This relationship is illustrated in Fig. 16.13.

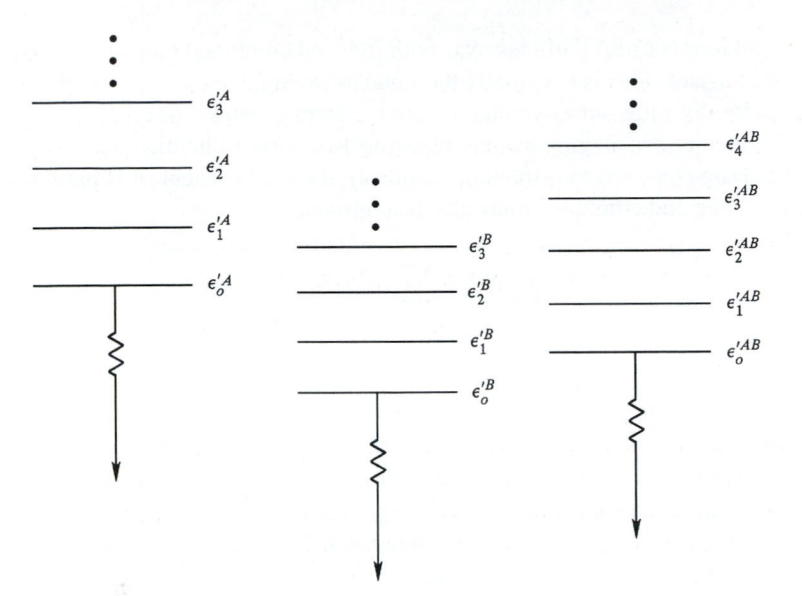

Figure 16.12 | Schematic of energy levels for three different chemical species.

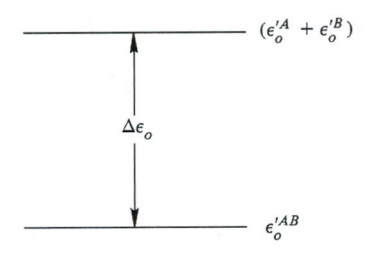

Figure 16.13 | Illustration of the meaning of change in zero-point energy.

The equilibrium mixture of A, B, and AB particles has two constraints:

1. The total energy E is constant:

$$E^A = \sum_j N_j^A \varepsilon_j'^A = \sum_j N_j^A \left(\varepsilon_j^A + \varepsilon_o^A \right)$$

$$E^B = \sum_j N_j^B \varepsilon_j'^B = \sum_j N_j^B \left(\varepsilon_j^B + \varepsilon_o^B \right)$$

$$E^{AB} = \sum_j N_j^{AB} \varepsilon_j'^{AB} = \sum_j N_j^{AB} \left(\varepsilon_j^{AB} + \varepsilon_o^{AB} \right)$$

$$\boxed{E = E^A + E^B + E^{AB} = \text{const}} \qquad (16.52)$$

2. Total number of A particles, N_A, both free and combined (such as in AB), must be constant. This is essentially the same as saying that the total number of A nuclei stays the same, whether it is in the form of pure A or combined in AB. We are not considering nuclear reactions here—only chemical reactions which rearrange the electron structure. Similarly, the total number of B particles, N_B, both free and combined must also be constant:

$$\begin{aligned}
\sum_j N_j^A + \sum_j N_j^{AB} &= N_A = \text{const} \\
\sum_j N_j^B + \sum_j N_j^{AB} &= N_B = \text{const}
\end{aligned}$$

(16.53)

To obtain the properties of the system in chemical equilibrium, we must find the most probable macrostate of the system, much the same way as we proceeded in Secs. 16.3 and 16.4 for a single species. The theme is the same; only the details are different. Consult Refs. 59 and 62 for those details. From this statistical thermodynamic treatment of the mixture, we find

$$N_j^A = N^A \frac{g_j^A e^{-\varepsilon_j^A/kT}}{Q^A}$$

(16.54a)

$$N_j^B = N^B \frac{g_j^B e^{-\varepsilon_j^B/kT}}{Q^B}$$

(16.54b)

$$N_j^{AB} = N^{AB} \frac{g_j^{AB} e^{-\varepsilon_j^{AB}/kT}}{Q^{AB}}$$

(16.54c)

and

$$\frac{N^A N^B}{N^{AB}} = e^{-\Delta\varepsilon_o/kT} \frac{Q^A Q^B}{Q^{AB}}$$

(16.55)

Recall that N^A, N^B, and N^{AB} are the actual number of A, B, and AB particles present in the mixture; do not confuse these with N_A and N_B, which were defined as the number of A and B nuclei.

Equations (16.54a) through (16.54c) demonstrate that a Boltzmann distribution exists independently for each one of the three chemical species. More important, however, Eq. (16.55) gives some information on the relative *amounts* of A, B, and AB in the mixture. Equation (16.55) is called the *law of mass action,* and it relates the amounts of different species to the change in zero-point energy, $\Delta\varepsilon_o$, and to the ratio of partition functions for each species.

For gasdynamic calculations, there is a more useful form of Eq. (16.55), as follows. From Sec. 1.4, we can write the perfect gas equation of state for the mixture as

$$pV = NkT \qquad (16.56)$$

For each species i, the *partial pressure* p_i, can be written as

$$p_i V = N_i kT \qquad (16.57)$$

The partial pressure is defined by Eq. (16.57); it is the pressure that would exist if N_i particles of species i were the only matter filling the volume V. Letting N_i equal N^A, N^B, and N^{AB}, respectively, and defining the corresponding partial pressures p_A, p_B, and p_{AB}, Eq. (16.57) yields

$$\frac{N^A N^B}{N^{AB}} = \frac{p_A p_B}{p_{AB}} \frac{V}{kT} \qquad (16.58)$$

Combining Eqs. (16.58) and (16.55), we have

$$\frac{p_A p_B}{p_{AB}} = \frac{kT}{V} e^{-\Delta\varepsilon_o/kT} \frac{Q^A Q^B}{Q^{AB}} \qquad (16.59)$$

Recall from Eqs. (16.37) and (16.38) that Q is proportional to the volume V. Therefore, in Eq. (16.59) the V's cancel, and we obtain

$$\frac{p_A p_B}{p_{AB}} = f(T)$$

This function of temperature is defined as the *equilibrium constant* for the reaction $AB \rightleftharpoons A + B$, $K_p(T)$:

$$\boxed{\frac{p_A p_B}{p_{AB}} = K_p(T)} \qquad (16.60)$$

From Eq. (16.60), the equilibrium constant for the reaction $AB \rightleftharpoons A + B$ can be defined as the ratio of the partial pressures of the products of reaction to the partial pressures of the reactants.

Generalizing this idea, consider the general chemical equation

$$v_1 A_1 + v_2 A_2 + v_3 A_3 \rightleftharpoons v_4 A_4 + v_5 A_5$$

or

$$0 = \sum_i v_i A_i \qquad (16.61)$$

where v_i is the stoichiometric mole number for species i and A_i is the chemical symbol for species i. In Eq. (16.61) v_i is positive for products and negative for reactants. Then the equilibrium constant is defined as

$$K_p(T) \equiv \frac{\left(p_{A_4}\right)^{v_4} \left(p_{A_5}\right)^{v_5}}{\left(p_{A_1}\right)^{v_1} \left(p_{A_2}\right)^{v_2} \left(p_{A_3}\right)^{v_3}}$$

or

$$\boxed{K_p(T) \equiv \prod_i p_i^{v_i}} \qquad (16.62)$$

Equation (16.62) is another form of the law of mass action, and it is extremely useful in the calculation of the composition of an equilibrium chemically reacting mixture. Some typical reactions, with their associated equilibrium constants are

$N_2 \rightleftarrows 2N$:
$$K_{p,N_2} = \frac{(p_N)^2}{p_{N_2}}$$

$H_2O_2 \rightleftarrows 2H + 2O$:
$$K_{p,H_2O_2} = \frac{(p_H)^2(p_O)^2}{p_{H_2O_2}}$$

In summary, we have made three important accomplishments in this section:

1. We have defined the equilibrium constant, Eqs. (16.60) or (16.62).
2. We have shown it to be a function of temperature only, Eq. (16.60).
3. We have demonstrated a formula from which it may be calculated based on a knowledge of the partition functions, Eq. (16.59). Indeed, tables of equilibrium constants for many basic chemical reactions have been calculated, and are given in Refs. 63 and 64.

In perspective, the first part of this chapter has developed the high-temperature properties of a single species. Now, in order to focus on the properties of a chemically reacting mixture (such as high-temperature air), we must know *what* chemical species are present in the mixture, and in what *quantity*. After these questions are answered, we can sum over all the species and find the thermodynamic properties of the mixture. These matters are the subjects of the next few sections.

16.10 | CHEMICAL EQUILIBRIUM— QUALITATIVE DISCUSSION

Consider air at normal room temperature and pressure. The chemical composition under these conditions is approximately 79% N_2, 20% O_2, and 1 percent trace species such as Ar, He, CO_2, H_2O, etc., by volume. Ignoring these trace species, we can consider that normal air consists of two species, N_2 and O_2. However, if we heat this air to a high temperature, where 2500 K $< T <$ 9000 K, chemical reactions will occur among the nitrogen and oxygen. Some of the important reactions in this temperature range are

$$O_2 \rightleftarrows 2O \qquad\qquad (16.63a)$$

$$N_2 \rightleftarrows 2N \qquad\qquad (16.63b)$$

$$N + O \rightleftarrows NO \qquad\qquad (16.63c)$$

$$N + O \rightleftarrows NO^+ + e^- \qquad\qquad (16.63d)$$

That is, at high temperatures, we have present in the air mixture not only O_2 and N_2, but O, N, NO, NO^+ and e^- as well. Moreover, if the air is brought to a given T and

p, and then left for a period of time until the above reactions are occurring an equal amount in both the forward and reverse directions, we approach the condition of *chemical equilibrium.* For air in chemical equilibrium at a given p and T, the species O_2, O, N_2, N, NO, NO^+, and e^- are present in specific, fixed amounts, which are unique functions of p and T. Indeed, for any equilibrium chemically reacting gas, the chemical composition (the types and amounts of each species) is determined uniquely by p and T, as we will learn in Sec. 16.11.

16.11 | PRACTICAL CALCULATION OF THE EQUILIBRIUM COMPOSITION

The method discussed in this section is applicable to any equilibrium chemically reacting mixture. However, because a large number of high-speed, compressible flow problems deal with air, we will illustrate the method by treating the case of high-temperature air.

To begin with, there are several different ways of specifying the composition of a gas mixture. For example, the quantity of different gases in a mixture can be specified by means of

1. *The partial pressures* p_i. For air, we have p_{O_2}, p_O, p_{N_2}, p_N, p_{NO}, p_{NO^+}, and p_{e^-}.
2. *The concentrations,* i.e., the number of moles of species i per unit volume of the mixture, denoted by $[X_i]$. For air, we have $[O_2]$, $[O]$, $[N_2]$, etc.
3. The mole-mass ratios (see Sec. 1.4), i.e., the number of moles of i per unit mass of mixture, denoted by η_i. For air, we have η_{O_2}, η_O, η_{N_2}, etc.
4. *The mole fractions,* i.e., the number of moles of species i per unit mole of mixture, denoted by X_i. For air, we have X_{O_2}, X_O, X_{N_2}, etc.
5. *The mass fraction,* i.e., the mass of species i per unit mass of mixture, denoted by c_i. For air, we have c_{O_2}, c_O, c_{N_2}, etc.

Each of these is equally definitive for specifying the composition of a chemically reacting mixture—if we know the composition in terms of p_i, for example, then we can immediately convert to X_i, c_i, etc. (Try deriving the conversion formulas yourself.) However, for gasdynamic problems, the use of partial pressures is particularly convenient; therefore, the following development will deal with p_i.

Consider again a system of high-temperature air at a given T and p, and assume that the above seven species are present. We want to solve for p_{O_2}, p_O, p_{N_2}, p_N, p_{NO}, p_{NO^+}, and p_{e^-} at the given mixture temperature and pressure. We have seven unknowns, hence we need seven independent equations. The first equation is Dalton's law of partial pressures, which states that the total pressure of the mixture is the sum of the partial pressures (Dalton's law holds only for perfect gases, i.e., gases wherein intermolecular forces are negligible):

I. $$p = p_{O_2} + p_O + p_{N_2} + p_N + p_{NO} + p_{NO^+} + p_{e^-} \qquad (16.64)$$

In addition, using Eq. (16.62) we can define the equilibrium constants for the chemical reactions (16.63a) through (16.63d) as

II.
$$\frac{(p_O)^2}{p_{O_2}} = K_{p,O_2}(T) \tag{16.65}$$

III.
$$\frac{(p_N)^2}{p_{N_2}} = K_{p,N_2}(T) \tag{16.66}$$

IV.
$$\frac{p_{NO}}{p_N p_O} = K_{p,NO}(T) \tag{16.67}$$

V.
$$\frac{p_{NO^+} p_e}{p_N p_O} = K_{p,NO^+}(T) \tag{16.68}$$

In Eqs. (16.65) through (16.68), the equilibrium constants K_p are known values, calculated from statistical mechanics as previously described, or obtained from thermodynamic measurements. They can be found in established tables, such as the JANAF Tables (Ref. 63). However, Eqs. (16.64) through (16.68) constitute only five equations—we still need two more. The other equations come from the indestructibility of matter, as follows.

Fact. The number of O nuclei, both in the free and combined state, must remain constant. Let N_O denote the number of oxygen nuclei per unit mass of mixture.

Fact. The number of N nuclei, both in the free and combined state, must remain constant. Let N_N denote the number of nitrogen nuclei per unit mass of mixture.

Then, from the definition of Avogadro's number N_A, and the mole-mass ratios η_i,

$$N_A(2\eta_{O_2} + \eta_O + \eta_{NO} + \eta_{NO^+}) = N_O \tag{16.69}$$

$$N_A(2\eta_{N_2} + \eta_N + \eta_{NO} + \eta_{NO^+}) = N_N \tag{16.70}$$

However, from Eq. (1.12),

$$p_i v = \eta_i \mathcal{R} T$$

Hence
$$\eta_i = p_i \frac{v}{\mathcal{R} T} \tag{16.71}$$

Dividing Eqs. (16.69) and (16.70), and substituting Eq. (16.71) into the result, we have

VI.
$$\frac{2p_{O_2} + p_O + p_{NO} + p_{NO^+}}{2p_{N_2} + p_N + p_{NO} + p_{NO^+}} = \frac{N_O}{N_N} \tag{16.72}$$

Equation (16.72) is called the *mass-balance equation*. Here, the ratio N_O/N_N is *known* from the original mixture at low temperature. For example, assuming at normal conditions that air consists of 80% N_2 and 20% O_2,

$$\frac{N_O}{N_N} = \frac{0.2}{0.8} = 0.25$$

Finally, to obtain our last remaining equation, we state the fact that electric charge must be conserved, and hence

$$\eta_{NO^+} = \eta_{e^-} \tag{16.73}$$

Substituting Eq. (16.71) into (16.73), we have

VII. $$p_{NO^+} = p_{e^-} \tag{16.74}$$

In summary, Eqs. (16.64) through (16.68), (16.72), and (16.74) are seven nonlinear, simultaneous, algebraic equations that can be solved for the seven unknown partial pressures. Furthermore, Eq. (16.64) requires the pressure p as input, and Eqs. (16.65) through (16.68) require the temperature T in order to evaluate the equilibrium constants. Hence, these equations clearly demonstrate that, for a given chemically reacting mixture, the equilibrium composition is a function of T and p.

This procedure, carried out for high-temperature air, is an example of a general procedure that applies to any chemically reacting mixture in chemical equilibrium. In general, if the mixture has $\sum$ species and ϕ elements, then we need $\sum - \phi$ independent chemical equations [such as Eqs. (16.63a) through (16.63d)] with the appropriate equilibrium constants. The remaining equations are obtained from the mass-balance equations and Dalton's law of partial pressures. In our earlier example for air, $\sum = 7$ and $\phi = 3$ (the elements are O, N, and e^-). Therefore, we needed $\sum - \phi = 4$ independent chemical equations with four different equilibrium constants. These four equations were Eqs. (16.63a) through (16.63d).

The calculation of a chemical equilibrium composition is conceptually straightforward, as indicated in this section. However, the solution of a system of many nonlinear, simultaneous algebraic equations is not a trivial undertaking by hand, and today such calculations are almost always performed on a high-speed digital computer using custom-designed algorithms.

Also, the reader should note that the specific chemical species to be solved are chosen at the beginning of the problem. This choice is important; if a major species is not considered (for example, if N had been left out of our above calculations), the final results for chemical equilibrium will not be accurate. The proper choice of the type of species in the mixture is a matter of experience and common sense. If there is any doubt, it is always safe to assume all possible combinations of the atoms and molecules as potential species; then, if many of the choices turn out to be trace species, the results of the calculation will state so. At least in this manner, the possibility of overlooking a major species is minimized.

16.12 | EQUILIBRIUM GAS MIXTURE THERMODYNAMIC PROPERTIES

In perspective, to this point in our discussion of the properties of high-temperature gases we have accomplished two major goals:

1. From Secs. 16.1 through 16.8, we have obtained formulas for calculating the thermodynamic properties of a given single species.

2. From Secs. 16.9 through 16.11, we have seen how to calculate the *amount* of each species in an equilibrium chemically reacting mixture.

In this section, we now combine the above knowledge to obtain the thermodynamic properties of an equilibrium chemically reacting mixture. Because of its importance to gas dynamics, we will concentrate on the enthalpy of the mixture.

From Eq. (1.10),

$$p_i \mathscr{V} = \mathscr{N}_i \mathscr{R} T \tag{16.75}$$

and

$$p \mathscr{V} = \mathscr{N} \mathscr{R} T \tag{16.76}$$

where $\mathscr{V}$ is the volume of the system, $\mathscr{N}_i$ is the number of moles of species i, and $\mathscr{N}$ is the total number of moles of the mixture. Dividing Eq. (16.75) by (16.76):

$$\boxed{\frac{p_i}{p} = \frac{\mathscr{N}_i}{\mathscr{N}} \equiv X_i} \tag{16.77}$$

where X_i is the mole fraction defined in Sec. 16.11. Let H_i be the enthalpy of species i per mole of species i, and H be the enthalpy of the mixture per mole of mixture. Then

$$\boxed{H = \sum_i X_i H_i = \sum_i \frac{p_i}{p} H_i} \tag{16.78}$$

where the summation is taken over all species in the mixture.

In gasdynamics, we are more concerned with unit masses than with moles. Let

$h =$ enthalpy per unit mass of mixture

$\mathscr{M} =$ molecular weight (more properly called the molecular mass) of the mixture; it is the mass of mixture per mole of mixture

$\mathscr{M}_i =$ molecular weight of species i; it is the mass of i per mole of i.

Hence, from the definitions, we have

$$\boxed{\mathscr{M} = \sum_i X_i \mathscr{M}_i = \sum_i \frac{p_i}{p} \mathscr{M}_i} \tag{16.79}$$

and therefore,

$$\boxed{h = \frac{H}{\mathscr{M}} = \frac{\displaystyle\sum_i \frac{p_i}{p} H_i}{\displaystyle\sum_i \frac{p_i}{p} \mathscr{M}_i}} \tag{16.80}$$

Equation (16.80) provides an equation for obtaining the enthalpy per unit mass of mixture from molar quantities. There are two alternative expressions for h. Recalling the definition of the mole-mass ratio η_i, from Secs. 1.4 and 16.11, we have

$$\boxed{h = \sum_i \eta_i H_i} \tag{16.81}$$

Also, denoting the enthalpy of species i per unit mass of i by h_i, where $h_i = H_i/\mathcal{M}_i$, and recalling the definition of mass fraction c_i from Sec. 16.11, we have

$$h = \sum_i c_i h_i \tag{16.82}$$

Note from the definitions that

$$c_i = X_i \frac{\mathcal{M}_i}{\mathcal{M}} = \frac{p_i}{p} \frac{\mathcal{M}_i}{\sum_i (p_i/p)\mathcal{M}_i} \tag{16.83}$$

Let us now examine the meaning of H_i more closely:

$$\underbrace{H_i}_{\substack{\text{Absolute enthalpy} \\ \text{of species } i \text{ per} \\ \text{mole of } i}} = \underbrace{(H - E_o)_i}_{\substack{\text{Sensible enthalpy of} \\ \text{species } i \text{ per mole of } i}} + \underbrace{E_{o_i}}_{\substack{\text{Zero-point energy} \\ \text{of species } i \text{ per} \\ \text{mole of } i}} \tag{16.84}$$

The sensible enthalpy is obtained from statistical mechanics, as we have already seen:

$$(H - E_o)_i = (E - E_o)_i + \mathcal{R}T$$

$$(H - E_o)_i = \underbrace{\tfrac{3}{2}\mathcal{R}T}_{\text{Translation}} + \underbrace{\mathcal{R}T}_{\text{Rotation}} + \underbrace{\frac{h\nu/kT}{e^{h\nu/kT} - 1}\mathcal{R}T}_{\text{Vibration}} + \mathcal{R}T + \text{ electronic energy} \tag{16.85}$$

Note that $(H - E_o)_i$ is a function of T only. Also, E_{o_i} is the zero-point energy of species i, that is, the energy of the species at $T = 0$ K; it is a constant for a given chemical species. The relationship is schematically shown in Fig. 16.14. As discussed in Secs. 16.2 and 16.7, the *absolute value* of E_{o_i} usually cannot be calculated or measured; nevertheless it is an important theoretical quantity. For example, in a complex chemically reacting mixture, we should establish some reference level from

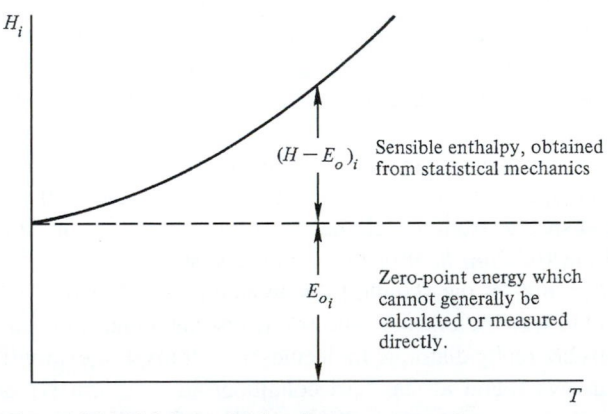

Figure 16.14 | Schematic showing the contrast between sensible enthalpy and zero-point energy.

which all the energies of the given species can be measured. Many times there is some difficulty and confusion in establishing what this level should be. However, by carrying through our concept of the absolute zero-point energy E_{o_i}, the choice of a proper reference level will soon become apparent.

Since the absolute value of E_{o_i} generally cannot be obtained, how can we calculate a number for h from Eq. (16.80), (16.81), or (16.82)? The answer lies in the fact that we never need an absolute number for h. In all thermodynamic and gasdynamic problems, we deal with *changes* in enthalpy and internal energy. For example, in Chap. 3 dealing with shock waves, we were always interested in the *change $h_2 - h_1$* across the shock. In the general conservation equations from Chap. 6, we dealt with the derivatives $\partial h/\partial x$, $\partial h/\partial y$, $\partial h/\partial z$, $\partial h/\partial t$, which are *changes* in enthalpy. Letting points 1 and 2 denote two different locations in a flowfield, we have from Eq. (16.81)

$$h_1 = \sum_i (\eta_i H_i)_1 = \sum [\eta_i (H - E_o)_i]_1 + \sum (\eta_i E_{o_i})_1$$

or
$$h_1 = h_{\text{sens}_1} + e_{o_1} \tag{16.86}$$

where h_{sens_1} and e_{o_1} are the sensible enthalpy and zero-point energy, respectively, per unit mass of mixture at point 1. Similarly, at point 2,

$$h_2 = \sum_i (\eta_i H_i)_2 = \sum_i [\eta_i (H - E_o)_i]_2 + \sum_i (\eta_i E_{o_i})_2$$

or
$$h_2 = h_{\text{sens}_2} + e_{o_2} \tag{16.87}$$

Subtracting Eq. (16.86) from (16.87), we have

$$\underbrace{h_2 - h_1}_{\substack{\text{Change in}\\\text{enthalpy}}} = \underbrace{\left(h_{\text{sens}_2} - h_{\text{sens}_1}\right)}_{\text{Change in sensible enthalpy}} + \underbrace{\left(e_{o_2} - e_{o_1}\right)}_{\substack{\text{Change in zero-}\\\text{point energy}}}$$

or
$$\boxed{\Delta h = \Delta h_{\text{sens}} + \Delta e_o} \tag{16.88}$$

It is important to note that in Eq. (16.88) we have circumvented the need to know the absolute value of the zero-point energy; rather, what we need now is a value for the *change* in zero-point energy, Δe_o. The value *can* be obtained from measurement, as discussed below.

The change in zero-point energy is related to the concept of the *heat of formation* for a given species. When a chemical reaction represents the formation of a single chemical species from its "elements" at standard conditions, the heat of reaction is called the *standard heat of formation*. The standard conditions are those of the stable "elements" at the standard temperature, $T_s = 298.16$ K. (The quotation marks around the word "elements" above reflects that some "elements" at the standard conditions are really diatomic molecules, not atoms. For example, nitrogen and oxygen are always found at standard conditions in the form N_2 and O_2, *not* N and O.) To illustrate, consider the formation of H_2O from its "elements" at standard conditions:

$$\underbrace{H_2 + \tfrac{1}{2}O_2}_{\text{At } T_s} \rightarrow \underbrace{H_2O}_{\text{At } T_s}$$

Then, by definition,

$$\underbrace{(\Delta H_f)^{T_s}_{H_2O}}_{\substack{\text{Standard heat} \\ \text{of formation} \\ \text{of } H_2O}} \equiv \underbrace{H^{T_s}_{H_2O} - H^{T_s}_{H_2} - \tfrac{1}{2}H^{T_s}_{O_2}}_{\substack{\text{Enthalpy of the product minus} \\ \text{the enthalpy of the reactants,} \\ \text{all at } T_s}}$$

In an analogous fashion, let us define the *heat of formation at absolute zero*. Here, both the product and reactants are assumed to be at absolute zero. For example,

$$\underbrace{H_2 + \tfrac{1}{2}O_2}_{\text{At } T=0\,K} \rightarrow \underbrace{H_2O}_{\text{At } T=0\,K}$$

Letting $(\Delta H_f)^o_{H_2O}$ denote the heat of formation of H_2O at absolute zero, we have

$$(\Delta H_f)^o_{H_2O} \equiv H^o_{H_2O} - H^o_{H_2} - \tfrac{1}{2}H^o_{O_2} \qquad (16.89)$$

However, the enthalpy of any species at absolute zero is, by definition, its zero-point energy. Hence, Eq. (16.89) becomes

$$(\Delta H_f)^o_{H_2O} \equiv (E_o)_{H_2O} - (E_o)_{H_2} - \tfrac{1}{2}(E_o)_{O_2} \qquad (16.90)$$

Note that these expressions are couched in terms of energy per mole. However, the heat of formation of species *i per unit mass*, $(\Delta h_f)_i$, is easily obtained as

$$(\Delta h_f)_i = \frac{(\Delta H_f)_i}{\mathscr{M}_i}$$

Also, the heats of formation for many species have been measured, and are tabulated in such references as NBS Circular 500, the JANAF Tables, and NASA SP-3001 (see Refs. 65, 63, and 64, respectively).

We now state this theorem:

Theorem
In a chemical reaction, the change in zero-point energy (zero-point energy of the products minus the zero-point energy of the reactants) is equal to the difference between the heats of formation of the products at $T = 0$ K and the heats of formation of the reactants at $T = 0$ K.

Proof of this theorem is obtained by induction from examples. For example, consider the water-gas reaction:

$$CO_2 + H_2 \rightarrow H_2O + CO$$

By definition of the change in zero-point energy,

$$\Delta E_o \equiv (E_o)_{H_2O} + (E_o)_{CO} - (E_o)_{CO_2} - (E_o)_{H_2} \qquad (16.91)$$

By definition of the heat of formation at absolute zero, we have

$$H_2 + \tfrac{1}{2}O_2 \rightarrow H_2O: \qquad (\Delta H_f)^o_{H_2O} = (E_o)_{H_2O} - (E_o)_{H_2} - \tfrac{1}{2}(E_o)_{O_2} \qquad (16.92)$$

$$C + \tfrac{1}{2}O_2 \rightarrow CO: \qquad (\Delta H_f)^o_{CO} = (E_o)_{CO} - (E_o)_C - \tfrac{1}{2}(E_o)_{O_2} \qquad (16.93)$$

$$C + O_2 \rightarrow CO_2: \qquad (\Delta H_f)^o_{CO_2} = (E_o)_{CO_2} - (E_o)_C - (E_o)_{O_2} \qquad (16.94)$$

$$H_2 \rightarrow H_2: \qquad (\Delta H_f)^o_{H_2} = 0 \qquad (16.95)$$

Adding Eqs. (16.92) and (16.93), and subtracting (16.94) and (16.95), we have

$$(\Delta H_f)^o_{H_2O} + (\Delta H_f)^o_{CO} - (\Delta H_f)^o_{CO_2} - (\Delta H_f)^o_{H_2}$$

$$= (E_o)_{H_2O} + (E_o)_{CO} - (E_o)_{CO_2} - (E_o)_{H_2} \equiv \Delta E_o$$

Thus, for the water-gas reaction, we have just shown that

$$\Delta E_o = (\Delta H_f)^o_{H_2O} + (\Delta H_f)^o_{CO} - (\Delta H_f)^o_{CO_2} - (\Delta H_f)^o_{H_2} \qquad (16.96)$$

This is precisely the statement of the theorem!

Compare Eqs. (16.91) and (16.96). It appears that the terms $(E_o)_{H_2O}$, $(E_o)_{CO}$, $(E_o)_{CO_2}$, and $(E_o)_{H_2}$ can be replaced in a one-to-one correspondence by $(\Delta H_f)^o_{H_2O}$, $(\Delta H_f)^o_{CO}$, $(\Delta H_f)^o_{CO_2}$, and $(\Delta H_f)^o_{H_2}$. Therefore, let us reorient our thinking about the enthalpy of a gas mixture. We have been writing

$$h = \sum_i \eta_i H_i = \underbrace{\sum_i \eta_i(H - E_o)_i}_{\substack{\text{Sensible enthalpy} \\ \text{of the mixture}}} + \underbrace{\sum_i \eta_i E_{o_i}}_{\substack{\text{Zero-point} \\ \text{energy of the} \\ \text{mixture}}} \qquad (16.97)$$

Let us replace this with

$$h = \underbrace{\sum_i \eta_i(H - E_o)_i}_{\substack{\text{Sensible enthalpy,} \\ \text{obtained for example} \\ \text{from statistical} \\ \text{mechanics}}} + \underbrace{\sum_i \eta_i(\Delta H_f)^o_i}_{\substack{\text{"Effective" zero-point} \\ \text{energy, obtained from tables}}} \qquad (16.98)$$

Equations (16.97) and (16.98) yield different absolute numbers for h; however, from the above theorem the values for *changes in enthalpy,* Δh, will be the same whether Eq. (16.97) or (16.98) is used. *Therefore, we are led to an important change in our interpretation of enthalpy; namely, from now on we will think of enthalpy as given by Eq. (16.98) with the term involving the heat of formation at absolute zero as an "effective" zero-point energy.* In terms of enthalpy per unit mass, we write

$$h = \sum_i c_i h_i$$

where

$$h_i = (h - e_o)_i + (\Delta h_f)^o_i$$

Thus

$$h = \sum_i c_i(h - e_o)_i + \sum_i c_i(\Delta h_f)^o_i \qquad (16.99)$$

[Note that in Eqs. (16.98) and (16.99), the effective zero-point energy $\sum_i \eta_i (\Delta H_f)_i^o = \sum_i c_i (\Delta h_f)_i^o$ is sometimes called the "chemical enthalpy" in the literature.]

With the above, we end our discussions on the thermodynamic properties of an equilibrium chemically reacting mixture. In summary, we have shown that

1. The sensible enthalpy of a mixture can be obtained from this:
   a. The sensible enthalpy for each species as given by the formulas for statistical mechanics, for example, Eqs. (16.48), (16.49), and (16.85).
   b. Knowledge of the equilibrium composition described in terms of p_i, X_i, η_i, or c_i.
2. The zero-point energy can be treated as an "effective" value by using the heats of formation at absolute zero in its place. Therefore, Eq. (16.98) or (16.99) can be construed as the enthalpy of a gas mixture.

Also, as a final note, a chemically reacting mixture that is commonly encountered in many high-speed compressible flow problems is high-temperature air. The equilibrium thermodynamic properties of high-temperature air have been calculated in detail, and are available from many sources, such as the reports by Hansen (Ref. 66) and Hilsenrath and Klein (Ref. 67). These calculations use essentially the same techniques as described in the previous sections. Also, high-temperature air properties are available on large Mollier diagrams (a plot of enthalpy versus entropy) available from the government and some commercial firms. An example of an abbreviated Mollier diagram for high-temperature air is given in Fig. 16.15a. Also, the variation of the equilibrium composition of air at 1 atm as a function of T is given in Fig. 16.15b.

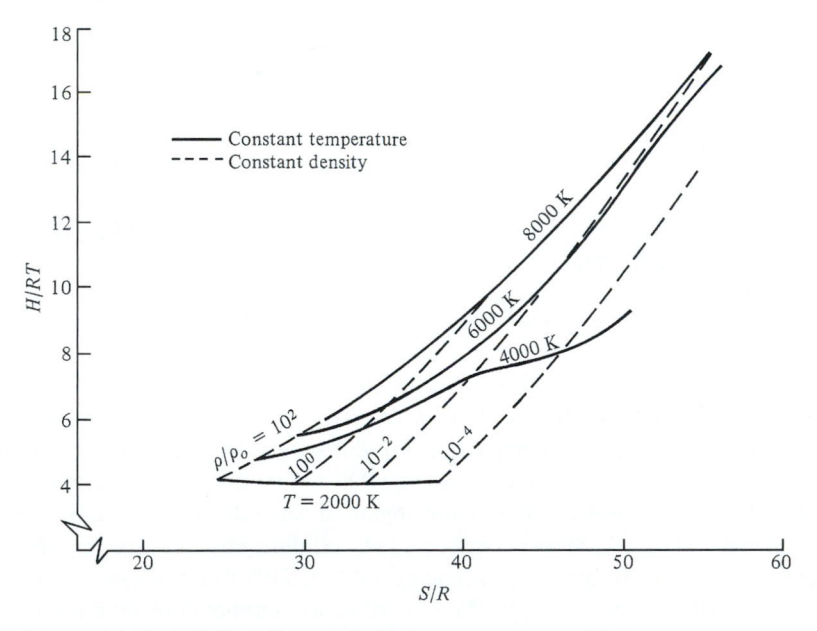

Figure 16.15a | Mollier diagram for high-temperature equilibrium air.

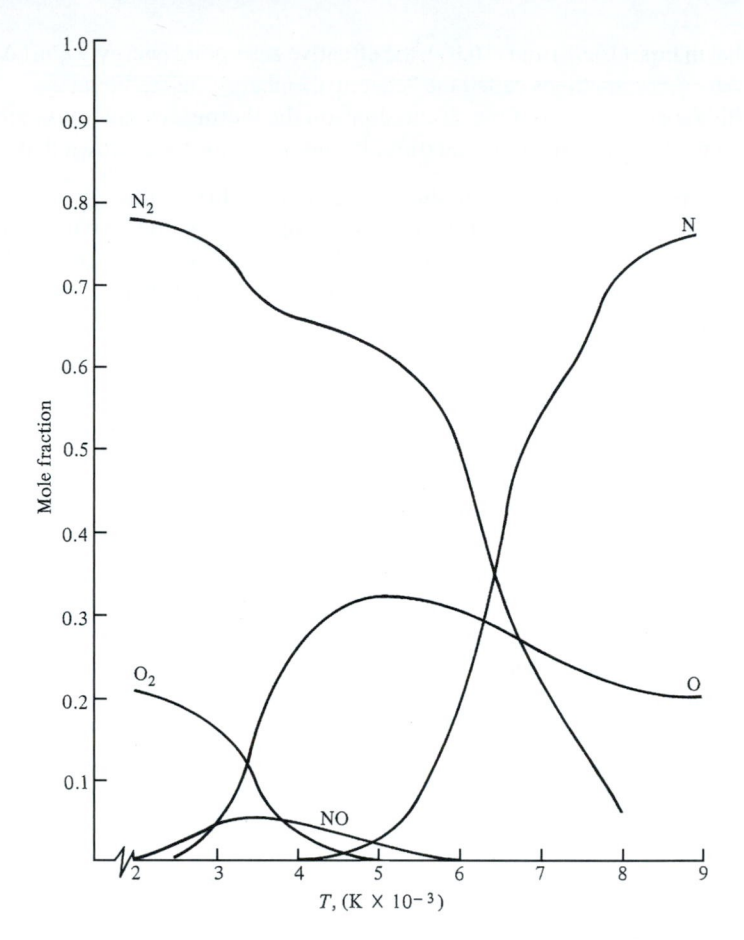

Figure 16.15*b* | Composition of equilibrium air versus temperature at 1 atm.

16.13 | INTRODUCTION TO NONEQUILIBRIUM SYSTEMS

All vibrational and chemical processes take place by molecular collisions and/or radiative interactions. Considering just molecular collisions, visualize for example an O_2 molecule colliding with other molecules in the system. If the O_2 vibrational energy is in the ground level before collision, it may or may not be vibrationally excited after the collision. Indeed, in general the O_2 molecule must experience a large number of collisions, typically on the order of 20,000, before it will become vibrationally excited. The actual number of collisions required depends on the type of molecule and the relative kinetic energy between the two colliding particles—the higher the kinetic energy (hence the higher the gas temperature), the fewer collisions are required for vibrational energy exchange. Moreover, as the temperature of the gas is increased, and hence the molecular collisions become more violent, it is probable that the O_2 molecule will be torn apart (dissociated) by collisions with other particles.

However, this requires a large number of collisions, on the order of 200,000. The important point to note here is that vibrational and chemical changes take place due to collisions. In turn, collisions take time to occur. Hence, vibrational and chemical changes in a gas take time to occur. The precise amount of time depends on the molecular collision frequency Z, which is the number of collisions a single particle makes with its neighboring particles per second. The results of kinetic theory show that $Z \propto p/\sqrt{T}$; hence the collision frequency is low for low pressures and very high temperatures.

The equilibrium systems considered in the previous sections assumed that the gas has had enough time for the necessary collisions to occur, and that the properties of the system at a fixed p and T are constant, independent of time. However, there are many problems in high-speed gasdynamics where the gas is not given the luxury of the necessary time to come to equilibrium. A typical example is the flow across a shock wave, where the pressure and temperature are rapidly increased within the shock front. Consider a fluid element passing through this shock front. When its p and T are suddenly increased, its equilibrium vibrational and chemical properties will change. The fluid element will start to seek these new equilibrium properties, but this requires molecular collisions, and hence time. By the time enough collisions have occurred and equilibrium properties have been approached, the fluid element has moved a certain distance downstream of the shock front. Hence, there will be a certain region immediately behind the shock wave where equilibrium conditions do not prevail—there will be a *nonequilibrium* region. To study the nonequilibrium region, additional techniques must be developed that take into account the time required for molecular collisions. Such techniques are the subject of the remaining sections of this chapter. The detailed study of both equilibrium and nonequilibrium flows through shock waves, as well as many other types of flows, will be made in Chap. 17.

16.14 | VIBRATIONAL RATE EQUATION

In this section we will derive an equation for the time rate of change of vibrational energy of a gas due to molecular collisions—the vibrational rate equation. In turn, this equation will be coupled with the continuity, momentum, and energy equations in Chap. 17 for the study of certain types of nonequilibrium flows.

Consider a diatomic molecule with a vibrational energy level diagram as illustrated in Fig. 16.16. Focus on the ith level. The population of this level, N_i, is *increased* by particles jumping up from the $i - 1$ level [transition (a) shown in Fig. 16.16] and by particles dropping down from the $i + 1$ level [transition (b)]. The population N_i is *decreased* by particles jumping up to the $i + 1$ level [transition (c)] and dropping down to the $i - 1$ level [transition (d)]. For the time being, consider just transition (c). Let $P_{i,i+1}$ be the *probability* that a molecule in the ith level, upon collision with another molecule, will jump up to the $i + 1$ level, $P_{i,i+1}$ is called the *transition probability*, and can be interpreted on a dimensional basis as the "number of transitions per collision per particle" (of course keeping in mind that a single transition requires many collisions). The value of $P_{i,i+1}$ is always *less* than unity. Also, let Z be the collision frequency as discussed above, where Z is the number of collisions per particle per second. Hence, the product $P_{i,i+1}Z$ is physically the

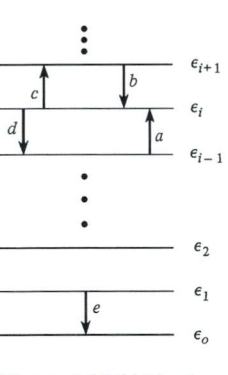

Figure 16.16 | Single quantum transitions for vibrational energy exchange.

number of transitions per particle per second. If there are N_i particles in level i, then $P_{i,i+1}ZN_i$ is the total number of transitions per second for the gas from the ith to the $i+1$ energy level. Similar definitions can be made for transitions (a), (b), and (d) in Fig. 16.16. Therefore, on purely physical grounds, using the above definitions, we can write the net rate of change of the population of the ith level as

$$\frac{dN_i}{dt} = \underbrace{P_{i+1,i}ZN_{i+1} + P_{i-1,i}ZN_{i-1}}_{\text{Rate of increase of } N_i} - \underbrace{P_{i,i+1}ZN_i - P_{i,i-1}ZN_i}_{\text{Rate of decrease of } N_i}$$

To simplify the above equation, define a vibrational *rate constant* $k_{i+1,i}$ such that $P_{i+1,i}Z \equiv k_{i+1,i}$; similarly for the other transitions. Then the above equation becomes

$$\boxed{\frac{dN_i}{dt} = k_{i+1,i}N_{i+1} + k_{i-1,i}N_{i-1} - k_{i,i+1}N_i - k_{i,i-1}N_i} \qquad (16.100)$$

Equation (16.100) is called the *master equation* for vibrational relaxation.

For a moment, consider that the gas is in equilibrium. Hence, from the Boltzmann distribution, Eq. (16.25), and the quantum mechanical expression for vibrational energy, $hv(n + \frac{1}{2})$, given in Sec. 16.7,

$$\frac{N_i^*}{N_{i-1}^*} = \frac{e^{-\varepsilon_i/kT}}{e^{-\varepsilon_{i-1}/kT}} = e^{-hv/kT} \qquad (16.101)$$

Moreover, in equilibrium, each transition in a given direction is exactly balanced by its counterpart in the opposite direction—this is called the *principle of detailed balancing*. That is, the number of transitions (a) per second must exactly equal the number of transitions (d) per second:

$$k_{i-1,i}N_{i-1}^* = k_{i,i-1}N_i^*$$

or
$$k_{i-1,i} = k_{i,i-1}\frac{N_i^*}{N_{i-1}^*} \qquad (16.102)$$

Combining Eqs. (16.101) and (16.102), we have

$$\boxed{k_{i-1,i} = k_{i,i-1}e^{-hv/kT}} \qquad (16.103)$$

Equation (16.103) is simply a relation between reciprocal rate constants; hence it holds for nonequilibrium as well as equilibrium conditions. Taking a result from quantum mechanics, it can also be shown that all the rate constants for higher-lying energy levels can be expressed in terms of the rate constant for transition (e) in Fig. 16.16, i.e., the transition from $i = 1$ to $i = 0$:

$$\boxed{k_{i,i-1} = ik_{1,0}} \qquad (16.104)$$

From Eq. (16.104), we can also write

$$k_{i+1,i} = (i + 1)k_{1,0} \qquad (16.105)$$

Combining Eqs. (16.103) and (16.104), we have

$$k_{i-1,i} = i k_{1,0} e^{-h\nu/kT} \qquad (16.106)$$

and from Eqs. (16.103), (16.104), and (16.105), we have

$$k_{i,i+1} = k_{i+1,i} e^{-h\nu/kT} = (i+1) k_{1,0} e^{-h\nu/kT} \qquad (16.107)$$

Substituting Eqs. (16.104) through (16.107) into (16.100), we have

$$\frac{dN_i}{dt} = (i+1)k_{1,0}N_{i+1} + i k_{1,0} e^{-h\nu/kT} N_{i-1} - (i+1)k_{1,0} e^{-h\nu/kT} N_i - i k_{1,0} N_i$$

or

$$\frac{dN_i}{dt} = k_{1,0}\{-i N_i + (i+1)N_{i+1} + e^{-h\nu/kT}[-(i+1)N_i + i N_{i-1}]\} \qquad (16.108)$$

In many gasdynamic problems, we are more interested in energies than populations. Let us convert Eq. (16.108) into a rate equation for e_{vib}. Assume that we are dealing with a unit mass of gas. From Secs. 16.2 and 16.7,

$$e_{\text{vib}} = \sum_{i=0}^{\infty} \varepsilon_i N_i = \sum_{i=0}^{\infty} (ih\nu)N_i = h\nu \sum_{i=1}^{\infty} i N_i$$

Hence

$$\frac{de_{\text{vib}}}{dt} = h\nu \sum_{i=1}^{\infty} i \frac{dN_i}{dt} \qquad (16.109)$$

Substitute Eq. (16.108) into (16.109):

$$\frac{de_{\text{vib}}}{dt} = h\nu k_{1,0} \sum_{i=1}^{\infty}$$

$$\times \{-i^2 N_i + i(i+1)N_{i+1} + e^{-h\nu/kT}[-i(i+1)N_i + i^2 N_{i-1}]\} \qquad (16.110)$$

Considering the first two terms in Eq. (16.110), and letting $s = i + 1$,

$$\sum_{i=1}^{\infty} [-i^2 N_i + i(i+1)N_{i+1}] = -\sum_{i=1}^{\infty} i^2 N_i + \sum_{s=2}^{\infty} (s-1)s N_s$$

$$= \sum_{i=1}^{\infty} i^2 N_i + \sum_{s=2}^{\infty} s^2 N_s - \sum_{s=2}^{\infty} s N_s$$

$$= -N_1 - \sum_{i=2}^{\infty} i^2 N_i + \sum_{s=2}^{\infty} s^2 N_s - \sum_{s=2}^{\infty} s N_s$$

$$= -N_i - \sum_{s=2}^{\infty} s N_s = -\sum_{i=1}^{\infty} i N_i = -\sum_{i=0}^{\infty} i N_i$$

Also, a similar reduction for the last two terms in Eq. (16.110) leads to

$$\sum_{i=1}^{\infty} [-i(i+1)N_i + i^2 N_{i-1}] = \sum_{i=0}^{\infty} (i+1)N_i$$

Thus, Eq. (16.110) becomes

$$\frac{de_{vib}}{dt} = h\nu k_{1,0} \sum_{i=0}^{\infty} [-iN_i + e^{-h\nu/kT}(i+1)N_i]$$

$$= h\nu k_{1,0}\left[e^{-h\nu/kT} \sum_{i=0}^{\infty} N_i - (1 - e^{-h\nu/kT}) \sum_{i=0}^{\infty} iN_i \right] \qquad (16.111)$$

However,

$$\sum_{i=0}^{\infty} N_i = N$$

and

$$e_{vib} = \sum_{i=0}^{\infty} \varepsilon_i N_i = h\nu \sum_{i=0}^{\infty} iN_i$$

Therefore,

$$\sum_{i=0}^{\infty} iN_i = \frac{e_{vib}}{h\nu}$$

Thus, Eq. (16.111) can be written as

$$\frac{de_{vib}}{dt} = h\nu k_{1,0}\left[e^{-h\nu/kT}N - (1 - e^{-h\nu/kT})\frac{e_{vib}}{h\nu} \right]$$

or

$$\frac{de_{vib}}{dt} = k_{1,0}(1 - e^{-h\nu/kT})\left[\frac{h\nu N}{e^{h\nu/kT} - 1} - e_{vib} \right] \qquad (16.112)$$

However, recalling that we are dealing with a unit mass, and hence N is the number of particles per unit mass, we have from Sec. 1.4 that $Nk = R$, the specific gas constant. Then, considering one of the expressions in Eq. (16.112),

$$\frac{h\nu N}{e^{h\nu/kT} - 1} = \frac{h\nu/kT}{e^{h\nu/kT} - 1}(NkT) = \frac{h\nu/kT}{e^{h\nu/kT} - 1}RT \qquad (16.113)$$

The right-hand side of Eq. (16.113) is simply the *equilibrium* vibrational energy from Eq. (16.47); we denote it by e_{vib}^{eq}. Hence, from Eq. (16.113)

$$\frac{h\nu N}{e^{h\nu/kT} - 1} = e_{vib}^{eq} \qquad (16.114)$$

Substituting Eq. (16.114) into Eq. (16.112),

$$\frac{de_{vib}}{dt} = k_{1,0}(1 - e^{-h\nu/kT})\left(e_{vib}^{eq} - e_{vib} \right) \qquad (16.115)$$

In Eq. (16.115), the factor $k_{1,0}(1 - e^{-h\nu/kT})$ has units of s^{-1}. Therefore, we define a vibrational relaxation time τ as

$$\tau \equiv \frac{1}{k_{1,0}(1 - e^{-h\nu/kT})}$$

Thus, Eq. (16.115) becomes

$$\frac{de_{\text{vib}}}{dt} = \frac{1}{\tau}\left(e_{\text{vib}}^{\text{eq}} - e_{\text{vib}}\right) \tag{16.116}$$

Equation (16.116) is called the *vibrational rate equation,* and it is the main result of this section. Equation (16.116) is a simple differential equation which relates the time rate of change of e_{vib} to the difference between the equilibrium value it is seeking and its local instantaneous nonequilibrium value.

The physical implications of Eq. (16.116) can be seen as follows. Consider a unit mass of gas in equilibrium at a given temperature T. Hence,

$$e_{\text{vib}} = e_{\text{vib}}^{\text{eq}} = \frac{h\nu/kT}{e^{h\nu/kT} - 1} RT \tag{16.117}$$

Now let us instantaneously excite the vibrational mode above its equilibrium value (say, by the absorption of radiation of the proper wavelength, e.g., we "zap" the gas with a laser). Let e_{vib_o} denote the instantaneous value of e_{vib} immediately after excitation, at time $t = 0$. This is illustrated in Fig. 16.17. Note that $e_{\text{vib}_o} > e_{\text{vib}}^{\text{eq}}$. Due to molecular collisions, the excited particles will exchange this "excess" vibrational energy with the translational and rotational energy of the gas, and after a period of time e_{vib} will decrease and approach its equilibrium value. This is illustrated by the solid curve in Fig. 16.17. However, note that, as the vibrational energy drains away, it reappears in part as an increase in translational energy. Since the temperature of the gas is proportional to the translational energy [see Eq. (16.43)], T increases. In turn, the equilibrium value of vibrational energy, from Eq. (16.117), will also increase. This is shown by the dashed line in Fig. 16.17. At large times, e_{vib} and $e_{\text{vib}}^{\text{eq}}$ will asymptotically approach the same value.

The relaxation time τ in Eq. (16.116) is a function of both local pressure and temperature. This is easily recognized because τ is a combination of the transition

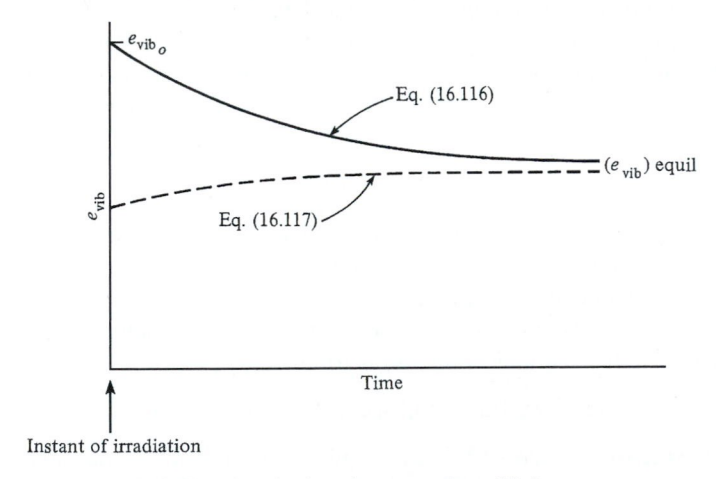

Figure 16.17 | Vibrational relaxation toward equilibrium.

probability P and the collision frequency Z, both defined earlier. In turn, P depends on T (on the relative kinetic energy between colliding particles), and $Z \propto p/\sqrt{T}$. For most diatomic gases, the variation of τ is given by the form

$$\tau p = c_1 e^{(C_2/T)^{1/3}}$$

or
$$\ln \tau p = \ln c_1 + \left(\frac{C_2}{T}\right)^{1/3} \tag{16.118}$$

In summary, the nonequilibrium variation of vibrational energy is given by the vibrational rate equation expressed as Eq. (16.116). Note that in Eq. (16.116) both τ and e_{vib}^{eq} are variables, with $\tau = (p, T)$ from Eq. (16.118) and $e_{vib}^{eq} = e(T)$ from Eq. (16.117). However, a word of caution is given. Equation (16.116) has certain limitations that have not been stressed during this derivation, namely, it holds only for diatomic molecules that are harmonic oscillators. The use of $\varepsilon_{vib} = h\nu n$, obtained from Sec. 16.7, is valid only if the molecule is a harmonic oscillator. Moreover, from Fig. 16.16, we have considered only single quantum jumps between energy levels, i.e., we did not consider transitions say from the ith directly to the $i + 2$ level. Such multiple quantum jumps can occur for anharmonic molecules, but their transition probabilities are very small. In spite of these restrictions, experience has proven that Eq. (16.116) is reasonably valid for real problems dealing with diatomic gases, and it is employed in almost all nonequilibrium analyses of such gases.

Recent developments in the study of vibrational nonequilibrium flows have highlighted a further limitation of Eq. (16.116), as follows. The energy level transitions included in the master equation, Eq. (16.100), are so-called "translation-vibration" (T-V) transfers. Here, a molecule upon collision with another will gain or lose vibrational energy, which then reappears as a decrease or increase in translational kinetic energy of the molecules. For example, a T-V transfer in CO can be given as

$$CO(n) + CO(n) \rightleftarrows CO(n - 1) + CO(n) + KE$$

where a CO molecule in the nth vibrational level drops to the $(n - 1)$ level after collision, with the consequent release of kinetic energy, KE. However, "vibration-vibration" (V-V) transfers also occur, where the vibrational quantum lost by one molecule is gained by its collision partner. For example, a V-V transfer in CO can be given as

$$CO(n) + CO(n) \rightleftarrows CO(n + 1) + CO(n - 1)$$

The above equation assumes a harmonic oscillator, where the spacings between all energy levels are the same. However, all molecules are in reality anharmonic oscillators, which results in unequal spacings between vibrational energy levels. Thus, in a V-V transfer involving anharmonic molecules, there is a small amount of translational energy exchanged in the process, as follows.

$$CO(n) + CO(n) \rightleftarrows CO(n + 1) + CO(n - 1) + KE$$

During an expansion process (decreasing temperature), the V-V transfers among anharmonic molecules result in an overpopulation of some of the higher energy levels than would be the case of a harmonic oscillator. This is called *anharmonic pumping,* and is particularly important in several types of gasdynamic and chemical lasers. The reverse effect occurs in a compression process (increasing temperature). In cases where anharmonic pumping is important, Eq. (16.116) is not valid, and the analysis must start from a master rate equation [such as Eq. (16.100)] expanded to include V-V transfers. For a fundamental discussion of the anharmonic pumping effect at an introductory level, see pages 112–120 of Ref. 21.

16.15 | CHEMICAL RATE EQUATIONS

Consider a system of oxygen in chemical equilibrium at $p = 1$ atm and $T = 3000$ K. Although Fig. 16.15b is for air, it clearly demonstrates that the oxygen under these conditions should be partially dissociated. Thus, in our system, both O_2 and O will be present in their proper equilibrium amounts. Now, assume that somehow T is instantaneously increased to, say, 4000 K. Equilibrium conditions at this higher temperature demand that the amount of O_2 decrease and the amount of O increase. However, as explained in Sec. 16.13, this change in composition takes place via molecular collisions, and hence it takes time to adjust to the new equilibrium conditions. During this nonequilibrium adjustment period, chemical reactions are taking place at a definite net rate. The purpose of this section is to establish relations for the finite time rate of change of each chemical species present in the mixture—the chemical rate equations.

Continuing with our example of a system of oxygen, the only chemical reaction taking place is

$$O_2 + M \rightarrow 2O + M \tag{16.119}$$

where M is a collision partner; it can be either O_2 or O. Using the bracket notation for concentration given in Sec. 16.11, we denote the number of moles of O_2 and O per unit volume of the mixture by $[O_2]$ and $[O]$, respectively. Empirical results have shown that the time rate of formation of O atoms via Eq. (16.119) is given by

$$\frac{d[O]}{dt} = 2k[O_2][M] \tag{16.120}$$

where $d[O]/dt$ is the *reaction rate*, k is the *reaction rate constant,* and Eq. (16.120) is called a *reaction rate equation.* The reaction rate constant k is a function of T only. Equation (16.120) gives the rate at which the reaction given in Eq. (16.119) goes from left to right; this is called the *forward rate,* and k is really the *forward rate constant k_f:*

$$O_2 + M \xrightarrow{k_f} 2O + M$$

Hence, Eq. (16.120) is more precisely written as

Forward rate: $$\frac{d[O]}{dt} = 2k_f[O_2][M] \tag{16.121}$$

The reaction in Eq. (16.119) that would proceed from right to left is called the *reverse reaction,* or *backward reaction,*

$$O_2 + M \xleftarrow[k_b]{} 2O + M$$

with an associated *reverse* or *backward rate constant k_b*, and a *reverse or backward rate* given by

Reverse rate:
$$\frac{d[O]}{dt} = -2k_b[O]^2[M]$$
(16.122)

Note that in both Eqs. (16.121) and (16.122), the right-hand side is the product of the concentrations of those particular colliding molecules that produce the chemical change, raised to the power equal to their stoichiometric mole number in the chemical equation. Equation (16.121) gives the time rate of *increase* of O atoms due to the forward rate, and Eq. (16.122) gives the time rate of *decrease* of O atoms due to the reverse rate. However, what we would actually observe in the laboratory is the *net* time rate of change of O atoms due to the *combined* forward and reverse reactions,

$$O_2 + M \underset{k_b}{\overset{k_f}{\rightleftarrows}} 2O + M$$

and this net reaction rate is given by

Net rate:
$$\boxed{\frac{d[O]}{dt} = 2k_f[O_2][M] - 2k_b[O]^2[M]}$$
(16.123)

Now consider our system to again be in chemical equilibrium; hence the composition is fixed with time. Then $d[O]/dt \equiv 0$, $[O_2] \equiv [O_2]^*$, and $[O] \equiv [O^*]$ where the asterisk denotes equilibrium conditions. In this case, Eq. (16.123) becomes

$$0 = 2k_f[O_2]^*[M^*] - 2k_b[O]^{*2}[M]^*$$

or
$$k_f = k_b\frac{[O]^{*2}}{[O_2]^*}$$
(16.124)

Examining the chemical equation given above, and recalling the substance of Sec. 16.9, we can define the ratio $[O]^{*2}/[O_2]^*$ in Eq. (16.124) as an equilibrium constant *based on concentrations, K_c*. This is related to the equilibrium constant based on partial pressures, K_p, defined in Sec. 16.9. From Eq. (1.13), it directly follows for this oxygen reaction that

$$K_c = \frac{1}{\mathscr{R}T}K_p$$

Hence, Eq. (16.124) can be written as

$$\boxed{\frac{k_f}{k_b} = K_c}$$
(16.125)

Equation (16.125), although derived by assuming equilibrium, is simply a relation between the forward and reverse rate constants, and therefore it holds in general for nonequilibrium conditions. Therefore, the net rate, Eq. (16.123), can be expressed as

$$\boxed{\frac{d[O]}{dt} = 2k_f[M]\left\{[O_2] - \frac{1}{K_c}[O]^2\right\}}$$
(16.126)

In practice, values for k_f are found from experiment, and then k_b can be directly obtained from Eq. (16.125). Keep in mind that k_f, k_b, K_c, and K_p for a given reaction are all functions of temperature only. Also, k_f in Eq. (16.126) is generally different depending on whether the collision partner M is chosen to be O_2 or O.

This example has been a special application of the more general case of a reacting mixture of n different species. Consider the general chemical reaction (but it must be an elementary reaction, as defined later)

$$\sum_{i=1}^{n} v_i' X_i \underset{k_b}{\overset{k_f}{\rightleftharpoons}} \sum_{i=1}^{n} v_i'' X_i \tag{16.127}$$

where v_i' and v_i'' represent the stoichiometric mole numbers of the reactants and products, respectively. (Note that in our above example for oxygen where the chemical reaction was $O_2 + M \rightleftharpoons 2O + M$, $v_{O_2}' = 1$, $v_O' = 0$, $v_{O_2}'' = 0$, $v_M' = 1$, $v_M'' = 1$, and $v_O'' = 2$.) For the above general reaction, Eq. (16.127), we can write

Forward rate:
$$\frac{d[X_i]}{dt} = (v_i'' - v_i')k_f \prod_i [X_i]^{v_i'} \tag{16.128}$$

Reverse rate:
$$\frac{d[X_i]}{dt} = -(v_i'' - v_i')k_b \prod_i [X_i]^{v_i''} \tag{16.129}$$

Net rate:
$$\frac{d[X_i]}{dt} = (v_i'' - v_i') \left\{ k_f \prod_i [X_i]^{v_i'} - k_b \prod_i [X_i]^{v_i''} \right\} \tag{16.130}$$

Equation (16.130) is a generalized net rate equation; it is a general form of the law of mass action first introduced in Sec. 16.9. In addition, the relation between k_f and k_b given by Eq. (16.125) holds for the general reaction given in Eq. (16.127).

The chemical rate constants are generally measured experimentally. Although methods from kinetic theory exist for their theoretical estimation, such results are sometimes uncertain by orders of magnitude. The empirical results for many reactions can be correlated in the form

$$k = Ce^{-\varepsilon_a/kT} \tag{16.131}$$

where ε_a is defined as the *activation energy* and C is a constant. Equation (16.131) is called the *Arrhenius equation*. An improved formula includes a "preexponential" temperature factor

$$k = c_1 T^\alpha e^{-\varepsilon_o/kT} \tag{16.132}$$

where c_1, α, and ε_o are all found from experimental data.

Returning to the special case of a dissociation reaction such as for diatomic nitrogen,

$$N_2 + M \overset{k_f}{\longrightarrow} 2N + M$$

the dissociation energy ε_d is defined as the difference between the zero-point energies,

$$\varepsilon_d \equiv \Delta\varepsilon_o = 2(\varepsilon_o)_N - (\varepsilon_o)_{N_2}$$

For this reaction, the rate constant is expressed as

$$k_f = c_f T^\alpha e^{-\varepsilon_d/kT} \tag{16.133}$$

where the activation energy $\varepsilon_a = \varepsilon_d$. Physically, the dissociation energy is the energy required to dissociate the molecule at $T = 0$ K. It is obviously a finite number: It takes energy—sometimes a considerable amount of energy—to tear a molecule apart. In contrast, consider the recombination reaction,

$$N_2 + M \xleftarrow[k_b]{} 2N + M$$

Here, no relative kinetic energy between the two colliding N atoms is necessary to bring about a change; indeed, the role of the third body M is to carry away some of the energy that must be given up by the two colliding N atoms before they can recombine. Hence, for recombination, there is no activation energy; $\varepsilon_a = 0$. Thus, the recombination rate constant is written as

$$k_b = c_b T^{\eta_b} \tag{16.134}$$

with no exponential factor.

Finally, it is important to note that all of the above formalism applies only to *elementary reactions*. An elementary chemical reaction is one that takes place in a single step. For example, a dissociation reaction such as

$$O_2 + M \rightarrow 2O + M$$

is an elementary reaction because it literally takes place by a collision of an O_2 molecule with another collision partner, yielding directly two oxygen atoms. On the other hand, the reaction

$$2H_2 + O_2 \rightarrow 2H_2O \tag{16.135}$$

is *not* an elementary reaction. Two hydrogen molecules do not come together with one oxygen molecule to directly yield two water molecules, even though if we mixed the hydrogen and oxygen together in the laboratory, our naked eye would observe what would appear to be the direct formation of water. Reaction (16.135) does not take place in a single step. Instead, Eq. (16.135) is a statement of an overall reaction that actually takes place through a series of elementary steps:

$$H_2 \rightarrow 2H \tag{16.136a}$$

$$O_2 \rightarrow 2O \tag{16.136b}$$

$$H + O_2 \rightarrow OH + O \tag{16.136c}$$

$$O + H_2 \rightarrow OH + H \tag{16.136d}$$

$$OH + H_2 \rightarrow H_2O + H \tag{16.136e}$$

Equations (16.136*a*) through (16.136*e*) constitute the *reaction mechanism* for the overall reaction (16.135). Each of Eqs. (16.136*a*) through (16.136*e*) is an elementary reaction.

We again emphasize that Eqs. (16.120) through (16.134) apply only for elementary reactions. In particular, the law of mass action given by Eq. (16.130) is valid for elementary reactions only. We *cannot* write Eq. (16.130) for reaction (16.135), but we can apply Eq. (16.130) to *each one* of the elementary reactions that constitute the reaction mechanism (16.136*a*) through (16.136*e*).

16.16 | CHEMICAL NONEQUILIBRIUM IN HIGH-TEMPERATURE AIR

We again highlight the importance of air in high-speed compressible flow problems. For the analysis of chemical nonequilibrium effects in high-temperature air, the following reaction mechanism occurs, valid below 9000 K:

$$O_2 + M \underset{k_{b_1}}{\overset{k_{f_1}}{\rightleftarrows}} 2O + M \tag{16.137}$$

$$N_2 + M \underset{k_{b_2}}{\overset{k_{f_2}}{\rightleftarrows}} 2N + M \tag{16.138}$$

$$NO + M \underset{k_{b_3}}{\overset{k_{f_3}}{\rightleftarrows}} N + O + M \tag{16.139}$$

$$O_2 + N \underset{k_{b_4}}{\overset{k_{f_4}}{\rightleftarrows}} NO + O \tag{16.140}$$

$$N_2 + O \underset{k_{b_5}}{\overset{k_{f_5}}{\rightleftarrows}} NO + N \tag{16.141}$$

$$N_2 + O_2 \underset{k_{b_6}}{\overset{k_{f_6}}{\rightleftarrows}} 2NO \tag{16.142}$$

$$N + O \underset{k_{b_7}}{\overset{k_{f_7}}{\rightleftarrows}} NO^+ + e^- \tag{16.143}$$

Equations (16.137) through (16.139) are dissociation reactions. Equations (16.140) and (16.141) are bimolecular exchange reactions (sometimes called the "shuffle" reactions); they are the two most important reactions for the formation of nitric oxide, NO, in air. Equation (16.143) is called a *dissociative-recombination reaction* because the recombination of the NO^+ ion with an electron produces not NO but rather a dissociated product $N + O$. Note that the above reactions are not all independent; for example, Eq. (16.142) can be obtained by adding Eqs. (16.140) and (16.141). However, in contrast to the calculation of an equilibrium composition as discussed in Sec. 16.11, for a nonequilibrium reaction mechanism the chemical equations do *not* have to be independent.

From this reaction mechanism, let us construct the rate equation for NO. Reactions (16.139) through (16.142) involve the production and extinction of NO. Moreover, in reaction (16.139), the collision partner M can be any of the different species, each requiring a different rate constant. That is, Eq. (16.139) is really these equations:

$$NO + O_2 \underset{k_{b3a}}{\overset{k_{f3a}}{\rightleftarrows}} N + O + O_2 \qquad (16.139a)$$

$$NO + N_2 \underset{k_{b3b}}{\overset{k_{f3b}}{\rightleftarrows}} N + O + N_2 \qquad (16.139b)$$

$$NO + NO \underset{k_{b3c}}{\overset{k_{f3c}}{\rightleftarrows}} N + O + NO \qquad (16.139c)$$

$$NO + O \underset{k_{b3d}}{\overset{k_{f3d}}{\rightleftarrows}} N + O + O \qquad (16.139d)$$

$$NO + N \underset{k_{b3e}}{\overset{k_{f3e}}{\rightleftarrows}} N + O + N \qquad (16.139e)$$

$$NO + NO^+ \underset{k_{b3f}}{\overset{k_{f3f}}{\rightleftarrows}} N + O + NO^+ \qquad (16.139f)$$

$$NO + e^- \underset{k_{b3g}}{\overset{k_{f3g}}{\rightleftarrows}} N + O + e^- \qquad (16.139g)$$

Thus, the chemical rate equation for NO is

$$\begin{aligned}
\frac{d[NO]}{dt} = &-k_{f3a}[NO][O_2] + k_{b3a}[N][O][O_2] \\
&- k_{f3b}[NO][N_2] + k_{b3b}[N][O][N_2] \\
&- k_{f3c}[NO]^2 + k_{b3c}[N][O][NO] \\
&- k_{f3d}[NO][O] + k_{b3d}[N][O]^2 \\
&- k_{f3e}[NO][N] + k_{b3e}[N]^2[O] \\
&- k_{f3f}[NO][NO^+] + k_{b3f}[N][O][NO^+] \\
&- k_{f3g}[NO][e^-] + k_{b3g}[N][O][e^-] \\
&+ k_{f4}[O_2][N] - k_{b4}[NO][O] \\
&+ k_{f5}[N_2][O] - k_{b5}[NO][N] \\
&+ 2k_{f6}[N_2][O_2] - 2k_{b6}[NO]^2
\end{aligned} \qquad (16.144)$$

There are rate equations similar to Eq. (16.144) for O_2, N_2, O, N, NO^+, and e^-. Clearly, you can see that a major aspect of such a nonequilibrium analysis is simply bookkeeping, making certain to keep track of all the terms in the equations.

Values of the rate constants for high-temperature air are readily available in the literature. See, for example, Ref. 68. Again, keep in mind that there is always some uncertainty in the published rate constants; they are difficult to measure experimentally, and very difficult to calculate accurately. Hence, any nonequilibrium analysis is a slave to the existing rate data.

16.17 | SUMMARY OF CHEMICAL NONEQUILIBRIUM

To analyze and compute the finite-rate chemical kinetic processes in any gas mixture, it is necessary to

1. Define the reaction mechanism [such as reactions (16.137) through (16.143) above].

2. Obtain the rate constants from the literature, usually in the form of Eq. (16.132).

3. Write all the appropriate rate equations, such as Eq. (16.144).

4. Solve the rate equations simultaneously to obtain the time variation of the species concentrations, i.e., $[O_2] = f_1(t)$, $[O] = f_2(t)$, etc. This is a job for a high-speed digital computer. Indeed, most modern analyses of chemical nonequilibrium systems would not be practically possible without computers.

Finally, we will see how these considerations are used in the analysis of high-temperature flowfields in Chap. 17.

16.18 | CHAPTER SUMMARY

High-temperature effects are an important aspect of modern compressible flow. This chapter presents the basic fundamentals of chemical thermodynamics and statistical thermodynamics necessary for the understanding of such effects.

For high-temperature gases, the internal energy is given by the sum of the translational, rotational, vibrational, and electronic energy modes. If these energies are measured above the zero-point energy, then the internal energy of a given chemical species i is given by

$$e_i = (e_{\text{trans}} + e_{\text{rot}} + e_{\text{vib}} + e_{\text{el}} + e_o)_i$$

The zero-point energy e_{o_i} is a property of the given chemical species and for purposes of analysis can be replaced by the heat of formation of species i at absolute zero $(\Delta h_f)_i^o$:

$$e_i = (e_{\text{trans}} + e_{\text{rot}} + e_{\text{vib}} + e_{\text{el}})_i + (\Delta h_f)_i^o$$

Because only a few molecular collisions are needed to transfer translational and rotational energy among molecules, e_{trans} and e_{rot} are almost always in equilibrium, and hence are given by

$$e_{trans} = \tfrac{3}{2}RT$$

$$e_{rot} = RT \qquad \left(\begin{array}{l}\text{for a diatomic molecule, or a linear} \\ \text{polyatomic molecule}\end{array}\right)$$

or

$$e_{rot} = \tfrac{3}{2}RT \qquad \text{(for a nonlinear polyatomic molecule)}$$

If the vibrational energy is in equilibrium, it is given by

$$e_{vib} = \frac{h\nu/kT}{e^{h\nu/kT} - 1}RT \qquad (16.47)$$

for a diatomic molecule. For a polyatomic molecule, there can be a number of different fundamental vibrational frequencies and the vibrational energy is given by a sum of terms, one each of the form of Eq. (16.47) for each of the fundamental frequencies, with each term multiplied by the degeneracy of each fundamental vibrational mode. If the vibrational energy is not in equilibrium, it is given by the master equation for vibrational relaxation, Eq. (16.100), which can be approximated by

$$\frac{de_{vib}}{dt} = \frac{1}{\tau}\left(e_{vib}^{eq} - e_{vib}\right) \qquad (16.116)$$

The electronic energy, when in equilibrium, can be obtained from the sum $e_{el} = \sum_j (\varepsilon_j N_j^*)_{el}$ with N_j given by the Boltzmann distribution

$$N_j^* = N\frac{g_j e^{-\epsilon_j/kT}}{\displaystyle\sum_j g_j e^{-\epsilon_j/kT}} \qquad (16.24)$$

and where ϵ_j denotes the energy of the jth electronic energy level measured above the zero point. The Boltzmann distribution describes how particles are distributed over their energy levels in equilibrium at the temperature T. For electronic nonequilibrium, rate equations beyond the scope of this chapter must be employed to obtain e_{el}. For many applications in high temperature gas dynamics, e_{el} is small and can be neglected.

The internal energy for the chemically reacting mixture can then be obtained from

$$e = \sum_i c_i e_i$$

where c_i is the mass fraction of species i. If the gas is in chemical equilibrium, then c_i is obtained from an equilibrium analysis using the equilibrium constants as described in Sec. 16.11. If the gas is not in chemical equilibrium, the c_i, must be obtained as a function of time from the chemical rate equations described in Sec. 16.15.

PROBLEMS

16.1   Starting with Eq. (16.4), derive the most probable population distribution for Fermions, namely, Eq. (16.19).

16.2   Derive Eqs. (16.34) and (16.35). (*Note:* You will have to search some references on statistical thermodynamics to set up these derivations.)

16.3   Starting with Eq. (16.35), derive the perfect gas equation of state, $p = \rho RT$. (This demonstrates that the perfect gas equation of state, which historically was first obtained empirically, falls out directly from the fundamentals of statistical thermodynamics.)

16.4   Starting with the quantum mechanical expression for the quantized translational energy levels as a function of the quantum numbers n_1, n_2, and n_3, derive in detail the translational partition function given by Eq. (16.38).

16.5   In a similar vein as Prob. 16.4, derive in detail the rotational partition function given by Eq. (16.39).

16.6   Consider 1 kg of pure diatomic N_2 in thermodynamic equilibrium. The fundamental vibrational frequency of N_2 is $v = 7.06 \times 10^{13}$/s, the molecular weight $\mathcal{M}_{N_2} = 28$, Planck's constant $h = 6.625 \times 10^{-34}$ J $\cdot$ s, and the Boltzmann constant is $k = 1.38 \times 10^{-23}$ J/K.

   a.   Calculate and plot on graph paper the number of N_2 molecules in each of the first three vibrational energy levels, ε_o, ε_1, and ε_2 as a function of temperature from $T = 300$ to 3500 K, using 400 K increments.

   b.   Calculate and plot on graph paper the sensible enthalpy (including translation, rotation, and vibration) in joules per kilogram as a function of temperature from $T = 300$ to 3500 K.

   c.   Calculate and plot on graph paper the specific heat at constant pressure as a function of temperature from $T = 300$ to 3500 K.

16.7   Frequently in the literature, a characteristic temperature for vibration is defined as $\theta_{\text{vib}} = hv/k$. Express e and c_v for a diatomic molecule [Eqs. (16.49) and (16.51)] in terms of θ_{vib}.

16.8   Consider an equilibrium chemically reacting mixture of three general species denoted by A, B, and AB. In detail, derive Eqs. (16.54) and (16.55) for such a mixture.

16.9   Consider an equilibrium chemically reacting mixture of oxygen at $p = 1$ atm and $T = 3200$ K. The only species present are O_2 and O. $K_{p,O_2} = 0.04575$ atm. Calculate the partial pressures, mole fractions, mole-mass ratios, and mass fractions for this mixture.

16.10 For the conditions of the Prob. 16.9, calculate the internal energy of the mixture in joules per kilogram, including the translational, rotational, vibrational, and electronic energies. Note the following physical data: For O_2, $\Delta H_f^0 = 0$, $v = 4.73 \times 10^{13}$/s, $\varepsilon_{\text{el}_1}/k = 11{,}390$ K, $g_{\text{el}_o} = 3$, $g_{\text{el}_1} = 2$ (ignore higher electronic levels); for O, $\Delta H_f^0 = 2.47 \times 10^8$ J/(kg $\cdot$ mol),

$\varepsilon_{el_1}/k = 228$ K, $\varepsilon_{el_2}/k = 326$ K, $g_{el_o} = 5$, $g_{el_1} = 3$, $g_{el_2} = 1$ (ignore higher electronic levels).

16.11 Consider air at $p = 0.5$ atm and $T = 4500$ K. Assume the chemical species present are O_2, O, N_2, and N. (Ignore NO.) Calculate the enthalpy in joules per kilogram. In addition to that already given in Prob. 16.10, note these physical data: $K_{p,O_2} = 12.19$ atm, $K_{p,N_2} = 0.7899 \times 10^{-4}$ atm; for N_2, $\Delta H_f^0 = 0$, $\nu = 7.06 \times 10^{13}$/s (ignore electronic levels of N_2 because the first excited level is very high, $\varepsilon_{el_1}/k = 100,000$ K, and hence its population is very low); for N, $\Delta H_f^0 = 4.714 \times 10^8$ J/(kg · mol) (again, ignore electronic levels of N because the first excited level is high, $\varepsilon_{el_1}/k = 23,000$ K).

16.12 Consider a unit mass of N_2 in equilibrium at $p = 1$ atm and $T = 300$ K. For these conditions, the vibrational relaxation time is 190 s. Assume that, by some mechanism, the vibrational energy is instantaneously increased by a factor of 100, with all other properties remaining unchanged. At the end of 1 min after excitation, what is the value of the vibrational energy relative to the equilibrium value? Assume that p and T remain constant at 1 atm and 300 K, respectively.

CHAPTER 17

High-Temperature Flows: Basic Examples

With the advent of jet propulsion it became necessary to broaden the field of aerodynamics to include problems which before were treated mostly by physical chemists.

Theodore von Karman, 1958

PREVIEW BOX

High-temperature effects can be dramatic. Witness the two separate results for the inviscid flow of air over a blunt body at Mach 20 at an altitude of 20 km shown in Fig. 17.1; we see here the bow shock wave shape as well as the temperature contours in the flow between the shock and the body. The results shown in Fig. 17.1*a* are for a perfect gas with constant $\gamma = 1.4$; those in Fig. 17.1*b* take into account equilibrium chemically reacting flow. What a dramatic difference! For the chemically reacting case in Fig. 17.1*b*, the shock is much closer to the body, and the temperature levels in the shock layer are much lower than those predicted by the calorically perfect gas case. The amount of nitrogen dissociation in the shock layer is shown in Fig. 17.2*a*, and the

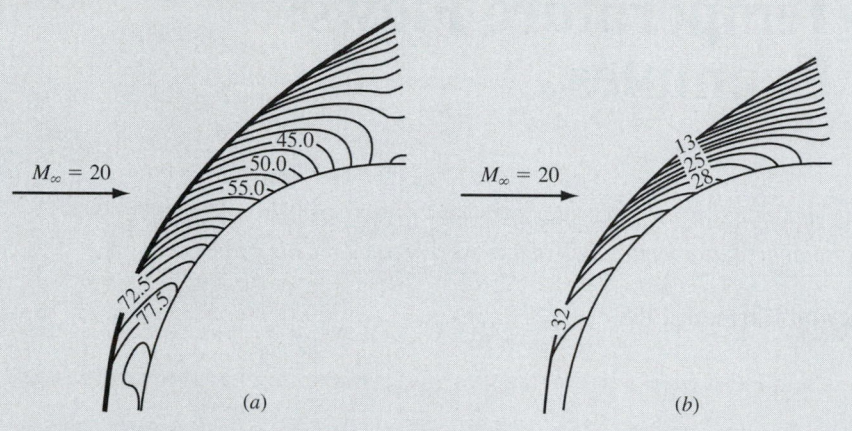

Figure 17.1 | Normalized temperature contours for blunt-body flow. (*a*) Calorically perfect gas; (*b*) equilibrium chemically reacting air. $M_\infty = 20$, altitude = 20 km. (From Palmer, Ref. 143.)

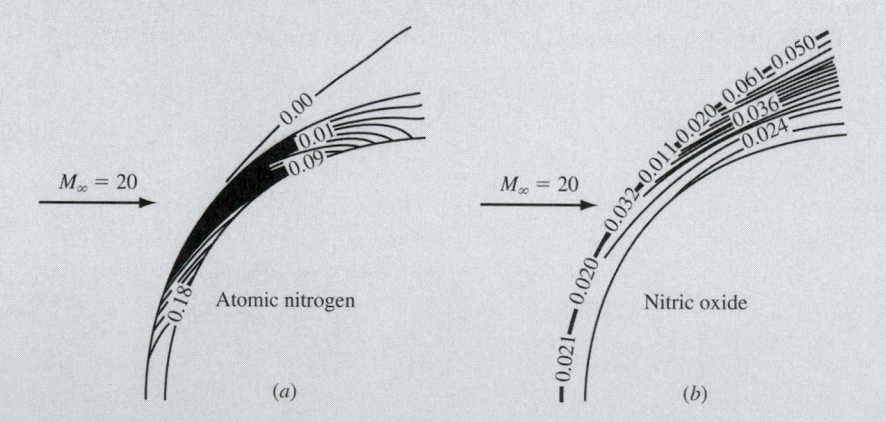

Figure 17.2 | Species mole fraction contours for blunt-body flow; equilibrium chemically reacting air. $M_\infty = 20$, altitude = 20 km. (*a*) X_N; (*b*) X_{NO}. (From Palmer, Ref. 143.)

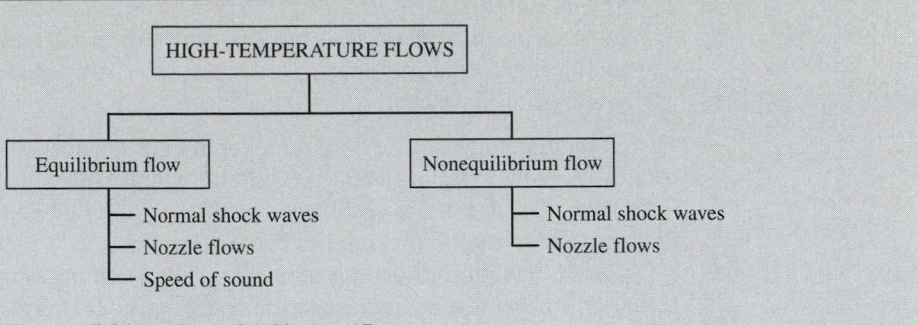

Figure 17.3 | Roadmap for Chapter 17.

amount of nitric oxide is shown in Fig. 17.2*b*. Clearly, the effect of chemical reactions is not trivial here; indeed, all aspects of the flow are dominated by high-temperature effects.

The purpose of this (and final) chapter is to illustrate some of the fundamental differences brought about by high-temperature effects in a flow. Rather than deal with more complex flows such as the blunt body shown in Figs. 17.1 and 17.2, we choose to examine high-temperature effects on relatively simple flows—normal shock waves and quasi-one-dimensional nozzle flows.

These flows are relatively uncomplicated, and the effects of high temperatures are easily demonstrated.

The roadmap for this chapter is given in Fig. 17.3. Our discussion of high-temperature flows is divided into equilibrium flow and nonequilibrium flow, listed on the left and right sides respectively in Fig. 17.3. We will discuss normal shock waves and nozzle flows for both cases. In addition, under equilibrium flow we will examine the speed of sound in a chemically reacting gas, a fundamental property that is changed considerably by high-temperature effects.

17.1 | INTRODUCTION TO LOCAL THERMODYNAMIC AND CHEMICAL EQUILIBRIUM

It is common in classical thermodynamics to define a system in *complete thermodynamic equilibrium* as one with these characteristics:

1. No gradients of pressure, temperature, velocity, or concentration exist anywhere in the system.
2. If the system is a mixture of gases, there is no tendency to undergo a spontaneous change in chemical composition, no matter how slowly. This characteristic is called *chemical equilibrium,* and is necessary for the overall concept of complete thermodynamic equilibrium.

In terms of our discussion in Chap. 16, these characteristics can be stated as:

1. If there are no pressure, temperature, velocity, or concentration gradients present in the system, the particles of the system are distributed over their allowed energy levels according to the *Boltzmann distribution* at the temperature T of the system.

2. If there is no tendency to undergo a spontaneous change in chemical composition, the composition itself is, of course, fixed, and the values for the equilibrium partial pressures p_i (or X_i, η_i, or c_i) are determined by the equilibrium constants.

Therefore, thermodynamic equilibrium implies a Boltzmann distribution at the temperature T of the system, chemical equilibrium implies that the composition is fixed for a given T and p and is calculated from the equilibrium constants, and *complete* thermodynamic equilibrium implies both of the above.

This discussion considers a stationary system, such as gas in a box left for a long period of time so that no gradients exist inside the box. However, what happens for the case of a *flowing* gas, say for flow through a nozzle or over a missile or reentry vehicle? Clearly, gradients in p and T exist in such flows. Therefore, any flowing gas in which properties are changing with location and/or time is *not* in complete thermodynamic equilibrium as defined above. In turn, can the equilibrium relations obtained in Chap. 16 be applied to such a flowing gas? The answer is, strictly speaking, no; however, in many practical cases, the answer is a qualified yes. The qualification is that if the gradients in the flow are small enough, we can shrink our "system" (our box) to an infinitesimal size around a given point in the flow and see a relatively constant property region in the immediate neighborhood of the point. Let us assume that the equilibrium thermodynamic relations apply *locally* at the local values of T and p at a point in the flow, and hence a local Boltzmann distribution applies at the local temperature. This is defined as the case of *local thermodynamic equilibrium*. Similarly, let us assume that the local chemical composition at a point in the flow is the same as that determined by equilibrium calculations (using the equilibrium constant) at the local T and p. This is defined as the case of *local chemical equilibrium*.

There are many practical situations where the flow gradients are moderate enough and the molecular collision frequencies (hence, the vibrational and chemical rates, for example) are large enough that both local thermodynamic and chemical equilibrium conditions hold at each point in the flow. In other situations, local thermodynamic equilibrium will hold, but the chemical composition may not be in equilibrium (chemical changes require more molecular collisions than vibrational or electronic energy changes). There are yet other cases where flows are neither in local thermodynamic nor chemical equilibrium. In this chapter, we will examine some basic high-temperature flow problems, such as normal shock waves and nozzle flows. In the first part, we will develop solutions assuming local thermodynamic and chemical equilibrium. This follows the left-hand column of our roadmap in Fig. 17.3. In the second half, we will emphasize techniques for solving nonequilibrium flows, which follows the right-hand column in Fig. 17.3.

17.2 | EQUILIBRIUM NORMAL SHOCK WAVE FLOWS

Consider a stationary normal shock wave as sketched in Fig. 3.4. Assume that the shock is strong enough, hence T_2 is high enough, such that vibrational excitation and chemical reactions occur behind the shock front. Moreover, assume that local thermodynamic and chemical equilibrium hold behind the shock. All conditions ahead of the shock wave (region 1) are known. Our objective is to calculate properties behind the shock.

The governing flow equations for steady one-dimensional flow were derived in Sec. 3.2, and were specialized to the case of a normal shock wave in Sec. 3.6. This resulted in Eqs. (3.38) through (3.40), which were general and hence apply to our present high-temperature case. They are repeated and renumbered here for convenience:

Continuity:
$$\rho_1 u_1 = \rho_2 u_2 \tag{17.1}$$

Momentum:
$$p_1 + \rho_1 u_1^2 = p_2 + \rho_2 u_2^2 \tag{17.2}$$

Energy:
$$h_1 + \frac{u_1^2}{2} = h_2 + \frac{u_2^2}{2} \tag{17.3}$$

In addition, the equilibrium thermodynamic properties for the high-temperature gas are assumed known from the techniques discussed in Chap. 16. These may take the form of tables or graphs, or may be calculated directly from the equations developed in Chap. 16. In any event, we can consider these properties in terms of these functional relations ("equations of state," if you will):

$$\rho_2 = \rho(p_2, h_2) \tag{17.4}$$

$$T_2 = T(p_2, h_2) \tag{17.5}$$

Recall from Chap. 3 that for a calorically perfect gas, these equations yield a series of closed-form algebraic relations for p_2/p_1, T_2/T_1, M_2, etc., as functions of M_1 [see, for example, Eqs. (3.51), (3.53), (3.57), and (3.59)]. Unfortunately, no simple formulas can be obtained when the gas is vibrationally excited and/or chemically reacting. For such high-temperature cases, Eqs. (17.1) through (17.5) must be solved numerically. To set up a numerical solution, let us first rearrange Eqs. (17.1) through (17.3). From Eq. (17.1),

$$u_2 = \frac{\rho_1 u_1}{\rho_2} \tag{17.6}$$

Substitute Eq. (17.6) into (17.2):

$$p_1 + \rho_1 u_1^2 = p_2 + \rho_2 \left(\frac{\rho_1 u_1}{\rho_2} \right)^2 \tag{17.7}$$

Solving Eq. (17.7) for p_2, we have

$$\boxed{p_2 = p_1 + \rho_1 u_1^2 \left(1 - \frac{\rho_1}{\rho_2} \right)} \tag{17.8}$$

In addition, substituting Eq. (17.6) into (17.3), we have

$$h_1 + \frac{u_1^2}{2} = h_2 + \left(\frac{\rho_1 u_1}{\rho_2} \right)^2 \Big/ 2 \tag{17.9}$$

Solving Eq. (17.9) for h_2,

$$h_2 = h_1 + \frac{u_1^2}{2}\left[1 - \left(\frac{\rho_1}{\rho_2}\right)^2\right] \tag{17.10}$$

Since all the upstream conditions ρ_1, u_1, p_1, h_1, etc., are known, Eqs. (17.8) and (17.10) express p_2 and h_2, respectively, in terms of only one unknown, namely, ρ_1/ρ_2. This establishes the basis for an iterative numerical solution:

1. *Assume* a value for ρ_1/ρ_2. (A value of 0.1 is usually good for a starter.)
2. Calculate p_2 from Eq. (17.8) and h_2 from Eq. (17.10).
3. With the values of p_2 and h_2 just obtained, calculate ρ_2 from Eq. (17.4).
4. Form a *new* value of ρ_1/ρ_2 using the value of ρ_2 obtained from Step 3.
5. Use this new value of ρ_1/ρ_2 in Eqs. (17.8) and (17.10) to obtain new values of p_2 and h_2, respectively. Then repeat steps 3 through 5 until convergence is obtained, i.e., until there is only a negligible change in ρ_1/ρ_2 from one iteration to the next. (This convergence is usually very fast, typically requiring less than five iterations.)
6. At this stage, we now have the correct values of p_2, h_2, and ρ_2. Obtain the correct value of T_2 from Eq. (17.5).
7. Obtain the correct value of u_2 from Eq. (17.6).

By means of steps 1 through 7 above, we can obtain all properties behind the shock wave for given properties in front of the wave.

There is a basic practical difference between the shock results for a calorically perfect gas and those for a chemically reacting gas. For a calorically perfect gas, we demonstrated in Sec. 3.6 that

$$\frac{p_2}{p_1} = f_1(M_1)$$

$$\frac{\rho_2}{\rho_1} = f_2(M_1)$$

$$\frac{h_2}{h_1} = f_3(M_1)$$

Note that in this case only M_1 is required to obtain the ratios of properties across a normal shock wave; such properties are tabulated in Table A.2 at the back of this book. In contrast, for an *equilibrium chemically reacting gas,* we have already seen that

$$\frac{p_2}{p_1} = g_1(u_1, p_1, T_1)$$

$$\frac{\rho_2}{\rho_1} = g_2(u_1, p_1, T_1)$$

$$\frac{h_2}{h_1} = g_3(u_1, p_1, T_1)$$

Note that in this case *three* free-stream parameters are necessary to obtain the ratios of properties across a normal shock wave. This makes plenty of sense—the equilibrium composition behind the shock depends on p_2 and T_2, which in turn are governed in part by p_1 and T_1. Hence, in addition to the upstream velocity u_1, the normal shock properties must depend also on p_1 and T_1. By this same reasoning, if no chemical reactions take place, but the vibrational and electronic energies are excited (a thermally perfect gas), then the downstream normal shock properties depend on two upstream conditions, namely, u_1 and T_1.

Also note that, in contrast to a calorically perfect gas, the Mach number no longer plays a pivotal role in the results for normal shock waves in a high-temperature gas. In fact, for most high-temperature flows in general, the Mach number is not a particularly useful quantity. The flow of a chemically reacting gas is mainly governed by the primitive variables of velocity, temperature, and pressure. For an equilibrium gas, the Mach number is still uniquely defined as V/a, and it can be used along with other determining variables—it just does not hold a dominant position as in the case of a calorically perfect gas considered in Chaps. 1 through 12. For a nonequilibrium gas, however, there is some ambiguity even in the definition of Mach number (to be discussed in Sec. 17.11), and hence the Mach number further loses significance for such cases.

For high-temperature air, a comparison between calorically perfect gas and equilibrium chemically reacting gas results was shown in Fig. 16.2. Here, the temperature behind a normal shock wave is plotted versus upstream velocity for conditions at a standard altitude of 52 km. The equilibrium results are plotted directly from normal shock tables prepared by the Cornell Aeronautical Laboratory (now CALSPAN Corporation), and published in Refs. 69 and 70. These reports should be consulted for equilibrium normal shock properties associated with air in the standard atmosphere. From Fig. 16.2, the calorically perfect results considerably overpredict the temperature, and for obvious reasons. For a calorically perfect gas, the directed kinetic energy of the flow ahead of the shock is mostly converted to translational and rotational molecular energy behind the shock. On the other hand, for a thermally perfect and/or chemically reacting gas, the directed kinetic energy of the flow, when converted across the shock wave, is shared across all molecular modes of energy, and/or goes into zero-point energy of the products of chemical reaction. Hence, the temperature (which is a measure of translational energy only) is less for such a case.

For further comparison, consider a reentry vehicle at 170,000-ft standard altitude with a velocity of 36,000 ft/s. The properties across a normal shock wave for this case are tabulated in Table 17.1. Note from that tabulation that chemical

Table 17.1

	For calorically perfect air, $\gamma = 1.4$ (see Table A.2)	For equilibrium chemically reacting air (CAL Report AG-1729-A-2)
p_2/p_1	1233	1387
ρ_2/ρ_1	5.972	15.19
h_2/h_1	206.35	212.8
T_2/T_1	206.35	41.64

reactions have the strongest effect on temperature, for the reasons given earlier. This is generally true for all types of chemically reacting flows—the temperature is by far the most sensitive variable. In contrast, the pressure ratio is affected only by a small amount. Pressure is a "mechanically" oriented variable; it is governed mainly by the fluid mechanics of the flow, and not so much by the thermodynamics. This is substantiated by examining the momentum equation, namely, Eq. (17.2). For high-speed flow, $u_2 \ll u_1$, and $p_2 \gg p_1$. Hence, from Eq. (17.2),

$$p_2 \approx \rho_1 u_1^2$$

This is a common hypersonic approximation; note that p_2 is mainly governed by the free-stream velocity, and that thermodynamic effects are secondary.

In an equilibrium dissociating and ionizing gas, increasing the pressure at constant temperature tends to decrease the atom and ion mass fractions, i.e., increasing the pressure tends to inhibit dissociation and ionization. The consequences of this effect on equilibrium normal shock properties are shown in Fig. 17.4, where the

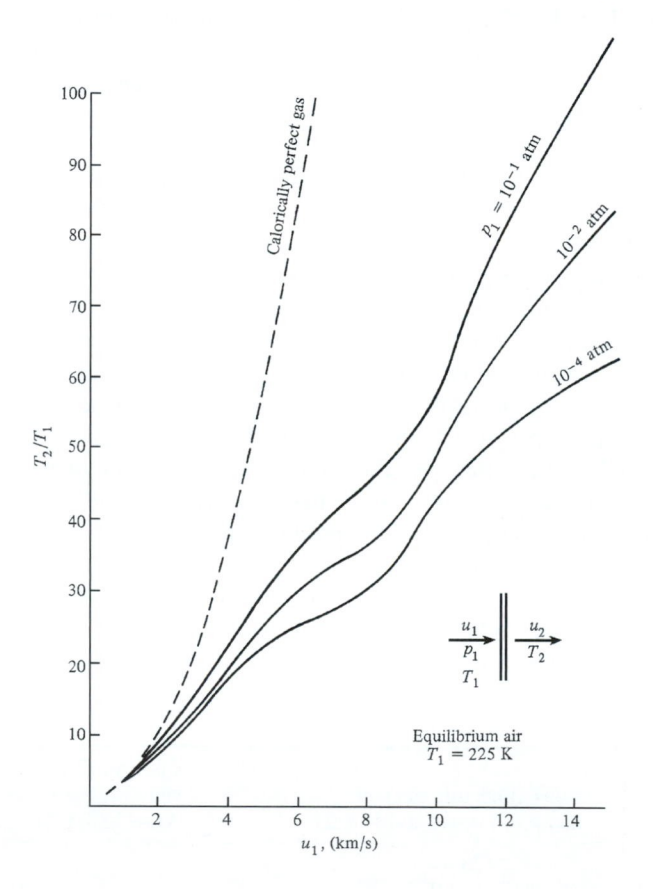

Figure 17.4 | Influence of pressure on the normal shock temperature in equilibrium air.

temperature ratio across the shock is plotted versus upstream velocity for three different values of upstream pressure. Note that T_2/T_1 is higher at higher pressures; the gas is less dissociated and ionized at higher pressure, and hence more energy goes into translational molecular motion behind the shock rather than into the zero-point energy of the products of dissociation.

17.3 | EQUILIBRIUM QUASI-ONE-DIMENSIONAL NOZZLE FLOWS

Consider the inviscid, adiabatic high-temperature flow through a convergent-divergent Laval nozzle, as sketched at the top of Fig. 17.5. As usual, the reservoir pressure and temperature are denoted by p_o and T_o, respectively. The throat conditions are denoted by an asterisk, and exit conditions by a subscript e. This nozzle could be a high-temperature wind tunnel, where air is heated in the reservoir, for example, by an electric arc (an "arc tunnel"), or by shock waves (a "shock tunnel"). In a shock tunnel, the nozzle is placed at the end of a shock tube, and the reservoir is

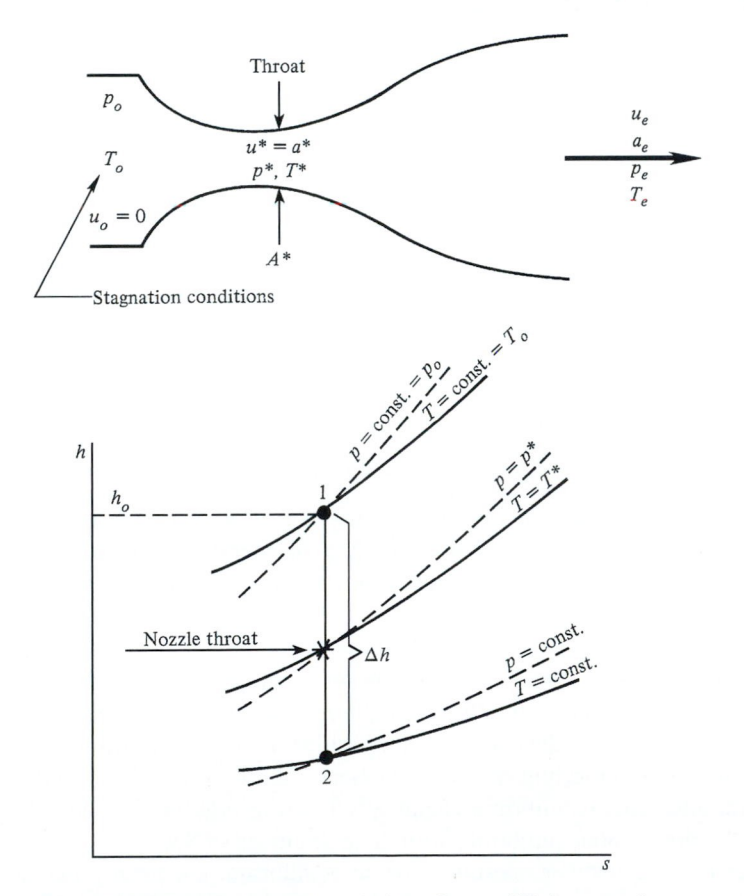

Figure 17.5 | Illustration of the solution of an equilibrium nozzle flow on a Mollier diagram.

essentially the hot, high-pressure gas behind a reflected shock wave (see Sec. 7.3). The nozzle in Fig. 17.5 could also be a rocket engine, where the reservoir conditions are determined by the burning of fuel and oxidizer in the combustion chamber. In either case—the high-temperature wind tunnel or the rocket engine—the flow through the nozzle is chemically reacting. Assuming local chemical equilibrium throughout the flow, let us examine the properties of the nozzle expansion.

First, let us pose the question: Is the chemically reacting flow isentropic? On a physical basis, the flow is both inviscid and adiabatic. However, this does not guarantee, in general, that the chemically reacting flow is irreversible. If we deal with an *equilibrium* chemically reacting flow, we can write the combined first and second laws of thermodynamics in the form of Eq. (1.32), repeated here:

$$T\,ds = dh - v\,dp \tag{1.32}$$

In Sec. 5.2, the governing equations for quasi-one-dimensional flow were derived in both algebraic and differential form. Moreover, in Sec. 5.2 no assumption was made about the type of gas; hence all the equations in that section hold in general. In particular, a form of the energy equation was obtained as Eq. (5.10):

$$dh + u\,du = 0 \tag{5.10}$$

In addition, Eq. (5.9) gave the momentum equation in the form

$$dp = -\rho u\,du \tag{5.9}$$

This can be rearranged as

$$u\,du = -\frac{dp}{\rho} = -v\,dp \tag{17.11}$$

Combining Eqs. (17.11) and Eq. (5.10), we have

$$dh - v\,dp = 0 \tag{17.12}$$

Substituting Eq. (17.12) into Eq. (1.32), we have

$$T\,ds = 0 \tag{17.13}$$

Hence, the equilibrium chemically reacting nozzle flow is isentropic. Moreover, since Eq. (17.13) was obtained by combining the energy and momentum equations, the assumption of isentropic flow can be used *in place of* either the momentum or energy equations in the analysis of the flow.

It is a general result that equilibrium chemical reactions do not introduce irreversibilities into the system; if an equilibrium reacting system starts at some conditions p_1 and T_1, deviates from these conditions for some reason, but then returns to the original p_1 and T_1, the chemical composition at the end returns to what it was at the beginning. Equilibrium chemical reactions are reversible. Hence, any shockless, inviscid, adiabatic, equilibrium chemically reacting flow is isentropic. This is *not* true if the flow is nonequilibrium, as will be discussed in Sec. 17.11.

Let us pose another question: For an equilibrium chemically reacting nozzle flow, does sonic flow exist at the throat? We have already established that the flow is isentropic. This was the only necessary condition for the derivation of the area-velocity relation, Eq. (5.15). Hence, the equation

$$\frac{dA}{A} = (M^2 - 1)\frac{du}{u} \tag{5.15}$$

holds for a general gas. In turn, when $M = 1$, $dA/A = 0$, and therefore sonic flow *does* exist at the throat of an equilibrium chemically reacting nozzle flow. The same is *not* true for a nonequilibrium flow, as will be discussed in Sec. 17.11.

We are now in a position to solve the equilibrium chemically reacting nozzle flow. A graphical solution is the easiest to visualize. Consider that we have the equilibrium gas properties on a Mollier diagram, as sketched in Fig. 17.5. Recall from Fig. 16.15*a* that a Mollier diagram is a plot of h versus s, and lines of constant p and constant T can be traced on the diagram. Hence, referring to Fig. 17.5, a given point on the Mollier diagram gives not only h and s, but p and T at that point as well (and any other equilibrium thermodynamic property, since the state of an equilibrium system is completely specified by any two-state variables). Let point 1 in Fig. 17.5 denote the known reservoir conditions in the nozzle. Since the flow is isentropic, conditions at all other locations throughout the nozzle must fall somewhere on the vertical line through point 1 in Fig. 17.5. In particular, choose a value of $u = u_2 \neq 0$. The point in Fig. 17.5 which corresponds to this velocity (point 2) can be found from Eqs. (5.5) and (5.6) as

$$h_1 + \frac{u_1^2}{2} = h_2 + \frac{u_2^2}{2} = h_o \tag{17.14}$$

Hence,

$$\Delta h = h_o - h_2 = \frac{u_2^2}{2} \tag{17.15}$$

Thus, for a given velocity u_2, Eq. (17.15) locates the appropriate point on the Mollier diagram. In turn, the constant-pressure and -temperature lines that run through point 2 define the pressure p_2 and temperature T_2 associated with the chosen velocity u_2. In this fashion, the variation of the thermodynamic properties through the nozzle expansion can be calculated as a function of velocity u for given reservoir conditions.

For an equilibrium gas, the speed of sound, $a \equiv (\partial p/\partial \rho)_s$, is also a unique function of the thermodynamic state. This will be discussed in more detail in Sec. 17.5. For example,

$$a = a(h, s) \tag{17.16}$$

Thus, at each point on the Mollier diagram in Fig. 17.5, there exists a definite value of a. Moreover, at some point along the vertical line through point 1, the speed of sound a will equal the velocity u at that point. Such a point is marked by an asterisk in Fig. 17.5. At this point, $u = a = u^* = a^*$. Since we demonstrated earlier that sonic flow corresponds to the throat in an equilibrium nozzle flow, then this point in

Fig. 17.5 must correspond to the throat. The pressure, temperature, and density at this point are p^*, T^*, and ρ^*, respectively. Thus, from the continuity equation (5.1), we have

$$\rho u A = \rho^* u^* A^* \tag{17.17}$$

or

$$\frac{A}{A^*} = \frac{\rho^* u^*}{\rho u} \tag{17.18}$$

Therefore, Eq. (17.18) allows the calculation of the nozzle area ratio as a function of velocity through the nozzle.

In summary, using the Mollier diagram in Fig. 17.5, we can compute the appropriate values of u, p, T, and A/A^* through an equilibrium nozzle flow for given reservoir conditions. An alternative to this graphical approach is a straight-forward numerical integration of Eqs. (5.7), (5.9), and (5.10) along with tabulated values of the equilibrium thermodynamic properties. The integration starts from known conditions in the reservoir and marches downstream. Such a numerical integration solution is left for the reader to construct.

In either case, numerical or graphical, it is clear that closed-form algebraic relations such as those obtained in Sec. 5.4 for a calorically perfect gas are not obtainable for chemically reacting nozzle flows. This is analogous to the case of chemically reacting flow through a shock wave discussed in Sec. 17.2. In fact, by now the reader should suspect, and correctly so, that closed-form algebraic relations cannot be obtained for any high-temperature chemically reacting flow of interest. Numerical or graphical solutions are necessary for such cases.

Recall from Chap. 5 that, for a calorically perfect gas, the nozzle flow characteristics were governed by the local Mach number only. For example, from Eqs. (5.20), (3.28), and (3.30), for a *calorically perfect gas*,

$$\frac{A}{A^*} = f_1(M)$$

$$\frac{T}{T_o} = f_2(M)$$

$$\frac{p}{p_o} = f_3(M)$$

In contrast, for an *equilibrium chemically reacting gas*,

$$\frac{A}{A^*} = g_1(p_o, T_o, u)$$

$$\frac{T}{T_o} = g_2(p_o, T_o, u)$$

$$\frac{p}{p_o} = g_3(p_o, T_o, u)$$

Note, as in the case of a normal shock, that the nozzle flow properties depend on three parameters. Also, once again we see that Mach number is not the pivotal parameter for a chemically reacting flow.

Some results for the equilibrium supersonic expansion of high-temperature air are shown in Fig. 17.6. Here the mole-mass ratios for N_2, O_2, N, O, and NO are given as a function of area ratio for $T_o = 8000$ K and $p_o = 100$ atm. At these conditions, the air is highly dissociated in the reservoir. However, as the gas expands through the nozzle, the temperature decreases, and as a result the oxygen

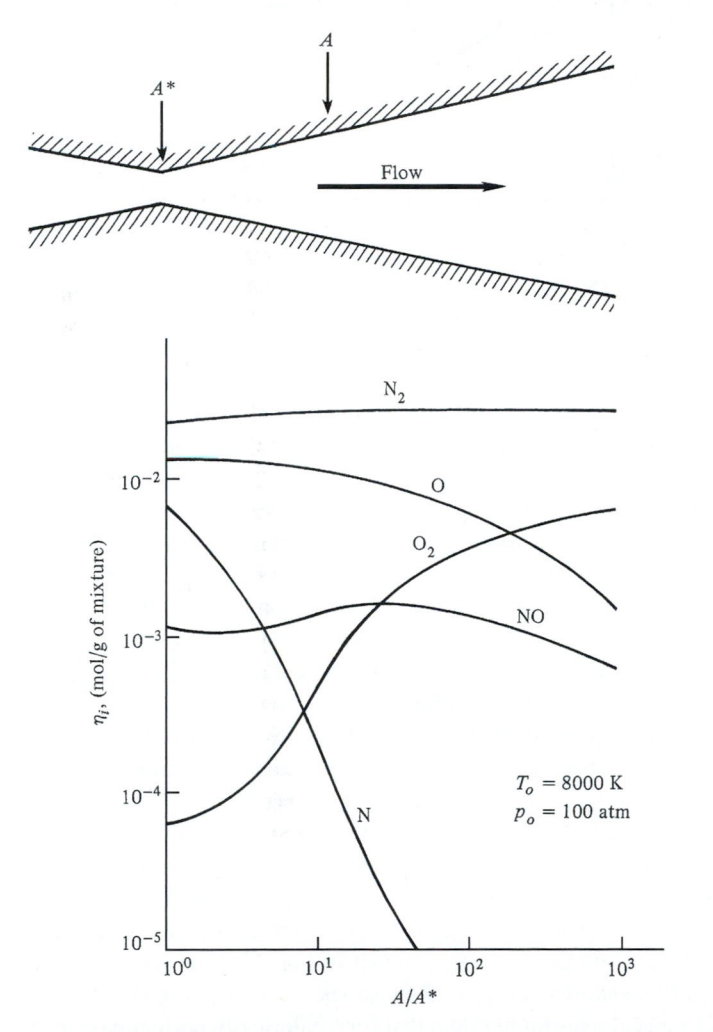

Figure 17.6 | Chemical composition for the equilibrium supersonic nozzle expansion of high-temperature air. (After Eschenroeder et al., "Shock Tunnel Studies of High Enthalpy Ionized Airflows," Cornell Aeronautical Lab. Report No. AF-1500 A1, 1962.)

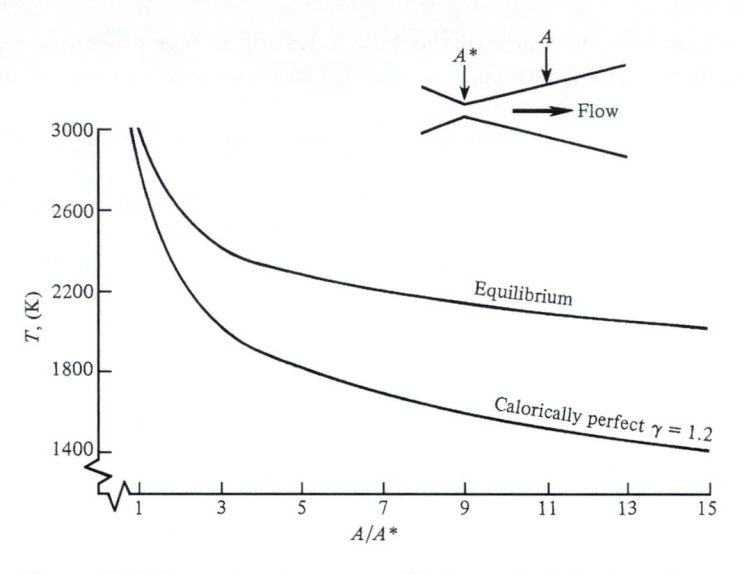

Figure 17.7 | Comparison between equilibrium and calorically perfect results for the flow through a rocket engine.

and nitrogen recombine. This is reflected in Fig. 17.6, which shows η_O and η_N decreasing and η_{O_2} and η_{N_2} increasing as the gas expands supersonically from $A/A^* = 1$ to 1000.

A typical result from equilibrium chemically reacting flow through a rocket nozzle is shown in Fig. 17.7. Here, the equilibrium temperature distribution is compared with that for a calorically perfect gas as a function of area ratio. The reservoir conditions are produced by the equilibrium combustion of an oxidizer (N_2O_2) with a fuel (half N_2H_4 and half unsymmetrical dimethyl hydrazine) at an oxidizer-to-fuel ratio of 2.25 and a chamber pressure of 4 atm. The calorically perfect gas is assumed to have a constant $\gamma = 1.20$. It is important to note from Fig. 17.7 that the equilibrium temperature is *higher* than that for the calorically perfect gas. This is because, as the gas expands and becomes cooler, the chemical composition changes from a high percentage of atomic species (O and H) in the reservoir with an attendant high zero-point energy to a high percentage of molecular products (H_2O, CO, etc.) in the nozzle expansion with an attendant lower zero-point energy. That is, the gas recombines, giving up chemical energy which serves to increase the translational energy of the molecules, hence resulting in a higher static temperature than would exist in the non-reacting case. Note that the trend shown in Fig. 17.7 for nozzle flow is exactly the opposite of that shown in Fig. 16.2 for shock waves. For nozzle flow, the equilibrium temperature is always higher than that for a calorically perfect gas; for flow behind a shock wave, the equilibrium temperature is always lower than that for a calorically perfect gas. In the former case, the reactions are exothermic, and energy is dumped into the translational molecular motion; in the latter, the reactions are endothermic and energy is taken from the translational mode.

17.4 | FROZEN AND EQUILIBRIUM FLOWS: SPECIFIC HEATS

In Secs. 17.1 through 17.3 we have discussed flows that are in local chemical equilibrium, i.e., flows where the local chemical composition at each point is dictated by the local temperature and pressure via equilibrium relations employing the equilibrium constants (see Sec. 16.11). However, we discussed in Secs. 16.13 through 16.16 that in reality all chemical reactions and vibrational energy exchanges take a finite time to occur. Therefore, in the case of a flow in local chemical and/or thermodynamic equilibrium, where the equilibrium properties of a moving fluid element demand instantaneous adjustments to the local T and p as the element moves through the field, the reaction rates have to be infinitely large. Therefore, equilibrium flow implies infinite chemical and vibrational rates.

The opposite of this situation is a flow where the reaction rates are precisely zero—so-called *frozen flow*. As a result, the chemical composition of a frozen flow remains constant throughout space and time. (This is true for an inviscid flow; for a viscous flow the composition of a given fluid element may change via diffusion, even though the flow is chemically frozen.)

The qualitative difference between chemical equilibrium and frozen nozzle flows is sketched in Fig. 17.8 for a case of fully dissociated oxygen in the reservoir. Examining Fig. 17.8c, the flow starts out with oxygen atoms in the reservoir ($c_O = 1$, $c_{O_2} = 0$). If we have equilibrium flow, as the temperature decreases throughout the expansion, the oxygen atoms will recombine; hence c_O decreases and c_{O_2} increases as a function of distance through the nozzle. If the expansion (area ratio) is large enough such that the exit temperature is near room temperature, equilibrium conditions demand that virtually all the oxygen atoms recombine, and for all practical purposes $c_{O_2} = 1$ and $c_O = 0$ at the exit. These equilibrium distributions are shown by the solid curves in Fig. 17.8. In contrast, if the flow is chemically frozen, then by definition the mass fractions are constant as a function of distance through the nozzle (the dashed lines in Fig. 17.8c). Recombination is an exothermic reaction; hence the equilibrium expansion results in the chemical zero-point energy of the atomic species being transferred into the translational, rotational, and vibrational modes of molecular energy. (The zero-point energy of two O atoms is much higher than the zero-point energy of one O_2 molecule. When two O atoms recombine into one O_2 molecule, the decrease in zero-point energy results in an increase in the internal molecular energy modes.) As a result, temperature distribution for equilibrium flow is higher than that for frozen flow, as sketched in Fig. 17.8b.

For *vibrationally* frozen flow, the vibrational energy remains constant throughout the flow. Consider a nonreacting vibrationally excited nozzle expansion as sketched in Fig. 17.9. Assume that we have diatomic oxygen in the reservoir at a temperature high enough to excite the vibrational energy, but low enough such that dissociation does not occur. If the flow is in local thermodynamic equilibrium, the translational, rotational, and vibrational energies are given by Eqs. (16.43), (16.45), and (16.47), respectively. The energies decrease through the nozzle, as shown by the solid curves in Fig. 17.9c. However, if the flow is vibrationally frozen, then e_{vib} is

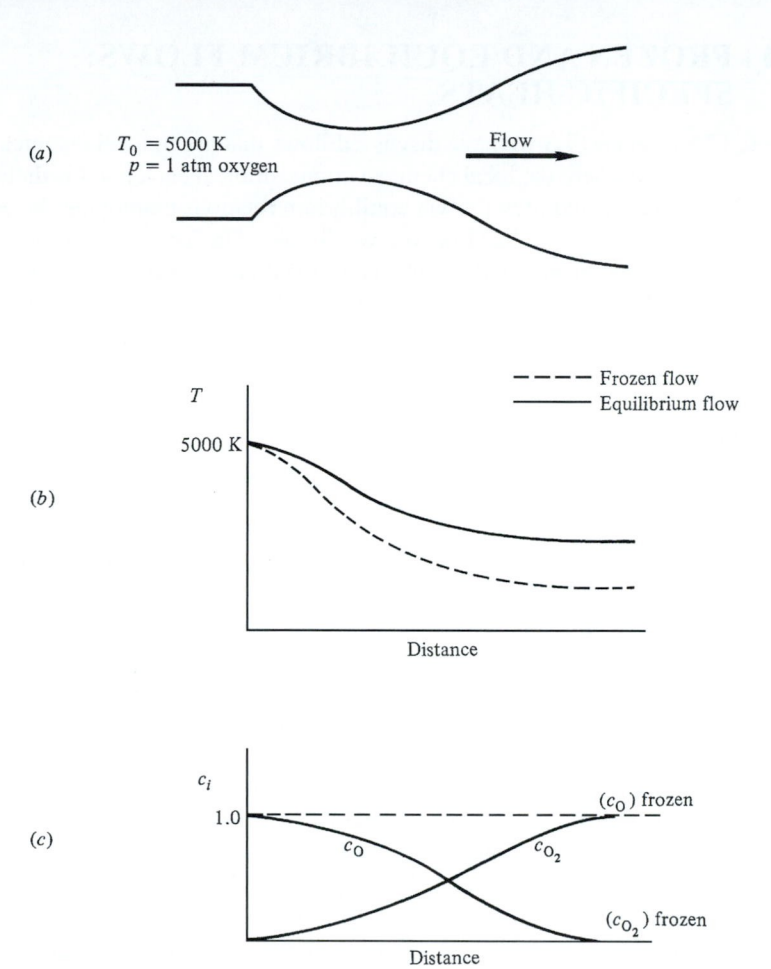

Figure 17.8 | A schematic comparing equilibrium and frozen chemically reacting flows through a nozzle.

constant throughout the nozzle, and is equal to its reservoir value. This is shown by the horizontal dashed line in Fig. 17.9c. In turn, because energy is permanently sealed in the frozen vibrational mode, less energy is available for the translational and rotational modes. Thus, because T is proportional to the translational energy, the frozen flow temperature distribution is less than that for equilibrium flow, as shown in Fig. 17.9b. In turn, the distributions of e_{trans} and e_{rot} will be lower for vibrationally frozen flow, as shown in Fig. 17.9c.

It is left as an exercise for the reader to compare the equilibrium and frozen flows across a normal shock wave.

Note that a flow which is both chemically and vibrationally frozen has constant specific heats. This is nothing more than the flow of a *calorically perfect gas* as we have treated the topic in Chaps. 1 through 12. Let us examine the specific heat in more detail.

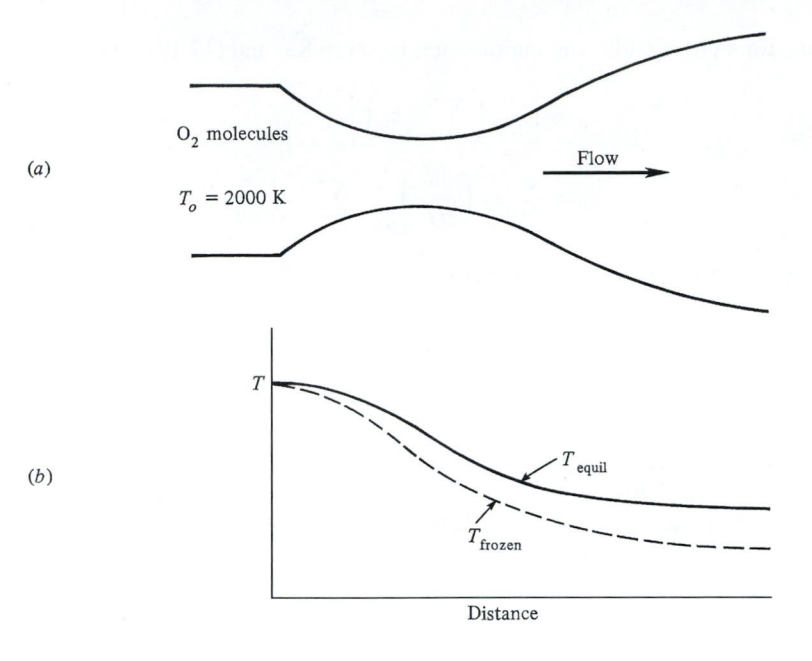

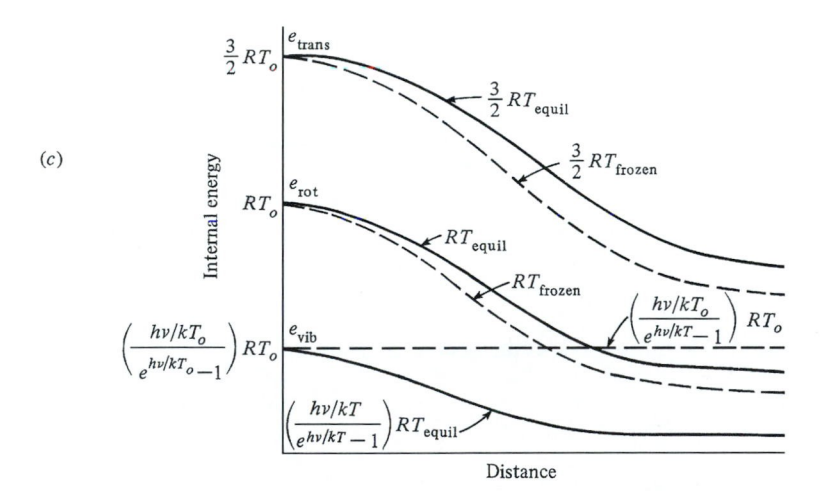

Figure 17.9 | A schematic comparing equilibrium and frozen vibrationally relaxing flows through a nozzle.

The enthalpy of a chemically reacting mixture can be obtained from Eq. (16.82), repeated here:

$$h = \sum_i c_i h_i \qquad (16.82)$$

By definition, the specific heat at constant pressure, c_p, is

$$c_p = \left(\frac{\partial h}{\partial T} \right)_p \qquad (17.19)$$

Thus, for a chemically reacting mixture, Eqs. (16.82) and (17.19) give

$$c_p = \left[\frac{\partial}{\partial T} \left(\sum_i c_i h_i \right) \right]_p$$

$$c_p = \sum_i c_i \left(\frac{\partial h_i}{\partial T} \right)_p + \sum_i h_i \left(\frac{\partial c_i}{\partial T} \right)_p \tag{17.20}$$

In Eq. (17.20) $(\partial h_i / \partial T)_p$ is the specific heat per unit mass for the pure species i, c_{p_i}. Hence, Eq. (17.20) becomes

$$\boxed{c_p = \sum_i c_i c_{p_i} + \sum_i h_i \left(\frac{\partial c_i}{\partial T} \right)_p} \tag{17.21}$$

Equation (17.21) is an expression for the specific heat of a chemically reacting mixture. If the flow is frozen, by definition there are no chemical reactions, and therefore in Eq. (17.21) the term $(\partial c_i / \partial T)_p = 0$. Thus, for a frozen flow, the specific heat becomes, from Eq. (17.21),

$$c_p = c_{p_f} = \sum_i c_i c_{p_i} \tag{17.22}$$

In turn, the frozen flow specific heat, denoted in Eq. (17.22) by c_{p_f}, can be inserted into Eq. (17.21), yielding for a chemically reacting gas

$$\boxed{\underbrace{c_p}_{\substack{\text{Specific heat at} \\ \text{constant pressure for} \\ \text{the reacting mixture}}} = \underbrace{c_{p_f}}_{\substack{\text{Frozen specific} \\ \text{heat}}} + \underbrace{\sum_i h_i \left(\frac{\partial c_i}{\partial T} \right)_p}_{\substack{\text{Contribution due to} \\ \text{chemical reaction}}}} \tag{17.23}$$

Considering the internal energy of the chemically reacting gas given by

$$e = \sum_i c_i e_i$$

and using the definition of specific heat at constant volume,

$$c_v = \left(\frac{\partial e}{\partial T} \right)_v$$

we obtain in a similar fashion

$$\boxed{c_v = c_{v_f} + \sum_i e_i \left(\frac{\partial c_i}{\partial T} \right)_v} \tag{17.24}$$

where

$$c_{v_f} = \sum_i c_i c_{v_i} \tag{17.25}$$

Equations (17.23) and (17.24) are conceptually important. Throughout our calorically perfect gas discussions in Chaps. 1 through 12, we were employing c_p and c_v as expressed by Eqs. (17.22) and (17.25). Now, for the case of a chemically reacting gas, we see from Eqs. (17.23) and (17.24) that an extra contribution, namely,

$$\sum_i h_i \left(\frac{\partial c_i}{\partial T}\right)_p \quad \text{or} \quad \sum_i e_i \left(\frac{\partial c_i}{\partial T}\right)_v$$

is made to the specific heats purely because of the reactions themselves. The magnitude of this extra contribution can be very large, and usually dominates the value of c_p and c_v.

For practical cases, it is not possible to find analytic expressions for $(\partial c_i/\partial T)_p$ or $(\partial c_i/\partial T)_v$. For an equilibrium mixture, they can be evaluated numerically by differentiating the data from an equilibrium calculation, such as was described in Sec. 16.11. Such evaluations have been made, for example, by Frederick Hansen in NASA TR-50 (see Ref. 66). Figure 17.10 is taken directly from Hansen's work, and shows the variation of c_v for air with temperature at several different pressures. The humps in each curve reflect the reaction term in Eq. (17.24),

$$\sum_i e_i \left(\frac{\partial c_i}{\partial T}\right)_v$$

and are due consecutively to dissociation of oxygen, dissociation of nitrogen, and then at very high temperatures the ionization of both O and N. (Note that the ordinate of Fig. 17.10 is a nondimensionalized specific heat, where $\mathscr{R}$ is the universal gas

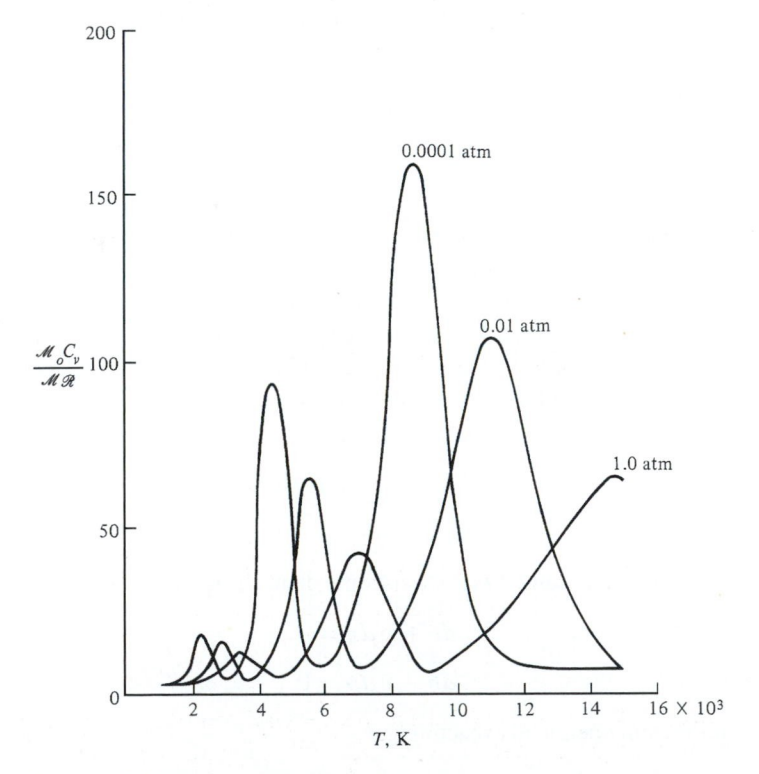

Figure 17.10 | Specific heat of equilibrium air at constant density as a function of temperature. (After Hansen, Ref. 66.)

constant, $\mathscr{M}_o$ is the initial molecular weight of undissociated air, $\mathscr{M}$ is molecular weight at the given T and p and C_v is the molar specific heat.)

Because c_p and c_v for a chemically reacting mixture are functions of both T and p (or T and v), and because they exhibit such wild variations as seen in Fig. 17.10, they are not usually employed directly in calculations of high-temperature flows. Note that, in our previous discussions on shock waves (Sec. 17.2) and nozzle flows (Sec. 17.3), h or e were used for a solution rather than c_p or c_v. However, it is important for an overall understanding of high-temperature flows to know how and why the specific heats vary. This has been the purpose of this discussion.

17.5 | EQUILIBRIUM SPEED OF SOUND

In both Secs. 3.3 and 7.5 we showed conclusively that the speed of sound in a gas is

$$a = \sqrt{\left(\frac{\partial p}{\partial \rho}\right)_s}$$

This is a physical fact, and is not changed by the presence of chemical reactions. Furthermore, in Sec. 3.3 we found for a calorically perfect gas that $a = \sqrt{\gamma RT}$. But what is the value of speed of sound in an equilibrium reacting mixture? How do we calculate it? Is it equal to $\sqrt{\gamma RT}$? The purpose of this section is to address these questions.

Consider an equilibrium chemically reacting mixture at a fixed p and T. Therefore, the chemical composition is uniquely fixed by p and T. Imagine a sound wave passing through this equilibrium mixture. Inside the wave, p and T will change slightly. If the gas remains in local chemical equilibrium through the internal structure of the sound wave, the gas composition is changed locally within the wave according to the local variations of p and T. For this situation, the speed of the sound wave is called the *equilibrium speed of sound,* denoted by a_e. In turn, if the gas is in motion at the velocity V, then V/a_e is defined as the equilibrium Mach number M_e.

To obtain a quantitative relation for the equilibrium speed of sound, consider the combined first and second laws of thermodynamics from Eqs. (1.30) and (1.32), repeated here:

$$T\,ds = de + p\,dv \qquad (1.30)$$

$$T\,ds = dh - v\,dp \qquad (1.32)$$

The process through a sound wave is isentropic; hence Eqs. (1.30) and (1.32) become

$$de + p\,dv = 0 \qquad (17.26)$$

and
$$dh - v\,dp = 0 \qquad (17.27)$$

For an equilibrium chemically reacting gas,

$$e = e(v, T)$$

Thus, the total differential is

$$de = \left(\frac{\partial e}{\partial v}\right)_T dv + \left(\frac{\partial e}{\partial T}\right)_v dT$$

$$de = \left(\frac{\partial e}{\partial v}\right)_T dv + c_v\, dT \qquad (17.28)$$

Similarly,

$$h = h(p, T)$$

$$dh = \left(\frac{\partial h}{\partial p}\right)_T dp + \left(\frac{\partial h}{\partial T}\right)_p dT$$

$$dh = \left(\frac{\partial h}{\partial p}\right)_T dp + c_p\, dT \qquad (17.29)$$

Note that, in Eqs. (17.28) and (17.29), c_v and c_p are given by Eqs. (17.24) and (17.21), respectively. Substituting Eq. (17.28) into (17.26),

$$\left(\frac{\partial e}{\partial v}\right)_T dv + c_v\, dT + p\, dv = 0$$

$$c_v\, dT + \left[p + \left(\frac{\partial e}{\partial v}\right)_T\right] dv = 0 \qquad (17.30)$$

Substituting Eq. (17.29) into (17.27),

$$\left(\frac{\partial h}{\partial p}\right)_T dp + c_p\, dT - v\, dp = 0$$

$$c_p\, dT + \left[\left(\frac{\partial h}{\partial p}\right)_T - v\right] dp = 0 \qquad (17.31)$$

Dividing Eq. (17.31) by (17.30),

$$\frac{c_p}{c_v} = \frac{[(\partial h/\partial p)_T - v]\, dp}{[(\partial e/\partial v)_T + p]\, dv} \qquad (17.32)$$

However, $v = 1/\rho$; hence $dv = -d\rho/\rho^2$. Thus, Eq. (17.32) becomes

$$\frac{c_p}{c_v} = \frac{[(\partial h/\partial p)_T - v]}{[(\partial e/\partial v)_T + p]}(-\rho^2)\frac{dp}{d\rho} \qquad (17.33)$$

Since we are dealing with isentropic conditions within the sound wave, any changes dp and $d\rho$ within the wave must take place isentropically. Thus, $dp/d\rho \equiv (\partial p/\partial \rho)_s \equiv a_e^2$.

Hence, Eq. (17.33) becomes

$$\left(\frac{\partial p}{\partial \rho}\right)_s = \frac{c_p}{c_v} \frac{1}{\rho^2} \frac{[(\partial e/\partial v)_T + p]}{[1/\rho - (\partial h/\partial p)_T]}$$

or

$$a_e^2 = \frac{c_p}{c_v} \frac{p}{\rho} \frac{[1 + (1/p)(\partial e/\partial v)_T]}{[1 - \rho(\partial h/\partial p)_T]} \tag{17.34}$$

As usual, let $\gamma \equiv c_p/c_v$. Also, note from the equation of state that $p/\rho = RT$. Thus, Eq. (17.34) becomes

$$\boxed{a_e^2 = \gamma RT \frac{[1 + (1/p)(\partial e/\partial v)_T]}{[1 - \rho(\partial h/\partial p)_T]}} \tag{17.35}$$

Equation (17.35) gives the equilibrium speed of sound in a chemically reacting mixture.

Equation (17.35) gives an immediate answer to one of the questions asked at the beginning of this section. The speed of sound in an equilibrium reacting mixture is *not* equal to the simple result $\sqrt{\gamma RT}$ obtained in Sec. 3.3 for a calorically perfect gas. However, if the gas is calorically perfect, then $h = c_p T$ and $e = c_v T$ (see Sec. 1.4). In turn, $(\partial h/\partial p)_T = 0$ and $(\partial e/\partial v)_T = 0$, and Eq. (17.35) reduces to the familiar result

$$a_f = \sqrt{\gamma RT} \tag{17.36}$$

The symbol a_f is used here to denote the *frozen speed of sound,* because a calorically perfect gas assumes no reactions. Equation (17.36) is the speed at which a sound wave will propagate when no chemical reactions take place internally within the wave, i.e., when the flow *inside* the wave is frozen.

For a thermally perfect gas, $h = h(T)$ and $e(T)$. Hence, again Eq. (17.35) reduces to Eq. (17.36).

Clearly, the full Eq. (17.35) must be used whenever $(\partial e/\partial v)_T$ and $(\partial h/\partial p)_T$ are finite. This occurs for two cases:

1. When the gas is chemically reacting
2. When intermolecular forces are important, i.e., when we are dealing with a *real gas* (see Sec. 1.4)

In both of these cases, $h = h(T, p)$ and $e = e(T, v)$ and hence Eq. (17.35) must be used.

Note from Eq. (17.35) that the equilibrium speed of sound is a function of *both* T and p, unlike the case for a calorically or thermally perfect gas where it depends on T only. This is emphasized in Fig. 17.11, which gives the equilibrium speed of sound for high-temperature air as a function of both T and p. In addition, note in Fig. 17.11 that the frozen speed of sound is given by a constant horizontal line at $a^2\rho/p = 1.4$, and that the difference between the frozen equilibrium speed of sound in air can be as large as 20 percent under practical conditions. In turn, this once again underscores the ambiguity in the definition of Mach number for high-temperature

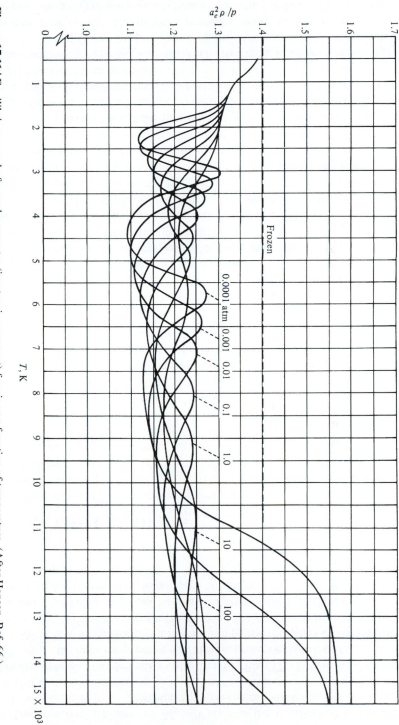

Figure 17.11 | Equilibrium speed of sound parameter (isentropic exponent) for air as a function of temperature. (After Hansen, Ref. 66.)

flows. The frozen Mach number $M_f = V/a_f$ and the equilibrium Mach number $M_e = V/a_e$ can differ by a substantial amount. Hence, Mach number is not particularly useful in this context.

Finally, note that the derivatives of e and h in Eq. (17.35) must be obtained numerically from the high-temperature equilibrium properties of the mixture. Although Eq. (17.35) is in a useful form to illustrate the physical aspects of the equilibrium speed of sound, it does not constitute a closed-form formula from which, given the local p and T, a value of a_e can be immediately obtained. Rather, the derivatives must be evaluated numerically, as has been carried out by Hansen (Ref. 66) and others and as is reported in Fig. 17.11.

17.6 | ON THE USE OF $\gamma = c_p/c_v$

As a corollary to Sec. 17.5, we emphasize that $\gamma \equiv c_p/c_v$ is a function of T and p for a chemically reacting gas. Hence, γ, which is so useful for the analysis of a calorically perfect gas as described in Chaps. 1 through 12, is virtually useless for the analysis of a high-temperature flow.

In spite of this, the temptation to use simple closed-form results such as Eqs. (3.28) through (3.31) has resulted in many approximate analyses of chemically reacting flows employing a constant "effective γ." This is particularly prevalent in the preliminary analysis and design of rocket engines. For example, consider the chemically reacting flow through a rocket nozzle, where T_o and T_e are the combustion chamber and exit temperatures, respectively. These temperatures can be used in Eq. (3.28) to define an effective value of γ, γ_{eff}, such that

$$\frac{T_o}{T_e} = 1 + \frac{\gamma_{\text{eff}} - 1}{2} M_e^2 \tag{17.37}$$

In Eq. (17.37), T_o, T_e, and M_e are all known quantities for the reacting flow, and γ_{eff} is solved from this equation. Of course, in turn γ_{eff} depends on p_o, T_o, and the gas composition. Moreover, this value of γ_{eff} holds only for Eq. (17.37), which is its definition. If the same value of γ_{eff} were used in Eqs. (3.30) and (3.31) for the pressure and density, respectively, the results would not be exact. Nevertheless, experience has shown that effective values of γ inserted into the closed-form results for a calorically perfect gas can be used in *approximate* analyses of chemically reacting flows. Choosing the appropriate value of γ_{eff} for a given problem is a matter of experience; in general, an answer to the problem, either from exact numerical calculations or experimental measurement, is usually necessary in order to estimate a reasonable value of γ_{eff}, say from Eq. (17.37). Note that, in terms of our discussion in Sec. 17.4, the effective γ approach is nothing more than assuming frozen flow with a proper value of γ_{eff} to give reasonably close results.

The reader is cautioned to approach such effective γ analyses with strong reservations. It is simply a "back-of-the-envelope" technique for estimating high-temperature flow results. The proper techniques, using numerical solutions of the proper governing equations, as described for example in Secs. 17.2 through 17.5, should be invoked whenever exact analyses of equilibrium high-temperature flows are desired.

17.7 | NONEQUILIBRIUM FLOWS: SPECIES CONTINUITY EQUATION

In the remainder of this chapter, we consider the flow of a high-temperature gas where the chemical and/or vibrational rates are finite, i.e., we consider nonequilibrium flows. This is in contrast to equilibrium (infinite-rate) or frozen (zero-rate) flows (see Sec. 17.4). In this capacity, we will need to incorporate into our flow analyses the finite-rate processes discussed in Secs. 16.13 through 16.16. In regard to our roadmap in Fig. 17.3, we now move to the right-hand column.

The analysis of nonequilibrium flows is inherently different from equilibrium or frozen flows in these fundamental ways:

1. The finite rates force the use of differential relationships for the governing equations. In contrast, note from our preceding discussions that equilibrium or frozen flows through nozzles or across shock waves can be treated with strictly algebraic relations. For nonequilibrium flow, the differential form of the governing equations (see Chap. 6) must be used.

2. For a nonequilibrium chemically reacting flow, the composition is no longer a unique function of the local p and T, but rather depends on the speed at which the finite-rate reactions are taking place, the speed of the fluid elements themselves, and the actual geometric scale of the flow problem. This will become clearer as our discussions progress. Similarly, the vibrational energy for a nonequilibrium vibrationally excited flow is no longer a function of the local T. Hence, for the analysis of such nonequilibrium flows, the unknown chemical composition and/or vibrational energies introduce the requirement for additional governing equations. The derivation of such equations for chemically reacting flows is given in the present section; analogous equations for vibrationally excited flows will be treated in Sec. 17.8.

In Sec. 2.3, we derived the integral form of the continuity equation, which states that mass can be neither created nor destroyed. This is a universal principle in classical physics, and therefore holds for a nonequilibrium chemically reacting gas. In the analysis of such flows, Eq. (2.2), as well as its differential counterparts, Eqs. (6.5) and (6.22), is called the *global continuity equation*. It deals with the overall conservation of mass for the reacting *mixture*. However, for a nonequilibrium chemically reacting flow, we must also utilize a continuity equation for each species separately. Such an equation is called the *species continuity equation,* and is derived as follows.

Consider a fixed, finite control volume in the nonequilibrium, inviscid flow of a chemically reacting gas; such a control volume is sketched in Fig. 2.4. Let ρ_i be the mass of species i per unit volume of mixture. Hence

$$\rho = \sum_i \rho_i$$

Examining Fig. 2.4, the mass flow of species i through the elemental surface area dS is $\rho_i \mathbf{V} \cdot d\mathbf{S}$, where $\mathbf{V}$ is the local flow velocity. Hence, the net mass flow of species i out of the control volume is

$$\oiint_S \rho_i \mathbf{V} \cdot d\mathbf{S}$$

The mass of species i inside the control volume is

$$\oiiint_\gamma \rho_i \, dV$$

Let $\dot{w}_i$ be the local rate of change of ρ_i *due to chemical reactions inside the control volume*. Therefore, the net time rate of change of the mass of species i inside the control volume is due to

1. The net flux of species i through the surface
2. The creation or extinction of species i inside the control volume due to chemical reactions

Writing this physical principle in terms of integrals over the control volume, we have

$$\boxed{\frac{\partial}{\partial t} \oiiint_\gamma \rho_i \, dV = -\oiint_S \rho_i \mathbf{V} \cdot d\mathbf{S} + \oiiint_\gamma \dot{w}_i \, dV} \qquad (17.38)$$

Equation (17.38) is the integral form of the species continuity equation; you will note that its derivation is quite similar to the global continuity equation given in Sec. 2.3. In turn, similar to the development given in Sec. 6.2, the differential form of the species continuity equation is obtained directly from Eq. (17.38) as

$$\boxed{\frac{\partial \rho_i}{\partial t} + \nabla \cdot (\rho_i \mathbf{V}) = \dot{w}_i} \qquad (17.39)$$

[Recall that we are dealing with an inviscid flow. If the flow were viscous, Eqs. (17.38) and (17.39) would each have an additional term for the transport of species i by mass diffusion, and the velocity would be the mass motion of species i, which is not necessarily the same as the mass motion of the mixture, $\mathbf{V}$.]

In Eqs. (17.38) and (17.39) an expression for $\dot{w}_i$ comes from the chemical rate equation (16.130), couched in suitable dimensions. For example, assume that we are dealing with chemically reacting air, and we write Eqs. (17.38) and (17.39) for NO, i.e., $\rho_i = \rho_{NO}$. The rate equation for NO is given by Eq. (16.144) in terms of

$$\frac{d[\text{NO}]}{dt} = -k_{f_{3a}}[\text{NO}][\text{O}_2] + \cdots$$

The dimensions of this equation are moles per unit volume per unit time. However, the dimensions of $\dot{w}_{NO}$ in Eqs. (17.38) and (17.39) are the mass of NO per unit volume per unit time. Recalling that molecular weight is defined as the mass of species i per mole of i, we can write

$$\dot{w}_{NO} = \mathcal{M}_{NO}\frac{d[NO]}{dt}$$

where $\mathcal{M}_{NO}$ is the molecular weight of NO. Therefore, Eq. (17.39) written for NO is

$$\frac{\partial \rho_{NO}}{\partial t} + \nabla \cdot (\rho_{NO}\mathbf{V}) = \mathcal{M}_{NO}\frac{d[NO]}{dt}$$

where $d[NO]/dt$ is obtained from Eq. (16.144).

For a nonequilibrium chemically reacting mixture with n different species, we need $n - 1$ species continuity equations of the form of Eq. (17.39). These, along with the additional result that

$$\sum_i \rho_i = \rho$$

provide n equations for the solution of the instantaneous composition of a nonequilibrium mixture of n chemical species.

An alternative form of the species continuity equation can be obtained as follows. The mass fraction of species i, c_i, is defined as $c_i = \rho_i/\rho$. Substituting this relation into Eq. (17.39),

$$\frac{\partial(\rho c_i)}{\partial t} + \nabla \cdot (\rho c_i \mathbf{V}) = \dot{w}_i \qquad (17.40)$$

Expanding Eq. (17.40), we have

$$\rho\left(\frac{\partial c_i}{\partial t} + \mathbf{V} \cdot \nabla c_i\right) + c_i\left[\frac{\partial \rho}{\partial t} + \nabla \cdot (\rho\mathbf{V})\right] = \dot{w}_i \qquad (17.41)$$

The first two terms of Eq. (17.41) constitute the substantial derivative of c_i (see Sec. 6.3). The second two terms (in brackets) result in zero from the global continuity equation (6.5). Hence, Eq. (17.41) can be written as

$$\boxed{\frac{Dc_i}{Dt} = \frac{\dot{w}_i}{\rho}} \qquad (17.42)$$

In terms of the mole-mass ratio, $\eta_i = c_i/\mathcal{M}_i$, Eq. (17.42) becomes

$$\boxed{\frac{D\eta_i}{Dt} = \frac{\dot{w}_i}{\mathcal{M}_i\rho}} \qquad (17.43)$$

Equations (17.42) and (17.43) are alternative forms of the species continuity equation, couched in terms of the substantial derivative.

Recall from Sec. 6.3 that the substantial derivative of a quantity is physically the time rate of change of that quantity as we follow a fluid element *moving with the flow*. Therefore, from Eqs. (17.42) and (17.43) as we follow a fluid element of fixed mass moving through the flowfield, we see that changes of c_i or η_i of the fluid element are due *only* to the finite-rate chemical kinetic changes taking place within the element. This makes common sense, and in hindsight, therefore, Eqs. (17.42) and (17.43) could have been written directly by inspection. We emphasize that in Eqs. (17.42) and (17.43) the flow variable inside the substantial derivative, c_i or η_i, is written *per unit mass*. As long as the nonequilibrium variable inside the substantial derivative is per unit mass of mixture, then the right-hand side of the conservation equation is simply due to finite-rate kinetics, such as shown in Eqs. (17.42) and (17.43). In contrast, Eq. (17.39) can also be written as

$$\frac{D\rho_i}{Dt} = \dot{w}_i - \rho_i(\nabla \cdot \mathbf{V}) \tag{17.44}$$

The derivation of Eq. (17.44) is left to the reader. In it, the nonequilibrium variable inside the substantial derivative, ρ_i, is per unit volume. Because it is *not* per unit mass, an extra term in addition to the finite-rate kinetics appears on the right-hand side to take into account the dilation effect of the changing specific volume of the flow. (Recall from basic fluid mechanics that $\nabla \cdot \mathbf{V}$ is physically the volume efflux of fluid from a point.) The distinction made here will be important in Sec. 17.8.

17.8 | RATE EQUATION FOR VIBRATIONALLY NONEQUILIBRIUM FLOW

Consider the nonequilibrium inviscid flow of a vibrationally excited diatomic gas. The finite-rate kinetics for vibrational energy exchange were discussed in Sec. 16.14, leading to Eq. (16.116) as the vibrational rate equation. Based on the discussion at the end of Sec. 17.7, if we follow a moving fluid element of fixed mass, the rate of change of e_{vib} for this element is equal to the rate of molecular energy exchange inside the element. Therefore, we can write the vibrational rate equation for a moving fluid element as

$$\boxed{\frac{De_{\text{vib}}}{Dt} = \frac{1}{\tau}\left(e_{\text{vib}}^{\text{eq}} - e_{\text{vib}}\right)} \tag{17.45}$$

Note in Eq. (17.45) that e_{vib} is the local nonequilibrium value of vibrational energy per unit mass of gas.

Flow with vibrational nonequilibrium is of particular practical interest in the analysis of modern gasdynamic and chemical lasers, and has been an important aspect of hypersonic wind tunnels since the mid-1950s.

17.9 | SUMMARY OF GOVERNING EQUATIONS FOR NONEQUILIBRIUM FLOWS

In a nonequilibrium flowfield, we wish to solve for p, ρ, T, $\mathbf{V}$, h, e_{vib}, and c_i as functions of space and time. For an inviscid, adiabatic nonequilibrium flow, the governing equations are summarized below. With the addition of the equations derived in Secs. 17.7 and 17.8, the governing equations are the same as developed in Chap. 6

(recall that in Chap. 6 we made no special assumptions regarding the type of gas):

Global continuity:
$$\frac{\partial \rho}{\partial t} + \nabla \cdot (\rho \mathbf{V}) = 0$$

Species continuity:
$$\frac{\partial \rho_i}{\partial t} + \nabla \cdot (\rho_i \mathbf{V}) = \dot{w}_i$$

or
$$\frac{Dc_i}{Dt} = \frac{\dot{w}_i}{\rho}$$

or
$$\frac{D\eta_i}{Dt} = \frac{\dot{w}_i}{\mathcal{M}_i \rho}$$

(Note that for a mixture of n species, we need $n - 1$ species continuity equations; the nth equation is given by $\sum_i \rho_i = \rho$, or $\sum_i c_i = 1$, or $\sum_i \eta_i = \eta$.)

Momentum:
$$\rho \frac{D\mathbf{V}}{Dt} = -\nabla p$$

Energy:
$$\frac{Dh}{Dt} - v\frac{Dp}{Dt} = 0$$

or for steady flow
$$h_o = h + \frac{V^2}{2} = \text{const along a streamline}$$

or any of Eqs. (6.17), (6.31), (6.36), (6.40), (6.43), (6.44), or (6.48). [Note that in the forms of the energy equation obtained in Chap. 6, a heat-addition term $\dot{q}$ was carried along. However, in the present chapter we are dealing with adiabatic flows; hence, $\dot{q} = 0$. The term $\dot{q}$ does *not* have anything to do with chemical reactions; it is simply an effect due to energy addition across the boundaries of the flow, such as absorption of radiant energy. The energy release or absorption due to chemical reactions is *not* included in $\dot{q}$; rather, these chemical energy changes are already naturally accounted for by the heats of formation appearing in the enthalpy terms, e.g., Eq. (16.99).]

Equation of state:
$$p = \rho R T$$

where
$$R = \frac{\mathcal{R}}{\mathcal{M}}$$

$$\mathcal{M} = \left(\sum_i \frac{c_i}{\mathcal{M}_i} \right)^{-1}$$

Enthalpy:
$$h = \sum_i c_i h_i$$

where
$$h_i = (e_{\text{trans}} + e_{\text{rot}} + e_{\text{vib}} + e_{\text{el}})_i + R_i T + \left(\Delta h_f^0 \right)_i$$

or
$$h_i = \frac{7}{2} R_i T + e_{\text{vib}_i} + e_{\text{el}_i} + \left(\Delta h_f^0 \right)_i$$

Vibrational energy:
$$\frac{D\left(c_i e_{\text{vib}_i}\right)}{Dt} = \frac{c_i}{\tau_i}\left(e_{\text{vib}_i}^{\text{eq}} - e_{\text{vib}_i} \right)$$

17.10 | NONEQUILIBRIUM NORMAL SHOCK WAVE FLOWS

Consider a strong normal shock wave in a gas. Moreover, assume the temperature within the shock wave is high enough to cause chemical reactions within the gas. In this situation, we need to reexamine the qualitative aspects of a shock wave, as sketched in Fig. 17.12. The thin region where large gradients in temperature, pressure, and velocity occur, and where the transport phenomena of viscosity and thermal conduction are important, is called the *shock front*. For all of our previous considerations of a calorically perfect gas, or equilibrium flow of a chemically reacting or vibrationally excited gas, this thin region *is* the shock wave. For these previous situations, the flow in front of and behind the shock front was uniform, and the only gradients in flow properties took place almost discontinuously within a thin region of no more than a few mean-free-paths thickness. However, in a nonequilibrium flow, all chemical reactions and/or vibrational excitations take place at a finite rate. Since the shock front is only a few mean-free-paths thick, the molecules in a fluid element can experience only a few collisions as the fluid element traverses the front. Consequently, the flow through the shock front itself is essentially *frozen*. In turn, the flow

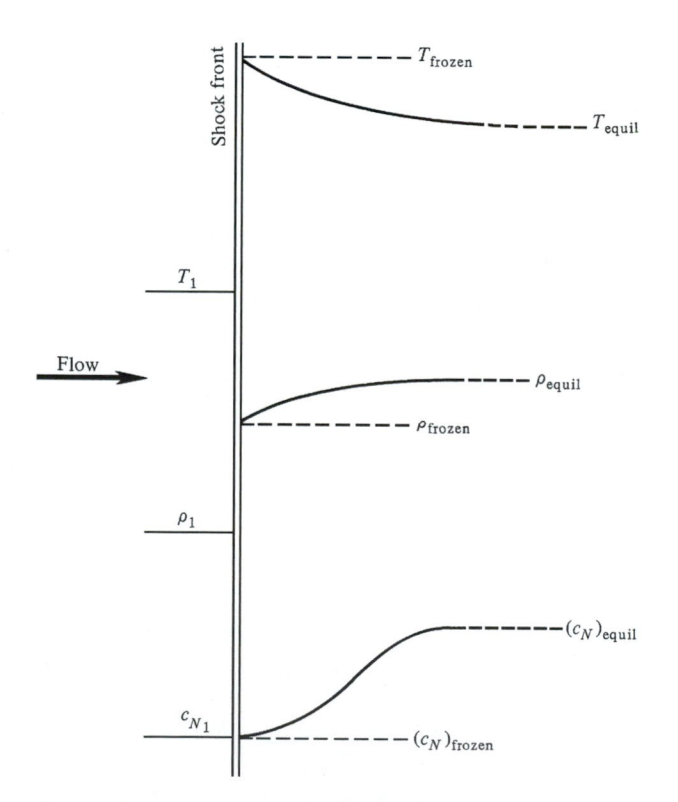

Figure 17.12 | Schematic of chemically reacting nonequilibrium flow behind a normal shock wave.

properties immediately behind the shock front are frozen flow properties, as discussed in Sec. 17.4 and as sketched in Fig. 17.12. Then, as the fluid element moves downstream, the finite-rate reactions take place, and the flow properties relax toward their equilibrium values, as also sketched in Fig. 17.12. With this picture in mind, the *shock wave* now encompasses both the shock front and the nonequilibrium region behind the front where the flow properties are changing due to the finite-rate reactions. For purposes of illustration, assume that the gas is pure diatomic nitrogen in front of the shock wave, i.e., $(C_N)_1 = 0$ in Fig. 17.12. The properties immediately behind the shock front are obtained from frozen flow results, i.e., the constant $\gamma = 1.4$ results from Sec. 3.6. Hence, the values of T_{frozen} and ρ_{frozen} shown in Fig. 17.12 can be obtained directly from Table A.2 at the back of this book. In addition, c_N immediately behind the shock front is still zero, since the flow is frozen. Downstream of the shock front, the nonequilibrium flow must be analyzed using the equations summarized in Sec. 17.9. In this region, the nitrogen becomes either partially or totally dissociated (depending on the strength of the shock wave), and c_N increases as sketched in Fig. 17.12. In turn, because this reaction is endothermic, the static temperature behind the shock front decreases, and the density increases. Finally, the downstream flow properties will approach their equilibrium values, as calculated from the technique described in Sec. 17.2.

A numerical calculation of the nonequilibrium region behind the shock front can be established as follows. Since the flow is one-dimensional and steady, the equations of Sec. 17.9 become

Global continuity:
$$\rho \, du + u \, d\rho = 0 \tag{17.46}$$

Momentum:
$$dp = -\rho u \, du \tag{17.47}$$

Energy:
$$dh - \frac{1}{\rho} dp = 0 \tag{17.48}$$

Species continuity:
$$u \, dc_i = \frac{\dot{w}_i}{\rho} \, dx \tag{17.49}$$

In Eq. (17.49) the x distance is measured from the shock front, extending downstream as shown in Fig. 17.13. Note that Eq. (17.49) explicitly involves the finite-rate chemical reaction term $\dot{w}_i$, and that a distance dx multiplies this term. Hence, Eq. (17.49) introduces a scale effect into the solution of the flowfield—a scale effect that is present solely because of the nonequilibrium phenomena. In turn, all flowfield properties become a function of distance behind the shock front, as sketched in Fig. 17.12. Continuing with the above equations, if Eq. (17.46) is multiplied by u,

$$\rho u \, du + u^2 \, d\rho = 0 \tag{17.50}$$

and Eq. (17.47) substituted into (17.50), we have

$$\boxed{dp = u^2 \, d\rho} \tag{17.51}$$

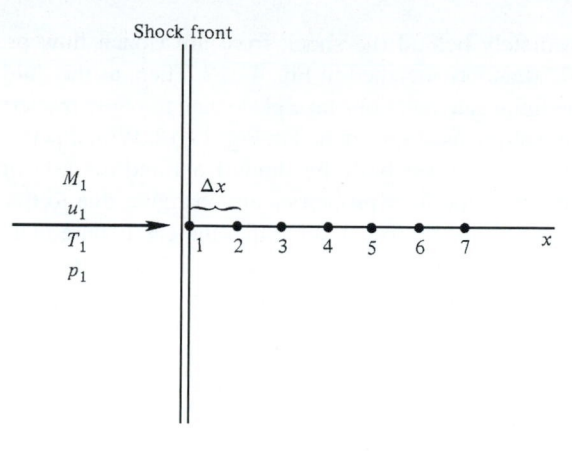

Figure 17.13 | Schematic of grid points for the numerical solution of nonequilibrium normal shock flows.

From the equation of state, $p = \rho R T$ where R is a variable, since $R = \mathcal{R}/\mathcal{M}$ and $\mathcal{M}$ varies due to chemical reactions. Hence,

$$\frac{dp}{p} = \frac{d\rho}{\rho} + \frac{dR}{R} + \frac{dT}{T}$$

or

$$dp = \frac{p}{\rho}\,d\rho + \frac{p}{R}\,dR + \frac{p}{T}\,dT$$

or

$$dp = RT\,d\rho + \rho T\,dR + \rho R\,dT \qquad (17.52)$$

Substituting Eq. (17.51) into (17.52), we have

$$u^2\,d\rho = RT\,d\rho + \rho T\,dR + \rho R\,dT \qquad (17.53)$$

Solving Eq. (17.53) for $d\rho$, we obtain

$$\boxed{d\rho = \frac{\rho T\,dR + \rho R\,dT}{u^2 - RT}} \qquad (17.54)$$

The mixture enthalpy is given by

$$h = \sum_i c_i h_i$$

where

$$h_i = \int_0^T c_{p_i}\,dT + \left(\Delta h_f^0\right)_i$$

Hence,

$$dh = \sum_i c_i\,dh_i + \sum_i h_i\,dc_i$$

$$= \sum_i c_i c_{p_i}\,dT + \sum_i h_i\,dc_i$$

$$= c_{p_f}\,dT + \sum_i h_i\,dc_i \qquad (17.55)$$

where c_{p_f} is the frozen specific heat introduced in Sec. 17.4. Substituting Eq. (17.55) into Eq. (17.48), we have

$$c_{p_f} \, dT + \sum_i h_i \, dc_i - \frac{dp}{\rho} = 0 \tag{17.56}$$

Substituting Eq. (17.51) into Eq. (17.56), we have

$$c_{p_f} \, dT + \sum_i h_i \, dc_i - \frac{u^2}{\rho} \, d\rho = 0 \tag{17.57}$$

Substituting Eq. (17.54) into Eq. (17.57), we obtain

$$c_{p_f} \, dT + \sum_i h_i \, dc_i - u^2 \left[\frac{T \, dR + R \, dT}{u^2 - RT} \right] = 0$$

or $$\left(c_{p_f} - \frac{u^2}{u^2 - RT} R \right) dT = \frac{u^2}{u^2 - RT} T \, dR - \sum_i h_i \, dc_i \tag{17.58}$$

Since $R = \mathscr{R}/\mathscr{M}$ and

$$\mathscr{M} = \left(\sum_i \frac{c_i}{\mathscr{M}_i} \right)^{-1}$$

then $$R = \mathscr{R} \sum_i \frac{c_i}{\mathscr{M}_i}$$

and $$dR = \mathscr{R} \sum_i \frac{dc_i}{\mathscr{M}_i} \tag{17.59}$$

Also, from Eq. (17.49),

$$dc_i = \frac{\dot{w}_i}{\rho u} \, dx \tag{17.60}$$

Substituting Eqs. (17.59) and (17.60) into Eq. (17.58), and solving for dT, we have

$$dT = \frac{\left(\dfrac{u^2}{u^2 - RT} \right) \mathscr{R} T \left(\dfrac{dx}{\rho u} \right) \sum_i \dfrac{\dot{w}_i}{\mathscr{M}_i} - \left(\dfrac{dx}{\rho u} \right) \sum_i h_i \dot{w}_i}{c_{p_f} - \dfrac{u^2}{u^2 - RT} R} \tag{17.61}$$

Equations (17.51), (17.54), (17.60), and (17.61) give the infinitesimal changes in p, ρ, c_i, and T, respectively, corresponding to the infinitesimal distance dx behind the shock front. These equations are in a convenient form for numerical solution. Consider the one-dimensional flowfield behind the shock front to be divided into a large number of grid points separated by an equal distance Δx, as sketched in Fig. 17.13. Because the flow is frozen across the shock front, all conditions at point 1 immediately behind the shock front are known. For purposes of illustration, let us use one-sided forward differences (see Sec. 11.11) in Eqs. (17.51), (17.54), (17.60), and (17.61). Hence, Eq. (17.60) is replaced by

$$\underbrace{c_{i_2}}_{\text{To be calculated}} - \underbrace{c_{i_1}}_{\text{Known}} = \underbrace{\left(\frac{\dot{w}_i}{\rho u}\right)_1 \Delta x}_{\text{Known}}$$

or

$$\boxed{c_{i_2} = c_{i_1} + \left(\frac{\dot{w}_i}{\rho u}\right)_1 \Delta x} \tag{17.62}$$

From Eq. (17.62), all the species mass fractions can be calculated at point 2. From Eq. (17.61) we have

$$\boxed{T_2 = T_1 + \Delta x \left[\frac{\left(\dfrac{u^2}{u^2 - RT}\right) \dfrac{\mathscr{R}T}{\rho u} \displaystyle\sum_i \dfrac{\dot{w}_i}{\mathscr{M}_i} - \dfrac{1}{\rho u} \displaystyle\sum_i h_i \dot{w}_i}{c_{p_f} - \dfrac{u^2}{u^2 - RT} R} \right]_1} \tag{17.63}$$

In Eq. (17.63) all terms on the right-hand side are known; hence, T_2 can be calculated directly. In turn, from Eq. (17.54)

$$\boxed{\rho_2 = \rho_1 + \frac{\rho_1 T_1 (R_2 - R_1) + \rho_1 R_1 (T_2 - T_1)}{u^2 - R_1 T_1}} \tag{17.64}$$

In Eq. (17.64), all terms on the right-hand side are known; hence ρ_2 can be calculated directly. [Note that $R_2 = \mathscr{R} \sum_i (c_{i_2}/\mathscr{M}_i)$.] In turn, from Eq. (17.51),

$$\boxed{p_2 = p_1 + u_1^2 (\rho_2 - \rho_1)} \tag{17.65}$$

Everything is known on the right-hand side of Eq. (17.65), thus directly yielding p_2. Finally, from Eq. (17.47),

$$\boxed{u_2 = u_1 - \frac{1}{\rho_1 u_1}(p_2 - p_1)} \tag{17.66}$$

and we can calculate u_2. Consequently, from Eqs. (17.62) through (17.66), all the flowfield variables at point 2 behind the shock front can be computed. Repeating these equations, we can march on to points 3, 4, 5, etc. In this fashion, the complete nonequilibrium flowfield can be obtained.

For simplicity in this illustration, we have used a simple first-order-accurate finite-difference for the derivatives. For such a method, however, the distance Δx must be made so small to maintain reasonable accuracy as to be totally impractical. In practice, Eqs. (17.51), (17.54), (17.60), and (17.61) would be solved by a higher-order method, such as the standard Runge–Kutta technique for ordinary differential equations. To complicate matters, if one or more of the finite-rate chemical reactions are very fast [if $\dot{w}_i$ in Eq. (17.60) is very large], then Δx must still be chosen very small even when a higher-order numerical method is used. The species continuity equations for such very fast reactions are called "stiff" equations, and readily lead to instabilities in the solution. Special methods for treating the solution of stiff ordinary differential equations have been reviewed by Hall and Treanor (see Ref. 71).

Typical results for the nonequilibrium flowfield behind a normal shock wave in air are given in Figs. 17.14 and 17.15, taken from the work of Marrone (Ref. 72). The Mach number ahead of the shock wave is 12.28, strong enough to produce major dissociation of O_2, but only slight dissociation of N_2. The variation of chemical composition with distance behind the shock front is given in Fig. 17.14. Note the expected increase in the concentration of O and N, rising from their frozen values (essentially zero) immediately behind the shock front, and monotonically approaching their equilibrium values about 10 cm downstream of the shock front. For the most part, the nonequilibrium flow variables will range between the two extremes of frozen and equilibrium values. However, in some cases, due to the complexities of the chemical kinetic mechanism, a species may exceed these two extremes. A case in point is the variation of NO concentration shown in Fig. 17.14. Note that it first increases from essentially zero behind the shock front, and overshoots its equilibrium value at about 0.1 cm. Further downstream, the NO concentration approaches its equilibrium value *from above*. This is a common behavior of NO when it is formed behind a shock front in air; it is not just a peculiarity of the given upstream conditions in Fig. 17.14. The

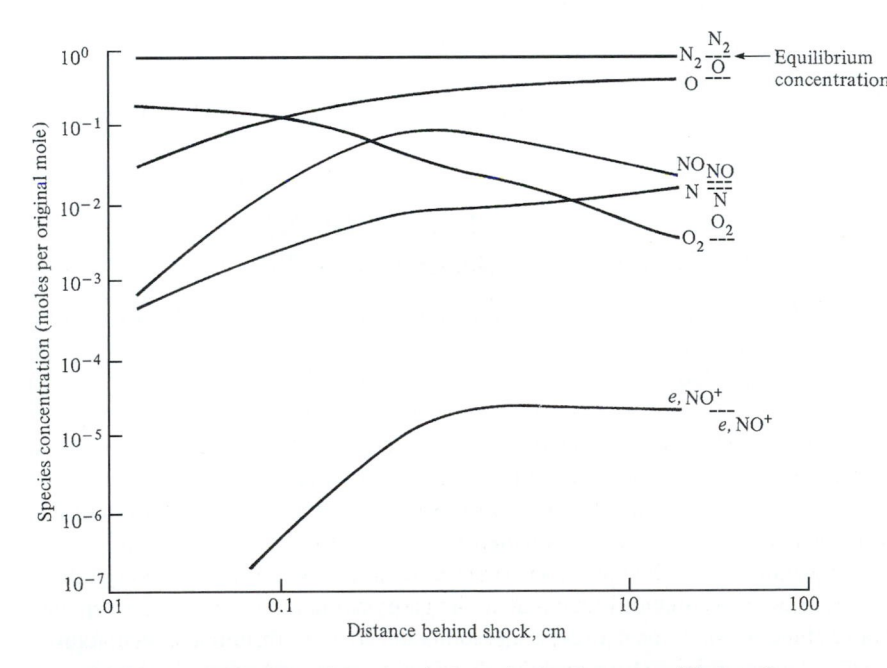

Figure 17.14 | Distributions of the chemical species for the nonequilibrium flow through a normal shock wave in air. $M_1 = 12.28$, $T_1 = 300$ K, $p_1 = 1.0$ mmHg. (After Marrone, Ref. 72.)

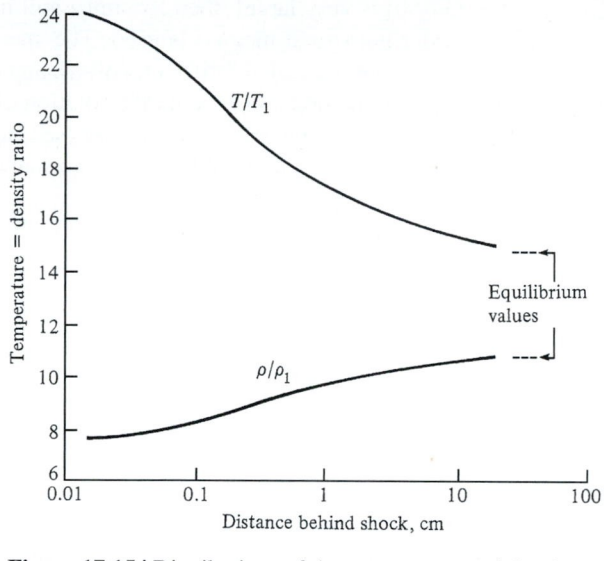

Figure 17.15 | Distributions of the temperature and density
for the nonequilibrium flow through a normal shock
wave in air. $M_1 = 12.28$, $T_1 = 300\,\text{K}$, $p_1 = 1.0\,\text{mmHg}$.
(After Marrone, Ref. 72.)

variations in temperature and density behind the shock front are shown in Fig. 17.15.
As noted earlier, the chemical reactions in air behind a shock front are predominantly
dissociation reactions, which are endothermic. Hence, T decreases and ρ increases
with distance behind the front—both by almost a factor of 2.

17.11 | NONEQUILIBRIUM QUASI-ONE-DIMENSIONAL NOZZLE FLOWS

Because of the practical importance of high-temperature flows through rocket nozzles
and high-enthalpy aerodynamic testing facilities, intensive efforts were made after
1950 to obtain relatively exact numerical solutions for the expansion of a high-
temperature gas through a nozzle when vibrational and/or chemical nonequilibrium
conditions prevail within the gas. In a rocket nozzle, nonequilibrium effects decrease
the thrust and specific impulse. In a high-temperature wind tunnel, the nonequilibrium
effects make the flow conditions in the test section somewhat uncertain. Both of these
are adverse effects, and hence rocket nozzles and wind tunnels are usually designed to
minimize the nonequilibrium effects; indeed, engineers strive to obtain equilibrium
conditions in such situations. In contrast, the gasdynamic laser (see Ref. 21) creates a
laser medium by intentionally fostering vibrational nonequilibrium in a supersonic expansion; here, engineers strive to obtain the highest degree of nonequilibrium possible.
In any event, the study of nonequilibrium nozzle flows is clearly important.

Until 1969, all solutions of nonequilibrium nozzle flows involved steady state
analyses. Such techniques were developed to a high degree, and are nicely reviewed

by Hall and Treanor (see Ref. 71). However, such steady state analyses were not straightforward. Complicated by the presence of stiff chemical rate equations (see Sec. 17.10), such solutions encountered a saddle-point singularity in the vicinity of the nozzle throat, and this made it very difficult to integrate from the subsonic to the supersonic sections of the nozzle. Moreover, for nonequilibrium nozzle flows the throat conditions and hence the mass flow are not known *a priori;* the nozzle mass flow must be obtained as part of the solution of the problem. Therefore, in 1969 a new technique for solving nonequilibrium nozzle flows was advanced by Anderson (see Refs. 73 and 74) using the time-marching finite-difference method discussed in Chap. 12. This time-marching approach circumvents the above problems encountered with steady-state analyses, and also has the virtue of being relatively easy and straightforward to program on the computer. Since its first introduction in 1969, the time-marching solution of nonequilibrium nozzle flows has gained wide acceptance.

Such a solution for a calorically perfect gas has already been introduced in Sec. 12.1. For this reason, the present section will highlight the time-marching solution of nonequilibrium nozzle flows. In this sense, our discussion here will be an extension of the ideas first presented in Sec. 12.1. Therefore, the reader is encouraged to review that section before proceeding further. Also, the reader is urged to study AGARD-ograph 124 by Hall and Treanor (see Ref. 71) for a broad outline of steady-state solutions for nonequilibrium nozzle flows.

Consider again the nozzle and grid-point distribution sketched in Fig. 12.5. The time-marching solution of nonequilibrium nozzle flows closely follows the technique described in Sec. 12.1, with the consideration of vibrational energy and chemical species concentrations as additional dependent variables. In this context, at the first grid point in Fig. 12.5, which represents the reservoir conditions, equilibrium conditions for e_{vib} and c_i at the given p_o and T_o are calculated, and held fixed, invariant with time. Guessed values of e_{vib} and c_i are then arbitrarily specified at all other grid points (along with guessed values of all other flow variables); these guessed values represent initial conditions for the time-marching solution. For the initial values of e_{vib} and c_i, it is recommended that equilibrium values be assumed from the reservoir to the throat, and then frozen values be prescribed downstream of the throat. Such an initial distribution of nonequilibrium variables is qualitatively similar to typical results obtained for nonequilibrium nozzle flows, as we will soon see.

The governing continuity, momentum, and energy equations for unsteady quasi-one-dimensional flow have been given as Eqs. (12.5), (12.6), and (12.7) respectively. In addition to these equations, for a nonequilibrium flow the appropriate vibrational rate and species continuity equations are

$$\frac{\partial e_{vib}}{\partial t} = \frac{1}{\tau} \left(e_{vib}^{eq} - e_{vib} \right) - u \frac{\partial e_{vib}}{\partial x} \tag{17.67}$$

and

$$\frac{\partial c_i}{\partial t} = -u \frac{\partial c_i}{\partial x} + \frac{\dot{w}_i}{\rho} \tag{17.68}$$

These equations are solved step by step in time using the finite-difference predictor-corrector approach described in Sec. 12.1, and of course are fully coupled with the other governing equations [Eqs. (12.5) through (12.7)] at each time step. Along with the other flow variables, e_{vib} and c_i at each grid point will vary with time; but after many time steps all flow variables will approach a steady state. As emphasized in Sec. 12.1, it is this steady flowfield we are interested in as our solution—the time-marching technique is simply a means to achieve this end.

The nonequilibrium phenomena introduce an important new stability criterion for Δt in addition to the CFL criterion discussed in Sec. 12.2. The value chosen for Δt must be geared to the speed of the nonequilibrium relaxation process, and must not exceed the characteristic time for the fastest finite rate taking place in the system. That is,

$$\Delta t < B\Gamma \tag{17.69}$$

where $\Gamma = \tau$ for vibrational nonequilibrium, $\Gamma = \rho(\partial \dot{w}_i/\partial c_i)^{-1}$ for chemical non-equilibrium, and B is a dimensionless proportionality constant found by experience to be less than unity, and sometimes as low as 0.1. The value chosen for Δt in a non-equilibrium flow must satisfy both Eqs. (17.69) and (12.14). Which of the two stability criteria is the smaller, and hence governs the time step, depends on the nature of the case being calculated. If the local pressure and temperature are low enough everywhere in the flow, the rates will be slow, and Eq. (12.14) generally dictates the value of Δt. On the other hand, if some of the rates have particularly high transition probabilities and/or the local p and T are very high, then Eq. (17.69) generally dictates Δt. This is almost always encountered in rocket nozzle flows of hydrocarbon gases, where some of the chemical reactions involving hydrogen are very fast and combustion chamber pressures and temperatures are reasonably high.

The nature of the time-marching solution of a vibrational nonequilibrium expansion of pure N_2 is shown in Fig. 17.16. Here, the transient e_{vib} profiles at

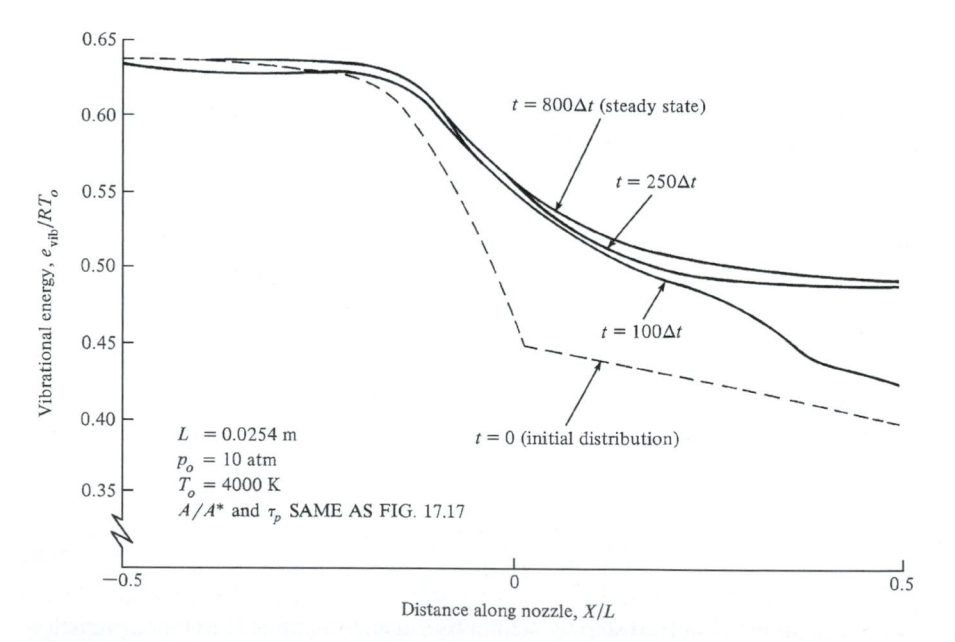

Figure 17.16 | Transient and final steady state e_{vib} distributions for the nonequilibrium expansion of N_2 obtained from the time-marching analysis. (After Anderson, Ref. 73.)

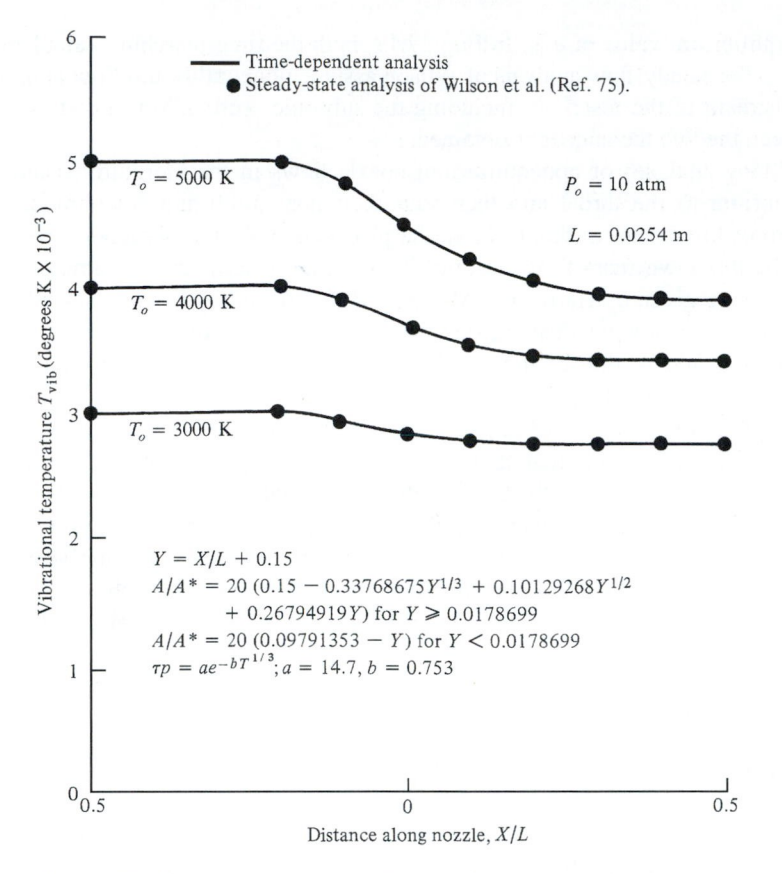

Figure 17.17 | Steady state T_{vib} distributions for the nonequilibrium expansion of N_2; comparison of the time-marching analysis with the steady-flow analysis of Wilson et al. (After Anderson, Ref. 73.)

various time steps are shown; the dashed curve represents the guessed initial distribution. Note that during the first 250 time steps, the proper steady state distribution is rapidly approached, and is reasonably attained after 800 time steps. Beyond this time, the time-marching solution produces virtually no change in the results from one time step to the next. This steady-state distribution agrees with the results of a steady-flow analysis after Wilson (see Ref. 75), as shown in Fig. 17.17. Here, a local "vibrational temperature" is defined from the local nonequilibrium value of e_{vib} using the relation

$$e_{vib} = \left[\frac{h\nu/kT_{vib}}{e^{h\nu/kT_{vib}} - 1} \right] R T_{vib} \qquad (17.70)$$

patterned after the equilibrium expression given by Eq. (16.47). Note that Eq. (17.70) is *not* a valid *physical* relationship for nonequilibrium flow; it is simply an equation that *defines* the vibrational temperature T_{vib} and that allows the calculation of a value of T_{vib} from the known value of e_{vib}. Hence, T_{vib} is simply an index for the local nonequilibrium value of e_{vib}. In Fig. 17.17, both the time-marching calculations as well as the steady-flow analysis of Wilson assume nonequilibrium flow at all points downstream of the reservoir, including the subsonic section. Very good agreement between the two techniques is obtained.

Many analyses of nonequilibrium nozzle flows in the literature assume local equilibrium to the throat and then start their nonequilibrium calculations downstream of the throat. In this fashion, the problems with the saddlepoint singularity and the unknown mass flow, described earlier, are sidestepped. Examples of such analyses are given by Harris and Albacete (Ref. 76), and by Erickson (see Ref. 77). However, for many practical nozzle flows, nonequilibrium effects become important in the subsonic section of the nozzle, and hence a fully nonequilibrium solution throughout the complete nozzle is required.

Figures 17.16 and 17.17 illustrate an important qualitative aspect of nonequilibrium nozzle flows. Note that, as the expansion proceeds and the static temperature (T_{trans}) decreases through the nozzle, the vibrational temperature and energy also decrease to begin with. However, in the throat region, e_{vib} and T_{vib} tend to "freeze," and are reasonably constant downstream of the throat. This is a qualitative comment only; the actual distributions depend on pressure, temperature, and nozzle length. It is generally true that equilibrium flow is reasonably obtained throughout large nozzles at high pressures. Reducing both the size of the nozzle and the reservoir pressure tends to encourage nonequilibrium flow.

Results for a chemical nonequilibrium nozzle flow are given in Fig. 17.18, where the transient mechanism of the time-marching technique is illustrated. Here,

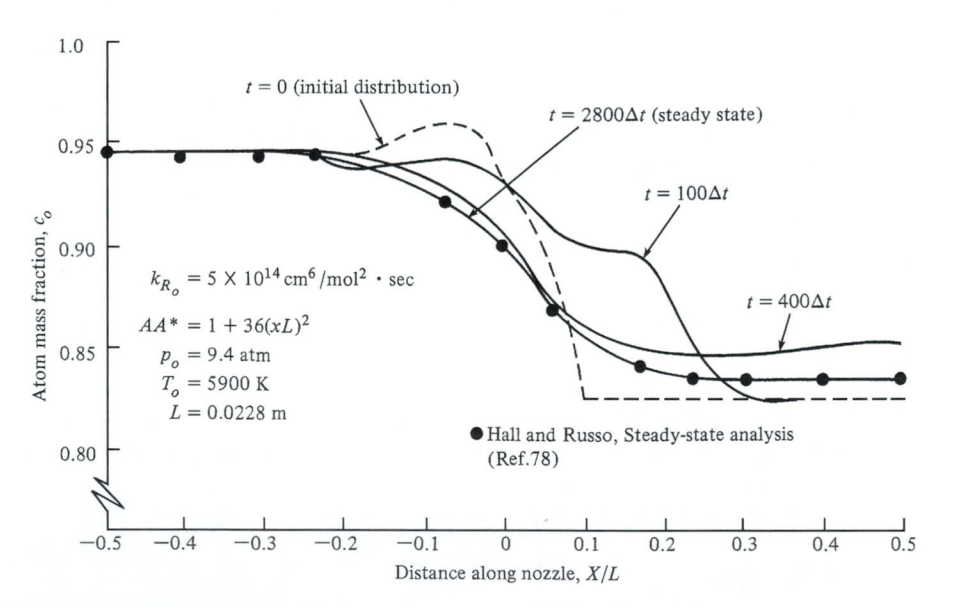

Figure 17.18 | Transient and final steady-state atom mass-fraction distributions for the nonequilibrium expansion of dissociated oxygen; comparison of the time-marching method with the steady-state approach of Hall and Russo. (After Anderson, Ref. 73.)

the nonequilibrium expansion of partially dissociated oxygen is calculated where the only chemical reaction is

$$O_2 + M \leftrightarrows 2O + M$$

In Fig. 17.18, the dashed line gives the initially assumed distribution for the atomic oxygen mass fraction, c_O. Note the rapid approach toward the steady state distribution during the first 400 time steps. The final steady-state distribution is obtained after 2800 time steps. This steady-state distribution compares favorably with the results of Hall and Russo (solid circles), who performed a steady-flow analysis of the complete nonequilibrium nozzle flow (see Ref. 78). Again, note the tendency of the oxygen mass fraction to freeze downstream of the throat.

A more complex chemically reacting nonequilibrium nozzle flow is illustrated by the expansion of a hydrocarbon mixture through a rocket engine. The configuration of a rocket nozzle is given in Fig. 17.19. Here, for the time-marching numerical solution two grids are used along the nozzle axis: a fine grid of closely spaced points through the subsonic section and slightly downstream of the throat, and a coarse grid of widely spaced points further downstream. Since most of the nonequilibrium behavior and the fastest reactions are occurring in the throat region, a fine grid is chosen here to maintain accuracy. In contrast, far downstream in the cooler supersonic region, the reactions are slower, the chemical composition is tending to freeze, and the grid spacing can be larger. (Parenthetically, we note that, for any of the finite-difference solutions discussed in this book, the grid spacings do not have to be constant. Indeed, the concept of *adaptive grids,* i.e., putting grid points only where you want them as dictated by the gradients in the flow, is a current state-of-the-art research problem of computational fluid dynamics.)

In Fig. 17.19, the reservoir conditions are formed by the equilibrium combustion of N_2O_4, N_2H_4, and unsymmetrical dimethyl hydrazine, with an oxydizer-to-fuel ratio of 2.25 and a chamber pressure of 4 atm. Results for the subsequent nonequilibrium expansion are shown in Figs. 17.20 through 17.23. In Fig. 17.20, the transient variation of the hydrogen atom mass fraction through the nozzle is shown. For

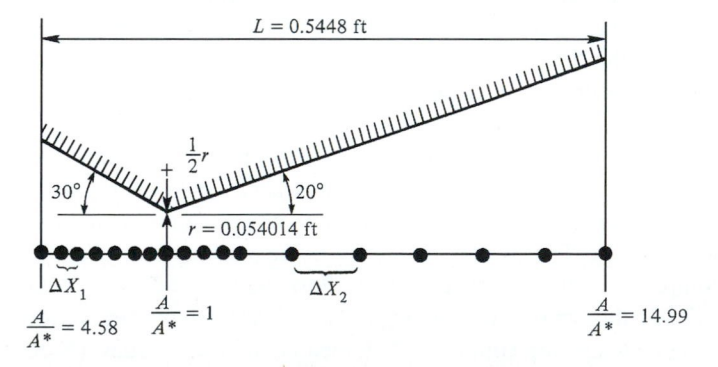

Figure 17.19 | Schematic representation of the rocket engine nozzle and grid-point system used by Vamos and Anderson, Ref. 80.

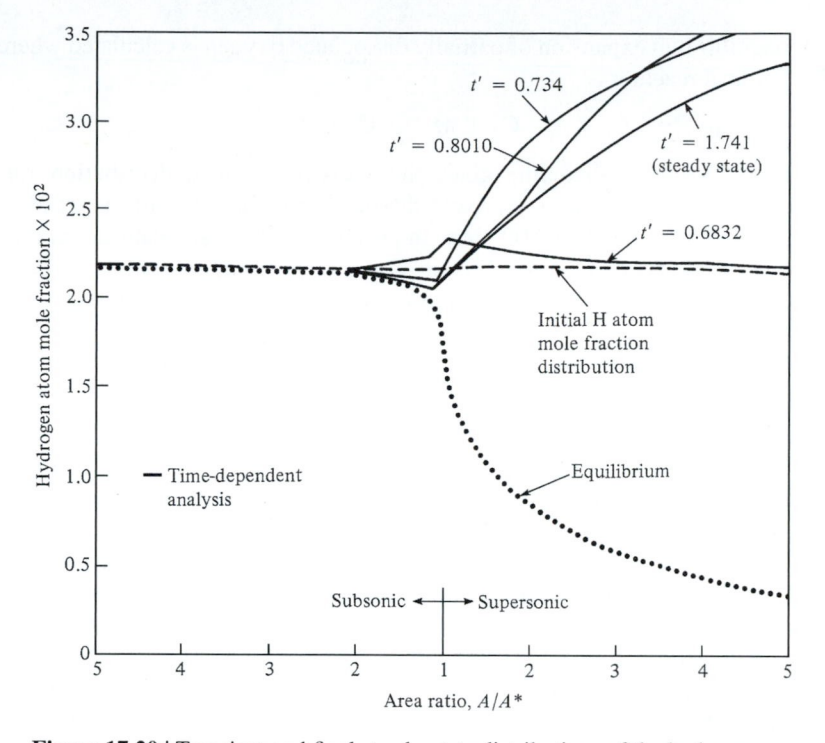

Figure 17.20 | Transient and final steady-state distributions of the hydrogen atom mole fraction through a rocket nozzle; nonequilibrium flow. (After Vamos and Anderson, Ref. 80.)

convenience, the initial distribution is assumed to be completely frozen from the reservoir (the dashed horizontal line). Several intermediate distributions obtained during the time-marching calculations are shown, with the final steady state being achieved at a dimensionless time of 1.741. Note that, if the flow were in local chemical equilibrium, X_H would decrease continuously as T decreases, as shown in Fig. 17.20. In contrast, however, due to the complexities of the H-C-O-N chemical kinetic mechanism, X_H actually *increases* with distance along the nozzle. Here is another example (the first was given in Sec. 17.10) where a nonequilibrium variable falls outside the bounds of equilibrium and frozen flows. The variation of static temperature is given in Fig. 17.21; note that for nonequilibrium flow the temperature distribution is *lower* than the equilibrium value. This is because the nonequilibrium flow tends to freeze some of the dissociated products, hence locking up some of the chemical zero-point energy which would otherwise be converted to random molecular translational energy. The steady-state temperature distribution in Fig. 17.21 (at $t' = 1.741$) compares favorably with the steady-flow analysis of Sarli et al. (see Ref. 79). In Fig. 17.22, the steady-state nonequilibrium distributions of various chemical species are given, and are compared with their equilibrium values. (Note that inconsistent units are used for the concentrations.) Clearly, a substantial degree of nonequilibrium exists in the nozzle expansion. A practical consequence of this nonequilibrium

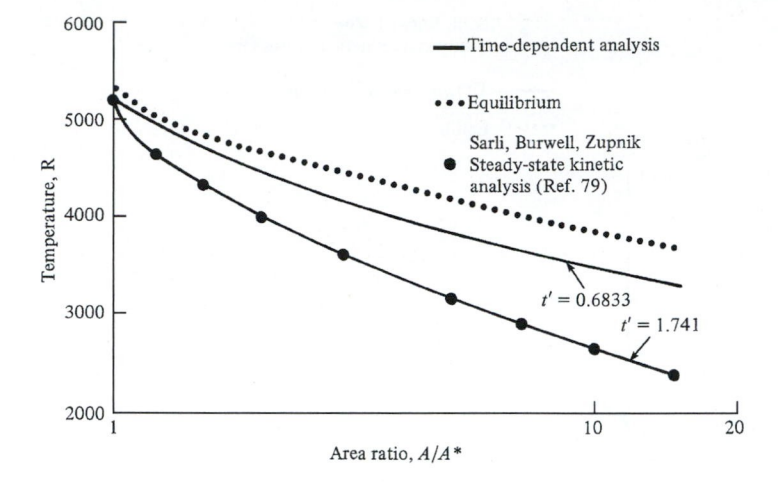

Figure 17.21 | Temperature distributions for the nonequilibrium flow through a rocket nozzle. (After Vamos and Anderson, Ref. 80.)

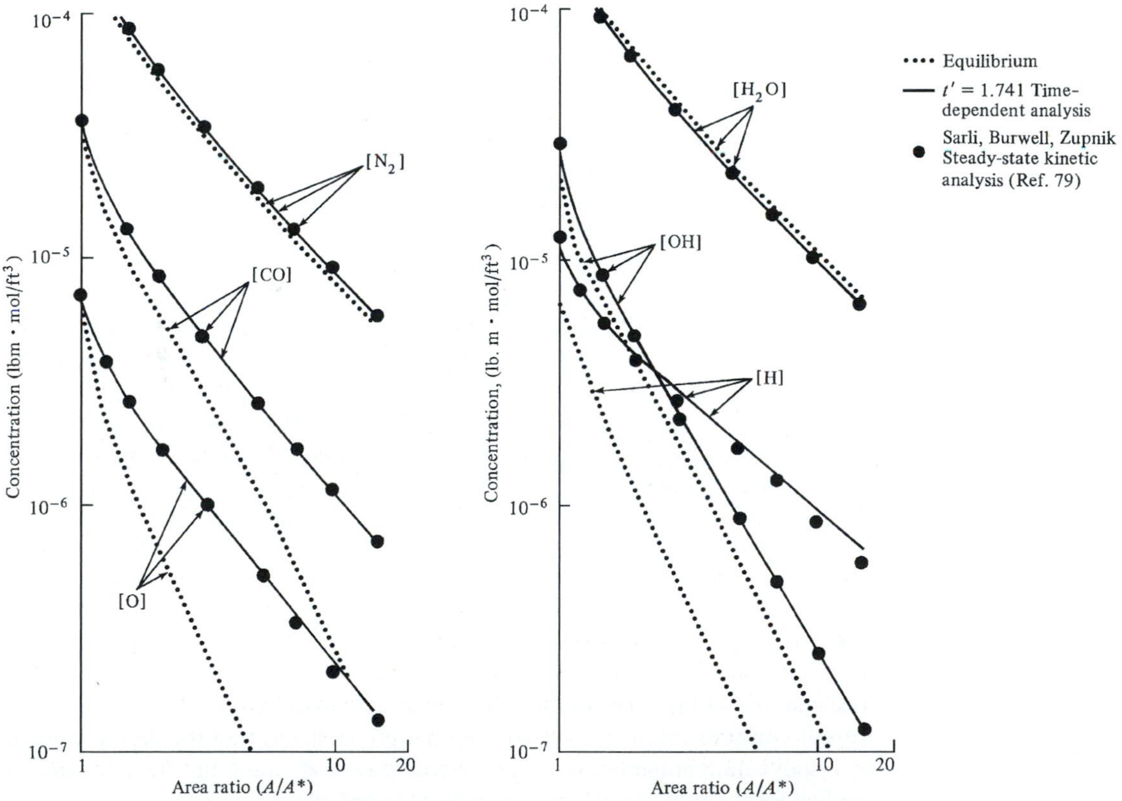

Figure 17.22 | Molecular and atomic species concentration profiles for the nonequilibrium flow through a rocket nozzle. (After Vamos and Anderson, Ref. 80.)

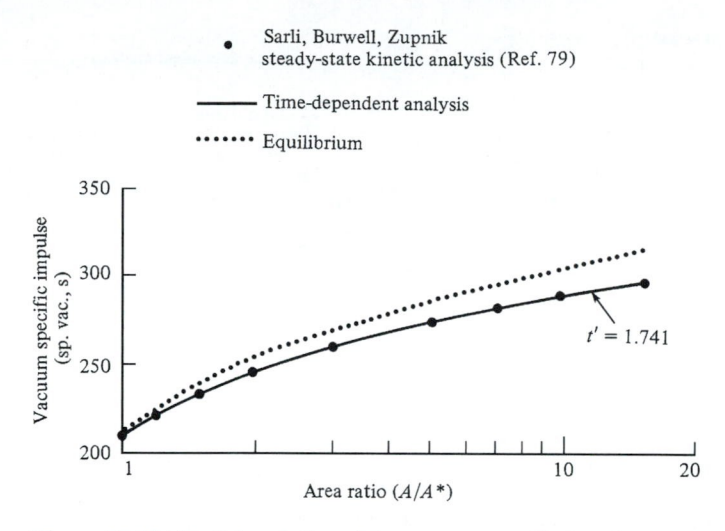

Figure 17.23 | Spatial variation of the vacuum specific impulse; comparison of equilibrium and nonequilibrium results. (After Vamos and Anderson, Ref. 80.)

flow is reflected in Fig. 17.23, which gives the variation of local specific impulse through the nozzle. The specific impulse (pounds of thrust per pound per second of mass flow) of the actual engine is given by the local value at the nozzle exit. Clearly, nonequilibrium flow throughout the nozzle expansion reduces the thrust and efficiency in comparison to equilibrium flow. See Ref. 80 for more details.

As a final point concerning nonequilibrium nozzle flows, note that any finite-rate phenomena are irreversible. Hence, an adiabatic, inviscid nonequilibrium nozzle flow is *non*isentropic. Because the entropy of a fluid element increases as it moves through the nozzle, a simple analysis shows that the local velocity at the nozzle throat is *not* sonic. Indeed, in a nonequilibrium flow, the speed of sound itself is not unique, and depends on the frequency of the sound wave. However, if either the frozen or equilibrium speeds of sound (see Sec. 17.5) are used to define the frozen or equilibrium Mach numbers at the nozzle throat, both Mach numbers will be less than unity. Sonic flow in a nonequilibrium nozzle expansion occurs slightly downstream of the throat.

17.12 | SUMMARY

The analysis of high-temperature flows is an important part of the modern application of compressible flow. This is why the principles behind the high-temperature thermodynamic properties of a gas were discussed at length in Chap. 16, and were applied for the analysis of some basic flows in the present chapter. Shock waves and nozzle flows are classic problems in the study of compressible flow; moreover, they occur so frequently in practice that such studies are of immense practical use.

Consequently, these basic flows were chosen in this chapter to illustrate the high-temperature effects of vibrational excitation and chemical reactions. However, the trends and physical results discussed here are characteristic of most high-temperature flows of interest. Therefore, a careful study of this chapter will prepare the reader for virtually any foray into more complex flows where high-temperature phenomena are of importance.

PROBLEMS

Note: Universal gas constant $\mathscr{R} = 8314$ J/(kg $\cdot$ mol $\cdot$ K), $\nu_{N_2} = 7.06 \times 10^{13}$/s, $k = 1.38 \times 10^{-23}$ J/K, $h = 6.625 \times 10^{-34}$ J $\cdot$ s, $\mathscr{M}_{N_2} = 28$, 1 atm $= 1.01 \times 10^5$ N/m^2.

17.1 Consider a normal shock wave in pure N_2. The upstream pressure, temperature, and velocity are 0.1 atm, 300 K, and 3500 m/s, respectively. Calculate T_2, p_2, and u_2 behind the shock assuming local thermodynamic equilibrium but no chemical reactions. Ignore the electronic energy.

17.2 The total temperature T_o is defined in Chap. 3 as that temperature that would exist at a point in the flow if the fluid elements were brought to rest adiabatically at that point. For each of these chemically reacting flows, is T_o constant or variable throughout the flow? Explain your answer.

    a.   Equilibrium flow across a shock wave

    b.   Nonequilibrium flow across a shock wave

    c.   Inviscid, adiabatic, equilibrium flow through nozzles

    d.   Inviscid, adiabatic nonequilibrium flow through nozzles

Table A.1 | Isentropic flow properties

M	$\dfrac{p_o}{p}$	$\dfrac{\rho_o}{\rho}$	$\dfrac{T_o}{T}$	$\dfrac{A}{A^*}$
0.2000 − 01	0.1000 + 01	0.1000 + 01	0.1000 + 01	0.2894 + 02
0.4000 − 01	0.1001 + 01	0.1001 + 01	0.1000 + 01	0.1448 + 02
0.6000 − 01	0.1003 + 01	0.1002 + 01	0.1001 + 01	0.9666 + 01
0.8000 − 01	0.1004 + 01	0.1003 + 01	0.1001 + 01	0.7262 + 01
0.1000 + 00	0.1007 + 01	0.1005 + 01	0.1002 + 01	0.5822 + 01
0.1200 + 00	0.1010 + 01	0.1007 + 01	0.1003 + 01	0.4864 + 01
0.1400 + 00	0.1014 + 01	0.1010 + 01	0.1004 + 01	0.4182 + 01
0.1600 + 00	0.1018 + 01	0.1013 + 01	0.1005 + 01	0.3673 + 01
0.1800 + 00	0.1023 + 01	0.1016 + 01	0.1006 + 01	0.3278 + 01
0.2000 + 00	0.1028 + 01	0.1020 + 01	0.1008 + 01	0.2964 + 01
0.2200 + 00	0.1034 + 01	0.1024 + 01	0.1010 + 01	0.2708 + 01
0.2400 + 00	0.1041 + 01	0.1029 + 01	0.1012 + 01	0.2496 + 01
0.2600 + 00	0.1048 + 01	0.1034 + 01	0.1014 + 01	0.2317 + 01
0.2800 + 00	0.1056 + 01	0.1040 + 01	0.1016 + 01	0.2166 + 01
0.3000 + 00	0.1064 + 01	0.1046 + 01	0.1018 + 01	0.2035 + 01
0.3200 + 00	0.1074 + 01	0.1052 + 01	0.1020 + 01	0.1922 + 01
0.3400 + 00	0.1083 + 01	0.1059 + 01	0.1023 + 01	0.1823 + 01
0.3600 + 00	0.1094 + 01	0.1066 + 01	0.1026 + 01	0.1736 + 01
0.3800 + 00	0.1105 + 01	0.1074 + 01	0.1029 + 01	0.1659 + 01
0.4000 + 00	0.1117 + 01	0.1082 + 01	0.1032 + 01	0.1590 + 01
0.4200 + 00	0.1129 + 01	0.1091 + 01	0.1035 + 01	0.1529 + 01
0.4400 + 00	0.1142 + 01	0.1100 + 01	0.1039 + 01	0.1474 + 01
0.4600 + 00	0.1156 + 01	0.1109 + 01	0.1042 + 01	0.1425 + 01
0.4800 + 00	0.1171 + 01	0.1119 + 01	0.1046 + 01	0.1380 + 01
0.5000 + 00	0.1186 + 01	0.1130 + 01	0.1050 + 01	0.1340 + 01
0.5200 + 00	0.1202 + 01	0.1141 + 01	0.1054 + 01	0.1303 + 01
0.5400 + 00	0.1219 + 01	0.1152 + 01	0.1058 + 01	0.1270 + 01
0.5600 + 00	0.1237 + 01	0.1164 + 01	0.1063 + 01	0.1240 + 01
0.5800 + 00	0.1256 + 01	0.1177 + 01	0.1067 + 01	0.1213 + 01
0.6000 + 00	0.1276 + 01	0.1190 + 01	0.1072 + 01	0.1188 + 01

Continued

Table A.1 | *Continued*

M	$\dfrac{p_o}{p}$	$\dfrac{\rho_o}{\rho}$	$\dfrac{T_o}{T}$	$\dfrac{A}{A^*}$
0.6200 + 00	0.1296 + 01	0.1203 + 01	0.1077 + 01	0.1166 + 01
0.6400 + 00	0.1317 + 01	0.1218 + 01	0.1082 + 01	0.1145 + 01
0.6600 + 00	0.1340 + 01	0.1232 + 01	0.1087 + 01	0.1127 + 01
0.6800 + 00	0.1363 + 01	0.1247 + 01	0.1092 + 01	0.1110 + 01
0.7000 + 00	0.1387 + 01	0.1263 + 01	0.1098 + 01	0.1094 + 01
0.7200 + 00	0.1412 + 01	0.1280 + 01	0.1104 + 01	0.1081 + 01
0.7400 + 00	0.1439 + 01	0.1297 + 01	0.1110 + 01	0.1068 + 01
0.7600 + 00	0.1466 + 01	0.1314 + 01	0.1116 + 01	0.1057 + 01
0.7800 + 00	0.1495 + 01	0.1333 + 01	0.1122 + 01	0.1047 + 01
0.8000 + 00	0.1524 + 01	0.1351 + 01	0.1128 + 01	0.1038 + 01
0.8200 + 00	0.1555 + 01	0.1371 + 01	0.1134 + 01	0.1030 + 01
0.8400 + 00	0.1587 + 01	0.1391 + 01	0.1141 + 01	0.1024 + 01
0.8600 + 00	0.1621 + 01	0.1412 + 01	0.1148 + 01	0.1018 + 01
0.8800 + 00	0.1655 + 01	0.1433 + 01	0.1155 + 01	0.1013 + 01
0.9000 + 00	0.1691 + 01	0.1456 + 01	0.1162 + 01	0.1009 + 01
0.9200 + 00	0.1729 + 01	0.1478 + 01	0.1169 + 01	0.1006 + 01
0.9400 + 00	0.1767 + 01	0.1502 + 01	0.1177 + 01	0.1003 + 01
0.9600 + 00	0.1808 + 01	0.1526 + 01	0.1184 + 01	0.1001 + 01
0.9800 + 00	0.1850 + 01	0.1552 + 01	0.1192 + 01	0.1000 + 01
0.1000 + 01	0.1893 + 01	0.1577 + 01	0.1200 + 01	0.1000 + 01
0.1020 + 01	0.1938 + 01	0.1604 + 01	0.1208 + 01	0.1000 + 01
0.1040 + 01	0.1985 + 01	0.1632 + 01	0.1216 + 01	0.1001 + 01
0.1060 + 01	0.2033 + 01	0.1660 + 01	0.1225 + 01	0.1003 + 01
0.1080 + 01	0.2083 + 01	0.1689 + 01	0.1233 + 01	0.1005 + 01
0.1100 + 01	0.2135 + 01	0.1719 + 01	0.1242 + 01	0.1008 + 01
0.1120 + 01	0.2189 + 01	0.1750 + 01	0.1251 + 01	0.1011 + 01
0.1140 + 01	0.2245 + 01	0.1782 + 01	0.1260 + 01	0.1015 + 01
0.1160 + 01	0.2303 + 01	0.1814 + 01	0.1269 + 01	0.1020 + 01
0.1180 + 01	0.2363 + 01	0.1848 + 01	0.1278 + 01	0.1025 + 01
0.1200 + 01	0.2425 + 01	0.1883 + 01	0.1288 + 01	0.1030 + 01
0.1220 + 01	0.2489 + 01	0.1918 + 01	0.1298 + 01	0.1037 + 01
0.1240 + 01	0.2556 + 01	0.1955 + 01	0.1308 + 01	0.1043 + 01
0.1260 + 01	0.2625 + 01	0.1992 + 01	0.1318 + 01	0.1050 + 01
0.1280 + 01	0.2697 + 01	0.2031 + 01	0.1328 + 01	0.1058 + 01
0.1300 + 01	0.2771 + 01	0.2071 + 01	0.1338 + 01	0.1066 + 01
0.1320 + 01	0.2847 + 01	0.2112 + 01	0.1348 + 01	0.1075 + 01
0.1340 + 01	0.2927 + 01	0.2153 + 01	0.1359 + 01	0.1084 + 01
0.1360 + 01	0.3009 + 01	0.2197 + 01	0.1370 + 01	0.1094 + 01
0.1380 + 01	0.3094 + 01	0.2241 + 01	0.1381 + 01	0.1104 + 01
0.1400 + 01	0.3182 + 01	0.2286 + 01	0.1392 + 01	0.1115 + 01
0.1420 + 01	0.3273 + 01	0.2333 + 01	0.1403 + 01	0.1126 + 01
0.1440 + 01	0.3368 + 01	0.2381 + 01	0.1415 + 01	0.1138 + 01
0.1460 + 01	0.3465 + 01	0.2430 + 01	0.1426 + 01	0.1150 + 01
0.1480 + 01	0.3566 + 01	0.2480 + 01	0.1438 + 01	0.1163 + 01
0.1500 + 01	0.3671 + 01	0.2532 + 01	0.1450 + 01	0.1176 + 01
0.1520 + 01	0.3779 + 01	0.2585 + 01	0.1462 + 01	0.1190 + 01
0.1540 + 01	0.3891 + 01	0.2639 + 01	0.1474 + 01	0.1204 + 01
0.1560 + 01	0.4007 + 01	0.2695 + 01	0.1487 + 01	0.1219 + 01
0.1580 + 01	0.4127 + 01	0.2752 + 01	0.1499 + 01	0.1234 + 01
0.1600 + 01	0.4250 + 01	0.2811 + 01	0.1512 + 01	0.1250 + 01

Table A.1 | *Continued*

M	$\dfrac{p_o}{p}$	$\dfrac{\rho_o}{\rho}$	$\dfrac{T_o}{T}$	$\dfrac{A}{A^*}$
0.1620 + 01	0.4378 + 01	0.2871 + 01	0.1525 + 01	0.1267 + 01
0.1640 + 01	0.4511 + 01	0.2933 + 01	0.1538 + 01	0.1284 + 01
0.1660 + 01	0.4648 + 01	0.2996 + 01	0.1551 + 01	0.1301 + 01
0.1680 + 01	0.4790 + 01	0.3061 + 01	0.1564 + 01	0.1319 + 01
0.1700 + 01	0.4936 + 01	0.3128 + 01	0.1578 + 01	0.1338 + 01
0.1720 + 01	0.5087 + 01	0.3196 + 01	0.1592 + 01	0.1357 + 01
0.1740 + 01	0.5244 + 01	0.3266 + 01	0.1606 + 01	0.1376 + 01
0.1760 + 01	0.5406 + 01	0.3338 + 01	0.1620 + 01	0.1397 + 01
0.1780 + 01	0.5573 + 01	0.3411 + 01	0.1634 + 01	0.1418 + 01
0.1800 + 01	0.5746 + 01	0.3487 + 01	0.1648 + 01	0.1439 + 01
0.1820 + 01	0.5924 + 01	0.3564 + 01	0.1662 + 01	0.1461 + 01
0.1840 + 01	0.6109 + 01	0.3643 + 01	0.1677 + 01	0.1484 + 01
0.1860 + 01	0.6300 + 01	0.3723 + 01	0.1692 + 01	0.1507 + 01
0.1880 + 01	0.6497 + 01	0.3806 + 01	0.1707 + 01	0.1531 + 01
0.1900 + 01	0.6701 + 01	0.3891 + 01	0.1722 + 01	0.1555 + 01
0.1920 + 01	0.6911 + 01	0.3978 + 01	0.1737 + 01	0.1580 + 01
0.1940 + 01	0.7128 + 01	0.4067 + 01	0.1753 + 01	0.1606 + 01
0.1960 + 01	0.7353 + 01	0.4158 + 01	0.1768 + 01	0.1633 + 01
0.1980 + 01	0.7585 + 01	0.4251 + 01	0.1784 + 01	0.1660 + 01
0.2000 + 01	0.7824 + 01	0.4347 + 01	0.1800 + 01	0.1687 + 01
0.2050 + 01	0.8458 + 01	0.4596 + 01	0.1840 + 01	0.1760 + 01
0.2100 + 01	0.9145 + 01	0.4859 + 01	0.1882 + 01	0.1837 + 01
0.2150 + 01	0.9888 + 01	0.5138 + 01	0.1924 + 01	0.1919 + 01
0.2200 + 01	0.1069 + 02	0.5433 + 01	0.1968 + 01	0.2005 + 01
0.2250 + 01	0.1156 + 02	0.5746 + 01	0.2012 + 01	0.2096 + 01
0.2300 + 01	0.1250 + 02	0.6076 + 01	0.2058 + 01	0.2193 + 01
0.2350 + 01	0.1352 + 02	0.6425 + 01	0.2104 + 01	0.2295 + 01
0.2400 + 01	0.1462 + 02	0.6794 + 01	0.2152 + 01	0.2403 + 01
0.2450 + 01	0.1581 + 02	0.7183 + 01	0.2200 + 01	0.2517 + 01
0.2500 + 01	0.1709 + 02	0.7594 + 01	0.2250 + 01	0.2637 + 01
0.2550 + 01	0.1847 + 02	0.8027 + 01	0.2300 + 01	0.2763 + 01
0.2600 + 01	0.1995 + 02	0.8484 + 01	0.2352 + 01	0.2896 + 01
0.2650 + 01	0.2156 + 02	0.8965 + 01	0.2404 + 01	0.3036 + 01
0.2700 + 01	0.2328 + 02	0.9472 + 01	0.2458 + 01	0.3183 + 01
0.2750 + 01	0.2514 + 02	0.1001 + 02	0.2512 + 01	0.3338 + 01
0.2800 + 01	0.2714 + 02	0.1057 + 02	0.2568 + 01	0.3500 + 01
0.2850 + 01	0.2929 + 02	0.1116 + 02	0.2624 + 01	0.3671 + 01
0.2900 + 01	0.3159 + 02	0.1178 + 02	0.2682 + 01	0.3850 + 01
0.2950 + 01	0.3407 + 02	0.1243 + 02	0.2740 + 01	0.4038 + 01
0.3000 + 01	0.3673 + 02	0.1312 + 02	0.2800 + 01	0.4235 + 01
0.3050 + 01	0.3959 + 02	0.1384 + 02	0.2860 + 01	0.4441 + 01
0.3100 + 01	0.4265 + 02	0.1459 + 02	0.2922 + 01	0.4657 + 01
0.3150 + 01	0.4593 + 02	0.1539 + 02	0.2984 + 01	0.4884 + 01
0.3200 + 01	0.4944 + 02	0.1622 + 02	0.3048 + 01	0.5121 + 01
0.3250 + 01	0.5320 + 02	0.1709 + 02	0.3112 + 01	0.5369 + 01
0.3300 + 01	0.5722 + 02	0.1800 + 02	0.3178 + 01	0.5629 + 01
0.3350 + 01	0.6152 + 02	0.1896 + 02	0.3244 + 01	0.5900 + 01
0.3400 + 01	0.6612 + 02	0.1996 + 02	0.3312 + 01	0.6184 + 01
0.3450 + 01	0.7103 + 02	0.2101 + 02	0.3380 + 01	0.6480 + 01
0.3500 + 01	0.7627 + 02	0.2211 + 02	0.3450 + 01	0.6790 + 01

Continued

Table A.1 | *Continued*

M	$\dfrac{p_o}{p}$	$\dfrac{\rho_o}{\rho}$	$\dfrac{T_o}{T}$	$\dfrac{A}{A^*}$
0.3550 + 01	0.8187 + 02	0.2325 + 02	0.3520 + 01	0.7113 + 01
0.3600 + 01	0.8784 + 02	0.2445 + 02	0.3592 + 01	0.7450 + 01
0.3650 + 01	0.9420 + 02	0.2571 + 02	0.3664 + 01	0.7802 + 01
0.3700 + 01	0.1010 + 03	0.2701 + 02	0.3738 + 01	0.8169 + 01
0.3750 + 01	0.1082 + 03	0.2838 + 02	0.3812 + 01	0.8552 + 01
0.3800 + 01	0.1159 + 03	0.2981 + 02	0.3888 + 01	0.8951 + 01
0.3850 + 01	0.1241 + 03	0.3129 + 02	0.3964 + 01	0.9366 + 01
0.3900 + 01	0.1328 + 03	0.3285 + 02	0.4042 + 01	0.9799 + 01
0.3950 + 01	0.1420 + 03	0.3446 + 02	0.4120 + 01	0.1025 + 02
0.4000 + 01	0.1518 + 03	0.3615 + 02	0.4200 + 01	0.1072 + 02
0.4050 + 01	0.1623 + 03	0.3791 + 02	0.4280 + 01	0.1121 + 02
0.4100 + 01	0.1733 + 03	0.3974 + 02	0.4362 + 01	0.1171 + 02
0.4150 + 01	0.1851 + 03	0.4164 + 02	0.4444 + 01	0.1224 + 02
0.4200 + 01	0.1975 + 03	0.4363 + 02	0.4528 + 01	0.1279 + 02
0.4250 + 01	0.2108 + 03	0.4569 + 02	0.4612 + 01	0.1336 + 02
0.4300 + 01	0.2247 + 03	0.4784 + 02	0.4698 + 01	0.1395 + 02
0.4350 + 01	0.2396 + 03	0.5007 + 02	0.4784 + 01	0.1457 + 02
0.4400 + 01	0.2553 + 03	0.5239 + 02	0.4872 + 01	0.1521 + 02
0.4450 + 01	0.2719 + 03	0.5480 + 02	0.4960 + 01	0.1587 + 02
0.4500 + 01	0.2894 + 03	0.5731 + 02	0.5050 + 01	0.1656 + 02
0.4550 + 01	0.3080 + 03	0.5991 + 02	0.5140 + 01	0.1728 + 02
0.4600 + 01	0.3276 + 03	0.6261 + 02	0.5232 + 01	0.1802 + 02
0.4650 + 01	0.3483 + 03	0.6542 + 02	0.5324 + 01	0.1879 + 02
0.4700 + 01	0.3702 + 03	0.6833 + 02	0.5418 + 01	0.1958 + 02
0.4750 + 01	0.3933 + 03	0.7135 + 02	0.5512 + 01	0.2041 + 02
0.4800 + 01	0.4177 + 03	0.7448 + 02	0.5608 + 01	0.2126 + 02
0.4850 + 01	0.4434 + 03	0.7772 + 02	0.5704 + 01	0.2215 + 02
0.4900 + 01	0.4705 + 03	0.8109 + 02	0.5802 + 01	0.2307 + 02
0.4950 + 01	0.4990 + 03	0.8457 + 02	0.5900 + 01	0.2402 + 02
0.5000 + 01	0.5291 + 03	0.8818 + 02	0.6000 + 01	0.2500 + 02
0.5100 + 01	0.5941 + 03	0.9579 + 02	0.6202 + 01	0.2707 + 02
0.5200 + 01	0.6661 + 03	0.1039 + 03	0.6408 + 01	0.2928 + 02
0.5300 + 01	0.7457 + 03	0.1127 + 03	0.6618 + 01	0.3165 + 02
0.5400 + 01	0.8335 + 03	0.1220 + 03	0.6832 + 01	0.3417 + 02
0.5500 + 01	0.9304 + 03	0.1320 + 03	0.7050 + 01	0.3687 + 02
0.5600 + 01	0.1037 + 04	0.1426 + 03	0.7272 + 01	0.3974 + 02
0.5700 + 01	0.1154 + 04	0.1539 + 03	0.7498 + 01	0.4280 + 02
0.5800 + 01	0.1283 + 04	0.1660 + 03	0.7728 + 01	0.4605 + 02
0.5900 + 01	0.1424 + 04	0.1789 + 03	0.7962 + 01	0.4951 + 02
0.6000 + 01	0.1579 + 04	0.1925 + 03	0.8200 + 01	0.5318 + 02
0.6100 + 01	0.1748 + 04	0.2071 + 03	0.8442 + 01	0.5708 + 02
0.6200 + 01	0.1933 + 04	0.2225 + 03	0.8688 + 01	0.6121 + 02
0.6300 + 01	0.2135 + 04	0.2388 + 03	0.8938 + 01	0.6559 + 02
0.6400 + 01	0.2355 + 04	0.2562 + 03	0.9192 + 01	0.7023 + 02
0.6500 + 01	0.2594 + 04	0.2745 + 03	0.9450 + 01	0.7513 + 02
0.6600 + 01	0.2855 + 04	0.2939 + 03	0.9712 + 01	0.8032 + 02
0.6700 + 01	0.3138 + 04	0.3145 + 03	0.9978 + 01	0.8580 + 02
0.6800 + 01	0.3445 + 04	0.3362 + 03	0.1025 + 02	0.9159 + 02
0.6900 + 01	0.3779 + 04	0.3591 + 03	0.1052 + 02	0.9770 + 02
0.7000 + 01	0.4140 + 04	0.3833 + 03	0.1080 + 02	0.1041 + 03

Table A.1 | *Continued*

M	$\dfrac{p_o}{p}$	$\dfrac{\rho_o}{\rho}$	$\dfrac{T_o}{T}$	$\dfrac{A}{A^*}$
0.7100 + 01	0.4531 + 04	0.4088 + 03	0.1108 + 02	0.1109 + 03
0.7200 + 01	0.4953 + 04	0.4357 + 03	0.1137 + 02	0.1181 + 03
0.7300 + 01	0.5410 + 04	0.4640 + 03	0.1166 + 02	0.1256 + 03
0.7400 + 01	0.5903 + 04	0.4939 + 03	0.1195 + 02	0.1335 + 03
0.7500 + 01	0.6434 + 04	0.5252 + 03	0.1225 + 02	0.1418 + 03
0.7600 + 01	0.7006 + 04	0.5582 + 03	0.1255 + 02	0.1506 + 03
0.7700 + 01	0.7623 + 04	0.5928 + 03	0.1286 + 02	0.1598 + 03
0.7800 + 01	0.8285 + 04	0.6292 + 03	0.1317 + 02	0.1694 + 03
0.7900 + 01	0.8998 + 04	0.6674 + 03	0.1348 + 02	0.1795 + 03
0.8000 + 01	0.9763 + 04	0.7075 + 03	0.1380 + 02	0.1901 + 03
0.9000 + 01	0.2110 + 05	0.1227 + 04	0.1720 + 02	0.3272 + 03
0.1000 + 02	0.4244 + 05	0.2021 + 04	0.2100 + 02	0.5359 + 03
0.1100 + 02	0.8033 + 05	0.3188 + 04	0.2520 + 02	0.8419 + 03
0.1200 + 02	0.1445 + 06	0.4848 + 04	0.2980 + 02	0.1276 + 04
0.1300 + 02	0.2486 + 06	0.7144 + 04	0.3480 + 02	0.1876 + 04
0.1400 + 02	0.4119 + 06	0.1025 + 05	0.4020 + 02	0.2685 + 04
0.1500 + 02	0.6602 + 06	0.1435 + 05	0.4600 + 02	0.3755 + 04
0.1600 + 02	0.1028 + 07	0.1969 + 05	0.5220 + 02	0.5145 + 04
0.1700 + 02	0.1559 + 07	0.2651 + 05	0.5880 + 02	0.6921 + 04
0.1800 + 02	0.2311 + 07	0.3512 + 05	0.6580 + 02	0.9159 + 04
0.1900 + 02	0.3356 + 07	0.4584 + 05	0.7320 + 02	0.1195 + 05
0.2000 + 02	0.4783 + 07	0.5905 + 05	0.8100 + 02	0.1538 + 05
0.2200 + 02	0.9251 + 07	0.9459 + 05	0.9780 + 02	0.2461 + 05
0.2400 + 02	0.1691 + 08	0.1456 + 06	0.1162 + 03	0.3783 + 05
0.2600 + 02	0.2949 + 08	0.2165 + 06	0.1362 + 03	0.5624 + 05
0.2800 + 02	0.4936 + 08	0.3128 + 06	0.1578 + 03	0.8121 + 05
0.3000 + 02	0.7978 + 08	0.4408 + 06	0.1810 + 03	0.1144 + 06
0.3200 + 02	0.1250 + 09	0.6076 + 06	0.2058 + 03	0.1576 + 06
0.3400 + 02	0.1908 + 09	0.8216 + 06	0.2322 + 03	0.2131 + 06
0.3600 + 02	0.2842 + 09	0.1092 + 07	0.2602 + 03	0.2832 + 06
0.3800 + 02	0.4143 + 09	0.1430 + 07	0.2898 + 03	0.3707 + 06
0.4000 + 02	0.5926 + 09	0.1846 + 07	0.3210 + 03	0.4785 + 06
0.4200 + 02	0.8330 + 09	0.2354 + 07	0.3538 + 03	0.6102 + 06
0.4400 + 02	0.1153 + 10	0.2969 + 07	0.3882 + 03	0.7694 + 06
0.4600 + 02	0.1572 + 10	0.3706 + 07	0.4242 + 03	0.9603 + 06
0.4800 + 02	0.2116 + 10	0.4583 + 07	0.4618 + 03	0.1187 + 07
0.5000 + 02	0.2815 + 10	0.5618 + 07	0.5010 + 03	0.1455 + 07

Table A.2 | Normal shock properties

M	$\dfrac{p_2}{p_1}$	$\dfrac{\rho_2}{\rho_1}$	$\dfrac{T_2}{T_1}$	$\dfrac{p_{o_2}}{p_{o_1}}$	$\dfrac{p_{o_2}}{p_1}$	M_2
0.1000 + 01	0.1000 + 01	0.1000 + 01	0.1000 + 01	0.1000 + 01	0.1893 + 01	0.1000 + 01
0.1020 + 01	0.1047 + 01	0.1033 + 01	0.1013 + 01	0.1000 + 01	0.1938 + 01	0.9805 + 00
0.1040 + 01	0.1095 + 01	0.1067 + 01	0.1026 + 01	0.9999 + 00	0.1984 + 01	0.9620 + 00
0.1060 + 01	0.1144 + 01	0.1101 + 01	0.1039 + 01	0.9998 + 00	0.2032 + 01	0.9444 + 00
0.1080 + 01	0.1194 + 01	0.1135 + 01	0.1052 + 01	0.9994 + 00	0.2082 + 01	0.9277 + 00
0.1100 + 01	0.1245 + 01	0.1169 + 01	0.1065 + 01	0.9989 + 00	0.2133 + 01	0.9118 + 00
0.1120 + 01	0.1297 + 01	0.1203 + 01	0.1078 + 01	0.9982 + 00	0.2185 + 01	0.8966 + 00
0.1140 + 01	0.1350 + 01	0.1238 + 01	0.1090 + 01	0.9973 + 00	0.2239 + 01	0.8820 + 00
0.1160 + 01	0.1403 + 01	0.1272 + 01	0.1103 + 01	0.9961 + 00	0.2294 + 01	0.8682 + 00
0.1180 + 01	0.1458 + 01	0.1307 + 01	0.1115 + 01	0.9946 + 00	0.2350 + 01	0.8549 + 00
0.1200 + 01	0.1513 + 01	0.1342 + 01	0.1128 + 01	0.9928 + 00	0.2408 + 01	0.8422 + 00
0.1220 + 01	0.1570 + 01	0.1376 + 01	0.1141 + 01	0.9907 + 00	0.2466 + 01	0.8300 + 00
0.1240 + 01	0.1627 + 01	0.1411 + 01	0.1153 + 01	0.9884 + 00	0.2526 + 01	0.8183 + 00
0.1260 + 01	0.1686 + 01	0.1446 + 01	0.1166 + 01	0.9857 + 00	0.2588 + 01	0.8071 + 00
0.1280 + 01	0.1745 + 01	0.1481 + 01	0.1178 + 01	0.9827 + 00	0.2650 + 01	0.7963 + 00
0.1300 + 01	0.1805 + 01	0.1516 + 01	0.1191 + 01	0.9794 + 00	0.2714 + 01	0.7860 + 00
0.1320 + 01	0.1866 + 01	0.1551 + 01	0.1204 + 01	0.9758 + 00	0.2778 + 01	0.7760 + 00
0.1340 + 01	0.1928 + 01	0.1585 + 01	0.1216 + 01	0.9718 + 00	0.2844 + 01	0.7664 + 00
0.1360 + 01	0.1991 + 01	0.1620 + 01	0.1229 + 01	0.9676 + 00	0.2912 + 01	0.7572 + 00
0.1380 + 01	0.2055 + 01	0.1655 + 01	0.1242 + 01	0.9630 + 00	0.2980 + 01	0.7483 + 00
0.1400 + 01	0.2120 + 01	0.1690 + 01	0.1255 + 01	0.9582 + 00	0.3049 + 01	0.7397 + 00
0.1420 + 01	0.2186 + 01	0.1724 + 01	0.1268 + 01	0.9531 + 00	0.3120 + 01	0.7314 + 00
0.1440 + 01	0.2253 + 01	0.1759 + 01	0.1281 + 01	0.9476 + 00	0.3191 + 01	0.7235 + 00
0.1460 + 01	0.2320 + 01	0.1793 + 01	0.1294 + 01	0.9420 + 00	0.3264 + 01	0.7157 + 00
0.1480 + 01	0.2389 + 01	0.1828 + 01	0.1307 + 01	0.9360 + 00	0.3338 + 01	0.7083 + 00
0.1500 + 01	0.2458 + 01	0.1862 + 01	0.1320 + 01	0.9298 + 00	0.3413 + 01	0.7011 + 00
0.1520 + 01	0.2529 + 01	0.1896 + 01	0.1334 + 01	0.9233 + 00	0.3489 + 01	0.6941 + 00
0.1540 + 01	0.2600 + 01	0.1930 + 01	0.1347 + 01	0.9166 + 00	0.3567 + 01	0.6874 + 00
0.1560 + 01	0.2673 + 01	0.1964 + 01	0.1361 + 01	0.9097 + 00	0.3645 + 01	0.6809 + 00
0.1580 + 01	0.2746 + 01	0.1998 + 01	0.1374 + 01	0.9026 + 00	0.3724 + 01	0.6746 + 00
0.1600 + 01	0.2820 + 01	0.2032 + 01	0.1388 + 01	0.8952 + 00	0.3805 + 01	0.6684 + 00
0.1620 + 01	0.2895 + 01	0.2065 + 01	0.1402 + 01	0.8877 + 00	0.3887 + 01	0.6625 + 00
0.1640 + 01	0.2971 + 01	0.2099 + 01	0.1416 + 01	0.8799 + 00	0.3969 + 01	0.6568 + 00
0.1660 + 01	0.3048 + 01	0.2132 + 01	0.1430 + 01	0.8720 + 00	0.4053 + 01	0.6512 + 00
0.1680 + 01	0.3126 + 01	0.2165 + 01	0.1444 + 01	0.8639 + 00	0.4138 + 01	0.6458 + 00
0.1700 + 01	0.3205 + 01	0.2198 + 01	0.1458 + 01	0.8557 + 00	0.4224 + 01	0.6405 + 00
0.1720 + 01	0.3285 + 01	0.2230 + 01	0.1473 + 01	0.8474 + 00	0.4311 + 01	0.6355 + 00
0.1740 + 01	0.3366 + 01	0.2263 + 01	0.1487 + 01	0.8389 + 00	0.4399 + 01	0.6305 + 00
0.1760 + 01	0.3447 + 01	0.2295 + 01	0.1502 + 01	0.8302 + 00	0.4488 + 01	0.6257 + 00
0.1780 + 01	0.3530 + 01	0.2327 + 01	0.1517 + 01	0.8215 + 00	0.4578 + 01	0.6210 + 00
0.1800 + 01	0.3613 + 01	0.2359 + 01	0.1532 + 01	0.8127 + 00	0.4670 + 01	0.6165 + 00
0.1820 + 01	0.3698 + 01	0.2391 + 01	0.1547 + 01	0.8038 + 00	0.4762 + 01	0.6121 + 00
0.1840 + 01	0.3783 + 01	0.2422 + 01	0.1562 + 01	0.7948 + 00	0.4855 + 01	0.6078 + 00
0.1860 + 01	0.3870 + 01	0.2454 + 01	0.1577 + 01	0.7857 + 00	0.4950 + 01	0.6036 + 00
0.1880 + 01	0.3957 + 01	0.2485 + 01	0.1592 + 01	0.7765 + 00	0.5045 + 01	0.5996 + 00
0.1900 + 01	0.4045 + 01	0.2516 + 01	0.1608 + 01	0.7674 + 00	0.5142 + 01	0.5956 + 00
0.1920 + 01	0.4134 + 01	0.2546 + 01	0.1624 + 01	0.7581 + 00	0.5239 + 01	0.5918 + 00
0.1940 + 01	0.4224 + 01	0.2577 + 01	0.1639 + 01	0.7488 + 00	0.5338 + 01	0.5880 + 00
0.1960 + 01	0.4315 + 01	0.2607 + 01	0.1655 + 01	0.7395 + 00	0.5438 + 01	0.5844 + 00
0.1980 + 01	0.4407 + 01	0.2637 + 01	0.1671 + 01	0.7302 + 00	0.5539 + 01	0.5808 + 00

Table A.2 | *Continued*

M	$\dfrac{p_2}{p_1}$	$\dfrac{\rho_2}{\rho_1}$	$\dfrac{T_2}{T_1}$	$\dfrac{p_{o_2}}{p_{o_1}}$	$\dfrac{p_{o_2}}{p_1}$	M_2
0.2000 + 01	0.4500 + 01	0.2667 + 01	0.1687 + 01	0.7209 + 00	0.5640 + 01	0.5774 + 00
0.2050 + 01	0.4736 + 01	0.2740 + 01	0.1729 + 01	0.6975 + 00	0.5900 + 01	0.5691 + 00
0.2100 + 01	0.4978 + 01	0.2812 + 01	0.1770 + 01	0.6742 + 00	0.6165 + 01	0.5613 + 00
0.2150 + 01	0.5226 + 01	0.2882 + 01	0.1813 + 01	0.6511 + 00	0.6438 + 01	0.5540 + 00
0.2200 + 01	0.5480 + 01	0.2951 + 01	0.1857 + 01	0.6281 + 00	0.6716 + 01	0.5471 + 00
0.2250 + 01	0.5740 + 01	0.3019 + 01	0.1901 + 01	0.6055 + 00	0.7002 + 01	0.5406 + 00
0.2300 + 01	0.6005 + 01	0.3085 + 01	0.1947 + 01	0.5833 + 00	0.7294 + 01	0.5344 + 00
0.2350 + 01	0.6276 + 01	0.3149 + 01	0.1993 + 01	0.5615 + 00	0.7592 + 01	0.5286 + 00
0.2400 + 01	0.6553 + 01	0.3212 + 01	0.2040 + 01	0.5401 + 00	0.7897 + 01	0.5231 + 00
0.2450 + 01	0.6836 + 01	0.3273 + 01	0.2088 + 01	0.5193 + 00	0.8208 + 01	0.5179 + 00
0.2500 + 01	0.7125 + 01	0.3333 + 01	0.2137 + 01	0.4990 + 00	0.8526 + 01	0.5130 + 00
0.2550 + 01	0.7420 + 01	0.3392 + 01	0.2187 + 01	0.4793 + 00	0.8850 + 01	0.5083 + 00
0.2600 + 01	0.7720 + 01	0.3449 + 01	0.2238 + 01	0.4601 + 00	0.9181 + 01	0.5039 + 00
0.2650 + 01	0.8026 + 01	0.3505 + 01	0.2290 + 01	0.4416 + 00	0.9519 + 01	0.4996 + 00
0.2700 + 01	0.8338 + 01	0.3559 + 01	0.2343 + 01	0.4236 + 00	0.9862 + 01	0.4956 + 00
0.2750 + 01	0.8656 + 01	0.3612 + 01	0.2397 + 01	0.4062 + 00	0.1021 + 02	0.4918 + 00
0.2800 + 01	0.8980 + 01	0.3664 + 01	0.2451 + 01	0.3895 + 00	0.1057 + 02	0.4882 + 00
0.2850 + 01	0.9310 + 01	0.3714 + 01	0.2507 + 01	0.3733 + 00	0.1093 + 02	0.4847 + 00
0.2900 + 01	0.9645 + 01	0.3763 + 01	0.2563 + 01	0.3577 + 00	0.1130 + 02	0.4814 + 00
0.2950 + 01	0.9986 + 01	0.3811 + 01	0.2621 + 01	0.3428 + 00	0.1168 + 02	0.4782 + 00
0.3000 + 01	0.1033 + 02	0.3857 + 01	0.2679 + 01	0.3283 + 00	0.1206 + 02	0.4752 + 00
0.3050 + 01	0.1069 + 02	0.3902 + 01	0.2738 + 01	0.3145 + 00	0.1245 + 02	0.4723 + 00
0.3100 + 01	0.1104 + 02	0.3947 + 01	0.2799 + 01	0.3012 + 00	0.1285 + 02	0.4695 + 00
0.3150 + 01	0.1141 + 02	0.3990 + 01	0.2860 + 01	0.2885 + 00	0.1325 + 02	0.4669 + 00
0.3200 + 01	0.1178 + 02	0.4031 + 01	0.2922 + 01	0.2762 + 00	0.1366 + 02	0.4643 + 00
0.3250 + 01	0.1216 + 02	0.4072 + 01	0.2985 + 01	0.2645 + 00	0.1407 + 02	0.4619 + 00
0.3300 + 01	0.1254 + 02	0.4112 + 01	0.3049 + 01	0.2533 + 00	0.1449 + 02	0.4596 + 00
0.3350 + 01	0.1293 + 02	0.4151 + 01	0.3114 + 01	0.2425 + 00	0.1492 + 02	0.4573 + 00
0.3400 + 01	0.1332 + 02	0.4188 + 01	0.3180 + 01	0.2322 + 00	0.1535 + 02	0.4552 + 00
0.3450 + 01	0.1372 + 02	0.4225 + 01	0.3247 + 01	0.2224 + 00	0.1579 + 02	0.4531 + 00
0.3500 + 01	0.1412 + 02	0.4261 + 01	0.3315 + 01	0.2129 + 00	0.1624 + 02	0.4512 + 00
0.3550 + 01	0.1454 + 02	0.4296 + 01	0.3384 + 01	0.2039 + 00	0.1670 + 02	0.4492 + 00
0.3600 + 01	0.1495 + 02	0.4330 + 01	0.3454 + 01	0.1953 + 00	0.1716 + 02	0.4474 + 00
0.3650 + 01	0.1538 + 02	0.4363 + 01	0.3525 + 01	0.1871 + 00	0.1762 + 02	0.4456 + 00
0.3700 + 01	0.1580 + 02	0.4395 + 01	0.3596 + 01	0.1792 + 00	0.1810 + 02	0.4439 + 00
0.3750 + 01	0.1624 + 02	0.4426 + 01	0.3669 + 01	0.1717 + 00	0.1857 + 02	0.4423 + 00
0.3800 + 01	0.1668 + 02	0.4457 + 01	0.3743 + 01	0.1645 + 00	0.1906 + 02	0.4407 + 00
0.3850 + 01	0.1713 + 02	0.4487 + 01	0.3817 + 01	0.1576 + 00	0.1955 + 02	0.4392 + 00
0.3900 + 01	0.1758 + 02	0.4516 + 01	0.3893 + 01	0.1510 + 00	0.2005 + 02	0.4377 + 00
0.3950 + 01	0.1804 + 02	0.4544 + 01	0.3969 + 01	0.1448 + 00	0.2056 + 02	0.4363 + 00
0.4000 + 01	0.1850 + 02	0.4571 + 01	0.4047 + 01	0.1388 + 00	0.2107 + 02	0.4350 + 00
0.4050 + 01	0.1897 + 02	0.4598 + 01	0.4125 + 01	0.1330 + 00	0.2159 + 02	0.4336 + 00
0.4100 + 01	0.1944 + 02	0.4624 + 01	0.4205 + 01	0.1276 + 00	0.2211 + 02	0.4324 + 00
0.4150 + 01	0.1993 + 02	0.4650 + 01	0.4285 + 01	0.1223 + 00	0.2264 + 02	0.4311 + 00
0.4200 + 01	0.2041 + 02	0.4675 + 01	0.4367 + 01	0.1173 + 00	0.2318 + 02	0.4299 + 00
0.4250 + 01	0.2091 + 02	0.4699 + 01	0.4449 + 01	0.1126 + 00	0.2372 + 02	0.4288 + 00
0.4300 + 01	0.2140 + 02	0.4723 + 01	0.4532 + 01	0.1080 + 00	0.2427 + 02	0.4277 + 00
0.4350 + 01	0.2191 + 02	0.4746 + 01	0.4616 + 01	0.1036 + 00	0.2483 + 02	0.4266 + 00
0.4400 + 01	0.2242 + 02	0.4768 + 01	0.4702 + 01	0.9948 − 01	0.2539 + 02	0.4255 + 00
0.4450 + 01	0.2294 + 02	0.4790 + 01	0.4788 + 01	0.9550 − 01	0.2596 + 02	0.4245 + 00

Continued

Table A.2 | *Continued*

M	$\dfrac{p_2}{p_1}$	$\dfrac{\rho_2}{\rho_1}$	$\dfrac{T_2}{T_1}$	$\dfrac{p_{o_2}}{p_{o_1}}$	$\dfrac{p_{o_2}}{p_1}$	M_2
0.4500 + 01	0.2346 + 02	0.4812 + 01	0.4875 + 01	0.9170 − 01	0.2654 + 02	0.4236 + 00
0.4550 + 01	0.2399 + 02	0.4833 + 01	0.4963 + 01	0.8806 − 01	0.2712 + 02	0.4226 + 00
0.4600 + 01	0.2452 + 02	0.4853 + 01	0.5052 + 01	0.8459 − 01	0.2771 + 02	0.4217 + 00
0.4650 + 01	0.2506 + 02	0.4873 + 01	0.5142 + 01	0.8126 − 01	0.2831 + 02	0.4208 + 00
0.4700 + 01	0.2560 + 02	0.4893 + 01	0.5233 + 01	0.7809 − 01	0.2891 + 02	0.4199 + 00
0.4750 + 01	0.2616 + 02	0.4912 + 01	0.5325 + 01	0.7505 − 01	0.2952 + 02	0.4191 + 00
0.4800 + 01	0.2671 + 02	0.4930 + 01	0.5418 + 01	0.7214 − 01	0.3013 + 02	0.4183 + 00
0.4850 + 01	0.2728 + 02	0.4948 + 01	0.5512 + 01	0.6936 − 01	0.3075 + 02	0.4175 + 00
0.4900 + 01	0.2784 + 02	0.4966 + 01	0.5607 + 01	0.6670 − 01	0.3138 + 02	0.4167 + 00
0.4950 + 01	0.2842 + 02	0.4983 + 01	0.5703 + 01	0.6415 − 01	0.3201 + 02	0.4160 + 00
0.5000 + 01	0.2900 + 02	0.5000 + 01	0.5800 + 01	0.6172 − 01	0.3265 + 02	0.4152 + 00
0.5100 + 01	0.3018 + 02	0.5033 + 01	0.5997 + 01	0.5715 − 01	0.3395 + 02	0.4138 + 00
0.5200 + 01	0.3138 + 02	0.5064 + 01	0.6197 + 01	0.5297 − 01	0.3528 + 02	0.4125 + 00
0.5300 + 01	0.3260 + 02	0.5093 + 01	0.6401 + 01	0.4913 − 01	0.3663 + 02	0.4113 + 00
0.5400 + 01	0.3385 + 02	0.5122 + 01	0.6610 + 01	0.4560 − 01	0.3801 + 02	0.4101 + 00
0.5500 + 01	0.3512 + 02	0.5149 + 01	0.6822 + 01	0.4236 − 01	0.3941 + 02	0.4090 + 00
0.5600 + 01	0.3642 + 02	0.5175 + 01	0.7038 + 01	0.3938 − 01	0.4084 + 02	0.4079 + 00
0.5700 + 01	0.3774 + 02	0.5200 + 01	0.7258 + 01	0.3664 − 01	0.4230 + 02	0.4069 + 00
0.5800 + 01	0.3908 + 02	0.5224 + 01	0.7481 + 01	0.3412 − 01	0.4378 + 02	0.4059 + 00
0.5900 + 01	0.4044 + 02	0.5246 + 01	0.7709 + 01	0.3180 − 01	0.4528 + 02	0.4050 + 00
0.6000 + 01	0.4183 + 02	0.5268 + 01	0.7941 + 01	0.2965 − 01	0.4682 + 02	0.4042 + 00
0.6100 + 01	0.4324 + 02	0.5289 + 01	0.8176 + 01	0.2767 − 01	0.4837 + 02	0.4033 + 00
0.6200 + 01	0.4468 + 02	0.5309 + 01	0.8415 + 01	0.2584 − 01	0.4996 + 02	0.4025 + 00
0.6300 + 01	0.4614 + 02	0.5329 + 01	0.8658 + 01	0.2416 − 01	0.5157 + 02	0.4018 + 00
0.6400 + 01	0.4762 + 02	0.5347 + 01	0.8905 + 01	0.2259 − 01	0.5320 + 02	0.4011 + 00
0.6500 + 01	0.4912 + 02	0.5365 + 01	0.9156 + 01	0.2115 − 01	0.5486 + 02	0.4004 + 00
0.6600 + 01	0.5065 + 02	0.5382 + 01	0.9411 + 01	0.1981 − 01	0.5655 + 02	0.3997 + 00
0.6700 + 01	0.5220 + 02	0.5399 + 01	0.9670 + 01	0.1857 − 01	0.5826 + 02	0.3991 + 00
0.6800 + 01	0.5378 + 02	0.5415 + 01	0.9933 + 01	0.1741 − 01	0.6000 + 02	0.3985 + 00
0.6900 + 01	0.5538 + 02	0.5430 + 01	0.1020 + 02	0.1635 − 01	0.6176 + 02	0.3979 + 00
0.7000 + 01	0.5700 + 02	0.5444 + 01	0.1047 + 02	0.1535 − 01	0.6355 + 02	0.3974 + 00
0.7100 + 01	0.5864 + 02	0.5459 + 01	0.1074 + 02	0.1443 − 01	0.6537 + 02	0.3968 + 00
0.7200 + 01	0.6031 + 02	0.5472 + 01	0.1102 + 02	0.1357 − 01	0.6721 + 02	0.3963 + 00
0.7300 + 01	0.6200 + 02	0.5485 + 01	0.1130 + 02	0.1277 − 01	0.6908 + 02	0.3958 + 00
0.7400 + 01	0.6372 + 02	0.5498 + 01	0.1159 + 02	0.1202 − 01	0.7097 + 02	0.3954 + 00
0.7500 + 01	0.6546 + 02	0.5510 + 01	0.1188 + 02	0.1133 − 01	0.7289 + 02	0.3949 + 00
0.7600 + 01	0.6722 + 02	0.5522 + 01	0.1217 + 02	0.1068 − 01	0.7483 + 02	0.3945 + 00
0.7700 + 01	0.6900 + 02	0.5533 + 01	0.1247 + 02	0.1008 − 01	0.7680 + 02	0.3941 + 00
0.7800 + 01	0.7081 + 02	0.5544 + 01	0.1277 + 02	0.9510 − 02	0.7880 + 02	0.3937 + 00
0.7900 + 01	0.7264 + 02	0.5555 + 01	0.1308 + 02	0.8982 − 02	0.8082 + 02	0.3933 + 00
0.8000 + 01	0.7450 + 02	0.5565 + 01	0.1339 + 02	0.8488 − 02	0.8287 + 02	0.3929 + 00
0.9000 + 01	0.9433 + 02	0.5651 + 01	0.1669 + 02	0.4964 − 02	0.1048 + 03	0.3898 + 00
0.1000 + 02	0.1165 + 03	0.5714 + 01	0.2039 + 02	0.3045 − 02	0.1292 + 03	0.3876 + 00
0.1100 + 02	0.1410 + 03	0.5762 + 01	0.2447 + 02	0.1945 − 02	0.1563 + 03	0.3859 + 00
0.1200 + 02	0.1678 + 03	0.5799 + 01	0.2894 + 02	0.1287 − 02	0.1859 + 03	0.3847 + 00
0.1300 + 02	0.1970 + 03	0.5828 + 01	0.3380 + 02	0.8771 − 03	0.2181 + 03	0.3837 + 00
0.1400 + 02	0.2285 + 03	0.5851 + 01	0.3905 + 02	0.6138 − 03	0.2528 + 03	0.3829 + 00
0.1500 + 02	0.2623 + 03	0.5870 + 01	0.4469 + 02	0.4395 − 03	0.2902 + 03	0.3823 + 00
0.1600 + 02	0.2985 + 03	0.5885 + 01	0.5072 + 02	0.3212 − 03	0.3301 + 03	0.3817 + 00
0.1700 + 02	0.3370 + 03	0.5898 + 01	0.5714 + 02	0.2390 − 03	0.3726 + 03	0.3813 + 00

Table A.2 | *Continued*

M	$\dfrac{p_2}{p_1}$	$\dfrac{\rho_2}{\rho_1}$	$\dfrac{T_2}{T_1}$	$\dfrac{p_{o_2}}{p_{o_1}}$	$\dfrac{p_{o_2}}{p_1}$	M_2
0.1800 + 02	0.3778 + 03	0.5909 + 01	0.6394 + 02	0.1807 − 03	0.4176 + 03	0.3810 + 00
0.1900 + 02	0.4210 + 03	0.5918 + 01	0.7114 + 02	0.1386 − 03	0.4653 + 03	0.3806 + 00
0.2000 + 02	0.4665 + 03	0.5926 + 01	0.7872 + 02	0.1078 − 03	0.5155 + 03	0.3804 + 00
0.2200 + 02	0.5645 + 03	0.5939 + 01	0.9506 + 02	0.6741 − 04	0.6236 + 03	0.3800 + 00
0.2400 + 02	0.6718 + 03	0.5948 + 01	0.1129 + 03	0.4388 − 04	0.7421 + 03	0.3796 + 00
0.2600 + 02	0.7885 + 03	0.5956 + 01	0.1324 + 03	0.2953 − 04	0.8709 + 03	0.3794 + 00
0.2800 + 02	0.9145 + 03	0.5962 + 01	0.1534 + 03	0.2046 − 04	0.1010 + 04	0.3792 + 00
0.3000 + 02	0.1050 + 04	0.5967 + 01	0.1759 + 03	0.1453 − 04	0.1159 + 04	0.3790 + 00
0.3200 + 02	0.1194 + 04	0.5971 + 01	0.2001 + 03	0.1055 − 04	0.1319 + 04	0.3789 + 00
0.3400 + 02	0.1348 + 04	0.5974 + 01	0.2257 + 03	0.7804 − 05	0.1489 + 04	0.3788 + 00
0.3600 + 02	0.1512 + 04	0.5977 + 01	0.2529 + 03	0.5874 − 05	0.1669 + 04	0.3787 + 00
0.3800 + 02	0.1684 + 04	0.5979 + 01	0.2817 + 03	0.4488 − 05	0.1860 + 04	0.3786 + 00
0.4000 + 02	0.1866 + 04	0.5981 + 01	0.3121 + 03	0.3477 − 05	0.2061 + 04	0.3786 + 00
0.4200 + 02	0.2058 + 04	0.5983 + 01	0.3439 + 03	0.2727 − 05	0.2272 + 04	0.3785 + 00
0.4400 + 02	0.2258 + 04	0.5985 + 01	0.3774 + 03	0.2163 − 05	0.2493 + 04	0.3785 + 00
0.4600 + 02	0.2468 + 04	0.5986 + 01	0.4124 + 03	0.1733 − 05	0.2725 + 04	0.3784 + 00
0.4800 + 02	0.2688 + 04	0.5987 + 01	0.4489 + 03	0.1402 − 05	0.2967 + 04	0.3784 + 00
0.5000 + 02	0.2916 + 04	0.5988 + 01	0.4871 + 03	0.1144 − 05	0.3219 + 04	0.3784 + 00

Reyleigh

Table A.3 | One-dimensional flow with heat addition

M	$\dfrac{p}{p^*}$	$\dfrac{T}{T^*}$	$\dfrac{\rho}{\rho^*}$	$\dfrac{p_o}{p_o^*}$	$\dfrac{T_o}{T_o^*}$
0.2000 − 01	0.2399 + 01	0.2301 − 02	0.1042 + 04	0.1268 + 01	0.1918 − 02
0.4000 − 01	0.2395 + 01	0.9175 − 02	0.2610 + 03	0.1266 + 01	0.7648 − 02
0.6000 − 01	0.2388 + 01	0.2053 − 01	0.1163 + 03	0.1265 + 01	0.1712 − 01
0.8000 − 01	0.2379 + 01	0.3621 − 01	0.6569 + 02	0.1262 + 01	0.3022 − 01
0.1000 + 00	0.2367 + 01	0.5602 − 01	0.4225 + 02	0.1259 + 01	0.4678 − 01
0.1200 + 00	0.2353 + 01	0.7970 − 01	0.2952 + 02	0.1255 + 01	0.6661 − 01
0.1400 + 00	0.2336 + 01	0.1069 + 00	0.2184 + 02	0.1251 + 01	0.8947 − 01
0.1600 + 00	0.2317 + 01	0.1374 + 00	0.1686 + 02	0.1246 + 01	0.1151 + 00
0.1800 + 00	0.2296 + 01	0.1708 + 00	0.1344 + 02	0.1241 + 01	0.1432 + 00
0.2000 + 00	0.2273 + 01	0.2066 + 00	0.1100 + 02	0.1235 + 01	0.1736 + 00
0.2200 + 00	0.2248 + 01	0.2445 + 00	0.9192 + 01	0.1228 + 01	0.2057 + 00
0.2400 + 00	0.2221 + 01	0.2841 + 00	0.7817 + 01	0.1221 + 01	0.2395 + 00
0.2600 + 00	0.2193 + 01	0.3250 + 00	0.6747 + 01	0.1214 + 01	0.2745 + 00
0.2800 + 00	0.2163 + 01	0.3667 + 00	0.5898 + 01	0.1206 + 01	0.3104 + 00
0.3000 + 00	0.2131 + 01	0.4089 + 00	0.5213 + 01	0.1199 + 01	0.3469 + 00
0.3200 + 00	0.2099 + 01	0.4512 + 00	0.4652 + 01	0.1190 + 01	0.3837 + 00
0.3400 + 00	0.2066 + 01	0.4933 + 00	0.4188 + 01	0.1182 + 01	0.4206 + 00
0.3600 + 00	0.2031 + 01	0.5348 + 00	0.3798 + 01	0.1174 + 01	0.4572 + 00
0.3800 + 00	0.1996 + 01	0.5755 + 00	0.3469 + 01	0.1165 + 01	0.4935 + 00
0.4000 + 00	0.1961 + 01	0.6151 + 00	0.3188 + 01	0.1157 + 01	0.5290 + 00
0.4200 + 00	0.1925 + 01	0.6535 + 00	0.2945 + 01	0.1148 + 01	0.5638 + 00
0.4400 + 00	0.1888 + 01	0.6903 + 00	0.2736 + 01	0.1139 + 01	0.5975 + 00
0.4600 + 00	0.1852 + 01	0.7254 + 00	0.2552 + 01	0.1131 + 01	0.6301 + 00
0.4800 + 00	0.1815 + 01	0.7587 + 00	0.2392 + 01	0.1122 + 01	0.6614 + 00
0.5000 + 00	0.1778 + 01	0.7901 + 00	0.2250 + 01	0.1114 + 01	0.6914 + 00
0.5200 + 00	0.1741 + 01	0.8196 + 00	0.2124 + 01	0.1106 + 01	0.7199 + 00
0.5400 + 00	0.1704 + 01	0.8469 + 00	0.2012 + 01	0.1098 + 01	0.7470 + 00
0.5600 + 00	0.1668 + 01	0.8723 + 00	0.1912 + 01	0.1090 + 01	0.7725 + 00
0.5800 + 00	0.1632 + 01	0.8955 + 00	0.1822 + 01	0.1083 + 01	0.7965 + 00
0.6000 + 00	0.1596 + 01	0.9167 + 00	0.1741 + 01	0.1075 + 01	0.8189 + 00
0.6200 + 00	0.1560 + 01	0.9358 + 00	0.1667 + 01	0.1068 + 01	0.8398 + 00
0.6400 + 00	0.1525 + 01	0.9530 + 00	0.1601 + 01	0.1061 + 01	0.8592 + 00
0.6600 + 00	0.1491 + 01	0.9682 + 00	0.1540 + 01	0.1055 + 01	0.8771 + 00
0.6800 + 00	0.1457 + 01	0.9814 + 00	0.1484 + 01	0.1049 + 01	0.8935 + 00
0.7000 + 00	0.1423 + 01	0.9929 + 00	0.1434 + 01	0.1043 + 01	0.9085 + 00
0.7200 + 00	0.1391 + 01	0.1003 + 01	0.1387 + 01	0.1038 + 01	0.9221 + 00
0.7400 + 00	0.1359 + 01	0.1011 + 01	0.1344 + 01	0.1033 + 01	0.9344 + 00
0.7600 + 00	0.1327 + 01	0.1017 + 01	0.1305 + 01	0.1028 + 01	0.9455 + 00
0.7800 + 00	0.1296 + 01	0.1022 + 01	0.1268 + 01	0.1023 + 01	0.9553 + 00
0.8000 + 00	0.1266 + 01	0.1025 + 01	0.1234 + 01	0.1019 + 01	0.9639 + 00
0.8200 + 00	0.1236 + 01	0.1028 + 01	0.1203 + 01	0.1016 + 01	0.9715 + 00
0.8400 + 00	0.1207 + 01	0.1029 + 01	0.1174 + 01	0.1012 + 01	0.9781 + 00
0.8600 + 00	0.1179 + 01	0.1028 + 01	0.1147 + 01	0.1010 + 01	0.9836 + 00
0.8800 + 00	0.1152 + 01	0.1027 + 01	0.1121 + 01	0.1007 + 01	0.9883 + 00
0.9000 + 00	0.1125 + 01	0.1025 + 01	0.1098 + 01	0.1005 + 01	0.9921 + 00
0.9200 + 00	0.1098 + 01	0.1021 + 01	0.1076 + 01	0.1003 + 01	0.9951 + 00
0.9400 + 00	0.1073 + 01	0.1017 + 01	0.1055 + 01	0.1002 + 01	0.9973 + 00
0.9600 + 00	0.1048 + 01	0.1012 + 01	0.1035 + 01	0.1001 + 01	0.9988 + 00
0.9800 + 00	0.1024 + 01	0.1006 + 01	0.1017 + 01	0.1000 + 01	0.9997 + 00
0.1000 + 01	0.1000 + 01	0.1000 + 01	0.1000 + 01	0.1000 + 01	0.1000 + 01

$M = 1$

Table A.3 | *Continued*

M	$\dfrac{p}{p^*}$	$\dfrac{T}{T^*}$	$\dfrac{\rho}{\rho^*}$	$\dfrac{p_o}{p_o^*}$	$\dfrac{T_o}{T_o^*}$
0.1020 + 01	0.9770 + 00	0.9930 + 00	0.9838 + 00	0.1000 + 01	0.9997 + 00
0.1040 + 01	0.9546 + 00	0.9855 + 00	0.9686 + 00	0.1001 + 01	0.9989 + 00
0.1060 + 01	0.9327 + 00	0.9776 + 00	0.9542 + 00	0.1002 + 01	0.9977 + 00
0.1080 + 01	0.9115 + 00	0.9691 + 00	0.9406 + 00	0.1003 + 01	0.9960 + 00
0.1100 + 01	0.8909 + 00	0.9603 + 00	0.9277 + 00	0.1005 + 01	0.9939 + 00
0.1120 + 01	0.8708 + 00	0.9512 + 00	0.9155 + 00	0.1007 + 01	0.9915 + 00
0.1140 + 01	0.8512 + 00	0.9417 + 00	0.9039 + 00	0.1010 + 01	0.9887 + 00
0.1160 + 01	0.8322 + 00	0.9320 + 00	0.8930 + 00	0.1012 + 01	0.9856 + 00
0.1180 + 01	0.8137 + 00	0.9220 + 00	0.8826 + 00	0.1016 + 01	0.9823 + 00
0.1200 + 01	0.7958 + 00	0.9118 + 00	0.8727 + 00	0.1019 + 01	0.9787 + 00
0.1220 + 01	0.7783 + 00	0.9015 + 00	0.8633 + 00	0.1023 + 01	0.9749 + 00
0.1240 + 01	0.7613 + 00	0.8911 + 00	0.8543 + 00	0.1028 + 01	0.9709 + 00
0.1260 + 01	0.7447 + 00	0.8805 + 00	0.8458 + 00	0.1033 + 01	0.9668 + 00
0.1280 + 01	0.7287 + 00	0.8699 + 00	0.8376 + 00	0.1038 + 01	0.9624 + 00
0.1300 + 01	0.7130 + 00	0.8592 + 00	0.8299 + 00	0.1044 + 01	0.9580 + 00
0.1320 + 01	0.6978 + 00	0.8484 + 00	0.8225 + 00	0.1050 + 01	0.9534 + 00
0.1340 + 01	0.6830 + 00	0.8377 + 00	0.8154 + 00	0.1056 + 01	0.9487 + 00
0.1360 + 01	0.6686 + 00	0.8269 + 00	0.8086 + 00	0.1063 + 01	0.9440 + 00
0.1380 + 01	0.6546 + 00	0.8161 + 00	0.8021 + 00	0.1070 + 01	0.9391 + 00
0.1400 + 01	0.6410 + 00	0.8054 + 00	0.7959 + 00	0.1078 + 01	0.9343 + 00
0.1420 + 01	0.6278 + 00	0.7947 + 00	0.7900 + 00	0.1086 + 01	0.9293 + 00
0.1440 + 01	0.6149 + 00	0.7840 + 00	0.7843 + 00	0.1094 + 01	0.9243 + 00
0.1460 + 01	0.6024 + 00	0.7735 + 00	0.7788 + 00	0.1103 + 01	0.9193 + 00
0.1480 + 01	0.5902 + 00	0.7629 + 00	0.7736 + 00	0.1112 + 01	0.9143 + 00
0.1500 + 01	0.5783 + 00	0.7525 + 00	0.7685 + 00	0.1122 + 01	0.9093 + 00
0.1520 + 01	0.5668 + 00	0.7422 + 00	0.7637 + 00	0.1132 + 01	0.9042 + 00
0.1540 + 01	0.5555 + 00	0.7319 + 00	0.7590 + 00	0.1142 + 01	0.8992 + 00
0.1560 + 01	0.5446 + 00	0.7217 + 00	0.7545 + 00	0.1153 + 01	0.8942 + 00
0.1580 + 01	0.5339 + 00	0.7117 + 00	0.7502 + 00	0.1164 + 01	0.8892 + 00
0.1600 + 01	0.5236 + 00	0.7017 + 00	0.7461 + 00	0.1176 + 01	0.8842 + 00
0.1620 + 01	0.5135 + 00	0.6919 + 00	0.7421 + 00	0.1188 + 01	0.8792 + 00
0.1640 + 01	0.5036 + 00	0.6822 + 00	0.7383 + 00	0.1200 + 01	0.8743 + 00
0.1660 + 01	0.4940 + 00	0.6726 + 00	0.7345 + 00	0.1213 + 01	0.8694 + 00
0.1680 + 01	0.4847 + 00	0.6631 + 00	0.7310 + 00	0.1226 + 01	0.8645 + 00
0.1700 + 01	0.4756 + 00	0.6538 + 00	0.7275 + 00	0.1240 + 01	0.8597 + 00
0.1720 + 01	0.4668 + 00	0.6445 + 00	0.7242 + 00	0.1254 + 01	0.8549 + 00
0.1740 + 01	0.4581 + 00	0.6355 + 00	0.7210 + 00	0.1269 + 01	0.8502 + 00
0.1760 + 01	0.4497 + 00	0.6265 + 00	0.7178 + 00	0.1284 + 01	0.8455 + 00
0.1780 + 01	0.4415 + 00	0.6176 + 00	0.7148 + 00	0.1300 + 01	0.8409 + 00
0.1800 + 01	0.4335 + 00	0.6089 + 00	0.7119 + 00	0.1316 + 01	0.8363 + 00
0.1820 + 01	0.4257 + 00	0.6004 + 00	0.7091 + 00	0.1332 + 01	0.8317 + 00
0.1840 + 01	0.4181 + 00	0.5919 + 00	0.7064 + 00	0.1349 + 01	0.8273 + 00
0.1860 + 01	0.4107 + 00	0.5836 + 00	0.7038 + 00	0.1367 + 01	0.8228 + 00
0.1880 + 01	0.4035 + 00	0.5754 + 00	0.7012 + 00	0.1385 + 01	0.8185 + 00
0.1900 + 01	0.3964 + 00	0.5673 + 00	0.6988 + 00	0.1403 + 01	0.8141 + 00
0.1920 + 01	0.3895 + 00	0.5594 + 00	0.6964 + 00	0.1422 + 01	0.8099 + 00
0.1940 + 01	0.3828 + 00	0.5516 + 00	0.6940 + 00	0.1442 + 01	0.8057 + 00
0.1960 + 01	0.3763 + 00	0.5439 + 00	0.6918 + 00	0.1462 + 01	0.8015 + 00
0.1980 + 01	0.3699 + 00	0.5364 + 00	0.6896 + 00	0.1482 + 01	0.7974 + 00
0.2000 + 01	0.3636 + 00	0.5289 + 00	0.6875 + 00	0.1503 + 01	0.7934 + 00

$M = 2$

Continued

Table A.3 | *Continued*

M	$\dfrac{p}{p^*}$	$\dfrac{T}{T^*}$	$\dfrac{\rho}{\rho^*}$	$\dfrac{p_o}{p_o^*}$	$\dfrac{T_o}{T_o^*}$
0.2050 + 01	0.3487 + 00	0.5109 + 00	0.6825 + 00	0.1558 + 01	0.7835 + 00
0.2100 + 01	0.3345 + 00	0.4936 + 00	0.6778 + 00	0.1616 + 01	0.7741 + 00
0.2150 + 01	0.3212 + 00	0.4770 + 00	0.6735 + 00	0.1678 + 01	0.7649 + 00
0.2200 + 01	0.3086 + 00	0.4611 + 00	0.6694 + 00	0.1743 + 01	0.7561 + 00
0.2250 + 01	0.2968 + 00	0.4458 + 00	0.6656 + 00	0.1813 + 01	0.7477 + 00
0.2300 + 01	0.2855 + 00	0.4312 + 00	0.6621 + 00	0.1886 + 01	0.7395 + 00
0.2350 + 01	0.2749 + 00	0.4172 + 00	0.6588 + 00	0.1963 + 01	0.7317 + 00
0.2400 + 01	0.2648 + 00	0.4038 + 00	0.6557 + 00	0.2045 + 01	0.7242 + 00
0.2450 + 01	0.2552 + 00	0.3910 + 00	0.6527 + 00	0.2131 + 01	0.7170 + 00
0.2500 + 01	0.2462 + 00	0.3787 + 00	0.6500 + 00	0.2222 + 01	0.7101 + 00
0.2550 + 01	0.2375 + 00	0.3669 + 00	0.6474 + 00	0.2317 + 01	0.7034 + 00
0.2600 + 01	0.2294 + 00	0.3556 + 00	0.6450 + 00	0.2418 + 01	0.6970 + 00
0.2650 + 01	0.2216 + 00	0.3448 + 00	0.6427 + 00	0.2523 + 01	0.6908 + 00
0.2700 + 01	0.2142 + 00	0.3344 + 00	0.6405 + 00	0.2634 + 01	0.6849 + 00
0.2750 + 01	0.2071 + 00	0.3244 + 00	0.6384 + 00	0.2751 + 01	0.6793 + 00
0.2800 + 01	0.2004 + 00	0.3149 + 00	0.6365 + 00	0.2873 + 01	0.6738 + 00
0.2850 + 01	0.1940 + 00	0.3057 + 00	0.6346 + 00	0.3001 + 01	0.6685 + 00
0.2900 + 01	0.1879 + 00	0.2969 + 00	0.6329 + 00	0.3136 + 01	0.6635 + 00
0.2950 + 01	0.1820 + 00	0.2884 + 00	0.6312 + 00	0.3277 + 01	0.6586 + 00
0.3000 + 01	0.1765 + 00	0.2803 + 00	0.6296 + 00	0.3424 + 01	0.6540 + 00
0.3050 + 01	0.1711 + 00	0.2725 + 00	0.6281 + 00	0.3579 + 01	0.6495 + 00
0.3100 + 01	0.1660 + 00	0.2650 + 00	0.6267 + 00	0.3741 + 01	0.6452 + 00
0.3150 + 01	0.1612 + 00	0.2577 + 00	0.6253 + 00	0.3910 + 01	0.6410 + 00
0.3200 + 01	0.1565 + 00	0.2508 + 00	0.6240 + 00	0.4087 + 01	0.6370 + 00
0.3250 + 01	0.1520 + 00	0.2441 + 00	0.6228 + 00	0.4272 + 01	0.6331 + 00
0.3300 + 01	0.1477 + 00	0.2377 + 00	0.6216 + 00	0.4465 + 01	0.6294 + 00
0.3350 + 01	0.1436 + 00	0.2315 + 00	0.6205 + 00	0.4667 + 01	0.6258 + 00
0.3400 + 01	0.1397 + 00	0.2255 + 00	0.6194 + 00	0.4878 + 01	0.6224 + 00
0.3450 + 01	0.1359 + 00	0.2197 + 00	0.6183 + 00	0.5098 + 01	0.6190 + 00
0.3500 + 01	0.1322 + 00	0.2142 + 00	0.6173 + 00	0.5328 + 01	0.6158 + 00
0.3550 + 01	0.1287 + 00	0.2088 + 00	0.6164 + 00	0.5568 + 01	0.6127 + 00
0.3600 + 01	0.1254 + 00	0.2037 + 00	0.6155 + 00	0.5817 + 01	0.6097 + 00
0.3650 + 01	0.1221 + 00	0.1987 + 00	0.6146 + 00	0.6078 + 01	0.6068 + 00
0.3700 + 01	0.1190 + 00	0.1939 + 00	0.6138 + 00	0.6349 + 01	0.6040 + 00
0.3750 + 01	0.1160 + 00	0.1893 + 00	0.6130 + 00	0.6631 + 01	0.6013 + 00
0.3800 + 01	0.1131 + 00	0.1848 + 00	0.6122 + 00	0.6926 + 01	0.5987 + 00
0.3850 + 01	0.1103 + 00	0.1805 + 00	0.6114 + 00	0.7232 + 01	0.5962 + 00
0.3900 + 01	0.1077 + 00	0.1763 + 00	0.6107 + 00	0.7550 + 01	0.5937 + 00
0.3950 + 01	0.1051 + 00	0.1722 + 00	0.6100 + 00	0.7882 + 01	0.5914 + 00
0.4000 + 01	0.1026 + 00	0.1683 + 00	0.6094 + 00	0.8227 + 01	0.5891 + 00
0.4050 + 01	0.1002 + 00	0.1645 + 00	0.6087 + 00	0.8585 + 01	0.5869 + 00
0.4100 + 01	0.9782 − 01	0.1609 + 00	0.6081 + 00	0.8958 + 01	0.5847 + 00
0.4150 + 01	0.9557 − 01	0.1573 + 00	0.6075 + 00	0.9345 + 01	0.5827 + 00
0.4200 + 01	0.9340 − 01	0.1539 + 00	0.6070 + 00	0.9747 + 01	0.5807 + 00
0.4250 + 01	0.9130 − 01	0.1506 + 00	0.6064 + 00	0.1016 + 02	0.5787 + 00
0.4300 + 01	0.8927 − 01	0.1473 + 00	0.6059 + 00	0.1060 + 02	0.5768 + 00
0.4350 + 01	0.8730 − 01	0.1442 + 00	0.6054 + 00	0.1105 + 02	0.5750 + 00
0.4400 + 01	0.8540 − 01	0.1412 + 00	0.6049 + 00	0.1152 + 02	0.5732 + 00
0.4450 + 01	0.8356 − 01	0.1383 + 00	0.6044 + 00	0.1200 + 02	0.5715 + 00
0.4500 + 01	0.8177 − 01	0.1354 + 00	0.6039 + 00	0.1250 + 02	0.5698 + 00

Table A.3 | *Continued*

M	$\dfrac{p}{p^*}$	$\dfrac{T}{T^*}$	$\dfrac{\rho}{\rho^*}$	$\dfrac{p_o}{p_o^*}$	$\dfrac{T_o}{T_o^*}$
0.4550 + 01	0.8004 − 01	0.1326 + 00	0.6035 + 00	0.1302 + 02	0.5682 + 00
0.4600 + 01	0.7837 − 01	0.1300 + 00	0.6030 + 00	0.1356 + 02	0.5666 + 00
0.4650 + 01	0.7675 − 01	0.1274 + 00	0.6026 + 00	0.1412 + 02	0.5651 + 00
0.4700 + 01	0.7517 − 01	0.1248 + 00	0.6022 + 00	0.1470 + 02	0.5636 + 00
0.4750 + 01	0.7365 − 01	0.1224 + 00	0.6018 + 00	0.1530 + 02	0.5622 + 00
0.4800 + 01	0.7217 − 01	0.1200 + 00	0.6014 + 00	0.1592 + 02	0.5608 + 00
0.4850 + 01	0.7073 − 01	0.1177 + 00	0.6010 + 00	0.1657 + 02	0.5594 + 00
0.4900 + 01	0.6934 − 01	0.1154 + 00	0.6007 + 00	0.1723 + 02	0.5581 + 00
0.4950 + 01	0.6798 − 01	0.1132 + 00	0.6003 + 00	0.1792 + 02	0.5568 + 00
0.5000 + 01	0.6667 − 01	0.1111 + 00	0.6000 + 00	0.1863 + 02	0.5556 + 00
0.5100 + 01	0.6415 − 01	0.1070 + 00	0.5994 + 00	0.2013 + 02	0.5532 + 00
0.5200 + 01	0.6177 − 01	0.1032 + 00	0.5987 + 00	0.2173 + 02	0.5509 + 00
0.5300 + 01	0.5951 − 01	0.9950 − 01	0.5982 + 00	0.2344 + 02	0.5487 + 00
0.5400 + 01	0.5738 − 01	0.9602 − 01	0.5976 + 00	0.2527 + 02	0.5467 + 00
0.5500 + 01	0.5536 − 01	0.9272 − 01	0.5971 + 00	0.2721 + 02	0.5447 + 00
0.5600 + 01	0.5345 − 01	0.8958 − 01	0.5966 + 00	0.2928 + 02	0.5429 + 00
0.5700 + 01	0.5163 − 01	0.8660 − 01	0.5962 + 00	0.3148 + 02	0.5411 + 00
0.5800 + 01	0.4990 − 01	0.8376 − 01	0.5957 + 00	0.3382 + 02	0.5394 + 00
0.5900 + 01	0.4826 − 01	0.8106 − 01	0.5953 + 00	0.3631 + 02	0.5378 + 00
0.6000 + 01	0.4669 − 01	0.7849 − 01	0.5949 + 00	0.3895 + 02	0.5363 + 00
0.6100 + 01	0.4520 − 01	0.7603 − 01	0.5945 + 00	0.4174 + 02	0.5349 + 00
0.6200 + 01	0.4378 − 01	0.7369 − 01	0.5942 + 00	0.4471 + 02	0.5335 + 00
0.6300 + 01	0.4243 − 01	0.7145 − 01	0.5938 + 00	0.4785 + 02	0.5322 + 00
0.6400 + 01	0.4114 − 01	0.6931 − 01	0.5935 + 00	0.5117 + 02	0.5309 + 00
0.6500 + 01	0.3990 − 01	0.6726 − 01	0.5932 + 00	0.5468 + 02	0.5297 + 00
0.6600 + 01	0.3872 − 01	0.6531 − 01	0.5929 + 00	0.5840 + 02	0.5285 + 00
0.6700 + 01	0.3759 − 01	0.6343 − 01	0.5926 + 00	0.6232 + 02	0.5274 + 00
0.6800 + 01	0.3651 − 01	0.6164 − 01	0.5923 + 00	0.6645 + 02	0.5264 + 00
0.6900 + 01	0.3547 − 01	0.5991 − 01	0.5921 + 00	0.7082 + 02	0.5254 + 00
0.7000 + 01	0.3448 − 01	0.5826 − 01	0.5918 + 00	0.7541 + 02	0.5244 + 00
0.7100 + 01	0.3353 − 01	0.5668 − 01	0.5916 + 00	0.8026 + 02	0.5234 + 00
0.7200 + 01	0.3262 − 01	0.5516 − 01	0.5914 + 00	0.8536 + 02	0.5225 + 00
0.7300 + 01	0.3174 − 01	0.5370 − 01	0.5912 + 00	0.9072 + 02	0.5217 + 00
0.7400 + 01	0.3090 − 01	0.5229 − 01	0.5909 + 00	0.9636 + 02	0.5208 + 00
0.7500 + 01	0.3009 − 01	0.5094 − 01	0.5907 + 00	0.1023 + 03	0.5200 + 00
0.7600 + 01	0.2932 − 01	0.4964 − 01	0.5905 + 00	0.1085 + 03	0.5193 + 00
0.7700 + 01	0.2857 − 01	0.4839 − 01	0.5904 + 00	0.1150 + 03	0.5185 + 00
0.7800 + 01	0.2785 − 01	0.4719 − 01	0.5902 + 00	0.1219 + 03	0.5178 + 00
0.7900 + 01	0.2716 − 01	0.4603 − 01	0.5900 + 00	0.1291 + 03	0.5171 + 00
0.8000 + 01	0.2649 − 01	0.4491 − 01	0.5898 + 00	0.1366 + 03	0.5165 + 00
0.9000 + 01	0.2098 − 01	0.3565 − 01	0.5885 + 00	0.2339 + 03	0.5110 + 00
0.1000 + 02	0.1702 − 01	0.2897 − 01	0.5875 + 00	0.3816 + 03	0.5070 + 00
0.1100 + 02	0.1408 − 01	0.2400 − 01	0.5868 + 00	0.5977 + 03	0.5041 + 00
0.1200 + 02	0.1185 − 01	0.2021 − 01	0.5862 + 00	0.9041 + 03	0.5018 + 00
0.1300 + 02	0.1010 − 01	0.1724 − 01	0.5858 + 00	0.1327 + 04	0.5001 + 00
0.1400 + 02	0.8715 − 02	0.1489 − 01	0.5855 + 00	0.1896 + 04	0.4986 + 00
0.1500 + 02	0.7595 − 02	0.1298 − 01	0.5852 + 00	0.2649 + 04	0.4975 + 00
0.1600 + 02	0.6678 − 02	0.1142 − 01	0.5850 + 00	0.3625 + 04	0.4966 + 00
0.1700 + 02	0.5917 − 02	0.1012 − 01	0.5848 + 00	0.4873 + 04	0.4958 + 00
0.1800 + 02	0.5279 − 02	0.9030 − 02	0.5846 + 00	0.6445 + 04	0.4952 + 00

Continued

Table A.3 | *Continued*

M	$\dfrac{p}{p^*}$	$\dfrac{T}{T^*}$	$\dfrac{\rho}{\rho^*}$	$\dfrac{p_o}{p_o^*}$	$\dfrac{T_o}{T_o^*}$
0.1900 + 02	0.4739 − 02	0.8109 − 02	0.5845 + 00	0.8402 + 04	0.4946 + 00
0.2000 + 02	0.4278 − 02	0.7321 − 02	0.5844 + 00	0.1081 + 05	0.4942 + 00
0.2200 + 02	0.3537 − 02	0.6054 − 02	0.5842 + 00	0.1728 + 05	0.4934 + 00
0.2400 + 02	0.2973 − 02	0.5089 − 02	0.5841 + 00	0.2656 + 05	0.4928 + 00
0.2600 + 02	0.2533 − 02	0.4338 − 02	0.5839 + 00	0.3946 + 05	0.4924 + 00
0.2800 + 02	0.2185 − 02	0.3742 − 02	0.5839 + 00	0.5697 + 05	0.4920 + 00
0.3000 + 02	0.1903 − 02	0.3260 − 02	0.5838 + 00	0.8021 + 05	0.4917 + 00
0.3200 + 02	0.1673 − 02	0.2866 − 02	0.5837 + 00	0.1105 + 06	0.4915 + 00
0.3400 + 02	0.1482 − 02	0.2539 − 02	0.5837 + 00	0.1494 + 06	0.4913 + 00
0.3600 + 02	0.1322 − 02	0.2265 − 02	0.5837 + 00	0.1985 + 06	0.4911 + 00
0.3800 + 02	0.1187 − 02	0.2033 − 02	0.5836 + 00	0.2597 + 06	0.4910 + 00
0.4000 + 02	0.1071 − 02	0.1835 − 02	0.5836 + 00	0.3353 + 06	0.4909 + 00
0.4200 + 02	0.9714 − 03	0.1665 − 02	0.5836 + 00	0.4275 + 06	0.4908 + 00
0.4400 + 02	0.8852 − 03	0.1517 − 02	0.5835 + 00	0.5390 + 06	0.4907 + 00
0.4600 + 02	0.8099 − 03	0.1388 − 02	0.5835 + 00	0.6726 + 06	0.4906 + 00
0.4800 + 02	0.7438 − 03	0.1275 − 02	0.5835 + 00	0.8316 + 06	0.4906 + 00
0.5000 + 02	0.6855 − 03	0.1175 − 02	0.5835 + 00	0.1019 + 07	0.4905 + 00

Fanno Line

Table A.4 | One-dimensional flow with friction

M	$\dfrac{T}{T^*}$	$\dfrac{p}{p^*}$	$\dfrac{\rho}{\rho^*}$	$\dfrac{p_o}{p_o^*}$	$\dfrac{4fL^*}{D}$
0.2000 − 01	0.1200 + 01	0.5477 + 02	0.4565 + 02	0.2894 + 02	0.1778 + 04
0.4000 − 01	0.1200 + 01	0.2738 + 02	0.2283 + 02	0.1448 + 02	0.4404 + 03
0.6000 − 01	0.1199 + 01	0.1825 + 02	0.1522 + 02	0.9666 + 01	0.1930 + 03
0.8000 − 01	0.1198 + 01	0.1368 + 02	0.1142 + 02	0.7262 + 01	0.1067 + 03
0.1000 + 00	0.1198 + 01	0.1094 + 02	0.9138 + 01	0.5822 + 01	0.6692 + 02
0.1200 + 00	0.1197 + 01	0.9116 + 01	0.7618 + 01	0.4864 + 01	0.4541 + 02
0.1400 + 00	0.1195 + 01	0.7809 + 01	0.6533 + 01	0.4182 + 01	0.3251 + 02
0.1600 + 00	0.1194 + 01	0.6829 + 01	0.5720 + 01	0.3673 + 01	0.2420 + 02
0.1800 + 00	0.1192 + 01	0.6066 + 01	0.5088 + 01	0.3278 + 01	0.1854 + 02
0.2000 + 00	0.1190 + 01	0.5455 + 01	0.4583 + 01	0.2964 + 01	0.1453 + 02
0.2200 + 00	0.1188 + 01	0.4955 + 01	0.4169 + 01	0.2708 + 01	0.1160 + 02
0.2400 + 00	0.1186 + 01	0.4538 + 01	0.3825 + 01	0.2496 + 01	0.9386 + 01
0.2600 + 00	0.1184 + 01	0.4185 + 01	0.3535 + 01	0.2317 + 01	0.7688 + 01
0.2800 + 00	0.1181 + 01	0.3882 + 01	0.3286 + 01	0.2166 + 01	0.6357 + 01
0.3000 + 00	0.1179 + 01	0.3619 + 01	0.3070 + 01	0.2035 + 01	0.5299 + 01
0.3200 + 00	0.1176 + 01	0.3389 + 01	0.2882 + 01	0.1922 + 01	0.4447 + 01
0.3400 + 00	0.1173 + 01	0.3185 + 01	0.2716 + 01	0.1823 + 01	0.3752 + 01
0.3600 + 00	0.1170 + 01	0.3004 + 01	0.2568 + 01	0.1736 + 01	0.3180 + 01
0.3800 + 00	0.1166 + 01	0.2842 + 01	0.2437 + 01	0.1659 + 01	0.2705 + 01
0.4000 + 00	0.1163 + 01	0.2696 + 01	0.2318 + 01	0.1590 + 01	0.2308 + 01
0.4200 + 00	0.1159 + 01	0.2563 + 01	0.2212 + 01	0.1529 + 01	0.1974 + 01
0.4400 + 00	0.1155 + 01	0.2443 + 01	0.2114 + 01	0.1474 + 01	0.1692 + 01
0.4600 + 00	0.1151 + 01	0.2333 + 01	0.2026 + 01	0.1425 + 01	0.1451 + 01
0.4800 + 00	0.1147 + 01	0.2231 + 01	0.1945 + 01	0.1380 + 01	0.1245 + 01
0.5000 + 00	0.1143 + 01	0.2138 + 01	0.1871 + 01	0.1340 + 01	0.1069 + 01
0.5200 + 00	0.1138 + 01	0.2052 + 01	0.1802 + 01	0.1303 + 01	0.9174 + 00
0.5400 + 00	0.1134 + 01	0.1972 + 01	0.1739 + 01	0.1270 + 01	0.7866 + 00
0.5600 + 00	0.1129 + 01	0.1898 + 01	0.1680 + 01	0.1240 + 01	0.6736 + 00
0.5800 + 00	0.1124 + 01	0.1828 + 01	0.1626 + 01	0.1213 + 01	0.5757 + 00
0.6000 + 00	0.1119 + 01	0.1763 + 01	0.1575 + 01	0.1188 + 01	0.4908 + 00
0.6200 + 00	0.1114 + 01	0.1703 + 01	0.1528 + 01	0.1166 + 01	0.4172 + 00
0.6400 + 00	0.1109 + 01	0.1646 + 01	0.1484 + 01	0.1145 + 01	0.3533 + 00
0.6600 + 00	0.1104 + 01	0.1592 + 01	0.1442 + 01	0.1127 + 01	0.2979 + 00
0.6800 + 00	0.1098 + 01	0.1541 + 01	0.1403 + 01	0.1110 + 01	0.2498 + 00
0.7000 + 00	0.1093 + 01	0.1493 + 01	0.1367 + 01	0.1094 + 01	0.2081 + 00
0.7200 + 00	0.1087 + 01	0.1448 + 01	0.1332 + 01	0.1081 + 01	0.1721 + 00
0.7400 + 00	0.1082 + 01	0.1405 + 01	0.1299 + 01	0.1068 + 01	0.1411 + 00
0.7600 + 00	0.1076 + 01	0.1365 + 01	0.1269 + 01	0.1057 + 01	0.1145 + 00
0.7800 + 00	0.1070 + 01	0.1326 + 01	0.1240 + 01	0.1047 + 01	0.9167 − 01
0.8000 + 00	0.1064 + 01	0.1289 + 01	0.1212 + 01	0.1038 + 01	0.7229 − 01
0.8200 + 00	0.1058 + 01	0.1254 + 01	0.1186 + 01	0.1030 + 01	0.5593 − 01
0.8400 + 00	0.1052 + 01	0.1221 + 01	0.1161 + 01	0.1024 + 01	0.4226 − 01
0.8600 + 00	0.1045 + 01	0.1189 + 01	0.1137 + 01	0.1018 + 01	0.3097 − 01
0.8800 + 00	0.1039 + 01	0.1158 + 01	0.1115 + 01	0.1013 + 01	0.2179 − 01
0.9000 + 00	0.1033 + 01	0.1129 + 01	0.1093 + 01	0.1009 + 01	0.1451 − 01
0.9200 + 00	0.1026 + 01	0.1101 + 01	0.1073 + 01	0.1006 + 01	0.8913 − 02
0.9400 + 00	0.1020 + 01	0.1074 + 01	0.1053 + 01	0.1003 + 01	0.4815 − 02
0.9600 + 00	0.1013 + 01	0.1049 + 01	0.1035 + 01	0.1001 + 01	0.2057 − 02
0.9800 + 00	0.1007 + 01	0.1024 + 01	0.1017 + 01	0.1000 + 01	0.4947 − 03
0.1000 + 01	0.1000 + 01	0.1000 + 01	0.1000 + 01	0.1000 + 01	0.0000 + 00

Continued

Table A.4 | *Continued*

M	$\dfrac{T}{T^*}$	$\dfrac{p}{p^*}$	$\dfrac{\rho}{\rho^*}$	$\dfrac{p_o}{p_o^*}$	$\dfrac{4fL^*}{D}$
0.1020 + 01	0.9933 + 00	0.9771 + 00	0.9837 + 00	0.1000 + 01	0.4587 − 03
0.1040 + 01	0.9866 + 00	0.9551 + 00	0.9681 + 00	0.1001 + 01	0.1768 − 02
0.1060 + 01	0.9798 + 00	0.9338 + 00	0.9531 + 00	0.1003 + 01	0.3838 − 02
0.1080 + 01	0.9730 + 00	0.9133 + 00	0.9387 + 00	0.1005 + 01	0.6585 − 02
0.1100 + 01	0.9662 + 00	0.8936 + 00	0.9249 + 00	0.1008 + 01	0.9935 − 02
0.1120 + 01	0.9593 + 00	0.8745 + 00	0.9116 + 00	0.1011 + 01	0.1382 − 01
0.1140 + 01	0.9524 + 00	0.8561 + 00	0.8988 + 00	0.1015 + 01	0.1819 − 01
0.1160 + 01	0.9455 + 00	0.8383 + 00	0.8865 + 00	0.1020 + 01	0.2298 − 01
0.1180 + 01	0.9386 + 00	0.8210 + 00	0.8747 + 00	0.1025 + 01	0.2814 − 01
0.1200 + 01	0.9317 + 00	0.8044 + 00	0.8633 + 00	0.1030 + 01	0.3364 − 01
0.1220 + 01	0.9247 + 00	0.7882 + 00	0.8524 + 00	0.1037 + 01	0.3943 − 01
0.1240 + 01	0.9178 + 00	0.7726 + 00	0.8418 + 00	0.1043 + 01	0.4547 − 01
0.1260 + 01	0.9108 + 00	0.7574 + 00	0.8316 + 00	0.1050 + 01	0.5174 − 01
0.1280 + 01	0.9038 + 00	0.7427 + 00	0.8218 + 00	0.1058 + 01	0.5820 − 01
0.1300 + 01	0.8969 + 00	0.7285 + 00	0.8123 + 00	0.1066 + 01	0.6483 − 01
0.1320 + 01	0.8899 + 00	0.7147 + 00	0.8031 + 00	0.1075 + 01	0.7161 − 01
0.1340 + 01	0.8829 + 00	0.7012 + 00	0.7942 + 00	0.1084 + 01	0.7850 − 01
0.1360 + 01	0.8760 + 00	0.6882 + 00	0.7856 + 00	0.1094 + 01	0.8550 − 01
0.1380 + 01	0.8690 + 00	0.6755 + 00	0.7773 + 00	0.1104 + 01	0.9259 − 01
0.1400 + 01	0.8621 + 00	0.6632 + 00	0.7693 + 00	0.1115 + 01	0.9974 − 01
0.1420 + 01	0.8551 + 00	0.6512 + 00	0.7615 + 00	0.1126 + 01	0.1069 + 00
0.1440 + 01	0.8482 + 00	0.6396 + 00	0.7540 + 00	0.1138 + 01	0.1142 + 00
0.1460 + 01	0.8413 + 00	0.6282 + 00	0.7467 + 00	0.1150 + 01	0.1215 + 00
0.1480 + 01	0.8344 + 00	0.6172 + 00	0.7397 + 00	0.1163 + 01	0.1288 + 00
0.1500 + 01	0.8276 + 00	0.6065 + 00	0.7328 + 00	0.1176 + 01	0.1361 + 00
0.1520 + 01	0.8207 + 00	0.5960 + 00	0.7262 + 00	0.1190 + 01	0.1433 + 00
0.1540 + 01	0.8139 + 00	0.5858 + 00	0.7198 + 00	0.1204 + 01	0.1506 + 00
0.1560 + 01	0.8071 + 00	0.5759 + 00	0.7135 + 00	0.1219 + 01	0.1579 + 00
0.1580 + 01	0.8004 + 00	0.5662 + 00	0.7074 + 00	0.1234 + 01	0.1651 + 00
0.1600 + 01	0.7937 + 00	0.5568 + 00	0.7016 + 00	0.1250 + 01	0.1724 + 00
0.1620 + 01	0.7869 + 00	0.5476 + 00	0.6958 + 00	0.1267 + 01	0.1795 + 00
0.1640 + 01	0.7803 + 00	0.5386 + 00	0.6903 + 00	0.1284 + 01	0.1867 + 00
0.1660 + 01	0.7736 + 00	0.5299 + 00	0.6849 + 00	0.1301 + 01	0.1938 + 00
0.1680 + 01	0.7670 + 00	0.5213 + 00	0.6796 + 00	0.1319 + 01	0.2008 + 00
0.1700 + 01	0.7605 + 00	0.5130 + 00	0.6745 + 00	0.1338 + 01	0.2078 + 00
0.1720 + 01	0.7539 + 00	0.5048 + 00	0.6696 + 00	0.1357 + 01	0.2147 + 00
0.1740 + 01	0.7474 + 00	0.4969 + 00	0.6648 + 00	0.1376 + 01	0.2216 + 00
0.1760 + 01	0.7410 + 00	0.4891 + 00	0.6601 + 00	0.1397 + 01	0.2284 + 00
0.1780 + 01	0.7345 + 00	0.4815 + 00	0.6555 + 00	0.1418 + 01	0.2352 + 00
0.1800 + 01	0.7282 + 00	0.4741 + 00	0.6511 + 00	0.1439 + 01	0.2419 + 00
0.1820 + 01	0.7218 + 00	0.4668 + 00	0.6467 + 00	0.1461 + 01	0.2485 + 00
0.1840 + 01	0.7155 + 00	0.4597 + 00	0.6425 + 00	0.1484 + 01	0.2551 + 00
0.1860 + 01	0.7093 + 00	0.4528 + 00	0.6384 + 00	0.1507 + 01	0.2616 + 00
0.1880 + 01	0.7030 + 00	0.4460 + 00	0.6344 + 00	0.1531 + 01	0.2680 + 00
0.1900 + 01	0.6969 + 00	0.4394 + 00	0.6305 + 00	0.1555 + 01	0.2743 + 00
0.1920 + 01	0.6907 + 00	0.4329 + 00	0.6267 + 00	0.1580 + 01	0.2806 + 00
0.1940 + 01	0.6847 + 00	0.4265 + 00	0.6230 + 00	0.1606 + 01	0.2868 + 00
0.1960 + 01	0.6786 + 00	0.4203 + 00	0.6193 + 00	0.1633 + 01	0.2929 + 00
0.1980 + 01	0.6726 + 00	0.4142 + 00	0.6158 + 00	0.1660 + 01	0.2990 + 00
0.2000 + 01	0.6667 + 00	0.4082 + 00	0.6124 + 00	0.1687 + 01	0.3050 + 00

Table A.4 | *Continued*

M	$\dfrac{T}{T^*}$	$\dfrac{p}{p^*}$	$\dfrac{\rho}{\rho^*}$	$\dfrac{p_o}{p_o^*}$	$\dfrac{4fL^*}{D}$
0.2050 + 01	0.6520 + 00	0.3939 + 00	0.6041 + 00	0.1760 + 01	0.3197 + 00
0.2100 + 01	0.6376 + 00	0.3802 + 00	0.5963 + 00	0.1837 + 01	0.3339 + 00
0.2150 + 01	0.6235 + 00	0.3673 + 00	0.5890 + 00	0.1919 + 01	0.3476 + 00
0.2200 + 01	0.6098 + 00	0.3549 + 00	0.5821 + 00	0.2005 + 01	0.3609 + 00
0.2250 + 01	0.5963 + 00	0.3432 + 00	0.5756 + 00	0.2096 + 01	0.3738 + 00
0.2300 + 01	0.5831 + 00	0.3320 + 00	0.5694 + 00	0.2193 + 01	0.3862 + 00
0.2350 + 01	0.5702 + 00	0.3213 + 00	0.5635 + 00	0.2295 + 01	0.3983 + 00
0.2400 + 01	0.5576 + 00	0.3111 + 00	0.5580 + 00	0.2403 + 01	0.4099 + 00
0.2450 + 01	0.5453 + 00	0.3014 + 00	0.5527 + 00	0.2517 + 01	0.4211 + 00
0.2500 + 01	0.5333 + 00	0.2921 + 00	0.5477 + 00	0.2637 + 01	0.4320 + 00
0.2550 + 01	0.5216 + 00	0.2832 + 00	0.5430 + 00	0.2763 + 01	0.4425 + 00
0.2600 + 01	0.5102 + 00	0.2747 + 00	0.5385 + 00	0.2896 + 01	0.4526 + 00
0.2650 + 01	0.4991 + 00	0.2666 + 00	0.5342 + 00	0.3036 + 01	0.4624 + 00
0.2700 + 01	0.4882 + 00	0.2588 + 00	0.5301 + 00	0.3183 + 01	0.4718 + 00
0.2750 + 01	0.4776 + 00	0.2513 + 00	0.5262 + 00	0.3338 + 01	0.4809 + 00
0.2800 + 01	0.4673 + 00	0.2441 + 00	0.5225 + 00	0.3500 + 01	0.4898 + 00
0.2850 + 01	0.4572 + 00	0.2373 + 00	0.5189 + 00	0.3671 + 01	0.4983 + 00
0.2900 + 01	0.4474 + 00	0.2307 + 00	0.5155 + 00	0.3850 + 01	0.5065 + 00
0.2950 + 01	0.4379 + 00	0.2243 + 00	0.5123 + 00	0.4038 + 01	0.5145 + 00
0.3000 + 01	0.4286 + 00	0.2182 + 00	0.5092 + 00	0.4235 + 01	0.5222 + 00
0.3050 + 01	0.4195 + 00	0.2124 + 00	0.5062 + 00	0.4441 + 01	0.5296 + 00
0.3100 + 01	0.4107 + 00	0.2067 + 00	0.5034 + 00	0.4657 + 01	0.5368 + 00
0.3150 + 01	0.4021 + 00	0.2013 + 00	0.5007 + 00	0.4884 + 01	0.5437 + 00
0.3200 + 01	0.3937 + 00	0.1961 + 00	0.4980 + 00	0.5121 + 01	0.5504 + 00
0.3250 + 01	0.3855 + 00	0.1911 + 00	0.4955 + 00	0.5369 + 01	0.5569 + 00
0.3300 + 01	0.3776 + 00	0.1862 + 00	0.4931 + 00	0.5629 + 01	0.5632 + 00
0.3350 + 01	0.3699 + 00	0.1815 + 00	0.4908 + 00	0.5900 + 01	0.5693 + 00
0.3400 + 01	0.3623 + 00	0.1770 + 00	0.4886 + 00	0.6184 + 01	0.5752 + 00
0.3450 + 01	0.3550 + 00	0.1727 + 00	0.4865 + 00	0.6480 + 01	0.5809 + 00
0.3500 + 01	0.3478 + 00	0.1685 + 00	0.4845 + 00	0.6790 + 01	0.5864 + 00
0.3550 + 01	0.3409 + 00	0.1645 + 00	0.4825 + 00	0.7113 + 01	0.5918 + 00
0.3600 + 01	0.3341 + 00	0.1606 + 00	0.4806 + 00	0.7450 + 01	0.5970 + 00
0.3650 + 01	0.3275 + 00	0.1568 + 00	0.4788 + 00	0.7802 + 01	0.6020 + 00
0.3700 + 01	0.3210 + 00	0.1531 + 00	0.4770 + 00	0.8169 + 01	0.6068 + 00
0.3750 + 01	0.3148 + 00	0.1496 + 00	0.4753 + 00	0.8552 + 01	0.6115 + 00
0.3800 + 01	0.3086 + 00	0.1462 + 00	0.4737 + 00	0.8951 + 01	0.6161 + 00
0.3850 + 01	0.3027 + 00	0.1429 + 00	0.4721 + 00	0.9366 + 01	0.6206 + 00
0.3900 + 01	0.2969 + 00	0.1397 + 00	0.4706 + 00	0.9799 + 01	0.6248 + 00
0.3950 + 01	0.2912 + 00	0.1366 + 00	0.4691 + 00	0.1025 + 02	0.6290 + 00
0.4000 + 01	0.2857 + 00	0.1336 + 00	0.4677 + 00	0.1072 + 02	0.6331 + 00
0.4050 + 01	0.2803 + 00	0.1307 + 00	0.4663 + 00	0.1121 + 02	0.6370 + 00
0.4100 + 01	0.2751 + 00	0.1279 + 00	0.4650 + 00	0.1171 + 02	0.6408 + 00
0.4150 + 01	0.2700 + 00	0.1252 + 00	0.4637 + 00	0.1224 + 02	0.6445 + 00
0.4200 + 01	0.2650 + 00	0.1226 + 00	0.4625 + 00	0.1279 + 02	0.6481 + 00
0.4250 + 01	0.2602 + 00	0.1200 + 00	0.4613 + 00	0.1336 + 02	0.6516 + 00
0.4300 + 01	0.2554 + 00	0.1175 + 00	0.4601 + 00	0.1395 + 02	0.6550 + 00
0.4350 + 01	0.2508 + 00	0.1151 + 00	0.4590 + 00	0.1457 + 02	0.6583 + 00
0.4400 + 01	0.2463 + 00	0.1128 + 00	0.4579 + 00	0.1521 + 02	0.6615 + 00
0.4450 + 01	0.2419 + 00	0.1105 + 00	0.4569 + 00	0.1587 + 02	0.6646 + 00
0.4500 + 01	0.2376 + 00	0.1083 + 00	0.4559 + 00	0.1656 + 02	0.6676 + 00

Continued

Table A.4 | *Continued*

M	$\dfrac{T}{T^*}$	$\dfrac{p}{p^*}$	$\dfrac{\rho}{\rho^*}$	$\dfrac{p_o}{p_o^*}$	$\dfrac{4fL^*}{D}$
0.4550 + 01	0.2334 + 00	0.1062 + 00	0.4549 + 00	0.1728 + 02	0.6706 + 00
0.4600 + 01	0.2294 + 00	0.1041 + 00	0.4539 + 00	0.1802 + 02	0.6734 + 00
0.4650 + 01	0.2254 + 00	0.1021 + 00	0.4530 + 00	0.1879 + 02	0.6762 + 00
0.4700 + 01	0.2215 + 00	0.1001 + 00	0.4521 + 00	0.1958 + 02	0.6790 + 00
0.4750 + 01	0.2177 + 00	0.9823 − 01	0.4512 + 00	0.2041 + 02	0.6816 + 00
0.4800 + 01	0.2140 + 00	0.9637 − 01	0.4504 + 00	0.2126 + 02	0.6842 + 00
0.4850 + 01	0.2104 + 00	0.9457 − 01	0.4495 + 00	0.2215 + 02	0.6867 + 00
0.4900 + 01	0.2068 + 00	0.9281 − 01	0.4487 + 00	0.2307 + 02	0.6891 + 00
0.4950 + 01	0.2034 + 00	0.9110 − 01	0.4480 + 00	0.2402 + 02	0.6915 + 00
0.5000 + 01	0.2000 + 00	0.8944 − 01	0.4472 + 00	0.2500 + 02	0.6938 + 00
0.5100 + 01	0.1935 + 00	0.8625 − 01	0.4458 + 00	0.2707 + 02	0.6983 + 00
0.5200 + 01	0.1873 + 00	0.8322 − 01	0.4444 + 00	0.2928 + 02	0.7025 + 00
0.5300 + 01	0.1813 + 00	0.8034 − 01	0.4431 + 00	0.3165 + 02	0.7065 + 00
0.5400 + 01	0.1756 + 00	0.7761 − 01	0.4419 + 00	0.3417 + 02	0.7104 + 00
0.5500 + 01	0.1702 + 00	0.7501 − 01	0.4407 + 00	0.3687 + 02	0.7140 + 00
0.5600 + 01	0.1650 + 00	0.7254 − 01	0.4396 + 00	0.3974 + 02	0.7175 + 00
0.5700 + 01	0.1600 + 00	0.7018 − 01	0.4385 + 00	0.4280 + 02	0.7208 + 00
0.5800 + 01	0.1553 + 00	0.6794 − 01	0.4375 + 00	0.4605 + 02	0.7240 + 00
0.5900 + 01	0.1507 + 00	0.6580 − 01	0.4366 + 00	0.4951 + 02	0.7270 + 00
0.6000 + 01	0.1463 + 00	0.6376 − 01	0.4357 + 00	0.5318 + 02	0.7299 + 00
0.6100 + 01	0.1421 + 00	0.6181 − 01	0.4348 + 00	0.5708 + 02	0.7326 + 00
0.6200 + 01	0.1381 + 00	0.5994 − 01	0.4340 + 00	0.6121 + 02	0.7353 + 00
0.6300 + 01	0.1343 + 00	0.5816 − 01	0.4332 + 00	0.6559 + 02	0.7378 + 00
0.6400 + 01	0.1305 + 00	0.5646 − 01	0.4324 + 00	0.7023 + 02	0.7402 + 00
0.6500 + 01	0.1270 + 00	0.5482 − 01	0.4317 + 00	0.7513 + 02	0.7425 + 00
0.6600 + 01	0.1236 + 00	0.5326 − 01	0.4310 + 00	0.8032 + 02	0.7448 + 00
0.6700 + 01	0.1203 + 00	0.5176 − 01	0.4304 + 00	0.8580 + 02	0.7469 + 00
0.6800 + 01	0.1171 + 00	0.5032 − 01	0.4298 + 00	0.9159 + 02	0.7489 + 00
0.6900 + 01	0.1140 + 00	0.4894 − 01	0.4292 + 00	0.9770 + 02	0.7509 + 00
0.7000 + 01	0.1111 + 00	0.4762 − 01	0.4286 + 00	0.1041 + 03	0.7528 + 00
0.7100 + 01	0.1083 + 00	0.4635 − 01	0.4280 + 00	0.1109 + 03	0.7546 + 00
0.7200 + 01	0.1056 + 00	0.4512 − 01	0.4275 + 00	0.1181 + 03	0.7564 + 00
0.7300 + 01	0.1029 + 00	0.4395 − 01	0.4270 + 00	0.1256 + 03	0.7580 + 00
0.7400 + 01	0.1004 + 00	0.4282 − 01	0.4265 + 00	0.1335 + 03	0.7597 + 00
0.7500 + 01	0.9796 − 01	0.4173 − 01	0.4260 + 00	0.1418 + 03	0.7612 + 00
0.7600 + 01	0.9560 − 01	0.4068 − 01	0.4256 + 00	0.1506 + 03	0.7627 + 00
0.7700 + 01	0.9333 − 01	0.3967 − 01	0.4251 + 00	0.1598 + 03	0.7642 + 00
0.7800 + 01	0.9113 − 01	0.3870 − 01	0.4247 + 00	0.1694 + 03	0.7656 + 00
0.7900 + 01	0.8901 − 01	0.3776 − 01	0.4243 + 00	0.1795 + 03	0.7669 + 00
0.8000 + 01	0.8696 − 01	0.3686 − 01	0.4239 + 00	0.1901 + 03	0.7682 + 00
0.9000 + 01	0.6977 − 01	0.2935 − 01	0.4207 + 00	0.3272 + 03	0.7790 + 00
0.1000 + 02	0.5714 − 01	0.2390 − 01	0.4183 + 00	0.5359 + 03	0.7868 + 00
0.1100 + 02	0.4762 − 01	0.1984 − 01	0.4166 + 00	0.8419 + 03	0.7927 + 00
0.1200 + 02	0.4027 − 01	0.1672 − 01	0.4153 + 00	0.1276 + 04	0.7972 + 00
0.1300 + 02	0.3448 − 01	0.1428 − 01	0.4142 + 00	0.1876 + 04	0.8007 + 00
0.1400 + 02	0.2985 − 01	0.1234 − 01	0.4134 + 00	0.2685 + 04	0.8036 + 00
0.1500 + 02	0.2609 − 01	0.1077 − 01	0.4128 + 00	0.3755 + 04	0.8058 + 00
0.1600 + 02	0.2299 − 01	0.9476 − 02	0.4122 + 00	0.5145 + 04	0.8077 + 00
0.1700 + 02	0.2041 − 01	0.8403 − 02	0.4118 + 00	0.6921 + 04	0.8093 + 00
0.1800 + 02	0.1824 − 01	0.7502 − 02	0.4114 + 00	0.9159 + 04	0.8106 + 00

Table A.4 | *Continued*

M	$\dfrac{T}{T^*}$	$\dfrac{p}{p^*}$	$\dfrac{\rho}{\rho^*}$	$\dfrac{p_o}{p_o^*}$	$\dfrac{4fL^*}{D}$
0.1900 + 02	0.1639 − 01	0.6739 − 02	0.4111 + 00	0.1195 + 05	0.8117 + 00
0.2000 + 02	0.1481 − 01	0.6086 − 02	0.4108 + 00	0.1538 + 05	0.8126 + 00
0.2200 + 02	0.1227 − 01	0.5035 − 02	0.4104 + 00	0.2461 + 05	0.8142 + 00
0.2400 + 02	0.1033 − 01	0.4234 − 02	0.4100 + 00	0.3783 + 05	0.8153 + 00
0.2600 + 02	0.8811 − 02	0.3610 − 02	0.4098 + 00	0.5624 + 05	0.8162 + 00
0.2800 + 02	0.7605 − 02	0.3114 − 02	0.4095 + 00	0.8121 + 05	0.8170 + 00
0.3000 + 02	0.6630 − 02	0.2714 − 02	0.4094 + 00	0.1144 + 06	0.8176 + 00
0.3200 + 02	0.5831 − 02	0.2386 − 02	0.4092 + 00	0.1576 + 06	0.8180 + 00
0.3400 + 02	0.5168 − 02	0.2114 − 02	0.4091 + 00	0.2131 + 06	0.8184 + 00
0.3600 + 02	0.4612 − 02	0.1886 − 02	0.4090 + 00	0.2832 + 06	0.8188 + 00
0.3800 + 02	0.4141 − 02	0.1693 − 02	0.4090 + 00	0.3707 + 06	0.8190 + 00
0.4000 + 02	0.3738 − 02	0.1529 − 02	0.4089 + 00	0.4785 + 06	0.8193 + 00
0.4200 + 02	0.3392 − 02	0.1387 − 02	0.4088 + 00	0.6102 + 06	0.8195 + 00
0.4400 + 02	0.3091 − 02	0.1264 − 02	0.4088 + 00	0.7694 + 06	0.8197 + 00
0.4600 + 02	0.2829 − 02	0.1156 − 02	0.4087 + 00	0.9603 + 06	0.8198 + 00
0.4800 + 02	0.2599 − 02	0.1062 − 02	0.4087 + 00	0.1187 + 07	0.8200 + 00
0.5000 + 02	0.2395 − 02	0.9788 − 03	0.4087 + 00	0.1455 + 07	0.8201 + 00

Table A.5 | Prandtl–Meyer function and Mach angle

M	ν	μ	M	ν	μ
0.1000 + 01	0.0000	0.9000 + 02	0.2000 + 01	0.2638 + 02	0.3000 + 02
0.1020 + 01	0.1257 + 00	0.7864 + 02	0.2050 + 01	0.2775 + 02	0.2920 + 02
0.1040 + 01	0.3510 + 00	0.7406 + 02	0.2100 + 01	0.2910 + 02	0.2844 + 02
0.1060 + 01	0.6367 + 00	0.7063 + 02	0.2150 + 01	0.3043 + 02	0.2772 + 02
0.1080 + 01	0.9680 + 00	0.6781 + 02	0.2200 + 01	0.3173 + 02	0.2704 + 02
0.1100 + 01	0.1336 + 01	0.6538 + 02	0.2250 + 01	0.3302 + 02	0.2639 + 02
0.1120 + 01	0.1735 + 01	0.6323 + 02	0.2300 + 01	0.3428 + 02	0.2577 + 02
0.1140 + 01	0.2160 + 01	0.6131 + 02	0.2350 + 01	0.3553 + 02	0.2518 + 02
0.1160 + 01	0.2607 + 01	0.5955 + 02	0.2400 + 01	0.3675 + 02	0.2462 + 02
0.1180 + 01	0.3074 + 01	0.5794 + 02	0.2450 + 01	0.3795 + 02	0.2409 + 02
0.1200 + 01	0.3558 + 01	0.5644 + 02	0.2500 + 01	0.3912 + 02	0.2358 + 02
0.1220 + 01	0.4057 + 01	0.5505 + 02	0.2550 + 01	0.4028 + 02	0.2309 + 02
0.1240 + 01	0.4569 + 01	0.5375 + 02	0.2600 + 01	0.4141 + 02	0.2262 + 02
0.1260 + 01	0.5093 + 01	0.5253 + 02	0.2650 + 01	0.4253 + 02	0.2217 + 02
0.1280 + 01	0.5627 + 01	0.5138 + 02	0.2700 + 01	0.4362 + 02	0.2174 + 02
0.1300 + 01	0.6170 + 01	0.5028 + 02	0.2750 + 01	0.4469 + 02	0.2132 + 02
0.1320 + 01	0.6721 + 01	0.4925 + 02	0.2800 + 01	0.4575 + 02	0.2092 + 02
0.1340 + 01	0.7279 + 01	0.4827 + 02	0.2850 + 01	0.4678 + 02	0.2054 + 02
0.1360 + 01	0.7844 + 01	0.4733 + 02	0.2900 + 01	0.4779 + 02	0.2017 + 02
0.1380 + 01	0.8413 + 01	0.4644 + 02	0.2950 + 01	0.4878 + 02	0.1981 + 02
0.1400 + 01	0.8987 + 01	0.4558 + 02	0.3000 + 01	0.4976 + 02	0.1947 + 02
0.1420 + 01	0.9565 + 01	0.4477 + 02	0.3050 + 01	0.5071 + 02	0.1914 + 02
0.1440 + 01	0.1015 + 02	0.4398 + 02	0.3100 + 01	0.5165 + 02	0.1882 + 02
0.1460 + 01	0.1073 + 02	0.4323 + 02	0.3150 + 01	0.5257 + 02	0.1851 + 02
0.1480 + 01	0.1132 + 02	0.4251 + 02	0.3200 + 01	0.5347 + 02	0.1821 + 02
0.1500 + 01	0.1191 + 02	0.4181 + 02	0.3250 + 01	0.5435 + 02	0.1792 + 02
0.1520 + 01	0.1249 + 02	0.4114 + 02	0.3300 + 01	0.5522 + 02	0.1764 + 02
0.1540 + 01	0.1309 + 02	0.4049 + 02	0.3350 + 01	0.5607 + 02	0.1737 + 02
0.1560 + 01	0.1368 + 02	0.3987 + 02	0.3400 + 01	0.5691 + 02	0.1710 + 02
0.1580 + 01	0.1427 + 02	0.3927 + 02	0.3450 + 01	0.5773 + 02	0.1685 + 02
0.1600 + 01	0.1486 + 02	0.3868 + 02	0.3500 + 01	0.5853 + 02	0.1660 + 02
0.1620 + 01	0.1545 + 02	0.3812 + 02	0.3550 + 01	0.5932 + 02	0.1636 + 02
0.1640 + 01	0.1604 + 02	0.3757 + 02	0.3600 + 01	0.6009 + 02	0.1613 + 02
0.1660 + 01	0.1663 + 02	0.3704 + 02	0.3650 + 01	0.6085 + 02	0.1590 + 02
0.1680 + 01	0.1722 + 02	0.3653 + 02	0.3700 + 01	0.6160 + 02	0.1568 + 02
0.1700 + 01	0.1781 + 02	0.3603 + 02	0.3750 + 01	0.6233 + 02	0.1547 + 02
0.1720 + 01	0.1840 + 02	0.3555 + 02	0.3800 + 01	0.6304 + 02	0.1526 + 02
0.1740 + 01	0.1898 + 02	0.3508 + 02	0.3850 + 01	0.6375 + 02	0.1505 + 02
0.1760 + 01	0.1956 + 02	0.3462 + 02	0.3900 + 01	0.6444 + 02	0.1486 + 02
0.1780 + 01	0.2015 + 02	0.3418 + 02	0.3950 + 01	0.6512 + 02	0.1466 + 02
0.1800 + 01	0.2073 + 02	0.3375 + 02	0.4000 + 01	0.6578 + 02	0.1448 + 02
0.1820 + 01	0.2130 + 02	0.3333 + 02	0.4050 + 01	0.6644 + 02	0.1429 + 02
0.1840 + 01	0.2188 + 02	0.3292 + 02	0.4100 + 01	0.6708 + 02	0.1412 + 02
0.1860 + 01	0.2245 + 02	0.3252 + 02	0.4150 + 01	0.6771 + 02	0.1394 + 02
0.1880 + 01	0.2302 + 02	0.3213 + 02	0.4200 + 01	0.6833 + 02	0.1377 + 02
0.1900 + 01	0.2359 + 02	0.3176 + 02	0.4250 + 01	0.6894 + 02	0.1361 + 02
0.1920 + 01	0.2415 + 02	0.3139 + 02	0.4300 + 01	0.6954 + 02	0.1345 + 02
0.1940 + 01	0.2471 + 02	0.3103 + 02	0.4350 + 01	0.7013 + 02	0.1329 + 02
0.1960 + 01	0.2527 + 02	0.3068 + 02	0.4400 + 01	0.7071 + 02	0.1314 + 02
0.1980 + 01	0.2583 + 02	0.3033 + 02	0.4450 + 01	0.7127 + 02	0.1299 + 02

Table A.5 | *Continued*

M	v	μ	M	v	μ
0.4500 + 01	0.7183 + 02	0.1284 + 02	0.7400 + 01	0.9297 + 02	0.7766 + 01
0.4550 + 01	0.7238 + 02	0.1270 + 02	0.7500 + 01	0.9344 + 02	0.7662 + 01
0.4600 + 01	0.7292 + 02	0.1256 + 02	0.7600 + 01	0.9390 + 02	0.7561 + 01
0.4650 + 01	0.7345 + 02	0.1242 + 02	0.7700 + 01	0.9434 + 02	0.7462 + 01
0.4700 + 01	0.7397 + 02	0.1228 + 02	0.7800 + 01	0.9478 + 02	0.7366 + 01
0.4750 + 01	0.7448 + 02	0.1215 + 02	0.7900 + 01	0.9521 + 02	0.7272 + 01
0.4800 + 01	0.7499 + 02	0.1202 + 02	0.8000 + 01	0.9562 + 02	0.7181 + 01
0.4850 + 01	0.7548 + 02	0.1190 + 02	0.9000 + 01	0.9932 + 02	0.6379 + 01
0.4900 + 01	0.7597 + 02	0.1178 + 02	0.1000 + 02	0.1023 + 03	0.5739 + 01
0.4950 + 01	0.7645 + 02	0.1166 + 02	0.1100 + 02	0.1048 + 03	0.5216 + 01
0.5000 + 01	0.7692 + 02	0.1154 + 02	0.1200 + 02	0.1069 + 03	0.4780 + 01
0.5100 + 01	0.7784 + 02	0.1131 + 02	0.1300 + 02	0.1087 + 03	0.4412 + 01
0.5200 + 01	0.7873 + 02	0.1109 + 02	0.1400 + 02	0.1102 + 03	0.4096 + 01
0.5300 + 01	0.7960 + 02	0.1088 + 02	0.1500 + 02	0.1115 + 03	0.3823 + 01
0.5400 + 01	0.8043 + 02	0.1067 + 02	0.1600 + 02	0.1127 + 03	0.3583 + 01
0.5500 + 01	0.8124 + 02	0.1048 + 02	0.1700 + 02	0.1137 + 03	0.3372 + 01
0.5600 + 01	0.8203 + 02	0.1029 + 02	0.1800 + 02	0.1146 + 03	0.3185 + 01
0.5700 + 01	0.8280 + 02	0.1010 + 02	0.1900 + 02	0.1155 + 03	0.3017 + 01
0.5800 + 01	0.8354 + 02	0.9928 + 01	0.2000 + 02	0.1162 + 03	0.2866 + 01
0.5900 + 01	0.8426 + 02	0.9758 + 01	0.2200 + 02	0.1175 + 03	0.2605 + 01
0.6000 + 01	0.8496 + 02	0.9594 + 01	0.2400 + 02	0.1186 + 03	0.2388 + 01
0.6100 + 01	0.8563 + 02	0.9435 + 01	0.2600 + 02	0.1195 + 03	0.2204 + 01
0.6200 + 01	0.8629 + 02	0.9282 + 01	0.2800 + 02	0.1202 + 03	0.2047 + 01
0.6300 + 01	0.8694 + 02	0.9133 + 01	0.3000 + 02	0.1209 + 03	0.1910 + 01
0.6400 + 01	0.8756 + 02	0.8989 + 01	0.3200 + 02	0.1215 + 03	0.1791 + 01
0.6500 + 01	0.8817 + 02	0.8850 + 01	0.3400 + 02	0.1220 + 03	0.1685 + 01
0.6600 + 01	0.8876 + 02	0.8715 + 01	0.3600 + 02	0.1225 + 03	0.1592 + 01
0.6700 + 01	0.8933 + 02	0.8584 + 01	0.3800 + 02	0.1229 + 03	0.1508 + 01
0.6800 + 01	0.8989 + 02	0.8457 + 01	0.4000 + 02	0.1233 + 03	0.1433 + 01
0.6900 + 01	0.9044 + 02	0.8333 + 01	0.4200 + 02	0.1236 + 03	0.1364 + 01
0.7000 + 01	0.9097 + 02	0.8213 + 01	0.4400 + 02	0.1239 + 03	0.1302 + 01
0.7100 + 01	0.9149 + 02	0.8097 + 01	0.4600 + 02	0.1242 + 03	0.1246 + 01
0.7200 + 01	0.9200 + 02	0.7984 + 01	0.4800 + 02	0.1245 + 03	0.1194 + 01
0.7300 + 01	0.9249 + 02	0.7873 + 01	0.5000 + 02	0.1247 + 03	0.1146 + 01

B

APPENDIX

An Illustration and Exercise of Computational Fluid Dynamics

The purpose of this appendix is to give the interested reader an opportunity for a hands-on experience in computational fluid dynamics (CFD). In various chapters of this book, computational fluid dynamics is discussed in the context of *modern* compressible flow. It is not our purpose, however, to present CFD in any detail; rather, this book emphasizes the *physical* aspects of compressible flow. Indeed, computation fluid dynamics is a subject by itself, and the reader is encouraged to examine the number of texts devoted exclusively to CFD.

On the other hand, this appendix offers the opportunity to sample the essence of CFD through an example using MacCormack's time-marching explicit finite-difference technique—by far the most "student-friendly" CFD technique that can be found. The application is the subsonic-supersonic quasi-one-dimensional isentropic flow through a convergent-divergent nozzle—the CFD application discussed in Sec. 12.1. In this appendix we go into the details that produced the results given in Sec. 12.1. This example is the simplest possible exercise that reflects the essence of CFD, yet you will find its explanation requires a rather lengthy discussion. Any other example of CFD goes well beyond the scope of this book.

Finally, for a very basic introduction to CFD, you are encouraged to examine the author's book *Computational Fluid Dynamics: The Basics with Applications,* McGraw-Hill, 1995, which contains a number of worked examples in elementary CFD, including the one described in this appendix. What follows is excerpted from Chap. 7 of that book. This is the author's best attempt to provide you with a hands-on experience in CFD. The appendix ends with a FORTRAN code listing that the author wrote for this particular application.

THE EQUATIONS

Return to Sec. 12.1 and review the basic governing equations for unsteady quasi-one-dimensional flow, namely, Eqs. (12.5), (12.6), and (12.7)—the continuity, momentum, and energy equations, respectively. Assuming a calorically perfect gas, let us replace the internal energy in Eq. (12.7) with temperature. For a calorically perfect gas

$$e = c_v T$$

Hence, Eq. (12.7) becomes

$$\rho c_v \frac{\partial T}{\partial t} + \rho u c_v \frac{\partial T}{\partial x} = -p \frac{\partial u}{\partial x} - pu \frac{\partial (\ln A)}{\partial x} \qquad (B.1)$$

As an interim summary, our continuity, momentum, and energy equations for unsteady, quasi-one-dimensional flow are given by Eqs. (12.5), (12.6), and (B.1), respectively. Take the time to look at these equations; you see three equations with four unknown variables ρ, u, p, and T. The pressure can be eliminated from these equations by using the equation of state

$$p = \rho R T \qquad (B.2)$$

along with its derivative

$$\frac{\partial p}{\partial x} = R \left(\rho \frac{\partial T}{\partial x} + T \frac{\partial \rho}{\partial x} \right) \qquad (B.3)$$

With this, we expand Eq. (12.5) and rewrite Eqs. (12.6) and (B.1), respectively, as

Continuity:
$$\frac{\partial (\rho A)}{\partial t} + \rho A \frac{\partial u}{\partial x} + \rho u \frac{\partial A}{\partial x} + u A \frac{\partial \rho}{\partial x} = 0 \qquad (B.4)$$

Momentum:
$$\rho \frac{\partial u}{\partial t} + \rho u \frac{\partial u}{\partial x} = -R \left(\rho \frac{\partial T}{\partial x} + T \frac{\partial \rho}{\partial x} \right) \qquad (B.5)$$

Energy:
$$\rho c_v \frac{\partial T}{\partial t} + \rho u c_v \frac{\partial T}{\partial x} = -\rho R T \left[\frac{\partial u}{\partial x} + u \frac{\partial (\ln A)}{\partial x} \right] \qquad (B.6)$$

At this stage, we could readily proceed to set up our numerical solution of Eqs. (B.4) to (B.6). Note that these are written in terms of dimensional variables. This is fine, and many CFD solutions are carried out directly in terms of such dimensional variables. Indeed, this has an added engineering advantage because it gives you a feeling for the magnitudes of the real physical quantities as the solution progresses. However, for nozzle flows, the flowfield variables are frequently expressed in terms of nondimensional variables, where the flow variables are referenced to their reservoir values. The nondimensional variables p/p_o, ρ/ρ_o, and T/T_o vary between 0 and 1, which is an "aesthetic" advantage when presenting the results. Because fluid dynamicists dealing with nozzle flows so frequently use these nondimensional terms, we will follow suit here. (A number of CFD practitioners prefer to always deal with nondimensional variables, whereas others prefer dimensional variables; as far as the numerics are concerned, there should be no real difference, and the choice is really a matter of your personal preference.) Therefore, we define the nondimensional temperature and density, respectively, as

$$T' = \frac{T}{T_o} \qquad \rho' = \frac{\rho}{\rho_o}$$

where (for the time being) the prime denotes a dimensionless variable. Moreover, letting L denote the length of the nozzle, we define a dimensionless length as

$$x' = \frac{x}{L}$$

Denoting the speed of sound in the reservoir as a_o, where

$$a_o = \sqrt{\gamma R T_o}$$

we define a dimensionless velocity as

$$V' = \frac{u}{a_o}$$

Also, the quantity L/a_o has the dimension of time, and we define a dimensionless time as

$$t' = \frac{t}{L/a_o}$$

Finally, we ratio the local area A to the sonic throat area A^* and define a dimensionless area as

$$A' = \frac{A}{A^*}$$

Returning to Eq. (B.4) and introducing the nondimensional variables, we have

$$\frac{\partial(\rho'A')}{\partial t'}\left(\frac{\rho_o A^*}{L/a_o}\right) + \rho'A'\frac{\partial V'}{\partial x'}\left(\frac{\rho_o A^* a_o}{L}\right) + \rho'V'\frac{\partial A'}{\partial x'}\left(\frac{\rho_o a_o A^*}{L}\right)$$

$$+ V'A'\frac{\partial \rho'}{\partial x'}\left(\frac{a_o A^* \rho_o}{L}\right) = 0 \tag{B.7}$$

Note that A' is a function of x' only; it is *not* a function of time (the nozzle geometry is fixed, invariant with time). Hence, in Eq. (B.7) the time derivative can be written as

$$\frac{\partial(\rho'A')}{\partial t'} = A'\frac{\partial \rho'}{\partial t'}$$

With this, Eq. (B.7) becomes

Continuity:
$$\boxed{\frac{\partial \rho'}{\partial t'} = -\rho'\frac{\partial V'}{\partial x'} - \rho'V'\frac{\partial(\ln A')}{\partial x'} - V'\frac{\partial \rho'}{\partial x'}} \tag{B.8}$$

Returning to Eq. (B.5) and introducing the nondimensional variables, we have

$$\rho'\frac{\partial V'}{\partial t'}\left(\frac{\rho_o a_o}{L/a_o}\right) + \rho'V'\frac{\partial V'}{\partial x'}\left(\frac{\rho_o a_o^2}{L}\right) = -R\left(\rho'\frac{\partial T'}{\partial x'} + T'\frac{\partial \rho'}{\partial x'}\right)\left(\frac{\rho_o T_o}{L}\right)$$

or
$$\rho'\frac{\partial V'}{\partial t'} = -\rho'V'\frac{\partial V'}{\partial x'} - \left(\rho'\frac{\partial T'}{\partial x'} + T'\frac{\partial \rho'}{\partial x'}\right)\frac{RT_o}{a_o^2} \qquad (B.9)$$

In Eq. (B.9), note that

$$\frac{RT_o}{a_o^2} = \frac{\gamma R T_o}{\gamma a_o^2} = \frac{a_o^2}{\gamma a_o^2} = \frac{1}{\gamma}$$

Hence, Eq. (B.9) becomes

Momentum:
$$\boxed{\frac{\partial V'}{\partial t'} = -V'\frac{\partial V'}{\partial x'} - \frac{1}{\gamma}\left(\frac{\partial T'}{\partial x'} + \frac{T'}{\rho'}\frac{\partial \rho'}{\partial x'}\right)} \qquad (B.10)$$

Returning to Eq. (B.6) and introducing the nondimensional variables, we have

$$\rho'c_v\frac{\partial T'}{\partial t'}\left(\frac{\rho_o T_o}{L/a_o}\right) + \rho'V'c_v\frac{\partial T'}{\partial x'}\left(\frac{\rho_o a_o T_o}{L}\right)$$

$$= -\rho'RT'\left[\frac{\partial V'}{\partial x'} + V'\frac{\partial(\ln A')}{\partial x'}\right]\left(\frac{\rho_o T_o a_o}{L}\right) \qquad (B.11)$$

In Eq. (B.11), the factor R/c_v is given by

$$\frac{R}{c_v} = \frac{R}{R/(\gamma - 1)} = \gamma - 1$$

Hence, Eq. (B.11) becomes

Energy:
$$\boxed{\frac{\partial T'}{\partial t'} = -V'\frac{\partial T'}{\partial x'} - (\gamma - 1)T'\left[\frac{\partial V'}{\partial x'} + V'\frac{\partial(\ln A')}{\partial x'}\right]} \qquad (B.12)$$

That is it! After what may seem like an interminable manipulation of the governing equations, we have finally set up that particular form of the equations that will be most appropriate as well as convenient for the time-marching solution of quasi-one-dimensional nozzle flow, namely, Eqs. (B.8), (B.10), and (B.12).

The Finite-Difference Equations

We now proceed to the setting up of the finite-difference expressions using MacCormack's explicit technique for the numerical solution of Eqs. (B.8), (B.10), and (B.12). To implement a finite-difference solution, we divide the x axis along the nozzle into a number of discrete grid points, as shown in Fig. B.1. (Recall that in our quasi-one-dimensional nozzle assumption, the flow variables *across* the nozzle cross section at any particular grid point, say point i, are uniform.) In Fig. B.1, the first grid point, labeled point 1, is assumed to be in the reservoir. The points are evenly distributed along the x axis, with Δx denoting the spacing between grid points. The

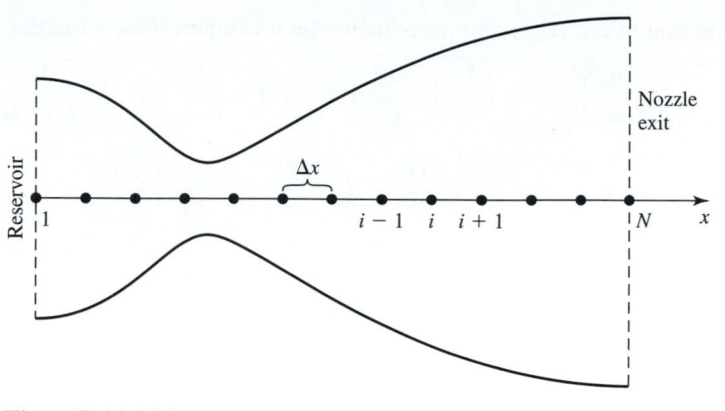

Figure B.1 | Grid point distribution along the nozzle.

last point, namely, that at the nozzle exit, is denoted by N; we have a total number of N grid points distributed along the axis. Point i is simply an arbitrary grid point, with points $i-1$ and $i+1$ as the adjacent points. Recall from Sec. 12.1 that MacCormack's technique is a predictor-corrector method. In the time-marching approach, remember that we know the flowfield variables at time t, and we use the difference equations to solve explicitly for the variables at time $t + \Delta t$.

First, consider the predictor step. Following the discussion in Sec. 12.1, we set up the spatial derivatives as forward differences. Also, to reduce the complexity of the notation, we will drop the use of the prime to denote a dimensionless variable. In what follows, *all* variables are the nondimensional variables, denoted earlier by the prime notation. From Eq. (B.8) we have

$$\left(\frac{\partial \rho}{\partial t}\right)_i^t = -\rho_i^t \frac{V_{i+1}^t - V_i^t}{\Delta x} - \rho_i^t V_i^t \frac{\ln A_{i+1} - \ln A_i}{\Delta x} - V_i^t \frac{\rho_{i+1}^t - \rho_i^t}{\Delta x} \qquad (B.13)$$

From Eq. (B.10), we have

$$\left(\frac{\partial V}{\partial t}\right)_i^t = -V_i^t \frac{V_{i+1}^t - V_i^t}{\Delta x} - \frac{1}{\gamma}\left(\frac{T_{i+1}^t - T_i^t}{\Delta x} + \frac{T_i^t}{\rho_i^t} \frac{\rho_{i+1}^t - \rho_i^t}{\Delta x}\right) \qquad (B.14)$$

From Eq. (B.12), we have

$$\left(\frac{\partial T}{\partial t}\right)_i^t = -V_i^t \frac{T_{i+1}^t - T_i^t}{\Delta x} - (\gamma - 1)T_i^t\left(\frac{V_{i+1}^t - V_i^t}{\Delta x} + V_i^t \frac{\ln A_{i+1} - \ln A_i}{\Delta x}\right)$$
$$(B.15)$$

We obtain predicted values of ρ, V, and T, denoted by barred quantities, from

$$\bar{\rho}_i^{t+\Delta t} = \rho_i^t + \left(\frac{\partial \rho}{\partial t}\right)_i^t \Delta t \qquad (B.16)$$

$$\bar{V}_i^{t+\Delta t} = V_i^t + \left(\frac{\partial V}{\partial t}\right)_i^t \Delta t \qquad (B.17)$$

$$\bar{T}_i^{t+\Delta t} = T_i^t + \left(\frac{\partial T}{\partial t}\right)_i^t \Delta t \qquad (B.18)$$

In Eqs. (B.16) to (B.18), p_i^t, V_i^t, and T_i^t are known values at time t. Numbers for the time derivatives in Eqs. (B.16) to (B.18) are supplied directly by Eqs. (B.13) to (B.15)

Moving to the corrector step, we return to Eqs. (B.8), (B.10), and (B.12) and replace the spatial derivatives with rearward differences, using the predicted (barred) quantities. We have from Eq. (B.8)

$$
\left(\frac{\overline{\partial \rho}}{\partial t} \right)_i^{t+\Delta t} = -\bar{\rho}_i^{t+\Delta t} \frac{\bar{V}_i^{t+\Delta t} - \bar{V}_{i-1}^{t+\Delta t}}{\Delta x} - \bar{\rho}_i^{t+\Delta t} \bar{V}_i^{t+\Delta t} \frac{\ln A_i - \ln A_{i-1}}{\Delta x}
$$
$$
- \bar{V}_i^{t+\Delta t} \frac{\bar{\rho}_i^{t+\Delta t} - \bar{\rho}_{i-1}^{t+\Delta t}}{\Delta x}
\tag{B.19}
$$

From Eq. (B.10), we have

$$
\left(\frac{\overline{\partial V}}{\partial t} \right)_i^{t+\Delta t} = -\bar{V}_i^{t+\Delta t} \frac{\bar{V}_i^{t+\Delta t} - \bar{V}_{i-1}^{t+\Delta t}}{\Delta x}
$$
$$
- \frac{1}{\gamma} \left(\frac{\bar{T}_i^{t+\Delta t} - \bar{T}_{i-1}^{t+\Delta t}}{\Delta x} + \frac{\bar{T}_i^{t+\Delta t}}{\bar{\rho}_i^{t+\Delta t}} \frac{\bar{\rho}_i^{t+\Delta t} - \bar{\rho}_{i-1}^{t+\Delta t}}{\Delta x} \right)
\tag{B.20}
$$

From Eq. (B.12), we have

$$
\left(\frac{\overline{\partial T}}{\partial t} \right)_i^{t+\Delta t} = -\bar{V}_i^{t+\Delta t} \frac{\bar{T}_i^{t+\Delta t} - \bar{T}_{i-1}^{t+\Delta t}}{\Delta x} - (\gamma - 1)\bar{T}_i^{t+\Delta t}
$$
$$
\times \left(\frac{\bar{V}_i^{t+\Delta t} - \bar{V}_{i-1}^{t+\Delta t}}{\Delta x} + \bar{V}_i^{t+\Delta t} \frac{\ln A_i - \ln A_{i-1}}{\Delta x} \right)
\tag{B.21}
$$

The average time derivatives are given by

$$
\left(\frac{\partial \rho}{\partial t} \right)_{\text{av}} = 0.5 \left[\underbrace{\left(\frac{\partial \rho}{\partial t} \right)_i^t}_{\substack{\text{From} \\ \text{Eq. (B.13)}}} + \underbrace{\left(\frac{\overline{\partial \rho}}{\partial t} \right)_i^{t+\Delta t}}_{\substack{\text{From} \\ \text{Eq. (B.19)}}} \right]
\tag{B.22}
$$

$$
\left(\frac{\partial V}{\partial t} \right)_{\text{av}} = 0.5 \left[\underbrace{\left(\frac{\partial V}{\partial t} \right)_i^t}_{\substack{\text{From} \\ \text{Eq. (B.14)}}} + \underbrace{\left(\frac{\overline{\partial V}}{\partial t} \right)_i^{t+\Delta t}}_{\substack{\text{From} \\ \text{Eq. (B.20)}}} \right]
\tag{B.23}
$$

$$
\left(\frac{\partial T}{\partial t} \right)_{\text{av}} = 0.5 \left[\underbrace{\left(\frac{\partial T}{\partial t} \right)_i^t}_{\substack{\text{From} \\ \text{Eq. (B.15)}}} + \underbrace{\left(\frac{\overline{\partial T}}{\partial t} \right)_i^{t+\Delta t}}_{\substack{\text{From} \\ \text{Eq. (B.21)}}} \right]
\tag{B.24}
$$

Finally, we have for the corrected values of the flowfield variables at time $t + \Delta t$

$$\rho_i^{t+\Delta t} = \rho_i^t + \left(\frac{\partial \rho}{\partial t}\right)_{\text{av}} \Delta t \tag{B.25}$$

$$V_i^{t+\Delta t} = V_i^t + \left(\frac{\partial V}{\partial t}\right)_{\text{av}} \Delta t \tag{B.26}$$

$$T_i^{t+\Delta t} = T_i^t + \left(\frac{\partial T}{\partial t}\right)_{\text{av}} \Delta t \tag{B.27}$$

Keep in mind that all the variables in Eqs. (B.13) to (B.27) are the *nondimensional* values. Also, Eqs. (B.13) to (B.27) constitute the finite-difference expressions of the governing equations in a form that pertains to MacCormack's technique.

Calculation of Time Step

We now proceed to the setting up of other details necessary for the numerical solution of the quasi-one-dimensional nozzle flow problem. First, we ask the question: What about the magnitude of Δt? The governing system of equations, Eqs. (B.4) to (B.6), is hyperbolic with respect to time. A stability constraint exists on this system, namely,

$$\Delta t = C \frac{\Delta x}{a + V} \tag{B.28}$$

where C is the *Courant number;* the simple stability analysis of a linear hyperbolic equation gives the result that $C \leq 1$ for an explicit numerical solution to be stable. The present application to subsonic-supersonic isentropic nozzle flow is governed by *nonlinear* partial differential equations, namely, Eqs. (B.8), (B.10), and (B.12). In this case, the exact stability criterion for a linear equation, namely, that $C \leq 1$, can only be viewed as general guidance for our present nonlinear problem. However, it turns out to be quite good guidance, as we shall see. Equation (B.28) is the *Courant-Friedrichs-Lowry (CFL) criterion* for a one-dimensional flow, where V is the local flow velocity at a point in the flow and a is the local speed of sound. Equation (B.28), along with $C \leq 1$, simply states that Δt must be less than, or at best equal to, the time it takes a sound wave to move from one grid point to the next. Note that t, x, a, and V are nondimensionalized. The nondimensional form of Eq. (B.28) is exactly the same form as the dimensional case. (Prove this to yourself.) Hence, we will hereafter treat the variables in Eq. (B.28) as our nondimensional variables defined earlier. That is, in Eq. (B.28), Δt is the increment in nondimensional time and Δx is the increment in nondimensional space; Δt and Δx in Eq. (B.28) are precisely the same as appear in the nondimensional equations (B.13) to (B.27). Examining Eq. (B.28) more carefully, we note that, although Δx is the same throughout the flow, both V and a are variables. Hence, at a given grid point at a given time step, Eq. (B.28) is written as

$$(\Delta t)_i^t = C \frac{\Delta x}{a_i^t + V_i^t} \tag{B.29}$$

At an adjacent grid point, we have from Eq. (B.28)

$$(\Delta t)_{i+1}^t = C \frac{\Delta x}{a_{i+1}^t + V_{i+1}^t} \tag{B.30}$$

Clearly, $(\Delta t)_i^t$ and $(\Delta t)_{i+1}^t$ obtained from Eqs. (B.29) and (B.30), respectively are, in general, different values. Hence, in the implementation of the time-marching solution, we have two choices:

1. In utilizing Eqs. (B.16) to (B.18) and (B.25) to (B.27), we can, at each grid point i, employ the *local* values of $(\Delta t)_i^t$ determined from Eq. (B.29). In this fashion, the flowfield variables at each grid point in Fig. B.1 will be advanced in time according to their own, local time step. Hence, the resulting flowfield at time $t + \Delta t$ will be in a type of artificial *"time warp,"* with the flowfield variables at a given grid point corresponding to some nonphysical time different from that of the variables at an adjacent grid point. Clearly, such a *local time-stepping* approach does not realistically follow the *actual, physical transients* in the flow and hence cannot be used for an accurate solution of the *unsteady* flow. However, if the final steady-state flowfield in the limit of large time is the only desired result, then the intermediate variation of the flowfield variables with time is irrelevant. Indeed, if such is the case, the *local* time-stepping will frequently lead to *faster* convergence to the steady state. This is why some practitioners use the local time-stepping approach. However, there is always a philosophical question that arises here, namely, does the *local* time-stepping method always lead to the *correct* steady state? Although the answer is usually yes, there is still some reason for a small feeling of discomfort in this regard.

2. The other choice is to calculate $(\Delta t)_i^t$ at all the grid points, $i = 1$ to $i = N$, and then choose the *minimum* value for use in Eqs. (B.16) to (B.18) and (B.12) to (B.27). That is,

$$\Delta t = \text{minimum}\left(\Delta t_1^t, \Delta t_2^t, \ldots, \Delta t_i^t, \ldots, \Delta t_N^t\right) \tag{B.31}$$

The resulting Δt obtained from Eq. (B.31) is then used in Eqs. (B.16) to (B.18) and (B.25) to (B.27). In this fashion, the flowfield variables at all the grid points at time $t + \Delta t$ all correspond to the *same* physical time. Hence, the time-marching solution is following the actual unsteady flow variations that would exist in nature; i.e., the solution gives a time-accurate solution of the actual transient flowfield, consistent with the unsteady continuity, momentum, and energy equations. This consistent time-marching is the approach we will use in the present example. Although it may require more time steps to approach the steady state in comparison to the "local" time stepping described earlier, we can feel comfortable that the consistent time-marching approach is giving us the physically meaningful transient variations—which frequently are of intrinsic value by themselves. Thus, in our subsequent calculations, we will use Eq. (B.31) to determine the value of Δt.

Boundary Conditions

Another aspect of the numerical solution is that of *boundary conditions*—an all-important aspect, because without the physically proper implementation of boundary conditions and their numerically proper representation, we have no hope whatsoever in obtaining a proper numerical solution to our flow problem. Returning to Fig. B.1, we note that grid points 1 and N represent the two boundary points on the x axis. Point 1 is essentially in the reservoir; it represents an *inflow* boundary, with flow coming from the reservoir and entering the nozzle. In contrast, point N is an *outflow* boundary, with flow leaving the nozzle at the nozzle exit. Moreover, the flow velocity at point 1 is a very low, subsonic value. (The flow velocity at point 1, which corresponds to a finite area ratio A_1/A^*, cannot be precisely zero; if it were, there would be no mass flow entering the nozzle. Hence, point 1 does not correspond *exactly* to the reservoir, where by definition the flow velocity is zero. That is, the area for the reservoir is theoretically infinite, and we are clearly starting our own calculation at point 1 where the cross-sectional area is finite.) Hence, not only is point 1 an *inflow* boundary, it is a *subsonic* inflow boundary. *Question:* Which flow quantities should be specified at this subsonic inflow boundary and which should be calculated as part of the solution (i.e., allowed to "float" as a function of time)? A formal answer can be obtained by using the method of characteristics for an unsteady, one-dimensional flow, as introduced in Chap. 7. We did not develop the method of characteristics in Chap. 7 to the extent necessary to precisely study this question about the boundary conditions; indeed, such a matter is beyond the scope of this book. However, we will mention the result of such a study, which you will find to be physically acceptable. Unsteady, inviscid flow is governed by hyperbolic equations, and therefore for one-dimensional unsteady flow there exist two real characteristic lines through any point in the xt plane. Physically, these two characteristics represent infinitely weak Mach waves that are propagating upstream and downstream, respectively. Both Mach waves are traveling at the speed of sound a. Now turn to Fig. B.2, which shows our convergent-divergent nozzle (Fig. B.2a) with an xt diagram sketched below it (Fig. B.2b). Concentrate on grid point 1 in the xt plane in Fig. B.2b. At point 1, the local flow velocity is subsonic, $V_1 < a_1$. Hence, the left-running characteristic at point 1 travels *upstream,* to the left in Fig. B.2; i.e., the left-running Mach wave, which is traveling toward the left (relative to a moving fluid element) at the speed of sound easily works its way *upstream* against the low-velocity subsonic flow, which is slowly moving from left to right. Hence, in Fig. B.2b, we show the left-running characteristic running to the left with a combined speed $a_1 - V_1$ (relative to the fixed nozzle in Fig. B.2a). Since the domain for the flowfield to be calculated is contained between grid points 1 and N, then at point 1 we see that the left-running characteristic is propagating *out of* the domain; it is propagating to the left, away from the domain. In contrast, the right-running characteristic, which is a Mach wave propagating to the right at the speed of sound relative to a fluid element, is clearly moving toward the right in Fig. B.2b. This is for two reasons: (1) the fluid element at point 1 is already moving toward the right, and (2) the right-running Mach wave (characteristic) is moving toward the right at the speed of sound relative to the fluid element. Hence, the right-running characteristic is propagating to the right (relative to the

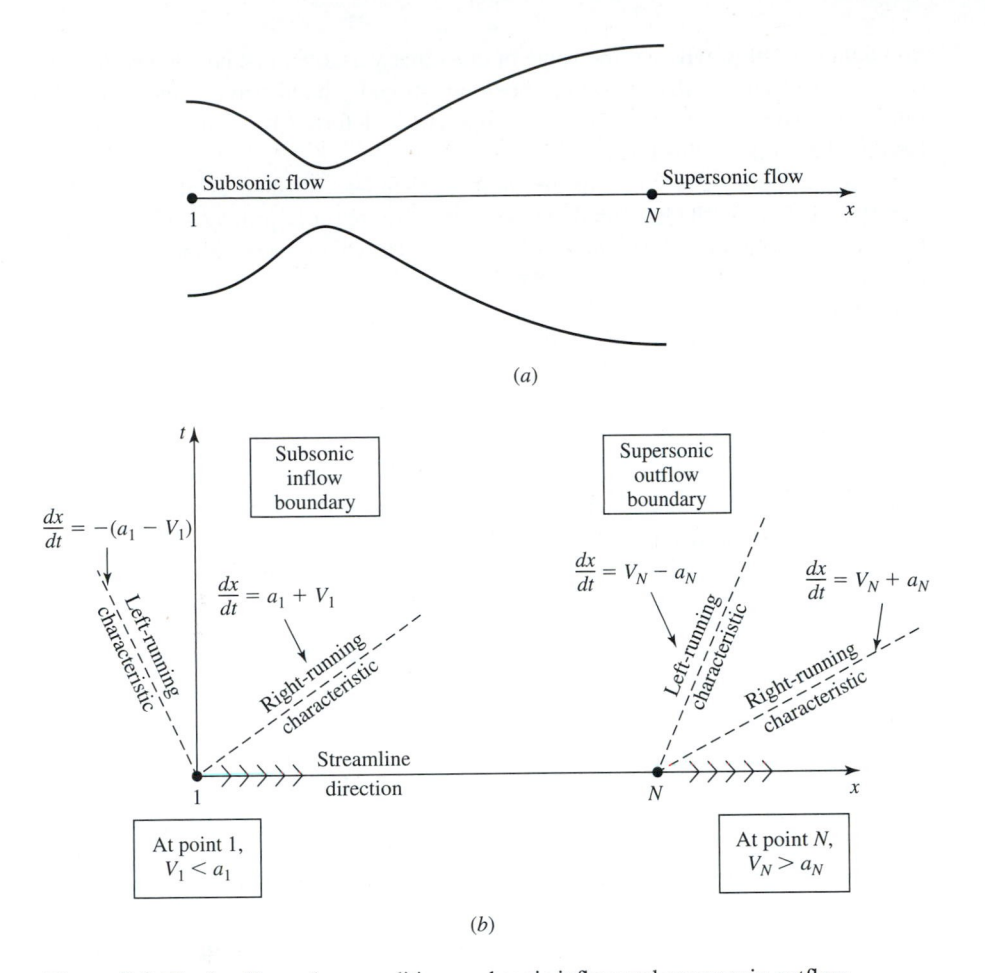

Figure B.2 | Study of boundary conditions: subsonic inflow and supersonic outflow.

nozzle) at a combined velocity of $V_1 + a_1$. What we see here is that the right-running characteristic is propagating from point 1 *into* the domain of the calculation.

What does all this have to do with boundary conditions? The method of characteristics tells us that at a boundary where one characteristic propagates *into* the domain, then the value of one dependent flowfield variable must be *specified* at that boundary, and if one characteristic line propagates *out* of the domain, then the value of another dependent flowfield variable must be allowed to *float* at the boundary; i.e., it must be calculated in steps of time as a function of the timewise solution of the flowfield. Also, note that at point 1 a streamline flows *into* the domain, across the inflow boundary. In terms of denoting what should and should not be specified at the boundary, the streamline *direction* plays the same role as the characteristic directions; i.e., the streamline moving *into* the domain at point 1 stipulates that the value of a second flowfield variable must be *specified* at the inflow boundary. *Conclusion:* At the *subsonic inflow boundary,* we must *stipulate* the values of *two* dependent flowfield variables, whereas the value of *one* other variable must be allowed to *float*. (Please note that this discussion has been intentionally hand-waving and somewhat intuitive; a rigorous mathematical development is deferred for your future studies, beyond the scope of this book.)

Let us apply these ideas to the *outflow* boundary, located at grid point N in Fig. B.2. As before, the left-running characteristic at point N propagates to the left at the speed of sound a *relative to a fluid element*. However, because the speed of the fluid element itself is supersonic, the left-running characteristic is carried *downstream* at the speed (relative to the nozzle) of $V_N - a_N$. The right-running characteristic at point N propagates to the right at the speed of sound a relative to the fluid element, and thus it is swept downstream at the speed (relative to the nozzle) of $V_N + a_N$. Hence, at the *supersonic outflow boundary,* we have both characteristics propagating *out* of the domain; so does the streamline at point N. Therefore, there are *no* flowfield variables that require their values to be stipulated at the supersonic outflow boundary; *all* variables must be allowed to *float* at this boundary.

This discussion details how the inflow and outflow boundary conditions are to be handled on an *analytical* basis. The *numerical* implementation of this discussion is carried out as follows.

Subsonic Inflow Boundary (Point 1).   Here, we must allow one variable to float; we choose the velocity V_1, because on a physical basis we know the mass flow through the nozzle must be allowed to adjust to the proper steady state, and allowing V_1 to float makes the most sense as part of this adjustment. The value of V_1 changes with time and is calculated from information provided by the flowfield solution over the internal points. (The *internal* points are those *not* on a boundary, i.e., points 2 through $N - 1$ in Fig. B.1). We use linear extrapolation from points 2 and 3 to calculate V_1. This is illustrated in Fig. B.3. Here, the slope of the linear extrapolation line is determined from points 2 and 3 as

$$\text{Slope} = \frac{V_3 - V_2}{\Delta x}$$

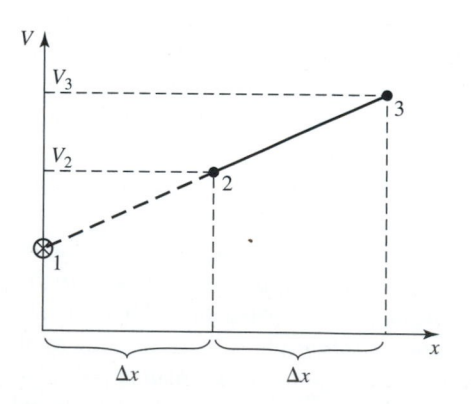

Figure B.3 | Sketch for linear extrapolation.

Using this slope to find V_1 by linear extrapolation, we have

$$V_1 = V_2 - \frac{V_3 - V_2}{\Delta x} \Delta x$$

or $\qquad\qquad\qquad\qquad V_1 = 2V_2 - V_3 \qquad\qquad\qquad\qquad$ (B.32)

All other flowfield variables are specified. Since point 1 is viewed as essentially the reservoir, we stipulate the density and temperature at point 1 to be their respective stagnation values, ρ_o and T_o, respectively. These are held *fixed,* independent of time. Hence, in terms of the *nondimensional* variables, we have

$$\left.\begin{array}{l} \rho_1 = 1 \\[2mm] T_1 = 1 \end{array}\right\} \text{ fixed, independent of time} \qquad\qquad \text{(B.33)}$$

Supersonic Outflow Boundary (Point N). Here, we must allow *all* flowfield variables to float. We again choose to use linear extrapolation based on the flowfield values at the internal points. Specifically, we have, for the *nondimensional* variables,

$$V_N = 2V_{N-1} - V_{N-2} \qquad\qquad\qquad \text{(B.34}a)$$

$$\rho_N = 2\rho_{N-1} - \rho_{N-2} \qquad\qquad\qquad \text{(B.34}b)$$

$$T_N = 2T_{N-1} - T_{N-2} \qquad\qquad\qquad \text{(B.34}c)$$

Nozzle Shape and Initial Conditions

The nozzle shape, $A = A(x)$, is specified and held fixed, independent of time. For the case illustrated in this appendix, we choose a parabolic area distribution given by

$$A = 1 + 2.2(x - 1.5)^2 \qquad 0 \leq x \leq 3 \qquad\qquad \text{(B.35)}$$

Note that $x = 1.5$ is the throat of the nozzle, that the convergent section occurs for $x < 1.5$, and that the divergent section occurs for $x > 1.5$. This nozzle shape is drawn to scale in Fig. B.4.

To start the time-marching calculations, we must stipulate *initial* conditions for ρ, T, and V as a function of x; that is, we must set up values of ρ, T, and V at time $t = 0$. In *theory,* these initial conditions can be purely arbitrary. In practice, there are two reasons why you want to choose the initial conditions *intelligently:*

1. The closer the initial conditions are to the final steady-state answer, the faster the time-marching procedure will converge, and hence the shorter will be the computer execution time.

2. If the initial conditions are too far away from reality, the initial timewise gradients at early time steps can become huge; i.e., the *time derivatives* themselves are initially very large. For a given time step Δt and a given spatial resolution Δx, it has been the author's experience that *inordinately* large gradients during the early part of the time-stepping procedure can cause the program to go unstable. In a sense, you can visualize the behavior of a

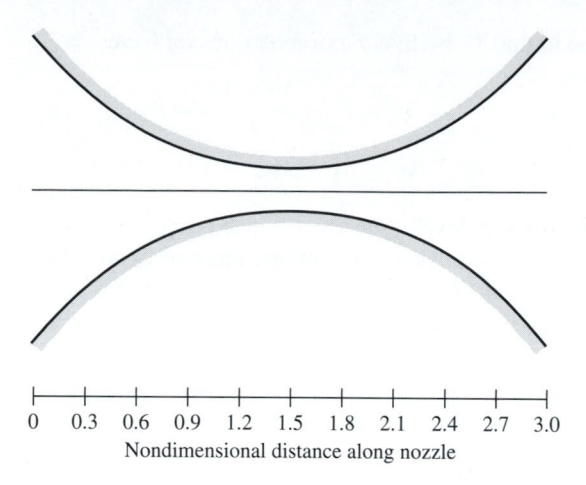

0   0.3   0.6   0.9   1.2   1.5   1.8   2.1   2.4   2.7   3.0
Nondimensional distance along nozzle

Figure B.4 | Shape of the nozzle used for the present calculations. This geometric picture is not unique; for a calorically perfect gas, what is germane is the area ratio distribution along the nozzle. Hence, assuming a two-dimensional nozzle, the ordinates of the shape shown here can be ratioed by any constant factor, and the nozzle solution would be the same.

time-marching solution as a stretched rubber band. At early times, the rubber band is highly stretched, thus providing a *strong* potential to push the flowfield *rapidly* toward the steady-state solution. As time progresses, the flowfield gets closer to the steady-state solution, and the rubber band progressively relaxes, hence slowing down the rate of approach [i.e., at larger times, the values of the time derivatives calculated from Eqs. (B.22) to (B.24) become progressively smaller]. At the beginning of the calculation, it is wise not to pick initial conditions which are so far off that the rubber band is "stretched too far," and may even break.

Therefore, in your choice of initial conditions, you are encouraged to use *any* knowledge you may have about a given problem in order to intelligently pick some initial conditions. For example, in the present problem, we know that ρ and T *decrease* and V *increases* as the flow expands through the nozzle. Hence, we choose initial conditions that *qualitatively* behave in the same fashion. For simplicity, let us assume linear variations of the flowfield variables, as a function of x. For the present case, we assume these values at time $t = 0$.

$$\rho = 1 - 0.3146x \qquad \qquad \text{(B.36a)}$$

$$T = 1 - 0.2314x \qquad \text{initial conditions at } t = 0 \qquad \text{(B.36b)}$$

$$V = (0.1 + 1.09x)T^{1/2} \qquad \qquad \text{(B.36c)}$$

INTERMEDIATE NUMERICAL RESULTS: THE FIRST FEW STEPS

In this section, we give a few numerical results that reflect the first stages of the calculation. This is to give you a more solid impression of what is going on and to provide some intermediate results for you to compare with when you write and run your own computer solution to this problem.

The first step is to feed the nozzle shape and the initial conditions into the program. These are given by Eqs. (B.35) and (B.36); the resulting numbers are tabulated in Table B.1. The values of ρ, V, and T given in this table are for $t = 0$.

The next step is to put these initial conditions into Eqs. (B.13) to (B.15) to initiate calculations pertaining to the predictor step. For purposes of illustration, let us return to the sketch shown in Fig. B.1 and focus on the calculations associated with

Table B.1 | Nozzle shape and initial conditions

$\dfrac{x}{L}$	$\dfrac{A}{A^*}$	$\dfrac{\rho}{\rho_o}$	$\dfrac{V}{a_o}$	$\dfrac{T}{T_o}$
0	5.950	1.000	0.100	1.000
0.1	5.312	0.969	0.207	0.977
0.2	4.718	0.937	0.311	0.954
0.3	4.168	0.906	0.412	0.931
0.4	3.662	0.874	0.511	0.907
0.5	3.200	0.843	0.607	0.884
0.6	2.782	0.811	0.700	0.861
0.7	2.408	0.780	0.790	0.838
0.8	2.078	0.748	0.877	0.815
0.9	1.792	0.717	0.962	0.792
1.0	1.550	0.685	1.043	0.769
1.1	1.352	0.654	1.122	0.745
1.2	1.198	0.622	1.197	0.722
1.3	1.088	0.591	1.268	0.699
1.4	1.022	0.560	1.337	0.676
1.5	1.000	0.528	1.402	0.653
1.6	1.022	0.497	1.463	0.630
1.7	1.088	0.465	1.521	0.607
1.8	1.198	0.434	1.575	0.583
1.9	1.352	0.402	1.625	0.560
2.0	1.550	0.371	1.671	0.537
2.1	1.792	0.339	1.713	0.514
2.2	2.078	0.308	1.750	0.491
2.3	2.408	0.276	1.783	0.468
2.4	2.782	0.245	1.811	0.445
2.5	3.200	0.214	1.834	0.422
2.6	3.662	0.182	1.852	0.398
2.7	4.168	0.151	1.864	0.375
2.8	4.718	0.119	1.870	0.352
2.9	5.312	0.088	1.870	0.329
3.0	5.950	0.056	1.864	0.306

grid point i. We will choose $i = 16$, which is the grid point at the throat of the nozzle drawn in Fig. B.4. From the initial data given in Table B.1, we have

$$\rho_i = \rho_{16} = 0.528$$

$$\rho_{i+1} = \rho_{17} = 0.497$$

$$V_i = V_{16} = 1.402$$

$$V_{i+1} = V_{17} = 1.463$$

$$T_i = T_{16} = 0.653$$

$$T_{i+1} = T_{17} = 0.630$$

$$\Delta x = 0.1$$

$$A_i = A_{16} = 1.0 \qquad \ln A_{16} = 0$$

$$A_{i+1} = A_{17} = 1.022 \qquad \ln A_{17} = 0.02176$$

Substitute these values into Eq. (B.13).

$$\left(\frac{\partial \rho}{\partial t}\right)_{16}^{t=0} = -0.528\left(\frac{1.463 - 1.402}{0.1}\right) - 0.528(1.402)\left(\frac{0.02176 - 0}{0.1}\right)$$

$$- 1.402\left(\frac{0.497 - 0.528}{0.1}\right)$$

$$= \boxed{-0.0445}$$

Substitute these values into Eq. (B.14).

$$\left(\frac{\partial V}{\partial t}\right)_{16}^{t=0} = -1.402\left(\frac{1.463 - 1.402}{0.1}\right)$$

$$- \frac{1}{1.4}\left[\frac{0.630 - 0.653}{0.1} + \frac{0.653}{0.528}\left(\frac{0.497 - 0.528}{0.1}\right)\right]$$

$$= \boxed{-0.418}$$

Substitute these values into Eq. (B.15).

$$\left(\frac{\partial T}{\partial t}\right)_{16}^{t=0} = -1.402\left(\frac{0.630 - 0.653}{0.1}\right) - (1.4 - 1)(0.653)$$

$$\times \left[\frac{1.463 - 1.402}{0.1} + 1.402\left(\frac{0.02176 - 0}{0.1}\right)\right]$$

$$= \boxed{0.0843}$$

Please note: The numbers shown in the *boxes* here are the precise numbers, rounded to three significant figures, that came out of the author's Macintosh computer. If you choose to run through these calculations with your hand calculator using all these entries, there will be slight differences because the numbers you feed into the calculator are *already* rounded to three significant figures, and hence the subsequent arithmetic operations on your calculator will lead to slight errors compared to the computer results. That is, your hand-calculator results may not always give you *precisely* the numbers you will find in the boxes, but they will certainly be close enough to check the results.

The next step is to calculate the *predicted* values (the "barred" quantities) from Eqs. (B.16) to (B.18). To do this, we first note that Δt is calculated from Eq. (B.31), which picks the minimum value of Δt_i from all those calculated from Eq. (B.29) evaluated for all internal points $i = 2, 3, \ldots, 30$. We do not have the space to show all these calculations here. As a sample calculation, let us calculate $(\Delta t)_{16}^{t=0}$ from Eq. (B.29). At present, we will assume a Courant number equal to 0.5; that is, $C = 0.5$. Also, in nondimensional terms, the speed of sound is given by

$$a = \sqrt{T} \tag{B.37}$$

where in Eq. (B.37) both a and T are the *nondimensional* values (a denotes the local speed of sound divided by a_0). Derive Eq. (B.37) for yourself. Thus, from Eq. (B.29), we have

$$(\Delta t)_{16}^{t=0} = C \left[\frac{\Delta x}{(T_{16})^{1/2} + V_{16}} \right] = 0.5 \left[\frac{0.1}{(0.653)^{1/2} + 1.402} \right] = 0.0226$$

This type of calculation is made at all the interior grid points, and the minimum value is chosen. The resulting minimum value is

$$\Delta t = 0.0201$$

With this, we can calculate $\bar{\rho}$, $\bar{V}$, and $\bar{T}$. From Eq. (B.16), noting that $t = 0 + \Delta t = \Delta t$,

$$\bar{\rho}_{16}^{t=\Delta t} = \rho_{16}^{t=0} + \left(\frac{\partial \rho}{\partial t} \right)_{16}^{t=0} \Delta t = 0.528 + (-0.0445)(0.0201)$$

$$= \boxed{0.527}$$

From Eq. (B.17),

$$\bar{V}_{16}^{t=\Delta t} = V_{16}^{t=0} + \left(\frac{\partial V}{\partial t} \right)_{16}^{t=0} \Delta t = 1.402 + (-0.418)(0.0201)$$

$$= \boxed{1.39}$$

From Eq. (B.18),

$$\bar{T}_{16}^{t=\Delta t} = T_{16}^{t=0} + \left(\frac{\partial T}{\partial t} \right)_{16}^{t=0} \Delta t = 0.653 + (0.0843)(0.0201)$$

$$= \boxed{0.655}$$

At this stage, we note that these calculations are carried out over *all* the internal grid points $i = 2$ to 30. The calculations are too repetitive to include here. Simply note that when the predictor step is completed, we have $\bar{\rho}$, $\bar{V}$, and $\bar{T}$ at all the internal grid points $i = 2$ to 30. This includes, of course, $\bar{\rho}_{15}^{t=\Delta t}$, $\bar{V}_{15}^{t=\Delta t}$, and $\bar{T}_{15}^{t=\Delta t}$. Focusing again on grid point 16, we now insert these *barred* quantities at grid points 15 and 16 into Eqs. (B.19) to (B.21). This is the beginning of the corrector step. From Eq. (B.19) we have

$$\left(\overline{\frac{\partial \rho}{\partial t}}\right)_{16}^{t=\Delta t} = -0.527(0.653) - 0.527(1.39)(-0.218) - 1.39(-0.368)$$

$$= \boxed{0.328}$$

From Eq. (B.20) we have

$$\left(\overline{\frac{\partial V}{\partial t}}\right)_{16}^{t=\Delta t} = -1.39(0.653) - \frac{1}{1.4}\left(-0.257 + \frac{0.655}{0.527}\right) = \boxed{-0.400}$$

From Eq. (B.21) we have

$$\left(\overline{\frac{\partial T}{\partial t}}\right)_{16}^{t=\Delta t} = -1.39(-0.257) - (1.4 - 1)(0.655)[0.653 + 1.39(-0.218)]$$

$$= \boxed{0.267}$$

With these values, we form the *average* time derivatives using Eqs. (B.22) to (B.24). From Eq. (B.22), we have at grid point $i = 16$,

$$\left(\frac{\partial \rho}{\partial t}\right)_{av} = 0.5(-0.0445 + 0.328) = \boxed{0.142}$$

From Eq. (B.23), we have at grid point $i = 16$,

$$\left(\frac{\partial V}{\partial t}\right)_{av} = 0.5(-0.418 - 0.400) = \boxed{-0.409}$$

From Eq. (B.24), we have at grid point $i = 16$,

$$\left(\frac{\partial T}{\partial t}\right)_{av} = 0.5(0.0843 + 0.267) = \boxed{0.176}$$

We now complete the corrector step by using Eqs. (B.25) to (B.27). From Eq. (B.25), we have at $i = 16$,

$$\rho_{16}^{t=\Delta t} = 0.528 + 0.142(0.0201) = \boxed{0.531}$$

From Eq. (B.26), we have at $i = 16$,

$$V_{16}^{t=\Delta t} = 1.402 + (-0.409)(0.0201) = \boxed{1.394}$$

From Eq. (B.27), we have at $i = 16$,

$$T_{16}^{t=\Delta t} = 0.653 + 0.176(0.0201) = \boxed{0.656}$$

Defining a nondimensional pressure as the local static pressure divided by the reservoir pressure p_o, the equation of state is given by

$$p = \rho T$$

where p, ρ, and T are *nondimensional* values. Thus, at grid point $i = 16$, we have

$$p_{16}^{t=\Delta t} = \rho_{16}^{t=\Delta t} T_{16}^{t=\Delta t} = 0.531(0.656) = \boxed{0.349}$$

*This now completes the corrector step for grid point $i = 16$. When the above corrector-step calculations are carried out for all grid points from $i = 2$ to 30, then we have completed the corrector step for all the *internal* grid points.*

It remains to calculate the flowfield variables at the boundary points. At the subsonic inflow boundary ($i = 1$), V_1 is calculated by linear extrapolation from grid points 2 and 3. At the end of the corrector step, from a calculation identical to that given above, the values of V_2 and V_3 at time $t = \Delta t$ are $V_2 = 0.212$ and $V_3 = 0.312$. Thus, from Eq. (B.32), we have

$$V_1 = 2V_2 - V_3 = 2(0.212) - 0.312 = \boxed{0.111}$$

At the supersonic outflow boundary ($i = 31$) all the flowfield variables are calculated by linear extrapolation from Eqs. (B.34a) to (B.34c). At the end of the corrector step, from a calculation identical to that given above, $V_{29} = 1.884$, $V_{30} = 1.890$, $\rho_{29} = 0.125$, $\rho_{30} = 0.095$, $T_{29} = 0.354$, and $T_{30} = 0.332$. When these values are inserted into Eqs. (B.34a) to (B.34c), we have

$$V_{31} = 2V_{30} - V_{29} = 2(1.890) - 1.884 = \boxed{1.895}$$

$$\rho_{31} = 2\rho_{30} - \rho_{29} = 2(0.095) - 0.125 = \boxed{0.066}$$

$$T_{31} = 2T_{30} - T_{29} = 2(0.332) - 0.354 = \boxed{0.309}$$

With this, we have completed the calculation of all the flowfield variables at all the grid points after the first time step, i.e., at time $t = \Delta t$. A tabulation of these variables is given in Table B.2. Note that the Mach number is included in this tabulation. In terms of the nondimensional velocity and temperature, the Mach number (which is already a dimensionless parameter defined as the local velocity divided by the local speed of sound) is given by

$$M = \frac{V}{\sqrt{T}} \tag{B.38}$$

Examine Table B.2 closely. By reading across the line labeled $I = 16$, you will find the familiar numbers that we have generated for grid point $i = 16$ in this discussion. Take the time to make this comparison. The entries for all other internal grid points

Table B.2 | Flowfield variables after the first time step

I	$\dfrac{x}{L}$	$\dfrac{A}{A^*}$	$\dfrac{\rho}{\rho_o}$	$\dfrac{V}{a_o}$	$\dfrac{T}{T_o}$	$\dfrac{p}{p_o}$	M
1	0.000	5.950	1.000	0.111	1.000	1.000	0.111
2	0.100	5.312	0.955	0.212	0.972	0.928	0.215
3	0.200	4.718	0.927	0.312	0.950	0.881	0.320
4	0.300	4.168	0.900	0.411	0.929	0.836	0.427
5	0.400	3.662	0.872	0.508	0.908	0.791	0.534
6	0.500	3.200	0.844	0.603	0.886	0.748	0.640
7	0.600	2.782	0.817	0.695	0.865	0.706	0.747
8	0.700	2.408	0.789	0.784	0.843	0.665	0.854
9	0.800	2.078	0.760	0.870	0.822	0.625	0.960
10	0.900	1.792	0.731	0.954	0.800	0.585	1.067
11	1.000	1.550	0.701	1.035	0.778	0.545	1.174
12	1.100	1.352	0.670	1.113	0.755	0.506	1.281
13	1.200	1.198	0.637	1.188	0.731	0.466	1.389
14	1.300	1.088	0.603	1.260	0.707	0.426	1.498
15	1.400	1.022	0.567	1.328	0.682	0.387	1.609
16	1.500	1.000	0.531	1.394	0.656	0.349	1.720
17	1.600	1.022	0.494	1.455	0.631	0.312	1.833
18	1.700	1.088	0.459	1.514	0.605	0.278	1.945
19	1.800	1.198	0.425	1.568	0.581	0.247	2.058
20	1.900	1.352	0.392	1.619	0.556	0.218	2.171
21	2.000	1.550	0.361	1.666	0.533	0.192	2.282
22	2.100	1.792	0.330	1.709	0.510	0.168	2.393
23	2.200	2.078	0.301	1.748	0.487	0.146	2.504
24	2.300	2.408	0.271	1.782	0.465	0.126	2.614
25	2.400	2.782	0.242	1.813	0.443	0.107	2.724
26	2.500	3.200	0.213	1.838	0.421	0.090	2.834
27	2.600	3.662	0.184	1.858	0.398	0.073	2.944
28	2.700	4.168	0.154	1.874	0.376	0.058	3.055
29	2.800	4.718	0.125	1.884	0.354	0.044	3.167
30	2.900	5.312	0.095	1.890	0.332	0.032	3.281
31	3.000	5.950	0.066	1.895	0.309	0.020	3.406

are calculated in a like manner. Also note the values at the boundary points, labeled $I = 1$ and $I = 31$ in Table B.2. You will find the numbers to be the same as discussed here.

FINAL NUMERICAL RESULTS: THE STEADY-STATE SOLUTION

Compare the flowfield results obtained after one time step (Table B.2) with the same quantities at the previous time (in this case the initial conditions given in Table B.1). Comparing these two tables, we see that the flowfield variables *have changed*. For example, the nondimensional density at the throat (where $A = 1$) has changed from 0.528 to 0.531, a 0.57 percent change over one time step. This is the natural behavior of a time-marching solution—the flowfield variables change from one time step to the next. However, in the approach toward the steady-state solution, at larger values of time (after a large number of time steps), the *changes* in the flowfield variables from one time step to the next become smaller and approach zero in the limit of large time. At this stage, the steady state (for all practical purposes) has been achieved, and the calculation can be stopped. This termination of the calculation can be done automatically by the computer program itself by having a test in the program to sense when the changes in the flowfield variables become smaller than some prescribed value (prescribed by you, depending on your desired accuracy of the final "steady-state" solution). Another option, and that preferred by the present author, is to simply stop the calculation after a prescribed number of time steps, look at the results, and see if they have approached the stage where the flowfield variables are not materially changing any more. If such is not the case, simply resume the calculations, and carry them out for the requisite number of time steps until you do see that the steady-state results have been reached.

What patterns do the timewise variations of the flowfield variables take? Some feeling for the answer is provided by Fig. B.5, which shows the variation of ρ, T, p, and M at the nozzle throat plotted versus the number of time steps. The abscissa starts at zero, which represents the initial conditions, and ends at time step 1000. Hence, the abscissa is essentially a time axis, with time increasing to the right. Note that the largest changes take place at early times, after which the final, steady-state value is approached almost asymptotically. Here is the "rubber band effect" mentioned previously; at early times the rubber band is "stretched" tightly, and therefore the flowfield variables are driven by a stronger potential and hence change rapidly. At later times, as the steady state is approached, the rubber band is less stretched; it becomes more "relaxed," and the changes become much smaller with time. The dashed lines to the right of the curves shown in Fig. B.5 represent the exact, analytical values as obtained from the equations discussed in Chap. 5. Note that the numerical time-marching procedure converges to the proper theoretical steady-state answer. We also note that no artificial viscosity has been explicitly added for these calculations; it is not needed.

It is interesting to examine the variation of the time derivatives as a function of time itself, or equivalently as a function of the number of time steps. Once again focusing on the nozzle throat (at grid point $i = 16$), Fig. B.6 gives the variation of the time derivatives of nondimensional density and velocity as a function of the number of time steps. These are the *average* time derivatives calculated from Eqs. (B.22) and (B.23), respectively. The *absolute value* of these time derivatives is shown in Fig. B.6. From these results, note two important aspects:

1. At early times, the time derivatives are large, and they oscillate in value. These oscillations are associated with various unsteady compression and expansion waves that propagate through the nozzle during the transient process. (See Chap. 7.)

2. At later times, the time derivatives rapidly grow small, changing by six orders of magnitude over a span of 1000 time steps. This is, of course, what we want to see happen. In the theoretical limit of the steady state (which is achieved at infinite time), the time derivatives should go to zero. However, numerically

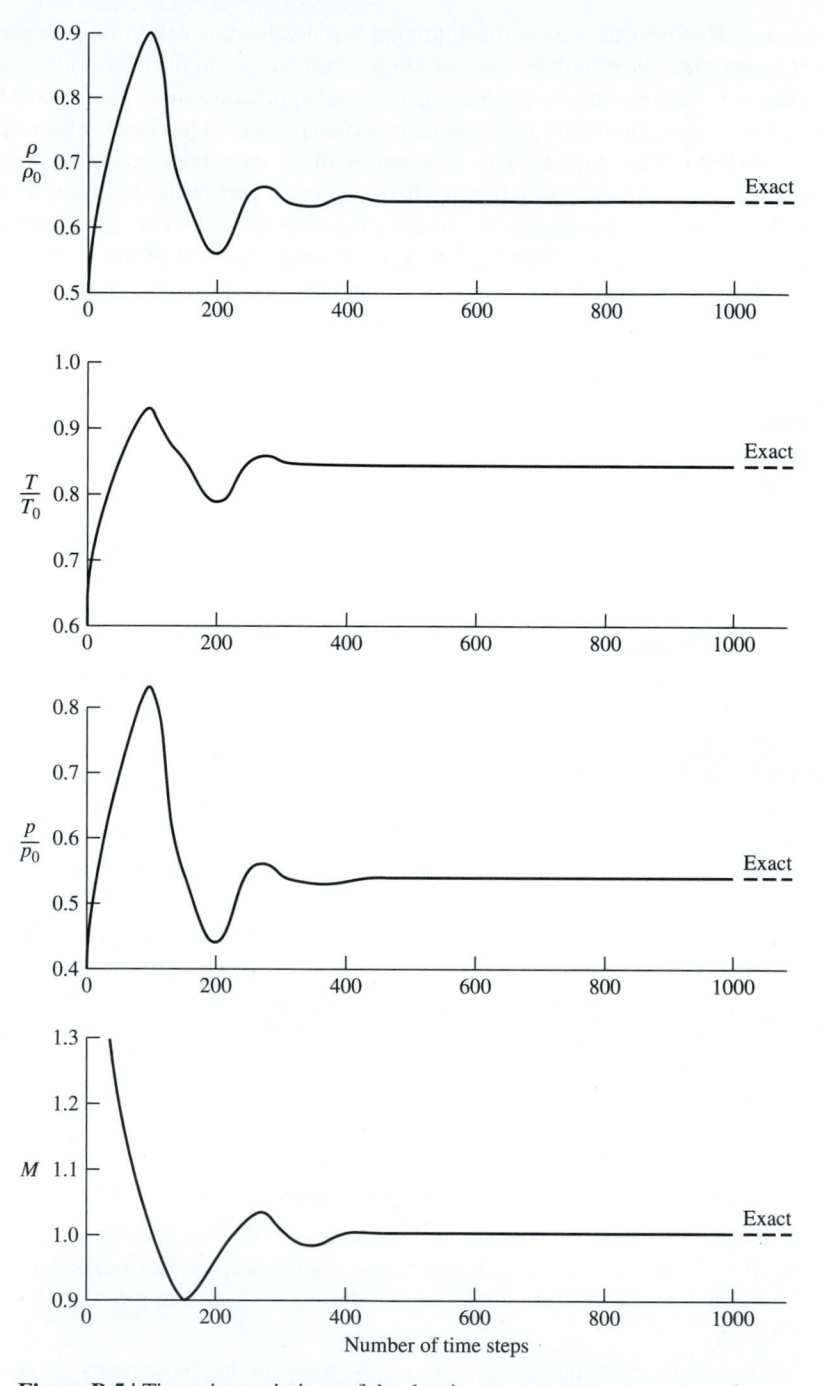

Figure B.5 | Timewise variations of the density, temperature, pressure, and Mach number at the nozzle throat (at grid point $i = 15$, where $A = 1$).

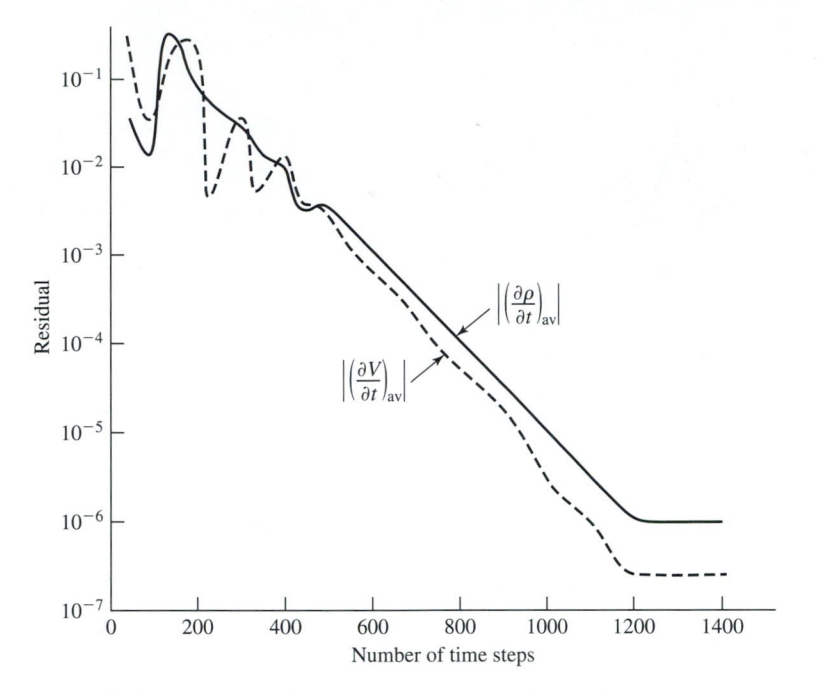

Figure B.6 | Timewise variations of the absolute values of the time derivatives of nondimensional density and velocity at the nozzle throat (at grid point $i = 16$).

this will never happen over a finite number of time steps. In fact, the results shown in Fig. B.6 indicate that the values of the time derivatives plateau after 1200 time steps. This seems to be a characteristic of MacCormack's technique. However, the values of the time derivatives at these plateaus are so small that, for all practical purposes, the numerical solution has arrived at the steady-state solution. Indeed, in terms of the values of the flowfield variables themselves, the results of Fig. B.5 indicate that the steady state is realistically achieved after 500 time steps, during which the time derivatives in Fig. B.6 have decreased only by two orders of magnitude.

Return to Eqs. (B.8) and (B.10) for a moment; we might visualize that what is being plotted in Fig. B.6 are the numerical values of the right-hand side of these equations. As time progresses and as the steady state is approached, the right-hand side of these equations should approach zero. Since the *numerical* values of the right-hand side are not precisely zero, they are called *residuals*. This is why the ordinate in Fig. B.6 is labeled as the residual. When CFD experts are comparing the relative merits of two or more different algorithms for a time-marching solution to the steady state, the magnitude of the residuals and their rate of decay are often used as figures of merit. That algorithm that gives the fastest decay of the residuals to the smallest value is usually looked upon most favorably.

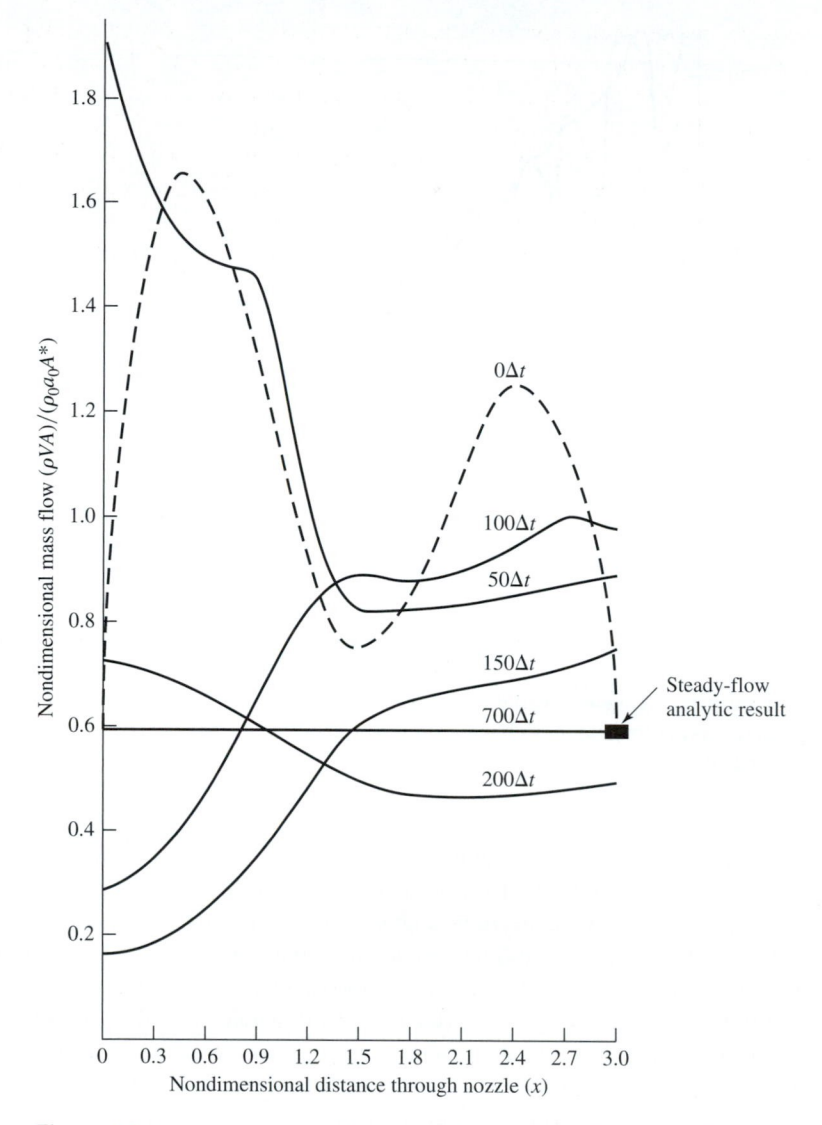

Figure B.7 | Instantaneous distributions of the nondimensional mass flow as a function of distance through the nozzle at six different times during the time-marching approach to the steady state.

Another insight to the mechanics of the timewise variation of the flow and its approach to the steady state is provided by the mass flow variations shown in Fig. B.7. Here, the nondimensional mass flow ρVA (where ρ, V, and A are the nondimensional values) is plotted as a function of nondimensional distance through the nozzle. Six different curves are shown, each for a different time during the course of the time-marching procedure. The dashed curve is the variation of ρVA, which pertains to the initial conditions, and hence it is labeled $0\Delta t$. The strange-looking, distorted sinelike variation of this dashed curve is simply the product of the assumed initial values for ρ and V combined with the specified parabolic variation of the nozzle area ratio A. After 50 time steps, the mass flow distribution through the nozzle has changed considerably; this is given by the curve labeled $50\Delta t$. After 100 time steps ($100\Delta t$), the mass flow distribution has changed radically; the mass flow variation is simply flopping around inside the nozzle due to the transient variation of the flowfield variables. However, after 200 time steps ($200\Delta t$), the mass flow distribution is beginning to settle down, and after 700 time steps ($700\Delta t$), the mass flow distribution is a straight, horizontal line across the graph. This says that the mass flow has converged to a *constant,* steady-state value throughout the nozzle. This agrees with our basic knowledge of steady-state nozzle flows, namely, that

$$\rho VA = \text{constant}$$

Moreover, it has converged to essentially the *correct value* of the steady mass flow, which in terms of the *nondimensional* variables evaluated at the nozzle throat is given by

$$\rho VA = \rho^* \sqrt{T^*} \qquad \text{(at throat)} \qquad (B.39)$$

where ρ^* and T^* are the nondimensional density and temperature at the throat, and where $M = 1$. [Derive Eq. (B.39) yourself—it is easy.] From the analytical equations discussed in Chap. 5, when $M = 1$ and $\gamma = 1.4$, we have $\rho^* = 0.634$ and $T^* = 0.833$. With these numbers, Eq. (B.39) yields

$$\rho VA = \text{constant} = 0.579$$

This value is given by the dark square in Fig. B.7; the mass flow result for $700\Delta t$ agrees reasonably well with the dark square.

Finally, let us examine the steady-state results. From our discussion and from examining Fig. B.5, the steady state is, for all practical purposes, reached after about 500 time steps. However, being very conservative, we will examine the results obtained after 1400 time steps; between 700 and 1400 time steps, there is no change in the results, at least to the three-decimal-place accuracy given in the tables herein.

A feeling for the graphical accuracy of the numerically obtained steady state is given by Fig. B.8. Here, the steady-state nondimensional density and Mach number distributions through the nozzle are plotted as a function of nondimensional distance along the nozzle. The numerical results, obtained after 1400 time steps, are given by the solid curves, and the exact analytical results are given by the circles. The analytical results are obtained from the equations discussed in Chap. 5; they can readily be obtained from the tables in App. A. They can also be obtained by writing your own short computer program to calculate numbers from the theoretically derived equations in Chap. 5. In any event, the comparison shown in Fig. B.8 clearly demonstrates that the numerical results agree very well with the exact analytical values, certainly to within graphical accuracy.

The detailed numerical results, to three decimal places, are tabulated in Table B.3. These are the results obtained after 1400 time steps. They are given here for you to

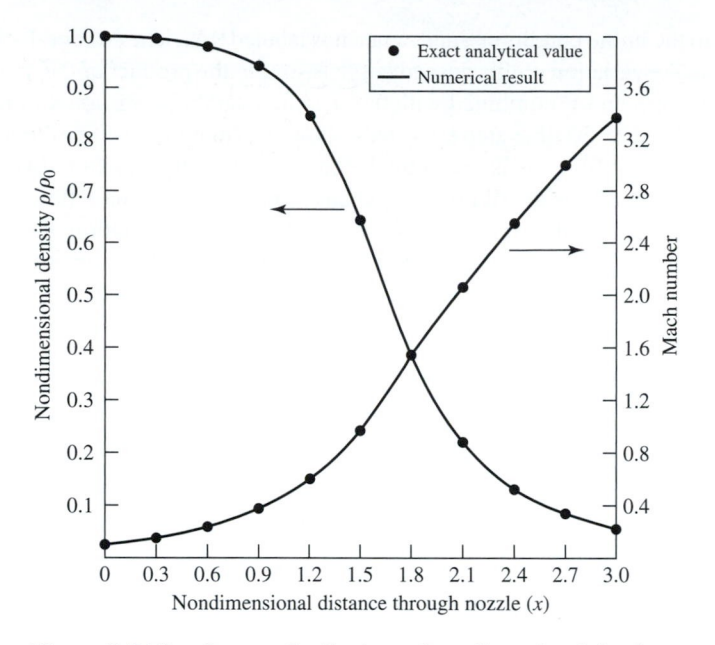

Figure B.8 | Steady-state distributions of nondimensional density and Mach number as a function of nondimensional distance through the nozzle. Comparison between the exact analytical values (circles) and the numerical results (solid curves).

compare numbers from your own computer program. It is interesting to note that the elapsed nondimensional time, starting at zero with the initial conditions, is, after 1400 time steps, a value of 28.952. Since time is nondimensionalized by the quantity L/a_o, let us assume a case where the length of the nozzle is 1 m and the reservoir temperature is the standard sea level value, $T = 288$ K. For this case, $L/a_o = (1 \text{ m})/(340.2 \text{ m/s}) = 2.94 \times 10^{-3}$ s. Hence, the total *real* time that has elapsed over the 1400 time steps is $(2.94 \times 10^{-3})(28.952) = 0.0851$ s. That is, the nozzle flow, starting from the assumed initial conditions, takes only 85.1 ms to reach steady-state conditions; in reality, since convergence is obtained for all practical purposes after about 500 time steps, the practical convergence time is more on the order of 30 ms.

A comparison between some of the numerical results and the corresponding exact analytical values is given in Table B.4; this provides you with a more detailed comparison than is given in Fig. B.8. Compared are the numerical and analytical results for the density ratio and Mach number. Note that the numerical results, to three decimal places, are not in precise agreement with the analytical values; there is a small percentage disagreement between the two sets of results, ranging from 0.3 to 3.29 percent. This amount of error is not discernable on the graphical display in Fig. B.8. At first thought, there might be three reasons for these small numerical inaccuracies: (1) a small inflow boundary condition error, (2) truncation errors associated with the

Table B.3 | Flowfield variables after 1400 time steps (nonconservation form of the governing equations)

I	$\dfrac{x}{L}$	$\dfrac{A}{A^*}$	$\dfrac{\rho}{\rho_o}$	$\dfrac{V}{a_o}$	$\dfrac{T}{T_o}$	$\dfrac{p}{p_o}$	M	$\dot{m}$
1	0.000	5.950	1.000	0.099	1.000	1.000	0.099	0.590
2	0.100	5.312	0.998	0.112	0.999	0.997	0.112	0.594
3	0.200	4.718	0.997	0.125	0.999	0.996	0.125	0.589
4	0.300	4.168	0.994	0.143	0.998	0.992	0.143	0.591
5	0.400	3.662	0.992	0.162	0.997	0.988	0.163	0.589
6	0.500	3.200	0.987	0.187	0.995	0.982	0.187	0.589
7	0.600	2.782	0.982	0.215	0.993	0.974	0.216	0.588
8	0.700	2.408	0.974	0.251	0.989	0.963	0.252	0.588
9	0.800	2.078	0.963	0.294	0.985	0.948	0.296	0.587
10	0.900	1.792	0.947	0.346	0.978	0.926	0.350	0.587
11	1.000	1.550	0.924	0.409	0.969	0.895	0.416	0.586
12	1.100	1.352	0.892	0.485	0.956	0.853	0.496	0.585
13	1.200	1.198	0.849	0.575	0.937	0.795	0.594	0.585
14	1.300	1.088	0.792	0.678	0.911	0.722	0.710	0.584
15	1.400	1.022	0.721	0.793	0.878	0.633	0.846	0.584
16	1.500	1.000	0.639	0.914	0.836	0.534	0.099	0.584
17	1.600	1.022	0.551	1.037	0.789	0.434	1.167	0.584
18	1.700	1.088	0.465	1.155	0.737	0.343	1.345	0.584
19	1.800	1.198	0.386	1.263	0.684	0.264	1.528	0.585
20	1.900	1.352	0.318	1.361	0.633	0.201	1.710	0.586
21	2.000	1.550	0.262	1.446	0.585	0.153	1.890	0.587
22	2.100	1.792	0.216	1.519	0.541	0.117	2.065	0.588
23	2.200	2.078	0.179	1.582	0.502	0.090	2.233	0.589
24	2.300	2.408	0.150	1.636	0.467	0.070	2.394	0.590
25	2.400	2.782	0.126	1.683	0.436	0.055	2.549	0.590
26	2.500	3.200	0.107	1.723	0.408	0.044	2.696	0.591
27	2.600	3.662	0.092	1.759	0.384	0.035	2.839	0.591
28	2.700	4.168	0.079	1.789	0.362	0.029	2.972	0.592
29	2.800	4.718	0.069	1.817	0.342	0.024	3.105	0.592
30	2.900	5.312	0.061	1.839	0.325	0.020	3.225	0.595
31	3.000	5.950	0.053	1.862	0.308	0.016	3.353	0.585

finite value of Δx, and (3) possible effects of the Courant number being substantially less than unity (recall that in the calculations discussed so far, the Courant number is chosen to be 0.5). Let us examine each of these reasons in turn.

Inflow Boundary Condition Error

There is a "built-in" error at the inflow boundary. At the first grid point, at $x = 0$, we *assume* that the density, pressure, and temperature are the reservoir properties ρ_o, p_o, and T_o, respectively. This is strictly true only if $M = 0$ at this point. In reality, there is a finite area ratio at $x = 0$, namely, $A/A^* = 5.95$, and hence a finite Mach number must exist at $x = 0$, both numerically and analytically (to allow a finite value of mass flow through the nozzle). Hence, in Table B.4, the numerical value of ρ/ρ_o at $x = 0$ is equal to 1.0—this is our prescribed boundary condition. On the other hand, the exact

Table B.4 | Density ratio and Mach number distributions through the nozzle

$\dfrac{x}{L}$	$\dfrac{A}{A^*}$	$\dfrac{\rho}{\rho_o}$ (numerical results)	$\dfrac{\rho}{\rho_o}$ (exact analytical results)	Difference, %	M (numerical results)	M (exact analytical results)	Difference, %
0.000	5.950	1.000	0.995	0.50	0.099	0.098	1.01
0.100	5.312	0.998	0.994	0.40	0.112	0.110	1.79
0.200	4.718	0.997	0.992	0.30	0.125	0.124	0.08
0.300	4.168	0.994	0.990	0.40	0.143	0.140	2.10
0.400	3.662	0.992	0.987	0.50	0.163	0.160	1.84
0.500	3.200	0.987	0.983	0.40	0.187	0.185	1.07
0.600	2.782	0.982	0.978	0.41	0.216	0.214	0.93
0.700	2.408	0.974	0.970	0.41	0.252	0.249	1.19
0.800	2.078	0.963	0.958	0.52	0.296	0.293	1.01
0.900	1.792	0.947	0.942	0.53	0.350	0.347	0.86
1.000	1.550	0.924	0.920	0.43	0.416	0.413	0.72
1.100	1.352	0.892	0.888	0.45	0.496	0.494	0.40
1.200	1.198	0.849	0.844	0.59	0.594	0.592	0.34
1.300	1.088	0.792	0.787	0.63	0.710	0.709	0.14
1.400	1.022	0.721	0.716	0.69	0.846	0.845	0.12
1.500	1.000	0.639	0.634	0.78	0.999	1.000	0.10
1.600	1.022	0.551	0.547	0.73	1.167	1.169	0.17
1.700	1.088	0.465	0.461	0.87	1.345	1.348	0.22
1.800	1.198	0.386	0.382	1.04	1.528	1.531	0.20
1.900	1.352	0.318	0.315	0.94	1.710	1.715	0.29
2.000	1.550	0.262	0.258	1.53	1.890	1.896	0.32
2.100	1.792	0.216	0.213	1.39	2.065	2.071	0.29
2.200	2.078	0.179	0.176	1.68	2.233	2.240	0.31
2.300	2.408	0.150	0.147	2.00	2.394	2.402	0.33
2.400	2.782	0.126	0.124	2.38	2.549	2.557	0.31
2.500	3.200	0.107	0.105	1.87	2.696	2.706	0.37
2.600	3.662	0.092	0.090	2.17	2.839	2.848	0.32
2.700	4.168	0.079	0.078	1.28	2.972	2.983	0.37
2.800	4.718	0.069	0.068	1.45	3.105	3.114	0.29
2.900	5.312	0.061	0.059	3.29	3.225	3.239	0.43
3.000	5.950	0.053	0.052	1.89	3.353	3.359	0.18

analytical value of ρ/ρ_o at $x = 0$ is 0.995, giving a 0.5 percent error. This built-in error is not viewed as serious, and we will not be concerned with it here.

Truncation Error: The Matter of Grid Independence

The matter of *grid independence* is a serious consideration in CFD, and this stage of our data analysis is a perfect time to introduce the concept. In general, when you solve a problem using CFD, you are employing a finite number of grid points (or a finite mesh) distributed over the flow field. Assume that you are using N grid points. If everything goes well during your solution, you will get some numbers out for the flowfield variables at these N grid points, and these numbers may look qualitatively good to you. However, assume that you rerun your solution, this time using twice as many grid points, $2N$, distributed over the same domain; i.e., you have decreased the value of the increment Δx (and also Δy in general, if you are dealing with a two-dimensional solution). You may find that the values of your flowfield variables are quite different for this second calculation. If this is the case, then your solution is a function of the number of grid points you are using—an untenable situation. You must, if at all practical, continue to increase the number of grid points until you reach a solution which is no longer sensitive to the number of points. When you reach this situation, then you have achieved *grid independence*.

Question: Do we have grid independence for the present calculation? Recall that we have used 31 grid points distributed evenly through the nozzle. To address this question, let us double the number of grid points; i.e., let us halve the value of Δx by using 61 grid points. Table B.5 compares the steady-state results for density, temperature, and pressure ratios, as well as for Mach numbers, at the throat for both the cases using 31 and 61 grid points. Also tabulated in Table B.5 are the exact analytical results. Note that although doubling the number of grid points did improve the numerical solution, it did so only marginally. The same is true for all locations within the nozzle. In other words, the two steady-state numerical solutions are essentially the same, and therefore we can conclude that our original calculations using 31 grid points is essentially *grid-independent*. This grid-independent solution does not agree *exactly* with the analytical results, but it is certainly close enough for our purposes. The degree of grid independence that you need to achieve in a given problem depends on what you want out of the solution. Do you need extreme accuracy? If so, you need to press the matter of grid independence in a very detailed fashion. Can you tolerate answers that can be a little less precise numerically (such as the 1 or 2 percent accuracy shown in the present calculations)? If so, you can slightly relax the criterion for extreme grid independence and use fewer grid points, thus saving computer time (which frequently means saving money). The proper decision depends on the circumstances. However, you should always be conscious of the question of grid independence and resolve the matter to your satisfaction for any CFD problem you solve. For example, in the present problem, do you think you can drive the numerical results shown in Table B.5 to agree exactly with the analytical results by using more and more grid points? If so, how many grid points will you need? You might want to experiment with this question by running your own program and seeing what happens.

Table B.5 | Demonstration of grid independence

	Conditions at the nozzle throat			
	$\dfrac{\rho^*}{\rho_o}$	$\dfrac{T^*}{T_o}$	$\dfrac{p^*}{p_o}$	M
Case 1: 31 points	0.639	0.836	0.534	0.999
Case 2: 61 points	0.638	0.835	0.533	1.000
Exact analytical solution	0.634	0.833	0.528	1.000

Courant Number Effects

There is the possibility that if the Courant number is too small, there might be problems in regard to the accuracy of the solution, albeit the solution will be very stable. Do we have such a problem with the present calculations? We have employed $C = 0.5$ for the present calculations. Is this too small, considering that the stability criterion for *linear* hyperbolic equations is $C \leq 1.0$? To examine this question, we can simply repeat the previous calculations but with progressively higher values of the Courant number. The resulting steady-state flowfield values at the nozzle throat are tabulated in Table B.6; the tabulations are given for six different values of C, starting at $C = 0.5$ and ranging to 1.2. For values ranging to as high as $C = 1.1$, the results were only marginally different, as seen in Table B.6. By increasing C to as high as 1.1, the numerical results do not agree any better with the exact analytical results (as shown in Table B.6) than the results at lower values of C. Hence, all our previous results obtained by using $C = 0.5$ are not tainted by any noticeable error due to the smaller-than-necessary value of C. Indeed, if anything, the numerical results for $C = 0.5$ in Table B.6 are marginally *closer* to the exact analytical solution than the results for higher Courant numbers. For the steady-state numerical results tabulated in Table B.6, the number of time steps was adjusted each time C was changed so that the nondimensional time at the end of each run was essentially the same. This adjustment is necessary because the value of Δt calculated from Eqs. (B.28) and (B.31) will obviously be different for different values of C. For example, when $C = 0.5$ as in our previous results, we carried out the time-marching procedure to 1400 time steps, which corresponded to a nondimensional time of 28.952. When C is increased to 0.7, the number of time steps carried out was $1400(\frac{5}{7}) = 1000$. This corresponded to a nondimensional time of 28.961—essentially the same as for the previous run. In the same manner, all the numerical data compared in Table B.6 pertain to the same nondimensional time.

It is interesting to note that for the present application, the CFL criterion, namely, that $C \leq 1$, does not hold exactly. In Table B.6, we show results where $C = 1.1$; a stable solution is obtained in spite of the fact that the CFL criterion is violated. However, as noted in Table B.6, when the Courant number is increased to 1.2, instabilities do occur, and the program blows up. Therefore, for the flow problem we

Table B.6 | Courant number effects

Courant number	$\dfrac{\rho^*}{\rho_o}$	$\dfrac{T^*}{T_o}$	$\dfrac{p^*}{p_o}$	M
0.5	0.639	0.836	0.534	0.999
0.7	0.639	0.837	0.535	0.999
0.9	0.639	0.837	0.535	0.999
1.0	0.640	0.837	0.535	0.999
1.1	0.640	0.837	0.535	0.999
1.2	Program went unstable and blew up			
Exact analytical solution	0.634	0.833	0.528	1.000

have been discussing in this appendix, which is governed by *nonlinear* hyperbolic partial differential equations, the CFL criterion (which is based on linear equations) does not hold exactly. However, from the results, we can see that the CFL criterion is certainly a good *estimate* for the value of Δt; it is the most reliable estimate for Δt that we can use, even though the governing equations are nonlinear.

SUMMARY

This appendix contains enough details for you to write your own computer program for the CFD solution of isentropic subsonic-supersonic quasi-one-dimensional flow. However, if you wish you can use the following FORTRAN program written by the author, who makes no claim of writing particularly efficient programs.

We note that this example has used the conservation form of the continuity equation, and the nonconservation form of the momentum and energy equations. These forms work fine for the application discussed here. For most modern applications in CFD, however, the conservation form of the equations is usually used, for reasons discussed in the author's book *Computational Fluid Dynamics: The Basics with Applications*. Also discussed in that book is the matter of artificial viscosity (numerical damping), which is important to many applications in CFD. These matters are beyond the scope of the present book, but you should be aware that most applications of CFD require considerations additional to those we have considered here. That is why CFD is a subject all by itself.

ISENTROPIC NOZZLE FLOW—SUBSONIC/ SUPERSONIC (NONCONSERVATION FORM)

```
      REAL A(31),RHO(31),T(31),U(31),DRHO(31),DT(31),PRHO(31),PU(31)
      REAL PT(31),ADRHO(31),ADU(31),P(31),XMACH(31),XR(31),DU(31)
      REAL ADT(31),XMFLOW(31)
      GAMMA=1.4
      COUR=0.2
      WRITE(6,200) COUR
      WRITE(*,200) COUR
      N=30
      N1=N+1
C     FEED IN NOZZLE AREA RATIO AND INITIAL CONDITIONS
      DX=3.0/FLOAT(N)
      X=0.0
      DO 1 I=1,N1
      A(I)=1.0+2.2*(X-1.5)**2
      RHO(I)=1.0-0.3146*X
      T(I)=1.0-0.2314*X
      U(I)=(0.1+1.09*X)*SQRT(T(I))
      XMFLOW(I)=RHO(I)*U(I)*A(I)
```

```
         XR(I)=X
         X=X+DX
1        CONTINUE
C        CALCULATION OF TIME STEP
         DELTY=1.0
         DO 2 I=2,N
          DELTX=DX/(U(I)+SQRT(T(I)))
          DELTIM=MIN(DELTX,DELTY)
          DELTY=DELTIM
          DELTIM=COUR*DELTIM
2        CONTINUE
         TIME=DELTIM
C        SOME ADDITIONAL VALUES TO BE INITIALIZED
         PRHO(1)=RHO(1)
         PT(1)=T(1)
         P(1)=1.0
          XMACH(1)=U(1)/SQRT(T(1))
         WRITE(6,100)
         WRITE(*,100)
          WRITE(6,101)(A(I),RHO(I),U(I),T(I),XMFLOW(I),I=1,N1)
          WRITE(*,101)(A(I),RHO(I),U(I),T(I),XMFLOW(I),I=1,N1)
         JMOD=3500
         JEND=3500
         DO 10 J=1,JEND
C        PREDICTED VALUES FOR INTERNAL POINTS
         DO 3 I=2,N
          DXLA=(ALOG(A(I+1))-ALOG(A(I)))/DX
          DXU=(U(I+1)-U(I))/DX
          DXRHO=(RHO(I+1)-RHO(I))/DX
          DXT=(T(I+1)-T(I))/DX
           DRHO(I)=-RHO(I)*U(I)*DXLA-RHO(I)*DXU-U(I)*DXRHO
           DU(I)=-U(I)*DXU-(1.0/GAMMA)*(DXT+T(I)/RHO(I)*DXRHO)
           DT(I)=-U(I)*DXT-(GAMMA-1.0)*(T(I)*DXU+T(I)*U(I)*DXLA)
          PRHO(I)=RHO(I)+DELTIM*DRHO(I)
          PU(I)=U(I)+DELTIM*DU(I)
          PT(I)=T(I)+DELTIM*DT(I)
3        CONTINUE
C        LINEAR EXTRAPOLATION FOR PU(1)
          PU(1)=2.0*PU(2)-PU(3)
C        CORRECTED VALUES FOR INTERNAL POINTS
         DO 4 I=2,N
          DXLA=(ALOG(A(I))-ALOG(A(I-1)))/DX
          DXRHO=(PRHO(I)-PRHO(I-1))/DX
          DXU=(PU(I)-PU(I-1))/DX
          DXT=(PT(I)-PT(I-1))/DX
```

```
      PDRHO=-PRHO(I)*PU(I)*DXLA-PRHO(I)*DXU-PU(I)*DXRHO
      PDU=-PU(I)*DXU-(1.0/GAMMA)*(DXT+PT(I)/PRHO(I)*DXRHO)
      PDT=-PU(I)*DXT-(GAMMA-1.0)*(PT(I)*DXU+PT(I)*PU(I)*DXLA)
    ADU(I)=0.5*(PDU+DU(I))
    ADRHO(I)=0.5*(PDRHO+DRHO(I))
    ADT(I)=0.5*(PDT+DT(I))
    RHO(I)=RHO(I)+ADRHO(I)*DELTIM
    U(I)=U(I)+ADU(I)*DELTIM
    T(I)=T(I)+ADT(I)*DELTIM
    P(I)=RHO(I)*T(I)
    XMACH(I)=U(I)/SQRT(T(I))
4   CONTINUE
C   EXTRAPOLATION TO END POINTS
    U(1)=2.0*U(2)-U(3)
    XMACH(1)=U(1)/SQRT(T(1))
    RHO(N1)=2.0*RHO(N)-RHO(N-1)
    U(N1)=2.0*U(N)-U(N-1)
    T(N1)=2.0*T(N)-T(N-1)
    P(N1)=RHO(N1)*T(N1)
    XMACH(N1)=U(N1)/SQRT(T(N1))
    DELTY=1.0
    DO 5 I=2,N
     DELTX=DX/(U(I)+SQRT(T(I)))
     DELTIM=MIN(DELTX,DELTY)
     DELTY=DELTIM
     DELTIM=COUR*DELTIM
5   CONTINUE
    DO 6 I=1,N1
     XMFLOW(I)=RHO(I)*U(I)*A(I)
6   CONTINUE
    TIME=TIME+DELTIM
    TEST=MOD(J,JMOD)
    IF(TEST.GT.0.01) GO TO 10
    WRITE(6,102) J,TIME
    WRITE(*,102) J,TIME
    WRITE(6,103)
    WRITE(*,103)
      WRITE(6,104)(I,XR(I),A(I),RHO(I),U(I),T(I),P(I),XMACH(I),
   +            XMFLOW(I),I=1,N1)
      WRITE(*,104)(I,XR(I),A(I),RHO(I),U(I),T(I),P(I),XMACH(I),
   +            XMFLOW(I),I=1,N1)
    WRITE(6,105)J,DELTIM
    WRITE(*,105)J,DELTIM
    WRITE(6,106)
    WRITE(*,106)
```

```
          WRITE(6,107)(I,ADRHO(I),ADU(I),ADT(I),I=2,N)
          WRITE(*,107)(I,ADRHO(I),ADU(I),ADT(I),I=2,N)
10    CONTINUE
100     FORMAT(3X,'INITIAL CONDITIONS'//12X,'A',8X,'RHO',8X,'U',8X,'T',
     +          8X,'MFLOW')
101     FORMAT(5X,5F10.3)
102      FORMAT(5X,'J=',I5,10X,'TIME=',F7.3//)
103      FORMAT(4X,'I',6X,'XD',6X,'A',3X,'RHO',6X,'U',6X,'T',6X,'P',
     +          6X,'M',6X,'MFLOW')
104     FORMAT(2X,I3,8F7.3)
105      FORMAT(5X,'J=',I5,10X,'DELTIM=',E10.3)
106      FORMAT(5X,'I',7X,'ADRHO',14X,'ADU',14X,'ADT')
107     FORMAT(2X,I3,3E15.3)
200     FORMAT(5X,'COURANT NUMBER =',F7.3)
      END
```

1. Anderson, J. D., Jr., *Introduction to Flight: Its Engineering and History,* 4th ed., McGraw-Hill, Boston, 2000.

2. Hallion, R. P., *Supersonic Flight: The Story of the Bell X-1 and Douglas D-558,* Macmillan, New York, 1972.

3. Courant, R. and K. O. Friedrichs, *Supersonic Flow and Shock Waves,* Interscience, New York, 1948.

4. Emmons, H. W. (ed.), *Foundations of Gas Dynamics,* Vol. III of *High Speed Aerodynamics and Jet Propulsion,* Princeton, Princeton, NJ, 1956.

5. Ferri, Antonio, *Elements of Aerodynamics of Supersonic Flows,* Macmillan, New York, 1949.

6. Hilton, W. F., *High Speed Aerodynamics,* Longmans, London, 1951.

7. Howarth, L. (ed.), *Modern Developments in Fluid Dynamics, High Speed Flow,* 2 vols., Oxford, London, 1953.

8. Johns, J. E. A., *Gas Dynamics,* Allyn and Bacon, Boston, 1969.

9. Liepmann, H. W. and A. Roshko, *Elements of Gasdynamics,* Wiley, New York, 1957.

10. Miles, E. R. C., *Supersonic Aerodynamics,* Dover, New York, 1950.

11. von Mises, Richard, *Mathematical Theory of Compressible Flow,* Academic, New York, 1958.

12. Oswatitsch, K., *Gasdynamik,* Springer, Vienna, 1952; Academic, New York, 1956.

13. Owczarek, J. A., *Fundamentals of Gas Dynamics,* International Textbook, Scranton, 1964.

14. Pope, Alan, *Aerodynamics of Supersonic Flight,* 2d ed., Pitman, New York, 1958.

15. Sears, W. R. (ed.), *Theory of High Speed Aerodynamics,* Vol. VI of *High Speed Aerodynamics and Jet Propulsion,* Princeton, Princeton, NJ, 1954.

16. Shapiro, A. H., *The Dynamics and Thermodynamics of Compressible Fluid Flow,* 2 vols., Ronald, New York, 1953.

17. Zucrow, M. J. and J. D. Hoffman, *Gas Dynamics,* 2 vols., Wiley, New York, 1976–77.

18. Anderson, J. D., Jr., *Computational Fluid Dynamics: The Basics with Applications,* McGraw-Hill, New York, 1995.

19. Hirschfelder, J. O., C. F. Curtiss, and R. B. Bird, *Molecular Theory of Gases and Liquids,* Wiley, New York, 1954.

20. Schlicting, H., *Boundary Layer Theory,* 6th ed., McGraw-Hill, New York, 1968.

21. Anderson, J. D., Jr., *Gasdynamic Lasers: An Introduction,* Academic, New York, 1976.

22. Durand, W. F. (ed.), *Aerodynamic Theory,* 6 vols., Springer, Berlin, 1934.

23. Laitone, E. V., "New Compressibility Correction for Two-Dimensional Subsonic Flow," *J. Aeronaut. Sci.* vol. 18, no. 5, 1951, p. 350.

24. Tsien. H. S., "Two-Dimensional Subsonic Flow of Compressible Fluids," *J. Aeronaut. Sci.* vol. 6, no. 10, 1939, p. 399.

25. von Karman, T. H., "Compressibility Effects in Aerodynamics," *J. Aeronaut. Sci.* vol. 8, no. 9, 1941, p. 337.

26. Busemann, A., "Drucke auf Kegelformige Spitzen bei Bewegung mit Uberschallgeschwindigkeit," *Z. Angew Math. Mech.* vol. 9, 1929, p. 496.

27. Taylor, G. I., and J. W. Maccoll, "The Air Pressure on a Cone Moving at High Speed" *Proc. Roy. Soc.* (London) ser. A, vol. 139, 1933, pp. 278–311.

28. Kopal, Z., "Tables of Supersonic Flow Around Cones," M.I.T. Center of Analysis Tech. Report No. 1, U.S. Govt. Printing Office, Washington, DC, 1947.

29. Sims, Joseph L., "Tables for Supersonic Flow Around Right Circular Cones at Zero Angle of Attack," NASA SP-3004, 1964.

30. Holt, Maurice, *Numerical Methods in Fluid Dynamics,* Springer-Verlag, Berlin, 1977.

31. Roache, Patrick J., *Computational Fluid Dynamics,* Hermosa, Albuquerque, 1972.

32. Taylor, T. D., *Numerical Methods for Predicting Subsonic, Transonic and Supersonic Flow,* AGARDograph No. 187, 1974.

33. Wirz, H. J. (ed.), *Progress in Numerical Fluid Dynamics,* Springer-Verlag, Berlin, 1975.

34. Saueriwein, H., "Numerical Calculation of Multidimensional and Unsteady Flows by the Method of Characteristics," *J. Comput. Phys.* vol. 1, no. 1, Feb. 1967, pp. 406–432.

35. Chushkin, P. I., *Numerical Method of Characteristics for Three-Dimensional Supersonic Flows,* vol. 9 in D. Kuchemann (ed.), *Progress in Aeronautical Sciences,* Pergamon, Elmsford, NY, 1968.

36. Butler, D. S., "The Numerical Solution of Hyperbolic Systems of Partial Differential Equations in Three Independent Variables," *Proc. Roy. Soc.* vol. A255, no. 1281, 1960, pp, 232–252.

37. Rakich, John V., "A Method of Characteristics for Steady Three-Dimensional Supersonic Flow with Application to Inclined Bodies of Revolution," NASA TN D-5341, 1969.

38. Rakich, John V. and Joseph W. Cleary, "Theoretical and Experimental Study of Supersonic Steady Flow Around Inclined Bodies of Revolution," *AIAA J.* vol. 8, no. 3, 1970, pp. 511–518.

39. MacCormack, R. W., "The Effect of Viscosity in Hypervelocity Impact Cratering," AIAA Paper No. 69-354, 1969.

40. Kutler, Paul and H. Lomax, "Shock-Capturing Finite-Difference Approach to Supersonic Flows," *J. Spacecraft Rockets* vol. 8, no. 12, 1971, pp. 1175–1182.

41. Thomas, P. D., M. Vinokur, R. A. Bastianon, and R. J. Conti, "Numerical Solutions for Three-Dimensional Inviscid Supersonic Flow," *AIAA J.* vol. 10, no. 7, 1972, pp. 887–894.

42. Warming, R. F., P. Kutler, and H. Lomax, "Second- and Third-Order Noncentered Difference Schemes for Nonlinear Hyperbolic Equations," *AIAA J.* vol. 11, no. 2, 1973, pp. 189–196.

43. Kutler, P., R. F. Warming, and H. Lomax, "Computation of Space Shuttle Flow Fields Using Noncentered Finite-Difference Schemes," *AIAA J.* vol. 11, no. 2, 1973, pp. 196–204.

44. Kutler, P., "Computation of Three-Dimensional, Inviscid Supersonic Flows," in H. J. Wirz (ed.), *Progress in Numerical Fluid Dynamics,* Springer-Verlag, Berlin, 1975, pp. 293–374.

45. Rakich, John V. and Paul Kutler, "Comparison of Characteristics and Shock Capturing Methods with Application to the Space Shuttle Vehicle," AIAA Paper No. 72-191, 1972.

46. Abbett, M. J., "Boundary Condition Calculation Procedures for Inviscid Supersonic Flow Fields," in *Proceedings of the 1st AIAA Computational Fluid Dynamics Conference,* Palm Springs, CA, 1973, pp. 153–172.

47. Moretti, G. and M. Abbett, "A Time-Dependent Computational Method for Blunt Body Flows," *AIAA J.* vol. 4, no. 12, 1966, pp. 2136–2141.

48. Anderson, J. D., Jr., "A Time-Dependent Analysis for Vibrational and Chemical Nonequilibrium Nozzle Flows," *AIAA J.* vol. 8, no. 3, 1970, pp. 545–550.

49. Anderson, J. D., Jr., "Time-Dependent Solutions of Nonequilibrium Nozzle Flows— A Sequel," *AIAA J.* vol. 5, no. 12, 1970, pp. 2280–2282.

50. Griffin, M. D. and J. D. Anderson, Jr., "On the Application of Boundary Conditions to Time-Dependent Computations for Quasi-One-Dimensional Fluid Flows," *Comput. Fluids* vol. 5, no. 3, 1977, pp. 127–237.

51. Courant, R., K. O. Friedrichs, and H. Lewy, "Uber die Differenzengleichungen der Mathematischen Physik," *Math. Ann.* vol. 100, 1928, p. 32.

52. Hayes, W. D. and R. F. Probstein, *Hypersonic Flow Theory,* 2d ed., Academic, 1966, vol. 1, Chap. 6.

53. Anderson, J. D., Jr., L. M. Albacete, and A. E. Winkelmann, "On Hypersonic Blunt Body Flow Fields Obtained with a Time-Dependent Technique," Naval Ordnance Laboratory NOLTR 68-129, 1968.

54. Lomax, H. and M. Inouye, "Numerical Analysis of Flow Properties About Blunt Bodies Moving at

Supersonic Speeds in an Equilibrium Gas," NASA TR-R-204, 1964.

55. Serra, R. A., "The Determination of Internal Gas Flows by a Transient Numerical Technique," AIAA Paper No. 71-45, 1971.

56. Lax, P., "Weak Solutions of Nonlinear Hyperbolic Equations and Their Numerical Computation," *Comm. Pure Appl. Math.* vol. VII, 1954, pp. 159–193.

57. Von Neumann, J. and R. D. Richtmyer, "A Method for the Numerical Calculation of Hydrodynamic Shocks," *J. Appl. Phys.* vol. 21, No. 3, 1950, pp. 232–237.

58. Griffin, M. D., J. D. Anderson, Jr., and E. Jones, "Computational Fluid Dynamics Applied to Three-Dimensional Nonreacting Inviscid Flows in an Internal Combustion Engine." *J. Fluids Eng.* vol. 101, no. 9, 1979, pp. 367–372.

59. Vincenti, W. G. and C. H. Kruger, *Introduction to Physical Gas Dynamics,* Wiley, New York, 1965.

60. Herzberg, G., *Atomic Spectra and Atomic Structure,* Dover, New York, 1944.

61. Herzberg, G., *Molecular Spectra and Molecular Structure,* D. Van Nostrand, New York, 1963.

62. Davidson, N., *Statistical Mechanics,* McGraw-Hill, New York, 1962.

63. Stull, D. R. et al., *JANAF Thermochemical Tables,* National Bureau of Standards NSRDS-NBS 37, 1971.

64. McBride, B. J., S. Heimel, J. G. Ehlers, and S. Gordon, "Thermodynamic Properties to 6000°K for 210 Substances Involving the First 18 Elements," NASA SP-3001, 1963.

65. Rossini, F. D., D. D. Wagman, W. H. Evans, S. Levine, and I. Jaffe, "Selected Values of Chemical Thermodynamic Properties," National Bureau of Standards Circular No. 500, 1952.

66. Hansen, C. F., "Approximations for the Thermodynamic and Transport Properties of High-Temperature Air," NASA TR-R-50, 1959.

67. Hilsenrath, J. and M. Klein, "Tables of Thermodynanic Properties of Air in Chemical Equilibrium Including Second Virial Corrections from 1500°K to 15,000°K," Arnold Engineering Development Center Report No. AEDC-TR-65-68, 1965.

68. Wray, K. L., "Chemical Kinetics of High Temperature Air," in R. F. Riddell (ed.), *Hypersonic Flow Research,* Academic, New York, 1961, pp. 181–204.

69. Wittliff, C. E. and J. T. Curtiss, "Normal Shock Wave Parameters in Equilibrium Air," Cornell Aeronautical Laboratory (now CALSPAN) Report No. CAL-111, 1961.

70. Marrone, P. V., "Normal Shock Waves in Air: Equilibrium Composition and Flow Parameters for Velocities from 26,000 to 50,000 ft/sec," Cornell Aeronautical Laboratory (now CALSPAN) Report No. AG-1729-A-2, 1962.

71. Hall, J. G. and C. E. Treanor, "Nonequilibrium Effects in Supersonic Nozzle Flows," AGAR-Dograph No. 124, 1968.

72. Marrone, P. V., "Inviscid Nonequilibrium Flow Behind Bow and Normal Shock Waves, Part I. General Analysis and Numerical Examples," Cornell Aeronautical Laboratory (now CALSPAN) Report No. QM-1626-A-12(I), 1963.

73. Anderson, J. D., Jr., "A Time-Dependent Analysis for Vibrational and Chemical Nonequilibrium Nozzle Flows," *AIAA J.* vol. 8, no. 3, 1970, pp. 545–550.

74. Anderson, J. D., Jr., "Time-Dependent Solutions of Nonequilibrium Nozzle Flows—A Sequel," *AIAA J.* vol. 8, no. 12, 1970, pp. 2280–2282.

75. Wilson, J. L., D. Schofield, and K. C. Lapworth, "A Computer Program for Nonequilibrium Convergent-Divergent Nozzle Flow," National Physical Laboratory Report No. 1250, 1967.

76. Harris, E. L. and L. M. Albacete, "Vibrational Relaxation of Nitrogen in the NOL Hypersonic Tunnel No. 4," Naval Ordnance Laboratory TR 63-221, 1964.

77. Erickson, W. D., "Vibrational Nonequilibrium Flow of Nitrogen in Hypersonic Nozzles," NASA TN D-1810, 1963.

78. Hall, J. G. and A. L. Russo, "Studies of Chemical Nonequilibrium in Hypersonic Nozzle Flows," Cornell Aeronautical Laboratory (now CALSPAN) Report No. AF-1118-A-6, 1959.

79. Sarli, V. J., W. G. Burwell, and T. F. Zupnik, "Investigation of Nonequilibrium Flow Effects in High Expansion Ratio Nozzles," NASA CR-54221, 1964.

80. Vamos, J. S. and J. D. Anderson, Jr., "Time-Dependent Analysis of Nonequilibrium Nozzle Flows with Complex Chemistry," *J. Spacecraft Rockets* vol. 10, no. 4, 1973, pp. 225–226.

81. Ferri, A., "Supersonic Flow around Circular Cones at Angles of Attack," NASA Technical Note 2236, 1950.

82. Ferri, A., "Supersonic Flows with Shock Waves," in W. R. Sears (ed.), *General Theory of High Speed Aerodynamics,* Princeton, Princeton, NJ, 1954, pp. 670–747.

83. Melnik, R. E., "Vortical Singularities in Conical Flow," *AIAA J.* vol. 5, no. 4, 1967, pp. 631–637.

84. Feldhuhn, R. H., A. E. Winkelmann, and L. Pasiuk, "An Experimental Investigation of the Flowfield around a Yawed Cone," *AIAA J.* vol. 9, no. 6, 1971, pp. 1074–1081.

85. Marconi, F., "Fully Three-Dimensional Separated Flows Computed with the Euler Equations," AIAA Reprint No. 87-0451, 1987.

86. Kopal, Z., "Tables of Supersonic Flow around Yawing Cones," M.I.T. Center of Analysis Tech. Report No. 3, 1947.

87. Sims, Joseph L., "Tables for Supersonic Flow around Right Circular Cones at Small Angle of Attack," NASA SP-3007, 1964.

88. Moretti, Gino, "Inviscid Flowfield about a Pointed Cone at an Angle of Attack," *AIAA J.* vol. 5, no. 4, 1967, pp. 789–791.

89. Tracy, R. R., "Hypersonic Flow over a Yawed Circular Cone," Graduate Aeronautical Labs., California Institute of Technology Memo 69, August 1, 1963.

90. Kutler, Paul and Howard Lomax, "The Computation of Supersonic Flow Fields about Wing-Body Combinations by Shock-Capturing Finite Difference Techniques," in *Proceedings of the Second International Conference on Numerical Methods in Fluid Dynamics,* Springer-Verlag, Berlin, 1971, pp. 24–29.

91. Fletcher, C. A. J., "GTT Method Applied to Cones at Large Angles of Attack," in *Proceedings of the Fourth International Conference on Numerical Methods in Fluid Dynamics,* Springer-Verlag, Berlin, 1975, pp. 161–166.

92. Moretti, Gino and Gary Bleich, "Three-Dimensional Flow around Blunt Bodies," *AIAA J.* vol. 5, no. 9, 1967, pp. 1557–1562.

93. Weilmuenser, K. J., "High Angle of Attack Inviscid Flow Calculations over a Shuttle-Like Vehicle with Comparisons to Flight Data, AIAA Paper No. 83-1798, 1983.

94. Newberry, C. F., H. S. Dresser, J. W. Byerly, and W. T. Riba, "The Evaluation of Forebody Compression at Hypersonic Mach Numbers," AIAA Paper No. 88-0479, 1988.

95. Chakravarthy, S. R. and K. Y. Szema, "An Euler Solver for Three-Dimensional Supersonic Flows with Subsonic Pockets," AIAA Paper No. 85-1703, 1985.

96. Chakravarthy, S. R., "The Versatility and Reliability of Euler Solvers Based on High-Accuracy TVD Foundations," AIAA Paper No. 86-0243, 1986.

97. Szema, K. Y., S. R. Chakravarthy, W. T. Riva, J. Byerly, and H. S. Dresser, "Multi-Zone Euler Marching Technique for Flow over Single and Multi-Body Configurations," AIAA Paper No. 87-0592, 1987.

98. Meyer, Richard E. (ed.), *Transonic, Shock, and Multidimensional Flows,* Academic, New York, 1982.

99. Becker, John V., *The High-Speed Frontier,* NASA SP-445, 1980.

100. Loftin, Lawrence K., Jr., *The Quest for Performance: The Evolution of Modern Aircraft,* NASA SP-468, 1985.

101. Murman, Earll M. and Julian D. Cole, "Calculation of Plane Steady Transonic Flows," *AIAA J.* vol. 9, no. 1, 1971, pp. 114–121.

102. Anderson, Dale A., John C. Tannehill, and R. H. Pletcher, *Computational Fluid Mechanics and Heat Transfer,* 2nd ed., Taylor and Francis, Washington, DC, 1997.

103. Knechtel, Earl D., "Experimental Investigation at Transonic Speeds of Pressure Distributions over Wedge and Circular-Arc Airfoil Sections and Evaluation of Perforated-Wall Interference," NASA TN D-15, 1959.

104. Anderson, John D., Jr., *Fundamentals of Aerodynamics,* McGraw-Hill, 3rd ed., Boston, 2001.

105. Caughey, David A., "The Computation of Transonic Potential Flows," in Milton van Dyke, J. V. Wehausen, and John L. Langley (eds.), *Annual Review of Fluid Mechanics,* Annual Reviews, Inc., Palo Alto, vol. 14, 1982, pp. 261–283.

106. Keyfitz, Barbara L., Robert E. Melnik, and Bernard Grossman, "Leading-Edge Singularity in Transonic Small-Disturbance Theory: Numerical Resolution," *AIAA J.* vol. 17, no. 3, 1979, pp. 296–298.

107. Magnus, R. and H. Yoshihara, "Inviscid Transonic Flow over Airfoils," *AIAA J.* vol. 8. no. 12, 1970, pp. 2157–2162.

108. Thompson, J. F., F. C. Thames, and C. W. Mastin, "Automatic Numerical Generation of Body-Fitted Curvilinear Coordinate System for Field Containing Any Number of Arbitrary Two-Dimensional Bodies," *J. Comput. Phys.,* vol. 15, 1974, 299–319.

109. Anderson, John D., "Introduction to Computational Fluid Dynamics," in *Introduction to Computation Fluid Dynamics,* von Karman Institute Lecture Notes, January 1989, pp. 1–208.

110. Kothari, Ajay P. and John D. Anderson, Jr., "Flow over Low Reynolds Number Airfoils—Compressible Navier-Stokes Numerical Solutions," AIAA Reprint 85-0107, 1985.

111. Ballhaus, W. F. and F. R. Bailey, "Numerical Calculation of Transonic Flow about Swept Wings," AIAA Paper No. 72-677, 1972.

112. Jameson, A. and D. A. Caughey, "Numerical Calculation of the Transonic Flow Past a Swept Wing," ERDA Report COO-3077-140, New York University, 1977.

113. Hinson, B. L. and K. P. Burdges, "An Evaluation of Three-Dimensional Transonic Codes Using New Correlation—Tailored Test Data," AIAA Paper No. 80-0003, 1980.

114. Stivers, L., "Effects of Subsonic Mach Number on the Forces and Pressure Distribution of Four NACA 64A-Series Airfoil Sections," NASA TN 3162, 1954.

115. Jameson, Antony, "Steady-State Solution of the Euler Equations for Transonic Flow," in Richard E. Meyer (ed.), *Transonic, Shock, and Multidimensional Flows,* Academic, New York, 1982, pp. 37–70.

116. Jameson, Antony, "Successes and Challenges in Computational Aerodynamics," in *Proceedings of the AIAA Eighth Computational Fluid Dynamics Conference,* American Institute of Aeronautics and Astronautics, June 9–11, 1987, pp. 1–35.

117. Reddy, K. C. and J. L. Jacobs, "A Locally Implicit Scheme for the Euler Equations," in *Proceedings of the AIAA Eighth Computational Fluid Dynamics Conference,* American Institute of Aeronautics and Astronautics, June 9–11, 1987, pp. 470–477.

118. Anderson, J. D., Jr., "A Survey of Modern Research in Hypersonic Aerodynamics," AIAA Paper No. 84-1578, 1984.

119. Anderson, John D., Jr., *Hypersonic and High Temperature Gas Dynamics,* McGraw-Hill, New York, 1989; reprinted by the American Institute of Aeronautics and Astronautics, Reston, VA, 2000.

120. Anderson, John D., Jr., "Hypersonic Viscous Flow over Cones at Nominal Mach 11 in Air," ARL Report 62-387, Aeronautical Research Laboratories, Wright-Patterson Air Force Base, Ohio, July 1962.

121. Charters, A. C. and R. N. Thomas, "The Aerodynamic Performance of Small Spheres from Subsonic to High Supersonic Velocities," *J. Aeronaut. Sci.* vol. 12, 1945, pp. 468–476.

122. Hodges, A. J., "The Drag Coefficient of Very High Velocity Spheres," *J. Aeronaut. Sci.* vol. 24, 1957, pp. 755–758.

123. Stevens, V. I., "Hypersonic Research Facilities at the Ames Aeronautical Laboratory," *J. Appl. Phys.* vol. 21, pp. 1150–1155.

124. Cox, R. N. and L. F. Crabtree, *Elements of Hypersonic Aerodynamics,* Academic, New York, 1965.

125. Neice, Stanford E. and Dorris M. Ehret, "Similarity Laws for Slender Bodies of Revolution in Hypersonic Flows," *J. Aeronautic. Sci.* vol. 18, no. 8, 1951, pp. 527–530, 568.

126. Tsien, H. S., "Similarity Laws of Hypersonic Flows," *J. Math. Phys.* vol. 25, 1946, pp. 247–251.

127. Hayes, Wallace D., "On Hypersonic Similitude," *Quart. Appl. Math.* vol. 5, no. 1, 1947, pp. 105–106.

128. Anderson, John D., Jr., Part I of *Computational Fluid Dynamics: An Introduction,* John F. Wendt (ed.), Springer, Berlin, 1996, pp. 1–147.

129. Rakich, John V. and Joseph W. Cleary, "Theoretical and Experimental Study of Supersonic Steady Flow around Inclined Bodies of Revolution," *AIAA J.* vol. 8, no. 3, 1970, pp. 511–518.

130. Thompson, M. J., "A Note on the Calculation of Oblique Shock-Wave Characteristics," *J. Aeronaut. Sci.* vol. 17, no. 11, 1950, p. 744.

131. Mascitti, V. R., "A Closed-Form Solution to Oblique Shock-Wave Properties," *J. Aircraft* vol. 6, no. 1, 1969, p. 66.

132. Wolf, T., "Comment on 'Approximate Formula of Weak Oblique Shock Wave Angle,'" *AIAA J.* vol. 31, no. 7, 1993, p. 1363.

133. Emanuel, George, *Analytical Fluid Dynamics,* 2nd ed., CRC Press, Boca Raton, FL, 2001, pp. 751–753.

134. Anderson, John D., Jr., *A History of Aerodynamics and Its Impact on Flying Machines,* Cambridge University Press, New York, 1997 (hardback), 1998 (paperback).

135. Kuchemann, D., *The Aerodynamic Design of Aircraft,* Pergamon Press, Oxford, 1978.

136. Bowcutt, K. G., J. D. Anderson, and D. Capriotti, "Numerical Optimization of Conical Flow Waveriders Including Detailed Viscous Effects," in *Aerodynamics of Hypersonic Lifting Vehicles, AGARD Conference Proceedings,* no. 428, November 1987, pp. 27.1–27.23.

137. Fletcher, C. A., *Computational Techniques for Fluid Dynamics,* vol. I: *Fundamental and General Techniques,* Springer-Verlag, Berlin, 1988.

138. Fletcher, C. A., *Computational Techniques for Fluid Dynamics,* vol. II: *Specific Techniques for Different Flow Categories,* Springer-Verlag, Berlin, 1988.

139. Hirsch, Charles, *Numerical Computation of Internal and External Flows,* vol. I: *Fundamentals of Numerical Discretization,* Wiley, New York, 1988.

140. Hirsch, Charles, *Numerical Computation of Internal and External Flows,* vol. II: *Methods for Inviscid and Viscous Flows,* Wiley, New York, 1990.

141. Hoffman, K. A., *Computational Fluid Dynamics for Engineers,* Engineering Education System, Austin, TX, 1989.

142. Laney, C. B., *Computational Gasdynamics,* Cambridge University Press, Cambridge, UK, 1998.

143. Palmer, Grant, "An Implicit Flux-Split Algorithm to Calculate Hypersonic Flowfields in Chemical Equilibrium," AIAA Paper No. 87-1580, June 1987.